D1731459

Springer Series in

OPTICAL SCIENCES 36

founded by H.K.V. Lotsch

Springer Series in
OPTICAL SCIENCES

The Springer Series in Optical Sciences, under the leadership of Editor-in-Chief *William T. Rhodes*, Georgia Institute of Technology, USA, provides an expanding selection of research monographs in all major areas of optics: lasers and quantum optics, ultrafast phenomena, optical spectroscopy techniques, optoelectronics, quantum information, information optics, applied laser technology, industrial applications, and other topics of contemporary interest.
With this broad coverage of topics, the series is of use to all research scientists and engineers who need up-to-date reference books.

The editors encourage prospective authors to correspond with them in advance of submitting a manuscript. Submission of manuscripts should be made to the Editor-in-Chief or one of the Editors. See also www.springeronline.com/series/624

L. Reimer   H. Kohl

Transmission Electron Microscopy

Physics of Image Formation

Fifth Edition

†Prof. Dr. L. Reimer

Prof. Dr. H. Kohl
Univ. Münster
Physikalisches Institut
Wilhelm-Klemm-Str. 10
48149 Münster
Germany
E-mail: kohl@uni-muenster.de

ISBN: 978-0-387-40093-8 e-ISBN: 978-0-387-34758-5
DOI: 10.1007/978-0-387-34758-5

Library of Congress Control Number: 2008933458

Printed on acid-free paper

9 8 7 6 5

springer.com

Preface to the Fifth Edition

The aim of this monograph is to outline the physics of image formation, electron–specimen interactions, and image interpretation in transmission electron microscopy. Since the last edition, transmission electron microscopy has undergone a rapid evolution. The introduction of monochromators and improved energy filters has allowed electron energy-loss spectra with an energy resolution down to about 0.1 eV to be obtained, and aberration correctors are now available that push the point-to-point resolution limit down below 0.1 nm.

After the untimely death of Ludwig Reimer, Dr. Koelsch from Springer-Verlag asked me if I would be willing to prepare a new edition of the book. As it had served me as a reference for more than 20 years, I agreed without hesitation. Distinct from more specialized books on specific topics and from books intended for classroom teaching, the Reimer book starts with the basic principles and gives a broad survey of the state-of-the-art methods, complemented by a list of references to allow the reader to find further details in the literature. The main objective of this revised edition was therefore to include the new developments but leave the character of the book intact.

The presentation of the material follows the format of the previous edition as outlined in the preface to that volume, which immediately follows. A few derivations have been modified to correspond more closely to modern textbooks on quantum mechanics, scattering theory, or solid state physics.

A special acknowledgement is due to M. Silder for preparing the new figures and helping with TeX and to all colleagues who gave permission to publish their results.

Münster, May 2007 *H. Kohl*

Preface

The aim of this monograph is to outline the physics of image formation, electron–specimen interactions, and image interpretation in transmission electron microscopy. The preparation of this fourth edition has made it possible to update the text and the bibliography. Meanwhile, the book *Energy-Filtering Transmission Electron Microscopy* has been published as Vol. 71 of the Springer Series in Optical Sciences. Discussion of this rapidly growing method has therefore been kept brief, and special aspects of energy filtering are discussed together with their conventional counterparts.

In the introductory chapter, the various electron–specimen interactions and their applications are summarized, the most important aspects of high-resolution, analytical, high-voltage, and energy-filtering electron microscopy are reviewed, and the different types of electron microscopes are compared. The optics of electron lenses are discussed in Chap. 2 in order to bring out electron-lens properties that are important for an understanding of the modes of operation of an electron microscope. In Chap. 3, the wave optics of electrons and the phase shifts caused by electrostatic and magnetic fields are introduced; Fresnel electron diffraction is treated using Huygens' principle. The recognition that the Fraunhofer diffraction pattern is the Fourier transform of the wave amplitude behind a specimen is important because the influence of the imaging process on the transfer of spatial frequencies can be described by introducing phase shifts and wave aberrations in the Fourier plane. In Chap. 4, the elements of an electron-optical column are described: the electron gun, the condenser, the imaging and recording system, and equipment for electron energy-loss spectroscopy and energy filtering.

A thorough understanding of electron–specimen interactions is essential to explain image contrast. Chapter 5 contains the most important facts about elastic, inelastic, and multiple scattering. The origin of scattering and phase contrast of noncrystalline specimens, the introduction of contrast-transfer functions, and the background of holographic and tomographic methods are described in Chap. 6. Chapter 7 introduces the most important laws about crystals and reciprocal lattices. The kinematical and dynamical theories of electron diffraction are then developed, and in Chap. 8 different modes and applications of electron diffraction are presented; convergent-beam electron diffraction (CBED) is of increasing interest. Electron diffraction is also the source of diffraction contrast. This type of contrast is important for the imaging of crystalline specimens and their defects and for the high-resolution study of crystal structure, treated in Chap. 9. Methods of elemental analysis and the formation of images representing the distribution of chemical elements by x-ray microanalysis and electron energy-loss spectroscopy are summarized in Chap. 10. The final chapter contains a brief account of the various specimen-damage processes caused by electron irradiation.

The author thanks Dr. P.W. Hawkes for thorough correction of the manuscript and many helpful comments.

Münster, January 1997 *L. Reimer*

Contents

1 **Introduction** ... 1
1.1 Transmission Electron Microscopy ... 1
1.1.1 Conventional Transmission Electron Microscopy ... 1
1.1.2 High-Resolution Electron Microscopy ... 3
1.1.3 Analytical Electron Microscopy ... 5
1.1.4 Energy-Filtering Electron Microscopy ... 7
1.1.5 High-Voltage Electron Microscopy ... 7
1.1.6 Dedicated Scanning Transmission Electron Microscopy ... 9
1.2 Alternative Types of Electron Microscopy ... 10
1.2.1 Emission Electron Microscopy ... 10
1.2.2 Reflection Electron Microscopy ... 11
1.2.3 Mirror Electron Microscopy ... 11
1.2.4 Scanning Electron Microscopy ... 12
1.2.5 X-ray and Auger-Electron Microanalysis ... 14
1.2.6 Scanning-Probe Microscopy ... 14

2 **Particle Optics of Electrons** ... 17
2.1 Acceleration and Deflection of Electrons ... 17
2.1.1 Relativistic Mechanics of Electron Acceleration ... 17
2.1.2 Deflection by Magnetic and Electric Fields ... 20
2.2 Electron Lenses ... 22
2.2.1 Electron Trajectories in a Magnetic Lens Field ... 22
2.2.2 Optics of an Electron Lens with a Bell-Shaped Field ... 25
2.2.3 Special Electron Lenses ... 29
2.3 Lens Aberrations ... 31
2.3.1 Classification of Lens Aberrations ... 31
2.3.2 Spherical Aberration ... 32
2.3.3 Astigmatism and Field Curvature ... 34
2.3.4 Distortion ... 36
2.3.5 Coma ... 37

2.3.6 Anisotropic Aberrations ... 38
2.3.7 Chromatic Aberration ... 38
2.4 Correction of Aberrations and Microscope Alignment ... 40
2.4.1 Correction of Astigmatism ... 40
2.4.2 Correction of Spherical and Chromatic Aberrations ... 42
2.4.3 Microscope Alignment ... 43

3 Wave Optics of Electrons ... 45
3.1 Electron Waves and Phase Shifts ... 45
3.1.1 De Broglie Waves ... 45
3.1.2 Probability Density and Wave Packets ... 49
3.1.3 Electron-Optical Refractive Index and the Schrödinger Equation ... 51
3.1.4 Electron Interferometry and Coherence ... 53
3.2 Fresnel and Fraunhofer Diffraction ... 55
3.2.1 Huygens' Principle and Fresnel Diffraction ... 55
3.2.2 Fresnel Fringes ... 59
3.2.3 Fraunhofer Diffraction ... 61
3.2.4 Mathematics of Fourier Transforms ... 63
3.3 Wave-Optical Formulation of Imaging ... 70
3.3.1 Wave Aberration of an Electron Lens ... 70
3.3.2 Wave-Optical Theory of Imaging ... 73

4 Elements of a Transmission Electron Microscope ... 77
4.1 Electron Guns ... 78
4.1.1 Physics of Electron Emission ... 78
4.1.2 Energy Spread ... 81
4.1.3 Gun Brightness ... 82
4.1.4 Thermionic Electron Guns ... 84
4.1.5 Schottky Emission Guns ... 88
4.1.6 Field-Emission Guns ... 89
4.2 The Illumination System of a TEM ... 90
4.2.1 Condenser-Lens System ... 90
4.2.2 Electron-Probe Formation ... 93
4.2.3 Illumination with an Objective Prefield Lens ... 96
4.3 Specimens ... 98
4.3.1 Useful Specimen Thickness ... 98
4.3.2 Specimen Mounting ... 99
4.3.3 Specimen Manipulation ... 100
4.4 The Imaging System of a TEM ... 103
4.4.1 Objective Lens ... 103
4.4.2 Imaging Modes of a TEM ... 104
4.4.3 Magnification and Calibration ... 107
4.4.4 Depth of Image and Depth of Focus ... 108
4.5 Scanning Transmission Electron Microscopy (STEM) ... 109

4.5.1 Scanning Transmission Mode of TEM 109
4.5.2 Dedicated STEM 112
4.5.3 Theorem of Reciprocity 113
4.6 Electron Spectrometers and Imaging Energy Filters 115
4.6.1 Postcolumn Prism Spectrometer 116
4.6.2 Wien Filter 119
4.6.3 Imaging Energy Filter 119
4.6.4 Operating Modes with Energy Filtering 124
4.7 Image Recording and Electron Detection 126
4.7.1 Fluorescent Screens 126
4.7.2 Photographic Emulsions 127
4.7.3 Imaging Plate 131
4.7.4 Detector Noise and Detection Quantum Efficiency 132
4.7.5 Low-Light-Level and Charge-Coupled-Device (CCD) Cameras 134
4.7.6 Semiconductor and Scintillation Detectors 138
4.7.7 Faraday Cages 139

5 Electron–Specimen Interactions 141
5.1 Elastic Scattering 141
5.1.1 Cross Section and Mean Free Path 141
5.1.2 Energy Transfer in an Electron–Nucleus Collision 143
5.1.3 Elastic Differential Cross Section for Small-Angle Scattering 146
5.1.4 Total Elastic Cross Section 152
5.2 Inelastic Scattering 153
5.2.1 Electron–Specimen Interactions with Energy Loss 153
5.2.2 Differential Cross Section for Single-Electron Excitation 156
5.2.3 Bethe Surface and Compton Scattering 158
5.2.4 Approximation for the Total Inelastic Cross Section 162
5.2.5 Dielectric Theory and Plasmon Losses in Solids 163
5.2.6 Surface-Plasmon Losses 171
5.3 Energy Losses by Inner-Shell Ionization 174
5.3.1 Position and Shape of Ionization Edges 174
5.3.2 Inner-Shell Ionization Cross Sections 177
5.3.3 Energy-Loss Near-Edge Structure (ELNES) 179
5.3.4 Extended Energy-Loss Fine Structure (EXELFS) 182
5.3.5 Linear and Circular Dichroism 183
5.4 Multiple-Scattering Effects 184
5.4.1 Angular Distribution of Scattered Electrons 184
5.4.2 Energy Distribution of Transmitted Electrons 186
5.4.3 Electron-Probe Broadening by Multiple Scattering 188
5.4.4 Electron Diffusion, Backscattering, and Secondary-Electron Emission 192

6 Scattering and Phase Contrast for Amorphous Specimens . . . 195
6.1 Scattering Contrast . . . 196
6.1.1 Transmission in the Bright-Field Mode . . . 196
6.1.2 Dark-Field Mode . . . 201
6.1.3 Examples of Scattering Contrast . . . 202
6.1.4 Improvement of Scattering Contrast by Energy Filtering . . . 205
6.1.5 Scattering Contrast in the STEM Mode . . . 208
6.1.6 Measurement of Mass Thickness and Total Mass . . . 209
6.2 Phase Contrast . . . 211
6.2.1 The Origin of Phase Contrast . . . 211
6.2.2 Defocusing Phase Contrast of Supporting Films . . . 212
6.2.3 Examples of Phase Contrast . . . 215
6.2.4 Theoretical Methods for Calculating Phase Contrast . . . 216
6.2.5 Imaging of a Scattering Point Object . . . 218
6.2.6 Relation between Phase and Scattering Contrast . . . 220
6.3 Imaging of Single Atoms . . . 221
6.3.1 Imaging of Single Atoms in TEM . . . 221
6.3.2 Imaging of Single Atoms in the STEM Mode . . . 225
6.4 Contrast-Transfer Function (CTF) . . . 228
6.4.1 The CTF for Amplitude and Phase Specimens . . . 228
6.4.2 Influence of Energy Spread and Illumination Aperture . . . 230
6.4.3 The CTF for Tilted-Beam and Hollow-Cone Illumination . . . 233
6.4.4 Contrast Transfer in STEM . . . 236
6.4.5 Phase Contrast by Inelastically Scattered Electrons . . . 237
6.4.6 Improvement of the CTF Inside the Microscope . . . 238
6.4.7 Control of the CTF by Optical or Digital Fourier Transform . . . 238
6.5 Electron Holography . . . 241
6.5.1 Fresnel and Fraunhofer In-Line Holography . . . 241
6.5.2 Single-Sideband Holography . . . 244
6.5.3 Off-Axis Holography . . . 245
6.5.4 Reconstruction of Off-Axis Holograms . . . 246
6.6 Image Restoration and Specimen Reconstruction . . . 249
6.6.1 General Aspects . . . 249
6.6.2 Methods of Optical Analog Filtering . . . 250
6.6.3 Digital Image Restoration . . . 252
6.6.4 Alignment by Cross-Correlation . . . 254
6.6.5 Averaging of Periodic and Aperiodic Structures . . . 255
6.7 Three-Dimensional Reconstruction . . . 258
6.7.1 Stereometry . . . 258
6.7.2 Electron Tomography . . . 259

6.8 Lorentz Microscopy ... 262
6.8.1 Lorentz Microscopy and Fresnel Diffraction ... 262
6.8.2 Imaging Modes of Lorentz Microscopy ... 264
6.8.3 Imaging of Electrostatic Specimen Fields ... 270

7 Theory of Electron Diffraction ... 273
7.1 Fundamentals of Crystallography ... 274
7.1.1 Bravais Lattice and Lattice Planes ... 274
7.1.2 The Reciprocal Lattice ... 279
7.1.3 Construction of Laue Zones ... 282
7.2 Kinematical Theory of Electron Diffraction ... 283
7.2.1 Bragg Condition and Ewald Sphere ... 283
7.2.2 Structure Amplitude and Lattice Amplitude ... 285
7.2.3 Column Approximation ... 289
7.3 Dynamical Theory of Electron Diffraction ... 292
7.3.1 Limitations of the Kinematical Theory ... 292
7.3.2 Formulation of the Dynamical Theory as a System of Differential Equations ... 293
7.3.3 Formulation of the Dynamical Theory as an Eigenvalue Problem ... 294
7.3.4 Discussion of the Two-Beam Case ... 298
7.4 Dynamical Theory Including Absorption ... 302
7.4.1 Inelastic-Scattering Processes in Crystals ... 302
7.4.2 Absorption of the Bloch-Wave Field ... 306
7.4.3 Dynamical n-Beam Theory ... 311
7.4.4 The Bethe Dynamical Potential and the Critical Voltage Effect ... 313
7.5 Intensity Distribution in Diffraction Patterns ... 317
7.5.1 Diffraction at Amorphous Specimens ... 317
7.5.2 Intensity of Debye–Scherrer Rings ... 318
7.5.3 Influence of Thermal Diffuse Scattering ... 321
7.5.4 Kikuchi Lines and Bands ... 323
7.5.5 Electron Spectroscopic Diffraction ... 326

8 Electron-Diffraction Modes and Applications ... 329
8.1 Electron-Diffraction Modes ... 329
8.1.1 Selected-Area Electron Diffraction (SAED) ... 329
8.1.2 Electron Diffraction Using a Rocking Beam ... 331
8.1.3 Electron Diffraction Using a Stationary Electron Probe ... 332
8.1.4 Electron Diffraction Using a Rocking Electron Probe ... 336
8.1.5 Further Diffraction Modes in TEM ... 338
8.2 Some Uses of Diffraction Patterns with Bragg Reflections ... 342
8.2.1 Lattice-Plane Spacings ... 342
8.2.2 Texture Diagrams ... 343

8.2.3 Crystal Structure . . . 345
8.2.4 Crystal Orientation . . . 347
8.2.5 Examples of Extra Spots and Streaks . . . 349
8.3 Convergent-Beam Electron Diffraction (CBED) . . . 352
8.3.1 Determination of Point and Space Groups . . . 352
8.3.2 Determination of Foil Thickness . . . 352
8.3.3 Charge-Density Distributions . . . 353
8.3.4 High-Order Laue Zone (HOLZ) Patterns . . . 354
8.3.5 HOLZ Lines . . . 355
8.3.6 Large-Angle CBED . . . 357

9 Imaging of Crystalline Specimens and Their Defects . . . 359
9.1 Diffraction Contrast of Crystals Free of Defects . . . 360
9.1.1 Edge and Bend Contours . . . 360
9.1.2 Dark-Field Imaging . . . 362
9.1.3 Moiré Fringes . . . 365
9.1.4 The STEM Mode and Multibeam Imaging . . . 367
9.1.5 Energy Filtering of Diffraction Contrast . . . 369
9.1.6 Transmission of Crystalline Specimens . . . 370
9.2 Calculation of Diffraction Contrast of Lattice Defects . . . 373
9.2.1 Kinematical Theory and the Howie–Whelan Equations . . . 373
9.2.2 Matrix-Multiplication Method . . . 375
9.2.3 Bloch-Wave Method . . . 376
9.3 Planar Lattice Faults . . . 378
9.3.1 Kinematical Theory of Stacking-Fault Contrast . . . 378
9.3.2 Dynamical Theory of Stacking-Fault Contrast . . . 379
9.3.3 Antiphase and Other Boundaries . . . 383
9.4 Dislocations . . . 385
9.4.1 Kinematical Theory of Dislocation Contrast . . . 385
9.4.2 Dynamical Effects in Dislocation Images . . . 390
9.4.3 Weak-Beam Imaging . . . 391
9.4.4 Determination of the Burgers Vector . . . 394
9.5 Lattice Defects of Small Dimensions . . . 396
9.5.1 Coherent and Incoherent Precipitates . . . 396
9.5.2 Defect Clusters . . . 398
9.6 High-Resolution Electron Microscopy (HREM) of Crystals . . . 400
9.6.1 Lattice-Plane Fringes . . . 400
9.6.2 General Aspects of Crystal-Structure Imaging . . . 402
9.6.3 Methods for Calculating Lattice-Image Contrast . . . 405
9.6.4 Simulation, Matching, and Reconstruction of Crystal Images . . . 407
9.6.5 Measurement of Atomic Displacements in HREM . . . 409
9.6.6 Crystal-Structure Imaging with a Scanning Transmission Electron Microscope . . . 411

9.7 Imaging of Atomic Surface Steps and Structures 412
9.7.1 Imaging of Surface Steps in Transmission 412
9.7.2 Reflection Electron Microscopy 416
9.7.3 Surface-Profile Imaging 418

10 Elemental Analysis by X-ray and Electron Energy-Loss Spectroscopy 419
10.1 X-ray and Auger-Electron Emission 419
10.1.1 X-ray Continuum 419
10.1.2 Characteristic X-ray and Auger-Electron Emission 421
10.2 X-ray Microanalysis in a Transmission Electron Microscope . . . 425
10.2.1 Wavelength-Dispersive Spectrometry 425
10.2.2 Energy-Dispersive Spectrometry (EDS) 427
10.2.3 X-ray Emission from Bulk Specimens and ZAF Correction 431
10.2.4 X-ray Microanalysis of Thin Specimens 434
10.2.5 X-ray Microanalysis of Organic Specimens 436
10.3 Electron Energy-Loss Spectroscopy 437
10.3.1 Recording of Electron Energy-Loss Spectra 437
10.3.2 Kramers–Kronig Relation 439
10.3.3 Background Fitting and Subtraction 441
10.3.4 Deconvolution 442
10.3.5 Elemental Analysis by Inner-Shell Ionizations 444
10.4 Element-Distribution Images 447
10.4.1 Elemental Mapping by X-Rays 447
10.4.2 Element-Distribution Images Formed by Electron Spectroscopic Imaging 448
10.4.3 Three-Window Method 449
10.4.4 White-Line Method 450
10.4.5 Correction of Scattering Contrast 450
10.5 Limitations of Elemental Analysis 452
10.5.1 Specimen Thickness 452
10.5.2 Radiation Damage and Loss of Elements 452
10.5.3 Counting Statistics and Sensitivity 453
10.5.4 Resolution and Detection Limits for Electron Spectroscopic Imaging 456

11 Specimen Damage by Electron Irradiation 459
11.1 Specimen Heating 459
11.1.1 Methods of Measuring Specimen Temperature 459
11.1.2 Generation of Heat by Electron Irradiation 461
11.1.3 Calculation of Specimen Temperature 463
11.2 Radiation Damage of Organic Specimens 466
11.2.1 Elementary Damage Processes in Organic Specimens . . . 466
11.2.2 Quantitative Methods of Measuring Damage Effects . . . 470

11.2.3 Methods of Reducing Radiation Damage 477
11.2.4 Radiation Damage and High Resolution 479
11.3 Radiation Damage of Inorganic Specimens 480
11.3.1 Damage by Electron Excitation 480
11.3.2 Radiation Damage by Knock-On Collisions 482
11.4 Contamination 484
11.4.1 Origin and Sources of Contamination 484
11.4.2 Methods for Decreasing Contamination 485
11.4.3 Dependence of Contamination on Irradiation Conditions 486

References 491

Index 575

1

Introduction

1.1 Transmission Electron Microscopy

1.1.1 Conventional Transmission Electron Microscopy

In a conventional transmission electron microscope (CTEM, or TEM for short) (Fig. 1.1), a thin specimen is irradiated with an electron beam of uniform current density. The acceleration voltage of routine instruments is 100–200 kV. Medium-voltage instruments work at 200–500 kV to provide better transmission and resolution, and in high-voltage electron microscopy (HVEM) the acceleration voltage reaches 500 kV–3 MV. Earlier books on the subject are listed as references [1.1–1.55]. The development of both theory and instrumentation as well as the different applications of TEM can be followed by consulting the proceedings of the International Conferences on Electron Microscopy [1.56–1.68].

Electrons are emitted in the electron gun by thermionic, Schottky, or field emission. The latter are used when high gun brightness and coherence are needed. A three- or four-stage condenser-lens system permits variation of the illumination aperture and the area of the specimen illuminated. The electron-intensity distribution behind the specimen is imaged with a lens system, composed of three to eight lenses, onto a fluorescent screen. The image can be recorded by direct exposure of a photographic emulsion or an image plate inside the vacuum, or digitally via a fluorescent screen coupled by a fiber-optic plate to a CCD camera.

Electrons interact strongly with atoms by elastic and inelastic scattering. The specimen must therefore be very thin, typically of the order of 5–100 nm for 100 keV electrons, depending on the density and elemental composition of the object and the resolution desired. Special preparation techniques are needed for this; electropolishing and ion-beam etching in materials science and ultramicrotomy of stained and embedded tissues or cryofixation in the biosciences.

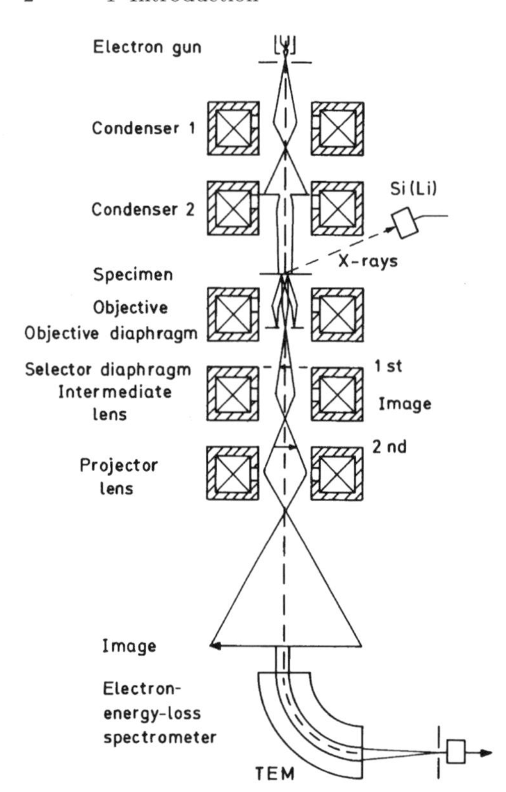

Fig. 1.1. Schematic ray path for a transmission electron microscope (TEM) equipped for additional x-ray and electron energy-loss spectroscopy.

The aberrations of the objective lens are so great that it is necessary to work with very small objective apertures, of the order of 10–25 mrad, to achieve a resolution of the order of 0.1–0.3 nm. Bright-field contrast is produced either by intercepting the electrons scattered through angles larger than the objective aperture (scattering contrast) or by interference between the scattered wave and the incident wave at the image point (phase contrast). The phase of the electron waves behind the specimen is modified by the wave aberration of the objective lens. This aberration and the energy spread of the electron gun, which is of the order of 0.3–2 eV, limit the contrast transfer of high spatial frequencies. Dark-field contrast is obtained by tilting the primary beam or by hollow-cone illumination so that the primary beam falls on the objective diaphragm.

In crystalline specimens, the use of the primary beam (bright field) or a Bragg-reflected beam on-axis (dark field) gives rise to diffraction contrast, which is important for the imaging of crystal defects. When Bragg-diffracted beams also pass through the aperture, crystal-structure imaging reveals projections of atomic rows. For the interpretation of these images, digital image simulation using the dynamical theory of electron diffraction is indispensable.

A further capability of modern TEM is the formation of nanometer-sized electron probes, 0.2–10 nm in diameter, by means of a three- or four-stage condenser-lens system, the last lens field of which is the objective prefield in front of the specimen. The main applications of such electron probes are in analytical electron microscopy (see below). This enables the instrument to operate in the scanning transmission (STEM) mode with a resolution determined by the electron-probe diameter; this has advantages for imaging thick specimens and for recording secondary electrons and backscattered electrons.

1.1.2 High-Resolution Electron Microscopy

The wave-optical theory of imaging is necessary to discuss high resolution. This theory can be expressed in terms of a two-stage Fourier transform. In the focal plane of the objective lens, the diffraction pattern of the specimen is formed; each scattering angle θ corresponds reciprocally to a periodic spacing Λ in the specimen, or in other words is proportional to a spatial frequency $q = 1/\Lambda$ since $\theta \simeq \lambda/\Lambda = \lambda q$ (λ : electron wavelength). The amplitude distribution $F(q)$ of the electron wave in the focal plane is the Fourier transform of the specimen transparency. The spherical aberration can be represented as a wave aberration, which is an additional phase shift that depends on scattering angle, the spherical-aberration constant C_{s}, and the defocusing Δz. This phase shift can be introduced as a phase factor applied to $F(q)$. The image amplitude is then the inverse Fourier transform of this weighted Fourier transform, in which the influences of the diaphragm, the finite illumination aperture (partial spatial coherence), and the energy spread of the electron gun (partial temporal coherence) can be included. The result may be expressed in terms of a contrast-transfer function for the different spatial frequencies. This transfer function is important because it characterizes the effect of the instrument on image formation and is independent of the particular specimen in question.

Transmission electron microscopy can provide high resolution [1.69, 1.70] because elastic scattering is an interaction process that is highly localized to the region occupied by the screened Coulomb potential of an atomic nucleus. The angular distribution of inelastically scattered electrons is concentrated within smaller scattering angles than that of elastically scattered electrons. Most of the inelastically scattered electrons normally pass through the objective diaphragm in the bright-field mode. Inelastically scattered electrons do not, however, contribute to high-resolution image details because the inelastic scattering is less localized. With increasing energy loss, the localization becomes narrower for inner-shell ionization, and resolutions of lattice periodicities of about 0.3–0.5 nm are possible with energy-filtering transmission electron microscopy.

The spherical-aberration coefficients C_{s} in present-day microscopes are about 0.5–2 mm. The optimum imaging condition in bright-field mode occurs at the Scherzer defocus $\Delta z = (C_{\mathrm{s}}\lambda)^{1/2}$, for which a broad band of spatial frequencies is imaged with positive phase contrast. This band has an upper limit at q_{max}. The value $\delta_{min} = 1/q_{max} = 0.67(C_s\lambda^3)^{1/4}$ is often used to define a limit of resolution, though it is not correct to characterize resolution by one number only. For $C_s = 1$ mm and $E = 100$ keV ($\lambda = 3.7$ pm), we find $\Delta z \simeq 60$ nm and $\delta_{min} = 0.32$ nm. Narrow bands of higher spatial frequencies can be imaged if the image is not blurred by imperfect spatial and temporal coherence. These effects limit the resolution of conventional microscopes to 0.15–0.3 nm and $\simeq$0.1 nm for crystal-structure imaging has been approached in a 1 MeV instrument.

The efforts of the last few years to increase resolution have been concentrated on using a Schottky or field-emission gun to decrease the damping of the contrast-transfer function at high spatial frequencies caused by partial spatial and temporal coherence. Normal TEMs equipped with thermionic cathodes work with illumination apertures α_i of about 0.1 mrad; with a Schottky or field-emission gun, apertures smaller than 10^{-2} mrad are possible. The energy spread $\Delta E = 1$–2 eV of a thermionic gun can be reduced to 0.3–0.6 eV with a Schottky or field-emission gun.

Using such guns, the resolution can be improved up to the information limit, which is determined by the spatial and temporal coherence rather than by the spherical-aberration constant. There are three routes to obtain a resolution at the information limit

1. Use a focal series combined with a reconstruction algorithm.
2. Improve holography, which was originally devised by Gabor in 1949 in the hope of overcoming the resolution limit imposed mainly by spherical aberration. With the development of the laser, a light source of high coherence, holography rapidly grew into a major branch of light optics. Holography has attracted renewed interest in electron optics with the development of field-emission or Schottky guns of high brightness and coherence. Apart from the attainment of better resolution than in the conventional bright-field mode, holography is becoming of increasing interest for quantitative studies of phase shifts [1.71–1.73].
3. Correct the spherical-aberration coefficient C_s by using multipole lens systems, so that the first zero of the phase-contrast transfer function is moved to spatial frequencies beyond 10 nm^{-1} [1.74].

High-resolution micrographs of specimens on supporting films are disturbed by a phase-contrast effect that creates defocus-dependent granularity. One way of reducing this granularity is to use hollow-cone illumination, which suppresses the granularity but does not destroy contrast arising from specimen structures containing heavy atoms. Furthermore, the contrast transfer does not show sign reversal with this type of illumination.

A further obstacle to obtaining high-resolution images of organic specimens is the radiation damage caused by ionization and subsequent breakage of chemical bonds and finally by a loss of mass. The radiation damage depends on the electron dose in C cm^{-2} (charge density) incident on the specimen. A dose of 1 C cm^{-2} corresponds to 6×10^4 electrons per nm^2, the value needed to form an image free of statistical noise at high magnification. Most amino-acid molecules are destroyed at doses of 10^{-2} C cm^{-2}, and only a few compounds, such as hexabromobenzene and phthalocyanine and related substances, can be observed at doses of the order of a few C cm^{-2}. The deterioration and mass loss can be reduced in various ways.

1. The specimen may be cooled to liquid-helium temperature. However, the ionization products are only frozen-in, and the primary ionization damage will be the same as at room temperature. Only those secondary radiation effects that are caused by loss of mass are appreciably reduced.
2. The electron dose may be kept very low, which produces a noisy image. The noise can be decreased by signal averaging, which is straightforward for periodic structures. Nonperiodic structures have to be aligned and superposed by correlation techniques. This technique is used especially for the tomography of biomacromolecules, where a resolution $\geq$1 nm can be reached reliably.

1.1.3 Analytical Electron Microscopy

The strength of TEM is that not only can it provide high-resolution images that contain information down to 0.1–0.2 nm but can also operate with small electron probes in various microanalytical modes with a spatial resolution of 0.2–100 nm [1.75–1.84].

X-Ray Microanalysis. X-ray microanalysis [1.75–1.77] in TEM mainly relies on energy-dispersive Si(Li) or highly pure germanium detectors, though instruments have been constructed with wavelength-dispersive spectrometers, as used in x-ray microanalyzers. The energy-dispersive Si(Li) detector with a resolution of $\Delta E_x \approx 150$ eV of x-ray quantum energy $E_x = h\nu$ has the disadvantages that neighboring characteristic lines are less well separated and the analytical sensitivity is poorer than in a wavelength-dispersive spectrometer; this is counterbalanced by the fact that all lines with quantum energies E_x greater than 0.2 keV can be recorded simultaneously, even at the low probe currents used in the TEM. Reliable quantitative information concerning elemental composition is provided because the x-ray signal generated by thin films needs only small corrections.

X-ray production in thin foils is confined to the small volume excited by the electron probe, only slightly broadened by multiple scattering. Better spatial resolution is therefore obtainable for segregation effects at crystal interfaces or precipitates, for example, than in an x-ray microanalyzer with bulk specimens, where the spatial resolution is limited to 0.1–1 μm by the diameter of the electron-diffusion cloud.

Electron Energy-Loss Spectroscopy. An electron energy-loss spectrum (EELS) can be recorded either with a magnetic prism spectrometer behind the final image or with an imaging energy filter inside the column of the microscope [1.78, 1.79]. With a CCD array, a large range of energy losses can be recorded in parallel. Because the inelastically scattered electrons are concentrated in small angles, a large fraction of the inner-shell ionizations can be collected by the spectrometer, whereas the collection efficiency of x-rays is much smaller due to the low fluorescence yield, the isotropic emission, and the small solid angle of the detector. Electron energy-loss spectrsocopy can therefore be superior for elemental analysis when recording in parallel by means of a CCD array; a disadvantage is that the background is larger than in x-ray spectra.

The low-loss region with energy losses $\Delta E \leq 50$ eV contains the plasmon losses and interband transitions, which are related by the dielectric theory to the optical constants. At higher energy losses, the inner-shell ionization processes result in sawtooth-like or delayed edges, which can be used for elemental analysis.

The ionization edges contain an energy-loss near-edge structure (ELNES) that contains information about the bonding and band structure of solids in a range of about 50 eV beyond the edge. An extended energy-loss fine structure (EXELFS) continuing to a few hundred electron volts beyond the edge furnishes information about the coordination of neighboring atoms.

Electron Diffraction. Information about crystal structure and orientation is provided by the electron-diffraction pattern [1.80–1.83]. The possibility of combining electron diffraction and the various imaging modes is the most powerful feature of TEM for the investigation of the crystal lattice and its defects in crystalline material. With the selected-area electron-diffraction technique, it is possible to switch from one mode to another simply by changing the excitation of the diffraction or intermediate lens and to select the diffraction pattern from areas 0.1–1 μm in diameter. Other modes of operation that permit electron-diffraction patterns to be obtained from small areas can be used when the instrument is capable of forming an electron probe 1–20 nm in diameter. In most cases, crystals are free of defects in such a small area and so convergent-beam electron diffraction (CBED) techniques can be applied. In particular, the appearance of Kikuchi lines in the convergent primary beam provides much additional information about crystal structure and defects. A high-order Laue zone pattern of large aperture, of the order of 10°, can be used for the three-dimensional reconstruction of the lattice because the Ewald sphere intersects high-order Laue zones in circles of large diameter. Convergent-beam electron diffraction patterns allow the determination of the space group of the crystal. Furthermore, the lattice constants and the Fourier coefficients V_g of the lattice potential can be measured accurately, and these can be used to calculate charge-density distributions inside the unit cell. Large-angle CBED patterns (LACBED) are used to investigate lattice defects or strains and misfits in multilayers.

1.1.4 Energy-Filtering Electron Microscopy

Energy-filtered images or diffraction patterns can be obtained either with an imaging spectrometer below the final screen or with an energy filter inside the column [1.84]. Zero-loss filtering allows us to remove the background of inelastically scattered electrons, which results in a considerable increase of contrast. Plasmon-loss filtering can be used for the analysis of different phases and precipitates. The contrast can also be enhanced or reversed by placing energy windows at a few hundred electron volts; for biological specimens, just below the carbon K edge at 285 eV. A three-window method with two images below and one image just beyond the ionization edge of an element allows us to extrapolate the background and to subtract the background in the third image pixel per pixel, which results in an element distribution image.

Zero-loss filtering of diffraction patterns allows a better comparison with the dynamical theory of electron diffraction to be made, and the structure amplitudes can be measured quantitatively in convergent-beam electron diffraction patterns. The Bragg-diffraction spots become diffuse with increasing energy loss, and at large energy losses the filtered pattern consists of excess or defect Kikuchi bands.

The method of angle-resolved EELS shows the intensity distribution as a function of scattering angle and energy loss along a stripe in the diffraction pattern. The recorded diagram contains the plasmon losses and their dispersion, the Compton scattering (Bethe ridge), and the ionization edges of the elements.

1.1.5 High-Voltage Electron Microscopy

For acceleration voltages higher than 500 kV, the high voltage must be generated in a tank on top of the microscope, typically filled with SF_6 at a pressure of a few bars, which decreases the critical distance for electrical breakdown. The high voltage is applied to a cascade of acceleration electrodes, with only 50–100 kV between neighboring rings. The structure occupies a considerable space, and the column of an HVEM is also large because the yokes of the electron lens must be scaled up to avoid magnetic saturation. A building some 10–15 m high is therefore needed to house an HVEM. For this reason, the present trend is more toward microscopes with acceleration voltages in the range 200–400 kV and with high resolution, which can be housed in normal rooms. In the following, we summarize some advantages of HVEM (for more details, see the review articles and special conferences on HVEM [1.85–1.91]).

Increased Useful Specimen Thickness. The investigation of thick specimens is limited by the full width ΔE of the energy-loss spectrum because the chromatic aberration of the objective lens blurs image points into image patches of width $C_c \alpha_o \Delta E/E$, where $C_c \simeq 0.5$–2 mm is the chromatic-aberration coefficient and α_o is the objective aperture. The decrease of the ratio $\Delta E/E$ markedly reduces the effect of chromatic aberration and allows thicker specimens to be investigated. Biological sections, for example, which can be observed in a 100 keV TEM only if their thicknesses are less than 200 nm, can be studied in a 1 MeV TEM with thicknesses as great as 1 μm. The investigation of whole cells and microorganisms by stereo pairs helps to establish the three-dimensional structure and the function of fibrillar and membranous cell components. At 100 kV, such large structures can be reconstructed only by analyzing serial sections.

Many ceramics and minerals are difficult to prepare in thin enough layers for 100 kV microscopy but can be studied by HVEM. An increase of useful thickness is also observed for metal foils, which can additionally show typical orientations for best transmission (10 μm silicon or 2 μm iron at 1 MV). This has two important advantages. Normally, areas thin enough at 100 kV are concentrated at edges; at 1 MV, the transparent area increases to nearly the whole specimen area. Secondly, the thicker parts of the specimen are more representative of the bulk material, an important point for dynamical experiments such as mechanical deformation, annealing, in situ precipitation, and environmental experiments.

Easier Specimen Manipulation. The polepiece gap of the objective lens is of the order of millimeters in 100 kV instruments and centimeters in HVEM; this extra space makes it a great deal easier to install complicated specimen stages or goniometers for heating, cooling, or stretching. Higher partial pressures of gases at the specimen controlled by using a differentially pumped system of diaphragms can be tolerated for environmental experiments. Similarly, organic specimens can be investigated in the native state with a partial pressure of water.

Radiation-Damage Experiments. For threshold energies of a few hundred keV, depending on the displacement energy $E_{\mathrm{d}} \simeq 20$–50 eV and the mass of the nuclei, energy losses greater than E_{d} can be transferred to the nuclei by elastic large-angle scattering; the nucleus is then knocked from its position in the crystal lattice to an interstitial site, for example. High-voltage electron microscopy thus becomes a powerful tool for the in situ study of irradiation processes and the kinetics of defect agglomeration. In normal operation, however, the current density can be kept low so that the specimen can be investigated over a reasonable time without damage.

Incorporation of Analytical Modes. At 100 kV, electron energy-loss spectroscopy is restricted to specimen thicknesses of the order of the mean free path for plasmon losses (10–30 nm) because the ionization edges are blurred by the low-energy part of the loss spectrum. Although the mean free path saturates at high energies, an increase of about a factor of 3 can be observed for 1 MV. In x-ray microanalysis, the x-ray continuum decreases owing to the pronounced forward bias of the emission of continuous x-ray quanta at higher energies.

Electron-diffraction analysis can be applied to thicker crystals because the dynamical absorption distance increases as the square of the velocity.

Many-beam dynamical theory has to be applied even for thin foils because the Ewald sphere is now large and many more Bragg reflections are excited simultaneously when a sample is irradiated near a low-index zone axis.

High Resolution. The relation between voltage (wavelength) and resolution was already discussed in Sect. 1.1.2. A notable feature is that many-beam imaging of the crystal structure is used to better advantage. Optimum results in crystal-lattice imaging are obtained with increasing acceleration voltage, and 0.1 nm resolution has been achieved with a 1 MV instrument. For organic material, the decrease of ionization probability (radiation damage) with increasing energy provides a gain of only a factor of 3 between 100 and 1000 kV. For thin specimens, however, the contrast in the image decreases by the same factor. The best images of organic crystals such as phthalocyanine have been obtained with an HVEM in the range 500–700 kV.

1.1.6 Dedicated Scanning Transmission Electron Microscopy

A dedicated STEM consists only of a field-emission gun, one probe-forming lens, and the electron-detection system, together with an electron spectrometer for electron energy-loss spectroscopy (EELS) and for separating the currents of unscattered and elastically scattered electron and inelastically scattered electrons (Fig. 1.2) [1.92–1.95]. The specimen is scanned by deflection coils in synchrony with the imaging TV tube. The whole column including the

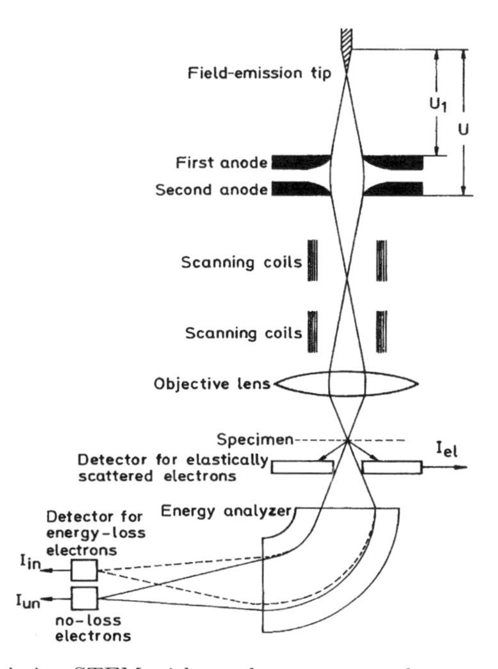

Fig. 1.2. Field-emission STEM with an electron energy-loss spectrometer.

specimen is under an ultrahigh vacuum. Electron-probe diameters of 0.2–0.5 nm can be formed, the spherical and chromatic aberrations of the lens being the limiting factors.

An advantage of STEM instruments is that the contrast can be enhanced by collecting several signals simultaneously and displaying differences and/or ratios of these by analog or digital processing. In particular, single atoms on a thin substrate can be imaged with a higher contrast than in the CTEM bright- or dark-field modes. An incoherent dark-field mode allows a high-resolution image of the crystal lattice to be formed, the contrast increasing with increasing atomic number. The irradiation of the specimen area can be reduced to a minimum in order to decrease radiation damage.

1.2 Alternative Types of Electron Microscopy

Although the main concern of this book is transmission electron microscopy, the function and limits of the other types of electron microscopes are also mentioned in this introductory chapter to show the advantages and disadvantages of their various imaging techniques. Several types of electron microscopes and analyzing instruments capable of furnishing an "image" can be distinguished. We now examine these briefly in turn, without considering the historical sequence in which these instruments were developed.

1.2.1 Emission Electron Microscopy

In an emission electron microscope [1.96–1.105], the cathode that emits the electrons is directly imaged by an electrostatic immersion lens, which accelerates the electrons and produces an intermediate image of the emission-intensity distribution at the cathode. This image can be magnified by further lenses and is observed on a fluorescent screen or with an image intensifier. The cathode (specimen) has to be planar and its surface should not be irregular. The electron emission can be stimulated by

1. heating the cathode (thermionic emission), which means that observation is possible only at elevated temperatures and for a limited number of materials, or alternatively, the cathode temperature need not be raised beyond 500°C–1000°C if a thin layer of barium is evaporated on the surface because this lowers the work function;
2. secondary-electron excitation by particle bombardment or by irradiating the cathode surface with a separate high-energy electron beam or an ion beam at grazing incidence; or
3. irradiation of the cathode with ultraviolet light to excite photoelectrons (using a photoelectron-emission microscope, PhEEM).

These instruments have a number of interesting applications, but their use is limited to particular specimens; at present, therefore, scanning electron microscopes and scanning tunneling microscopes and their variants are the most widely used instruments for imaging bulk specimens, especially because there is no need to limit the roughness of the specimen surface. The final restriction is the limited number of electrons emitted, which limits the image intensity at high magnification, and moreover the resolution of the immersion-lens system is only of the order of 10–30 nm. On the credit side, surfaces can be observed directly in situ, and each of the processes 1–3 generates a specific contrast. The photoelectron-emission microscope has the advantage of being applicable to nearly any flat specimen surface, including biological specimens. The image contrast is caused by differences of the emission intensity (material and crystal orientation contrast) and by angular selection with a diaphragm that intercepts electrons whose trajectories have been deflected by variations of the equipotentials near the surface caused by surface steps (topographic contrast), surface potentials (potential contrast), or magnetic stray fields (magnetic contrast). Investigations based on photoemission in combination with Auger-electron spectroscopy in an ultrahigh vacuum [1.100] are of special interest for surface physics. With improved access to synchrotron-radiation sources, PhEEM is developing into a versatile analytical tool in surface and materials science [1.104, 1.105].

1.2.2 Reflection Electron Microscopy

The electrons that emerge from a specimen as a result of primary-electron bombardment are either low-energy secondary electrons, which can be used in an emission microscope (see above) or a scanning electron microscope (see below), or primary (backscattered) electrons with large energy losses, which cannot be focused sharply by an electron lens because of the chromatic aberration. However, imaging of the surface is possible for a grazing electron incidence below 10°, the "reflected" electrons being imaged with an objective lens [1.106–1.109]. The energy-loss spectrum of the reflected electrons has a half-width of the order of 100–200 eV. With additional energy selection by means of an electrostatic filter lens, a resolution of 10–20 nm has been attained. Because the angle of incidence is so low, small image steps can be imaged with high contrast. The angular distribution of the electrons reflected at single crystals is a reflection high-energy electron diffraction (RHEED) pattern with Bragg-diffraction spots; images exhibiting crystallographic contrast can be formed by selecting individual Bragg spots. A TEM equipped with the appropriate specimen holder can be operated in this mode by tilting the incident beam and with the reflected electrons on the axis of the objective lens. This reflection electron microscopy (REM) mode in the TEM has become a powerful tool for the investigation of the surface structure of crystals, especially with additional energy filtering (Sect. 9.7.2) [1.109].

1.2.3 Mirror Electron Microscopy

An electron beam is deflected by a magnetic sector field and retarded and reflected at a flat specimen surface that is biased a few volts more negative than the cathode of the electron gun. The reflected electron trajectories are influenced by irregularities of the equipotential surfaces in front of the specimen, which may be caused by surface roughness or by potential differences and specimen charges; magnetic stray fields likewise act on the electron trajectories [1.110]. An advantage of this technique is that the electrons do not strike the specimen; it is the only method that permits surface charges to be imaged undisturbed. After passing through the magnetic sector field again, the electrons can be selected according to their angular deflection. A new design [1.111] of mirror electron microscope has a resolution of the order of 4 nm. Single surface steps, 5 nm in height, can produce discernible contrast. Such a mirror electron microscope can be combined with an electron interferometer, which offers the possibility of measuring phase shifts by the equipotentials or magnetic stray fields with high precision. There are types of scanning mirror electron microscopes [1.112, 1.113] that allow the relation between the observed image point and the local deflection to be established more quantitatively.

1.2.4 Scanning Electron Microscopy

The SEM is the most important electron-optical instrument for the investigation of bulk specimens [1.114–1.123]. An electron probe is produced by two- or three-stage demagnification of the smallest cross section of the electron beam after acceleration. This electron probe, 2–10 nm in diameter, is scanned in a raster over a region of the specimen (Fig. 1.3). The smallest diameter of the

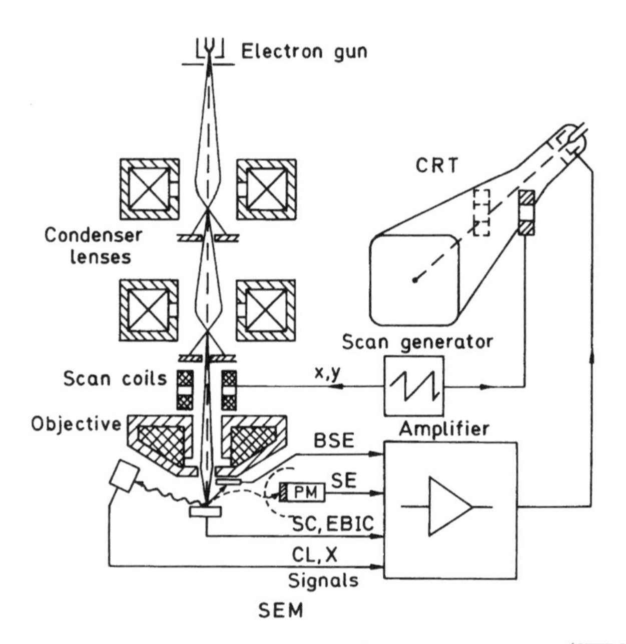

Fig. 1.3. Schematic ray path for a scanning electron microscope (SEM).

electron probe is limited by the minimum acceptable probe current, which lies in the range 10^{-12}–10^{-11} A. This value is determined by the need to generate an adequate signal-to-noise ratio and by the spherical and chromatic aberrations of the final probe-forming lens. The image is displayed on a cathode-ray tube (CRT) scanned in synchrony. The CRT beam intensity can be modulated by any of the different signals that result from the electron-specimen interactions.

The most important signals are those produced by secondary electrons (SE) with most probable exit energies of 2–5 eV and by backscattered electrons (BSE) with energies that range from the energy of the primary electrons to about 50 eV. The secondary-electron yield and the backscattering coefficient depend on the angle of electron incidence (topographic contrast), the mean atomic number (material contrast), the crystal orientation (channeling contrast), and electrostatic and magnetic fields near the surface (potential and magnetic contrast). A signal can also be produced by the specimen current and by electron-beam-induced currents in semiconductors. Analytical information is available from the x-ray spectrum and Auger electrons or from light quanta emitted by cathodoluminescence. The crystallographic structure and orientation can be obtained from electron channeling patterns, electron-backscattering patterns, and x-ray Kossel diagrams. An environmental SEM can work with a high partial pressure between the specimen and the objective-lens diaphragm.

The resolutions of the different modes of operation and types of contrast depend on the information volume that contributes to the signal. Secondary electrons provide the best resolution because the exit depth is very small, of the order of a few nanometers. The information depth of backscattered electrons is much greater, of the order of half the electron range, which is as much as 0.1–1 μm, depending on the density of the specimen and the electron energy. The secondary electron signal also contains a large contribution from the backscattered electrons when these penetrate the surface layer. At higher energies, the electron range and the diameter of the electron-diffusion region are greater. Conversely, higher energies are of interest for x-ray microanalysis if K shells of heavy elements are to be excited. The progress in Schottky and field-emission gun design has increased the gun brightness at low electron energies, too, so that low-voltage scanning electron microscopy (LVSEM) [1.123] in the range 0.5–5 keV is attracting interest because information can be extracted from a volume nearer to the surface.

Unlike in TEM, special specimen-preparation methods are rarely needed in SEM. Nevertheless, charging effects have to be avoided by coating a non-conductive specimen with a thin conductive film, for example, and organic specimens have to be protected from surface distortions by chemical fixation or cryo-fixation.

1.2.5 X-ray and Auger-Electron Microanalysis

By using a wavelength-dispersive x-ray spectrometer (Bragg reflection at a crystal), we can work with high x-ray excitation rates and electron-probe currents of the order of 10^{-8}–10^{-7} A, though the electron-probe diameter is then larger, about 0.1–1 μm. The main task of an x-ray microanalyzer [1.124–1.132] is to analyze the elemental composition of flat, polished surfaces at normal incidence with a high analytical sensitivity. The ray diagram of such an instrument is similar to that of an SEM, but two or three crystal spectrometers that can simultaneously record different characteristic x-ray wavelengths are attached to the column. The surface can be imaged by one of the SEM modes to select the specimen points to be analyzed.

An SEM or x-ray microanalyzer can be equipped with an Auger-electron spectrometer of the cylindrical mirror type, for example. It is then necessary to work with an ultrahigh vacuum in the specimen chamber because Auger electrons are extremely sensitive to the state of the surface: A few atomic layers are sufficient to halt them. Special Auger-electron microanalyzers have therefore been developed in which the 1–10 keV electron gun may, for example, be incorporated in the inner cylinder of a spectrometer. This type of instrument can also work in the scanning mode, or an element-distribution map can be generated using Auger electrons.

1.2.6 Scanning-Probe Microscopy

The scanning tunneling microscope (STM, Fig. 1.4) [1.133–1.137] uses a tungsten tip of small radius like that of a field-emission gun. When the tip, negatively biased by a few tenths of a volt (U_{T}), approaches the conductive surface

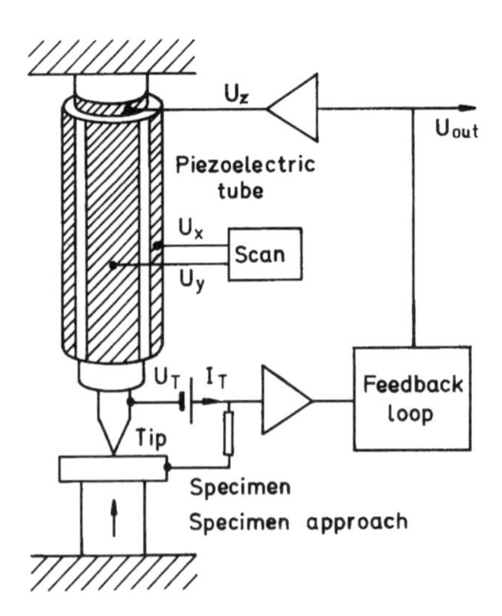

Fig. 1.4. Scanning tunneling microscope (STM) with a mechanical approach to the specimen, a piezo electric tube for x and y scanning, a z shift, and a feedback loop to keep the tunneling current I_{T} at a constant level.

in vacuum, air, or even a liquid at a distance below one nanometer, the quantum-mechanical tunneling effect causes a current I_{T} to flow through the barrier. The tunneling starts at the atom of the tip that is nearest to the surface, and it is possible to record the arrangement of single atoms and monoatomic steps on surfaces. The current depends on the distance between tip and surface, but it is convenient to maintain the current, and hence the distance, constant by moving the tip normal to the surface as it scans over the latter. This vertical movement is achieved by means of a piezoelectric transducer, that is, by the voltage U_z at the inner electrode of a piezoelectric tube. The scanning motion in the x and y directions is likewise effected by a crossed pair of outer electrodes. The voltage U_z is a measure of the local specimen height and can be used to modulate a CRT tube scanned in synchrony. This results in a very simple and compact microscope with atomic resolution.

Insulating specimens can be observed in the atomic force mode. The tip is mounted on an elastic ribbon (cantilever), which is deformed by the force between tip and specimen. The elastic deformation, on the order of nanometers, is recorded by a second tip or by reflection of a laser beam at the cantilever. Related scanning-probe methods are scanning near-field optical, acoustic, and thermal microscopies as well as capacitance, electrochemical, and micropipette scanning microscopies.

This wealth of additional modes, the atomic resolution of STM, and the possibility of direct surface profiling are the striking advantages of scanning-probe microscopy, which is, however, restricted to the imaging and analysis of surfaces. On the contrary, TEM mainly gives information about the bulk structure, including the high resolution of atomic rows in crystals. The analytical modes of x-ray microanalysis, electron energy-loss spectroscopy, and electron diffraction supplement this, though the specimens have to be prepared as thin films. Surface information can also be obtained by various surface-sensitive methods, though scanning-probe microscopy is superior. The two techniques, TEM and STM/AFM, should be regarded as complementary, and scanning-probe microscopists should take more notice of the advantages and results of TEM.

2

Particle Optics of Electrons

The acceleration of electrons in the electrostatic field between cathode and anode, the action of magnetic fields with axial symmetry as electron lenses, and the application of transverse magnetic and electrostatic fields for electron-beam deflection and electron spectroscopy can be analyzed by applying the laws of relativistic mechanics and hence calculating electron trajectories. Lens aberrations can likewise be introduced and evaluated by this kind of particle optics. In the case of spherical aberration, however, it will also be necessary to express this error in terms of a phase shift, known as the wave aberration, by using the wave-optical model introduced in the next chapter.

2.1 Acceleration and Deflection of Electrons

2.1.1 Relativistic Mechanics of Electron Acceleration

The relevant properties of an electron in particle optics are the rest mass m_0 and the charge $-e$ (Table 2.1). In an electric field $\boldsymbol{E}$ and a magnetic field $\boldsymbol{B}$, electrons experience the Lorentz force

$$\boldsymbol{F} = -e\,(\boldsymbol{E} + \boldsymbol{v} \times \boldsymbol{B}). \tag{2.1}$$

Inserting (2.1) in Newton's law

$$m\,\ddot{\boldsymbol{r}} = \boldsymbol{F} \tag{2.2}$$

yields the laws of particle optics.

We start with a discussion of the acceleration of an electron beam in an electron gun. Electrons leave the cathode of the latter as a result of thermionic or field emission (see Sect. 4.1 for details). The cathode is held at a negative potential $\Phi_\mathrm{C} = -U$ (U: acceleration voltage) relative to the anode, which is grounded, $\Phi_\mathrm{A} = 0$ (Fig. 2.1). The Wehnelt electrode of a thermionic gun, maintained at a potential $\Phi_\mathrm{W} = -(U + U_\mathrm{W})$, limits the emission to a small area around the cathode tip. Its action will be discussed in detail in Sect. 4.1.4.

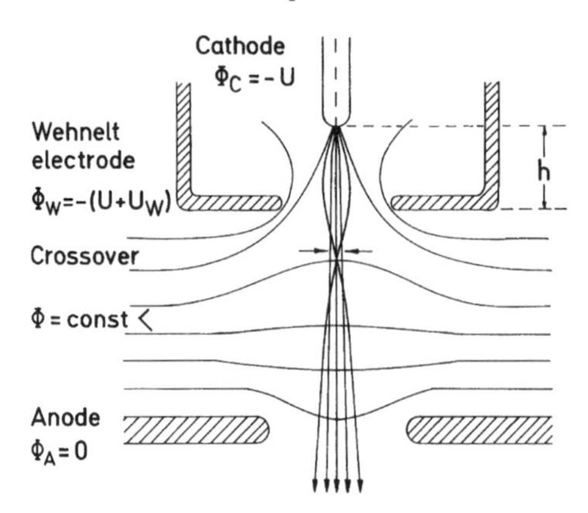

Fig. 2.1. Electron acceleration, trajectories, and equipotentials (Φ = const) in a triode electron gun.

The electrode potentials create an electric field $\boldsymbol{E}$ in the vacuum between cathode and anode, which can also be characterized by equipotentials Φ = const (Fig. 2.1). The electric field is the negative gradient of the potential

$$\boldsymbol{E} = -\nabla\Phi = -\left(\frac{\partial\Phi}{\partial x}, \frac{\partial\Phi}{\partial y}, \frac{\partial\Phi}{\partial z}\right). \tag{2.3}$$

The existence of a potential implies that the force $\boldsymbol{F} = -e\boldsymbol{E}$ is conservative and that the law of energy conservation

$$E + V = \text{const} \tag{2.4}$$

can be applied, as will be demonstrated by considering the electron acceleration in Fig. 2.1. The kinetic energy at the cathode is $E = 0$, whereas the potential energy V is zero at the anode. The potential energy at the cathode can be obtained from the work W that is needed to move an electron from the anode to the cathode against the force $\boldsymbol{F}$:

$$\begin{aligned} V = -W &= -\int_{\mathrm{A}}^{\mathrm{C}} \boldsymbol{F}\cdot \mathrm{d}\boldsymbol{s} = e\int_{\mathrm{A}}^{\mathrm{C}} \boldsymbol{E}\cdot \mathrm{d}\boldsymbol{s} = -e\int_{\mathrm{A}}^{\mathrm{C}} \nabla\Phi\cdot \mathrm{d}\boldsymbol{s} \\ &= -e(\Phi_{\mathrm{C}} - \Phi_{\mathrm{A}}) = eU. \end{aligned} \tag{2.5}$$

In the reverse direction, the electrons acquire this amount eU of kinetic energy at the anode. This implies that the gain of kinetic energy $E = eU$ of an accelerated electron depends only on the potential difference U, irrespective of the real trajectory between cathode and anode.

Relation (2.5) can also be used to define the potential energy $V(\boldsymbol{r})$ at each point $\boldsymbol{r}$ at which the potential is $\Phi(\boldsymbol{r})$:

$$V(\boldsymbol{r}) = -e\Phi(\boldsymbol{r}). \tag{2.6}$$

However, an arbitrary constant can be added to $V(\boldsymbol{r})$ or $\Phi(\boldsymbol{r})$ without changing the electric field $\boldsymbol{E}$ because the gradient of a constant in (2.3) is zero. We arbitrarily assumed $\Phi_A = 0$ in the special case discussed above, and the results do not change if we assume that $\Phi_C = 0$ and $\Phi_A = +U$, for example.

An electron acquires the kinetic energy $E = 1.602 \times 10^{-19}$ J if accelerated through a potential difference $U = 1$ V because in SI units

$$1\ \mathrm{C\ V} = 1\ \mathrm{A\ V\ s} = 1\ \mathrm{W\ s} = 1\ \mathrm{J}.$$

This energy of 1 eV = 1.602×10^{-19} J is used as a new unit and is called "one electron volt". Electrons accelerated through $U = 100$ kV have an energy of $E = 100$ keV.

Relativistic effects have to be considered at these energies particularly when acceleration voltages up to some megavolts (MV) are used in high-voltage electron microscopy. Table 2.1 therefore contains not only the classical (non-relativistic) formulas but also their relativistic counterparts.

Table 2.1. Properties of the electron.

Rest mass		m_0	=	9.1091×10^{-31} kg	
Charge		e	=	-1.602×10^{-19} C	
Kinetic energy		E	=	eU	
				1 eV = 1.602×10^{-19} J	
Velocity of light		c	=	2.9979×10^{8} m s^{-1}	
Rest energy		E_0	=	$m_0c^2 = 511$ keV	
Spin		s	=	$h/4\pi$	
Planck's constant		h	=	6.6256×10^{-34} J s	
Nonrelativistic	$(E \ll E_0)$			Relativistic $(E \sim E_0)$	
Newton's law	$\boldsymbol{F} = \frac{\mathrm{d}\boldsymbol{p}}{\mathrm{d}\tau}$	$\boldsymbol{F}$	=	$\frac{\mathrm{d}}{\mathrm{d}\tau}(m\boldsymbol{v})$	(2.7)
Mass	$m = m_0$	m	=	$m_0/\sqrt{1 - v^2/c^2}$	(2.8a)
Energy	$E = eU = \frac{1}{2}m_0v^2$	mc^2	=	$m_0c^2 + eU = E_0 + E$	(2.9)
		m	=	$m_0(1 + E/E_0)$	(2.8b)
Velocity	$v = \sqrt{2E/m_0}$	v	=	$c\sqrt{1 - \frac{1}{(1 + E/E_0)^2}}$	(2.10)
Momentum	$p = m_0v = \sqrt{2m_0E}$	p	=	$\sqrt{2m_0E(1 + E/2E_0)}$	(2.11)
			=	$\frac{1}{c}\sqrt{2EE_0 + E^2}$	
Wavelength	$\lambda = \frac{h}{p} = h/\sqrt{2m_0E}$	λ	=	$h/\sqrt{2m_0E(1 + E/2E_0)}$	(2.12)
			=	$hc/\sqrt{2EE_0 + E^2}$	

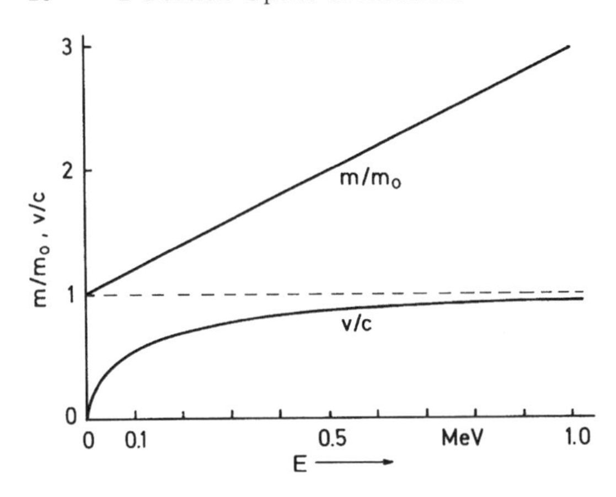

Fig. 2.2. Increase of electron mass m and velocity v with increasing electron energy $E = eU$.

The formula (2.8a) for the increase of the electron mass with increasing velocity v can be obtained from the invariance of the conservation of momentum under a Lorentz transformation, and Newton's law, in the form $\boldsymbol{F} = \mathrm{d}\boldsymbol{p}/\mathrm{d}\tau$, can also be used for relativistic energies.

The most important law of relativistic mechanics is the equivalence of energy and mass: $E = mc^2$. The total energy mc^2 of an accelerated electron is the sum of the rest energy $E_0 = m_0c^2$ and the kinetic energy $E = eU$ (2.9). $E_0 = m_0c^2$ corresponds to an energy of 0.511 MeV. The relativistic increase of the mass m can be formulated not only as in (2.8a) but also in terms of energy as in (2.8b), which follows directly from (2.9). The mass, therefore, increases linearly with increasing energy E; it reaches three times the rest mass m_0 at $E = 2E_0 \simeq 1$ MeV (Fig. 2.2).

The velocity v (2.10) cannot exceed the velocity of light c (Fig. 2.2) and can be obtained by comparing the right-hand sides of (2.8a) and (2.8b). At 100 keV, the electron velocity v reaches 1.64×10^8 m s^{-1}; that is, more than half of the velocity of light. The electron momentum $\boldsymbol{p}$ (2.11) is important because the conservation of both energy and momentum has to be considered in electron collisions (Sect. 5.1). The radius of an electron trajectory in a homogeneous magnetic field $\boldsymbol{B}$ and the de Broglie wavelength λ – (2.12) and Sect. 3.1.1 – also depend on the value of the momentum.

A further property of the electron is its spin (angular momentum) $s = h/4\pi$, and electrons can be polarized by scattering [2.1]. However, spin polarization does not occur in small-angle scattering, which is responsible for the image contrast in TEM.

2.1.2 Deflection by Magnetic and Electric Fields

The force generated by the magnetic part of the Lorentz force (2.1) is normal to both the velocity $\boldsymbol{v}$ and the magnetic field $\boldsymbol{B}$ and has a magnitude $|\boldsymbol{F}| = evB\sin\theta$, θ being the angle between $\boldsymbol{v}$ and $\boldsymbol{B}$. An electron entering a magnetic field with velocity $\boldsymbol{v}$ undergoes an acceleration that is everywhere normal to the local velocity vector. This causes no change in the magnitude of $\boldsymbol{v}$ but does alter its direction. In the magnetic field, therefore, energy is conserved.

In a homogeneous magnetic field, the continuous change of direction of $\boldsymbol{v}$ results in a circular trajectory if $\boldsymbol{v} \perp \boldsymbol{B}$ or $\theta = 90°$. On a circular trajectory, the centrifugal force $F = mv^2/r$ and the centripetal force $F = evB$ are equal, so that the radius of the circle can be calculated from

$$r = \frac{mv}{eB} = \frac{[2m_0 E(1 + E/2E_0)]^{1/2}}{eB}$$
$$= 3.37 \times 10^{-6}\,[U(1 + 0.9788 \times 10^{-6} U)]^{1/2} B^{-1} \tag{2.13}$$

with r (m), U (V), and B (T) (1 T = 1 Tesla = 1 V s m^{-2}).

Large beam deflections through angles of about 90° are used in magnetic prism spectrometers for electron energy-loss spectroscopy and magnetic imaging energy filters (Sect. 4.6).

Small beam deflections produced by transverse electric and magnetic fields are needed for the alignment of electron microscopes or for scanning and rocking electron beams (Sect. 4.2.1). An expression for small-angle deflection ϵ with $\sin\epsilon \simeq \epsilon$ can be obtained by the momentum method (Fig. 2.3). An electron moves in the z direction with an unchanged velocity $v = \mathrm{d}z/\mathrm{d}\tau$ and with a momentum $p_z = mv$. The electric deflection field is obtained by applying a voltage $\pm u$ to plates d apart. The momenta transferred during the time of flight $T = L/v$ are as follows:

electric field $\boldsymbol{E}$

$$p_x = \int_0^T F \mathrm{d}\tau = e \int_0^T |\boldsymbol{E}| \mathrm{d}\tau = \frac{e}{v} \int_0^L |\boldsymbol{E}| \mathrm{d}z = \frac{e|\boldsymbol{E}|L}{v}\,, \tag{2.14}$$

magnetic field $\boldsymbol{B}$

$$p_x = e \int_0^T vB \mathrm{d}\tau = e \int_0^L B \mathrm{d}z = eBL, \tag{2.15}$$

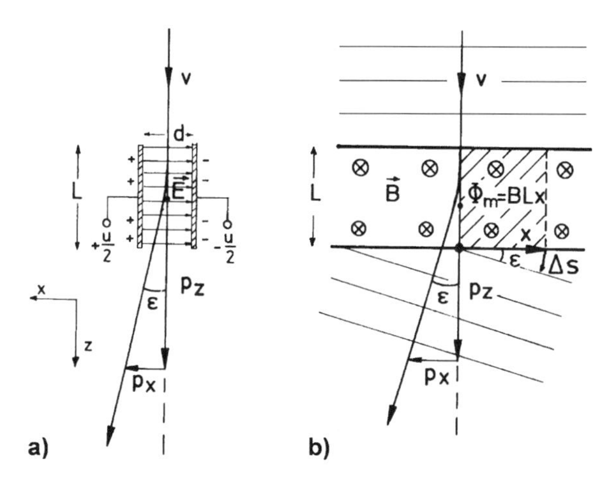

Fig. 2.3. Small-angle deflections ϵ in a transverse electric field (**a**) and magnetic field (**b**).

and the angles of deflection ϵ can be obtained from

$$\epsilon = \frac{p_x}{p_z} = \frac{e|\boldsymbol{E}|L}{mv^2} = \frac{euL}{2Ed}\frac{1+E/E_0}{1+E/2E_0}, \tag{2.16}$$

$$\epsilon = \frac{eBL}{mv} = \frac{eBL}{[2m_0E(1+E/2E_0)]^{1/2}}, \tag{2.17}$$

for the electric and magnetic fields, respectively. The formula (2.17) for the magnetic deflection will also be obtained in Sect. 3.1.5 by a wave-optical calculation. This formula is important for Lorentz microscopy (Sect. 6.8).

As an example, we calculate the field strengths needed to deflect 100 keV electrons through an angle $\epsilon = 5° \simeq 0.1$ rad in a field of length $L = 1$ cm with a plate or polepiece separation $d = 1$ mm. The electric field has to be $|\boldsymbol{E}| = 2 \times 10^6$ V m^{-1}, which implies a voltage u of $\pm$ 1000 V at the plates. The magnetic field $\boldsymbol{B}$ produced by an electromagnet with a slit width d is given approximately by $B = \mu_0 NI/d$ ($\mu_0 = 4\pi \times 10^{-7}$ Vs/(Am), N: number of turns, I: coil current). A deflection ϵ of 5° requires $B = 10^{-2}$ T and can be achieved with $NI = 10$ A; e.g. 100 turns and $I = 0.1$ A.

2.2 Electron Lenses

2.2.1 Electron Trajectories in a Magnetic Lens Field

The physical background of electron-lens optics will be described only briefly to give a quantitative understanding of the function of an electron lens (see [2.2, 2.3, 2.4, 2.5, 2.6, 2.7]).

Magnetic lenses with short focal lengths are obtained by concentrating the magnetic field by means of magnetic polepieces. Figure 2.4 shows the distribution of a magnetic field produced by a coil enclosed in an iron shield, apart from an open slit. The magnetic field has rotational symmetry; the distribution on the optic z axis can be represented approximately by Glaser's "Glockenfeld" (bell-shaped field)

$$B_z = \frac{B_0}{1+(z/a)^2}, \tag{2.18}$$

where B_0 denotes the maximum field in the lens center and $2a$ the full-width at half-maximum [2.9]. Other approximations for the field distribution $B_z(z)$ are also in use, but the Glaser field offers the advantage that the most important properties, the positions of foci and principal planes (Sect. 2.2.2), for example, can be calculated straightforwardly. A knowledge of the magnetic field B_z on the axis is sufficient for calculating the paraxial rays because the radial component B_r close to the axis can be calculated from $B_z(z)$. For stationary fields in a vacuum (no currents: $\boldsymbol{j} = 0$), we can use Maxwell's equation curl $\boldsymbol{B} = \boldsymbol{j} = 0$, which implies that $\boldsymbol{B}$ can be written as the gradient of a scalar

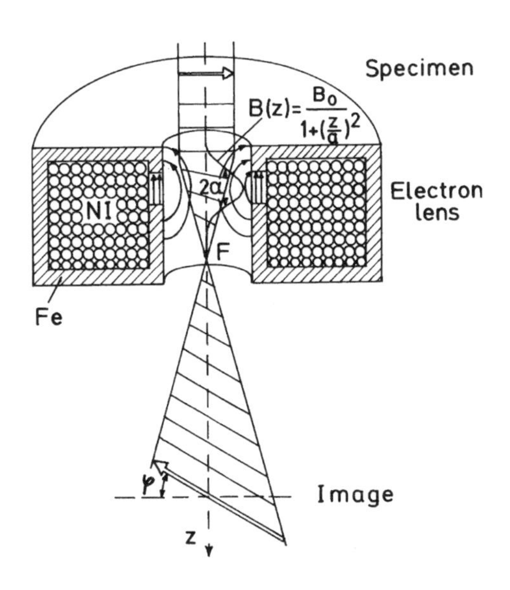

Fig. 2.4. Concentration of a rotationally symmetric magnetic field in the gap of an electron lens (φ: image rotation).

magnetic potential [2.8] $\Psi(\boldsymbol{r})$: $\boldsymbol{B}(\boldsymbol{r}) = -\mathrm{grad}\Psi(\boldsymbol{r})$. Inserting this expression into Gauss' law div $\boldsymbol{B} = 0$, we obtain Laplace's equation

$$\Delta\Psi(\boldsymbol{r}) = 0, \tag{2.19}$$

which can be written in cylindrical coordinates

$$\frac{1}{r}\frac{\partial}{\partial r}r\frac{\partial}{\partial r}\Psi + \frac{\partial^2\Psi}{\partial z^2} + \frac{1}{r^2}\frac{\partial^2\Psi}{\partial\varphi^2} = 0. \tag{2.20}$$

For cylinder symmetric setups, the solution of this equation can be expanded in a power series of r as

$$\Psi(r,z) = \sum_{n=0}^{\infty} a_n(z)r^{2n}. \tag{2.21}$$

Inserting this sum into (2.20), we obtain the recurrence relation

$$a_{n+1}(z) = -\frac{1}{4(n+1)^2}a_n''(z). \tag{2.22}$$

From the series, we obtain

$$B_z(z) = -\frac{\partial a_0(z)}{\partial z}, \tag{2.23}$$

and, for small r,

$$B_r \cong -2a_1(z)r = \frac{-r}{2}\frac{\partial B_z}{\partial z}. \tag{2.24}$$

From the recurrence relation, we see that the scalar magnetic potential, and thus the magnetic field, is determined by its values on the optic axis.

The system of differential equations (Newton's law) for the electron trajectories can be separated in a cylindrical coordinate system r, φ, z:

$$\text{radial component :} \qquad m\ddot{r} = F_r + mr\dot{\varphi}^2. \tag{2.25}$$

$$\text{circular component :} \qquad \frac{\mathrm{d}}{\mathrm{d}t}(mr^2\dot{\varphi}) = rF_\varphi. \tag{2.26}$$

$$\text{longitudinal component :} \qquad m\ddot{z} = F_z. \tag{2.27}$$

The last term in (2.25) can be interpreted as the centrifugal force. Equation (2.26) represents the change of angular momentum $\boldsymbol{L}$ caused by the torque $M = rF_\varphi$ ($\dot{\boldsymbol{L}} = \boldsymbol{M}$).

On substituting the Lorentz force $\boldsymbol{F} = -e\boldsymbol{v} \times \boldsymbol{B}$ with $\boldsymbol{v} = (\dot{r}, r\dot{\varphi}, \dot{z})$ and $B_\varphi = 0$ and using (2.24), we obtain

$$m\ddot{r} = -eB_z r\dot{\varphi} + mr\dot{\varphi}^2, \tag{2.28}$$

$$\frac{\mathrm{d}}{\mathrm{d}t}(mr^2\dot{\varphi}) = eB_z r\dot{r} + e\frac{r^2}{2}\dot{z}\frac{\partial B_z}{\partial z} = \frac{\mathrm{d}}{\mathrm{d}t}\left(\frac{e}{2}r^2 B_z\right), \tag{2.29}$$

$$m\ddot{z} = eB_r r\dot{\varphi}. \tag{2.30}$$

Integration of (2.29) results in

$$mr^2\dot{\varphi} = \frac{e}{2}r^2 B_z + C. \tag{2.31}$$

The constant of integration C becomes zero for meridional rays, and only a trajectory $r(z)$ need be considered in a meridional plane rotating at the angular velocity

$$\omega_\mathrm{L} = \dot{\varphi} = \frac{e}{2m}B_z. \tag{2.32}$$

This is known as the Larmor frequency, which is half the cyclotron frequency of an electron on a circular trajectory.

For paraxial rays (small values of r), equation (2.30) can be approximated by $\ddot{z} = 0$, which implies that v_z is constant. Substitution of (2.32) in (2.28) results in

$$m\ddot{r} = -eB_z r\frac{e}{2m}B_z + mr\left(\frac{e}{2m}B_z\right)^2 = -\frac{e^2}{4m}rB_z^2. \tag{2.33}$$

The time can be eliminated by writing $v_z = \mathrm{d}z/\mathrm{d}\tau \simeq v$. Using (2.9) and (2.10), we find

$$\frac{\mathrm{d}^2 r}{\mathrm{d}z^2} = -\frac{e}{8m_0 U^*}\, r\, B_z^2(z) \quad \text{with} \quad U^* = U\left(1 + \frac{E}{2E_0}\right). \tag{2.34}$$

This is the equation for the trajectory $r(z)$ in the meridional plane rotating at the angular velocity ω_L.

2.2.2 Optics of an Electron Lens with a Bell-Shaped Field

Let us now substitute the bell-shaped field (2.18) in (2.34). The solution of the differential equation can be simplified by introducing reduced coordinates $y = r/a$ and $x = z/a$ and a dimensionless lens parameter

$$k^2 = \frac{eB_0^2 a^2}{8m_0 U^*} , \tag{2.35}$$

resulting in

$$\frac{\mathrm{d}^2 y}{\mathrm{d}x^2} = y'' = -\frac{k^2}{(1+x^2)^2}\, y. \tag{2.36}$$

This equation can be further simplified by the substitution

$$x = \cot\phi; \quad \mathrm{d}x = -\mathrm{d}\phi/\sin^2\phi; \quad 1 + x^2 = \mathrm{cosec}^2\phi. \tag{2.37}$$

The meaning of the angle ϕ can be seen from Fig. 2.5. The variable ϕ varies from π for $z = -\infty$ to $\phi = \pi/2$ for $z = 0$ and then to $\phi = 0$ for $z = +\infty$. Equation (2.36) becomes

$$y''(\phi) + 2\cot\phi\, y'(\phi) + k^2 y(\phi) = 0. \tag{2.38}$$

The solution of (2.38) is a linear combination,

$$y(\phi) = C_1 u(\phi) + C_2 w(\phi), \tag{2.39}$$

of the two particular integrals

$$u(\phi) = \sin(\omega\phi)/\sin\phi,$$
$$w(\phi) = \cos(\omega\phi)/\sin\phi \quad \text{with} \quad \omega = \sqrt{1+k^2}\,. \tag{2.40}$$

The coefficients C_1 and C_2 can be determined from the initial conditions. Thus, for a parallel incident ray, the initial conditions are $r = r_0$ for $z = -\infty$

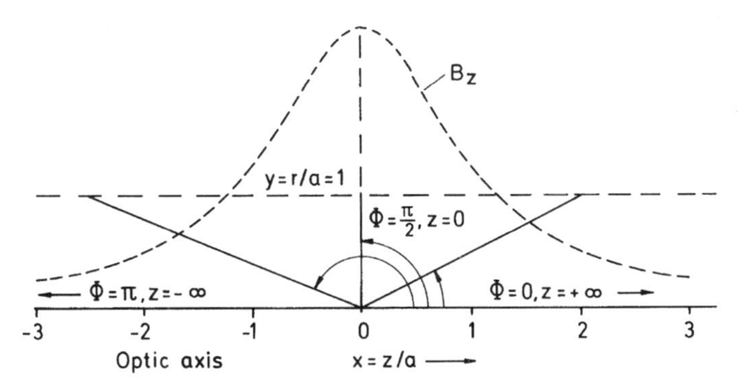

Fig. 2.5. Angular coordinate ϕ for the calculation of electron trajectories and lens parameters.

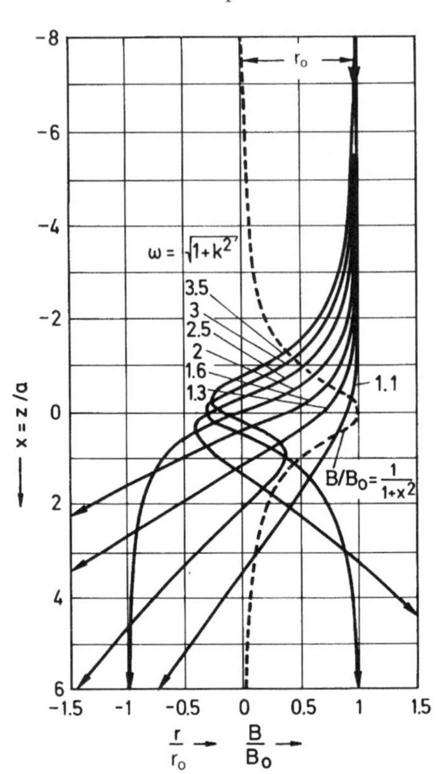

Fig. 2.6. Electron trajectories incident parallel to the axis for increasing values of lens strength $\omega = \sqrt{1+k^2}$ [2.2].

or $y(\pi) = r_0/a$ and $y'(\pi) = 0$, which results in $C_2 = 0$; the radial component of the trajectory becomes

$$y = \frac{r}{a} = -\frac{r_0}{a\omega}\,\frac{\sin(\omega\phi)}{\sin\phi}. \tag{2.41}$$

Such trajectories are plotted in Fig. 2.6 for increasing values of the strength parameter $\omega = \sqrt{1+k^2}$ of the lens.

For a more general discussion, we assume that the ray passes through a point $\mathrm{P}_0(y_0, \phi_0)$ in front of the lens. Substituting $y = y_0$ and $\phi = \phi_0$ in (2.39) and solving for C_1 yields

$$C_1 = \frac{y_0 \sin\phi_0}{\sin(\omega\phi_0)} - C_2\frac{\cos(\omega\phi_0)}{\sin(\omega\phi_0)}. \tag{2.42}$$

We substitute (2.42) in (2.39), giving

$$y(\phi) = \frac{\sin(\omega\phi)\,\sin\phi_0}{\sin(\omega\phi_0)\,\sin\phi}y_0 + \frac{C_2}{\sin\phi}\left[\cos(\omega\phi) - \frac{\cos(\omega\phi_0)}{\sin(\omega\phi_0)}\sin(\omega\phi)\right]. \tag{2.43}$$

The coefficient C_2 can be determined from the direction (slope) of the ray at the point P_0, and different values of C_2 will correspond to different directions. The image point $\mathrm{P}_1(y_1, \phi_1)$ conjugate to the object point P_0 can be obtained from the condition that the last square bracket in (2.43) becomes zero, which means that P_1 has the coordinate

$$y_1 = \frac{\sin(\omega\phi_1)\,\sin\phi_0}{\sin(\omega\phi_0)\,\sin\phi_1} y_0 = My_0, \tag{2.44}$$

independent of C_2 (M: magnification). Multiplying the bracket in (2.43) by $\sin(\omega\phi_0)$ yields the addition theorem for a sine function, and the condition for a zero bracket can be written

$$\sin[\omega(\phi_1 - \phi_0)] = 0, \tag{2.45}$$

which is satisfied by

$$\phi_{1n} = \phi_0 - n\frac{\pi}{\omega}, \quad n = 1, 2, \ldots . \tag{2.46}$$

This means that more than one image point can occur in strong lenses. However, $n = 2$ will not be possible until $\omega = \sqrt{1+k^2} \geq 2$ or $k^2 \geq 3$.

The positions of the object and image points are

$$z_0 = a\cot\phi_0\,; \quad z_{1n} = a\cot\phi_{1n}. \tag{2.47}$$

Substitution of (2.46) into (2.47) gives

$$z_0 = a\cot\left(\phi_{1n} + n\frac{\pi}{\omega}\right) = \frac{a\cot\phi_{1n}\cot\left(n\frac{\pi}{\omega}\right) - a}{\cot\phi_{1n} + \cot\left(n\frac{\pi}{\omega}\right)}. \tag{2.48}$$

This equation can be rewritten in the form

$$\left[z_0 - a\cot\left(n\frac{\pi}{\omega}\right)\right]\left[z_{1n} + a\cot\left(n\frac{\pi}{\omega}\right)\right] = -a^2\mathrm{cosec}^2\left(n\frac{\pi}{\omega}\right), \tag{2.49}$$

which is equivalent to Newton's lens equation of light optics

$$Z_0 Z_1 = f_0 f_1, \tag{2.50}$$

where f_0 and f_1 denote the focal lengths and the distances

$$Z_0 = z_0 - z(\mathrm{F}_0), \quad Z_1 = z_1 - z(\mathrm{F}_1), \tag{2.51}$$

separate the object and image points from the corresponding foci F_0 and F_1. Comparison of (2.49) and (2.50) shows that

$$f_0 = -f_1 = a\,\mathrm{cosec}\left(n\frac{\pi}{\omega}\right), \quad z(F_0) = -z(F_1) = a\cot\left(n\frac{\pi}{\omega}\right). \tag{2.52}$$

The focal lengths f are not the same as the distances $z(\mathrm{F})$ of the foci from the lens center at $z = 0$. This means that electron lenses cannot be treated as thin lenses. Principal planes can be introduced, as in light optics, to construct the position of the corresponding image. The positions of the principal planes are, for $n = 1$,

$$z(\mathrm{H}_0) = z(\mathrm{F}_0) + f_0 = a\frac{\cos\left(\frac{\pi}{\omega}\right) + 1}{\sin\left(\frac{\pi}{\omega}\right)} = a\cot\left(\frac{\pi}{2\omega}\right) = -z(\mathrm{H}_1). \tag{2.53}$$

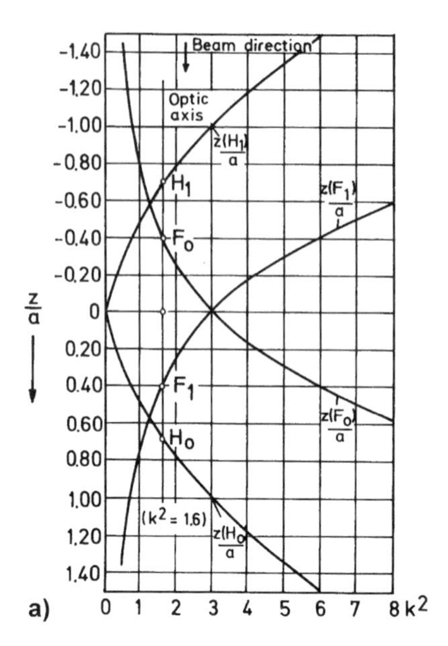

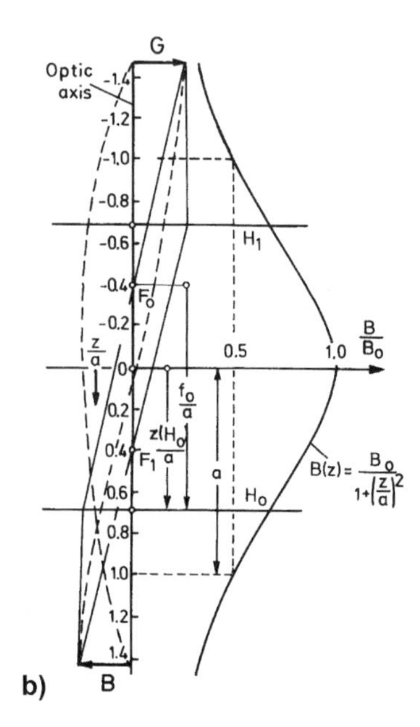

Fig. 2.7. (**a**) Positions of the foci $\mathrm{F_0}$, $\mathrm{F_1}$ and principal planes $\mathrm{H_0}$, $\mathrm{H_1}$ as the lens parameter k^2 is increased and (**b**) example of a geometrical construction for $k^2 = 1.6$ [2.2].

The positions $z(\mathrm{F})$ of the foci and $z(\mathrm{H})$ of the principal planes are plotted in Fig. 2.7a as a function of the lens parameter k^2 (2.35). Figure 2.7b also shows how the image point can be geometrically constructed for the particular case $k^2 = 1.6$. A ray parallel to the axis is refracted at $\mathrm{H_1}$ and continued as a straight line through the focus $\mathrm{F_1}$; a ray through $\mathrm{F_0}$ is refracted at $\mathrm{H_0}$, continuing parallel to the axis. The intersection of these two lines is the image point. Unlike in light-optical lenses, corresponding foci and principal planes are situated on opposite sides of the lens center.

The magnification M in (2.44) can be written in terms of f and Z by substituting $\phi = \phi_{1n}$ from (2.46) and using (2.49–2.52):

$$M = f_0/Z_0 = Z_1/f_1. \tag{2.54}$$

In reality, the trajectories are curved, and the coordinate system rotates with the angular velocity $\dot{\varphi}$ of (2.32). The total rotation angle φ between image and object (Fig. 2.4) can be calculated by using the substitution $\mathrm{d}z = v\,\mathrm{d}\tau$ and (2.35, 2.37, and 2.46):

$$\begin{aligned}\varphi &= \frac{e}{2m}\int_0^{\tau_1} B_z \mathrm{d}\tau = \frac{e}{2mv}\int_{z_0}^{z_1} B_z \mathrm{d}z = \sqrt{\frac{e}{8m_0U^*}}\int_{z_0}^{z_1}\frac{B_0\mathrm{d}z}{1+(z/a)^2} \\ &= -\sqrt{\frac{e}{8m_0U^*}}\,a\,B_0\int_{\phi_0}^{\phi_1}\mathrm{d}\phi = k(\phi_0-\phi_1) = k\,\frac{\pi}{\sqrt{1+k^2}}.\end{aligned} \tag{2.55}$$

The values of the focal lengths and the positions (2.52) of the foci do not depend on the direction of B_z, whereas the image-rotation angle φ is reversed when B_z or the lens current is reversed.

The image rotation in an electron microscope can therefore be partially compensated for by changing the sign of the currents in different lenses. The image rotation does not influence the quality of the image, but its magnitude has to be known if directions in the image have to be correlated with corresponding directions in the specimen or in an electron-diffraction pattern.

The formulas above are for lenses with symmetric polepieces. Lenses with asymmetric polepiece diameters are often used in practice. If the larger diameter is on the specimen side, more space is available for specimen translation with top-entry specimen stages. These lenses can be treated in a similar way by approximating the lens field on the axis by two Glaser fields (2.36) with different parameters a_1 and a_2 on the two sides [2.10].

2.2.3 Special Electron Lenses

Objective Lenses with $\mathbf{k^2 \geq 3}$. A lens with an excitation $k^2 = 3$ (single-field condenser-objective lens) will be optimal in the sense that the focal length is shortest (Fig. 2.11) [2.11, 2.12] and the spherical-aberration coefficient C_s is low (Sect. 2.3.2). Figures 2.6 and 2.7 show that the focus of such a lens is in the center of the lens field at $z = 0$. The specimen position is at the lens center, and the prefield of the lens acts as a condenser lens. Figure 2.8 shows the electron trajectories in such a single-field condenser-objective lens and Fig. 4.14 the corresponding ray diagram, with straight lines and two separate lenses representing the pre- and postfields. The front focal plane (FFP) and back focal plane (BFP) are conjugate. A parallel beam in the FFP is focused at the specimen and is again parallel in the BFP. The lens is thus operating in the "telefocal condition". The specimen area illuminated can be limited by placing a diaphragm in a plane conjugate to the specimen plane. By focusing the last condenser lens in front of the condenser-objective lens on this diaphragm plane, a demagnified electron probe of 1–5 nm in diameter is produced in the specimen plane. All modern microscopes work with such a lens.

The specimen position is shifted beyond the lens center in a second-zone lens with $k^2 > 3$ [2.12, 2.13].

Superconducting Lenses. The strength of a given magnetic lens with an iron core cannot be increased indefinitely owing to the saturation of magnetization M_s at about 2.1 T ($B = \mu_0 H + M$); strong lenses require an increase of size and power supply. Superconducting hollow cylinders or rings have the property of screening the inner space from external magnetic fields and can trap magnetic flux that penetrated the ring in the normal conducting state. The critical magnetic field that destroys superconductivity is very high

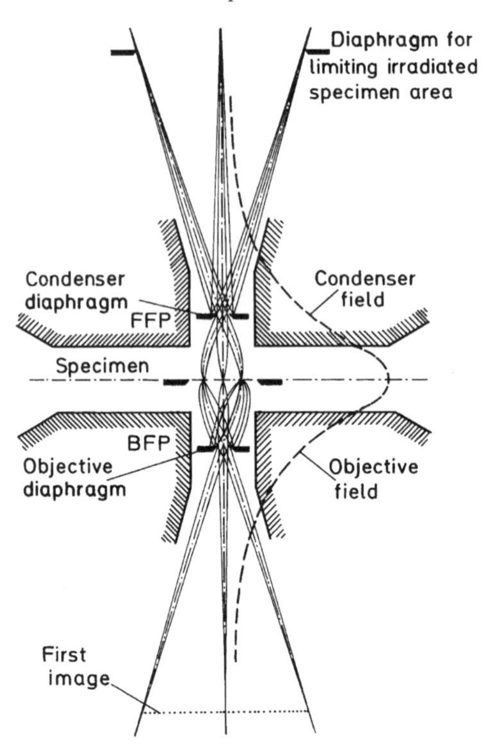

Fig. 2.8. Electron trajectories and conjugate planes in a single-field condenser-objective lens.

($B_c \simeq$ 5–10 T) in type II superconductors (e.g., Nb-Zr, Nb-Ti, Nb_3Sn). Superconducting lenses can be designed in three different ways [2.14, 2.15, 2.16, 2.17] (see also the review in [2.18]):

1. The lens still has ferromagnetic polepieces, which may be of dysprosium or holmium, for which M_s = 3–3.4 T at low temperatures; it is excited by a superconducting coil.
2. Superconductors are introduced into the bore of a conventional magnetic lens in the form of hollow cylinders, thus confining the magnetic flux to a smaller space by screening.
3. The flux trapped in superconducting rings or discs may be exploited.

Minilenses. Any decrease in the size of magnetic lenses will have the advantage of decreasing the length of the electron-optical column, thus reducing the influence of mechanical vibrations and a.c. magnetic stray fields. Small lenses (minilenses) are also useful in front of an objective lens to decrease and control the electron-probe diameter. One way of reducing the size is to use superconducting lenses; alternatively, a stronger excitation may be employed with a more efficient water-cooling system [2.19, 2.20].

Multipole Lenses. A quadrupole lens can be constructed from four polepieces of opposite polarity (Fig. 2.9). Because the magnetic field is normal

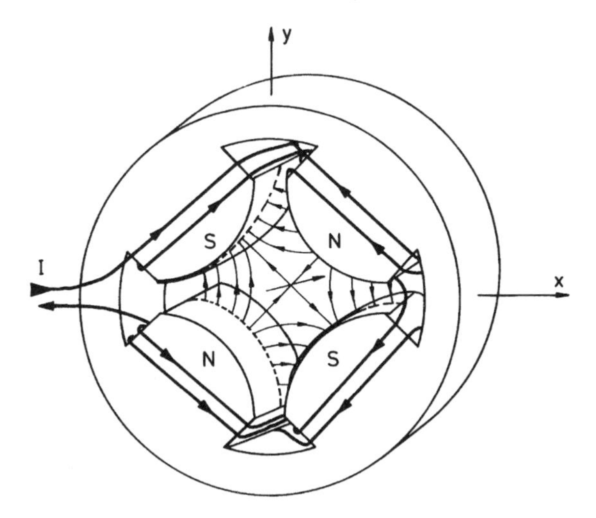

Fig. 2.9. Construction of a quadrupole lens.

to the electron beam, a stronger Lorentz force is exerted. A point object is focused as a line image, as with a cylindrical lens in light optics. In electron microscopes, quadrupole lenses are used as stigmators for compensating the axial astigmatism (Sect. 2.4.1) or to correct the focusing distance of an electron prism for energy analysis (Sect. 4.6.1).

Hexapole lenses consist of six polepieces and octopole lenses of eight polepieces, with alternating polarities. Combinations of hexapole or quadrupole and octopole lenses can be used to correct lens aberrations (Sects. 2.4.2 and 4.6).

2.3 Lens Aberrations

2.3.1 Classification of Lens Aberrations

There are five possible isotropic aberrations of third order in lenses with rotational symmetry, as in light optics:

1) spherical aberration (Sect. 2.3.2)
2) astigmatism (Sect. 2.3.3)
3) field curvature (Sect. 2.3.3)
4) distortion (Sect. 2.3.4)
5) coma (Sect. 2.3.5)

There are three further anisotropic aberrations (Sect. 2.3.6):

6) anisotropic coma
7) anisotropic astigmatism
8) anisotropic distortion

If the electron beam is not monochromatic, owing to

a) insufficient stabilization of the acceleration voltage,

b) the energy spread of the electron gun, and
c) energy losses in the specimen,

9) chromatic aberration (Sect. 2.3.7)

also has to be considered. Departure of the magnetic-lens field from exact rotational symmetry causes an

10) axial astigmatism (Sect. 2.3.3).

The spherical aberration, a distortion associated with this aberration, the axial astigmatism, the coma, and the chromatic aberration are the most important aberrations for electron microscopy, and only these on-axis errors will be discussed in detail. The other aberrations can normally be neglected because the electron beam necessarily remains close to the optic axis and small lens apertures are needed for high resolution. After compensation of axial astigmatism and coma-free alignment, a threefold astigmatism has to be considered at high resolution.

The aberrations can be calculated by the eikonal method [2.21, 2.22], for example, where

$$S(\mathrm{P}_0, \mathrm{P}_1) = \int_{\mathrm{P}_0}^{\mathrm{P}_1} n \mathrm{d}s \tag{2.56}$$

represents the point eikonal as the set of optical path lengths between two points P_0 and P_1. The true path makes the eikonal (2.56) an extremum, which is known as Fermat's principle in light optics.

The so-called diffraction error is not caused by the lens itself but is a consequence of the presence of diaphragms; this error will therefore be discussed not in this section but in Sects. 3.3.2 and 6.2, where the wave-optical theory of image formation is presented.

2.3.2 Spherical Aberration

The spherical aberration has the effect of reducing the focal length for electron rays passing through outer zones of the lens (Fig. 2.10). Electrons crossing the optic axis at different angles θ or scattered in the specimen through angles θ will intersect the Gaussian image plane at a distance

$$r'_{\mathrm{s}} = C_{\mathrm{s}} \theta^3 M \tag{2.57}$$

from the paraxial image point. The Gaussian image plane is the position of the image when very small apertures are used (paraxial rays). C_{s} is the spherical-aberration coefficient and M the magnification. We use coordinates x, y, or r in the specimen plane and the corresponding coordinates $x' = -Mx$ and y', r', respectively, in the image plane.

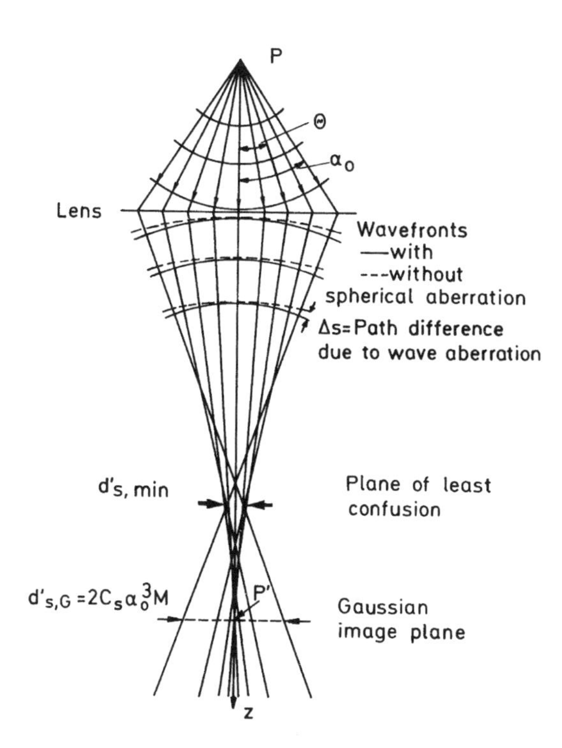

Fig. 2.10. Electron trajectories and wavefronts in a lens with spherical aberration.

A conical electron beam with angular aperture α_o defined by the objective diaphragm does not produce a sharp image point, but the beam diameter passes through a minimum, $d'_{s,\min}$, in a plane of least confusion; in the Gaussian image plane, the diameter is $d'_{s,G} = 2C_s\alpha_o^3 M$. The corresponding diameters referred back to the specimen plane are $d_{s,G} = d'_{s,G}/M$ and $d_{s,\min} = d'_{s,\min}/M$. It can be shown that the smallest diameter is given by

$$d_{s,\min} = 0.5\, C_s\alpha_o^3. \tag{2.58}$$

The spherical-aberration coefficients of objective lenses are normally of the order of 0.5–2 mm. Calculated values of the spherical-aberration coefficient C_s of magnetic lenses are plotted in Fig. 2.11 as a function of the lens parameter k^2. C_s decreases with increasing lens strength. The minimum focal length occurs at $k^2 = 3$, and C_s shows a flat minimum at $k^2 = 7$.

The spherical aberration of the objective lens not only influences the resolution but can also be observed when imaging crystalline specimens. The diffracted beams produce shifted twin images if the objective aperture diaphragm is removed or if the primary beam and the diffracted beam can both pass through the diaphragm. The bright bend contours of crystalline foils observable in the dark-field image are shifted relative to the corresponding dark contours in the bright-field image. This effect can be used for the measurement of C_s [2.23, 2.24, 2.25]. The same effect limits the useful area in selected-area electron diffraction (Sect. 8.1.1).

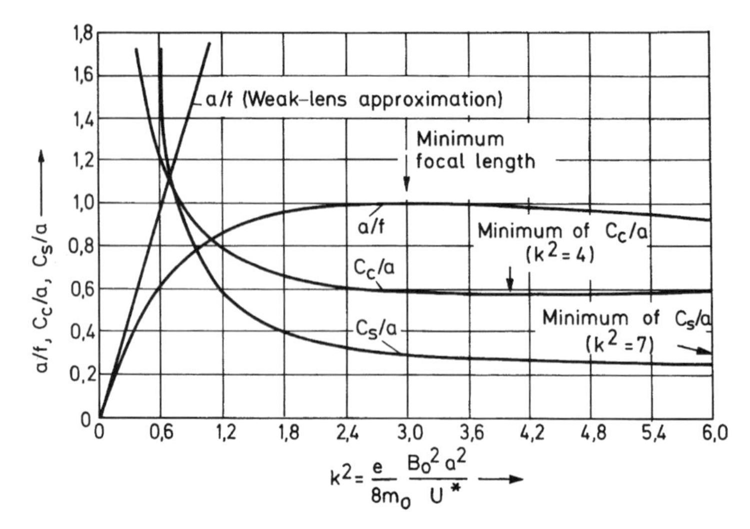

Fig. 2.11. Dependence of reciprocal focal length a/f, reduced spherical-aberration constant C_s/a, and chromatic-aberration constant C_c/a on the lens parameter k^2 [2.2].

A wave-optical formulation of the effect of spherical aberration, which is important for the discussion of phase contrast, will be presented in Sect. 3.3.1. Determination of the contrast-transfer functions by optical diffractometry (Sect. 6.4.7) or a digital Fourier transform and from a defocus series of crystal-lattice images (Sect. 9.6.4) also allows C_s to be evaluated.

2.3.3 Astigmatism and Field Curvature

A cone of rays of semiangle θ from a specimen point P at a distance x from the axis is focused in the Gaussian-image plane as an ellipse with its center at the Gaussian-image point x'. The principal axes of the ellipse are parallel to x' and y', and their lengths are proportional to x^2 and θ.

Rays passing through points around a circle of radius R in the lens and the corresponding points on the ellipse form an astigmatic bundle of rays that collapses to perpendicular focal lines F_s and F_m for rays in the sagittal and meridional planes. These foci lie on the curved sagittal and meridional image surfaces shown in Fig. 2.12. A circle of least confusion is formed in the curved mean image surface. This error disappears for on-axis specimen points ($x = 0$), and this type of astigmatism can in practice be neglected because small apertures are used to decrease the influence of spherical aberration and because the electron beam is necessarily adjusted on-axis to decrease the influence of coma and chromatic aberration.

However, astigmatism will be observed even for points on-axis if the lens field is not exactly rotationally symmetric, owing to inhomogeneity of the

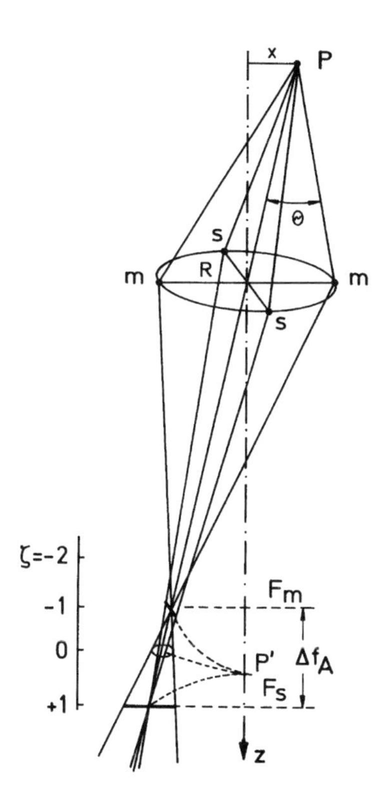

Fig. 2.12. Astigmatic focal differences between meridional and sagittal ray bundles.

magnetization of the polepiece, ellipticity of the polepiece bores, or electric charging of aperture diaphragms. This error is therefore called axial astigmatism. In consequence, a pair of diametrically opposite zones of a circular specimen will be focused sharply at one focal point F_s, and the two other diametrically opposite zones, 90° from the first, will be focused at the other focal point F_m. The difference Δf_A of the focal lengths (Fig. 2.12) will be small and is only of the order of 0.1 to 1 μm. Nevertheless, the resolution can be reduced, as is shown by the following estimate.

The diameter of the error disc at the specimen plane will be

$$d_A = \Delta f_A \alpha_o . \tag{2.59}$$

If a resolution $\delta = 0.5$ nm is wanted for an aperture α_o of 10 mrad, d_A should be smaller than δ and, therefore, $\Delta f_A < \delta/\alpha_o = 50$ nm. If we assume that the polepiece bore is elliptical with semiaxes $b_0 \pm \Delta b$, the relative focal difference becomes

$$\frac{\Delta f_A}{f} = 2\,\frac{\Delta b}{b_0} \tag{2.60}$$

because the focal length is of the order of the diameter b_0. It follows that Δb must be less than 25 nm with the estimated value of Δf_A. It is very difficult to obtain such precision in the diameter of the bore.

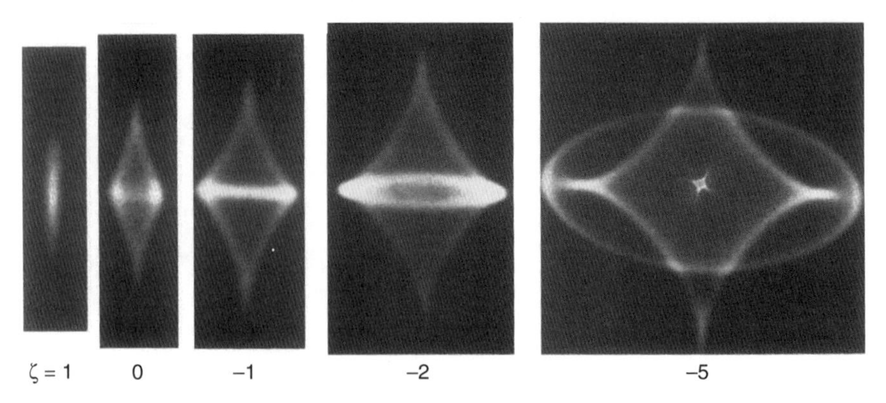

Fig. 2.13. Cross sections through the caustic at different values of the coordinate ζ of Fig. 2.12 [2.27].

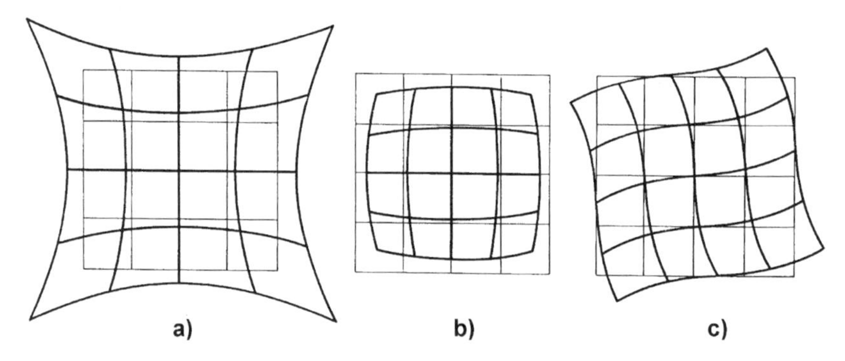

Fig. 2.14. (**a**) Pin cushion, (**b**) barrel, and (**c**) spiral distortion of a square grid.

The simple drawing of Fig. 2.12 is not adequate for calculating the cross section of the electron beam in an astigmatic image. If all rays, including those not in the sagittal or in the meridional plane, are considered, a complicated intensity distribution in the neighborhood of the focus results, the so-called caustic. Figure 2.13 shows observed intensity distributions [2.26, 2.27] corresponding to cross sections through the caustic at the positions ζ indicated in Fig. 2.12. The orthogonal focal lines have the coordinates $\zeta = \pm 1$.

2.3.4 Distortion

Distortion causes a displacement

$$\Delta r' = -C_{\mathrm{E}} r'^3 \tag{2.61}$$

in the Gaussian image plane for off-axis points. This results in a geometrical distortion of a square, which is known as pin cushion distortion for $C_{\mathrm{E}} > 0$ and barrel distortion for $C_{\mathrm{E}} < 0$ (Fig. 2.14a,b).

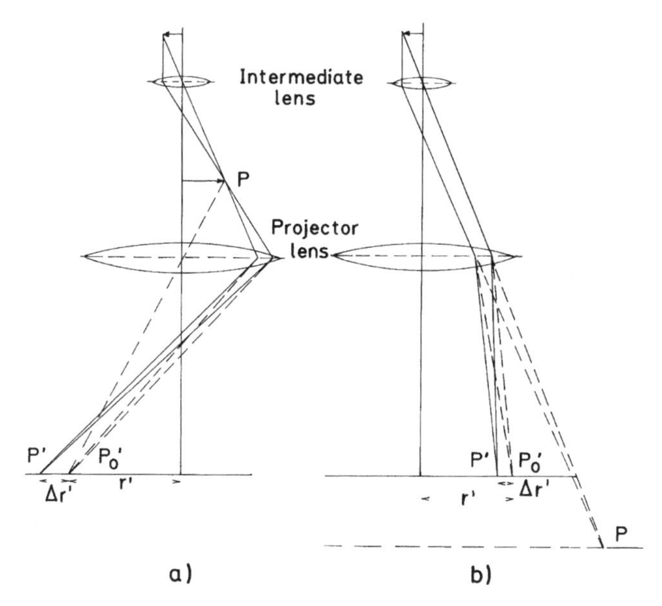

Fig. 2.15. (**a, b**) Examples of distortion caused indirectly by the spherical aberration of a projector lens.

The spherical aberration may be used to explain an image distortion found in intermediate and projector lenses operating at low magnifications. These large-bore lenses magnify an intermediate image in which the angular aperture at any image point is smaller by a factor $1/M$ than the objective aperture α_o. Therefore, no further decrease of image resolution by the spherical aberration is expected. However, Fig. 2.15 shows how a distortion of the image can be generated indirectly by the spherical aberration. A conical beam coming from P in the intermediate image of Fig. 2.15a converges to an image point P_0' in the absence of spherical aberration but to a point P' if it is present. The deviation $\Delta r'$ on the image screen increases with r' as r'^3, resulting in a pin cushion distortion of a square specimen area. The opposite situation is observed when the intermediate image lies beyond the second lens (Fig. 2.15b); the deviation $\Delta r' \propto r'^3$ is now directed toward the optical axis, resulting in a barrel distortion. It is possible to compensate for this type of distortion by suitably exciting the lens system, and a pin cushion distortion can be compensated for by a barrel distortion in another intermediate image step [2.20]. This compensation of distortion is very important for preliminary exploration of the specimen at low magnification.

2.3.5 Coma

Coma causes a cone of rays passing through the specimen point P at an off-axis distance r at angles θ to the axis to be imaged as a circle with radius proportional to θ^2 and r. The center of the circle does not coincide with the

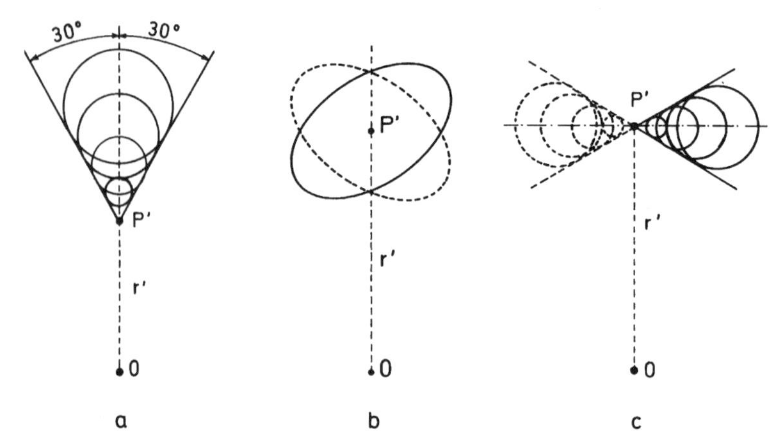

Fig. 2.16. (**a**) The effect of coma on the image P′ of a specimen point P at a distance $r = r'/M$ from the axis for increasing angular apertures. (**b**) Anisotropic astigmatism and (**c**) anisotropic coma [(- - -) magnetic field of the lens reversed].

Gaussian image point $r' = -rM$ but is shifted in the radial direction by twice the radius. Circles corresponding to different angles θ therefore lie within a sector of 60° (Fig. 2.16a). Coma-free alignment is necessary for high resolution (Sect. 2.5.3).

2.3.6 Anisotropic Aberrations

The anisotropic distortion is caused by the dependence of the image rotation on the off-axis distance r of the object point; the latter is imaged with an additional rotation angle φ proportional to r'^2. Straight lines in the specimen plane become cubic parabolas in the image plane (Fig. 2.14c). Reversal of the lens current changes the sense of rotation.

The anisotropic astigmatism together with the astigmatism discussed in Sect. 2.3.3 results in an ellipse, the principal axes of which are not parallel to the x' and y' axes (Fig. 2.16b).

The anisotropic coma differs from the coma (Sect. 2.3.5) in the direction of the coma sector, which is not perpendicular to the radius r' (Fig. 2.16c).

2.3.7 Chromatic Aberration

Variations of electron energy and lens current cause a variation of focal length

$$\frac{\Delta f_{\mathrm{c}}}{f} = \frac{\Delta E}{E} - 2\frac{\Delta I}{I} \tag{2.62}$$

because f is proportional to E and B^{-2} or I^{-2} (I: lens current). This means that chromatic aberration can be caused by fluctuations of the acceleration voltage, by the energy spread of the emitted beam, by energy losses inside the specimen, and by fluctuations of the lens current. An energy spread ΔE causes a point to be imaged as a chromatic-aberration disc of diameter

$$d_c' = d_c M = \frac{1}{2} C_c \frac{\Delta E}{E} \frac{1 + E/E_0}{1 + E/2E_0} \alpha_o M. \quad (2.63)$$

The chromatic-aberration coefficient C_c is of the order of the focal length f for weak lenses and decreases to a minimum of about $0.6f$ for stronger lenses (Fig. 2.11) [2.28].

The chromatic aberration caused by the energy spread ΔE of the electron beam limits the resolution. If a resolution $\delta = 0.2$ nm $< d_c$ is wanted for $\alpha_o =$ 20 mrad and $C_c = 2$ mm, we must ensure that $\Delta E/E < 10^{-5}$. Owing to the Boersch effect (Sect. 4.1.2), the half-width of the electron energy distribution from a thermionic cathode is of the order of 1–2 eV. If the focusing corresponds to the maximum of this energy distribution, only half of this value should be used for ΔE in (2.63). This means that Schottky or field-emission cathodes with $\Delta E < 1$ eV must be used for high-resolution work, that is, not worse than 0.1–0.2 nm. Furthermore, the acceleration voltage and the lens currents have to be stabilized to better than 10^{-5}. The influence of chromatic aberration on contrast transfer will be discussed in detail in Sect. 6.4.2. To minimize the influence of energy losses ΔE inside the specimen, the proportion of the beam scattered inelastically should be very much smaller than that scattered elastically or unscattered.

The number of unscattered and elastically scattered electrons is strongly reduced in thick films, and the energy-loss spectrum is broadened by multiple energy losses (Fig. 5.34b,c). An operator will focus on the most probable energy (the maximum of the energy-loss spectrum). The resolution will be limited by the half-width of the energy-loss spectrum [2.29]. The chromatic aberration associated with film thickness can be measured from the blurring of sharp edges [2.30].

Equation (2.63) describes the axial chromatic aberration, which is still present for electron beams entering the objective lens from the axial point of the specimen. When the electron beam passes the specimen at a distance r from the axis, a chromatic error streak $\delta_{r\varphi}$ with two components is observed: a radial component δ_r, due to changes of magnification as a function of electron energy, and an azimuthal component δ_φ, due to variation of the image rotation angle φ (2.55). Together, these give (Fig. 2.17a)

$$\delta_{r\varphi} = C_{r\varphi}\, r \frac{\Delta E}{E} \quad \text{with} \quad C_{r\varphi}^2 = C_r^2 + C_\varphi^2. \quad (2.64)$$

The constant C_r remains positive whatever the lens excitation, whereas C_φ can change sign [2.31]. This chromatic-error streak is illustrated in Fig. 2.17b, where several exposures of polystyrene spheres corresponding to different values of the lens current are superimposed.

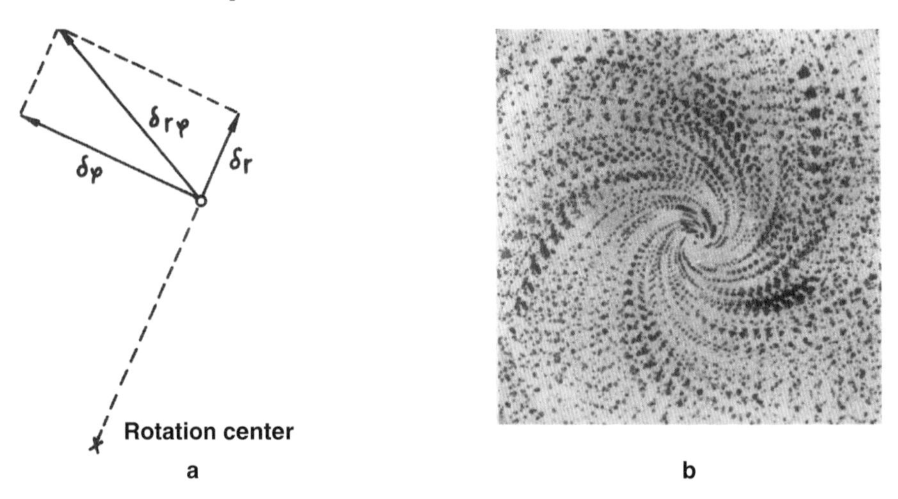

Fig. 2.17. (**a**) Chromatic error streak and (**b**) demonstration by superimposed focal series of polystyrene spheres.

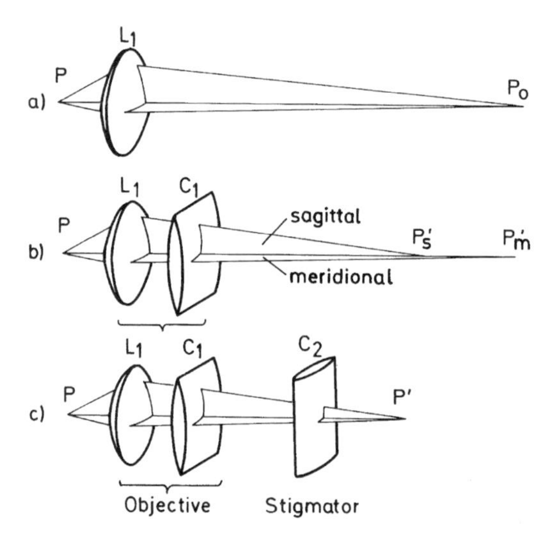

Fig. 2.18(a–c). Light-optical analogue of the action of astigmatism (L1 + C1) and a stigmator C2 [2.32].

2.4 Correction of Aberrations and Microscope Alignment

2.4.1 Correction of Astigmatism

Axial astigmatism can be compensated for by placing a simple stigmator in the polepiece bore of a lens. The function of this correction element can be understood from a light-optical analogue [2.32] (Fig. 2.18). Axial astigmatism can be simulated by adding a cylindrical lens C1 to the rotationally symmetric lens L1 (Fig. 2.18a). The lens C1 acts only on the sagittal bundle, resulting in a shorter focal length (Fig. 2.18b). The stigmator consists of a cylindrical lens C2 rotated through 90° relative to C1. It acts only on the meridional bundle (Fig. 2.18c), so that P'_s and P'_m coincide in P'. This means that the lens astigmatism is compensated for by a perpendicular astigmatism of the same magnitude. The orientation and strength of C2 therefore have to be adjustable.

In electron optics, toric rather than cylindrical lenses are employed in the form of very weak quadrupole lenses (Sect. 2.2.3). Two quadrupoles mounted with a relative rotation of 45° around the axis and excited by different currents allow the direction and strength of the quadrupole lens system to be varied.

For high resolution, the astigmatic focal difference Δf_A should be smaller than 10 nm. Sensitive methods of detecting such small focal differences are required to adjust the stigmator correctly. The following methods can be used.

Fresnel-Fringe Method. Defocusing causes Fresnel diffraction fringes to be seen at edges (Sect. 3.2.2). These fringes disappear in focus. The distance x_1 (3.35) of the first fringe from the edge is proportional to the square root of the defocus $\Delta z = R_0$. If a small hole of about 0.1 μm diameter in a supporting film is observed, the Fresnel fringes disappear in the presence of astigmatism only on opposite sides of the hole. They remain visible in a perpendicular direction as a result of the astigmatic focal difference. The astigmatism is compensated for when the fringe visibility for small defocusing is the same around the edge of a hole. This method is capable of revealing values of Δf_A greater than 0.1 μm by visual observation of the viewing screen and about half of this value on a micrograph.

Granularity of Supporting Films. Supporting films of carbon exhibit a granularity caused by phase contrast (Sect. 6.2.2) that is very sensitive to defocusing. In the presence of astigmatism, the granularity shows preferential directions that change through 90° if the focusing is changed from the sagittal to the meridional focus. A very high sensitivity can be obtained by recording the image with a CCD camera and observing the granularity on the TV screen. An improved contrast can be observed when using a thin evaporated amorphous germanium film.

Fraunhofer Diffraction. The spatial-frequency spectrum of the granularity can be observed by light-optical Fraunhofer diffraction on developed micrographs (Sect. 6.4.7) or by online digital Fourier analysis of images recorded by a CCD camera. Spherical aberration and defocusing lead to gaps in the transfer of spatial frequencies, which can be seen as a ring pattern in Fraunhofer diffractograms (Fig. 6.27). Astigmatism deforms the rings to ellipses or hyperbolas.

The latter two methods can detect values of Δf_A greater than 10 nm, which is sufficient for high resolution. All three methods are based on phase-contrast effects caused by defocusing. It is necessary to work with a nearly coherent electron beam to prevent blurring of the fringes and the granularity.

With an illumination aperture α_i of 1 mrad, for example, only one Fresnel fringe can be resolved, whereas with a very coherent beam produced by a Schottky emitter or a field-emission gun, hundreds of fringes may be seen (Fig. 3.9).

2.4.2 Correction of Spherical and Chromatic Aberrations

Because the third-order lens aberrations are observable only for nonparaxial rays, aberration correction will be necessary only for lenses working with larger apertures, $\alpha > 1$ mrad, such as a probe-forming condenser lens or an objective lens. In the intermediate and projector lenses, the angular aperture is decreased to α/M.

The resolution is limited by both the spherical and chromatic aberrations of the objective lens, and it will hence be of interest to correct both defects simultaneously. In light optics, spherical aberration is caused by the spherical shape of the glass-lens surfaces, and chromatic aberration is caused by the dispersion of the refractive index. Both errors can be corrected by using nonspherical surfaces and/or a suitable combination of lenses. The magnetic field of an electron lens cannot be "polished", and spherical aberration is a consequence of the structure of the rotationally symmetric magnetic field. Thus, in (2.24), we used the relation div $\boldsymbol{B} = 0$ to show that the radial component B_r cannot be independent of the axial component B_z. Starting from this relation and an equivalent one for electric fields, Scherzer has demonstrated that the spherical- and chromatic-aberration coefficients of a stationary, charge-free round lens are always positive [2.33].

Scherzer [2.34] proposed that correction of the third-order spherical aberration and first-order chromatic aberration should be possible by introducing an additional system of multipole lenses behind the objective lens. The spherical aberration can in principle be compensated for by a combination of magnetic quadrupole and octopole lenses, whereas a combination of electrostatic and magnetic quadrupoles is necessary for the chromatic aberration [2.35, 2.36, 2.37]. Koops et al. [2.38] showed experimentally that such a system works and that the sensitivity to misalignment can be decreased by additional trim coils.

When using Schottky or field-emission guns with $\Delta E \leq 0.3$ eV, the chromatic aberration at voltages $\geq$200 kV will be less than the spherical aberration. Correction of the spherical aberration thus permits the extension of the point-to-point resolution to the information limit, which is determined by the chromatic aberration and the mechanical and electrical stability of the instrument. Rose [2.39] has proposed such a C_s-corrector, composed of two sextupoles and two round lenses. This system has been built for a 200 kV microscope by Haider [2.40]. This corrector is now commercially available and has proven its usefulness for materials applications [2.41]. More recent ideas for corrector systems include a quadrupole-octupole corrector for a STEM [2.42] and an "ultracorrector" to correct all primary spherical and chromatic aberrations in TEM [2.43].

2.4.3 Microscope Alignment

When the acceleration voltage or the objective current is wobbled, specimen structures move on spirals around the corresponding voltage or current center (Fig. 2.17b). Both centers should coincide with the center of the final screen. This correction is sufficient for medium resolution. If the centers do not coincide, preference should be given to the voltage center. Owing to mechanical limitations, the condenser-, objective- and projector-lens systems are not perfectly aligned. An on-axis alignment of the electron beam in the objective lens is essential for high resolution (coma-free alignment). The microscope can be aligned with the aid of beam-tilt coils above and image-shift coils below the objective lens. Three different types of autoalignment methods are in use.

Diffractogram Method. When using an untilted beam and the compensation for astigmatism described in Sect. 2.4.1, the Fraunhofer diffractogram shows concentric circles from which the defocus and C_s can be determined; the axial coma and threefold astigmatism cannot be detected. On tilting the beam by $\pm\theta$, the latter aberrations cause noncentrosymmetric differences in the diffractograms. A useful procedure is to take micrographs of amorphous carbon or germanium films with a beam tilt $\theta \simeq$ 5–10 mrad at 6–20 azimuths between 0 and 2π and produce a tableau of diffractograms [2.44, 2.45, 2.46]. When the illumination direction is aligned, the tableau of diffractograms is centrosymmetric. Further alignment is necessary for the voltage and current centers.

Image-Contrast Method. The image shows minimum contrast when the beam is aligned, the image focused, and astigmatism corrected [2.47, 2.48]. Accurate settings are obtained by a deliberate variation of defocus, astigmatism, and alignment. The method only works efficiently when the parameters are close to their correct values, and the dose required is high.

Image-Shift Method. This method exploits the beam-tilt-induced displacements of an image [2.49, 2.50, 2.51] and is independent of the particular specimen structure, whereas the two previous methods use amorphous carbon or germanium test films. Coma-free alignment is based on the nonlinear relation between displacement and beam tilt. The displacements can be measured by seeking the maximum of the cross-correlation (Sect. 6.5). For coma-free alignment, five images have to be recorded: one without and four with equal but oppositely tilted beams, for example. The reproducibility of alignment on the coma-free axis is better than 0.1 mrad and that of focusing and stigmation better than 3 nm at $M = 500\,000$.

3

Wave Optics of Electrons

A de Broglie wavelength can be attributed to each accelerated particle, and the propagation of electrons can be described by means of the concept of a wave packet. The interaction with magnetic and electrostatic fields can be described in terms of a phase shift, or the notion of a refractive index can be employed, leading to the Schrödinger equation. The interaction with matter can similarly be reduced to an interaction with the Coulomb potentials of the atoms.

Many of the interference experiments of light optics can be transferred to electron optics. The most important are the Fresnel biprism experiment and Fresnel diffraction at edges. The diffraction pattern far from the specimen or in the focal plane of an objective lens can be described by means of Fraunhofer diffraction. As in light optics, the Fraunhofer-diffraction amplitude is the Fourier transform of the amplitude distribution of the wave leaving the specimen where the lens aberrations are incorporated in the wave-aberration function.

The image amplitude can be described in terms of an inverse Fourier transform, which does not, however, result in an aberration-free image owing to the phase shifts introduced by the electron lens and the use of a diaphragm in the focal plane.

3.1 Electron Waves and Phase Shifts

3.1.1 De Broglie Waves

In 1924, de Broglie showed that an electron can be treated as a quantum of an electron wave and that the relation $E = h\nu$ for light quanta should also be valid for electrons. As a consequence, he postulated that the momentum $\boldsymbol{p} = m\boldsymbol{v}$ is also related by $\boldsymbol{p} = h\boldsymbol{k}$ to the wave vector $\boldsymbol{k}$, the magnitude of which (the wave number) may be written $|\boldsymbol{k}| = 1/\lambda$ (λ: wavelength); this is analogous to $p = h\nu/c = hk$ for light quanta. This implies that $\lambda = h/p$ (2.12) with the relativistic momentum p (2.11). Substitution of the constants in (2.12) results in the formula

$$\lambda = \frac{h}{mv} = \frac{1.226}{[U(1 + 0.9788 \times 10^{-6}U]^{1/2}} \tag{3.1}$$

with λ (nm) and U (V) ($\lambda = 3.7$ pm for $U = 100$ kV and 0.8715 pm for $U =$ 1 MV).

A stationary plane wave that propagates in the z direction can be described by a wave function ψ that depends on space and time τ,

$$\psi = \psi_0 \exp[2\pi i(kz - \nu\tau)] = \psi_0 \exp\left(\frac{2\pi i}{\lambda}z - 2\pi i\nu\tau\right) = \psi_0 \exp(i\varphi), \tag{3.2}$$

where ψ_0 is called the amplitude and φ the phase of the wave. The phase changes by 2π for $\tau = \text{const}$ if the difference between two positions $(z_2 - z_1)$ is equal to λ (Fig. 3.1a).

When the electron moves in an electrostatic field, we have to distinguish between the kinetic energy E_{kin} and the total energy E_{tot}, which is given by the sum of the kinetic and the potential energies $E_{tot} = m_0c^2 + E_{kin} + V(\boldsymbol{r})$. Whereas the frequency is directly related to the energy, $E = h\nu$, the definition of a wavelength or a wave number is more complicated. If we assume that the potential varies only slowly, we can define a spatially varying wave number $\boldsymbol{k}$ by dividing the local momentum $\boldsymbol{p}$ by Planck's constant h. In the one-dimensional case, we then obtain a wave function

$$\psi(z) = \psi_0 \exp\left\{2\pi i\left[\int_{z_0}^{z} k(z)\mathrm{d}z - \nu\tau\right]\right\}. \tag{3.3}$$

Formally, this so-called WKB approximation can be obtained from the Schrödinger equation (3.21). Details can be found in many textbooks on

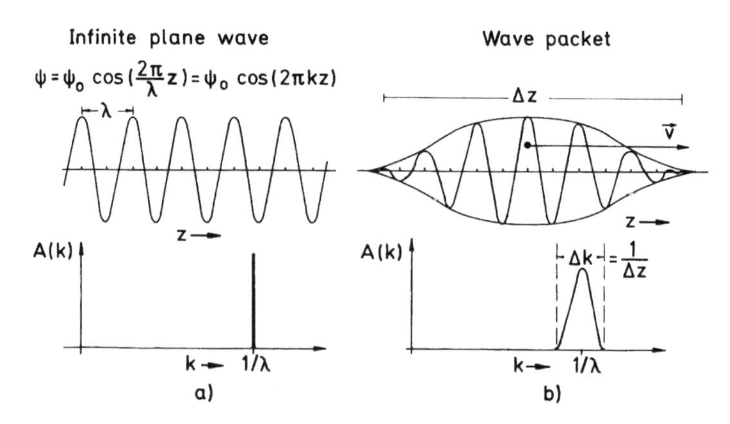

Fig. 3.1. (**a**) Infinite cosine wave with a discrete k-spectrum and (**b**) a wave packet of lateral width Δz and a broadened k-spectrum of width $\Delta k = 1/\Delta z$.

quantum mechanics [3.1, 3.2, 3.3]. Due to the fact that an arbitrary constant can be added to the potential Φ or the potential energy V as shown in (2.6), the frequency of an electron wave is not a clearly defined quantity. This does not matter because it is not an observable quantity.

In the presence of fields, the quantities $E' = mc^2 - e\Phi$ and $\boldsymbol{p} = m\boldsymbol{v}$ do not form a relativistic four-vector, whereas that is a necessary condition if the physical laws are to satisfy the invariance requirements of relativity. An electrostatic field is time dependent if it is seen from a frame of reference moving with a velocity $\boldsymbol{v}$ relative to the original frame. It is therefore associated with a magnetic field via Maxwell's equations. The correct value $\boldsymbol{p}'$ that must be used for $\boldsymbol{p}$ in $\boldsymbol{p} = h\boldsymbol{k}$ – the canonical momentum – is

$$\boldsymbol{p}' = m\boldsymbol{v} - e\boldsymbol{A} = h\boldsymbol{k}. \tag{3.4}$$

The magnetic vector potential $\boldsymbol{A}$ is related to the magnetic field by $\boldsymbol{B} = \nabla\times\boldsymbol{A}$. The vector $(p'_x, p'_y, p'_z, E'/c)$ thus becomes a relativistic four-vector.

In the relation (3.4), the vector potential $\boldsymbol{A}$ is not uniquely defined because an arbitrary field $\boldsymbol{A}'$ that satisfies the condition $\nabla \times \boldsymbol{A}' = 0$ can be added to $\boldsymbol{A}$ without affecting the value of $\boldsymbol{B}$ because $\boldsymbol{B} = \nabla \times (\boldsymbol{A} + \boldsymbol{A}') = \nabla \times \boldsymbol{A}$. The arbitrary field $\boldsymbol{A}'$ therefore has only to be curl-free. Just like the frequency, then, the wave number k and the wavelength λ are not uniquely defined quantities for electrons and therefore are not observable quantities.

This is a very strange conclusion for an electron microscopist, who daily sees electron-diffraction patterns and uses (3.1), but it transpires that electron-interference effects can be observed even though the wavelength of an electron is not a clearly defined quantity. Consider the following experiment, which can serve as a model for all interference and diffraction experiments. A wave from a source Q (Fig. 3.2) passes through a double slit, beyond which the two partial waves overlap at P. There will be constructive interference if the difference between the phases is an integral multiple of 2π or an even integral multiple

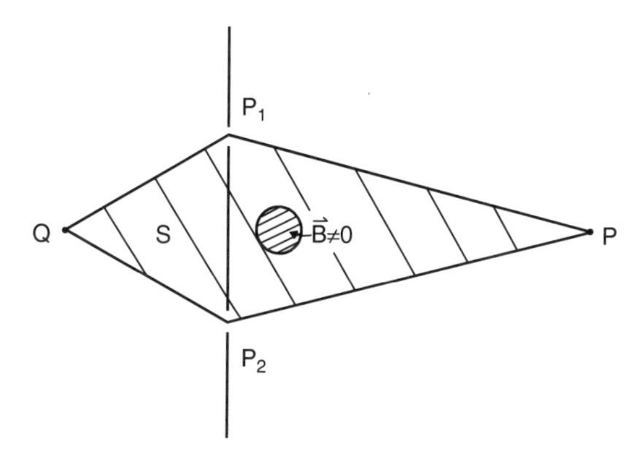

Fig. 3.2. Calculation of the phase difference between two partial waves passing the double slit at P_1 and P_2.

of π and destructive interference if the phase is an odd integral multiple of π. In the presence of an electrostatic and a magnetic field, the phase difference between QP_1P and QP_2P becomes

$$\begin{aligned}\varphi_2 - \varphi_1 &= 2\pi \left(\int_Q^{P_2} \boldsymbol{k} \cdot \mathrm{d}\boldsymbol{s} + \int_{P_2}^{P} \boldsymbol{k} \cdot \mathrm{d}\boldsymbol{s} - \int_Q^{P_1} \boldsymbol{k} \cdot \mathrm{d}\boldsymbol{s} - \int_{P_1}^{P} \boldsymbol{k} \cdot \mathrm{d}\boldsymbol{s} \right) \\ &= \frac{2\pi}{h} \oint (m\boldsymbol{v} - e\boldsymbol{A}) \cdot \mathrm{d}\boldsymbol{s}. \end{aligned} \tag{3.5}$$

The signs of the last integrals have been changed by interchanging the lower and upper integration limits, and the four integrals have thus been reduced to an integral over the closed loop QP_2PP_1Q in which (3.4) for $\boldsymbol{k}$ has been used. Stokes' law can be applied to the integral involving $\boldsymbol{A}$, and the result may be expressed in terms of the magnetic flux Φ_m enclosed within the loop of area S:

$$\begin{aligned}\varphi_2 - \varphi_1 &= \frac{2\pi}{h} \oint m\boldsymbol{v} \cdot \mathrm{d}\boldsymbol{s} - \frac{2\pi e}{h} \int_S (\nabla \times \boldsymbol{A}) \cdot \mathrm{d}\boldsymbol{S} \\ &= \frac{2\pi}{h} \oint m\boldsymbol{v} \cdot \mathrm{d}\boldsymbol{s} - \frac{2\pi e}{h} \Phi_\mathrm{m}. \end{aligned} \tag{3.6}$$

An arbitrary additional vector field $\boldsymbol{A}'$, such that $\nabla \times \boldsymbol{A}' = 0$, which caused the trouble in the definition of the wavelength, does not influence the difference between the phases in (3.6). Equation (3.6) therefore shows that the relation $k = 1/\lambda = mv/h$ can be used to calculate phase or wave number differences for interference and diffraction experiments in the absence of a magnetic field ($\Phi_\mathrm{m} = 0$) and that if magnetic flux does pass through the loop ($\Phi_\mathrm{m} \neq 0$), it causes an additional phase shift. This phase shift can be measured by means of a biprism. Consider two rays (Fig. 3.2) from the source Q to the point P of the interference pattern. The optical phase difference φ depends on the enclosed magnetic flux Φ_m. A phase shift occurs even if there is no magnetic field at the trajectory, and hence if no magnetic term of the Lorentz force acts on the electrons (Fig. 3.2). For the phase shift, only the magnetic flux through the enclosed area is important. This Aharonov-Bohm effect [3.4] has been verified experimentally by many authors. A magnetic flux $\Phi_\mathrm{m} = h/e = 4.135 \times 10^{-15}$ Vs is sufficient to cause a phase shift $\varphi = 2\pi$ corresponding to a path difference $\Delta s = \lambda$ and to a shift of the interference pattern by one fringe distance. Such a small flux can be created by an iron whisker with a cross section of 2000 nm^2 and a saturation magnetization $B_\mathrm{s} = 2.1$ T [3.5, 3.6] or by a 25 nm permalloy film evaporated on the biprism wire [3.7]. The theoretical value of the phase shift was confirmed from two exposures of the fringe system obtained with B_s in opposite directions. By using three biprism wires, a larger spatial separation of the electron rays can be achieved and the flux of a coil 20 μm in diameter can be enclosed [3.8, 3.9]. Lischke [3.10] verified the quantization of the enclosed flux in superconductors, which is a multiple of the flux quantum (fluxon) $h/2e$. One fluxon corresponds to a shift of the fringe pattern by one-half of the fringe distance. The fact that the effect occurs even if the electrons themselves do not experience any magnetic field has been proven by carefully shielding the magnetic field created in a toroidal magnet by embedding it in a superconductor [3.11].

The deflection of electrons in a transverse magnetic field of length L through an angle ϵ (Sect. 2.1.2 and Fig. 2.3b) means, in wave-optical terms, that the incident wave is tilted through an angle ϵ after passing the magnetic field. With an arbitrary origin at $x = 0$ in Fig. 2.3, rays at a distance x enclose a magnetic flux $\Phi_{\mathrm{m}} = BLx$. Equation (3.6) gives the same value for the deflection angle $\epsilon = \Delta s/x$, with $\Delta s = \lambda(\varphi_2 - \varphi_1)/2\pi$, as that given by (2.17), obtained by using classical mechanics.

Because only the time-independent term of the phase in (3.2) is important for interference experiments, we reduce the wave function (3.2) of a plane wave to

$$\psi(z) = \psi_0 \exp(2\pi i k z). \tag{3.7}$$

For many applications, it is also of interest to use spherical waves

$$\psi = A_{\mathrm{Q}} \frac{\mathrm{e}^{2\pi i k r}}{r}. \tag{3.8}$$

A_{Q} is a measure of the magnitude of the source Q, and r denotes the distance from the source. The plane-wave function (3.7) and the spherical-wave function (3.8) are special solutions of the time-independent wave equation (3.19).

The widely used terms "wavefront" or "wave surface" can be defined as surfaces of constant phase φ. The wavefronts of a plane wave are planes normal to the direction of propagation. In the case of a spherical wave, they are concentric spheres. For a vanishing magnetic vector potential $\boldsymbol{A} = 0$, the rays of particle optics are trajectories normal to the wavefronts.

3.1.2 Probability Density and Wave Packets

A parallel electron beam with N electrons per unit volume and velocity v represents a current density

$$j = Nev \tag{3.9}$$

in A m^{-2}, where Nv is the flux of particles; that is, the number of electrons traversing a unit area per unit time. In electron microscopy, we can measure the current or current density, proportional to Nv, or we can count the number N by single-particle detection. To combine these possibilities of measuring with the wave concept, we use the quantum-mechanical formula for a flux of particles

$$\boldsymbol{j} = e \frac{i\hbar}{2m} (\psi \nabla \psi^* - \psi^* \nabla \psi). \tag{3.10}$$

When this formula is applied to a plane wave (3.7), the operator ∇ becomes $\partial/\partial z$, and substitution of (3.7) in (3.10) results in

$$j = e\frac{\hbar}{m}2\pi k|\psi_0|^2 = ev|\psi_0|^2 \tag{3.11}$$

by using the relation $k = mv/h$. Comparison of (3.11) with (3.9) shows that $|\psi_0|^2 = N$, which corresponds to the interpretation of $|\psi|^2 = \psi\psi^*$ as a probability density or $\psi\psi^*\mathrm{d}V = N\mathrm{d}V$ as the probability of finding N electrons in the volume element $\mathrm{d}V$. We shall call the quantity

$$I = |\psi\,|^2 = \psi\psi^* \tag{3.12}$$

the intensity, which can be used to relate the wave amplitude to measurable quantities.

We have to be careful when substituting $|\psi_0| = N^{1/2}$ because we can describe only one electron by a de Broglie wave ($N = 1$); interference effects between electron waves can occur only within the wave field of one electron (see also the discussion in Sect. 3.1.4).

Because $\psi\psi^*$ means the probability of finding an electron, its integral over all space should be unity for one electron:

$$\int_V \psi\psi^*\mathrm{d}V = 1. \tag{3.13}$$

An infinite plane wave such as (3.7) cannot be normalized by means of (3.13). The concept of a wave packet is therefore introduced to combine the motion of a particle of velocity v with the concept of a wave. A monochromatic wave with a discrete wavelength λ or wave number $k = 1/\lambda$ represents a plane wave with an infinite extension (Fig. 3.1a). A limited wave packet moving with the particle velocity v (Fig. 3.1b) can be obtained by superposing a broad spectrum $A(k)$ of wavelengths or wave numbers:

$$\psi = \int_{-\infty}^{+\infty} A(k)\mathrm{e}^{2\pi\mathrm{i}kz}\mathrm{d}k. \tag{3.14}$$

The amplitudes in front of and behind a wave packet vanish by destructive interference. The amplitudes are summed up by constructive interference with the correct phase only inside the wave packet. The width Δk of the wave-number spectrum $A(k)$ and the spatial width Δz of the wave packet are related by the Heisenberg uncertainty principle $\Delta p\Delta z > h/(4\pi)$ or $\Delta k\Delta z > 1/(4\pi)$ (compare the Fourier transform of a finite cosine wave in Table 3.2).

In practice, it is inconvenient to use a broad spectrum $A(k)$, which corresponds to the superposition (3.14) of many partial waves. Therefore, we continue to use the expression (3.7). The results obtained for the center of the k spectrum are not appreciably different from the behavior of the wave packet, provided that $\Delta k \ll k$.

3.1.3 Electron-Optical Refractive Index and the Schrödinger Equation

An electron-optical refractive index n can be introduced as in light optics, where it is defined as the ratio of the velocity c in a vacuum to the velocity c_m in matter or by the corresponding ratio of the wavelengths: $n = c/c_m = \lambda/\lambda_m$. The velocity of electrons in matter is influenced by the attractive Coulomb potential $V(r)$:

$$V(r) = -\frac{e^2 Z_{\text{eff}}(r)}{4\pi\epsilon_0 r}. \tag{3.15}$$

The effective number $Z_{\text{eff}}(r)$ takes into account the increased screening of the nuclear charge by the atomic electrons with increased r (Sect. 5.1.3). We have only to replace the energy in (2.12) by $E - V(r)$ to obtain the dependence of electron wavelength on r. The refractive index in the absence of a magnetic field becomes

$$n(r) = \frac{\lambda}{\lambda_m} = \frac{p_m}{p} = \left[\frac{2(E-V)E_0 + (E-V)^2}{2EE_0 + E^2}\right]^{1/2}. \tag{3.16}$$

This formula can be simplified if it is assumed that $V(r) \ll E$ and E_0,

$$n(r) = 1 - \frac{V(r)}{E}\,\frac{E_0 + E}{2E_0 + E} + \ldots; \tag{3.17}$$

$n \geq 1$ because $V(r)$ in matter is negative.

Figure 3.3 shows schematically the potential energy $V(r)$ along a row of atoms. The mean value $V_i = -eU_i$, which is the constant term of a Fourier expansion, is called the inner potential (Table 3.1). This inner potential causes

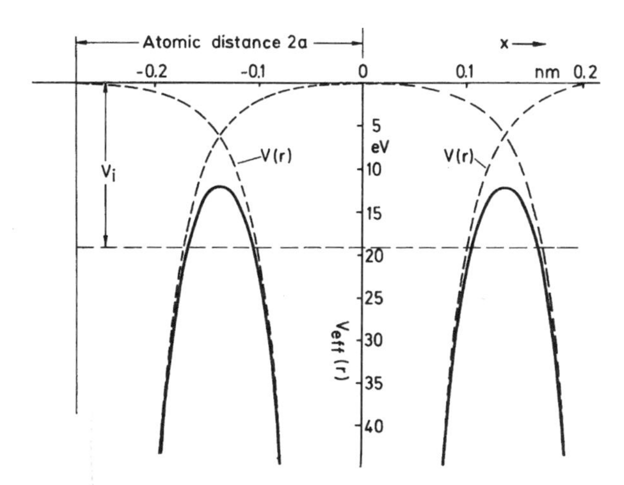

Fig. 3.3. Potential $V(r)$ of a crystal lattice along a row of Ge atoms with interatomic spacing $2a$ and definition of the inner potential $V_i = eU_i$.

Table 3.1. Values of the inner potential U_i (V) of various elements.

Be	7.8 ± 0.4	[3.12]	Au	21.1 ± 2	[3.14]
C	7.8 ± 0.6	[3.13]		$22.1 - 27.0$	[3.13]
Al	13.0 ± 0.4	[3.13]	Si	11.5	[3.16]
	12.4 ± 1	[3.14]	Ge	15.6 ± 0.8	[3.15]
	11.9 ± 0.7	[3.15]	W	23.4	[3.16]
Cu	23.5 ± 0.6	[3.13]	ZnS	10.2 ± 1	[3.14]
	20.1 ± 1.0	[3.15]			
Ag	20.7 ± 2	[3.14]			
	$17.0 - 21.8$	[3.14]			

a phase shift relative to a wave traveling in a vacuum. (The large local variations of $V(r)$ in the specimen produce elastic scattering of electrons at the nuclei; see Sect. 5.1.3.)

An optical path difference $\Delta s = (n-1)\,t$ and hence a phase shift φ that corresponds to a layer of thickness t can be introduced by writing

$$\varphi = \frac{2\pi}{\lambda}\Delta s = \frac{2\pi}{\lambda}(n-1)\,t = \frac{2\pi}{\lambda}\frac{eU_\mathrm{i}}{E}\frac{E_0+E}{2E_0+E}\,t. \tag{3.18}$$

Thus, for carbon films, for example, we have $U_\mathrm{i} = 8$ V, giving $n-1 = 4\times 10^{-5}$ for 100 keV electrons, for which $\lambda = 3.7$ pm. A film thickness t of 21 nm will be needed to obtain a phase shift φ of $\pi/2$.

The electron-optical refractive index or the inner potential U_i can be determined from the shift of single-crystal diffraction spots or Kikuchi lines in electron-diffraction patterns with oblique incidence (RHEED, Sect. 8.1.4) [3.16, 3.17]. An interference effect due to double refraction can be observed in small polyhedral crystals (e.g., MgO smoke); however, this can be explained completely only by the dynamical theory of electron diffraction [3.18, 3.19]. The phase shift also causes modifications of the Fresnel fringes at the edges of transparent foils (Sect. 3.2.2). However, the most accurate method of measuring U_i involves the use of electron interferometry (Sect. 3.1.4). Other methods of measuring U_i are discussed in [3.20]. The inner potential measured by surface-sensitive methods can be different from the value for bulk material.

Substitution of the wave number $k_\mathrm{m} = n(r)k$ in the time-independent wave equation

$$\nabla^2\psi + 4\pi k_\mathrm{m}^2\psi = 0, \tag{3.19}$$

which is also valid for electromagnetic waves and light quanta, yields the quantum-mechanical Schrödinger equation in a relativistically corrected form,

$$\begin{aligned}\nabla^2\psi &+ 4\pi^2 n^2(r)k^2\psi \\ &= \nabla^2\psi + 4\pi^2\left(1 - \frac{V}{E}\frac{2E_0+2E}{2E_0+E}\right)\frac{2EE_0+E^2}{h^2c^2}\psi = 0\end{aligned} \tag{3.20}$$

or with $\hbar = h/2\pi$,

$$\nabla^2\psi + \frac{2m_0}{\hbar^2}\left[E\left(1+\frac{E}{2E_0}\right) - V(r)\left(1+\frac{E}{E_0}\right)\right]\psi = 0, \tag{3.21}$$

or

$$\left[\frac{\hbar^2}{2m}\nabla^2 + E^* - V(r)\right]\psi = 0 \quad \text{with} \quad E^* = E\,\frac{2E_0+E}{2(E_0+E)}. \tag{3.22}$$

For energies well below the rest energy of the electron $E << E_0$, we obtain the conventional Schrödinger equation

$$\left[\frac{-\hbar^2}{2m}\nabla^2 + V(r)\right]\psi = E\psi. \tag{3.23}$$

3.1.4 Electron Interferometry and Coherence

An electron wave can be split into two coherent waves by an electron-optical biprism interferometer, the analogue of the Fresnel biprism of light optics as developed by Möllenstedt and Düker [3.21, 3.22].

Figure 3.4a shows the light-optical Fresnel biprism. The refracted waves behind the prism are generated by the virtual sources Q_1 and Q_2. The optical phase difference (3.18) can be calculated as a function of the coordinate x in the viewing plane from the path difference Δs. From Fig. 3.4a, we see that

$$\Delta s = \left[L^2 + \left(x+\frac{a}{2}\right)^2\right]^{1/2} - \left[L^2 + \left(x-\frac{a}{2}\right)^2\right]^{1/2} \simeq \frac{ax}{L} \tag{3.24}$$

if $x, a \ll L$.

Constructive interference maxima are obtained if $\Delta s = n\lambda$ or $\varphi = 2\pi n$, n being an integer. The distance between the maxima becomes

$$\Delta x = \frac{L}{a}\lambda = \frac{\lambda}{2\beta}. \tag{3.25}$$

The electric field between a thin wire (diameter $\simeq 1\mu$m) and grounded plates can form such a biprism for electrons (Fig. 3.4b). A shadow of the wire is seen at the viewing plane if the wire is grounded. As the positive bias of the wire is increased, the two waves with wave vectors $\boldsymbol{k}_1$ and $\boldsymbol{k}_2$ can overlap, resulting in an amplitude distribution

$$\begin{aligned}\psi &= \psi_0[\exp(2\pi i\boldsymbol{k}_1\cdot\boldsymbol{r}) + \exp(2\pi i\boldsymbol{k}_2\cdot\boldsymbol{r})]\\ &= \psi_0\{\exp[\pi i(\boldsymbol{k}_1-\boldsymbol{k}_2)\cdot\boldsymbol{r}] + \exp[-\pi i(\boldsymbol{k}_1-\boldsymbol{k}_2)\cdot\boldsymbol{r}]\}\exp[\pi i(\boldsymbol{k}_1+\boldsymbol{k}_2)\cdot\boldsymbol{r}]\\ &= 2\psi_0\cos[\pi(\boldsymbol{k}_1-\boldsymbol{k}_2)\cdot\boldsymbol{r}]\exp(2\pi ikz).\end{aligned} \tag{3.26}$$

The intensity distribution $I(x) = \psi\psi^*$ becomes

$$I(x) = 4I_0\cos^2(2\pi\beta x/\lambda), \tag{3.27}$$

in which we have written $I_0 = |\psi_0|^2$, $(\boldsymbol{k}_1+\boldsymbol{k}_2)\cdot\boldsymbol{r} \simeq 2kz$, and $(\boldsymbol{k}_1-\boldsymbol{k}_2)\cdot\boldsymbol{r} = 2kx\sin\beta \simeq 2\beta x/\lambda$ (see Curve 1 in Fig. 3.4c).

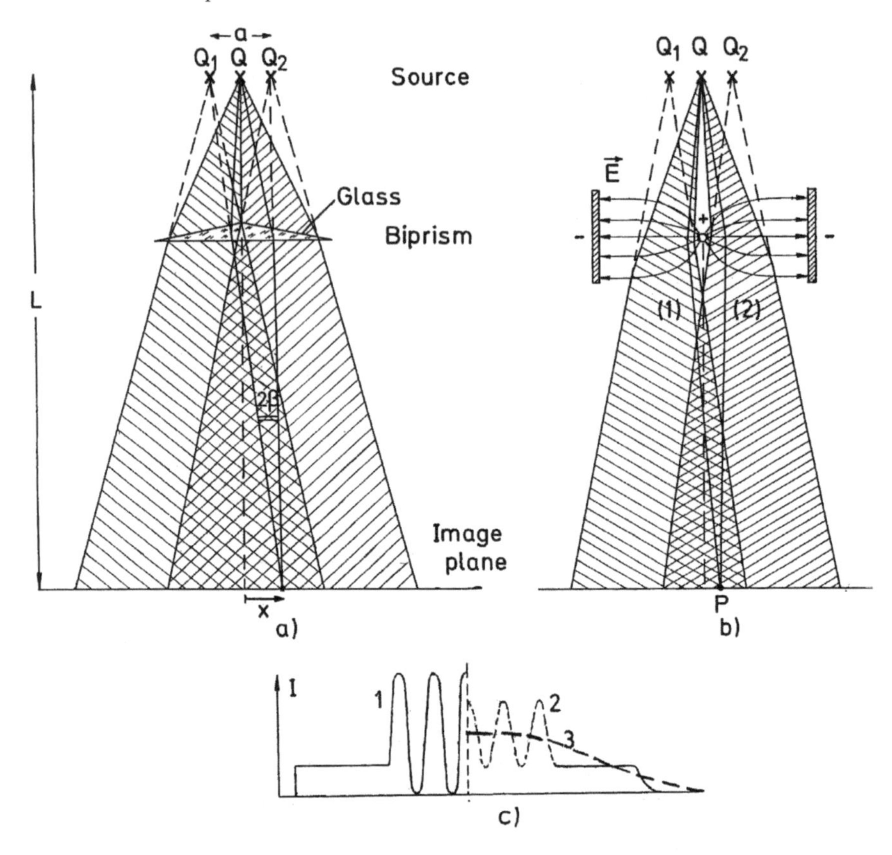

Fig. 3.4. (**a**) Fresnel-biprism experiment of light optics. (**b**) Electron-optical realization of a biprism experiment with a positively biased wire (**c**) Curve 1 (*left*); intensity distribution of interference fringes for coherent illumination and curves 2 and 3 (*right*) for partially coherent and incoherent illumination.

If an extended source is used rather than a point source, as assumed above, the probability (3.27) of observing an electron at any point x will not be changed. However, electrons from other points of the extended source will produce shifted interference patterns. The maxima and minima are totally blurred if the extension Δa of the source is larger than the distance Δx between the maxima. The illumination is said to be "incoherent" (see Curve 3 in Fig. 3.4c). In the center of the overlap, the intensity becomes $2I_0$, the value expected if no interference effects occur. Partially coherent illumination, with $\Delta a < \Delta x$, leads to a decrease of the maxima, and the minima no longer fall to zero (Curve 2 in Fig. 3.4c).

When the source size is sufficiently small, $\Delta a \ll \Delta x = L\lambda/a$, the radiation is said to be spatially coherent. The angle, $\alpha_{\rm i} = \Delta a/2L$, can be interpreted as the illumination angle; that is, the cone angle of the rays from different points of the source at the point P in the observation plane. The condition $\Delta a \ll \Delta x$ for spatial coherence is thus equivalent to $\Delta a \alpha_{\rm i} \ll \lambda/2$, which is also used as a coherence condition in light optics.

A coherence condition for temporal coherence results from the finite coherence length Δz of the wave packet (Fig. 3.1b). The path difference Δs between two interfering waves has to be much smaller than Δz. The value of $\Delta z = c\Delta\tau$ is related to the emission time $\Delta\tau$. In light emission, a normal dipole transition has an emission time $\Delta\tau \simeq 10^{-8}$ s so that with $v = c$ we have $\Delta z \simeq 3$ m $\simeq 6 \times 10^6\lambda$ for $\lambda = 0.5$ μm. For electrons, $\Delta\tau$ can be estimated from the Heisenberg uncertainty relation $\Delta E \Delta\tau \simeq h$, where $\Delta E \simeq$ 1 eV is the energy spread at the electron gun; this gives $\Delta\tau \simeq 4 \times 10^{-15}$ s. Thus 100 keV electrons, for which $v = 1.64 \times 10^8$ m s^{-1}, have a coherence length $\Delta z = v\Delta\tau = 600$ nm $\simeq 2 \times 10^5\lambda$. Möllenstedt and Wohland [3.23] produced path differences Δs of the order of Δz using a biprism combined with a Wien filter and confirmed that the biprism interference pattern decreases in amplitude if $\Delta s \simeq \Delta z$.

The influence of spatial and temporal coherence on phase contrast is discussed in Sects. 6.4.2 and 6.4.3. For further discussion of coherence and the introduction of coherence functions, see [3.24, 3.25].

The biprism experiments shed light on another important aspect of wave optics. In particle optics, the concept of a trajectory is used. In our example, the particle can pass either side of the wire. In wave optics, the wave of a single electron passes on both sides of the wire simultaneously and we can observe only the probability of detecting the electron at some position x. It is therefore nonsense to ask on which side the electron has passed. If we put a detector on one side of the wire, half of the total number of electrons will be detected, but we thereby suppress all wave amplitudes on this side and will observe no interference pattern.

Introduction of a thin foil on one side of the wire causes a phase shift (3.18) given by the inner potential $U_{\rm i}$ (Sect. 3.1.3). The phase shift can be measured accurately by using an electron interference microscope, which images both the specimen and an interference pattern. It is advisable to use a simple geometry for measuring, such as evaporated stripes [3.12] or circular areas [3.14].

3.2 Fresnel and Fraunhofer Diffraction

3.2.1 Huygens' Principle and Fresnel Diffraction

In wave optics, all other wavefronts can be calculated once the shape of one of them is known by using the Kirchhoff diffraction theory based on the wave equation (3.19). However, the simpler treatment offered by Huygens' principle can also be used in electron optics; this states that each surface element dS

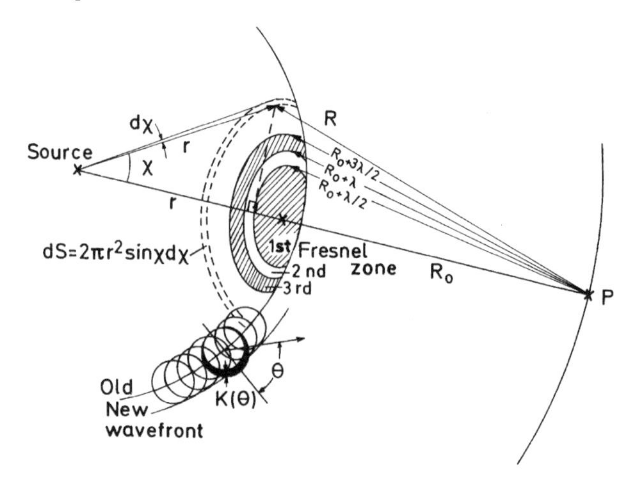

Fig. 3.5. Illustration of Huygens' principle and Fresnel zones showing how the wave amplitude at the point P is obtained by summing the amplitudes of the Huygens wavelets from a spherical wavefront of radius r.

of a wavefront generates a secondary spherical wave with amplitude

$$\mathrm{d}\psi = \frac{A(\theta)}{\mathrm{i}\lambda}\,\psi\,\frac{\mathrm{e}^{2\pi\mathrm{i}kR}}{R}\mathrm{d}S, \quad A(\theta) = (1+\cos\theta)/2, \tag{3.28}$$

where ψ denotes the amplitude of the incident wave at dS and θ the angle of emission to the normal of the wavefront. A new wavefront is generated by the superposition of all of the secondary waves (Fig. 3.5). At a point P in front of the wavefront, the amplitudes of all the secondary waves have to be summed, considering their phase shifts. The factor $A(\theta)$ is unity in the direction of the propagating wave and decreases with increasing θ. For the reverse direction, A is zero. The exact form of $A(\theta)$ is not important for the following calculation. The factor $1/\mathrm{i} = \exp(-\mathrm{i}\pi/2)$ in (3.28) represents a phase shift of $-\pi/2$ relative to the incident wave.

We apply Huygens' principle to the propagation of a spherical wave (Fig. 3.5). The known wavefront is thus spherical with radius r and amplitude ψ (3.8). The surface element $\mathrm{d}S = r\mathrm{d}\chi \cdot 2\pi r \sin\chi$ by using the spherical polar coordinates r and χ. The distance R to the point P can be calculated from $R^2 = r^2 + (r+R_0)^2 - 2r(r+R_0)\cos\chi$, from which we obtain $2R\mathrm{d}R = 2r(r+R_0)\sin\chi\mathrm{d}\chi$, and it follows that $\mathrm{d}S = 2\pi[r/(r+R_0)]R\mathrm{d}R$. The amplitude ψ_P at P is obtained by integration over all the secondary waves,

$$\begin{aligned}\psi_\mathrm{P} &= \int_S \frac{A(\theta)}{\mathrm{i}\lambda}A_\mathrm{Q}\frac{\mathrm{e}^{2\pi\mathrm{i}kr}}{r}\frac{\mathrm{e}^{2\pi\mathrm{i}kR}}{R}\mathrm{d}S \\ &= \frac{2\pi A_\mathrm{Q}\,\mathrm{e}^{2\pi\mathrm{i}kr}}{\mathrm{i}\lambda(r+R_0)}\int_{R_0}^{R_\mathrm{max}} A(\theta)\mathrm{e}^{2\pi\mathrm{i}kR}\mathrm{d}R.\end{aligned} \tag{3.29}$$

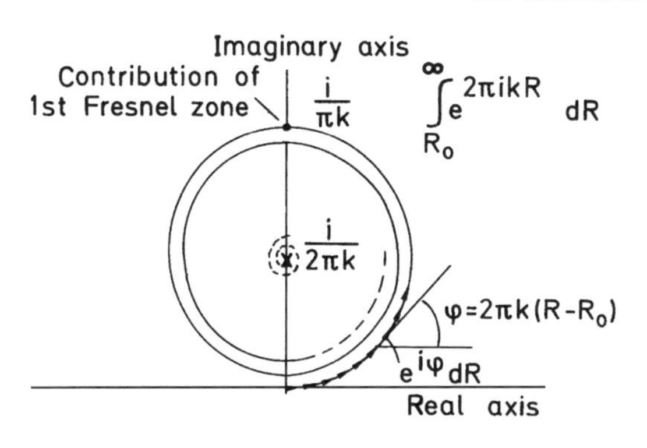

Fig. 3.6. Amplitude-phase diagram for the integral in (3.29).

The result can be established by means of an amplitude-phase diagram (APD) (Fig. 3.6). The term $\exp(\mathrm{i}\varphi)\mathrm{dR}$ can be represented in the complex number plane by a line element $\mathrm{d}R$ inclined at an angle φ to the real axis. The integration in (3.29) means adding infinitesimal line elements $\mathrm{d}R$ with increasing $\varphi = 2\pi kR$, resulting in a circle. The radius of the circle decreases because of the decrease of $A(\theta)$ with increasing θ. The result is a spiral that converges to the center of the circle. Starting from the lower limit of integration, $R = R_0$, the integral reaches its greatest value when $\varphi = 2\pi k(R - R_0) = \pi$ or $R - R_0 = \lambda/2$. The value of the integral then decreases again because the phase shift of the secondary wave becomes greater than π. This is the basic idea of the Fresnel-zone construction. If a sphere of radius $R = R_0 + \lambda/2$ centered at P is drawn, as in Fig. 3.5, the first Fresnel zone is obtained, as indicated by the hatched area on the wavefront, which contributes to the amplitude ψ_P with a positive value. The second Fresnel zone, between the corresponding radii $R_0 + \lambda/2$ and $R_0 + \lambda$, results in a negative contribution, the next Fresnel zone again gives a positive contribution, and so on with alternating signs. The convergence of the APD to the center of the circle means that the integral in (3.29) becomes only half of the value ψ_P of the first Fresnel zone. This results in the following value for the integral in (3.29):

$$\int_{R_0}^{R_{\max}} A(\theta)\mathrm{e}^{2\pi\mathrm{i}kR}\mathrm{d}R = \frac{1}{2}\int_{R_0}^{R_0+\lambda/2} \mathrm{e}^{2\pi\mathrm{i}kR}\mathrm{d}R = -\frac{1}{2\pi\mathrm{i}k}\mathrm{e}^{2\pi\mathrm{i}kR_0} . \tag{3.30}$$

Substituting this value in (3.29) gives

$$\psi_\mathrm{P} = A_\mathrm{Q}\exp[2\pi\mathrm{i}k(r + R_0)]/(r + R_0) . \tag{3.31}$$

This is the expected formula for the wavefront at a distance $r + R_0$ from the point source. This simple example demonstrates the power of Huygens' principle.

We now use Huygens' principle with another choice of coordinates, which directly yields Fresnel diffraction at an edge (Sect. 3.2.2). The surface element $\mathrm{d}S$ is placed in an x-y plane normal to the line from the source Q to the point P (Fig. 3.7). The distance r becomes

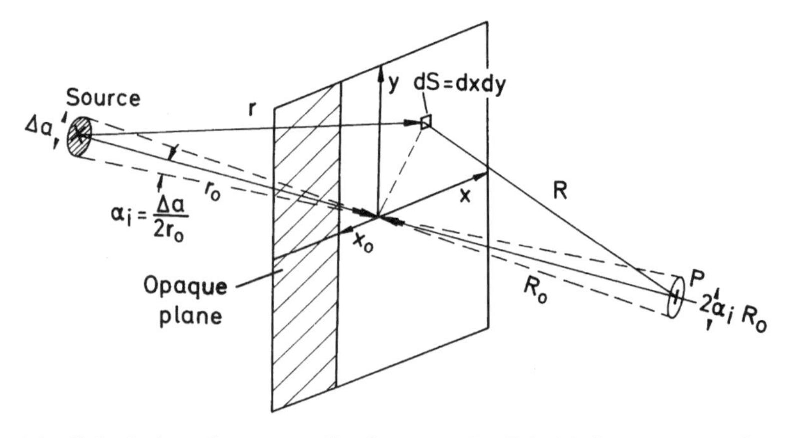

Fig. 3.7. Calculation of wave amplitude at a point P behind an opaque edge.

$$r = (r_0^2 + x^2 + y^2)^{1/2} = r_0\left(1 + \frac{x^2+y^2}{r_0^2}\right)^{1/2}$$

$$= r_0\left(1 + \frac{x^2+y^2}{2r_0^2} + \ldots\right), \tag{3.32}$$

with a corresponding formula for R if r_0 is replaced by R_0. Substitution of (3.32) into (3.29) results in (3.33); $A(\theta)$ has been omitted because the integral already converges for small values of x and y, for which $A(\theta) = 1$:

$$\psi_P = \frac{A_Q \exp[2\pi i k(r_0 + R_0)]}{i\lambda r_0 R_0} \int\limits_{-x_0}^{+\infty}\int\limits_{-\infty}^{+\infty} \exp\left(2\pi i k x^2 \frac{r_0+R_0}{2r_0R_0}\right)$$

$$\times \exp\left(2\pi i k y^2 \frac{r_0+R_0}{2r_0R_0}\right) dx dy. \tag{3.33}$$

The substitutions $u = x\left(\frac{2(r_0+R_0)}{\lambda r_0 R_0}\right)^{1/2}$ and $v = y\left(\frac{2(r_0+R_0)}{\lambda r_0 R_0}\right)^{1/2}$ give

$$\psi_P = \frac{A_Q \exp[2\pi i(r_0 + R_0)]}{2i(r_0+R_0)} \int\limits_{-u_0}^{+\infty} \exp(i\pi u^2/2) du \int\limits_{-\infty}^{+\infty} \exp(i\pi v^2/2) dv$$

$$= \frac{A_Q \exp[2\pi i(r_0 + R_0)]}{i(r_0+R_0)} \frac{1}{2}[C(u) + iS(u)]_{-u_0}^{+\infty}[C(v) + iS(v)]_{-\infty}^{+\infty}, \tag{3.34}$$

where $C(u)$ and $S(u)$ are the tabulated Fresnel integrals and x_0 replaces u_0. $C(u) + iS(u)$ produces the Cornu spiral in the APD of Fig. 3.8a. The more complicated shape of the APD as compared with Fig. 3.6 results only from the different choice of coordinate system. The point of convergence is –0.5(1+i) for the limit of integration $u \to -\infty$ and 0.5(1+i) for $u \to +\infty$. The total amplitude is obtained by connecting the two points of convergence and is

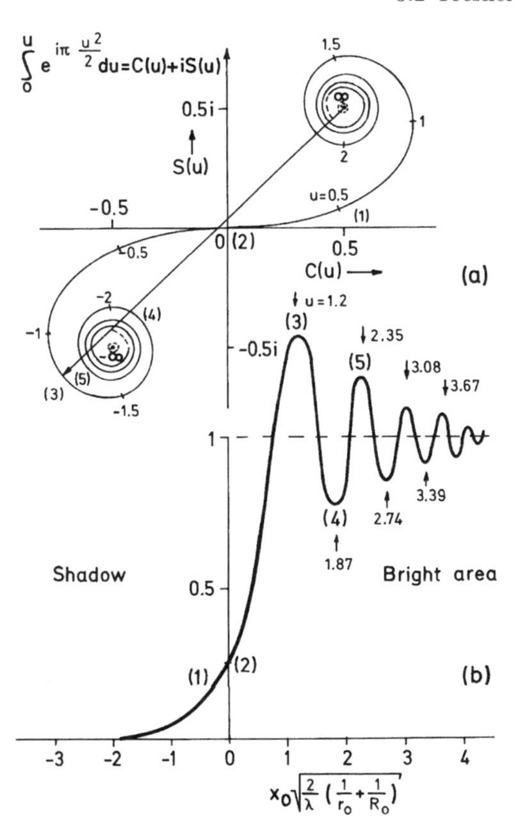

Fig. 3.8. (**a**) Amplitude-phase diagram *(Cornu spiral)* for one of the integrals in (3.34). (**b**) Intensity distribution of Fresnel fringes in the Fresnel diffraction pattern of an opaque edge.

hence 1 + i. Because we have to consider the product of two integrations in the x and y directions, the two quantities in square brackets in (3.34) give $(1 + \mathrm{i})^2 = 2\mathrm{i}$, and again we obtain the expected value ψ_{P} of the wave excitation at P at a distance $r_0 + R_0$ from the source Q.

3.2.2 Fresnel Fringes

The last section has shown that the wave amplitude at a point of a wavefront and the wave propagation can both be described by Huygens' elementary waves and Fresnel integrals. This formalism will now be applied to electron-opaque obstacles. One important example is the appearance of Fresnel diffraction fringes at opaque half-planes (Fig. 3.9).

At a distance x_0 from the shadow of a half-plane (Fig. 3.7), the intensity is obtained by integration in the x direction from x_0 to $+\infty$. In the y direction, we again consider stripes of width dy from $-\infty$ to $+\infty$, and the integral in the y direction has the value 1 + i, as before. The amplitude contribution from the x direction is obtained by connecting the point of convergence 0.5 + 0.5 i for $x_0 = u = +\infty$ to the corresponding point $u = -x_0[2(r_0+R_0)/\lambda r_0 R_0]^{1/2}$ on the

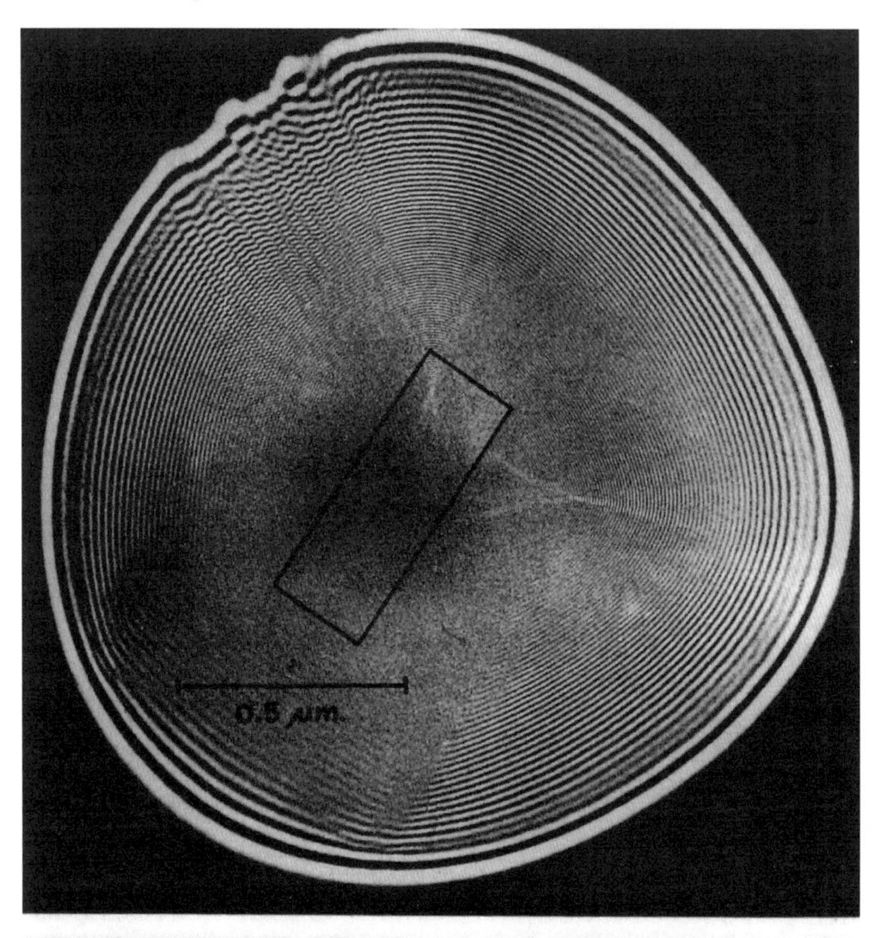

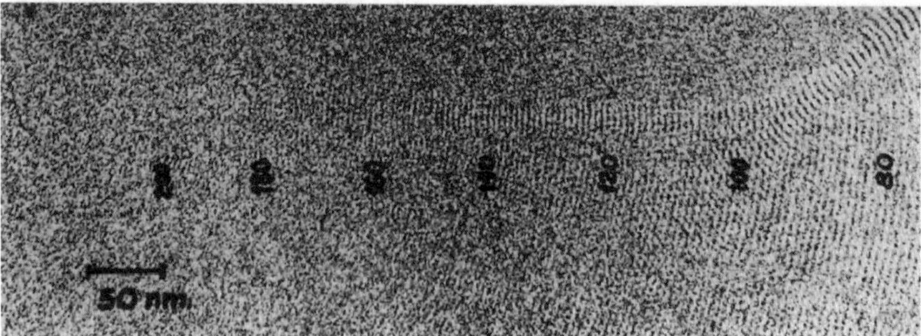

Fig. 3.9. Numerous Fresnel fringes around a hole in a carbon foil obtained with a highly coherent field-emission gun.

Cornu spiral. The coordinate u is the arc length along the Cornu spiral. At the point (3) of the spiral farthest from the positive-convergence point, we obtain a maximum of amplitude and intensity. In Fig. 3.8a,b, further corresponding points on the Cornu spiral and in the intensity distribution are numbered. In all practical cases, $r_0 \gg R_0$, a relatively accurate formula can be obtained for the positions of the maxima u_n or x_n from the condition that the phase in the integral of (3.34) will be $\pi u^2/2 = \pi(2n - 5/4)$ for $n = 1, 2, \ldots$ or, in other words, that the tangent to the Cornu spiral is inclined at an angle of –45° to the real axis. This condition results in

$$u_n = \sqrt{(8n-5)/2}; \quad x_n = \sqrt{\lambda R_0 (8n-5)/4}. \tag{3.35}$$

The intensity distribution of these Fresnel fringes in a plane at a distance R_0 below an edge can be imaged by defocusing the objective lens [3.26]. Observation of these Fresnel fringes is important for the recognition and correction of astigmatism (Sect. 2.3.4). The number of Fresnel fringes visible is a measure of the spatial coherence of the electron beam. Figure 3.9 shows many Fresnel fringes around a hole in a supporting film illuminated with a field-emission gun. The influence of the spherical aberration of the objective lens on the intensity and position of Fresnel fringes has to be considered [3.27] for transparent supporting films. The decrease of wave amplitude caused by scattering and the phase shift due to the inner potential U_{i} also have an effect [3.28, 3.29]. Fresnel fringes can likewise be used to characterize grain boundaries [3.30].

If the source has a finite size Δa, the specimen is irradiated with an angular aperture $\alpha_{\mathrm{i}} = \Delta a/2r_0$, which causes blurring of the Fresnel diffraction pattern proportional to $2\alpha_{\mathrm{i}} R_0$. The intensity distribution with such partially coherent illumination is given by the convolution of the coherent distribution with the geometric shadow distribution of the source at a distance R_0. Because the distances between the diffraction maxima $x_{n+1} - x_n$ (3.35) decrease with increasing n, the diffraction maxima of high order disappear first, and only one Fresnel maximum can be observed with a thermionic cathode when a larger illumination aperture necessary for visual observation of the viewing screen at high magnification is used. A small illumination aperture, use of an image intensifier or CCD camera, and observation on a TV screen are necessary to see more than one fringe. All of the maxima are blurred when the angular aperture α_{i} is very large, as in the case of incoherent illumination; the intensity distribution is then the same as that of the purely geometric shadow of the edge thrown by an extended source.

3.2.3 Fraunhofer Diffraction

Fresnel diffraction goes over into Fraunhofer diffraction if a plane incident wave is used and if the diffraction pattern is observed at an infinite distance. Alternatively, this pattern can be observed in the focal (diffraction) plane of a lens (Fig. 3.10). A parallel beam inclined at a small angle θ to the optical axis converges to a point in this plane at a distance $f\theta$ from the optic axis. No further phase shifts occur if the lens is aberration-free. The exit wave amplitude after passing the specimen can be described by

$$\psi = \psi_0 a_{\mathrm{s}}(\boldsymbol{r}) \exp[\mathrm{i}\varphi_{\mathrm{s}}(\boldsymbol{r})] \exp(2\pi \mathrm{i} kz) = \psi_{\mathrm{s}}(\boldsymbol{r}) \exp(2\pi \mathrm{i} kz) \tag{3.36}$$

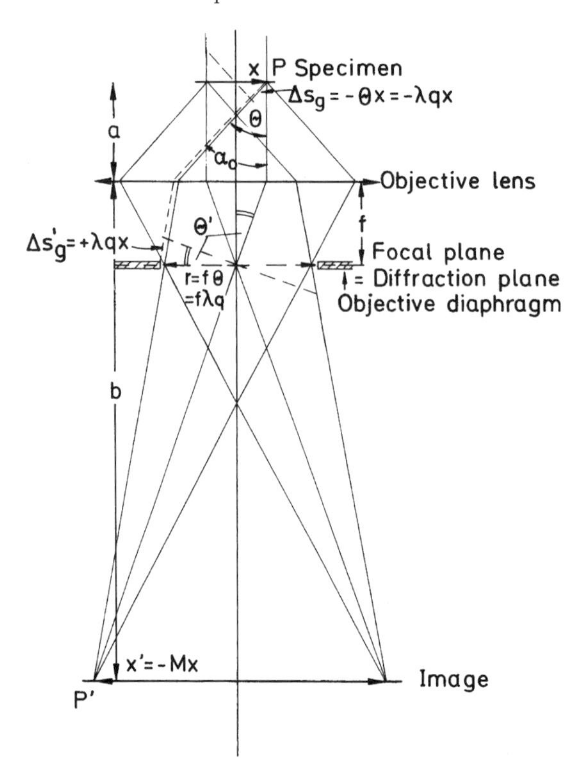

Fig. 3.10. Ray diagram for image formation by an objective lens; from the wavefronts (– – –), the path differences Δs_{g} and $\Delta s'_{\mathrm{g}}$ can be derived.

($\boldsymbol{r}$: radius vector in the specimen plane from the origin on the optic axis, $a_{\mathrm{s}}(\boldsymbol{r}) \leq 1$: local decrease of amplitude (absorption), and $\varphi_{\mathrm{s}}(\boldsymbol{r})$: phase shift caused by the specimen).

Furthermore, a phase shift φ_{g} has to be introduced that results from the geometric path difference of the plane wavefront in the direction θ. Figure 3.11 shows that the two points O and P separated by a distance $\boldsymbol{r}$ correspond to an optical path difference

$$\Delta s_{\mathrm{g}} = \boldsymbol{u}_0 \cdot \boldsymbol{r} - \boldsymbol{u} \cdot \boldsymbol{r} \tag{3.37}$$

($\boldsymbol{u}_0 = \lambda \boldsymbol{k}_0$ and $\boldsymbol{u} = \lambda \boldsymbol{k}$ are unit vectors in the direction of the incident and scattered waves, respectively). The phase difference is given by

$$\varphi_{\mathrm{g}} = \frac{2\pi}{\lambda} \Delta s_{\mathrm{g}} = -2\pi(\boldsymbol{k} - \boldsymbol{k}_0) \cdot \boldsymbol{r} = -2\pi \boldsymbol{q} \cdot \boldsymbol{r}. \tag{3.38}$$

Figure 3.11 shows that

$$|\boldsymbol{k} - \boldsymbol{k}_0| = |\boldsymbol{q}| = 2k \sin\frac{\theta}{2} \simeq \frac{\theta}{\lambda}. \tag{3.39}$$

A diffraction grating with period Λ (lattice spacing) generates a diffraction maximum at an angle $\sin\theta \simeq \theta = \lambda/\Lambda$. This implies that q in (3.37)–(3.39) is equal to Λ^{-1}; it is known as the *spatial frequency* by analogy with the relation

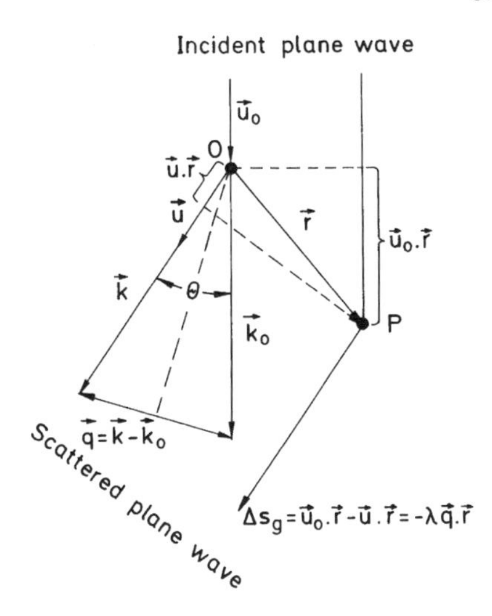

Fig. 3.11. Demonstration of the path-length difference Δs_{g} caused by scattering at two points O and P.

$\nu = T^{-1}$ between the temporal frequency and the period T. From now on, we shall use $\boldsymbol{q}$ as a coordinate in the diffraction plane.

The amplitude $F(\boldsymbol{q})$ in the diffraction plane can be obtained by integration over all of the surface elements $\mathrm{d}\boldsymbol{S} = \mathrm{d}^2\boldsymbol{r}$ of the specimen plane,

$$F(\boldsymbol{q}) = \int_S \psi_{\mathrm{s}}(\boldsymbol{r}) \exp(\mathrm{i}\varphi_{\mathrm{g}})\mathrm{d}\boldsymbol{S} = \int_S \psi_{\mathrm{s}}(\boldsymbol{r}) \exp(-2\pi\mathrm{i}\boldsymbol{q}\cdot\boldsymbol{r})\mathrm{d}^2\boldsymbol{r}. \tag{3.40}$$

This shows that $F(\boldsymbol{q})$ is the Fourier transform of $\psi_{\mathrm{s}}(\boldsymbol{r})$.

3.2.4 Mathematics of Fourier Transforms

This section contains a short review of the mathematics of Fourier transforms, which are important not only in the description of Fraunhofer diffraction and electron diffraction at crystal lattices but also in the electron-optical theory of image formation. For simplicity, we normally discuss one-dimensional functions $f(x)$. There is no difficulty in extending this to two- and three-dimensional Fourier transforms (see the examples in Table 3.2).

Let $f(x)$ be a real or complex function of the real variable x. The Fourier transform of $f(x)$ is defined by the mathematical operation $\mathbf{F}$:

$$\mathbf{F}\{f(x)\} = F(q) = \int_{-\infty}^{+\infty} f(x)\mathrm{e}^{-2\pi\mathrm{i}qx}\mathrm{d}x. \tag{3.41}$$

$f(x)$ can be obtained from $F(q)$ by the inverse Fourier transform $\mathbf{F}^{-1}$, which has the opposite sign in the exponent:

$$\mathbf{F}^{-1}\{F(q)\} = f(x) = \int_{-\infty}^{+\infty} F(q)\mathrm{e}^{+2\pi\mathrm{i}qx}\mathrm{d}q. \tag{3.42}$$

The following relations can be obtained from the definition (3.41) of a Fourier transform:

1) *Linearity*

$$\mathbf{F}\{af(x) + bg(x)\} = aF(q) + bG(q). \quad (3.43)$$

2) *Translation theorem*

$$\mathbf{F}\{f(x - x')\} = F(q)e^{-2\pi i q x'} \quad \text{or} \quad \mathbf{F}^{-1}\{F(q - q')\} = f(x)e^{+2\pi i q' x}. \quad (3.44)$$

3) *Scale change*

$$\mathbf{F}\{f(ax)\} = \frac{1}{|a|}F\left(\frac{q}{a}\right) \quad \text{or} \quad \mathbf{F}^{-1}\left\{F\left(\frac{q}{a}\right)\right\} = |a|f(ax). \quad (3.45)$$

Table 3.2 contains concrete examples of Fourier transforms. Example 1a, a rectangular function (the slit in a diffraction experiment), will be calculated in detail as an example. Because of the Euler relation $\exp(2\pi i qx) = \cos(2\pi qx)$ + i $\sin(2\pi qx)$, the last sine term can be omitted in the integration of (3.41) owing to the antisymmetry of this term:

$$F_1(q) = \int_{-\infty}^{+\infty} f_1(x)e^{2\pi i qx}dx = \int_{-a/2}^{+a/2} \cos(2\pi qx)dx = a\,\frac{\sin(\pi qa)}{\pi qa}\,. \quad (3.46)$$

If the width of the slit a tends to zero, a δ-function results (the point source in Example 1b). The width of the diffraction maximum in $F_1(q)$ then goes to infinity. This means that the diffraction amplitude $F(q)$ of a point source is isotropic in all directions q. The Fourier transform of a δ-function at the position b relative to the origin (Example 1c) is obtained by using the translation theorem (3.44).

Further examples are shown in Table 3.2 - a rectangular slit (2), a parallelepiped (3), a circular diaphragm (4) and a sphere (5) - illustrating Fourier transforms of functions in more than one dimension.

A Gaussian function (Example 6) is again a Gaussian function after a Fourier transform but with the reciprocal half-width.

The Fourier transform $F_7(q)$ of a one-dimensional point lattice with N points distance d apart shows principal maxima with spacing $1/d$ if the numerator and denominator of $F_7(q)$ simultaneously become zero. There are $N - 1$ zeros between these principal maxima, where only the numerator is zero. The widths of the principal maxima decrease in proportion to $1/Nd$, and the amplitudes of the subsidiary maxima are further decreased. Example 7 becomes Example 11 for $N \to \infty$.

Convolution Theorem for Fourier Transforms. This theorem is of interest for calculating Fourier transforms of a product or a convolution of two functions each of whose Fourier transforms is known. We first introduce the

Table 3.2. Examples of Fourier transforms.

Specimen function $f(x)$	Fourier Transform $\mathbf{F}\{f(x)\} = F(q)$
1 a) One-dimensional slit	
$f_1(x) = \begin{cases} 1 & \text{if } \lvert x\rvert < a/2 \\ 0 & \text{if } \lvert x\rvert > a/2 \end{cases}$	$F_1(q) = a\dfrac{\sin \pi a q}{\pi a q}$
	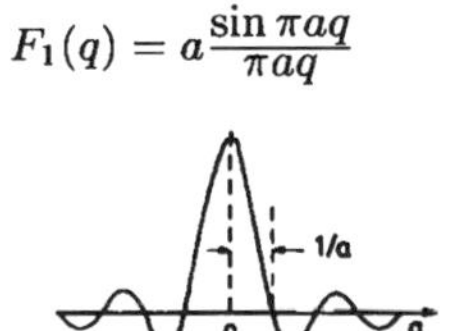
1 b) Point source ($a \to 0$) one- $f_1(x) = \delta(x)$, $\int \delta(x)\mathrm{d}x = 1$ two-dimensional $f_1(\boldsymbol{r}) = \delta(\boldsymbol{r})$, $\int\int \delta(\boldsymbol{r})\mathrm{d}^2\boldsymbol{r} = 1$	$F_1(q) = 1$ (isotropic scattering) 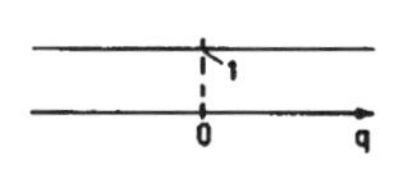
1 c) $f_1(x) = \delta(x - b)$	$F_1(q) = \exp(2\pi i b q)$; $\lvert F_1(q)\rvert = 1$ 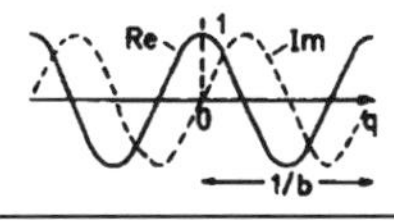
2) Transparent rectangle $f_2(x, y)$ 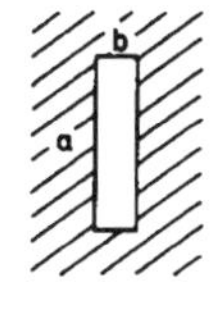	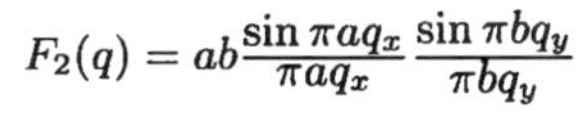 $F_2(q) = ab\dfrac{\sin \pi a q_x}{\pi a q_x}\dfrac{\sin \pi b q_y}{\pi b q_y}$ 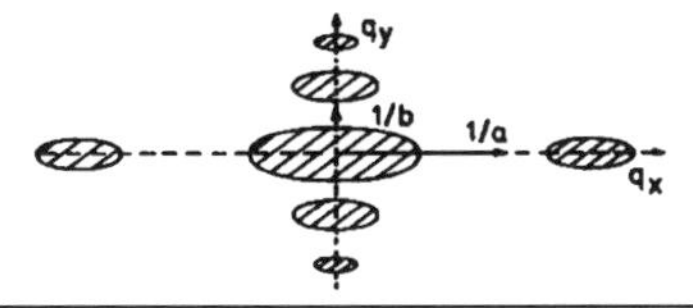
3) Parallelepiped $f_3(x, y, z)$ 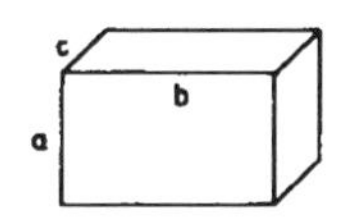	$F_3(\boldsymbol{q}) = abc\dfrac{\sin \pi a q_x}{\pi a q_x}\dfrac{\sin \pi b q_y}{\pi b q_y}\dfrac{\sin \pi c q_z}{\pi c q_z}$
4) Circular hole 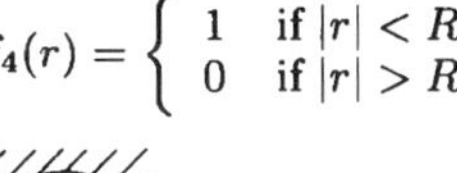 $f_4(r) = \begin{cases} 1 & \text{if } \lvert r\rvert < R \\ 0 & \text{if } \lvert r\rvert > R \end{cases}$ 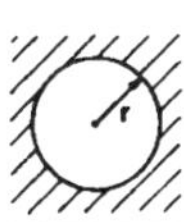	$F_4(q) = \pi R^2 \dfrac{J_1(2\pi q R)}{2\pi q R}$ (Airy distribution) 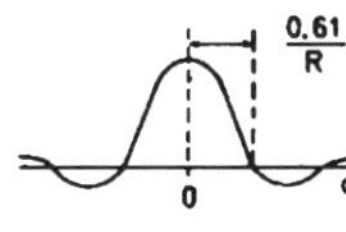

Table 3.2. (continued)

Specimen function $f(x)$	Fourier transform $\mathbf{F}\{f(x)\} = F(q)$
5) Sphere of radius R: $f_5(r)$	$F_5(q) = \frac{4\pi}{3} R^3 3 \frac{\sin u - u\cos u}{u^3}; \quad u = 2\pi q R$
6) Gaussian function $f_6(x) = \exp[-(x/a)^2]$	$F_6(q) = \sqrt{\pi} a \exp[-(\pi q a)^2]$
$h_x = \sqrt{\ln 2}\, a$; 0; x	$h_q = \sqrt{\ln 2}\, \frac{1}{\pi a}$; 0; q
7) One-dimensional point lattice $f_7(x) = \sum_{n=1}^{N} \delta(x - x_n)$	$F_7(q) = \frac{\sin \pi q N d}{\sin \pi q d}$
d; 0; x	1/d; 0; q
8) N slits of width a $f_8(x) = f_1(x) \otimes f_7(x)$	$F_8(q) = F_1(q) \cdot F_7(q) = a \frac{\sin \pi a q}{\pi a q} \frac{\sin \pi q N d}{\sin \pi q d}$
0; x	$F_1(q)$; 0; q
9) Infinite wave $f_9(x) = \cos(2\pi x/\Lambda)$	$F_9(q) = \frac{1}{2}\left[\delta\left(q + \frac{1}{\Lambda}\right) + \delta\left(q - \frac{1}{\Lambda}\right)\right]$
Λ; x	-1/Λ; 0; +1/Λ; q
10) Wave packet of width a $f_{10}(x) = f_1(x) \cdot f_9(x)$	$F_{10}(q) = F_1(q) \otimes F_9(q)$
Λ; a; x	1/a; -1/Λ; 0; +1/Λ; q

Table 3.2 (continued)

Specimen function $f(x)$	Fourier transform $\mathbf{F}\{f(x)\} = F(q)$
11) Infinite point row $N \to \infty$	
$f_{11} = \sum_{n=-\infty}^{+\infty} \delta(x - nd)$	$F_{11}(q) = \sum_{n=-\infty}^{+\infty} \delta\left(q - \frac{n}{d}\right)$
 0 d x	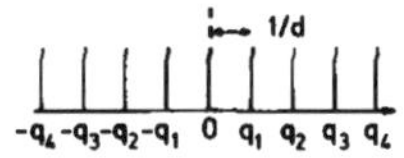 1/d $-q_4$ $-q_3$ $-q_2$ $-q_1$ 0 q_1 q_2 q_3 q_4
12) Infinite periodic function	
$f_{12}(x) = f_{11}(x) \otimes f_1(x)$	$F_{11}(q) = F_{12}(q) \cdot F_1(q) = \sum_{n=-\infty}^{+\infty} F_1(q_n)\delta(q - q_n)$
 a 0 d x	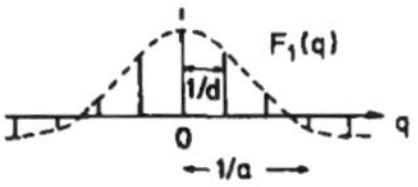 $F_1(q)$ 1/d 0 q 1/a

concept of convolution of two functions $f(x)$ and $g(x)$. Let us consider, for example, measurements of the intensity distribution $f(x)$ of a photographic emulsion with a densitometer. Let $g(x)$ be the slit function $f_1(x)$ introduced in Table 3.2 describing the transmission of a slit. The slit moves across the function $f(x)$, and at a position x all values of the function $f(x)$ between the limits $x - a/2$ and $x + a/2$ will be integrated. The resulting intensity curve is a convolution of the functions $f(x)$ and $g(x)$, which can be described mathematically by

$$C(x) = \int_{-\infty}^{+\infty} f(\xi)g(x - \xi)\mathrm{d}\xi = f(x) \otimes g(x). \tag{3.47}$$

The symbol $\otimes$ stands for a convolution. If $F(q)$ and $G(q)$ are the Fourier transforms of $f(x)$ and $g(x)$, respectively, then the convolution theorem states that

$$\mathbf{F}\{f \otimes g\} = F(q) \cdot G(q), \tag{3.48}$$

$$\mathbf{F}\{f \cdot g\} = F(q) \otimes G(q). \tag{3.49}$$

The proof of the first relation passes through the stages of reversal of the order of integration, use of the translation theorem (3.44), and withdrawal of G from the integral because it no longer depends on ξ:

$$\mathbf{F}\{f(x) \otimes g(x)\} = \int_{-\infty}^{+\infty} \left[\int_{-\infty}^{+\infty} f(\xi)g(x - \xi)\mathrm{d}\xi \right] \mathrm{e}^{-2\pi \mathrm{i} q x} \mathrm{d}x$$

$$= \int_{-\infty}^{+\infty} \left[\int_{-\infty}^{+\infty} g(x-\xi) e^{-2\pi i q x} dx \right] f(\xi) d\xi$$

$$= \int_{-\infty}^{+\infty} G(q) e^{-2\pi i q \xi} f(\xi) d\xi = F(q) \cdot G(q). \tag{3.50}$$

Some applications of this convolution theorem will now be discussed in detail. The diffraction grating consisting of N slits with spacing a (Example 8) can be described as the convolution of a discrete point function $f_7(x)$ that coincides with the centers of the slits with the function $f_1(x)$ of a single slit. By using (3.48), the Fourier transform $F_8(q)$ is equal to the product of the Fourier transforms $F_1(q)$ of the single slit and $F_7(q)$ of the lattice function. The slit function $F_1(q)$ therefore acts as an envelope, modulating the amplitudes of the principal maxima.

Examples 9 and 10 contain an application of (3.49) to a wave of infinite extent, as already discussed in Sect. 3.1.2 and illustrated in Fig. 3.1b. The Fourier transform $F_9(q)$ of the infinite sine wave $f_9(x)$ has nonzero values only for $q = \pm 1/\Lambda$. A finite wave can be described by the product $f_{10}(x) = f_1(x) \cdot f_9(x)$, which is zero for $x < -a/2$ and $x > a/2$. The Fourier transform is obtained from (3.49) by a convolution of the Fourier transforms of the individual functions: $F_{10}(q) = F_9(q) \otimes F_1(q)$. This means that the δ-functions at the positions $q = \pm 1/\Lambda$ will be broadened by the function $F_1(q)$. If the sine wave does not decrease abruptly to zero at $x = \pm a/2$ but is multiplied by a Gaussian function $f_6(x)$, the example shown schematically in Fig. 3.1b is obtained.

An infinite row of points (Example 11), with spacings d, has a Fourier transform consisting of an infinite number of δ-functions at the positions $q_n = n/d$ (n integer), whereas only a first-order maximum appears for the function $f_9(x)$. Each δ-function at q_n corresponds to a function $\exp(2\pi i q_n x)$ (Example 1c). An infinite periodic function $f_P(x)$ with the period d (Example 12) can be described by a convolution of the infinite point row (Example 11) with a function $f(x)$ defined in the interval $-d/2 < x < +d/2$. The Fourier transform of $f(x)$ is therefore the envelope of the maxima of $\delta(q - q_n)$. At the positions q_n, the Fourier amplitudes of the periodic function f_P become

$$F_n = \frac{1}{d} \int_{-\infty}^{+\infty} f(x) \exp(-2\pi i q_n x) dx. \tag{3.51}$$

The inverse Fourier transform of $F_P(x) = \sum_{-\infty}^{+\infty} F_n \delta(q - q_n)$ gives the description of a periodic function in terms of a sum of sine and cosine terms (Fourier sum):,

$$f_P(x) = \sum_{-\infty}^{+\infty} F_n \exp(2\pi i q_n x)$$

$$= \frac{a_0}{2} + \sum_{n=1}^{+\infty} [a_n \cos(2\pi q_n x) + b_n \sin(2\pi q_n x)], \tag{3.52}$$

where the coefficients a_n and b_n can be calculated from

$$a_n = \frac{2}{d}\int_{-d/2}^{+d/2} f(x)\cos\left(2\pi\frac{n}{d}x\right)\mathrm{d}x; \quad b_n = \frac{2}{d}\int_{-d/2}^{+d/2} f(x)\sin\left(2\pi\frac{n}{d}x\right)\mathrm{d}x. \tag{3.53}$$

If this formula is applied to a periodic rectangular function, a diffraction grating of slit width $a = d/2$, for example, all of the b_n become zero because $f(x)$ is a symmetric function and

$$a_0 = \frac{1}{2}, \quad a_n = \frac{2}{\pi n}\sin\frac{n\pi}{2},$$

$$\text{which means} \quad a_n = \begin{cases} 0 & \text{for } n \text{ even} \\ \frac{2}{\pi n}(-1)^{(n-1)/2} & \text{for } n \text{ odd}. \end{cases} \tag{3.54}$$

Figure 3.12 shows how the rectangular function can be approximated successively by an increasing number of cosine functions of the Fourier sum (3.52) with the coefficients (3.54) up to the fifth order. The last curve in Fig. 3.12 can be observed experimentally as the intensity distribution of a grid in the image plane if all diffraction maxima with $n > 5$ are removed by a diaphragm in the focal plane of the objective. A pure cosine wave will be observed when

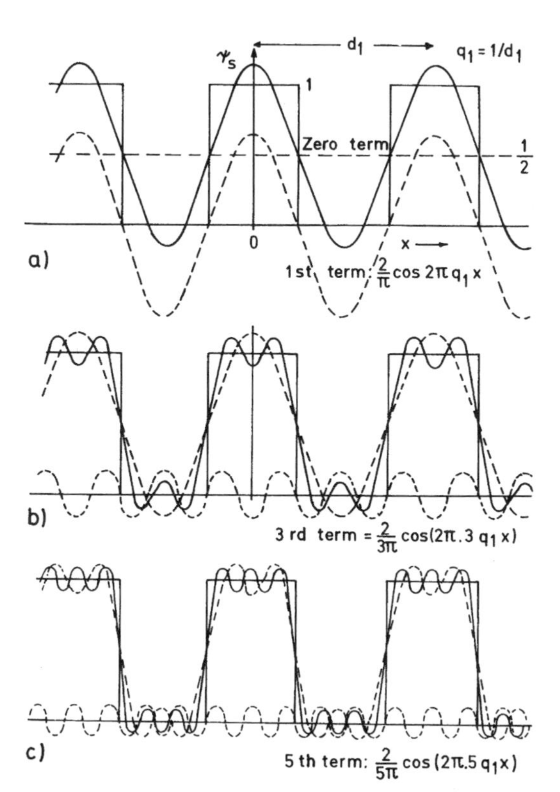

Fig. 3.12(a–c). Approximation of a step function by an increasing number of terms of a Fourier series.

only the first maximum is transmitted by the diaphragm (the first curve in Fig. 3.12). The image will then contain only the information that there is a periodicity d in the specimen but no information about the detailed form of the periodic function.

Fourier transforms of electron-microscope images can be obtained by means of an optical diffractometer (Sect. 6.3.6) or by digital computation using a fast Fourier transform (FFT) algorithm. In the latter case, the Fourier transform or its inverse will be calculated using a limited number of discrete image points. The intensity distribution $\sigma(x, y)$ in the image can be obtained by digital recording over a square of side length L. The smallest spatial frequency will be $q = 1/L$. All higher spatial frequencies are multiples of this frequency: spectral points with the coordinates $q_{xn} = n/L$ and $q_{ym} = m/L$ (m, n integer). The Fourier transform $S(q)$ of $\sigma(x, y)$ inside the square of area L^2 can be calculated from the sum

$$\begin{aligned} S(q_x, q_y) &= \mathbf{F}\{\sigma(x, y)\} \\ &= \sum_{n=-N}^{+N} \sum_{m=-N}^{+N} F_{nm} \frac{\sin[\pi L(q_{xn} - q_x)]}{\pi L(q_{xn} - q_x)} \frac{\sin[\pi L(q_{ym} - q_y)]}{\pi L(q_{ym} - q_y)} \end{aligned} \tag{3.55}$$

with the coefficients

$$F_{nm} = \frac{1}{N^2} \sum_{i=0}^{N-1} \sum_{j=0}^{N-1} \sigma(x_i, y_i) \exp[-2\pi i(q_{xn}x_i + q_{ym}y_j)]. \tag{3.56}$$

If δ denotes the resolution of the electron-microscope image, then $q_{max} = 1/\delta = N/L$ will be the highest spatial frequency when the number of sampling points is N^2 (sampling theorem of Shannon). The area of the densitometer slit or the electron detector should be of the order of δ^2 to ensure good averaging and to reduce the noise.

3.3 Wave-Optical Formulation of Imaging

3.3.1 Wave Aberration of an Electron Lens

The spherical aberration can be treated in wave optics in the following manner. An object point P emits a spherical, scattered wave with concentric wavefronts of equal phase (Fig. 2.10). An ideal lens would introduce thc phase shifts necessary to create a spherical wave beyond the lens, converging onto the image point P′. The rays of geometric optics are trajectories orthogonal to the wavefronts, and the wave amplitudes scattered into different angles θ of the cone with the aperture α_o are summed in the image point with equal phase. A radial decrease is observed in the intensity distribution of the blurred image point only because of the finite aperture (Airy disc, see Fig. 3.16b). The spherical aberration reduces the focal length for rays at larger θ. Because the rays and wavefronts are orthogonal, the wavefronts beyond a lens with spherical aberration are more strongly curved in the outer zones of the lens;

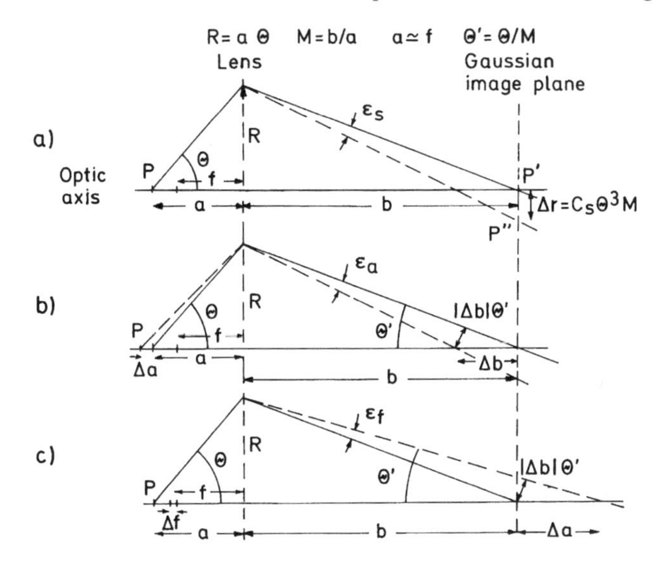

Fig. 3.13. Ray diagram for evaluating the angular deviations ϵ caused by (**a**) spherical aberration, (**b**) change Δa of specimen position, and (**c**) change Δf of focal length.

there is a difference Δs of optical path relative to the spherical wavefronts of an ideal lens (Fig. 2.10). The wave amplitudes are therefore not all in phase at the Gaussian image point. The smallest diameter of the intensity distribution, similar to an Airy disc, will be observed in front of the Gaussian image plane (Fig. 3.16c). A phase shift caused by defocusing of the lens also has to be considered; this can be generated either by a displacement Δa of the specimen or by a change Δf of focal length.

Figure 3.13 will be used to calculate the dependence of the phase shift $W(\theta) = 2\pi\Delta s/\lambda$, which is known as the *wave aberration*, on the scattering angle θ. First, however, a comment on its sign should be added, because different conventions are found in the literature. By using $\exp(2\pi i k z)$ instead of $\exp(-2\pi i k z)$ for a plane wave in Sect. 3.1.1, the convention is made that the phase increases with increasing z, that is, the direction of wave propagation. Because the optical path length along a trajectory in Fig. 2.10 is decreased by Δs, the phase shift $W(\theta)$ is also decreased. This phase shift therefore has to be represented by a phase factor $\exp[-iW(\theta)]$ with a negative sign in the exponent (see also comments in [3.31, 3.32]).

A ray that leaves the specimen point P at a scattering angle θ reaches the lens at a distance $R \simeq a\theta$ ($\theta \simeq$ a few tens of mrad) from the optic axis (Fig. 3.13a). The ray intersects the optic axis in the Gaussian image plane at P$'$ if there is no spherical aberration and at P$''$, a distance $\Delta r = C_s\theta^3 M$ from P$'$ in this plane, if the spherical aberration does not vanish; see (2.57). This causes a small angular deviation,

$$\epsilon_s \simeq \Delta r/b = C_s\theta^3 M/b. \tag{3.57}$$

By using the relation $\theta = R/a$, $M = b/a$, and $a \simeq f$, (3.57) becomes

$$\epsilon_s = C_s R^3/f^4. \tag{3.58}$$

We now assume that there is no spherical aberration and that the specimen distance a is increased by Δa (Fig. 3.13b). The focal length f of the lens is unchanged. The variation Δb of the image distance can be calculated from the well-known lens equation

$$\frac{1}{f} = \frac{1}{a+\Delta a} + \frac{1}{b+\Delta b} = \frac{1}{a}\left(1 - \frac{\Delta a}{a} + \ldots\right) + \frac{1}{b}\left(1 - \frac{\Delta b}{b} + \ldots\right). \tag{3.59}$$

Solving for Δb and using $1/f = 1/a + 1/b$, we obtain

$$\Delta b = -\Delta a\, b^2/a^2. \tag{3.60}$$

The corresponding angular deviation is obtained from

$$\epsilon_a = |\Delta b|\theta'/b = \Delta a\, R/f^2 \tag{3.61}$$

by using $\theta' \simeq R/b$.

A third case (Fig. 3.13c), where the focal length is changed to $f + \Delta f$, can be treated in a similar way. The lens equation $1/(f+\Delta f) = 1/a + 1/(b+\Delta b)$ gives $\Delta b = \Delta f\, b^2/f^2$ and so

$$\epsilon_f = -|\Delta b|\theta'/b = -\Delta f\, R/f^2. \tag{3.62}$$

Adding the three angular deviations of the geometric optical trajectories, we obtain the total angular deviation

$$\epsilon = \epsilon_s + \epsilon_a + \epsilon_f = C_s(R^3/f^4) - (\Delta f - \Delta a)R/f^2. \tag{3.63}$$

Figure 3.14 shows an enlargement of part of the lens between two trajectories and their orthogonal wavefronts, which reach the lens at distances R and $R + \mathrm{d}R$ from the optic axis. The angular deviation causes an optical path difference $\mathrm{d}s = \epsilon \mathrm{d}R$ between the two trajectories. These path differences $\mathrm{d}s$ have to be summed (integrated) to get the total path difference Δs or the phase shift $W(\theta)$ relative to the optic axis:

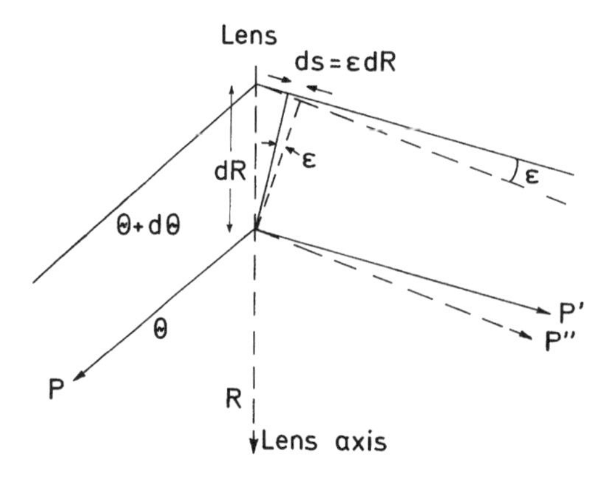

Fig. 3.14. Part of the outer zone of a lens at a distance R from the optic axis showing the relation between the angular deviation ϵ and the optical path difference $\mathrm{d}s = \epsilon \mathrm{d}R$.

$$W(\theta) = \frac{2\pi}{\lambda}\Delta s = \frac{2\pi}{\lambda}\int_0^R \mathrm{d}s = \frac{2\pi}{\lambda}\int_0^R \epsilon \mathrm{d}R$$
$$= \frac{2\pi}{\lambda}\left[\frac{1}{4}C_s\frac{R^4}{f^4} - \frac{1}{2}(\Delta f - \Delta a)\frac{R^2}{f^2}\right]. \quad (3.64)$$

With $R/f \simeq \theta$ and the defocusing $\Delta z = \Delta f - \Delta a$, the so-called Scherzer formula [3.33] is obtained,

$$W(\theta) = \frac{\pi}{2\lambda}(C_s\theta^4 - 2\Delta z\theta^2), \quad (3.65)$$

or by introducing the spatial frequency $q = \theta/\lambda$ (3.39),

$$W(q) = \frac{\pi}{2}(C_s\lambda^3 q^4 - 2\Delta z\lambda q^2). \quad (3.66)$$

In more accurate calculations, the change of the positions of the principal planes of the lens and the variation of C_s when the lens excitation is changed also have to be considered. Axial astigmatism (Sect. 2.3.3) can be included in (3.65) by introducing an additional term

$$W_A = \frac{\pi}{2\lambda}\Delta f_A \sin[2(\chi - \chi_0)], \quad (3.67)$$

which depends on an azimuthal angle χ.

The relation (3.65) is important for the discussion of phase contrast and for the study, in wave-optical terms, of the formation of a small electron probe for scanning transmission electron microscopy (STEM). Because the wave aberration depends on the two parameters C_s and λ, it is convenient to discuss the wave aberration in terms of reduced coordinates [3.34, 3.35],

$$\theta^* = \theta\,(C_s/\lambda)^{1/4} \quad \text{and} \quad \Delta z^* = \Delta z\,(C_s\lambda)^{-1/2}. \quad (3.68)$$

This results in the reduced wave aberration

$$\frac{W(\theta^*)}{2\pi} = \frac{\theta^{*4}}{4} - \frac{\theta^{*2}}{2}\Delta z^*. \quad (3.69)$$

Figure 3.15 shows $W(\theta^*)$ for different values of reduced defocus $\Delta z^* = \sqrt{n}$ (n an integer), for which the minima of $W(\theta^*)$ are $-n\pi/2$.

Defocusing with positive Δz is called underfocusing and defocusing with negative Δz is called overfocusing. Scattered electron waves are shifted with a phase $\pi/2$ relative to the unscattered wave (Sect. 6.2.1). For this reason, a reduced defocusing $\Delta z^* = 1$ (Scherzer focus) is advantageous for phase contrast because $W(\theta)$ has the value $-\pi/2$ over a relatively broad range of scattering angles or spatial frequencies in the vicinity of the minimum for $\Delta z^* = 1$ in Fig. 3.15.

3.3.2 Wave-Optical Theory of Imaging

The rays from an object point P are reunited by the lens at the image point P′ (Fig. 3.10), a distance $x' = -Mx$ from the optic axis, where x is the off-axis

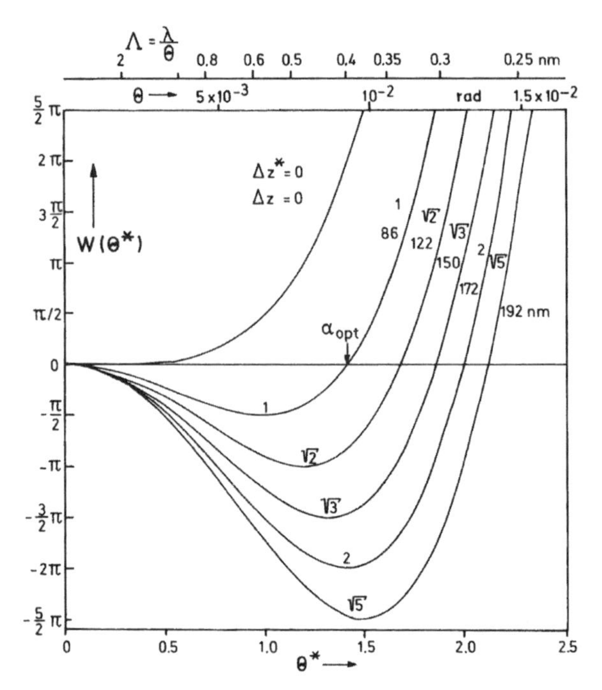

Fig. 3.15. Wave aberration $W(\theta^*)$ as a function of the reduced scattering angle θ^* for various reduced focusing distances Δz^*. The upper scale shows the values of $\Lambda = \lambda/\theta$ for the special case $E = 100$ keV and $C_s = 2$ mm.

distance of P ($M = b/a$: magnification). Rays with equal scattering angles from different points of the specimen intersect in the focal plane of the lens. It has already been shown in Sect. 3.2.3 that the wave-amplitude $F(\boldsymbol{q})$ in this plane is obtained from the exit wave-amplitude distribution $\psi_s(\boldsymbol{r})$ behind the specimen by a Fourier transform. The wave amplitudes from different points of this $\boldsymbol{q}$ plane have to be summed at the image point P′, taking into account differences of the optical path. Just as we defined the geometrical path difference Δs_g in Sect. 3.2.3, we now formulate the path difference $\Delta s'_g$ from Fig. 3.10 using $r = f\theta$, $\theta' = x/f$,

$$\Delta s'_g = r\theta' = +\lambda q x = -\Delta s_g \quad \text{or} \quad \Delta s'_g = +\lambda \boldsymbol{q} \cdot \boldsymbol{r}, \tag{3.70}$$

for the two-dimensional $\boldsymbol{q}$ plane. This corresponds to a phase shift $\varphi'_g = 2\pi \boldsymbol{q}\cdot\boldsymbol{r}$. As in (3.40), the wave amplitude ψ_m at the image point P′ is obtained by integrating over all elements of area $d^2\boldsymbol{q}$ of the focal plane

$$\psi_m(\boldsymbol{r}) = \frac{1}{M} \int\int F(\boldsymbol{q}) e^{2\pi i \boldsymbol{q}\cdot\boldsymbol{r}} d^2\boldsymbol{q} = \frac{1}{M}\psi_s(\boldsymbol{r}). \tag{3.71}$$

Thus, ψ_m is obtained as the inverse Fourier transform of $F(\boldsymbol{q})$. The image intensity $I = \psi_m \psi_m^*$ decreases as M^{-2} because the electrons are spread over an area M^2 times as large as the corresponding specimen area.

For aberration-free imaging, there will be no further phase shift, apart from φ'_g, and the integration in (3.71) will be taken over the whole range of spatial frequencies $\boldsymbol{q}$ that appear in the specimen. In practice, a maximum scattering angle $\theta_{max} = \alpha_o$ (objective aperture) corresponding to a maximum spatial frequency q_{max} is used. This limitation on spatial frequencies by an objective diaphragm can be expressed in terms of a multiplicative masking function $M(q)$, which would have the values $M(q) = 1$ for $q = |\boldsymbol{q}| < q_{max}$ and $M(q) = 0$ for $|\boldsymbol{q}| > q_{max}$ in the normal bright-field mode. Because the wave aberration $W(q)$ in (3.66) representing the spherical aberration and defocusing depends only on q, the action of this contribution can be represented by a multiplication of the amplitudes at the focal plane by the phase factor $\exp[-iW(q)]$. Equation (3.71) therefore has to be modified to

$$\psi_m(\boldsymbol{r}') = \frac{1}{M} \int\int F(\boldsymbol{q}) \underbrace{[e^{-iW(q)} M(q)]}_{H(q)} e^{2\pi i \boldsymbol{q}\cdot\boldsymbol{r}} d^2\boldsymbol{q}. \tag{3.72}$$

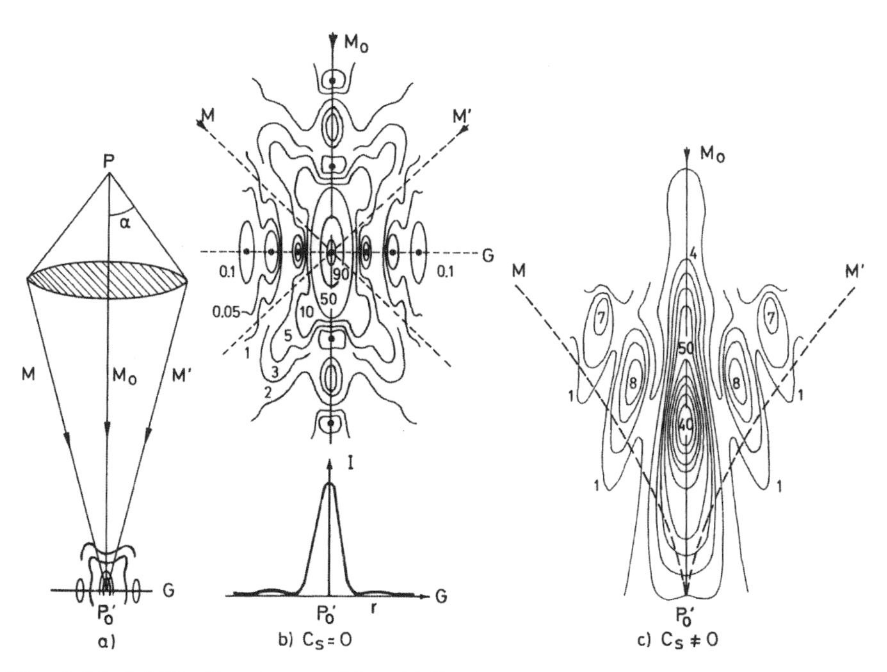

Fig. 3.16. (**a**) The limiting rays M and M′, which produce an image of the point P at the image point P'_0 in the Gaussian image plane G. (**b**) and (**c**) show the enlarged intensity distribution near P'_0 and lines of equal intensity for an aberration-free lens (**b**) and for a lens with spherical aberration (**c**). The curve at the bottom of (**b**) represents the cross section $I(r)$ through the Gaussian focus (Airy disc). The horizontal axis corresponds to the distance to the optical axis and the vertical axis to the defocus.

$H(q)$ is known as the pupil function. The convolution theorem (3.49) can be applied to (3.72) as

$$\psi_m(\boldsymbol{r}') = \frac{1}{M}\psi_s(\boldsymbol{r}) \otimes h(\boldsymbol{r}) = \frac{1}{M}\int\int \psi_s(\boldsymbol{r}_1)h(\boldsymbol{r}-\boldsymbol{r}_1)\mathrm{d}^2\boldsymbol{r}_1, \tag{3.73}$$

where $h(\boldsymbol{r}) = \mathbf{F}^{-1}\{H(\boldsymbol{q})\}$ is the inverse Fourier transform of the pupil function $H(\boldsymbol{q})$. This means that sharp image points will not be obtained. Instead, each image point will be blurred (convolved) with the point-spread function $h(\boldsymbol{r})$. The image of a point source that scatters in all scattering angles with uniform amplitude $[F(q) = \text{const and } \psi_s(r) = \delta(0)]$ would be the function $h(r)/M$.

We consider now the case $W(q) = 0$ (no spherical aberration or defocusing). The image amplitude of a point source can be calculated as the Fourier transform of $M(q)$ (see Example 4 in Table 3.2),

$$\psi_m(\boldsymbol{r}) \propto h(\boldsymbol{r}) \propto \frac{J_1(x)}{x} \quad \text{with} \quad x = \frac{2\pi}{\lambda}\alpha_o r. \tag{3.74}$$

This amplitude distribution, which corresponds to the intensity distribution $I(x) \propto [J_1(x)/x]^2$, of the blurring function of a point source with an aberration-free but aperture-limited objective lens is called the Airy distribution (Fig. 3.16b). The intensity distribution for $\Delta z \neq 0$ is symmetrical in Δz and is plotted in Fig. 3.16b as lines of equal intensity. The same situation is shown in Fig. 3.16c with spherical aberration present. The distribution is asymmetric in Δz; the smallest error disc occurs at underfocus, in agreement with the geometrical-optical construction of Fig. 2.10. The dashed line is the caustic of geometrical optics. Further discussion of the wave-optical imaging theory will be found in Sect. 6.2.

4

Elements of a Transmission Electron Microscope

Not only does the electron gun of an electron microscope emit electrons into the vacuum and accelerate them between cathode and anode, but it is also required to produce an electron beam of high brightness and high temporal and spatial coherence. The conventional thermionic emission from a tungsten wire is limited in temporal coherence by an energy spread of the emitted electrons of the order of a few electron volts and in spatial coherence by the gun brightness. Schottky-emission and field-emission guns are newer alternatives for which the energy spread is less and the gun brightness higher.

The condenser-lens system of the microscope controls the specimen illumination, which ranges from uniform illumination of a large area at low magnification, through a stronger focusing for high magnification, to the production of an electron probe of the order of a few nanometers or even less than a nanometer in diameter for scanning transmission electron microscopy or for microanalytical methods.

The useful specimen thickness depends on the operation mode used and the information desired. Specimen manipulation methods inside the microscope are of increasing interest but are restricted by the size of the specimen and by the free space inside the polepiece system of the objective lens.

The different imaging modes of a TEM can be described by ray diagrams, as in light optics, which can also be used to evaluate the depth of focus or to establish a theorem of reciprocity between conventional and scanning transmission electron microscopy. Electron prism spectrometers or imaging energy filters allow electron energy-loss spectra (EELS) to be recorded and various operating modes of electron spectroscopic imaging (ESI) and diffraction (ESD) to be used.

Observation of the image on a fluorescent screen and image recording on photographic emulsions can be replaced by techniques that allow digital, parallel, and quantitative recording of the image intensity.

4.1 Electron Guns

4.1.1 Physics of Electron Emission

Thermionic Emission. The conduction electrons in metals or compounds have to overcome the work function ϕ_w if they are to be emitted from the cathode into the vacuum. Figure 4.1 shows the dependence of potential energy on a coordinate z normal to the surface. The potential energy $V(z)$ of an electron in front of a conducting surface at a distance z larger than the atomic diameter can be calculated by considering the effect of a mirror charge with opposite sign behind the surface; with an electric field $\boldsymbol{E}$, the potential energy $V = -e|\boldsymbol{E}|z$ is superposed on that of the mirror charge, giving

$$V(z) = \phi_w - \frac{e^2}{16\pi\epsilon_0}\frac{1}{z} - e|\boldsymbol{E}|z. \tag{4.1}$$

Increasing the cathode temperature leads to a broadening of the Fermi distribution $f(E)$ at the Fermi level E_F, and for high temperatures, electrons in the tail of the Fermi distribution acquire enough kinetic energy to overcome the work function ϕ_w. The current density j_c (A m^{-2}) of the cathode emission can be estimated from Richardson's law [4.1],

$$j_c = AT_c^2 \exp(-\phi_w/kT_c), \tag{4.2}$$

where $k = 1.38 \times 10^{-23}$ J K^{-1} is Boltzmann's constant, T_c is the cathode temperature, and $A \simeq 12 \times 10^5$ A K^{-2} m^{-2} is a constant that depends on the cathode material.

Most metals melt before they reach a sufficiently high temperature for thermionic emission. An exception is tungsten, which is widely used at a working temperature T_c of 2500–3000 K (melting point $T_m = 3650$ K). Lanthanum

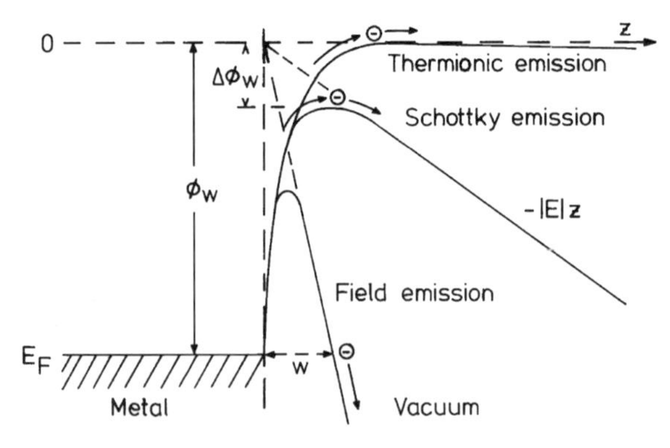

Fig. 4.1. Potential energy $V(z)$ of electrons at the metal–vacuum boundary. Electrons with energies beyond the Fermi energy E_F have to overcome the barriers ϕ_w and $\phi_w - \Delta\phi_w$ for thermionic or Schottky emission or can tunnel through the barrier of width w for field emission.

Table 4.1. Parameters of thermionic, Schottky, and field-emission cathodes at $E = 100$ keV.

Characteristic parameters:	
Cathode temperature T_c	Tip radius r of pointed cathodes
Work function ϕ_w	Diameter d of source
Emission current density j_c	Operating vacuum p
Gun brightness β at $E = 100$ keV	Field strength $\|\boldsymbol{E}\|$ at cathode
Energy spread ΔE	
Thermionic cathodes (field at cathode reduced by Wehnelt electrode)	
Tungsten hairpin	**Pointed LaB_6 rod**
$T_c = 2500$–3000 K	$T_c = 1400$–2000 K
$\phi_w = 4.5$ eV	$\phi_w = 2.7$ eV
$j_c \simeq (1-3) \times 10^4$ A/m^2	$j_c \simeq (2-5) \times 10^5$ A/m^2
$\beta = (1-5) \times 10^9$ A/m^2 sr	$\beta = (1-5) \times 10^{10}$ A/m^2 sr
$\Delta E = 1.5$–3 eV	$\Delta E = 1$–2 eV
$d = 20$–50 μm	$d = 10$–20 μm
$p \leq 10^{-3}$ Pa (1 Pa $= 10^{-5}$ bar)	$p \leq 10^{-4}$ Pa
$\|\boldsymbol{E}\| \simeq 10^6$ V/m	
Point-source cathodes	
Schottky emission	**Field emission**
(Thermal emission from ZrO/W tip at 1800 K with high electric field)	(Tunneling from cold or heated tungsten tips)
$T_c = 1800$ K	$T_c = 300$ K or $\simeq 1500$ K
$\phi_w = 2.7$ eV	$\phi_w = 4.5$ eV
$j_c \simeq 5 \times 10^6$ A/m^2	$j_c \simeq 10^9 - 10^{10}$ A/m^2
	$\beta = 2 \times 10^{12} - 2 \times 10^{13}$ A/m^2 sr
$\Delta E = 0.3$–0.7 eV	$\Delta E = 0.2$–0.7 eV
$r = 0.5$–1 μm	$r \leq 0.1$ μm
$d \simeq 15$ nm	$d \simeq 2.5$ nm
$p \leq 10^{-6}$ Pa	$p \leq 10^{-8}$ Pa
$\|\boldsymbol{E}\| \simeq 2 \times 10^8$ V/m	$\|\boldsymbol{E}\| \simeq 5 \times 10^9$ V/m

hexaboride (LaB_6) cathodes with $T_c = 1400$–2000 K are also employed because their work function is lower (Table 4.1). The tungsten metal evaporates continuously during operation, limiting the lifetime of the filament, which decreases from $\simeq 200$ h to 5 h if T_c increases from 2500 K to 2900 K [4.2]. Also, CeB_6 cathodes are now offered commercially.

Schottky Emission. When the field strength $\boldsymbol{E}$ at the cathode is increased, the overlap of potential energies in (4.1) results in a decrease $\Delta\phi_w$ of the work function (Schottky effect). At the maximum of (4.1), the effective work function is lowered to [4.1]

$$\phi_{w,\text{eff}} = \phi_w - \Delta\phi_w = \phi_w - e\sqrt{\frac{e|\boldsymbol{E}|}{4\pi\epsilon_0}}. \tag{4.3}$$

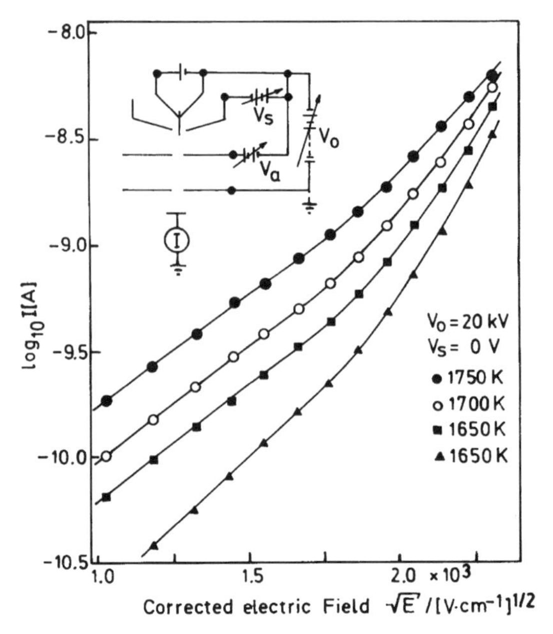

Fig. 4.2. Schottky plot of emission current from a ZrO/W tip at different tip temperatures [4.2].

This decrease can be neglected in normal thermionic cathodes. When using a Schottky cathode with a radius $r \leq 1\ \mu$m at the tip and a field strength 10^8 V/m, the decrease is $\Delta\phi_{\mathrm{w}} \simeq 0.4$ eV. In contrast to field emission, discussed below, the electrons still have to overcome the now lowered work function $\phi_{\mathrm{w,eff}}$ by their kinetic energy, which is furnished by heating the cathode. This can be confirmed experimentally by a Schottky plot (Fig. 4.2). Substitution of (4.3) in (4.2) shows that a semilogarithmic plot of the emission current log I versus the square root of the electric field strength $|\boldsymbol{E}|$ results in a straight line when the temperature of the tip is constant and $|\boldsymbol{E}|$ is increased by increasing the extraction voltage [4.3]. As in a thermionic cathode, the emission increases with increasing temperature of the tip. Beyond $|\boldsymbol{E}| = 4 \times 10^8$ V/m, the stronger increase of emission indicates the onset of field emission; the latter becomes independent of cathode temperature at higher $|\boldsymbol{E}|$ (4.4) and a Fowler-Nordheim plot of log I versus $1/|\boldsymbol{E}|$ then results in a straight line.

Field Emission. The width b of the potential barrier at the metal–vacuum boundary decreases with increasing $\boldsymbol{E}$; for $|\boldsymbol{E}| \geq 10^9$ V m^{-1}, using a tip radius $r \leq 0.1\ \mu$m, the width b becomes less than 10 nm (Fig. 4.1) and electrons at the Fermi level can penetrate the potential barrier by the quantum-mechanical tunneling effect. This means that the electron waves near the Fermi energy are reflected at the potential barrier but penetrate with an exponential decrease of their amplitude ψ into the barrier. When the width b of the barrier is small, the amplitude at the vacuum side of the barrier is still appreciable and the probability of tunneling across the barrier is proportional to $\psi\psi^*$. The emitted electrons do not need to overcome the potential barrier and it is not necessary to heat the cathode, whereas this is essential for thermionic and Schottky emissions. If a field-emission source is heated, it is mainly to prevent the adsorption of gas molecules.

The current density of the field emission can be estimated from the Fowler–Nordheim formula (see also [4.4])

$$j = \frac{k_1|\boldsymbol{E}|^2}{\phi_{\mathrm{w}}} \exp\left(-\frac{k_2\phi_{\mathrm{w}}^{3/2}}{|\boldsymbol{E}|}\right). \tag{4.4}$$

The constants k_1 and k_2 depend only weakly on $|\boldsymbol{E}|$ and ϕ_{w}.

4.1.2 Energy Spread

The tail of the Fermi distribution $f(E) = \{1 + \exp[(E - E_{\mathrm{F}})/kT_{\mathrm{c}}]\}^{-1}$, which can overcome the work function ϕ_{w}, results for $E > E_{\mathrm{F}}$ in a Maxwell–Boltzmann distribution of the exit momenta p or energies $E = p^2/2m$,

$$f(E) \propto \exp(-E/kT_{\mathrm{c}}). \tag{4.5}$$

Electrons are emitted in all directions within the half-space; the electron motion is characterized by the tangential (t) and normal (n) components of $\boldsymbol{p}$, so that (4.5) has to be multiplied by the volume element (density of states) $2\pi p^2 \mathrm{d}p$ of the momentum space to get the number of electrons with momenta between p and $p + \mathrm{d}p$ or energies between $E = (p_{\mathrm{t}}^2 + p_{\mathrm{n}}^2)/2m$ and $E + \mathrm{d}E$. This yields the normalized total energy distribution (Fig. 4.3)

$$N(E)\mathrm{d}E = \frac{E}{(kT_{\mathrm{c}})^2} \exp(-E/kT_{\mathrm{c}})\mathrm{d}E, \tag{4.6}$$

$$\begin{aligned} &\text{with a most probable energy } & E_p &= kT_{\mathrm{c}}, \\ &\text{a mean energy} & \langle E\rangle &= 2kT_{\mathrm{c}}, \\ &\text{and a half-width} & \Delta E &= 2.45kT_{\mathrm{c}}. \end{aligned} \tag{4.7}$$

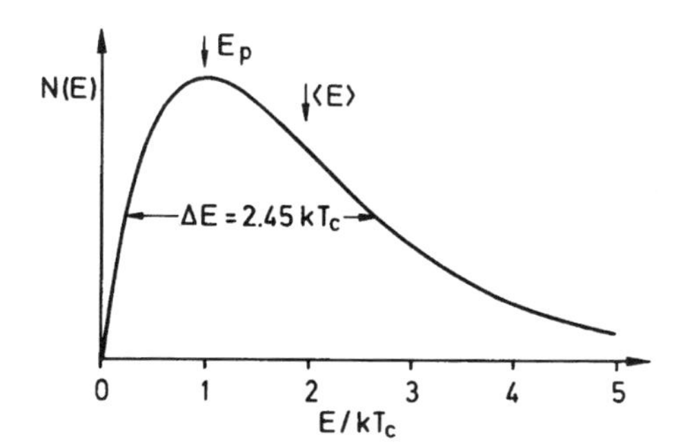

Fig. 4.3. Maxwellian distribution of electron energies emitted from a thermionic cathode ($\langle E\rangle$: mean energy, E_{p}: most probable energy, ΔE: energy half-width).

Thus, for a cathode temperature T_c of 2500 K, the half-width ΔE will be 0.5 eV. This energy spread is superposed on the accelerating energy $E = eU$. This theoretical value will occur only when the cathode is operated in the saturation mode with low current density. In the normal operating mode with a triode gun (Sect. 4.1.4), an anomalous energy spread is observed (the Boersch effect [4.5]), with the result that $\Delta E \simeq 1$–2 eV and even worse. This can be explained by Coulomb interactions of the electrons in the crossover [4.6, 4.7, 4.8, 4.9]. The energy spread of a thermionic cathode increases with increasing emission current and depends on the shape of the Wehnelt electrode [4.10]. The energy spread of Schottky and field-emission guns is of the order of $\Delta E = 0.2$–0.7 eV; for the dependence on tip orientation and temperature, see [4.11].

4.1.3 Gun Brightness

The components p_t of the initial exit momenta tangential to the exit surface result in an angular spread of the electron beam and limit the value of the *gun brightness* β. This quantity is defined as the current density $j = \Delta I/\Delta S$ per solid angle $\Delta\Omega = \pi\alpha^2$, where α denotes the half-aperture of the cone of electrons that pass through the surface element ΔS,

$$\beta = \frac{\Delta I}{\Delta S \Delta\Omega} = \frac{j}{\pi\alpha^2}. \tag{4.8}$$

The maximum possible value β_{max} for a thermionic cathode can be estimated from the following simplified model (Fig. 4.4) (see [4.12, 4.13] for details). The components p_t and p_n are each described by a Maxwell–Boltzmann distribution (4.5) with mean-square values

$$\langle p_t^2 \rangle = \langle p_n^2 \rangle = 2m_0 kT_c. \tag{4.9}$$

The electron acceleration contributes an additional kinetic energy $E = eU$ so that, in all, using (2.11), we find

$$\langle p_n^2 \rangle = 2m_0 kT_c + 2m_0 E(1 + E/2E_0). \tag{4.10}$$

The angular aperture α of a virtual electron source behind the cathode surface can be obtained from the vector sum of $\boldsymbol{p}_t$ and $\boldsymbol{p}_n$ (Fig. 4.4): $\alpha = p_t/p_n$ or $\langle\alpha^2\rangle = \langle p_t^2\rangle/\langle p_n^2\rangle$. Substituting (4.9) and (4.10) into (4.8) gives

$$\beta_{max} = \frac{j_c}{\pi}\left[1 + \frac{E}{kT_c}(1 + E/2E_0)\right] \simeq \frac{j_c E}{\pi k T_c}. \tag{4.11}$$

This formula is valid even for nonuniform fields in front of the cathode.

Numerical values of the gun brightness are listed in Table 4.1. With thermionic cathodes, the maximum value β_{max} can be attained by using optimum operating conditions (Sect. 4.1.4). Otherwise, lower values, which also depend on the Wehnelt shape, are found, ranging from 0.1 to 0.5 β_{max}. The angular spread is also increased by the Coulomb interactions in the crossover (the lateral Boersch effect), again decreasing β.

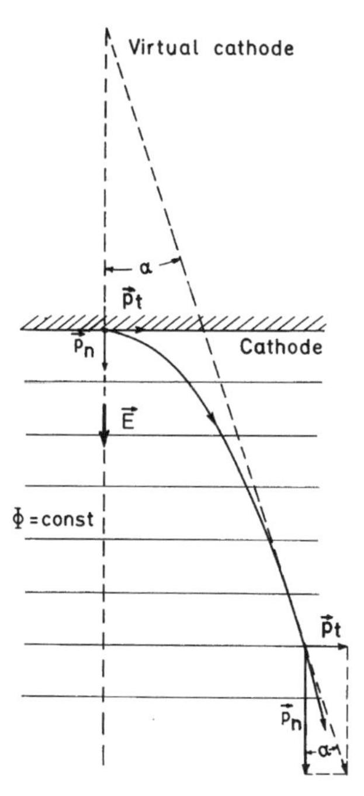

Fig. 4.4. Evaluation of the angular spread α of electrons emitted; with a transverse exit momentum p_t and a uniform electric field in front of the cathode, the trajectories are parabolic.

This axial gun brightness β (that is, the brightness for points on the axis of an electron-optical column) remains constant for all points on the axis, from the anode to the final image. This invariance of axial gun brightness along the optic axis will now be demonstrated by considering an aberration-free lens with a diaphragm in front of it, though the result is true for real lenses with aberrations. Lenses and diaphragms are typical elements of any electron-optical system. We assume that an intermediate image of the source is formed in the plane indicated by the suffix 1 (Fig. 4.5). The electron current density in this intermediate image may have a Gaussian distribution (4.14). We consider only the center of this distribution because we are interested only in the axial brightness. A fraction ΔI_1 of the total current passes through the area ΔS_1 with an angular aperture α_1 corresponding to a solid angle $\Delta\Omega_1 = \pi\alpha_1^2$. The gun brightness in this plane is

$$\beta_1 = \frac{\Delta I_1}{\Delta S_1 \Delta\Omega_1} = \frac{\Delta I_1}{\Delta S_1 \pi\alpha_1^2}. \tag{4.12}$$

The diaphragm in front of the lens cuts off a fraction of the current ΔI_1, and only a fraction

$$\Delta I_2 = \Delta I_1 \frac{\pi\alpha^2}{\pi\alpha_1^2} \tag{4.13}$$

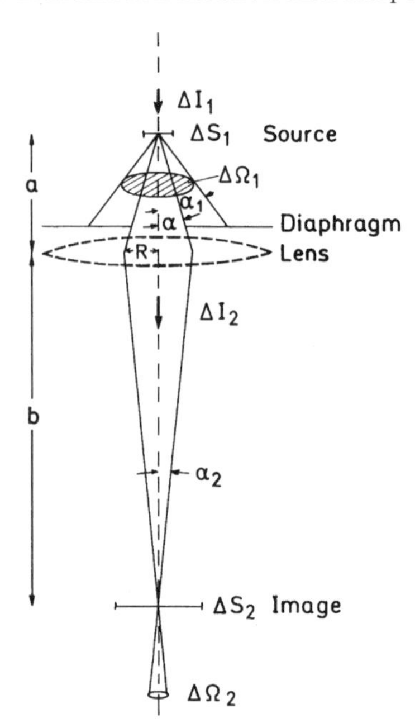

Fig. 4.5. Demonstration of the conservation of gun brightness on the axis of an electron-optical system in the presence of apertures and lenses.

will pass the diaphragm. This current is concentrated in an image area $\Delta S_2 = \Delta S_1 M^2$, where $M = b/a$ is the magnification, which can be smaller than unity if the lens is demagnifying. The aperture is decreased to $\alpha_2 = \alpha/M$ because $\tan\alpha \simeq \alpha = R/a$ and $\alpha_2 \simeq R/b$ so that $\alpha_2/\alpha = a/b = 1/M$. The gun brightness in the image plane is $\beta_2 = \Delta I_2/\Delta S_2 \Delta\Omega_2$. Substituting for the quantities with the suffix 2 gives $\beta_1 = \beta_2$, which demonstrates the invariance of β for this special case.

The invariance of β means that high values of the current density j at the specimen can be obtained only by using large apertures of the convergent electron probe or beam. If it is essential to use very small apertures, for Lorentz microscopy (Sect. 6.8) and small-angle electron diffraction (Sect. 8.1.5), for example, correspondingly low values of j must be expected. The gun brightness is therefore an important characteristic of an electron gun. The need for high gun brightness has stimulated the development of LaB_6 thermionic cathodes and Schottky and field-emission guns.

4.1.4 Thermionic Electron Guns

The most widely used thermionic cathodes consist of a tungsten wire 0.1–0.2 mm in diameter bent like a hairpin and soldered on contacts. The wire is directly heated by a current of a few amperes.

LaB_6 cathodes consist of small, pointed crystals [4.14, 4.15, 4.16, 4.17, 4.18]. They require indirect heating because their electrical resistance is too high for direct-current heating. The heating power can be decreased by supporting a small crystal between carbon rods or fibers or binding it to refractory metals (rhenium or tantalum) that have a low rate of reaction with LaB_6. These cathodes need a better vacuum than tungsten cathodes to reduce the damage caused by positive-ion bombardment. They provide a higher gun brightness, and the value of the energy spread is lower (Table 4.1). The emission current is greatest for (100)-oriented tips, ten times higher than for the (510) orientation [4.19].

A thermionic electron gun consists of three electrodes (triode structure):

1. the heated filament, which forms the cathode, at the potential $\Phi_C = -U$;
2. the Wehnelt electrode, at a potential Φ_W some hundreds of volts more negative than the cathode; and
3. the grounded anode ($\Phi_A = 0$).

The electron optics of a triode electron gun is reviewed in [4.21]. Figure 2.1 shows the equipotentials Φ = const in a cross section through a triode gun and Fig. 4.6 those near the cathode tip. In Fig. 4.6a, the negative bias of the Wehnelt electrode is not great enough to decrease $|\boldsymbol{E}|$ at the cathode surface.

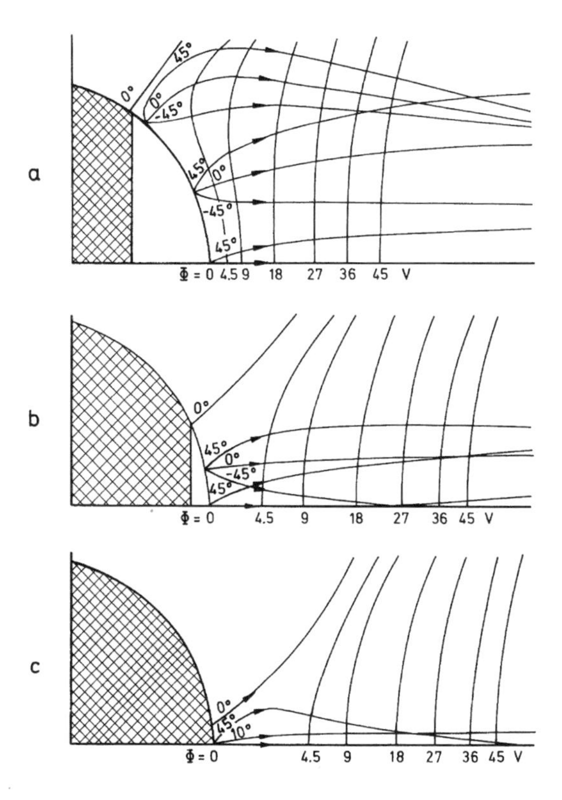

Fig. 4.6. Equipotentials ϕ = const in front of the cathode tip for (from **a** to **c**) increasing negative bias $-U_W$ of the Wehnelt electrode; electron trajectories are shown with an exit energy of 0.3 eV and various angles of emission [4.20].

The zero equipotential intersects the tip around a circle. All of the electrons emitted from a large cathode area (nonshaded) are accelerated. Beyond the circle, the electric field strength is of opposite sign, and no electrons can leave the shaded area. In Fig. 4.6b, the negative bias is further increased and the area of emission is thus reduced. In Fig. 4.6c, the zero equipotential reaches the tip of the cathode. No electrons will leave the cathode if the Wehnelt bias is increased further.

Figure 4.6 also shows some electron trajectories, with an initial exit energy of 0.3 eV and different angles of emission. In Fig. 4.6b, the electrons enter a more or less uniform electric field, which exerts small additive radial force components on the electron trajectories; the cross section of the electron beam passes through a minimum, known as the *crossover*, between the cathode and the anode. This crossover acts as an effective electron source for the electron optical system of the microscope. Large radial components of velocity (momentum) are produced near the zero equipotentials in Fig. 4.6a. The corresponding electrons cross the axis and result in a *hollow-beam* cross section. Radial components are also produced in Fig. 4.6c near the cutoff bias. No further decrease of the crossover is observed, but the emission current falls. Figure 4.7 shows enlarged images of the crossover and the transition from hollow beam to optimum cross section as the gun filament current is increased in an autobiased gun discussed below. A lower energy spread due to the Boersch effect can be observed at moderate underheating corresponding to the crossover profile of Fig. 4.7b.

The minimum diameter of the crossover is limited not only by the lens-like action of the electric field in front of the cathode but also by the radial components of the electron exit momenta. The Maxwellian distribution of exit velocities gives the radial current-density distribution in the crossover an approximately Gaussian shape:

$$j(r) = j_0 \exp[-(r^2/r_0^2)]. \tag{4.14}$$

In practice, the Wehnelt electrode is biased not by a separate voltage supply but by the voltage drop $U_W = I_c R_W$ across the resistor R_W in the

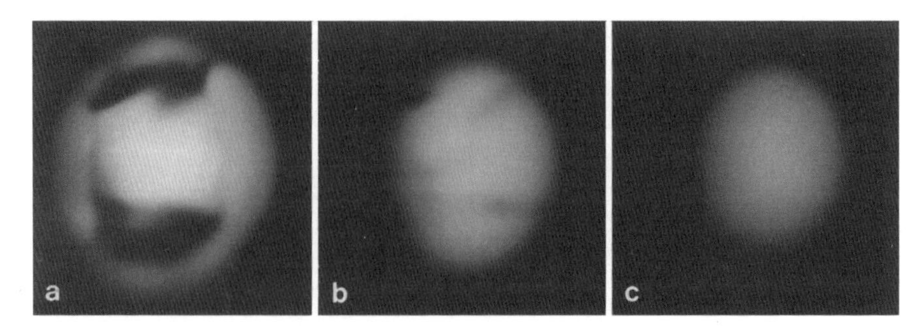

Fig. 4.7. Enlarged images of the crossover of an autobiased tungsten-hairpin cathode. From (**a**) to (**c**), the heating current of the gun is increased.

high-tension supply line (Fig. 4.8) produced by the emission current I_c. The resistance R_W can be altered by means of a mechanical potentiometer or a vacuum diode, the filament heating of which is varied. It will now be shown that this system is autobiasing. When U_W is generated by an independent voltage supply, the dependence of the emission current I_c on U_W shown in Fig. 4.9a

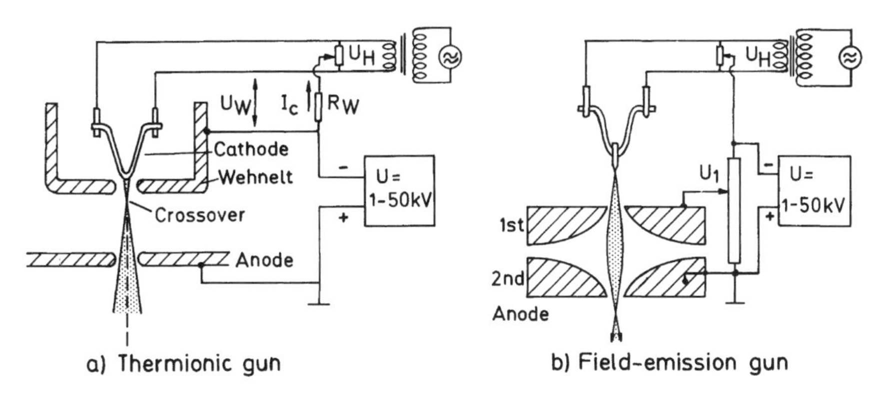

Fig. 4.8(a,b). Generation of Wehnelt bias U_W as a voltage drop across a resistance R_W by the total beam current I_c in an electron gun with autobias.

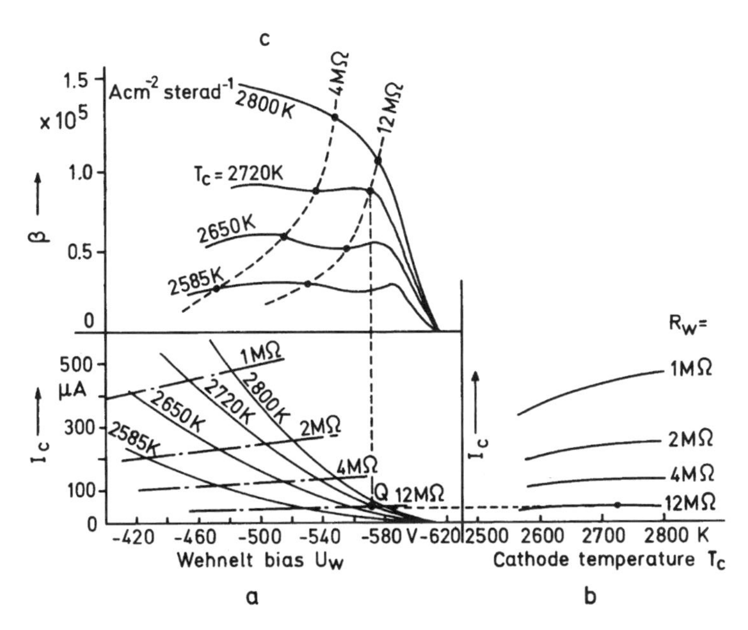

Fig. 4.9. (**a**) Dependence of the total beam current I_c, (— · — · —) $I_c = U_W/R_W$ for different R_W on Wehnelt bias. (**b**) Construction of the dependence of I_c on cathode temperature T_c from the intersection of the dash-dotted line in (a) with the $I_c - U_W$ curves. (**c**) Dependence of gun brightness β on Wehnelt bias U_W for different cathode temperatures T_c, (— — —) working points for constant R_W. [4.20].

will be observed as the cathode temperature T_c is increased. When Wehnelt biasing is produced by the voltage drop across R_W, the working points shown in Fig. 4.9a are obtained; these are the points of intersection of the straight lines $I_c = U_W/R_W$, plotted for different values of R_W (1–12 MΩ), with the I_c vs. U_W curves. From this diagram, the dependence of I_c on T_c for constant R_W (Fig. 4.9b) can be constructed. This plot shows an emission current I_c that increases as T_c is raised until it reaches a saturation value beyond which any further increase of T_c produces little increase of I_c. This has been attributed, in older publications, to the fact that the gun is running into space-charge-limited conditions. That this is not so can be seen when this type of biasing is replaced by a variable, independent bias. The saturation effect does not result from space-charge limitation but from the shape of the $I_c - U_W$ curves. Space-charge effects can therefore be neglected at normal temperatures, $T_c \simeq$ 2560 K, but can occur at higher values of T_c [4.20].

The optimum value of R_W for a given cathode geometry and temperature can be constructed by dropping a vertical line from the maximum of β in Fig. 4.9c onto the corresponding $I_c - U_W$ curve. The slope of the straight line from the intersection Q to the origin of Fig. 4.9a determines the optimum resistance R_W. These optimum points are situated just at the onset (*knee*) of the saturation of the emission current in Fig. 4.9b. Decreasing R_W and increasing T_c produces higher brightness and larger saturation currents but shortens the lifetime of the cathode.

The gun-brightness values shown in Fig. 4.9 are for a particular gun geometry. The shape of the Wehnelt electrode (flat or conical, with the cone apex turned toward or away from the anode) has a large influence on the brightness and other gun parameters.

4.1.5 Schottky Emission Guns

So-called Schottky emission cathodes are of the ZrO/W(100) type with a tip radius $r \simeq 0.1$–$1\ \mu$m [4.22, 4.23]. Just after etching, the middle of the rod is coated with ZrH_2 powder, which dissociates at $\simeq$1800 K in UHV and lets metallic Zr diffuse to the tip. Zirconium oxide is formed at 1600 K for a few hours at a partial oxygen pressure of 10^{-4}–10^{-5} Pa and flashed for a few seconds at 2000 K [4.3]. The work function is lowered by the ZrO coating from $\phi_w = 4.5$ eV (W) to 2.7 eV. This allows the electrons to overcome the work function at a temperature of 1800 K. The cathode is surrounded by a negatively biased suppressor (Wehnelt) electrode beyond which the tip apex protrudes $\simeq$0.3 mm. The electrons are extracted by a voltage of 4–8 kV at an extractor electrode. The field strength at the cathode is much higher than in thermionic cathodes but is still ten times lower than in field-emission sources. This means that the field strength is not sufficient for quantum-mechanical tunneling. Although the potential barrier is lowered by $\Delta\phi_w$ (Fig. 4.1), the Schottky effect, the electrons have to overcome the barrier with their thermal energy. It is therefore confusing to call this type of source a field-emission gun. The only common feature is emission from a pointed cathode, and it is more reasonable to regard the Schottky emitter as a field-assisted thermionic emitter. In contrast to a thermionic cathode, the Schottky emission gun has an energy spread $\Delta E \simeq 0.5$ eV not widened by the Boersch effect, and its emission current density $j_c \simeq 5 \times 10^6$ A/m^2 is higher by two orders of magnitude. The size of the virtual source as defined by the intersection of extrapolated trajectories behind the tip is $\simeq$15 nm, much smaller than the tip radius $r \simeq$ 0.5–1 μm. Thanks to these properties and to the high gun brightness, this type of cathode is coming into widespread use.

4.1.6 Field-Emission Guns

Field-emission guns also consist of a pointed cathode tip and at least two anodes. Tungsten is normally used as the tip material because etching is easy, but it has the disadvantage of sensitivity to surface layers. Wires of 0.1 mm diameter are spot-welded on a tungsten hairpin cathode and electrolytically etched to a radius of curvature of about 0.1 μm. The hairpin can be heated to eliminate absorbed gas atoms from the tip, to work at higher temperatures (of the order of 1500 K), or to raise the temperature when the tip requires remolding. So-called cold field-emission guns work with the cathode at room temperature. In both cases, the temperature is too low for the work function to be overcome; the electric field strength at the tip is so high that the emission occurs by the quantum-mechanical tunneling effect. $\langle 310 \rangle$-oriented tips are mainly used for cold field emitters and $\langle 100 \rangle$- and $\langle 111 \rangle$-oriented tips for heated ones.

The electron optics of a field-emission source are discussed in [4.24]. The positive voltage U_1 of a few kV at the first (extraction) anode (Fig. 1.2) generates a field strength $|\boldsymbol{E}| \simeq U_1/r$ of about 5×10^9 V m^{-1} at the cathode tip; this produces a field-emission current of the order of 1–10 μA. The electrons are postaccelerated to the final energy $E = eU$ by the voltage U between the cathode tip and the grounded second anode. The field-emission current (4.4) depends on the work function ϕ_w and on $|\boldsymbol{E}|$. Both quantities vary during operation of the gun. The work function changes owing to diffusion of impurities from within the tip material or to surface reactions or the adsorption of gases. The electric field strength changes as a result of damage to the tip by ion bombardment. This damage is unacceptable unless an ultrahigh vacuum $\leq 5 \times 10^{-8}$ Pa is maintained in the field-emission system. Even with a constant emission current, these factors can alter the solid angle of emission. The current emitted by a field-emission gun therefore drifts over long periods, and the tip has to be reactivated and remolded from time to time to concentrate the emitted current within a smaller angular cone.

A focused electron probe with a diameter of about 10 nm is formed as an image of the source by the action of anodes 1 and 2, which behave as an electrostatic lens. The diameter and the position of the focused probe and aberration constants C_s and C_c depend on the shape and dimensions of the anodes and on the ratio U_1/U [4.25, 4.26, 4.27, 4.28, 4.29]. The strong dependence of the position of the probe on the ratio U_1/U for a constant geometry is a disadvantage when the field-emission gun is combined with the condenser system of a TEM, whereas the dependence of the position of the crossover of thermionic cathodes on the operating parameters can be neglected. Anode 1 can be replaced by an electrostatic lens to overcome this problem [4.30, 4.31, 4.32, 4.33]; the electron-probe position can then be adjusted independently of the necessary voltage U_1. Other authors have proposed that a magnetic-lens field should be superimposed to provide a fully controllable field-emission gun [4.34, 4.35].

Field-emission guns have the advantage of high brightness and low energy spread (Table 4.1). They are of interest in all work that needs high coherence, which means low beam apertures and high current densities (high-resolution phase contrast, electron holography and interferometry, Lorentz microscopy, and STEM), though these modes can also function satisfactorily with a Schottky emitter. The high coherence of a field-emission gun is demonstrated in Fig. 3.9 by the large number of resolvable Fresnel fringes.

4.2 The Illumination System of a TEM

4.2.1 Condenser-Lens System

The condenser-lens system of a TEM (Fig. 4.10) performs the following tasks:

1. focusing of the electron beam on the specimen in such a way that sufficient image intensity is obtainable even at high magnification;
2. irradiation of a specimen area that corresponds as closely as possible to the viewing screen with a uniform current density, whatever the magnification, thereby reducing specimen drift by heating and limiting the radiation damage and contamination in nonirradiated areas;
3. variation of the illumination aperture α_i, which is of the order of 1 mrad for medium magnifications and must be $\leq$0.1 mrad for high-resolution and phase-contrast microscopy and $\leq 10^{-2}$ mrad for Lorentz microscopy, small-angle electron diffraction, and holographic experiments;
4. production of a small electron probe (0.2–100 nm in diameter) for x-ray microanalysis, electron-energy-loss spectroscopy, microbeam electron-diffraction methods and the scanning mode, and simple switching from the probe mode to area illumination.

Transmission electron microscopes are equipped with at least two condenser lenses to satisfy these requirements; the prefield of a strongly excited objective lens can act as an additional condenser lens, especially for point 4 (Sect. 4.2.3).

Figure 4.10 shows the most important modes of operation of a two-lens condenser system. In cases a–c, only the condenser lens C2 is excited. When focusing (Fig. 4.10b), the familiar lens formula $1/f_2 = 1/s_2 + 1/s_2'$ can be

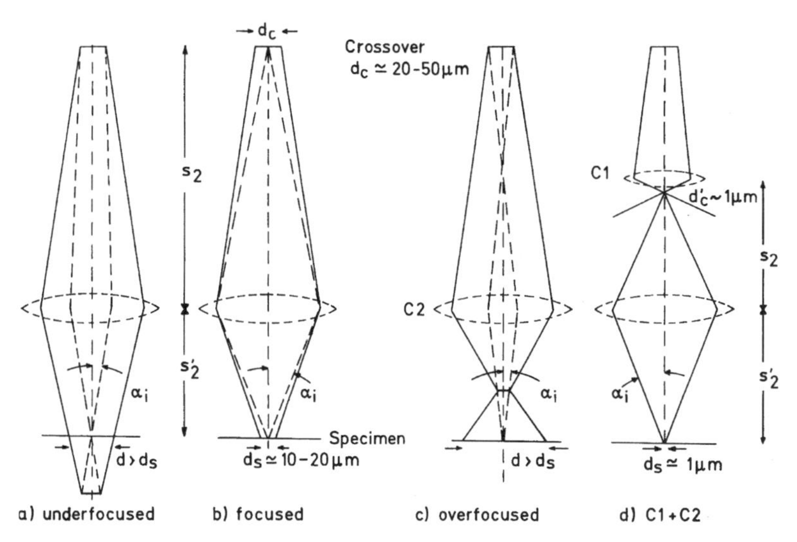

Fig. 4.10. Operation of a two-lens condenser system for illuminating the specimen. (**a**) Underfocused and (**c**) overfocused and (**b**) in-focus operation with condenser lens C2. (**d**) Additional use of condenser lens C1 to demagnify the crossover, the demagnified image then being focused on the specimen with condenser lens C2.

applied, and the crossover or virtual source is demagnified by the factor $d_s/d_c = s_2'/s_2$. The current density j_s at the specimen and the illumination aperture α_i reach a maximum and the diameter d_s of the irradiated area a minimum (Fig. 4.11). For underfocus (Fig 4.10a) and overfocus (Fig 4.10c) j_s and α_i increase. A condenser diaphragm (100–200 μm diameter) near the center of the condenser lens selects only the center of the beam. In focus, d_s has the same value as with no diaphragm because the crossover is imaged in both cases; the maximum current density j_s in the center of the beam and the illumination aperture α_i are, however, decreased as the diaphragm is made smaller. The current density and the aperture are related via the gun brightness (4.8) $\beta = j_s/\pi\alpha_i^2$.

The size of the final fluorescent screen corresponds to a specimen diameter of 1 μm at $M = 100\,000$. It is therefore sufficient to illuminate specimen areas as small as this. This can be achieved by fully exciting the condenser lens C1 (Fig. 4.10d); the strongly demagnified intermediate image of the crossover, with a diameter $d_c' \simeq 1\,\mu$m in the case of a thermionic cathode, can then be imaged on the specimen by condenser C2 with $M = s_2'/s_2$, resulting in $d_s \simeq$ 0.5–1 μm.

Very small values of α_i (for Lorentz microscopy or small-angle electron diffraction, for example) can be obtained by exciting condenser lens C1 and using the demagnified image of the crossover as the electron source (C2 switched off). The smallest obtainable illumination aperture with thermionic cathodes will be $\alpha_i = r_c'/(s_2 + s_2') \simeq 10^{-2}$ mrad. In view of (4.8), this operating mode

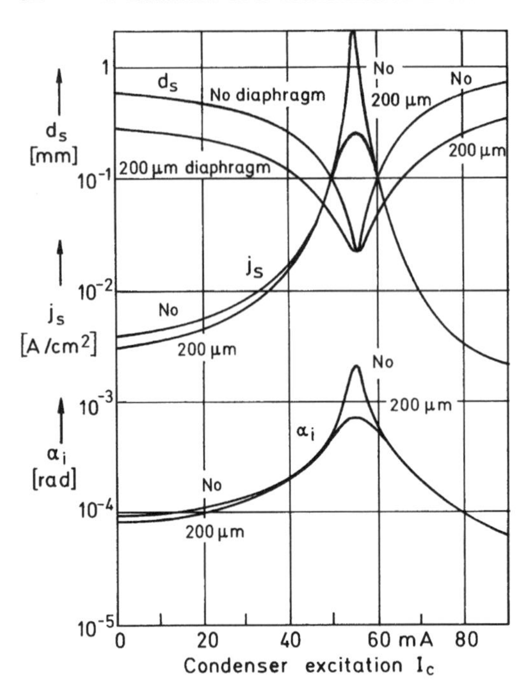

Fig. 4.11. Dependence of the diameter d_s of the irradiated area, the current density j_s, and the illumination aperture α_i in the specimen plane on the excitation of condenser lens C2 with no diaphragm and with a 200 μm diaphragm.

only works when very small current densities j_s are acceptable. These results for the case of thermionic cathodes can be improved by the use of Schottky or field-emission guns, which have a higher brightness and a virtual source of about 10–100 nm diameter.

The condenser lens C1 works with a relatively large entrance aperture and is therefore equipped with a stigmator to compensate for the astigmatism and to decrease the diameter d'_c of the crossover image. It is usually sufficient to observe caustic cross sections as in Fig. 2.13 for the compensation of astigmatism.

The electron-gun system of a microscope can be adjusted onto the axis of the condenser-lens system by tilting and shifting the gun system. Further adjustments are necessary to bring the electron beam onto the axis of the objective-lens and magnifying-lens system. The specimen structures spiral around the image-rotation center if the high voltage, or preferably the lens current of the objective lens, is varied periodically (wobbled). For easy observation of this spiral movement during alignment, holey formvar films or polystyrene spheres on supporting films can be used as specimens (Fig. 2.17b). The distance of the image-rotation center from the point of intersection of the objective axis can be calibrated by shifting the condenser-lens system relative to the objective lens. The point of intersection of the objective-field axis with the final screen does not necessarily coincide with the center of the final screen [4.36, 4.37]. Its position can be determined by reversing the objective-lens cur-

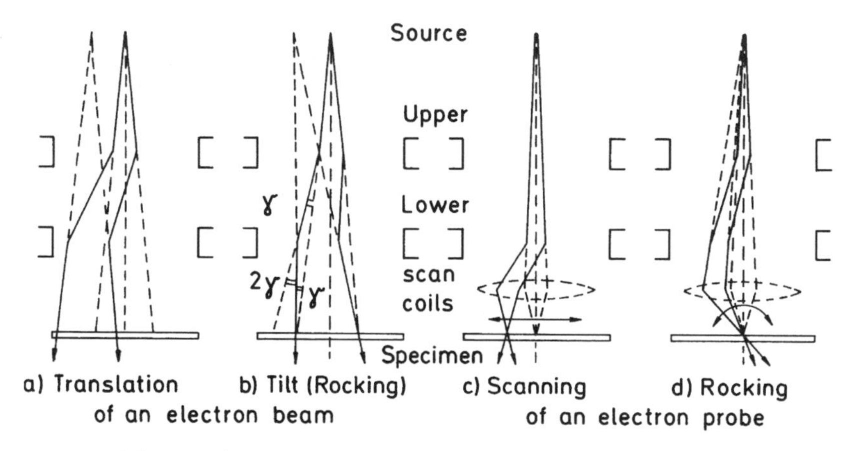

Fig. 4.12. (**a**) Shift (translation) and (**b**) tilt (rocking) of an electron beam by a double deflection-coil system when exciting the coils with a sawtooth current. (**c**) Scanning and (**d**) rocking when working with the prefield of an objective lens as an additional condenser lens.

rent. Specimen structures at the point of intersection will remain stationary. Coma-free alignment for high resolution is discussed in Sect. 2.4.3.

The alignment procedure needed to bring the electron beam on-axis involves a mechanical shift and tilt of the condenser-lens system or electromagnetic deflection of the electron beam by pairs of alignment coils (Fig. 4.12a,b). Such coils can also be used to generate the dark-field mode (Sects. 4.4.2 and 6.1.2). The incident electron beam is tilted (Fig. 4.12b) so that a cone of scattered electrons or Bragg-reflected electrons is on-axis and can pass the objective diaphragm. The transition from the bright- to the dark-field mode and back can easily be achieved by switching the alignment coils off and on, respectively. These coils can also be used for irradiation with a rocking beam at low and medium magnifications as an additional focusing aid (Sect. 4.4.4) or, together with the objective-lens prefield, for scanning and rocking (Fig. 4.12c,d) the electron probe in a scanning mode (Sect. 4.2.3) or for special diffraction techniques (Sect. 8.1).

A dark-field mode with hollow-cone illumination (Sect. 6.1.2) can be created by replacing the circular diaphragm in the condenser lens C2 by an annular diaphragm. Alternatively, the beam may be deflected successively around a circle (Sect. 6.4.3).

4.2.2 Electron-Probe Formation

The illumination of specimens with small electron probes of diameter less than 0.1 μm is important for the x-ray microanalysis and energy-loss spectroscopy of small specimen areas and for microbeam-diffraction methods in TEM as well as for scanning transmission electron microscopy (STEM). An electron probe is formed by two- to three-stage demagnification of the electron-gun crossover. If the geometrical diameter of the probe is d_0, the total probe current will be given by

$$I_p = \frac{\pi}{4} d_0^2 j_p. \tag{4.15}$$

In reality, the intensity is distributed more as a Gaussian distribution (4.14). This, however, will only change the results of our simple estimation by correction factors of the order of unity. The conservation of gun brightness (4.8) on the optic axis implies $j_p = \pi\alpha_p^2\beta$ (α_p is the electron-probe aperture). Substitution in (4.15) gives

$$I_p = \frac{\pi^2}{4} \beta d_0^2 \alpha_p^2. \tag{4.16}$$

Solving for d_0, we find

$$d_0 = \left(\frac{4I_p}{\pi^2\beta}\right)^{1/2} \frac{1}{\alpha_p} = \frac{C_0}{\alpha_p}, \tag{4.17}$$

which shows that for a given probe current I_p, small values of d_0 can be obtained only for large values of the gun brightness β and probe aperture α_p.

The geometrical diameter d_0 is broadened by the action of lens aberrations; chromatic aberration produces an error disc of diameter d_c (2.63) and spherical aberration of diameter d_s (2.58). The aperture limitation α_p causes a diffraction disc $d_d = 0.6\lambda/\alpha_p$; that is, the half-width of the Airy distribution in Fig. 3.16b.

To estimate the final probe size d_p, this blurring can be treated approximately as a quadratic superposition of the error-disc diameters [4.38], though this is strictly valid only when the error discs are all of Gaussian shape and independent from one another

$$\begin{aligned} d_p^2 &= d_0^2 + d_d^2 + d_s^2 + d_c^2 \\ &= [C_0^2 + (0.6\lambda)^2]\frac{1}{\alpha_p^2} + \frac{1}{4}C_s^2\alpha_p^6 + \left(C_c\frac{\Delta E}{E}\right)^2 \alpha_p^2. \end{aligned} \tag{4.18}$$

For a thermionic cathode, the constant C_0 (4.17) will be much greater than the wavelength. Then the chromatic-error and diffraction terms in (4.18) can be neglected.

Figure 4.13 shows how the diameters d_0 and d_s superpose and produce a minimum probe diameter d_{min} at an optimum aperture α_{opt} for a constant probe current I_p. The optimum aperture is obtained by writing $\partial d_p/\partial\alpha_p = 0$, giving

$$\alpha_{opt} = (4/3)^{1/8}(C_0/C_s)^{1/4}, \tag{4.19}$$

and substitution in (4.18) gives

$$d_{min} = (4/3)^{3/8}(C_0^3 C_s)^{1/4}. \tag{4.20}$$

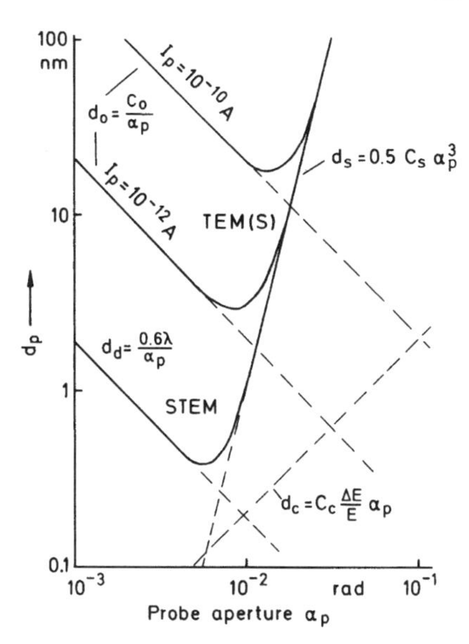

Fig. 4.13. Superposition of error discs shown as a function of probe aperture α_{p} in a double logarithmic plot. Upper two curves: superposition of d_0 and d_{s} for $I_{\mathrm{p}} = 10^{-10}$ and 10^{-12} A for the probe-forming and scanning transmission mode of a TEM. Lower curve: superposition of d_{d} and d_{s} for a field-emission STEM. $E =$ 100 keV, $\Delta E = 1$ eV, $C_{\mathrm{s}} = C_{\mathrm{c}} = 2$ mm, $\beta = 10^5$ A cm^{-2} sr^{-1}.

For a field-emission gun, the constant C_0 will be much smaller than the wavelength, and the energy spread ΔE is also smaller. Superposition of the largest terms in (4.18), now d_{d} and d_{s}, again yields a minimum (Fig. 4.13). For this case, a wave-optical calculation should strictly be used (see, e.g., Fig. 3.16c), but this has no influence on the position of the minimum [1.94]. The only difference is that the increase due to spherical aberration for $\alpha > \alpha_{\mathrm{opt}}$ is not so rapid.

It is of practical interest to express the maximum probe current I_{p} as a function of the probe diameter d_{p}. For thermionic cathodes, (4.20) can be solved for I_{p}, which is included in C_0 (4.17):

$$I_{\mathrm{p}} = (3\pi^2/16)\beta\, C_{\mathrm{s}}^{-2/3} d_{\mathrm{p}}^{8/3}. \tag{4.21}$$

To obtain a formula similar to (4.21) for a field-emission gun, the spherical-aberration constants C_{s1} and C_{s2} of the lens system of the field-emission gun and the objective lens, respectively, have to be considered [4.39, 4.40], resulting in

$$I_{\mathrm{p}} = \frac{1.3J}{C_{s1}^{1/2} C_{s2}^{1/6}} d_{\mathrm{p}}^{2/3} \tag{4.22}$$

with $J = \mathrm{d}I/\mathrm{d}\Omega$ as the emission current per solid angle. The probe current I_{p} increases only as $d_{\mathrm{p}}^{2/3}$ and reaches saturation at a relatively small value of d_{p} because I_{p} cannot become larger than the emission current. Although these calculations are somewhat oversimplified, they show that the field-emission gun can have disadvantages if large currents are needed. For the production of electron probes smaller than 0.1 μm, the field-emission gun has the advantage of providing larger beam currents for a constant probe diameter, which is important for increasing the signal-to-noise ratio in STEM.

The importance of the probe current I_p for achieving a good signal-to-noise ratio is shown by the following estimate. A signal S is produced by a number

$$n = fI_p\tau/e \tag{4.23}$$

of electrons, where f denotes the fraction of electrons recorded by the detector ($f < 1$) or the number of recorded x-ray quanta per electron ($f \ll 1$) and τ denotes the recording time for one image point (pixel); that is, the frame time (1/20–1000 s) divided by the number of pixels (10^5-10^6). As a result of statistical shot noise, the noise signal is $N = \sqrt{n}$. The signal-to-noise ratio must be larger than some value κ, which should be of the order of 3–5 to detect a signal in a noisy record. If a signal difference ΔS on a background S is to be detected, then

$$\frac{\Delta S}{S} \geq \kappa \frac{n^{1/2}}{n} = \frac{\kappa}{(fI_p\tau/e)^{1/2}} \tag{4.24}$$

and so I_p has to satisfy the inequality

$$I_p \geq \left(\frac{\kappa}{\Delta S/S}\right)^2 \frac{e}{f\tau}. \tag{4.25}$$

As a numerical example, for $\kappa = 3$, $\Delta S/S = 5\%$, $f = 0.1$, $\tau = 1$ ms (scanning of a frame with 10^6 pixels in 1000 s), and $e = 1.6 \times 10^{-19}$ C, we find $I_p \geq 3.6$ pA.

4.2.3 Illumination with an Objective Prefield Lens

A single-field condenser-objective lens (Figs. 2.8 and 4.14) with an excitation $k^2 = 3$ not only has the advantage of a low spherical-aberration coefficient C_s but also simplifies the transition from the extended illumination needed for the TEM bright- and dark-field modes (Fig. 4.14a) to the illumination required to form a small electron probe for the scanning transmission mode and for x-ray and energy-loss spectroscopy and electron diffraction of small specimen areas (Fig. 4.14b).

As discussed in Sect. 2.2.3, this type of lens operates in the telefocal condition with the specimen at the lens center. The action of the prefield condenser and postfield objective field can be represented in a ray diagram by two separate lenses. The optimum working condition for illumination with an extended beam (Fig. 4.14a) will be achieved by fully exciting condenser lens C1 and focusing the crossover on the front focal plane (FFP) with condenser lens C2. This can be checked by imaging the back focal plane (BFP) on the final viewing screen, the BFP being conjugate to the FFP. Furthermore, the specimen plane and the plane of the condenser C2 diaphragm of diameter d_2 are conjugate, and the diameter of the irradiated area is thus $d_s = Md_2$ with the demagnification $M = f_0/s_2'$. The illumination aperture $\alpha_i = d_c'/2f_0$ is limited by the diameter d_c' of the crossover image in the FFP. Thus, for $f_0 = 1$ mm,

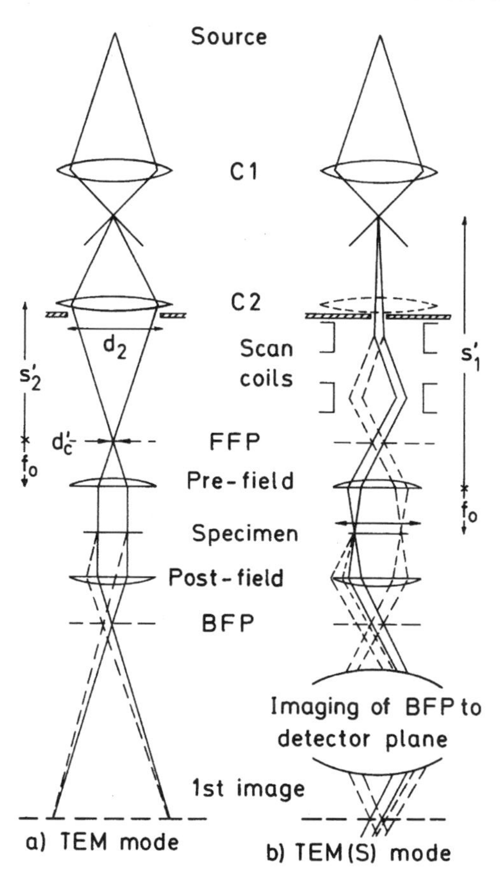

Fig. 4.14. Specimen illumination by an objective prefield: (**a**) large-area illumination for TEM, (**b**) electron-probe formation and scanning for the scanning transmission mode of a TEM.

$s_2' = 200$ mm, $d_2 = 100$ μm, and $d_c' = 0.5$ μm, we find $d_s = 0.5$ μm and $\alpha_i =$ 0.25 mrad, which are optimum operating conditions for high resolution.

A very small spot diameter d_s can be obtained for the scanning mode (Fig. 4.14b) by switching off C2 and using a small C2 diaphragm. The geometrical diameter d_0 of the electron probe in the specimen plane can be estimated with the demagnification factor $M = f_0/s_1' \simeq 1/250$; if the diameter of the crossover image in the focal plane of C1 is 0.5 μm, we obtain $d_0 \simeq 2$ nm. The aperture of the electron probe is determined by the projected diameter of the C2 diaphragm in the FFP. The aperture will be of the order of 5 mrad and therefore one order of magnitude larger than the illumination aperture α_i in the extended-beam-producing mode of Fig. 4.14a. As shown in Sect. 4.2.2, a large probe aperture α_p will be necessary to produce a small probe diameter d_p.

This principle, which has been illustrated with a two-lens condenser system, can be optimized by using a three-lens system, which allows illumination analogous to the Köhler illumination of light microscopes to be achieved; the condenser diaphragms can likewise be selected automatically for different diameters of the irradiated area [4.41].

The electron probe can be scanned across the specimen plane by means of pairs of scanning coils, as in Fig. 4.14b, which rock the incident electron beam. If the pivot point of this beam rocking is at the FFP, the pivot point of the rays behind the specimen will be at the BFP because these planes are conjugate. This means that the position of the first diffraction pattern in the BFP is stationary, and the BFP can be imaged on the detector plane in the scanning transmission mode (Sect. 4.5.1).

4.3 Specimens

4.3.1 Useful Specimen Thickness

The maximum useful specimen thickness depends on the type of electron–specimen interaction used to form the image and on the mode of operation. For high-resolution imaging ($\leq$1 nm) in the bright-field mode, phase-contrast effects are important. The image contrast is then due to the interference of the scattered waves with the unscattered primary incident wave. Phase-contrast effects therefore decrease with increasing specimen thickness owing to the attenuation of the incident-wave amplitude (Sect. 6.2). For irregular structures, this limits the useful thickness range to a few tens of nanometers. Typical specimens for this mode of operation are single atoms or clusters of heavy elements, organic macromolecules, viruses, phages, etc. This implies that the specimen must be mounted on a thin supporting film of thickness $t \leq 5$ nm.

The imaging of lattice planes of crystals results from the interference of the primary beam and one or more Bragg-reflected waves and can be observed for thicknesses of a few tens of nanometers for which sufficient wave amplitude remains. Directly interpretable high-resolution images of the crystal structure using several Bragg reflections can only be obtained for thicknesses less than 10 nm because the wave amplitudes of the Bragg reflections are changed by dynamical electron diffraction; false contrast results, which can be interpreted only by computer simulations.

For medium and low resolutions ($\geq$1 nm), most work on amorphous specimens relies on scattering contrast. In the bright-field mode, the image intensity depends on the number of electrons that pass the objective diaphragm. The decrease in image intensity is caused by the absence of those electrons that have been scattered outside the cone with aperture α_o (objective aperture). In biological sections, the scattering contrast is increased by staining the tissue or thin sections with heavy atoms. Quantitative examples of scattering contrast are reported in Sect. 6.1.3. Another example is the negative staining technique, where microorganisms or macromolecules are embedded in a layer of a heavy metal compound, such as phosphotungstic acid. The energy lost by electrons during inelastic scattering and the chromatic aberration of the objective lens limit the maximum useful specimen thickness to 100–300 nm for 100 kV and about 1 μm for 1 MV. This chromatic error can be avoided in the

STEM mode of TEM. However, the resolution is limited by the broadening of the electron probe due to multiple scattering (Sect. 5.4.3). The electron spectroscopic imaging mode of an energy-filtering TEM also avoids the chromatic error resulting from energy losses in thick specimens. The investigation is limited only by the decrease of transmission below $T = 10^{-3}$, which means $\simeq$75 μg/cm^2 for amorphous and 150 μg/cm^2 for crystalline specimens at 80 keV.

By using the primary beam in the bright-field mode or a Bragg-reflected beam in the dark-field mode, lattice defects in crystalline specimens can be imaged. The maximum thickness is limited by the intensity of the primary or Bragg-reflected beam and by the chromatic error for thick specimens. The intensities of the beams depend on the crystal orientation, and better penetration is observed in the case of anomalous transmission near a Bragg condition. At 100 keV, the useful thickness of metal foils and other crystalline material is of the order of 50–200 nm. The increase in the useful thickness when the accelerating voltage is increased from 100 kV to 1 MV is only of the order of three to five times (see also Sect. 9.1.6). However, a large number of specimens (electropolished metal foils, for example) are some 200–500 nm thick over most of the thinned specimen area and, in many cases, the only areas that can be used at 100 keV are the edges of holes in the center of electropolished or ion-beam thinned discs.

4.3.2 Specimen Mounting

Metals and other materials can be used directly as thin discs of 3 mm diameter and $\simeq$0.1 mm thickness if they can be thinned in the center by electropolishing or chemical or ion etching.

Other specimens for TEM (crystal flakes, surface replicas, evaporated films, biological sections) are mounted on copper grids with 100–200 μm meshes. Grids of 3 mm diameter are commercially available with different mesh sizes and orientation marks.

Small particles, microorganisms, viruses, macromolecules, and single molecules need a supporting film possessing the following properties:

1. low atomic number to reduce scattering
2. high mechanical strength
3. resistance to electron irradiation (and heating)
4. electrical conductivity to avoid charging
5. low granularity (caused by phase contrast) for high resolution
6. easy preparation

For medium magnifications, formvar films of 10–20 nm are in use that are produced by dipping a glass slide in a 0.3% solution of formvar in chloroform and floating the dried film on a water surface. A higher mechanical strength is obtained by evaporating an additional thin film of carbon ($\simeq$5 nm) on a formvar or collodion film. Pure carbon films are more brittle but can be used as 3–5 nm films on plastic supporting films with holes.

For high resolution, the granularity of carbon (or amorphous germanium) films is useful for investigating the contrast-transfer function of the TEM (Sect. 6.4.6), but it obscures the image of small particles, macromolecules, and single atoms. Numerous attempts have therefore been made to prepare supporting films with less phase-contrast granularity: amorphous aluminum oxide [4.42], boron [4.43], single-crystal films of graphite [4.44], or vermiculite [4.45].

The specimen grids or discs are mounted in a specimen cartridge, which can be transferred through an airlock system either into the bore of the upper polepiece of the objective lens (top entry) or, mounted on a rod, into the polepiece gap (side entry). The specimen position is near the center of the bell-shaped lens field for a strongly excited objective lens with $k^2 \simeq 3$. The polepiece gap also contains the objective diaphragms and the anti-contamination blades or *cold finger* (Sect. 11.4.2), which decrease the partial pressure of organic molecules near the specimen. This decreases the space available for special specimen manipulations. The gap is only of the order of a few millimeters for 100 kV TEMs and 1–2 cm in an HVEM.

4.3.3 Specimen Manipulation

The principal methods of specimen manipulation are summarized in [4.46, 4.47, 4.48] and the proceedings of the HVEM symposia [1.86, 1.87, 1.88, 1.89, 1.90, 1.91].

Specimen rotation about an axis parallel to the electron beam can be used to bring specimen structures into a convenient orientation in the final image. Tilting devices with one axis normal to the electron beam can produce stereo pairs for quantitative measurement and stereoscopic observation of the three-dimensional specimen structure. A goniometer can tilt the specimen with high precision in any desired direction up to $\pm 60°$ or even $\pm 70°$. Side-entry goniometers are available that cause a specimen shift less than 1 μm when tilting the specimen $\pm 30°$. The second degree of freedom for angular adjustment is often exploited as a specimen rotation about an axis normal to the specimen plane. Top-entry goniometers can often tilt the specimen and move the specimen normal over a cone around the optic axis. Goniometer stages can be useful when studying biological tissue sections to bring lamellar systems or other structures into favorable orientations or for tomography, for example. Crystalline specimens have to be tilted in a goniometer for

1. observation of lattice fringes and crystal structures,
2. observation of diffraction contrast of lattice defects with distinct Bragg reflections or known orientation,
3. determination of the Burgers vector of lattice defects, and
4. determination of crystal orientation by electron diffraction.

A large variety of specimen tilting, heating, and straining cartridges have been developed for 100 kV TEMs. The problem arises of whether the phenomena of deformation, annealing, and precipitation are the same in thin specimen areas that can be studied at 100 kV as in the bulk material. Such specimen manipulations are therefore of particular interest in HVEM [4.49] in which there is more space in the polepiece gap and the specimens will be more similar to bulk material because greater thicknesses can be penetrated.

Specimen-cooling devices operating at temperatures below $-150°C$ can be used to reduce the contamination of inorganic material associated with radiation-induced etching of carbon in the presence of oxygen molecules (Sect. 11.4.2). Such devices must not be confused with the cooled anti-contamination blades mentioned above. Specimens that melt at room temperature or due to electron-beam heating or that sublimate in the vacuum may be observable if the specimen is cooled. A special application is the direct observation of cryosections. The sections have to be transferred from the cryomicrotome to the cooled specimen cartridge of the microscope via a cooling chain.

Specimen-cooling devices operating at liquid-helium temperature need very careful design and construction [4.50, 4.51, 4.52]. The specimen and an additional storage tube for liquid helium have to be shielded against radiative heat loss by surrounding them in a liquid-nitrogen-cooled trap. Specimen structures or physical effects that are normally present only at very low temperatures can be observed such as the crystal structure of condensed gases, magnetic fields around superconducting domains, and ferromagnetic films of low Curie temperature. The mobility of radiation-induced lattice defects decreases at low temperatures. These defects can be generated directly in a cooled specimen by bombardment with α-particles or high-energy electrons beyond the threshold energy (Sect. 5.1.2); the coagulation of dislocation loops or stacking faults can then be observed when the temperature is raised. The suppression of the radiation damage of organic specimens is another application of liquid-helium-cooled stages. However, specimen cooling obviously only retards secondary radiation effects such as the distortion of the crystal lattice, which leads to fading of the electron-diffraction pattern, but cannot prevent primary damage of the individual organic molecules (Sect. 11.2).

Environmental cells in which the partial pressures of inert and reactive gases up to atmospheric pressure are maintained allow us to observe in situ reactive processes between a gas and the specimen; with a partial pressure of water, hydrated biological specimens can be observed [4.53]. Such studies are limited by electron scattering at the gas molecules. The large gas pressure in the specimen area can be obtained either by using differentially pumped systems of diaphragms or by confining the gas between diaphragms covered with thin films. High-voltage electron microscopy is more suitable for environmental experiments because much more space is available in the polepiece gap and the scattering in the gas atmosphere is less severe. Table 4.2 contains some further examples of in situ experiments.

Table 4.2. Specimen manipulations.

Procedure	Application
1. Specimen rotation	
Rotation on an axis parallel to the electron beam	Orientation of specimen structures or diffraction patterns relative to the edges of the final screen
2. Specimen tilt	
a) Tilt ($\pm 5^\circ$ – $\pm 10^\circ$) about an axis in the specimen plane	Stereo pairs
b) Tilt about an axis normal to the beam and rotation about an axis parallel to the beam	Lattice defects Determination of orientation
c) Double tilt ($\pm 25^\circ$ – $\pm 70^\circ$) about two perpendicular axes normal to the beam	Favorable orientation of biological sections
d) Specimen goniometer (± 25 – $\pm 50^\circ$) Small specimen shift by adjustment of tilt axis in height and position (accuracy: $\pm 0.1^\circ$)	Lattice defects Three-dimensional reconstruction by tomography
3. Straining devices	
Straining of the specimen by movement of two clamps by mechanical or piezoelectric effects	Straining of metals and high polymers
4. Specimen heating	
Direct heating of a grid	Recovery and recrystallization
Indirect heating ($\simeq$10 W for 1000°C)	Precipitation and transition phenomena
5. Specimen cooling	
a) Cooling to between –100°C and –150°C with liquid nitrogen	Temperature-sensitive specimens Decrease of specimen contamination Direct observation of cryosections
b) Cooling with liquid helium (4–10 K)	Structure of condensed gases Decrease of radiation damage Superconducting states Magnetic structure in low-Curie-point ferromagnetics
6. Environmental cells	
Gas pressure between diaphragms covered with foils or separated from the microscope vacuum by additional pumping stages	Biological specimens in wet atmosphere Gas reactions on the specimen Corrosion tests
Spraying the specimen with a gas jet	
7. Other in situ methods	
Evaporation in the specimen chamber	Investigation of film growth
Particle bombardment by an ion source or the beam of an HVEM	Radiation-damage experiments
Magnetization of the specimen by additional coils (Lorentz microscopy)	Direct observation of movement of ferromagnetic domain walls

4.4 The Imaging System of a TEM

4.4.1 Objective Lens

Alhough the first intermediate image formed by the objective lens has a magnification of only 20–50 times, it is from this lens that the highest performance will be demanded. The mechanical tolerances necessary have already been discussed in Sect. 2.3.3. The natural astigmatism of an objective lens must be so small that the main task of the stigmator is to compensate for the astigmatism caused by contamination of the diaphragm and other perturbing effects. The resolution-limiting errors such as the spherical and chromatic aberrations are important only for the objective lens because a magnification M decreases the apertures for the following lenses to $\alpha = \alpha_o/M$. The diameter of the spherical-aberration disc is proportional to α^3 (2.58). Even for a modest magnification M of 20–50 times at the first intermediate image, the aperture becomes so small that the spherical aberration of the intermediate and subsequent lenses can be neglected, even though these lenses normally have larger values of C_s and C_c than the objective lens. In projector lenses (Sect. 2.3.4), the dominant aberration will usually be distortion (pin cushion or barrel) at low magnifications, which distorts but does not impair the sharpness of the image.

Any of three or four diaphragms of 20–200 μm diameter can be inserted in the focal plane of the objective lens, thus permitting the objective aperture α_o to be changed (Fig. 4.15). We should distinguish between apertures (angles) and solid diaphragms and not use the word "aperture" for both. The trajectories in Fig. 4.15 show that a diaphragm introduced in this plane stops

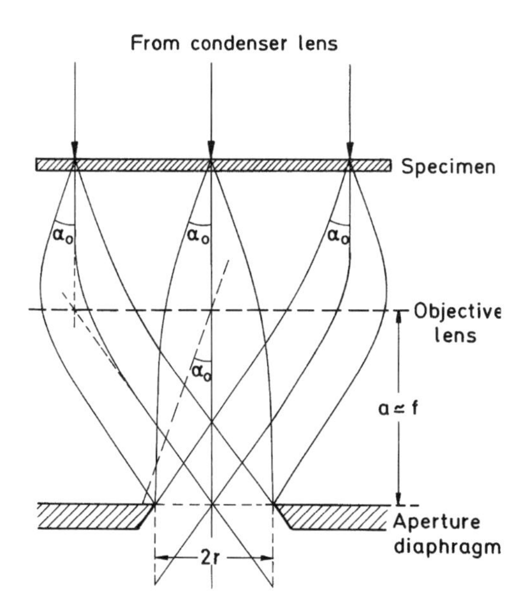

Fig. 4.15. Action of the objective diaphragm in the focal plane of the objective lens as an angle-selective diaphragm.

all electrons that have been scattered through angles $\theta \geq \alpha_o$. Decreasing the objective aperture increases the scattering contrast (Sect. 6.1). For high resolution, an aperture as large as possible is used so that high spatial frequencies can contribute to the image (Sect. 6.2) and so that any contamination and charging of the diaphragm do not disturb the image.

The objective aperture α_o will be given by r/f, where r is the radius of the diaphragm, only to a first approximation because the electron trajectories in the thick objective lens are curved (Fig. 4.15). The objective aperture can be measured accurately by selected-area electron diffraction (SAED, Sect. 8.1.1), in which the focal plane of the objective lens is imaged on the final image screen. For the measurement of α_o, exposures of the diffraction pattern (of an evaporated Au film, for example) are taken with and without the aperture diaphragm. The ratio of the objective aperture α_o to the Bragg–diffraction angle $2\theta_B$ of a Debye–Scherrer ring is related to the diameter d_0 of the shadow of the diaphragm and the diameter d_B of the Debye–Scherrer ring by

$$\alpha_o/2\theta_B = d_0/d_B. \tag{4.26}$$

The same procedure can be used to measure the illumination aperture α_i, which corresponds to the radius of the primary beam in a diffraction pattern. For small values of α_i below 0.1 mrad, the magnification (*camera length*) of the SAED pattern has to be increased.

The nature of the objective diaphragm is important for the quality of the image. The diaphragm has to be of a heat-resistant material (Pt, Pt-Ir, Mo, or Ta) capable of tolerating the largest possible current density in the focal plane; this may reach 10^5 A m^{-2}. Dust, fragments of the specimen, and contamination in general can cause local charging, which generates an additional astigmatism, especially if small apertures are used. Charging effects can be delayed by using diaphragms in the form of thin metal foils (1–2 μm) with circular holes [4.54, 4.55, 4.56].

4.4.2 Imaging Modes of a TEM

The imaging system of a TEM consists of at least three lenses (Fig. 4.16): the objective lens, the intermediate lens (or lenses), and the projector lens. The intermediate lens can magnify the first intermediate image, which is formed just in front of this lens (Fig. 4.16a), or the first diffraction pattern, which is formed in the focal plane of the objective lens (Fig. 4.16b), by reducing the excitation (selected-area electron diffraction, Sect. 8.1.1). In many microscopes, an additional diffraction lens is inserted between the objective and intermediate lenses to image the diffraction pattern and to enable the magnification to be varied in the range 10^2 to 10^6.

The *bright-field mode* (BF) (Figs. 4.16a and 4.17a) with a centered objective diaphragm is the typical TEM mode, with which scattering contrast (Sect. 6.1.1) and diffraction contrast (Sect. 9.1) can be produced with objective apertures α_o between 5 and 20 mrad. For high-resolution phase contrast

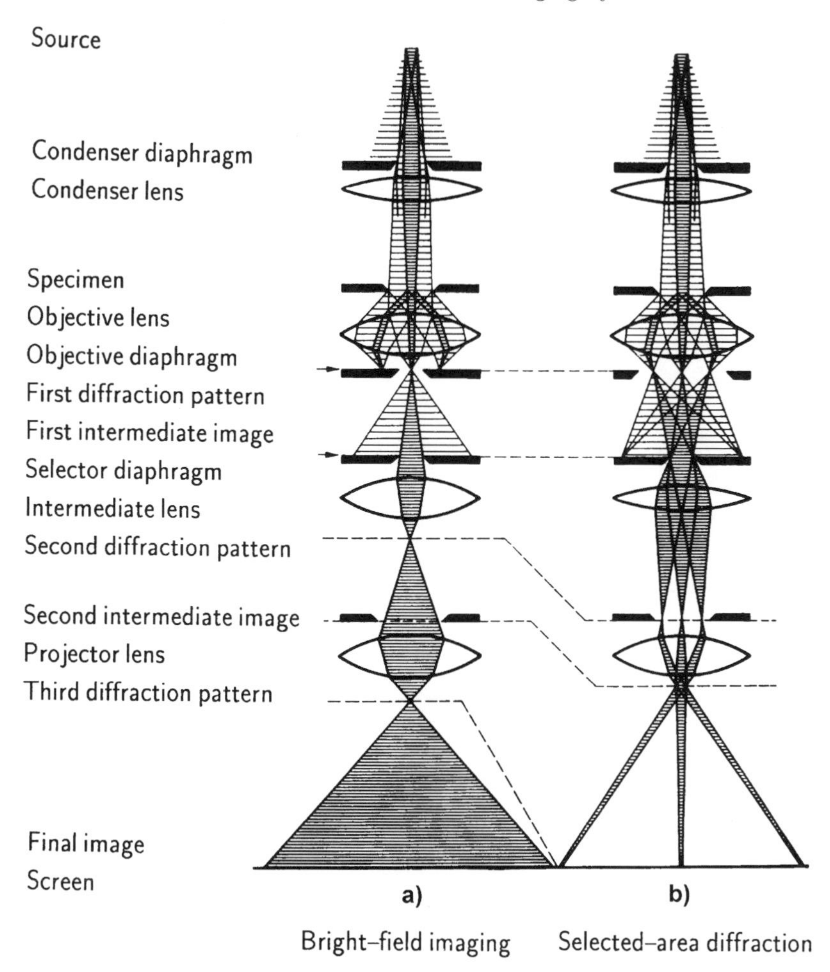

Fig. 4.16. Ray diagrams for a TEM in (**a**) the bright-field mode and (**b**) selected-area electron diffraction (SAED) mode.

(Sect. 6.2), the aperture should be larger ($\alpha_o \geq 20$ mrad) to transfer high spatial frequencies. The only purpose of the diaphragm in this mode is to decrease the background by absorbing electrons scattered at very large angles. The resolution is limited by the attenuation of the contrast-transfer function (CTF) caused by chromatic aberration (Sect. 6.4.2) and not by the objective aperture α_o. Normally, the specimen is irradiated with small illumination apertures $\alpha_i \leq 1$ mrad. For high resolution, an even smaller aperture $\alpha_i \leq 0.1$ mrad is necessary to avoid additional attenuation of the CTF by partial spatial coherence (Sect. 6.4.2). When unconventional types of contrast transfer are desired, it is often necessary to change the illumination condition by tilting the beam or using hollow-cone illumination, for example.

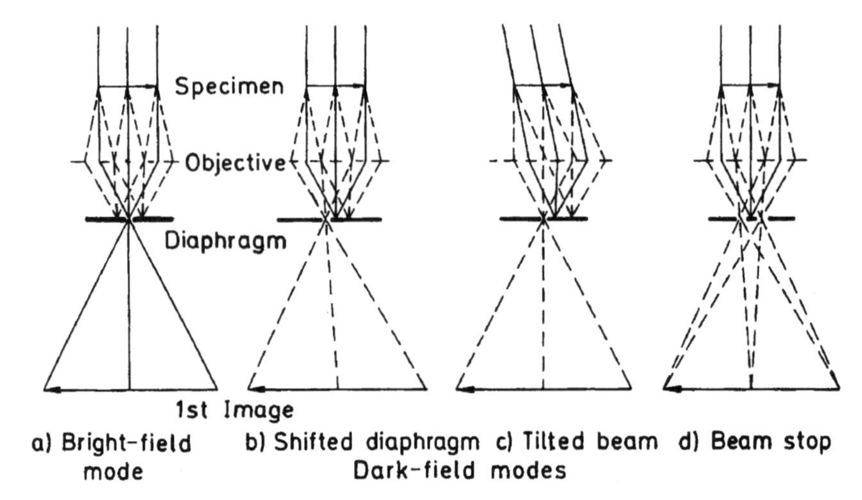

Fig. 4.17. (**a**) Bright-field mode with a centered objective diaphragm and production of a dark-field mode by (**b**) a shifted diaphragm, (**c**) a tilted beam, and (**d**) a central beam stop.

In the *dark-field mode* (DF), the primary beam is intercepted in the focal plane of the objective lens. Different ways of producing dark-field conditions are in use. The shifted-diaphragm method (Fig. 4.17b) has the disadvantage that the scattered electrons pass through the objective lens on off-axis trajectories, which worsens the chromatic aberration. The most common mode is therefore that in which the primary beam is tilted (Fig. 4.17c) so that the axis strikes the centered diaphragm. The image is produced by electrons scattered into an on-axis cone of aperture α_o. This mode has the advantage that off-axis aberrations are avoided. There is thus no increase of chromatic error. Asymmetries in the dark-field image can be avoided by swiveling the direction of tilt around a cone, or conical illumination can be produced by introducing an annular diaphragm in the condenser lens. Another possibility is to use a central beam stop that intercepts the primary beam in the back focal plane; for this, a thin wire stretched across a circular diaphragm may be employed (Fig. 4.17d). DF micrographs need a longer exposure time because there are fewer scattered electrons. For high resolution, the contrast-transfer function (CTF) of DF is nonlinear, whereas the CTF of the BF mode is linear for weak-phase specimens. The DF mode can also be employed to image crystalline specimens with selected Bragg-diffraction spots.

Increasing the objective aperture in the BF mode allows us to transfer the primary and one Bragg-reflected beam through the diaphragm. These beams can interfere in the final image. The fringe pattern is then an image of the crystal-lattice planes (Sect. 9.6.1). Optimum results are obtained for this mode when the primary beam is tilted by the Bragg angle $+\theta_B$. The Bragg-reflected beam that is deflected by $2\theta_B$ passes through the objective lens with an angle $-\theta_B$ relative to the axis.

In the *crystal-lattice imaging mode*, more than one Bragg reflection and the primary beam form a lattice image that consists of crossed lattice fringes, or an image of the lattice and its unit cells if a large number of Bragg reflections are used (Sect. 9.6). This mode is most successful for the imaging of large unit cells, which produce diffraction spots at low Bragg angles so that the phase shifts produced by spherical aberration and defocusing are not sufficiently different to cause imaging artifacts.

Further operating modes of a TEM are described in other sections: *scanning transmission mode* (Sect. 4.5), *Lorentz microscopy* (Sect. 6.8) and the analytical modes of *x-ray microanalysis* (Sect. 10.2), *electron energy-loss spectroscopy* (EELS, Sect. 10.3), and *electron diffraction* (Chap. 8).

4.4.3 Magnification and Calibration

If structures as small as 0.1 nm are to be resolved, the instrument must be capable of magnifying this distance until it is larger than the resolution of the photographic emulsion or the pixel size of the CCD camera (20–50 μm); this requires a magnification M of at least $250\,000 - 500\,000$ times, for which more than two imaging lenses are needed.

The accuracy of magnification depends on the excitation of the objective lens. If the magnification is to be constant to within about $\pm 1\%$, the following precautions have to be taken:

1. The height of the specimen in the specimen cartridge must be reproducible. Depending on the microscope used, a variation of the vertical position of ± 50 μm results in a variation of 2–5% in the magnification [4.57, 4.58].
2. The lens current and acceleration voltage must be highly stable. The lens current necessary for focusing is related to the height of the specimen. Differences of specimen height can therefore be compensated for by reading the lens current and using a calibration curve relating lens current and magnification. However, accuracies of reading the lens currents of the order of $\pm 1\%$ are needed [4.59].
3. Hysteresis effects in the iron parts of the objective lens must be avoided [4.58]. This can be achieved by setting the lens excitation at its maximum value and then reducing the lens current down to (but not below) that needed for focusing. This cycle of maximum excitation and focusing has to be repeated two or three times. It would be better to keep the excitation of the objective lens constant and move the specimen by means of a mechanical and/or piezoelectric specimen drive [4.60]; this would also minimize the effort required for microscope alignment (Sect. 2.4.3).

Up to values of about 20 000 times, the magnification can be calibrated by means of surface replicas of metal gratings, which are commercially available. Polystyrene spheres should be avoided for magnification calibration because their diameters are affected by the preparation, by radiation damage, and by contamination [4.61]. For magnifications of the order of 100 000 times, images of lattice planes (Sect. 9.6.1) can be used, provided that the lattice spacings are not altered by radiation damage. Dowell [4.62] discussed the procedure and the possible errors (±2%) if the lattice constant is calibrated with a TlCl standard by electron diffraction; these errors can mainly be attributed to the distortion of electron-diffraction patterns. For intermediate magnifications, catalase crystals can be employed; the measured lattice constants are 8.8 ± 0.3 nm [4.63] and 8.6 ± 0.2 nm [4.64].

A magnification standard covering the whole range of magnifications has been proposed that consists of molecular-beam-epitaxy-grown single-crystal layers of alternating Si and SiGe (two sets of layer distances at low and medium resolutions and the Si lattice for high resolution) [4.65].

4.4.4 Depth of Image and Depth of Focus

The depth of image S is defined in Fig. 4.18. A blurring of the image $\delta_s M$ will be observed at a distance $\pm S/2$ from the final image plane, where δ_s and S are related as

$$\delta_s M = \alpha' S, \qquad S = \frac{\delta_s M}{\alpha'} = \frac{\delta_s M^2}{\alpha_o}. \tag{4.27}$$

Here, $\alpha' = \alpha_o/M$ denotes the aperture in the final image. As a numerical example, for $M = 10\,000$, $\alpha_o = 10$ mrad, and $\delta_s = 5$ nm, we find that $S > 50$

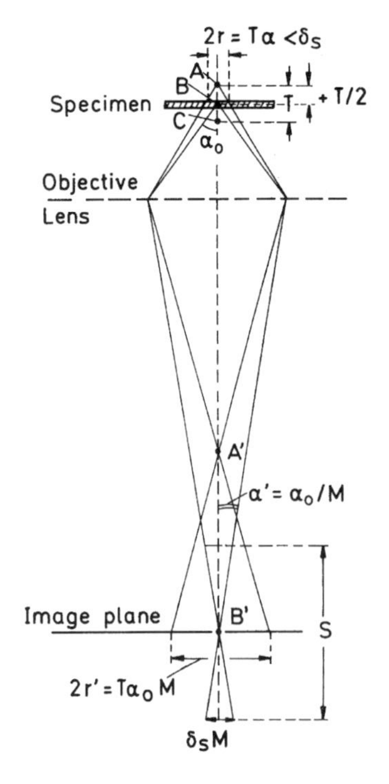

Fig. 4.18. Calculation of the depth of image S and the depth of focus T.

cm. Because of this large depth of image, a focused image will be obtained on the photographic plate or CCD camera even though these are some centimeters below the viewing screen, which is normally inclined for focusing; only the magnification will be different.

Another property of the instrument is the depth of focus. This is the axial distance $\pm T/2$ within which specimen details on the axis will be focused with a resolution δ_s. From Fig. 4.18, we have

$$T < \delta_s/\alpha_0. \tag{4.28}$$

As a numerical example, for a resolution $\delta_s M = 50\ \mu$m of the image recording system, a magnification $M = 10\,000$, and an aperture $\alpha = 1$ mrad, we find $T = 5\ \mu$m. Therefore, focusing at low magnifications sometimes becomes difficult owing to the large depth of focus. For thick specimens, a larger aperture can be used for focusing, which decreases the depth of focus; for thin specimens, the illumination aperture can be artificially increased by rocking (wobbling) the electron beam [4.66].

These geometrical estimates of the depth of focus are equivalent to wave-optical considerations at high resolution because defocusing differences Δz change the image-intensity distribution. If the specimen contains a periodicity Λ or a spatial frequency $q = 1/\Lambda$, a diffraction maximum will be formed at $\sin\theta \simeq \theta = \lambda/\Lambda$. The maxima and minima of the specimen periodicity will be reversed in contrast when the second term of the wave aberration $W(\theta)$ in (3.65) caused by the defocusing Δz changes the phase of the diffracted beam by π. Setting $W(\theta) = \pi\Delta z\theta^2/\lambda \leq \pi$ results in

$$\Delta z \leq \frac{\lambda}{\theta^2} = \frac{\Lambda}{\theta}. \tag{4.29}$$

With $\Delta z \leftrightarrow T$, $\Lambda \leftrightarrow \delta_s$, $\theta \leftrightarrow \alpha$, this formula corresponds to (4.28). As a numerical example, for $\alpha_o = 10$ mrad, and $\Lambda = 0.1$ nm, we find $\Delta z \leq 10$ nm.

4.5 Scanning Transmission Electron Microscopy (STEM)

4.5.1 Scanning Transmission Mode of TEM

Unlike the conventional transmission mode of TEM, in which the whole imaged specimen area is illuminated simultaneously, the specimen is scanned in a raster point-by-point with a small electron probe in the scanning transmission mode.

The prefield of the objective lens is used as an additional condenser lens (Sect. 4.2.3) to form a small electron probe at the specimen when operating in the STEM mode (Fig. 4.14b) [4.67]. The objective lens works near $k^2 = 3$ (condenser-objective lens, Sect. 2.2.3). An electron-probe diameter of the order of 0.2–5 nm can be produced. No further lenses are needed below the objective lens. Nevertheless, the later lenses may be excited to image the first diffraction pattern in the back focal plane (BFP) through the small polepiece bores of the subsequent lenses onto the electron-detector plane, above or below the final image plane. The generator that produces the saw-tooth currents for the x and y deflection coils simultaneously deflects, in synchrony, the electron beam of a cathode-ray tube (CRT). The intensity of the CRT beam can be modulated by any of the signals that can be obtained from the electron–specimen interactions. The transmitted electrons can be recorded in the bright- and dark-field modes with a semiconductor detector or a scintillator–photomultiplier combination. These modes can be selected by placing large circular or sector diaphragms in front of the detector; we recall that an enlarged far-field diffraction pattern is produced in the detector plane by each object element in turn and does not move during scanning if the pivot point of the primary-beam rocking is at the FFP (Fig. 4.14b).

In conventional TEM, small illumination apertures α_{i} are used in the bright- and dark-field modes (Sect. 4.4.2). In the STEM mode, a small electron probe can be obtained only with a large value of $\alpha_{\mathrm{i}} \simeq 10$ mrad (Sect. 4.2.2). The detector aperture α_{d} has to be matched to this illumination condition. Thus, in the BF mode it will be necessary to use a detector aperture $\alpha_{\mathrm{d}} \simeq \alpha_{\mathrm{i}}$. Otherwise, a large part of the unscattered electrons would not be recorded, and the signal-to-noise ratio would be correspondingly decreased. Details of contrast mechanisms and differences between STEM and the conventional TEM modes are discussed in Sect. 6.1.5 for amorphous specimens and in Sect. 9.1.4 for crystalline specimens.

An annular semiconductor detector or scintillator can be used below the specimen to record the electrons scattered through angles $\theta \geq 10°$ [4.68]. Another detector can be placed above the specimen to record the backscattered electrons (BSE). Secondary electrons (SE) with exit energies $\leq$50 eV will move around the axis in spiral trajectories owing to the strong axial magnetic field and can be detected by a scintillator–photomultiplier combination situated between the objective and condenser lenses (Fig. 4.19). This SE mode can be used to image the surface structure of the specimen.

The effect of chromatic aberration of the objective lens can be avoided in the STEM mode. This is of interest for thick specimens. However, the gain in resolution will be limited by the top–bottom effect (Sect. 5.4.3) caused by broadening the electron probe by multiple scattering. The main advantages of the STEM mode are the production and positioning of small electron probes $\leq$0.1 μm for the microbeam electron diffraction and convergent-beam diffraction techniques, x-ray microanalysis, and electron energy-loss spectroscopy (EELS) of small specimen areas.

The STEM mode can also be used to generate other signals, such as cathodoluminescence and electron-beam-induced current (EBIC) in semiconductors. The use of cathodoluminescence (CL) is a well-established technique in scanning electron microscopy (SEM). The CL signal can also be recorded in a transmission electron microscope equipped with a STEM attachment by collecting the light quanta emitted. An advantage of this mode is the possibility

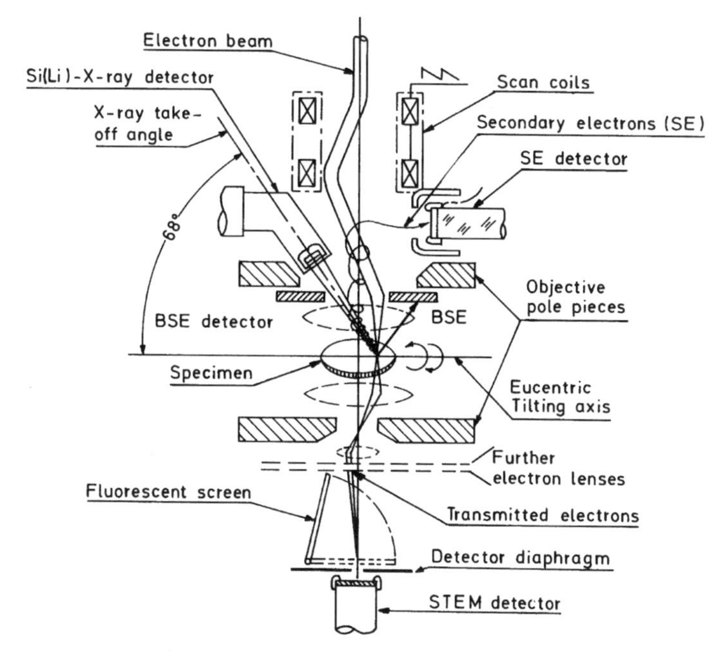

Fig. 4.19. Detectors for x-rays, secondary electrons (SE), backscattered electrons (BSE), and transmitted electrons (TE) in the scanning transmission mode of TEM.

of simultaneously imaging lattice defects in the STEM bright-field mode and examining their influence on CL; alternatively, additional information about variations in the concentration of dopants, which act as luminescence centers or nonradiative recombination centers, can be obtained. A disadvantage is that the CL intensity is accumulated only in a foil thickness much smaller than the electron range and, in addition, the film surface acts as a dead layer due to surface recombinations. The method is restricted to those semiconductors with a high luminescence yield. For these, resolutions of the order of a few tens of nanometers are obtainable thanks to the reduction of electron diffusion.

The low intensity requires an efficient light-collection system with a large solid angle and lateral selection of the irradiated area to shield the signal from CL contributed by diffusely scattered electrons. An obstacle to the collection of the light quanta is the narrowness of the polepiece gap. A tapered silver tube or an elliptical mirror is used to transmit the light to a quartz light pipe and a photomultiplier [4.69]. X-rays can cause CL in the quartz light pipe and this signal must be eliminated by placing additional lead-shielded mirrors between the collection system and the light pipe.

In diamond, for example, almost all of the luminescence is emitted from dislocations as a result of localized electron states near these defects [4.69, 4.70]. The CL depends on the crystal orientation and exhibits bend contours that are similar to channeling effects in energy-loss spectroscopy and x-ray emission; this has been shown for ZnS single-crystal foils [4.71]. In a Ga_{1-x} Al_xAs laser structure, CL can be employed to analyze radiative and nonradiative centers and to record the luminescence spectra from the different parts of the structure, which may be separated by wedge-shaped etching of the structure; the CL signal may be compared with that obtained in EBIC experiments [4.72].

The EBIC mode is also widely used in SEM [4.73]. An electric field must be present in a p-n junction, or a Schottky barrier must be present to separate the electron-hole pairs generated by the electron beam. A current can be recorded at zero bias or with a reverse bias, which increases the field strength and the width of the depletion layer. The EBIC signal consists not only of electron-hole pairs generated in the depletion layer but also minority carriers, which reach the layer by diffusion. The EBIC signal may decrease at lattice defects, such as dislocations or stacking faults, which act as recombination centers. It is thus of interest to image the lattice defects in the TEM or STEM mode. Combination of the SEM/EBIC and the TEM modes (in different instruments) [4.74] has the advantage that the EBIC mode can be applied first to the bulk semiconductor device, after which a TEM investigation of the same area after thinning gives information about the faults. Another possibility is to observe the thin sample in a scanning transmission electron microscope or a transmission electron microscope with a scanning attachment in the STEM and EBIC modes simultaneously [4.72, 4.75]. Because the active area and depletion layers are of the order of a few micrometers thick, HVEM offers better penetration of thick regions. Unlike SEM/EBIC experiments, in which the electron-hole pairs are generated in the whole volume of the electron-diffusion cloud a few micrometers in diameter, the generation in STEM/EBIC is concentrated in the volume irradiated by the electron probe, which is only slightly broadened by multiple scattering. This can result in better resolution, though the latter is ultimately limited by the diffusion of the minority carriers.

Another interesting method is scanning deep-level transient spectroscopy (SDLTS) [4.72, 4.76], which can provide the profile of the defect concentration in the direction normal to the junction. The electron probe is switched on and off so that the deep levels are filled by the injected carriers when the beam is on. They can be detected by observing the thermally stimulated current transient that flows when the levels empty during the off time of the beam. The depth of the levels (activation energy for emission) can be determined from the temperature dependence of the transient-time constant, which can be measured by opening two sampling-rate windows at times t_1 and t_2 after the electron-beam chopping pulse.

4.5.2 Dedicated STEM

This type of electron microscope is designed to work only in the scanning transmission mode. Figure 1.2 schematically shows a version introduced by

Crewe and coworkers [4.77, 4.78, 4.79, 4.80]. A field-emission gun is used to produce a very small electron source of the order of 10 nm. Only one magnetic lens with a short focal length, low spherical aberration, and equipped with a stigmator is needed to demagnify this source to an electron probe of 0.2–0.5 nm on the specimen. The scanning coils are arranged in front of the lens. A signal I_{el} of large-angle elastically scattered electrons can be detected by an annular detector. The cone of small-angle scattered electrons, which enters a prism spectrometer, can be separated into unscattered (I_{un}) and inelastically scattered (I_{in}) signals.

The diffraction pattern of the illuminated area of the specimen is formed in the detection plane, and various annular, semiannular, quadrant, or multichannel detectors can therefore be used to get optimum contrast. This offers new possibilities for contrast enhancement that are not available in a conventional transmission electron microscope:

1. Z-contrast of amorphous (especially biological) specimens (Sect. 6.1.5)
2. quantitative determination of mass thickness (Sect. 6.1.6)
3. imaging and contrast enhancement of single atoms (Sect. 6.3.2)
4. methods of differential phase contrast (Sect. 6.4.4)
5. differential phase contrast in Lorentz microscopy (Sect. 6.8.2)
6. imaging of lattice planes and atomic rows by high-angle Z-contrast dark-field imaging (Sect. 9.6.6).

The scanning transmission electron microscope can also be used for x-ray microanalysis and electron energy-loss spectroscopy of a selected area or for energy-filtering microscopy.

The field-emission gun, lens, and spectrometer occupy little space, and the whole STEM column can be kept at a UHV of $10^{-8} - 10^{-7}$ Pa. This allows the gun to operate satisfactorily and drastically reduces specimen contamination.

4.5.3 Theorem of Reciprocity

The reciprocity theorem was first discussed by Helmholtz (1860) in light optics. In geometrical optics, it is known as the reciprocity of ray diagrams. However, in wave optics it also implies that the excitation of a wave at a point P by a wave from a source Q is the same as that detected at Q with the source at P.

The ray diagram of STEM is the reciprocal of that of TEM [4.81, 4.82]. This will be demonstrated with the aid of the ray diagram of Fig. 4.20. The *source* in the ray diagram of TEM in Fig. 4.20a is already a demagnified image of the crossover produced by the condenser lenses. The intermediate image can be further enlarged by the subsequent lenses, not shown in the diagram. The specimen is illuminated with an illumination aperture α_{i} of the order of 0.1–1 mrad, which is much smaller than the objective aperture $\alpha_{\mathrm{o}} = 5$–20 mrad. The ray diagram of STEM (Fig. 4.20b) has to be read in the reverse direction. The

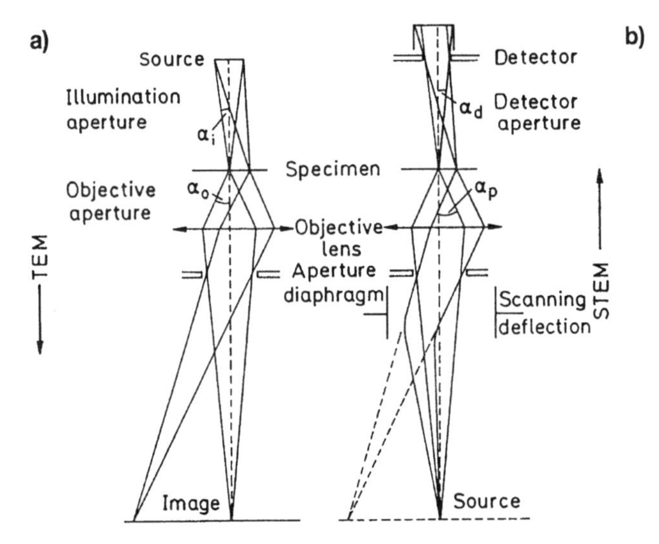

Fig. 4.20. Demonstration of the theorem of reciprocity for (**a**) TEM and (**b**) STEM in terms of ray diagrams connecting the intermediate source and image.

objective lens now demagnifies a source point on the specimen. A large probe aperture $\alpha_p \simeq \alpha_o$ is necessary to obtain the smallest possible spot size (Sect. 4.2.2). A fraction of the incident cone of the electron probe and scattered electrons are collected by the detector aperture α_d. If $\alpha_d = \alpha_i \ll \alpha_o = \alpha_p$, the same image contrast is obtained as in TEM. A scanning unit between the source and the objective lens deflects the electron probe in a raster across the specimen. Projected backwards, the rays scan over a virtual source plane, which corresponds to the image plane in TEM in Fig. 4.20a. We can argue that, in STEM, the CRT is needed to *image* this virtual plane by modulating the CRT with the detector signal.

By enlarging the ray diagram near the specimen in Fig. 4.21, we can demonstrate that the theorem of reciprocity can also be applied to wave-optical imaging, Fresnel fringes, and phase contrast, for example. The reciprocity theorem for the imaging of crystal lattices by STEM is discussed in Sect. 9.6.6. Here, we consider the case of Fresnel fringes. The source and the detector are assumed to be very distant so that the incident and exit waves can be regarded as plane waves. The objective lens in TEM enlarges the intensity distribution in the plane at a distance $\Delta z = z_0$ (defocusing) behind the specimen. At one point of this plane, the Huygens elementary wavelets from each point beside the specimen edge overlap with their corresponding geometrical phase shifts and form the Fresnel fringes of an edge (Fig. 3.8). When the diagram is reversed for discussion of the STEM mode, an electron probe is formed in the focal plane at a distance z_0 in front of the specimen edge. The same geometric phase shifts as in TEM will occur during the wave propagation to the detector. It should be mentioned that the distance of the

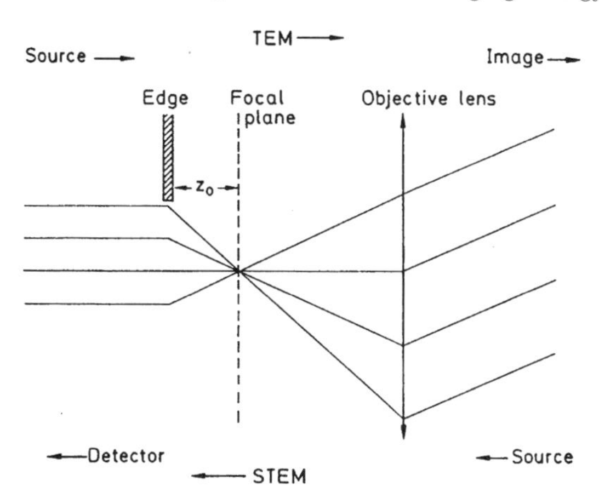

Fig. 4.21. Demonstration of the theorem of reciprocity of TEM and STEM for phase contrast (Fresnel fringes at an edge) (z_0 = defocusing).

first Fresnel fringe from the edge increases as $\sqrt{z_0}$ with increasing defocusing [4.83], whereas a fringe distance increasing as z_0 can be explained by refraction at the wedge-shaped edge [4.84].

The phase shifts caused by the spherical aberration of the lens also act in the same manner in TEM and STEM. It was shown in Sect. 3.2.2 that an increase of the illumination aperture α_i causes a blurring of the Fresnel fringes by $\pm\alpha_i \Delta z$, thus decreasing the number of observable fringes. The same effect would be obtained when recording a TEM image with a slit width of $2\alpha_i \Delta z$. An analogous blurring is observed in STEM if the detector area or the detector aperture α_i is increased. Therefore, if phase-contrast effects are to be observed in STEM, a small detector aperture has to be used ($\alpha_d \ll \alpha_p$). It will be shown in Sect. 6.1.5 that this is an unfavorable operating condition. Exposure of the specimen to damaging radiation has to be kept low, and all of the unscattered electrons have to be collected in order to image single atoms, for example. This means that α_d should be approximately equal to α_p. This corresponds to extremely incoherent illumination in TEM. Single atoms are imaged in STEM mainly by their scattering contrast. In TEM, the optimum condition for imaging atoms corresponds to phase-contrast operation, for which $\alpha_i \ll \alpha_o$.

4.6 Electron Spectrometers and Imaging Energy Filters

Electron spectrometers of high energy resolution are needed to resolve the relatively low energy losses between $\Delta E = 0$ and 3000 eV for electron energy-loss spectroscopy (EELS). The energy spread $\Delta E \simeq 1$–2 eV of a thermionic gun is normally narrow enough to record energy losses by the excitation of plasmons and inner shells, though in some cases a fine structure can be seen in the spectrum if the resolution is better. Higher resolution needs a field-emission gun ($\Delta E \simeq 0.2 - 0.5$ eV) or a monochromator ($\Delta E \geq 0.1$ eV). A widely used postlens spectrometer is the magnetic-prism spectrometer. The Wien filter is also described below because of its historical importance and its use as the monochromator for high-resolution electron energy-loss spectroscopy. A new trend in analytical microscopy is the use of imaging energy filters, which can record either energy spectra or energy-filtered images with an energy-selecting slit in the energy-dispersive plane (Sect. 4.6.3).

4.6.1 Postcolumn Prism Spectrometer

Electron prisms consist of transverse magnetic or electric fields. In a transverse magnetic field, the radius of the trajectories is proportional to the momentum (2.13). In a radial electric field, the radius is proportional to the electron energy. A spectrometer should have a high resolution and a large angle of acceptance, which means a large entrance aperture $\alpha = d/2p_r$ (d: diameter of the entrance diaphragm, p_r: distance PH in Fig. 4.22). The two aims can be reconciled only by designing the sector field to give additional focusing and by correcting the second-order aberrations. A point source P will then be imaged by the spectrometer as a sharp line or image point Q, at which a slit can be placed in front of an electron detector (Fig. 4.22).

The central beam in a magnetic sector field (Fig. 4.22) is bent into the form of a circle of radius $r_0 = mv/eB$ with center C. If the incident and exit directions are normal to the edges of the sector field with sector angle ϕ, focusing occurs for small α (paraxial rays), and the points P, C, and Q are collinear (Barber's rule). The distances p_r = PH and q_r = H′Q (focal lengths) are given by $p_r = q_r = r_0/\tan(\phi/2)$ for a symmetric prism. There is no focusing of the momentum components in the z direction parallel to the magnetic field.

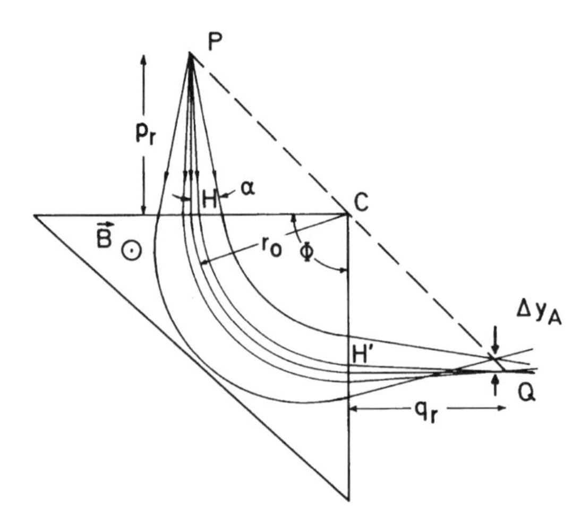

Fig. 4.22. Radial focusing property of a 90° magnetic electron prism with second-order aberration Δy_A for large α.

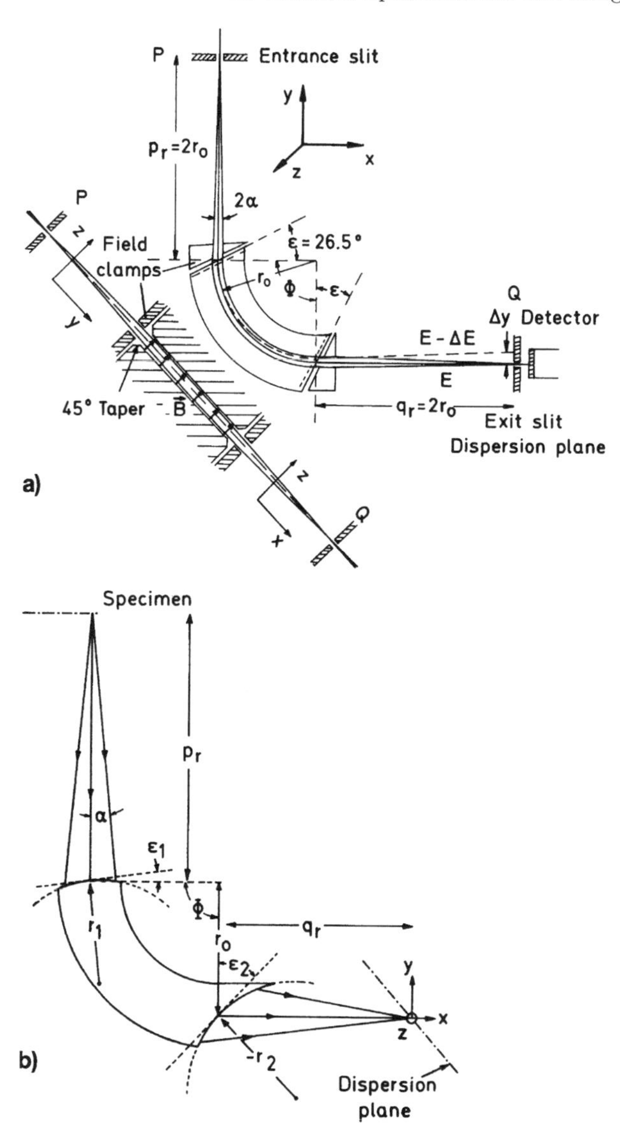

Fig. 4.23. (**a**) Double focusing in a 90° electron prism with a tilt angle $\epsilon = 26.5°$ of the edges, tapered polepieces, and field clamps. The deflection Δy of electrons with energy loss ΔE is responsible for the dispersion $\Delta y/\Delta E$. (**b**) Correction of second-order aberration by curvature of the edges.

Tilting the magnet edge ($\epsilon \neq 0$) (Fig. 4.23a) has the same effect (in first order) as adding a quadrupole lens with focal length $f = \pm r_0 \cot\epsilon$ for components of the momenta in the radial (+) and axial (–) directions, respectively. The focal lengths are therefore

$$\frac{1}{p_\mathrm{r}} = \frac{1}{q_\mathrm{r}} = \frac{1}{r_0}\left(\tan\frac{\phi}{2} - \tan\epsilon\right), \quad \frac{1}{p_\mathrm{z}} = \frac{1}{q_\mathrm{z}} = \frac{1}{r_0}\tan\epsilon. \tag{4.30}$$

So-called *double-stigmatic focusing* can be obtained if $p_\mathrm{r} = q_\mathrm{r} = p_\mathrm{z} = q_\mathrm{z} = r_0/\tan\epsilon$ and $\tan(\phi/2) = 2\tan\epsilon$. For $\phi = 90°$, this gives $\tan\epsilon = 0.5$ or $\epsilon = 26.5°$ (Fig. 4.23a).

Focusing in the z direction is not necessary when a slit is used to record a spectrum. However, the slit has to be aligned and the line focus may be curved. Double focusing will therefore be advantageous, though complete focusing in the z direction is not necessary; indeed, a small width in the z direction can be desirable to avoid damaging the detector.

It has been assumed in the foregoing that the magnetic field terminates abruptly at the edges (sharp cutoff fringe field, or SCOFF, approximation). The real fringe fields influence the focal lengths, and the effective prism angle ϕ becomes larger. This can be counteracted by finishing the edge of the magnetic polepiece plates with a 45° taper and by introducing field clamps, which are constructed from the same high-permeability material as the polepieces and placed at half the gap length in front of the prism, with a small hole for the incident and exit rays (Fig. 4.23a) [4.85, 4.86].

The *dispersion* $\Delta y/\Delta E$ is defined as the displacement Δy of electrons with energy $E - \Delta E$ in the dispersion plane (Fig. 4.23) and becomes (here p is the electron momentum)

$$\frac{\Delta y}{\Delta p} = \frac{4r_0}{p}, \qquad \frac{\Delta y}{\Delta E} = \frac{2r_0}{E}\frac{1+E/E_0}{1+E/2E_0}, \tag{4.31}$$

for a symmetric prism with $\phi = 90°$. We find $\Delta y/\Delta E = 1\ \mu\text{m/eV}$ for $E = 100$ keV and $r_0 = 5$ cm.

On increasing α, a second-order angular aberration $\Delta y_{\text{A}} = B\alpha^2$ becomes apparent (Fig. 4.22). This aberration can be corrected in the radial direction by curving the edges of the magnet (Fig. 4.23b) [4.85, 4.86, 4.87, 4.88]. The total width Δs of the zone occupied by the zero-loss electrons in the dispersion plane is determined by the size of the image of the entrance slit or diaphragm, which is blurred owing to the energy width of the electron gun, and also by the second-order aberrations. This width and the dispersion (4.31) limit the *resolution* $\Delta E_{\text{r}} = \Delta s/(\Delta y/\Delta E)$. With a thermionic electron gun at 100 keV, a resolution of 1–2 eV is obtainable.

A magnetic prism spectrometer is normally situated below the viewing screen of the transmission electron microscope. The lens system can be used to adapt the different operating modes of TEM to the spectrometer [4.89, 4.90]. Thus, any corrections needed to focus the beam on the exit slit can be made by means of a pre-spectrometer lens or a quadrupole lens between the spectrometer and the exit slit. The point source P is formed at the focus of the last projector lens by the demagnified diffraction pattern or the image of a selected area if the entrance plane of the spectrometer contains an image or a diffraction pattern, respectively.

The resolution of a prism spectrometer can be increased by decelerating the electrons in a retarding field to an energy of the order of 1 keV. Magnetic [4.91, 4.92] and electrostatic prisms are used in this way. Electrostatic prisms consist of radial electric fields between concentric cylindrical or spherical electrodes. An electrostatic prism-spectrometer without retardation is described in [4.93].

4.6.2 Wien Filter

The field strength $\boldsymbol{E}$ of a transverse electric field and the magnetic induction $\boldsymbol{B}$ of a crossed transverse magnetic field normal to $\boldsymbol{E}$ can be adjusted so that electrons of velocity v are not deflected. The condition for this is

$$F = e|\boldsymbol{E}| = ev|\boldsymbol{B}| \quad \rightarrow \quad v = |\boldsymbol{E}|/|\boldsymbol{B}|. \tag{4.32}$$

Electrons passing through the filter with other energies are spread out into a spectrum, or a spectrum is serially recorded by varying one of the field strengths. This type of filter has the advantage that it is situated on-axis and there is no overall deflection of the beam. Focusing conditions have to be found such that the entrance slit is focused on the exit slit. Wien filters with 1 eV resolution for a commercial transmission electron microscope are described in [4.94, 4.95].

A deceleration of 10–20 keV electrons to 20–300 eV by an electrostatic retarding lens yields a resolution of 2 meV [4.96]. In order to obtain an energy-loss spectrum with this resolution in the range ΔE = 0–10 eV (Fig. 5.7), the electron beam must be monochromatized by placing a further Wien filter in front of the specimen. This resolution can be realized for EELS experiments only. A resolution of 80 meV with monochromatizing and analyzing Wien filters has been realized in a scanning transmission electron microscope [4.97] and a transmission electron microscope [4.98, 4.99]. More recently, monochromators have been developed, that can be incorporated in a regular transmission electron micrcoscope without impeding its imaging capabilities [4.100, 4.101, 4.102].

4.6.3 Imaging Energy Filter

To understand the functioning of an imaging energy filter, we regard the filter as a black box (Fig. 4.24) with the following properties. In the source plane SP (focal plane of a projector lens), we find either a demagnified image of the

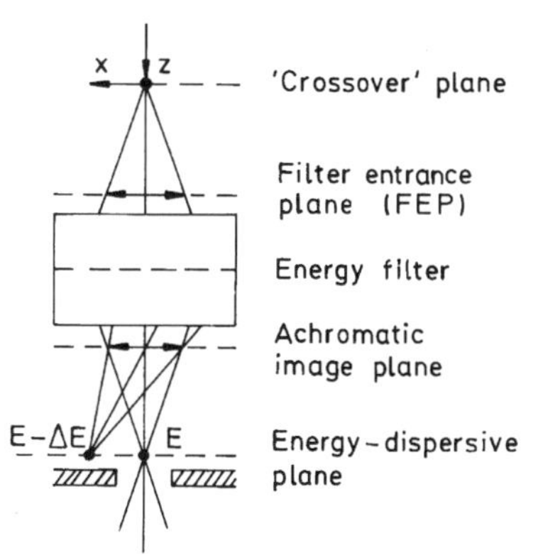

Fig. 4.24. Schematic action of an imaging energy filter and its important planes.

focal plane of the objective lens with the diffraction pattern or a demagnified image of the selector diaphragm in the intermediate lens. The filter entrance plane then contains an enlarged image or diffraction pattern, respectively. The imaging filter produces a 1:1 image in the achromatic image plane (AIP) with the difference that electrons that have lost energy now pass their image point with an angular deviation that increases with increasing energy loss. This means that the image is sharp, apart from the chromatic aberration of the objective lens, of course. Rays of equal energy loss from different points of the AIP intersect in the energy-dispersive plane (EDP) and form an energy-loss spectrum (EELS). A further projector lens can magnify either the EDP or the AIP. The former case results in a magnified EELS on the final image plane and the latter in an energy-filtered image or diffraction pattern when a slit in the EDP selects an energy window of width Δ. The final image can be observed on a fluorescent screen or recorded on a photographic emulsion or with a CCD camera.

Energy Filtering with a Prism Spectrometer. The 90° magnetic sector field spectrometer (Sect. 4.6.1) can be used as an energy-dispersive imaging filter (Fig. 4.25) with the properties shown in Fig. 4.24 [4.103, 10.156, 4.105]. The entrance diaphragm with a diameter of 0.6–5 mm contains a magnified image or diffraction pattern. The prespectrometer optics permit adjustment of the beam and compensation for some of the aberrations. The spectrum in

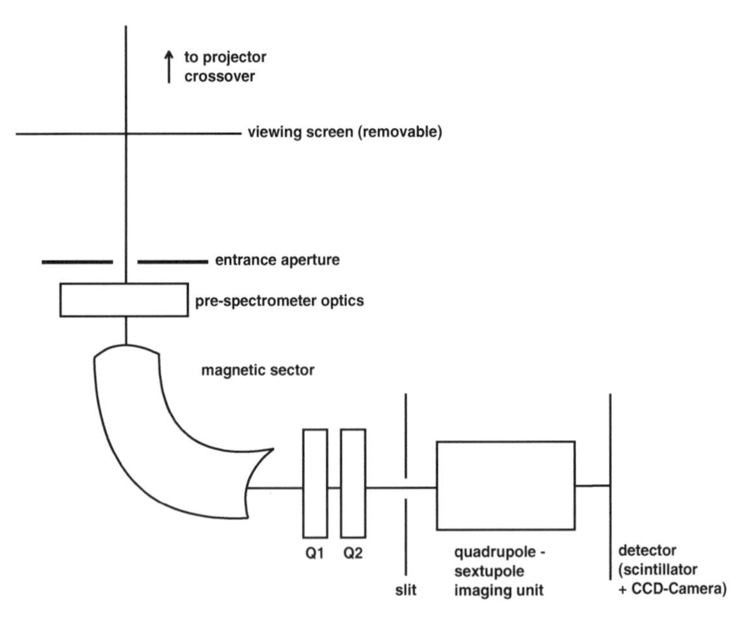

Fig. 4.25. Schematic diagram of the imaging-filter system consisting of a magnetic sector-field spectrometer and a quadrupole-octopole system for magnifying the energy-selected image.

the energy-dispersive plane can be enlarged using a pair of quadrupole lenses Q1 and Q2 just behind the spectrometer. The slit width for selecting the energy window can be adjusted piezoelectrically, and the slit system can be removed pneumatically when parallel recording the energy-loss spectrum. By introducing a quadrupole-sextupole imaging unit after the slit, it is possible to work in the electron spectroscopic imaging (ESI) or diffraction (ESD) mode or the parallel-recorded EELS mode. Electron spectroscopic imaging is realized by selecting the desired energy-loss window with the slit. The system of quadrupoles and sextupoles behind the slit produces a filtered image on the CCD array, corrected for the most important aberrations and magnified by a factor ranging from 8 to 20. One therefore has to operate the microscope itself with a corresponding reduction of the magnification on the viewing screen.

An advantage of this type of imaging filter is that it can be attached below conventional microscopes up to 1.25 MeV [4.106].

Castaing–Henry and Ω-Filters. Castaing and Henry [4.107, 4.108] combined a retarding-field electrode (electron mirror) and a double magnetic prism to form an imaging energy filter that can be incorporated in the column of a transmission electron microscope [4.109, 4.110] between the first and second projector lenses. The function of such an imaging energy filter for electron spectroscopic imaging (ESI) and diffraction (ESD) or electron energy-loss spectroscopy (EELS) can best be understood by considering the conjugate planes (Fig. 4.26) [4.111, 4.112].

In the ESI mode, the objective lens produces a first diffraction pattern in its focal plane and a magnified image in the first intermediate image plane. The primary spot in the diffraction pattern is an image of the crossover. Electrons scattered through larger angles are absorbed by the objective diaphragm, and that part of the pattern passing the diaphragm is demagnified by the first projector system P1 into its focal plane. This "crossover" plane acts as the source plane (SP) for the energy filter. The plane conjugate to the latter, after passing the energy filter with 1:1 magnification, is the energy-dispersive plane (EDP) containing the energy-loss spectrum. The filter entrance plane (FEP) is conjugate to the achromatic image plane (AIP). Electron spectroscopic imaging is now realized by magnifying the achromatic image plane with the second projector P2 onto the final image plane (FIP) and selecting an energy window of width $\Delta = 0.5$–50 eV by a slit in the energy-dispersive plane.

In the ESD mode, the objective diaphragm is withdrawn and a selector diaphragm in the intermediate image plane limits the area contributing to the "selected-area electron diffraction" (Sect. 8.1.1). P1 produces a conjugate diffraction pattern in the final image and the achromatic image plane. The source plane behind the first projector P1 now contains a demagnified image of the selector diaphragm, which becomes conjugate to the energy-dispersive plane. Magnifying the achromatic image with the diffraction pattern by P2 and selecting an energy window in the energy-dispersive plane now results in an energy-filtered diffraction pattern on the final image.

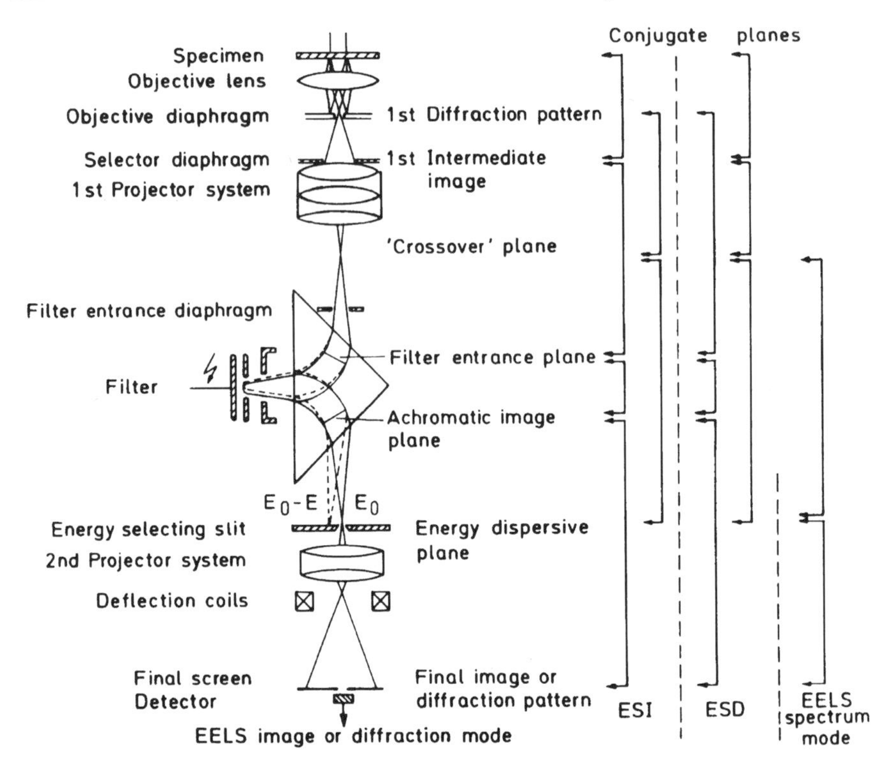

Fig. 4.26. Castaing–Henry filter between the first and second projector lens systems of a transmission electron microscope as incorporated in a Zeiss EM902. Conjugate planes for the electron spectroscopic imaging (ESI) and diffraction (ESD) mode and the EELS spectrum mode are shown at the right. [4.111].

In the EELS spectrum mode, P2 is more strongly excited to magnify the energy-loss spectrum in the energy-dispersive plane onto the final image plane by withdrawing the energy-selecting slit. The energy-loss spectrum can be recorded either serially, by shifting the spectrum across a slit in front of a scintillator–photomultiplier combination, or in parallel with a scintillator coupled by a fiber plate to a CCD camera. Because the energy-dispersive plane is conjugate to the source plane, the observed energy-loss spectrum is convoluted with the demagnified image of the diffraction pattern (objective diaphragm) when P1 is excited as in the ESI mode or with the demagnified image of the selector diaphragm in the ESD mode. Further modes of recording energy-loss spectra are described in Sect. 10.3.1.

An advantage of incorporating such an imaging energy filter in the TEM column is that energy-filtered images and the positions of diaphragms and slits can be observed directly on the fluorescent screen in the final image. However, the second-order aberration of the Castaing–Henry filter limits the diameter of the exactly filtered image to about 2–3 cm on the FIP. Furthermore, the

Castaing–Henry filter is limited to an acceleration voltage of 80–100 kV because the same voltage is applied to the mirror electrode and breakdown may occur for higher voltages.

Pure magnetic imaging energy filters have been proposed and built [4.113, 4.114] in order to extend the technique to higher voltages. Rose and Plies [4.115, 4.116] proposed the first symmetric magnetic equivalent of the Castaing–Henry filter, which is called an Ω-filter because of the shape of the trajectories. This filter can be equipped with a system of multipoles between the magnetic sector fields in order to correct the second-order aberration so that an energy-filtered image can be observed with $\Delta = 1$ eV over the whole final screen [4.117, 4.118]. Figure 4.27 shows such a fully corrected Ω-filter with the adjacent lenses in front of and behind the filter, which has been built at the Fritz-Haber-Institut in Berlin (see [4.119] for an extensive discussion of the theory and alignment procedures). Uhlemann and Rose

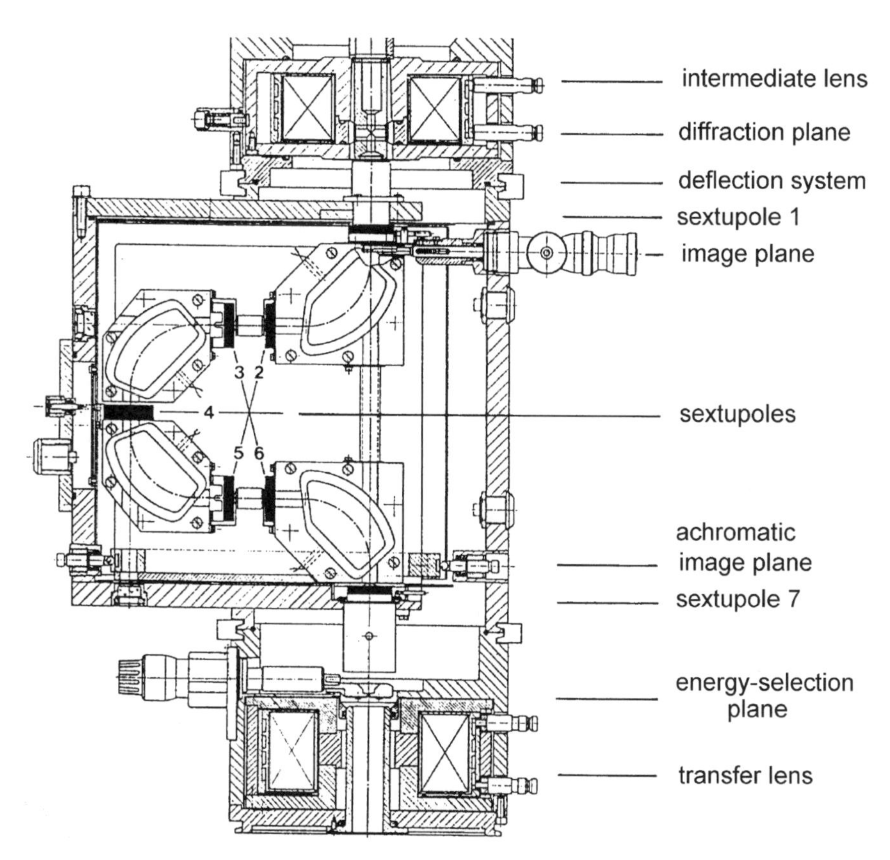

Fig. 4.27. Cross section through the sextupole-corrected energy filter of the Fritz-Haber-Institut (Berlin) and the adjacent lenses located in front of and behind the filter.

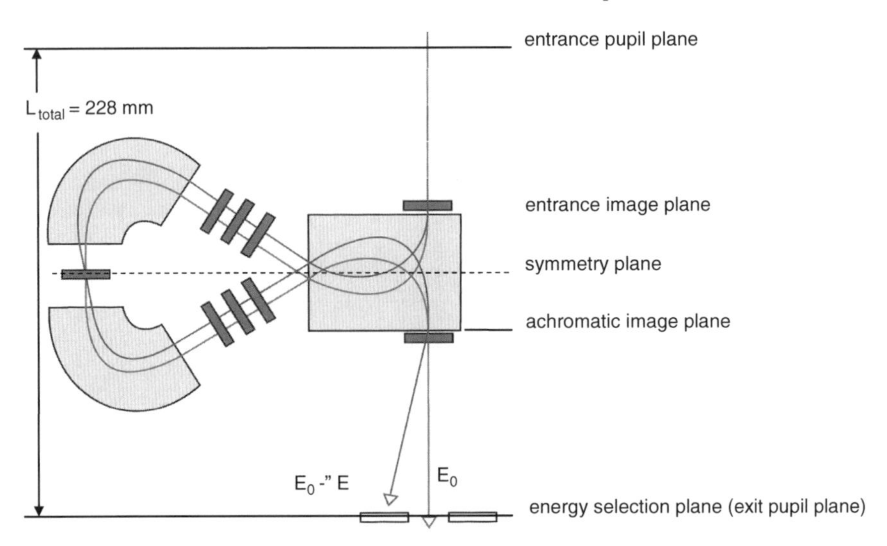

Fig. 4.28. Schematic diagram of the MANDOLINE filter (by courtesy of E. Essers).

[4.120] proposed to use inclined pole faces in the sector magnet to reduce the residual aberrations. Such a MANDOLINE filter (Fig. 4.28) has an energy dispersion of $\simeq$10 eV/μm at 200 keV. Together with an appropriate set of multipole elements for correcting aberrations, it is particularly suitable for 200–400 kV microscopes [4.121].

4.6.4 Operating Modes with Energy Filtering

Energy-filtering transmission electron microscopy (EFTEM) can be performed with a dedicated scanning transmission electron microscope (Sect. 4.5.2), or in a transmission electron microscope equipped with a postcolumn imaging prism spectrometer or an in-column imaging energy filter (Sect. 4.6.3). The method of EFTEM is extensively described in [4.112, 4.122]. A dedicated scanning transmission electron microscope can scan the specimen pixel by pixel and store the parallel-recorded EELS (PEELS) signal, whereas an imaging filter can record two-dimensional images at successively increased energy losses. For both techniques, an image series at many energy losses occupies about several tens of Mbytes of computer memory [4.123]. The complete information can be described as a data cube (Fig. 4.29) with the spatial coordinates x and y and the energy loss E as the third dimension. Whereas in the scanning transmission electron microscope the information is acquired column by column (Fig. 4.29a), EFTEM permits detection of the data slice by slice (Fig. 4.29b). When many energy losses are of interest, it is obvious from this scheme that a scanning transmission electron microscope can operate with a much lower irradiation dose, whereas for a large number of image points the EFTEM technique is less time-consuming.

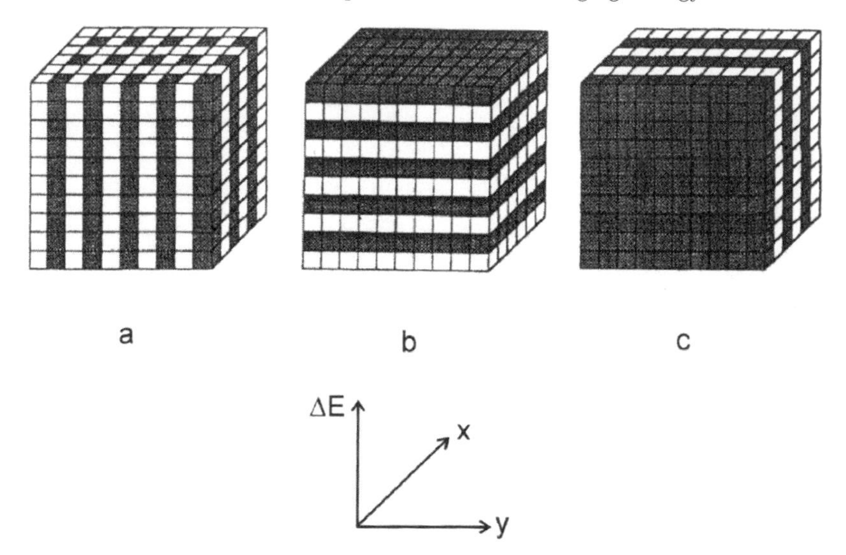

Fig. 4.29. The data cube depicting the complete information that can be explored in different ways. (**a**) A scanning transmission electron microscope acquires spectra point by point. (**b**) In EFTEM, the information is obtained energy slice by energy slice. (**c**) Using an energy filter, one can obtain a spectrum from a line selected in the image.

Figure 4.30 describes schematically a classification of the different modes of electron spectroscopic imaging (ESI) depending on the selected energy-loss range and the type of information available [4.111, 4.112, 4.124]:

1. zero-loss filtering to remove the inelastically scattered electrons in images of amorphous and crystalline specimens
2. plasmon-loss imaging for selectively imaging phases with a shift of plasmon losses and investigation of the preservation of phase and diffraction contrast
3. structure-sensitive contrast for biological sections
4. contrast tuning of optimum energy window for imaging biological specimens and polymers
5. elemental distribution images at inner-shell ionization edges
6. most-probable-loss imaging of amorphous and crystalline specimens

In electron spectroscopic diffraction (ESD), the following modes can be employed:

1. zero-loss filtering of whole diffraction patterns of amorphous, polycrystalline, and single-crystal specimens
2. plasmon-loss filtering for analyzing the anisotropy of energy losses
3. high-energy-loss filtering for the imaging of Compton scattering and the contributions of inelastically scattered electrons to single-crystal diffraction patterns

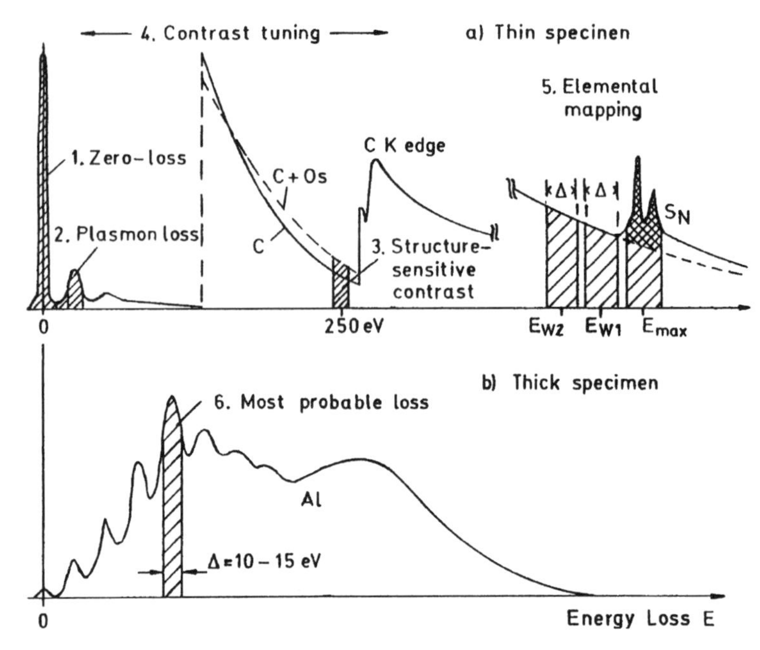

Fig. 4.30. Imaging modes of electron spectroscopic imaging (ESI) with selected energy windows at different parts of the electron energy-loss spectrum (EELS).

The following modes of electron energy-loss spectroscopy (EELS) can be used:

1. EELS image mode with an image at the entrance plane
2. EELS diffraction mode with a diffraction pattern at the entrance plane; shifting the pattern across the entrance diaphragm allows energy-loss spectra to be recorded at different scattering angles
3. angle-resolved EELS of a line through the diffraction pattern selected by a slit in the filter entrance plane with energy dispersion normal to the slit (see Fig. 5.13)
4. time-resolved EELS for recording radiation damage [4.123]
5. spatially resolved EELS by selecting a line through an image (Fig. 4.29c)
6. "Image EELS" by taking a series of ESIs, digitally selecting an area of interest, and plotting the integrated intensity versus the selected energy-loss windows

4.7 Image Recording and Electron Detection

4.7.1 Fluorescent Screens

The final image of a transmission electron microscope can be observed on a fluorescent screen consisting of ZnS or ZnS/CdS powder, which is excited by cathodoluminescence. The color can be varied by adding small concentrations of activator atoms, such as Cu or Mn. The maximum emission is normally in the green (550 nm), where the sensitivity of the human eye is at a maximum.

The light intensity of a fluorescent screen is proportional to the incident electron current density j, usually measured in A m^{-2}. For constant j, the intensity might be expected to increase in proportion to the electron energy because more light quanta are generated by high-energy electrons. In fact, a slower rate of increase is observed owing to the increasing depth of generation and the subsequent absorption and scattering of the light quanta. The light-generating efficiency likewise decreases when the electron range exceeds the thickness of the fluorescent layer; this can be a problem in high-voltage electron microscopy [4.125].

The decay of intensity with time proceeds in two stages: A fast decrease with a time constant of the order of $10^{-5} - 10^{-3}$ s is followed by an afterglow of the order of seconds. For a faster response, in STEM for example, fluorescent materials with time constants less than 1 μs are needed (Sect. 4.7.6).

4.7.2 Photographic Emulsions

Photographic emulsions are directly exposed to the electrons inside the microscope vacuum. The gelatin of the emulsion contains a considerable amount of water, and it is necessary to dehydrate the photographic material in a desiccator at 1 Pa and to load the microscope camera as quickly as possible [4.126].

The basic processes that occur in the exposure of photographic emulsions to electrons will now be discussed; for more details, see [4.127, 4.128, 4.129, 4.130, 4.131, 4.132]. The ionization probability of electrons is so large that each silver halide particle penetrated is rendered developable and can be reduced to a silver grain. High-energy electrons in the MeV range can probably penetrate some particles without ionization. For light, on the contrary, several quanta have to be absorbed in a single particle for it to be made developable. Therefore, unlike light exposure, there is no illumination threshold for exposure to electrons.

The following laws for the photographic density D can be derived with this exposure mechanism. The density D of a developed emulsion is defined as the logarithm of the ratio of the light transmission L_0 of an unexposed part and that of an exposed region (L):

$$D = \log_{10}(L_0/L). \tag{4.33}$$

A saturation density $D_{\max}$ is reached when all of the grains are developed. Owing to the statistical nature of silver-grain production, the density D of an unsaturated emulsion exposed to a charge density $J = j\tau = en$ in units C m^{-2} [j: current density (A m^{-2}), τ: exposure time, n: number of incident electrons per unit area] will be given by

$$D = D_{\max}(1 - e^{-cJ}). \tag{4.34}$$

The validity of this law means that a long exposure with low j produces the same density as a short exposure with high j if the product $j\tau$ is constant. This law of reciprocity is not true for light exposure. For the latter, the relation can be expressed in terms of the Schwarzschild exponent κ, different from unity, equal densities being obtained for $j\tau^{\kappa}$ = const. All experiments show that $\kappa = 1$ for exposure to electrons. However, for some emulsions, the results depend on the delay between exposure and development [4.133].

For small values of J, equation (4.34) leads to the proportionality

$$D = cD_{\max}J = \epsilon J, \tag{4.35}$$

where ϵ is known as the sensitivity. This is valid for $D \le 0.2D_{\max}$, which means, in practice, $D \le 0.6 - 1.5$ (Fig. 4.31a).

If N grains are developed per unit area with a mean projected area $\bar{a}$, the density D can be written for small J as

$$D = 0.46N\bar{a} = 0.46\,pJ\bar{a}/e, \tag{4.36}$$

where p denotes the number of particles exposed by one electron. This gives for the sensitivity

$$\epsilon = 0.46\,p\bar{a}/e. \tag{4.37}$$

The mean number of particles exposed depends on the electron energy and the following parameters of the emulsion: quantity of silver per unit area (0.4–0.6 mg cm^{-2}), mean density $\rho = 1 - 2$ g cm^{-3}, thickness of emulsion t = 1–50 μm, and grain diameter (0.5–2 μm). The electron energy and the mean density

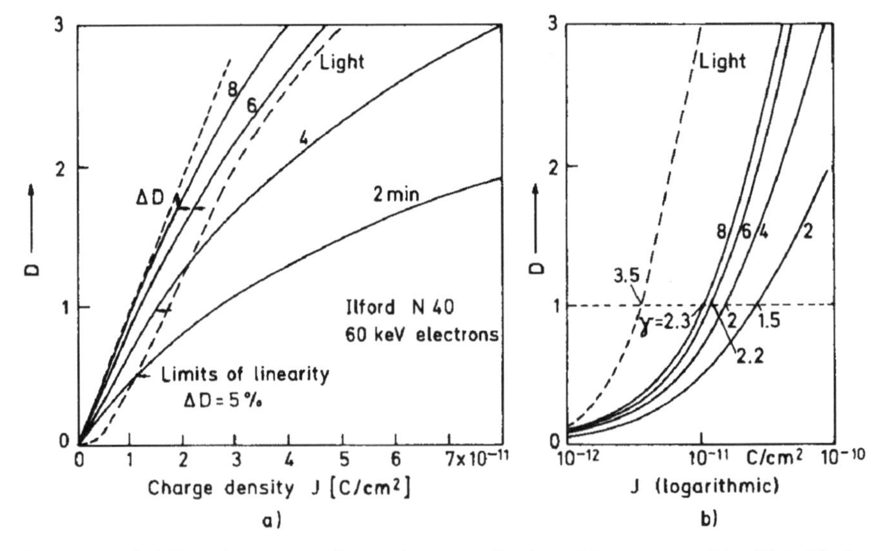

Fig. 4.31. (**a**) Density curves for a photographic emulsion exposed to 60 keV electrons (full lines) and to light (dashed lines) for the developing times indicated. (**b**) Double logarithmic plot for various values of $\gamma = dD/d(\log_{10}J)$ at $D = 1$.

determine the electron range R, which lies between 75 and 120 μm for 100 keV electrons. The sensitivity increases as E increases if $R < t$ and decreases if $R > t$ because the ionization probability per unit path length of an electron trajectory decreases with increasing energy. Photoemulsions therefore exhibit decreasing sensitivity with increasing energy in HVEM.

It is usual to plot D versus $\log_{10} J$ (Fig. 4.31b). This curve has a straight part with slope γ for medium densities. This slope is used to characterize the photographic emulsion because high values of γ correspond to high-contrast recording. A relative variation of current density $\Delta j/j$ or charge density $\Delta J/J$ produces a relative variation of light transmission $\Delta L/L = -\gamma \Delta J/J$. During exposure to light, the value of γ can be high even at low density owing to the existence of a threshold (see exposure to light with dotted lines in Fig. 4.31a). Because the density curve for electron exposure does not show a threshold, γ cannot increase beyond a certain limit. The proportionality (4.35) can be written as

$$D = \epsilon\, J = \epsilon\, 10^{\log_{10} J}, \tag{4.38}$$

and the maximum possible slope γ is given by

$$\gamma = \frac{\mathrm{d}D}{\mathrm{d}(\log_{10} J)} = \epsilon \ln 10 \cdot 10^{\log_{10} J} = 2.3D. \tag{4.39}$$

With electron exposure, it is therefore impossible to obtain a value of γ greater than 2.3 for a density of unity. γ can increase as long as the density increases with J. No further increase is observed when D approaches the saturation value D_{max}. A further increase of contrast can be obtained by a suitable choice of the photographic material used for printing the micrograph.

The resolution of an emulsion is limited by two effects: the diameter of the electron-diffusion cloud and the granularity of the emulsion. When exposed to light, a halo is formed by scattering at the silver halide grains, the radius of which depends on the grain size. The diffusion halo in electron exposure depends only on electron energy and the mean density of the emulsion. If a slit of width d is illuminated with unit intensity, a density distribution (edge spread function)

$$S(x) = \frac{2.3d}{x_{\mathrm{k}}} 10^{-2|x|/x_{\mathrm{k}}} \tag{4.40}$$

($d \ll x_{\mathrm{k}}$) is obtained [4.128]. The quantity x_{k}, typically 30–50 μm, is the width over which the intensity falls to 10% of the central value. Figure 4.32a shows the intensity recorded by an emulsion for which $x_{\mathrm{k}} = 50\mu$m exposed to a slit of width $d = 10\mu$m.

Suppose now that the density varies periodically with a spacing corresponding to a spatial frequency $q = 1/\Lambda : D = D_0 + \Delta D \cos(2\pi q x)$. This function has to be convoluted with $S(x)$, which results in a decrease of the

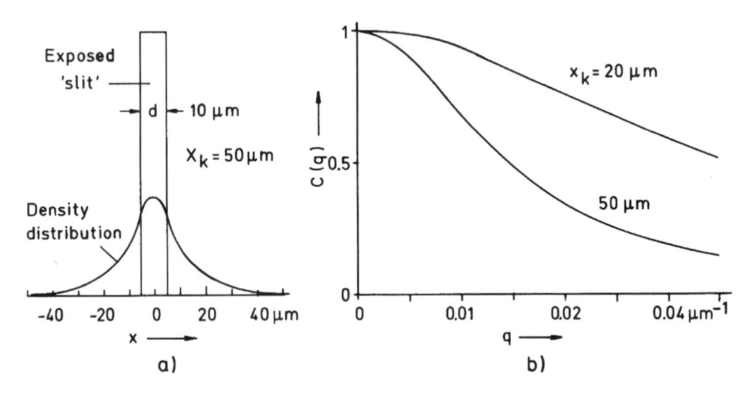

Fig. 4.32. (**a**) Edge-spread function and (**b**) contrast-transfer function $C(q)$ representing electron diffusion in the photographic emulsion.

density amplitude from ΔD to $\Delta D'$. The contrast-transfer function,

$$C(q) = \frac{\Delta D'}{\Delta D} = \left[1 + \left(\frac{\pi q x_\mathrm{k}}{\ln 10}\right)^2\right]^{-1} = \frac{1}{1 + (1.36 q x_\mathrm{k})^2}, \tag{4.41}$$

is plotted in Fig. 4.32b.

The granularity can be considered in the following manner. The number p of neighboring developed grains is greater when an electron passes through the emulsion than when it is stopped by it. Depending on the grain size, the thickness of the emulsion, and the electron energy, p lies between 6 and 50. For light ($p = 1$), the mean-square deviation of the density with a photometric slit of area A is

$$\overline{\Delta D_L^2} = \frac{1}{2.3}\frac{\overline{a}}{A} D. \tag{4.42}$$

An emulsion exposed to a homogeneous current density j appears more granular than one exposed to light because, during electron exposure, clusters of neighboring silver grains are rendered developable by single electrons. The observed mean-square deviation $\overline{\Delta D_E^2}$ will lie between the limits

$$\overline{\Delta D_L^2} < \overline{\Delta D_E^2} < (p+1)\overline{\Delta D_L^2}. \tag{4.43}$$

In order to detect a periodicity, the amplitude $\Delta D'$, already decreased by electron diffusion, must be approximately five times greater than the noise,

$$\Delta D' \geq 5\sqrt{\overline{\Delta D_E^2}}. \tag{4.44}$$

Furthermore, the shot noise caused by the statistical variation $\Delta N = N^{1/2}$ of the number

$$N = n\delta^2 = \frac{j\tau}{e}\delta^2 \tag{4.45}$$

of electrons incident on a small area δ^2 (δ: resolution of the emulsion) must be less than the noise caused by granularity. The necessary charge density

$j\tau$ for a density $D = 1$ in Fig. 4.31a is of the order of 10^{-11} C cm^{-2}. For a resolution δ of 30 μm, this results in $N = 350$ electrons and a noise-to-signal ratio $\Delta N/N = N^{-1/2} = 4\%$. The human eye can detect relative intensity variations of the order of 5%. This means that the sensitivity of photographic emulsions to electron exposure is of just the right order, and emulsions are optimum for the recording of electron micrographs. A film size of $A = 6 \times 9$ cm^2 contains $A/\delta^2 = 6 \times 10^6$ image points, which corresponds to a very high storage capability.

4.7.3 Imaging Plate

The imaging plate (IP) was first developed for x-ray radiography to have a higher sensitivity and better image quality than conventional x-ray films. The IP is also an interesting image recording system for TEM [4.134, 4.135, 4.136].

The IP is a flexible sheet with a thickness of 0.3–0.5 mm that is composed of a transparent protective layer, a phosphor layer (50–100 μm thick), and a plastic support. The main part is the photostimulatable phosphor BaFX:Eu (X = Cl, Br, I). The phosphor, with a grain size of $\simeq$5 μm, is spread over the plastic support together with an organic binder. The IP is directly exposed to electrons in the vacuum in the same way as for photographic emulsions. Part of the electron energy dissipated is stored in luminescence centers. When such an exposed IP is scanned in air with a small spot of He-Ne laser light, the stored energy is emitted as blue light with an emission maximum at $\lambda =$ 390 nm; the reading time is about 1000 s. About 100 photons are generated by a single electron. The emitted light is detected by a photomultiplier. This allows a digitized image with 3760 $\times$ 3000 pixels to be recorded directly with a reading pixel size of 25–50 μm and 2^{14} grey levels (14 bits). The image can equally well be printed directly on photographic printing paper. The dissipated energy remaining in the phosphor is erased by irradiating it with light, so that the plate is reusable.

The IP has a total reading range of charge density from 2×10^{-10} to 2×10^{-5} C/m^2; the sensitivity is about three orders of magnitude higher than for a Fuji FG film, which needs 5×10^{-7} C/m^2 for a density $S = 1$ when irradiated with 200 keV electrons. The thickness of the phosphor layer is such that the sensitivity is approximately constant over the range 100–400 keV.

The contrast-transfer function $C(q)$ of an IP with a 50 μm layer falls to 50% at $\Lambda = 1/q \simeq 200$ μm compared with $\simeq$60 μm for a 50 μm photographic emulsion (Fig. 4.32b). The resolution can be increased by decreasing the thickness of the phosphor layer, but this will of course also decrease the sensitivity.

The high dynamic range allows both the central and the outer parts of diffraction patterns to be recorded under conditions in which the central region would be overexposed on a photographic emulsion.

4.7.4 Detector Noise and Detection Quantum Efficiency

Not only do the primary electrons show shot noise, which means that the number of electrons hitting a detector during a given pixel time is statistically distributed, but the statistics also increase the noise in the different steps of the detector system; the signal-to-noise ratio $\mathrm{SNR_{out}}$ behind the detector system is lower than $\mathrm{SNR_{in}}$. The squared ratio of these is called the detection quantum efficiency,

$$\mathrm{DQE} = \frac{\mathrm{SNR}^2_{\mathrm{out}}}{\mathrm{SNR}^2_{\mathrm{in}}} \leq 1. \tag{4.46}$$

It will be unity for an ideal detector that produces no further noise. The choice of definition of this quantity for electronics is historical; the square, not very intuitive for particle detectors, is associated with the power spectrum of noise.

For the calculation of the DQE, we have to evaluate the variances in the different steps. We assume that one particle (electron or photon) can randomly generate $i = 0, 1, 2, \ldots$ particles in the subsequent step with a probability $P_x(i)$, where $\sum_i P_x(i) = 1$. The mean yield of this step will be

$$x = \sum_i i P_x(i), \tag{4.47}$$

and the variance of x, which is the square of the standard deviation σ, becomes

$$\begin{aligned}\mathrm{var}(x) = \sigma^2 &= \sum_i (i-x)^2 = \sum_i i^2 P_x(i) - 2x \underbrace{\sum_i i P_x(i)}_{=x} + x^2 \underbrace{\sum_i P_x(i)}_{=1} \\ &= \sum_i i^2 P_x(i) - x^2. \end{aligned} \tag{4.48}$$

Whereas $P_x(i)$ has been introduced as the probability that one incident particle generates i particles, the probability $P_x(n,i)$ is defined to be the probability that n incident particles will result in a total number i of particles:

$$P_x(n,i) = \sum_j P_x(j) P_x(n-1, i-j). \tag{4.49}$$

$P_x(i)$ is called a binomial or binary distribution if only the values $i = 0$, 1 are possible. This means that the particles are either absorbed ($i = 0$) or transmitted, backscattered or ejected ($i = 1$), and

$$P_x(0) = 1 - x, \qquad P_x(1) = x, \tag{4.50}$$

with the mean yield x and the variance

$$\mathrm{var}(x) = x(1-x). \tag{4.51}$$

For a binomial distribution, the relation (4.49) gives

$$P_x(n,i) = \frac{n!}{i!(n-i)!} x^i (1-x)^{n-i} \tag{4.52}$$

with a mean yield $y = nx$ and the variance

$$\mathrm{var}(nx) = nx(1-x). \tag{4.53}$$

For example, $P_x(2,0) = (1-x)^2$, $P_x(2,1) = 2x(1-x)$, $P_x(2,2) = x^2$.

The binomial distribution degenerates to a Poisson distribution for $y = nx$ if $x \ll 1$ and n is very large:

$$P_y(i) = \frac{y^i}{i!}e^{-y}, \qquad \mathrm{var}(y) = y. \tag{4.54}$$

The following equations are relevant to the statistics of cascade processes. If one particle generates i particles in the first stage with a probability $P_{x1}(i)$, and if these i particles enter the second stage and generate j particles with a probability $P_{x2}(i,j)$ per incident particle, then the probability $P_y(k)$ of generating k particles in the second stage per incident particle in the first stage is given by

$$P_y(k) = \sum_i P_{x1}(i)P_{x2}(i,k) \tag{4.55}$$

with mean yield y and variance

$$y = x_1 x_2, \qquad \mathrm{var}(y) = \mathrm{var}(x_1)x_2^2 + x_1\mathrm{var}(x_2). \tag{4.56}$$

For m statistical processes, the cascade has the mean yield $y = x_1 x_2 \ldots x_m$, and variance

$$\mathrm{var}(y) = y^2\left(\frac{\mathrm{var}(x_1)}{x_1^2} + \frac{\mathrm{var}(x_2)}{x_1 x_2^2} + \ldots \frac{\mathrm{var}(x_m)}{x_1 x_2 \ldots x_{m-1} x_m^2}\right). \tag{4.57}$$

A cascade of two binomial distributions again results in a binomial distribution with $y = x_1 x_2$ and $\mathrm{var}(y) = y(1-y)$. If a Poisson distribution with mean value x_1 is followed by a binomial distribution, we get a Poisson distribution with $y = x_1 x_2$ and $\mathrm{var}(y) = x$. However, two Poisson distributions in cascade do not result in a Poisson distribution.

We first apply these general laws to the statistics of the primary electrons. The mean number of incident electrons per pixel is

$$N = I_\mathrm{p}\tau/e. \tag{4.58}$$

The shot noise, may be analyzed by the following argument. The time τ for one pixel can be divided into a large number n of time intervals, so that the probability x of observing one electron in one of these time intervals is much less than unity and the probability of observing more than one electron per time interval is negligible. We then expect the mean value of the number of electrons y in the time interval τ to follow a Poisson distribution (4.54), $y = nx = N$ and $\mathrm{var}(N) = N$.

The electron current consists of pulses of charge e and can be Fourier analyzed, from which we obtain the rms current of the a.c. or noise component

$$I_{\mathrm{n,rms}} = \sqrt{\overline{I_n^2}} = \sqrt{2e\Delta f I_\mathrm{p}}, \tag{4.59}$$

where Δf is the bandwidth of the detection system. Furthermore, (4.58) gives the noise amplitude

$$I_{\mathrm{n,rms}} = \frac{e}{\tau}\sqrt{\mathrm{var}(N)} = \frac{e}{\tau}\sqrt{N} = \sqrt{eI_{\mathrm{p}}/\tau} \tag{4.60}$$

of the primary current, and the last two equations become identical for $\tau = 1/(2\Delta f)$. The signal-to-noise ratio of the primary electrons will be $\mathrm{SNR_{in}}$ in (4.46):

$$\mathrm{SNR_{in}} = I_{\mathrm{p}}/I_{\mathrm{n,rms}} = N/\sqrt{\mathrm{var}(N)} = \sqrt{N} = \sqrt{I_{\mathrm{p}}/2e\Delta f}. \tag{4.61}$$

These laws of statistics are used below to calculate the DQE of scintillators coupled by a fiber plate to a CCD.

4.7.5 Low-Light-Level and Charge-Coupled-Device (CCD) Cameras

For digital image processing and for electron microscope alignment (Sect. 2.4.3), it is of interest to record two-dimensional arrays of pixels directly, avoiding the darkroom work required for photographic emulsions or the read-out for imaging plates. In the first attempts, the image on a fluorescent screen was captured by a low-light-level TV camera. Several commercial TV tubes were tested and employed for electron microscopy; see [4.137] for a comparison of their sensitivities and DQEs. Of these, it was the SIT camera (silicon intensifier target) that was mainly used. However, CCD cameras have come into increasingly widespread use; see [4.138] for a comparison of a frame-transfer CCD and a SIT camera.

A CCD image sensor consists of an array of 1024^2 to 4096^2 silicon-based photodiodes (pixels), each typically about $20 \times 20\ \mu\mathrm{m}^2$. Absorbed light quanta generate electron-hole pairs, which are separated in the depletion layer of the diodes, and the electrons are accumulated in the potential wells of the diodes during storage. By applying sequences of different biases to neighboring diodes, the charges can be transferred to a serial shift register, after which a built-in amplifier and ADC transfers the signal to an external buffer store. Charge-coupled devices in light-optical TV cameras can be read at TV frequency, but the read-out (accumulation) time can be varied and increased to 100 s when the CCD is cooled to $-30°$C using a Peltier element. This decreases the background dark current of the diodes and considerably increases the signal-to-noise ratio.

The CCD cannot be irradiated with electrons directly owing to the generation of defects, which cause a long-time fading of the sensitivity. Furthermore, the large number of electron-hole pairs created by each incident electron would limit the number of recordable electrons per diode to only about a hundred. The electrons are therefore converted to photons in a thin scintillator (powder layer or YAG single crystal with a thickness of 50 μm) and transferred through

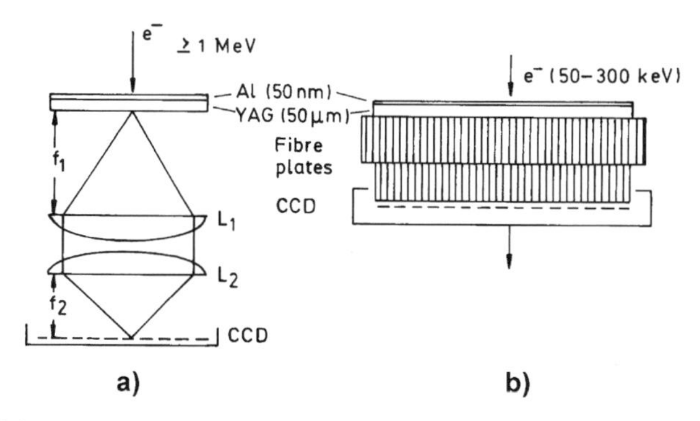

Fig. 4.33. (**a**) Tandem optics and (**b**) fiber plate for coupling the light excited in a fluorescent layer to a charge-coupled device (CCD).

a fiber plate with fiber diameters of 6 μm or a light-optical tandem objective to the CCD array (Fig. 4.33) [4.138, 4.139, 4.140, 4.141, 4.142, 4.143, 4.144].

The fiber-plate coupling has the advantage of better light-collection efficiency. A disadvantage is the honeycomb pattern, which can be seen as a Moiré pattern between the gratings of the diodes and fibers. Digital gain normalization of the recorded signal is hence necessary using an "image" recorded without any specimen with a uniform incident current density.

The light-collection efficiency and the DQE are less good with light-optical coupling. Otherwise, it has the advantages that the optics and cooled CCD are outside the microscope, there is no honeycomb structure, and only some shading from the center to the corner. It can be used for HVEM [4.145, 4.146], although the transparency of the fiber plate decreases because of formation of color centers. Also, the generation of bright spots excited by x-rays in the CCD can be reduced.

With $n = j/e$ incident electrons per unit area and $N = nd^2$ per diode area d^2, the three different conversion stages (scintillator, fiber plate, and CCD) result in the mean value

$$N_e = N n_e = N \underbrace{\epsilon \frac{\overline{E}}{E_{\text{ph}}}}_{n_{\text{ph}}} \cdot \underbrace{\eta_{\text{opt}} \eta_{\text{CCD}}}_{\eta} \tag{4.62}$$

of stored electrons per diode, where, for example,

$$\begin{aligned}
\overline{E} &= (1-\eta_c)E &&: \text{energy dissipated in phosphor,}\\
\eta_c &\simeq 0.1 &&: \text{energy fraction lost by backscattering,}\\
\epsilon &= 5\% &&: \text{energy conversion coefficient,}\\
E_{\text{ph}} &= 2.21\text{eV} &&: \text{mean photon energy,}\\
\eta_{\text{opt}} &= 0.06 &&: \text{optical efficiency of the fiber plate,}\\
\eta_{\text{CCD}} &= 35\% &&: \text{quantum efficiency of the CCD.}
\end{aligned}$$

This results in a mean number $n_{\text{ph}} \simeq 2000$ of emitted photons and $n_{\text{e}} \simeq 40$ accumulated electrons in the CCD per incident 100 keV electron on the phosphor layer. The saturation charge (full-well capacity) is $\simeq 2.5 \times 10^5$ e-/pixel, which means 5000 incident electrons can be recorded per pixel. This is far superior to a photographic emulsion and provides a dynamic range of 12 bits.

Application of (4.57) to (4.62) results in

$$\mathrm{SNR}^2_{\mathrm{out}} = \frac{\mathrm{var}(N_\mathrm{e})}{N_\mathrm{e}^2} = \frac{\mathrm{var}(N)}{N^2} + \frac{\mathrm{var}(n_\mathrm{ph})}{N n_\mathrm{ph}^2} + \frac{\eta(1-\eta)}{N n_\mathrm{ph}\eta^2}. \tag{4.63}$$

The third term is the variance of a binomial distribution (4.51), and the first term (Poisson distribution) becomes $1/N$ (4.54). Using the definition (4.46) of the DQE, (4.63), and (4.61) results in [4.142]

$$\mathrm{DQE} = \left[1 + \frac{\mathrm{var}(n_\mathrm{ph})}{n_\mathrm{ph}^2} + \frac{1-\eta}{n_e} + \frac{1}{N}\frac{\Delta n_\mathrm{r}^2}{n_e^2}\right]^{-1}. \tag{4.64}$$

The last term is included to represent the readout noise with $\Delta n_\mathrm{r} = 20e^-$. When the values above are inserted, the last two terms in (4.64) are very small so long as $n_e \gg 1$; the DQE is then determined by $\mathrm{var}(n_\mathrm{ph})$, which can be calculated from a measured or Monte Carlo simulated pulse-height distribution (Fig. 4.34a). With increasing electron energy E, the maxima of the distribution are broadened and shift to higher pulse heights. However, when the electrons penetrate the phosphor, the maxima decrease strongly and a new maximum appears at low pulse heights caused by electrons backscattered at the fiber plate. This results in the decrease of the DQE with increasing E shown in Fig. 4.34b for different thicknesses t of the single-crystal YAG disc.

As in Fig. 4.32 for a photographic emulsion, an edge- or point-spread function can be measured for a CCD (Fig. 4.34c); this shows that the signal is spread over more than one pixel as a consequence of electron diffusion and light scattering in the phosphor.

The relation between the point-spread function $p(x, y)$ and the edge-spread function $e(x, y)$ can easily be deduced from the equation

$$g(x,y) = \int f(x', y')p(x - x', y - y')dx'dy' \tag{4.65}$$

for the image intensity $g(x, y)$ for an object described by the transmission function $f(x, y)$. For a small slit, the transmission function is given by

$$f(x,y) = \delta(x), \tag{4.66}$$

yielding the line-spread function

$$l(x) = \int_{-\infty}^{\infty} \delta(x')p(x - x', y - y')dx'dy' = \int_{-\infty}^{\infty} p(x, y')dy'. \tag{4.67}$$

Correspondingly, for a sharp edge, we insert the transmission function

$$f(x,y) = \begin{cases} 1 & \text{for } x > 0 \\ 0 & \text{otherwise} \end{cases} \tag{4.68}$$

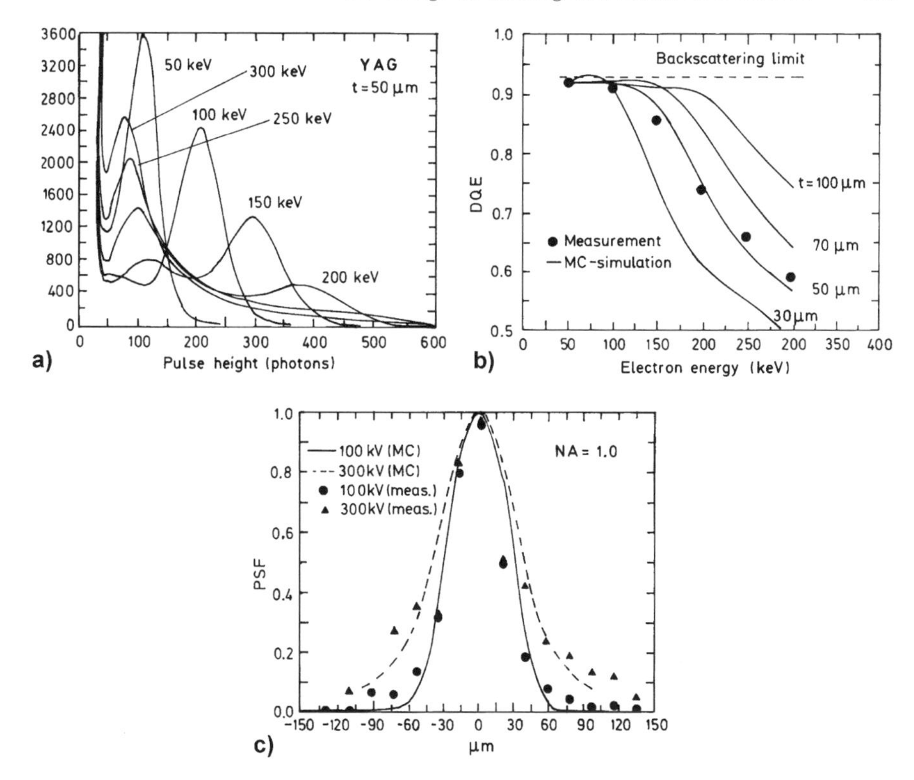

Fig. 4.34. (**a**) Pulse-height distribution of a 50 μm YAG single crystal irradiated with different electron energies, (**b**) the DQE calculated from the measured pulse-height distribution and Monte Carlo simulations as a function of electron energy and different thicknesses t of the YAG single crystal, and (**c**) point-spread function of the CCD for 100 and 300 keV electrons [4.142].

and obtain

$$e(x) = \int_{x'=0}^{\infty} \int_{-\infty}^{\infty} p(x - x', y - y')dx'dy' = \int_{-x}^{\infty} l(\tilde{x})d\tilde{x}. \tag{4.69}$$

The last identity shows that the line-spread function can be obtained from the edge-spread function by a simple differentiation,

$$l(x) = \frac{d}{dx}e(x). \tag{4.70}$$

Alternatively, the transfer properties of the camera can be described by a modulation transfer function defined by

$$M(\boldsymbol{q}) = \int p(\boldsymbol{r})e^{2\pi i \boldsymbol{q}\boldsymbol{r}}d^2\boldsymbol{r}. \tag{4.71}$$

Assuming rotational symmetry, we obtain

$$M(q) = \int_{r=0}^{\infty} \int_{\varphi=0}^{2\pi} p(r)e^{2\pi i q r \cos\varphi}d\varphi r dr = 2\pi \int_{0}^{\infty} p(r)Jo(2\pi q r)r dr. \tag{4.72}$$

The modulation transfer function can also be obtained from the line-spread function via

$$M(q) = \int_{-\infty}^{\infty} l(x) e^{2\pi i q x} dx. \quad (4.73)$$

In practice, one often measures the edge-spread function, from which the other functions can be easily obtained.

4.7.6 Semiconductor and Scintillation Detectors

The following detectors can be used for the recording of signals in STEM or for sequentially recording electron energy-loss spectra or diffraction patterns, for example.

Semiconductor Detector. This type of detector consists of a p-n junction diode below a conductive surface layer; optimally the thickness of the depletion layer should be of the order of the electron range R. High-energy electrons of energy E create $n = E/\overline{E_i}$ electron-hole pairs. The mean energy $\overline{E_i}$ for creating one pair is 3.6 eV in silicon. The electron-hole pairs created in the depletion layer are separated and produce a charge-collection current

$$I_{cc} = I_p(1 - \eta_c) \frac{E - E_{th}}{\overline{E_i}} \epsilon_c, \quad (4.74)$$

where η_c takes into account the loss of ionization by backscattering, which is only of the order of 10% for silicon; $E_{th} = 1$–5 keV is the threshold energy for the incident electrons arising from absorption in an evaporated gold contact layer and/or from an increased surface recombination rate (dead layer); and ϵ_c is the charge-collection efficiency of the depletion layer.

Because of the relatively large capacitance of the depletion layer, a low-impedance current amplifier has to be used to convert I_{cc} to a video voltage of a few hundred meV. The time constant $\tau_0 = RC$ decreases with decreasing area of the depletion layer and increasing current I_p. The capacitance and the background noise can be further decreased by employing reverse biasing of the p-n junction. Currents of 10^{-11} A can be recorded in about 10^{-5} s, which corresponds to a cutoff frequency of 100 kHz of the video signal. By decreasing C, it is possible to observe backscattered electrons at TV scan rates [4.147, 4.148]. With the high electron energies used in TEM, single-electron counting is also possible when the single pulses are higher than a threshold that exceeds the noise level.

Scintillation Detector. Scintillator materials emit light quanta (photons) under electron bombardment. Zinc sulfide (ZnS), which is used for fluorescent screens in TEM, has a high efficiency, but its light-intensity decay time is of the order of one millisecond and the afterglow persists for several seconds; it therefore cannot be used for fast recording. Plastic scintillators (NE102A of Nuclear Enterprises Ltd., for example), P-47 powder, or single crystals (yttrium silicate doped with 1% cerium) have become standard scintillator materials for TEM, STEM, and SEM because their time constants are of the order of 10^{-8} s and their efficiency is not worse than one-tenth that of ZnS.

A conductive and light-absorbing Al coating about 100 nm thick is evaporated on plastic scintillators. The light emission decreases with increasing irradiation time owing to radiation damage of the organic material. However, the thin damaged layer can be removed by polishing. P-47 powder layers exhibit a much longer radiation resistance. Methods of preparing P-47 layers with optimum thickness are reported in [4.149, 4.150]. Single-crystal scintillators consisting of cerium-doped yttrium aluminum garnet (YAG) can be used for the detection of transmitted and backscattered electrons in STEM [4.151] and as thin polished slices in front of fiber plates connected to a CCD.

The photons emitted are collected by a light pipe in front of the photomultiplier, which reflects the light by total reflection with a transmission T. The photons are converted to photoelectrons at the photocathode of the multiplier with a quantum efficiency q_c that is between 5% and 20%. The photoelectrons are accelerated by a potential of about +100 V to an electrode of high secondary-electron yield $\delta_\mathrm{PM} = 8$–15. The total gain of the multiplier is obtained by successive acceleration and secondary-electron emission at $n =$ 8–10 electrodes (dynodes), resulting in a total gain $g_\mathrm{PM} = \delta^n_\mathrm{PM}$. The pulse of g_PM electrons or the current induced by a higher rate of incident electrons causes a voltage drop U across a resistor $R = 100\ \mathrm{k}\Omega$, which can be amplified by operational amplifiers. For an incident probe current I_p and a detector collection efficiency f, which depends on the signal generated (transmitted, secondary, or backscattered electrons) and on the solid angle of collection, the signal is

$$U = fI_\mathrm{p}\frac{E}{\overline{E_\mathrm{ph}}}Tq_\mathrm{c}\delta^n_\mathrm{PM}R, \tag{4.75}$$

where $\overline{E_\mathrm{ph}}$ denotes the mean energy needed to produce one photon in the scintillator. Such a scintillator–photomultiplier combination can be operated with a large bandwidth Δf, up to some MHz, and low noise. It is possible to achieve an rms noise amplitude that is only a factor of 1–2 larger than the shot noise., The latter is the noise amplitude (4.59) associated with an electron current I_p caused by statistical fluctuations of the number of electrons incident during equal sampling times.

4.7.7 Faraday Cages

Direct measurement of electron currents is of interest for determination of electron-current densities and electron-beam currents. Quantitative measurements require a Faraday cage, which consists of a grounded shield that contains a hole somewhat smaller than that of the inner cage (Fig. 4.35). The hole has to be small enough to ensure that the solid angle of escape for electrons backscattered at the bottom of the cage is negligible. The backscattering coefficient of the bottom material must be low ($\eta = 6\%$ for carbon and 13%

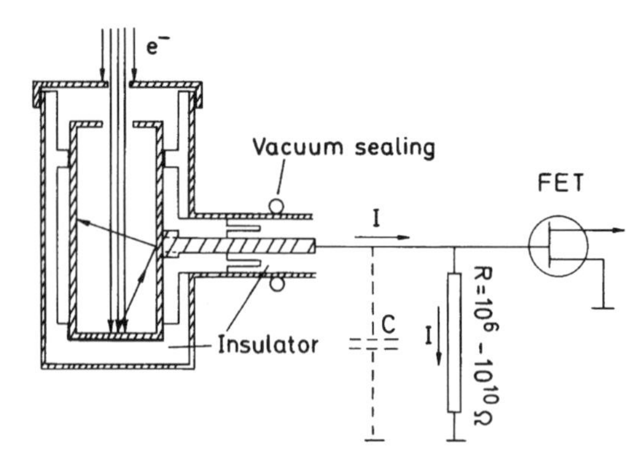

Fig. 4.35. Construction and input circuit of a Faraday cage for measuring electron currents.

for Al). Furthermore, the secondary electrons produced at the inner walls of the cage must remain inside the cage. The low currents can be measured with a commercial electrometer, which makes use of the voltage drop $U = RI$ of the order of 1 mV to 1 V across a high resistance $R = 10^6 - 10^{10}\,\Omega$. A low-impedance output signal can be obtained by using a field-effect transistor (FET) or a vibrating-reed electrometer. The high resistance R and the by no means negligible capacitance C of the cage, the cables, and the electrometer input result in a time constant $\tau_0 = RC$, which may reach a few seconds for very small currents. A Faraday cage therefore cannot be used to record fast variations of low electron currents.

Some microscopes are equipped with an insulated fluorescent screen to measure the incident current for an automatic exposure system. This cannot be used for quantitative measurements because the large backscattered fraction of the electrons travel like bouncing balls through the chamber and an unknown fraction hit the screen again.

5

Electron–Specimen Interactions

The elastic scattering of electrons by the Coulomb potential of a nucleus is the most important of the interactions that contribute to image contrast. Cross sections and mean-free-path lengths are used to describe the scattering process quantitatively. A knowledge of the screening of the Coulomb potential of the nuclei by the atomic electrons is important when calculating the differential cross sections at small scattering angles.

The inelastic scattering is concentrated within smaller scattering angles, and the excitation of energy states results in energy losses. The dominant mechanisms are plasmon and interband excitations, which can be described by the dielectric theory. These inelastic scattering processes are less localized than elastic scattering and cannot contribute to high resolution. Inner-shell ionizations result in edge-shaped structures in the electron energy-loss spectrum (EELS), on which are superposed a near-edge structure (ELNES) and an extended energy-loss fine structure (EXELFS), which can be used for analytical electron microscopy at high spatial resolution.

Even quite thin specimen layers, of the order of a few nanometers, do not show the angular or energy-loss distribution corresponding to a single scattering process. Multiple-scattering effects have to be considered as the specimen thickness is increased, and this can also result in electron-probe broadening.

5.1 Elastic Scattering

5.1.1 Cross Section and Mean Free Path

The most convenient quantity for characterizing the angular distribution of scattered particles is the differential cross section, which is introduced in Fig. 5.1a using the Coulomb model for the scattering of an electron by a nucleus. The electrons travel on hyperbolic trajectories due to the attractive Coulomb force (3.15) between the electron and the nucleus. If there were no

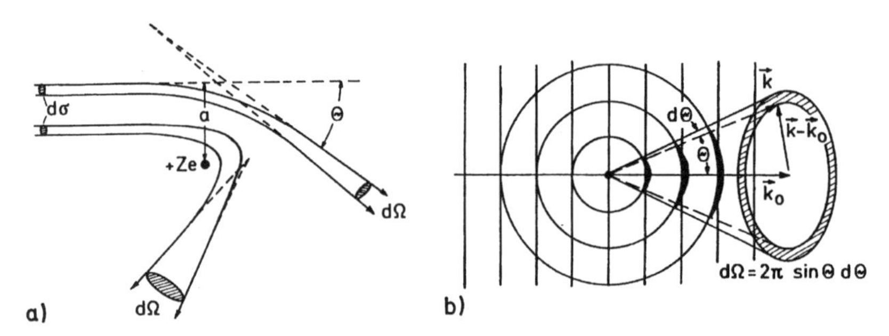

Fig. 5.1. (**a**) Elastic electron scattering in the particle model and explanation of the differential cross section $\mathrm{d}\sigma/\mathrm{d}\Omega$ (a: impact parameter). (**b**) Scattering in the wave model with the superposition of a plane incident wave of wave number $\boldsymbol{k}_0$ and a spherical scattered wave of amplitude $f(\theta)$, depending on the scattering angle θ.

interaction between them, the electron would travel straight past the nucleus; the shortest distance between them, the impact parameter, is denoted by a. Increasing a decreases the scattering angle θ. Electrons that pass through an element of area $\mathrm{d}\sigma$ of the parallel incident beam will be scattered into a cone of solid angle $\mathrm{d}\Omega$. The ratio $\mathrm{d}\sigma/\mathrm{d}\Omega$ is known as the differential cross section and is a function of the scattering angle θ.

This cross section $\mathrm{d}\sigma/\mathrm{d}\Omega$ cannot be calculated exactly from the classical particle model; quantum mechanics has to be used (Sect. 5.1.3) [3.1, 3.2, 3.3, 5.1]. Far from the nucleus, the total wave field can be expressed as the superposition of the undisturbed plane incident wave of amplitude $\psi = \psi_0 \exp(2\pi \mathrm{i} k_0 z)$ and a spherical scattered wave of amplitude

$$\psi_{\mathrm{sc}} = \psi_0 f(\theta) \frac{\mathrm{e}^{2\pi \mathrm{i} kr}}{r} \tag{5.1}$$

depending on the scattering angle θ (Fig. 5.1b).

The current density $j_0 = eNv$ of a parallel beam has been introduced in (3.9); Nv is the flux of particles that pass through a unit area per unit time. Scattering into the solid angle $\mathrm{d}\Omega$ is observed when the electron hits the fraction $\mathrm{d}\sigma$ of the unit area. The scattered current $\mathrm{d}I_{\mathrm{sc}}$ that passes through the area $\mathrm{d}S = r^2\,\mathrm{d}\Omega$ will be

$$\mathrm{d}I_{\mathrm{sc}} = j_{\mathrm{sc}} r^2 \mathrm{d}\Omega = j_0 \mathrm{d}\sigma, \quad \text{which implies} \quad j_{\mathrm{sc}} = \frac{j_0}{r^2} \frac{\mathrm{d}\sigma}{\mathrm{d}\Omega} \,. \tag{5.2}$$

Substituting the scattered-wave amplitude ψ_{sc} (5.1) in the quantum-mechanical expression for the current density (3.10) yields

$$j_{\mathrm{sc}} = ev|\psi_0|^2 \frac{|f(\theta)|^2}{r^2} = j_0 \frac{|f(\theta)|^2}{r^2} \,. \tag{5.3}$$

Comparing (5.2) and (5.3), we find

$$\frac{\mathrm{d}\sigma}{\mathrm{d}\Omega} = |f(\theta)|^2 \,. \tag{5.4}$$

The total number of scattered electrons can be calculated by dividing the corresponding solid angle into small segments $\mathrm{d}\Omega = 2\pi \sin\theta \, \mathrm{d}\theta$ (Fig. 5.1b) and integrating over θ from 0 to π. This gives the *total elastic cross section*

$$\sigma_{\mathrm{el}} = \int_0^{\pi} \frac{\mathrm{d}\sigma}{\mathrm{d}\Omega} 2\pi \sin\theta \mathrm{d}\theta. \tag{5.5}$$

This quantity can be used to calculate the number of unscattered electrons, for example. Whether or not scattering occurs is determined by the total (elastic and inelastic) cross section $\sigma_{\mathrm{t}} = \sigma_{\mathrm{el}} + \sigma_{\mathrm{inel}}$ (σ_{inel} is the total inelastic cross section, Sect. 5.2.2).

Suppose that n unscattered electrons are incident on a thin layer of a solid film with a mass thickness $\mathrm{d}x = \rho \mathrm{d}z$ in units g cm^{-2}. There will be $N\rho\mathrm{d}z$ atoms per unit area in a layer of thickness $\mathrm{d}z$ with $N = N_{\mathrm{A}}/A$ atoms per gram (N_{A} is Avogadro's number and A the atomic weight). Scattering occurs when the electrons hit a small area σ_{t} in the vicinity of each atom. A scattering event will be recorded when the electrons strike a fraction $N\sigma_{\mathrm{t}}\mathrm{d}x$ of the unit area, and a fraction

$$\frac{\mathrm{d}n}{n} = -N\sigma_{\mathrm{t}}\mathrm{d}x \tag{5.6}$$

will be scattered in the layer of thickness $\mathrm{d}x$. The negative sign indicates that n is decreased by scattering.

Integrating (5.6), we find

$$\ln n = -N\sigma_{\mathrm{t}}x + \ln n_0, \tag{5.7}$$

where the constant of integration, $\ln n_0$, is determined by the initial number $n = n_0$ of incident electrons per unit area at $x = 0$. This shows that the number of unscattered electrons decreases exponentially with increasing mass thickness,

$$n = n_0 \exp(-N\sigma_{\mathrm{t}}x) = n_0 \exp(-x/x_{\mathrm{t}}). \tag{5.8}$$

The lengths

$$x_{\mathrm{t}} = \rho\Lambda_{\mathrm{t}} = 1/N\sigma_{\mathrm{t}} \quad \text{and} \quad \Lambda_{\mathrm{t}} = x_{\mathrm{t}}/\rho \tag{5.9}$$

are both known as the total mean-free-path length (in units g cm^{-2} and cm, respectively) between scattering events.

5.1.2 Energy Transfer in an Electron–Nucleus Collision

An elastic collision is defined as a collision in which the total kinetic energy and momentum are conserved. The laws of conservation of energy and momentum before and after the collision can be written without any detailed knowledge of the interaction process between the particles. We characterize quantities after the collision by a dash and those of the nucleus by the suffix n. From Fig. 5.2, the conservation of momentum can be expressed as

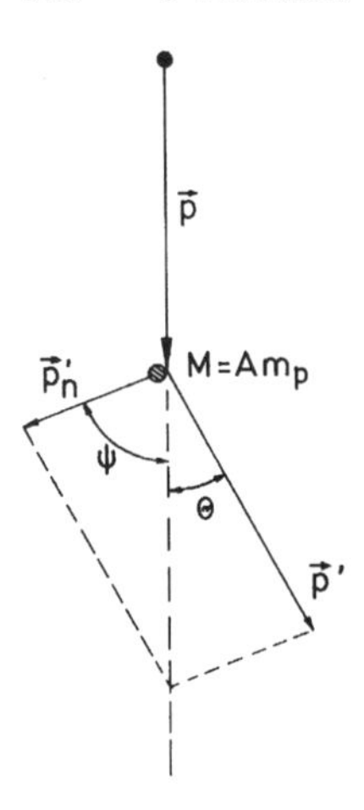

Fig. 5.2. Conservation of momentum in elastic scattering.

$$\boldsymbol{p} = \boldsymbol{p}' + \boldsymbol{p}'_{\rm n} = \begin{cases} p = p'\cos\theta + p'_n \cos\psi \\ 0 = p'\sin\theta - p'_{\rm n}\sin\psi , \end{cases} \tag{5.10}$$

and the conservation of kinetic energy requires that

$$E = E' + E'_{\rm n} . \tag{5.11}$$

Equation (2.11) has to be used for the relativistic momentum of the electron, whereas the nonrelativistic formula

$$p'_{\rm n} = (2ME'_{\rm n})^{1/2} \tag{5.12}$$

can be used for the momentum of the nucleus because its rest mass $M = Am_{\rm p}$ is very large ($m_{\rm p}$: atomic mass unit). Solving the lower equation in (5.10) for $\sin\psi$ and (5.11) for E' and substituting these quantities in the upper equation in (5.10), we obtain

$$\frac{1}{c}\left[E(E+2E_0)\right]^{1/2} = \frac{1}{c}\left[(E-E'_{\rm n})(E-E'_{\rm n}+2E_0)\right]^{1/2}\cos\theta + \left[2ME'_{\rm n}\left(1-\frac{(E-E'_{\rm n})(E-E'_{\rm n}+2E_0)}{2Mc^2E'_{\rm n}}\sin^2\theta\right)\right]^{1/2} . \tag{5.13}$$

The energy transfer $E'_{\rm n}$ to the nucleus will be small compared with E, so $E - E'_{\rm n} \simeq E$. Transferring the first term on the right-hand side of (5.13) to the left-hand side, squaring the equation, and using the relation $1 - \cos\theta = 2\sin^2(\theta/2)$, we find

$$E'_{\rm n} = \frac{2E(E+2E_0)}{Mc^2}\sin^2\frac{\theta}{2} = \frac{E(E+1.02)}{496A}\sin^2\frac{\theta}{2} , \tag{5.14}$$

with $E'_{\rm n}$, E, and E_0 in MeV.

From the conservation of energy (5.11), this energy $E'_{\rm n}$ transferred to the nucleus must be equal to the energy loss ΔE of the primary electron. Table 5.1 shows typical values of $E'_{\rm n}$. This energy loss is negligible for small scattering

Table 5.1. Energy transfer E'_{n} to a nucleus, which is equal to the energy loss ΔE of the primary electron of energy E in an elastic scattering process with a scattering angle θ for 100 keV and 1 MeV electrons.

E	100 keV			1 MeV		
	C	Cu	Au	C	Cu	Au
θ	$(A = 12)$	$(A = 63.5)$	$(A = 197)$			
0.5°	0.5 meV	0.1 meV	0.03 meV	9 meV	1.7 meV	0.54 meV
10°	0.15 eV	29 meV	9 meV	2.7 eV	0.5 eV	0.17 eV
90°	10 eV	1.9 eV	0.6 eV	179 eV	34 eV	11 eV
180°	20 eV	3.8 eV	1.2 eV	359 eV	68 eV	22 eV

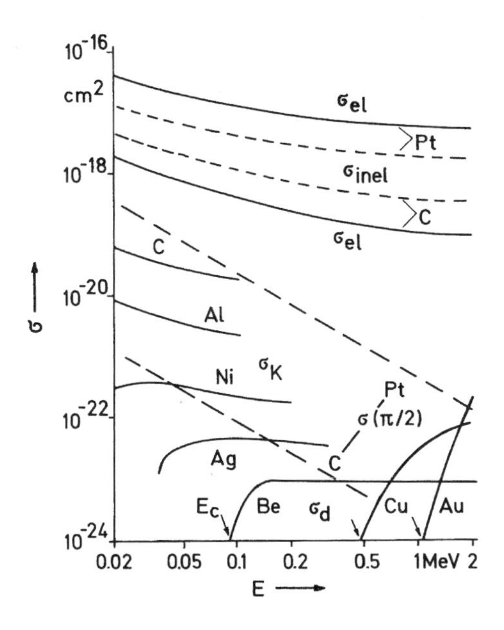

Fig. 5.3. Comparison of total cross sections σ for elastic scattering (σ_{el}), inelastic scattering (σ_{inel}), K-shell ionization (σ_{K}), backscattering into angles $\theta \geq \pi/2$ ($\sigma_{\pi/2}$), and an atomic displacement with a displacement energy $E_{\mathrm{d}} = 20$ eV (σ_{d}) as functions of electron energy E.

angles owing to the presence of the factor $\sin^2(\theta/2)$. Therefore, in elastic electron-nucleus small-angle scattering, we can say that effectively no energy is lost by the primary electron, though very small energy losses of the order of meV are possible in solids by electron–phonon scattering (Sect. 5.2.1).

However, the energy losses are not negligible for higher electron energies and scattering angles. If the energy transfer E'_{n} is greater than the displacement energy $E_{\mathrm{d}} \simeq 10$–30 eV, nuclei can be displaced from their lattice points to interstitial sites, resulting in radiation damage, which has to be considered in high-voltage electron microscopy (Sect. 11.3.2). Carbon atoms can also be knocked out of organic compounds by this direct transfer of momentum. However, the cross section σ_{d} for such knock-on processes is smaller by orders of magnitude than the cross sections σ_{el} and σ_{inel} (Sect. 5.1.4 and 5.2.2) for elastic and inelastic scattering (Fig. 5.3). The displacement cross section σ_{d} for an energy transfer $\Delta E = E'_{\mathrm{n}} \geq E_{\mathrm{d}}$ can be calculated by first determining the minimum scattering angle $\theta_{\min}$ for which $E'_{\mathrm{n}} = E_{\mathrm{d}}$ using (5.14). We expect $E'_{\mathrm{n}} \geq E_{\mathrm{d}}$ for all $\theta \geq \theta_{\min}$ and

$$\sigma_{\mathrm{d}} = \int_{\theta_{\min}}^{\pi} \frac{\mathrm{d}\sigma_{\mathrm{el}}}{\mathrm{d}\Omega} 2\pi \sin\theta \mathrm{d}\theta, \tag{5.15}$$

in which we use the differential elastic cross section for large-angle scattering and relativistic energies. The threshold primary electron energy E_{c} for transfer of the minimum energy E_{d} to a nucleus can be obtained by setting $E'_{\mathrm{n}} = E_{\mathrm{d}}$ and $\theta = 180°$ in (5.14).

5.1.3 Elastic Differential Cross Section for Small-Angle Scattering

The elastic cross section $\mathrm{d}\sigma_{\mathrm{el}}/\mathrm{d}\Omega$ or the scattering amplitude $f(\theta)$ can be calculated from the Schrödinger equation (3.21). The asymptotic solution far from the nucleus can be represented by a plane, unscattered wave and a scattered, spherical wave (5.1) with an amplitude $f(\theta)$ depending on the scattering angle θ (Fig. 5.1b),

$$\psi_{\mathrm{s}} = \psi_0 \left[\exp(2\pi \mathrm{i} kz) + f(\theta) \frac{\mathrm{e}^{2\pi \mathrm{i} kr}}{r} \right]. \tag{5.16}$$

The scattering amplitude

$$f(\theta) = |f(\theta)| \mathrm{e}^{\mathrm{i}\eta(\theta)} \tag{5.17}$$

is complex.

For scattering angles $\theta \leq 10°$, which are important for TEM, the scattering amplitude $f(\theta)$ can be calculated by the so-called WKB method (Wentzel, Kramer, Brillouin) in the small-angle approximation of Molière [5.2], also associated with the name of Glauber [5.3], and by the Born approximation. The latter only gives real values of $f(\theta)$ and fails for atoms of high atomic number. An exact solution of the Schrödinger equation (3.22) resulting in complex scattering amplitudes can be obtained by the partial-wave analysis. These three methods will be described in the following.

WKB Method. The scattering amplitude $f(\theta)$ is the amplitude of the spherical wave far from the scattering event (Fig. 5.1b) and is therefore identical with the diffraction amplitude in Fraunhofer diffraction. We need to calculate the wavefront behind the atom (Fig. 5.4) in the form (3.36). We can assume that $a_{\mathrm{s}}(\boldsymbol{r}) = 1$ because there is no absorption of electrons, and the phase shift $\varphi_{\mathrm{s}}(\boldsymbol{r})$ can be obtained from the optical path difference Δs relative to the wavefront in vacuum,

$$\varphi_{\mathrm{s}}(\boldsymbol{r}) = \frac{2\pi}{\lambda} \int_{-\infty}^{+\infty} [n(\boldsymbol{r}) - 1] \mathrm{d}z \approx -\frac{2\pi}{\lambda E} \frac{E + E_0}{E + 2E_0} \int_{-\infty}^{+\infty} V(r) \mathrm{d}z. \tag{5.18}$$

Equation (3.17) has been used for the electron-optical refractive index $n(\boldsymbol{r})$; this contains the potential energy, which involves not only the Coulomb

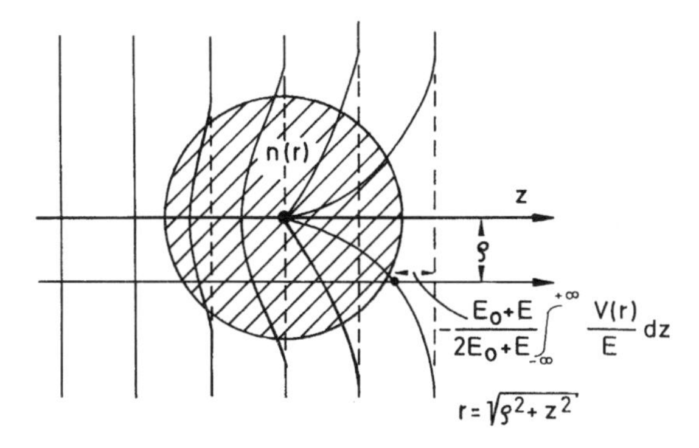

Fig. 5.4. Phase shift φ of a plane incident wavefront passing the Coulomb potential $V(r)$ of an atom.

potential of the nucleus but also that of the atomic electrons. The latter cause a screening of the nuclear charge $+Ze$. The charge distribution $\rho(\boldsymbol{r}_j)$ inside an atom can be described by a δ-function at the nucleus ($r = 0$) together with the charge density $-e\rho_e(\boldsymbol{r}_j)$ of the electron cloud with probability density $\rho_e(\boldsymbol{r}_j)$; ψ_{0s} denotes the wave amplitudes of the Z atomic electrons at the position $\boldsymbol{r}_j$,

$$e\rho(\boldsymbol{r}_j) = eZ\delta(0) - e\rho_e(\boldsymbol{r}_j) = eZ\delta(0) - e\sum_{s=1}^{Z}\psi_{0s}(\boldsymbol{r}_j)\psi_{0s}^*(\boldsymbol{r}_j). \tag{5.19}$$

The probability density $\rho_e(\boldsymbol{r}_j)$ can be calculated from the Thomas–Fermi model or by the Hartree–Fock method. An element of volume $\mathrm{d}^3\boldsymbol{r}_j$ at a distance $\boldsymbol{r}_j$ from the nucleus contributes $-e^2\rho(\boldsymbol{r}_j)(4\pi\epsilon_0|\boldsymbol{r}_i - \boldsymbol{r}_j|)^{-1}$ to the Coulomb energy of a beam electron at a distance $\boldsymbol{r}_i$. The total Coulomb energy becomes

$$V(\boldsymbol{r}_i) = -\frac{e^2}{4\pi\epsilon_0}\int\frac{\rho(\boldsymbol{r}_j)}{|\boldsymbol{r}_i - \boldsymbol{r}_j|}\mathrm{d}^3\boldsymbol{r}_j\ . \tag{5.20}$$

If $\rho_e(\boldsymbol{r}_j)$ is assumed to be rotationally symmetric (5.20), $V(r_i)$ can be derived from the elementary law of electrostatics

$$\boldsymbol{E}(r_i) = -\frac{Q(r_i)}{4\pi\epsilon_0 r_i^2}\boldsymbol{e}_r, \tag{5.21}$$

where Q_{r_i} is the charge inside a sphere of radius r_i. It is advantageous in the calculation that follows to approximate the screening action in (5.21) in various ways:

1. One exponential term (Wentzel atom model)

$$V(r) = -\frac{e^2 Z}{4\pi\epsilon_0 r}\mathrm{e}^{-r/R} \quad \text{with} \quad R = a_H Z^{-1/3} \tag{5.22}$$

$a_H = \epsilon_0 h^2/\pi m_0 e^2 = 0.0529$ nm is the Bohr radius.

2. A sum of exponentials [5.4, 5.5, 5.6]

$$V(r) = -\frac{e^2 Z}{4\pi\epsilon_0 r}\sum_{i=1}^{k} b_i \exp(-a_i r), \quad \sum_{i=1}^{k} b_i = 1. \tag{5.23}$$

To distinguish between the unscattered and the scattered parts of the wave, it is useful to rewrite (3.36) in the form

$$\psi_s(\boldsymbol{r}) = \psi_0 + \psi_0\{\exp[i\varphi_s(\boldsymbol{r}] - 1\}, \tag{5.24}$$

where the first term describes the unscattered part of the wave. From the second term one obtains the scattering amplitude [3.1, 3.2]

$$\begin{aligned} f(\theta) &= -ik\int\{\exp[i\varphi_s(\boldsymbol{r})] - 1\}\, e^{2\pi i\boldsymbol{q}\cdot\boldsymbol{r}} d^2\boldsymbol{r} \\ &= -ik\int\{\exp[i\varphi_s(\boldsymbol{r})] - 1\}e^{2\pi iqr\cos\chi} r dr d\chi \\ &= -2\pi ik\int_0^\infty\{\exp[i\varphi_s(r)] - 1\}J_0(2\pi qr) r dr. \end{aligned} \tag{5.25}$$

Here we have used $\boldsymbol{q}\cdot\boldsymbol{r} = qr\cos\chi$ (r, χ are polar coordinates in the specimen plane) and $d^2\boldsymbol{r} = r dr d\chi$. The integral over χ for constant r yields the Bessel function J_0. Figure 5.5 shows calculated scattering amplitudes $f(\theta)$ for C and Pt atoms with the muffin-tin model [5.8]. The value of $f(\theta)$ increases with increasing electron energy for small θ but decreases for large θ. In consequence, the total elastic cross section σ_{el} decreases with increasing energy (see Fig. 5.3 and Sect. 5.1.4). The additional phase shift $\eta(\theta)$ is very much less for C than for Pt. Complex scattering amplitudes have also been reported in [5.7, 5.9, 5.10].

Born Approximation. The Born approximation can be used only for *weak-phase specimens*, for which $\varphi_s \ll 1$ and the Taylor series $\exp[i\varphi_s(r)] = 1 + i\varphi_s(r) + \ldots$ can be truncated after the first two terms. Substituting this in (5.25), we obtain the Born approximation

$$f(\theta) = 2\pi k\int\varphi_s(\boldsymbol{r})e^{2\pi i\boldsymbol{q}\cdot\boldsymbol{r}} d^2\boldsymbol{r}. \tag{5.26}$$

By writing $\varphi_s = (2\pi/\lambda)\int(n-1)dz$ with the refractive index n (3.17), $f(q)$ in (5.26) becomes

$$f(\theta) = -\frac{2\pi}{\lambda^2 E}\frac{E_0 + E}{2E_0 + E}\int V(\boldsymbol{r})e^{2\pi i\boldsymbol{q}\cdot\boldsymbol{r}} d^3\boldsymbol{r}, \tag{5.27}$$

where $F(q)$ has been transformed to $f(\theta)$ by multiplying by the factor λ^{-1}. The factor in front of the integral in (5.27) is written in various ways in the literature, which are connected by the identities

$$\frac{\pi e^2}{\epsilon_0\lambda^2 E}\frac{E_0 + E}{2E_0 + E} \equiv \frac{e^2 m_0(1 + E/E_0)}{4\pi\epsilon_0\hbar^2} \equiv \frac{1 + E/E_0}{a_H}, \tag{5.28}$$

where $a_H = 0.0529$ nm is the Bohr radius.

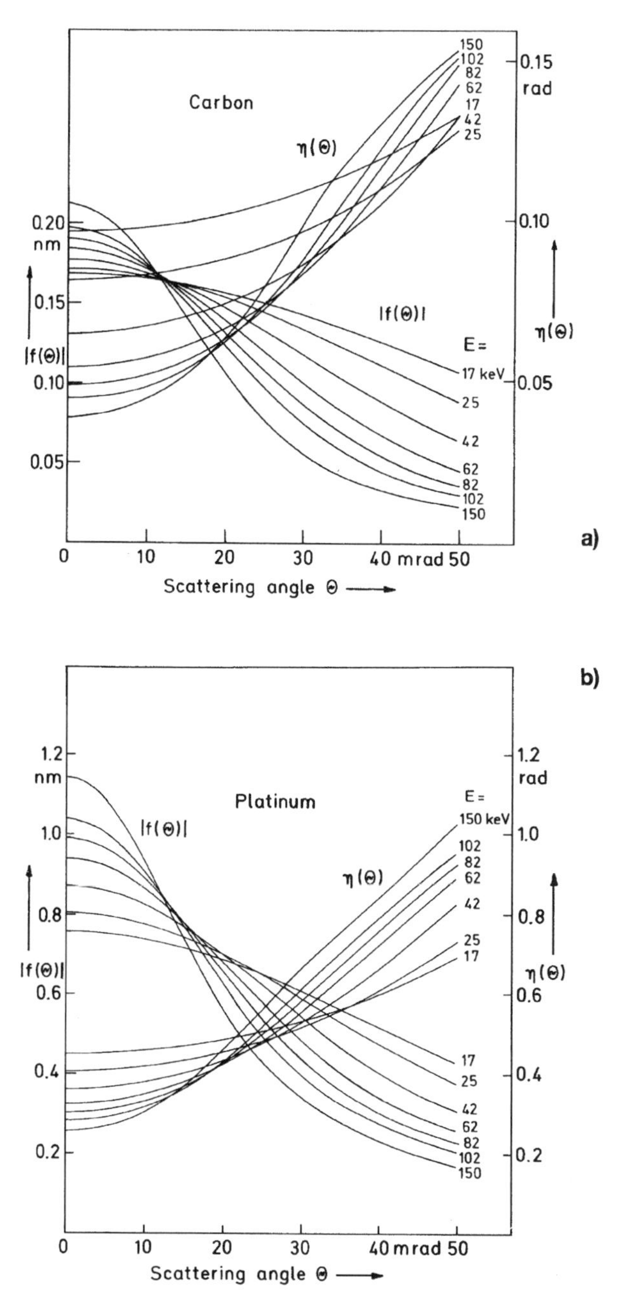

Fig. 5.5. Values of the scattering amplitude $|f(\theta)|$ and phase shift $\eta(\theta)$ of the complex scattering amplitude (5.17) calculated by the WKB method using a muffin-tin model for (**a**) carbon and (**b**) platinum.

The scattering amplitude for atoms $f(\theta)$ is a real quantity in the Born approximation, and the additional phase shift $\eta(\theta)$ is zero. The Born approximation therefore cannot be used for atoms of high atomic number because these are never "weak-phase objects". The difference between the WKB method and the Born approximation is shown in Fig. 6.3 for the total elastic cross section σ_{el} or $x_{\text{el}} = A/(N_A \sigma_{\text{el}})$ (see the discussion in Sect. 6.1.1). The Born approximation has the advantages that an analytical solution can be obtained for simple potential models and that the dependence of $f(\theta)$ on the various parameters can be comprehended more readily. Substitution of (5.20) in (5.27), with the coordinate $\boldsymbol{r}_j$ describing the atomic charge density and $\boldsymbol{r}_i$ the beam electrons, gives

$$f(\theta) = \frac{2\pi}{\lambda^2 E} \frac{E_0 + E}{2E_0 + E} \frac{e^2}{4\pi\epsilon_0} \cdot \underbrace{\int_0^\infty \rho(r_j) \mathrm{e}^{-2\pi \mathrm{i} \boldsymbol{q}\cdot\boldsymbol{r}_j} \mathrm{d}^3\boldsymbol{r}_j}_{Z - f_{\mathrm{x}}} \cdot \underbrace{\int_0^\infty \frac{\exp[-2\pi \mathrm{i}\boldsymbol{q}\cdot(\boldsymbol{r}_i - \boldsymbol{r}_j)]}{|\boldsymbol{r}_i - \boldsymbol{r}_j|} \mathrm{d}^3\boldsymbol{r}_i}_{1/\pi q^2}, \tag{5.29}$$

where f_{x} is the scattering amplitude for x-rays. This quantity is dimensionless, whereas $f(\theta)$ has the dimension of a length. This may be written

$$f(\theta) = \frac{\lambda^2(1 + E/E_0)}{8\pi^2 a_H}(Z - f_{\mathrm{x}})\frac{1}{\sin^2(\theta/2)}, \tag{5.30}$$

in which we have used (5.28) and (3.39). The differential cross section for large-angle scattering can be obtained by setting $f_{\mathrm{x}} = 0$ in (5.30). This yields the Rutherford cross section

$$\begin{aligned}\frac{\mathrm{d}\sigma_{\mathrm{R}}}{\mathrm{d}\Omega} &= \left(\frac{Ze^2}{8\pi\epsilon_0 E}\right)^2 \left(\frac{E_0 + E}{2E_0 + E}\right)^2 \frac{1}{\sin^4(\theta/2)} \\ &= \left(\frac{Ze^2}{8\pi\epsilon_0 m v^2}\right)^2 \operatorname{cosec}^4\frac{\theta}{2}.\end{aligned} \tag{5.31}$$

Writing $\sin(\theta/2) \simeq \theta/2$, we obtain the small-angle elastic cross section

$$\frac{\mathrm{d}\sigma_{\mathrm{el}}}{\mathrm{d}\Omega} = |f(\theta)|^2 = \frac{\lambda^4(1 + E/E_0)^2}{4\pi^4 a_H^2} \frac{(Z - f_{\mathrm{x}})^2}{\theta^4}. \tag{5.32}$$

This unscreened cross section and the Rutherford cross section (5.31) have a singularity at $\theta = 0$. However, the numerator also goes to zero as $\theta \to 0$ because f_{x} tends to $Z = \int \rho_{\mathrm{e}}(r_j)\mathrm{d}^3\boldsymbol{r}_j$ as θ or q tends to zero; $f(0)$ therefore takes a finite value, which is sensitive to the choice of the screening model.

This influence of screening on $f(0)$ can be better understood if we substitute (5.22) for $V(r)$ in (5.27) and consider small scattering angles for which $\sin\theta \simeq \theta$ [5.11, 5.12]. We have

$$f_{\mathrm{x}} = \frac{Z}{1 + 4\pi^2 q^2 R^2}. \tag{5.33}$$

Substitution in (5.30) gives

$$\frac{d\sigma_{el}}{d\Omega} = \frac{4Z^2R^4(1+E/E_0)^2}{a_H^2}\frac{1}{[1+(\theta/\theta_0)^2]^2}$$
$$\text{with } \theta_0 = \frac{\lambda}{2\pi R};\ R = a_H Z^{-1/3}. \tag{5.34}$$

At the *characteristic angle* θ_0, the differential cross section (5.34) falls to a quarter of the value at $\theta = 0$. Calculations of the elastic differential cross section or $f(\theta)$ in the Born approximation using electron-density distributions $\rho_e(r_j)$ given by relativistic Hartree–Fock calculations have been published in [5.13, 5.14, 5.15, 5.16].

Differential scattering cross sections can only be measured for gas targets if the concentration of atoms is so low that multiple scattering does not occur. Figure 5.6 shows measurements of the elastic and inelastic differential cross sections of argon atoms [5.17] that confirm the dependence (5.34) on the scattering angle θ predicted by the Wentzel model (5.22). The angular intensity distribution resulting from scattering in thin films is influenced by multiple scattering even when the films are very thin (Sect. 5.3.1). The results can be compared with calculated values of $d\sigma/d\Omega$ only after applying a deconvolution procedure to the measured data; alternatively, the transmission $T(\alpha)$ through thin films into a cone of aperture α (Sect. 6.1.1) may be compared with calculated partial cross sections $\sigma(\alpha)$ as defined in (6.1).

Partial Wave Analysis. For a spherically symmetric potential $V(r)$, the scattering amplitude can be expanded in an infinite series of Legendre polynomials,

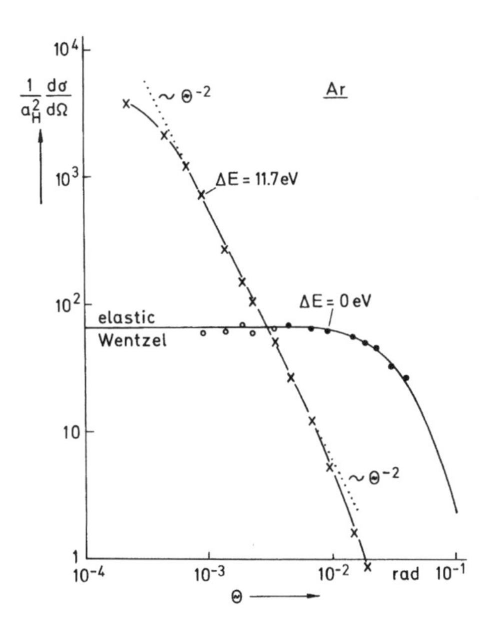

Fig. 5.6. Angular dependence of elastic and inelastic differential cross sections $d\sigma/d\Omega$ (resonance energy loss at $\Delta E = 11.7$ eV) for 25 keV electrons scattered at an argon gas target [5.17].

$$f(\theta) = \frac{1}{4\mathrm{i}\pi k}\sum_{l=0}^{\infty}(2l+1)[\exp(2\mathrm{i}\delta_l)-1]\mathrm{P}_l(\cos\theta), \tag{5.35}$$

where l denotes the quantum number of angular momentum. The phase shifts δ_l are positive for an attractive Coulomb potential, and these phase shifts can be calculated from [3.1, 3.2, 3.3, 5.1]

$$\sin\delta_l = -\int_0^{\infty}\mathrm{j}_{l+1/2}(2\pi kr)\frac{2m}{\hbar^2}V(r)u_l(r)\mathrm{d}r, \tag{5.36}$$

where the $j_{l+1/2}$ are spherical Bessel functions and the $u_l(r)$ are solutions of the radial Schrödinger equation

$$\frac{\mathrm{d}^2u_l}{\mathrm{d}r^2}+\left[4\pi^2k^2-\frac{2m}{\hbar^2}V(r)-\frac{l(l+1)}{r^2}\right]u_l(r)=0. \tag{5.37}$$

Of the order of a hundred partial waves have to be calculated to decrease the error at low scattering angles to less than 1% for 100 keV electrons (about a thousand partial waves are necessary for 1 MeV). A computer program is available that uses the muffin-tin model of $V(r)$ [Fig. 3.3] and can also calculate deviations from Rutherford cross sections (Mott cross sections) at large scattering angles [5.18].

5.1.4 Total Elastic Cross Section

The total elastic cross section σ_{el} can be calculated by applying (5.5) to the differential cross section. Because of the fast decrease of $f(\theta)$ with increasing θ, an exact knowledge of the large-angle scattering distribution is not necessary.

When a complex scattering amplitude (5.17) or (5.35) given by the WKB or partial-wave method is available, substitution of (5.35) in (5.5) using the orthogonality of the Legendre polynomials results in

$$\sigma_{\mathrm{el}} = \frac{1}{\pi k^2}\sum_{l=1}^{\infty}(2l+1)\sin^2\delta_l. \tag{5.38}$$

Comparison with (5.35) gives the *optical theorem* of quantum-mechanical scattering theory

$$\sigma_{\mathrm{el}} = \frac{2}{k}\,\mathrm{Im}\{f(0)\} = 2\lambda|f(0)|\sin\eta(0), \tag{5.39}$$

which shows that the imaginary part of the forward scattering amplitude for $\theta=0$ determines the total cross section.

Substituting the expression $\mathrm{d}\sigma_{\mathrm{el}}/\mathrm{d}\Omega$ given by (5.34) and $R=a_HZ^{-1/3}$ in (5.5), we obtain ($\beta=v/c$)

$$\sigma_{\mathrm{el}} = \frac{Z^2R^2\lambda^2(1+E/E_0)^2}{\pi a_H^2} = \frac{h^2Z^{4/3}}{\pi E_0^2\beta^2}\,. \tag{5.40}$$

The absolute value of σ_{el} or its reciprocal $x_{\mathrm{el}}=A/(N_A\sigma_{\mathrm{el}})$ does not agree well with experiments (Fig. 6.3) because the Wentzel screening model (5.22) is too simple and the Born approximation fails for high Z. However, the cross section σ_{el} is observed to be proportional to β^{-2} for low-Z material, and the predicted saturation for energies larger than 1 MeV is found for all elements (Figs. 5.3 and 6.3). Also, the proportionality with $Z^{4/3}$ is a typical result of the Wentzel model and the Born approximation. Calculations of the total elastic cross section for Hartree–Fock–Slater or Dirac–Slater atoms [5.19] lead to the approximate formula

$$\sigma_{el} = \frac{1.5 \times 10^{-6}}{\beta^2} Z^{3/2} \left(1 - 0.23 \frac{Z}{137\beta}\right) \quad \text{for} \quad \frac{Z}{137\beta} < 1.2, \tag{5.41}$$

where σ_{el} is measured in nm^2.

5.2 Inelastic Scattering

5.2.1 Electron–Specimen Interactions with Energy Loss

Whereas an elastic collision preserves kinetic energy and momentum, an inelastic collision conserves the total energy and momentum, a part of the kinetic energy being converted to atom–electron excitation. The primary electron is observed to lose energy even at small scattering angles. The following excitation mechanisms may be distinguished:

1. Excitation of oscillations in molecules [5.20] and phonon excitations in solids [5.21, 5.22]. These energy losses (Fig. 5.7) are of the order of 20

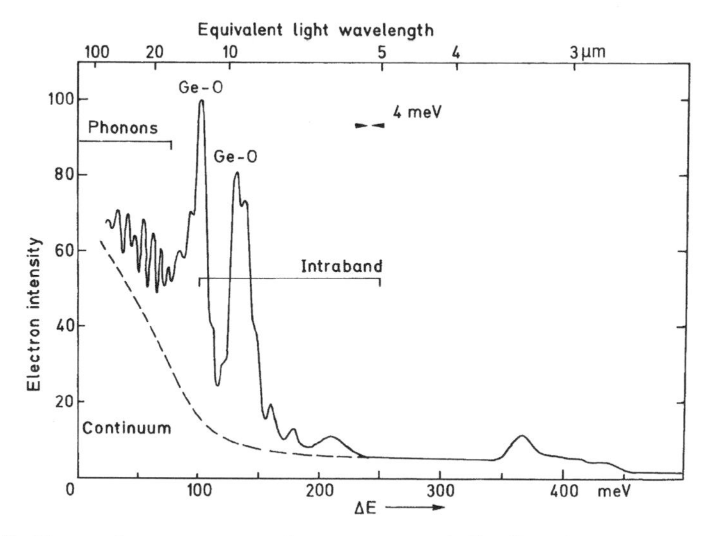

Fig. 5.7. Energy-loss spectrum of an evaporated Ge film due to phonon excitation, excitation of the GeO bonding and intraband transitions for energy losses $\Delta E \leq 500$ meV [5.22]

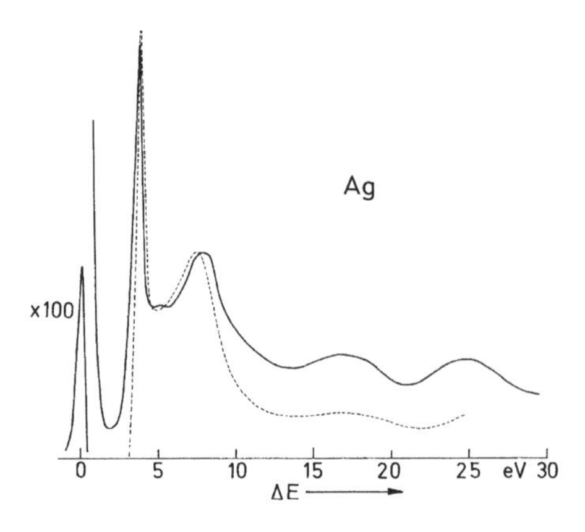

Fig. 5.8. Energy-loss spectrum of an Ag film and comparison with the dielectric theory (dotted line) [5.23].

meV–1 eV and can be observed only after monochromatization of the primary electron beam, which has an energy width of the order of 1 eV when a thermionic electron gun is used. These interaction processes are also excited by the infrared part of the electromagnetic spectrum. The observed energy losses would be of considerable interest for molecular and solid-state physics but, owing to the low intensity of a monochromatized beam, high spatial resolution is scarcely possible. At the moment these processes are therefore of little interest in electron microscopy.

2. Intra- and interband excitation of the outer atomic electrons and excitation of collective oscillations (plasmons) of the valence and conduction electrons (Sect. 5.2.4). Most of the plasma losses show relatively broad maxima in the energy-loss range of $\Delta E = 3$–25 eV (Figs. 5.8 and 5.20). The plasmon losses depend on the concentration of valence and conduction electrons and are influenced by chemical bonds and the electron-band structure in alloys. There are analogies with optical excitations in the visible and ultraviolet.
3. Ionization of core electrons in inner atomic shells (Sect. 5.3). Atomic electrons can be excited from an inner shell ($I = K, L, M, \ldots$) of ionization energy E_{I} to an unoccupied energy state above the Fermi level; such a transition needs an energy transfer (energy loss) $\geq E_{\mathrm{I}}$. The energy-loss spectrum $\mathrm{d}\sigma/\mathrm{d}E$ shows a steep increase for energy losses $\Delta E \geq E_{\mathrm{I}}$ (Fig. 5.9). A structure is observed in the loss spectrum a few eV beyond $\Delta E = E_{\mathrm{I}}$ caused by excitation into higher bound states. In organic molecules, the fine structure of the loss spectrum depends on the molecular structure [5.25]. This type of inelastic scattering is also concentrated within relatively small scattering angles $\theta < \theta_{\mathrm{E}} = E_{\mathrm{I}}/2E$, though part of the inelastic scattering extends to larger scattering angles. Energy-loss spectroscopy is therefore the best method for analyzing elements of low atomic number (e.g., C, N, O) in thin films with thicknesses smaller than the mean free path for inelastic scattering. When the electron gap in the inner shell is

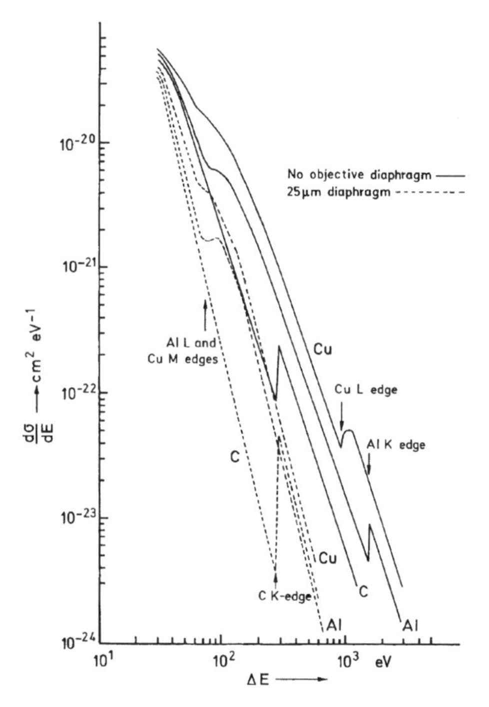

Fig. 5.9. Differential cross section $d\sigma/dE$ measured on 15 nm films of C, Al and Cu with 80 keV electrons (full curve: without an aperture diaphragm, dotted curve: using an objective aperture $\alpha_o = 4.5$ mrad). The arrows indicate the positions of different shell energies [5.24].

filled by an electron from an outer shell, the excess energy is emitted as an x-ray quantum or transferred to another atomic electron, which is emitted as an Auger electron (Sect. 10.1).

The inelastic interactions therefore form the basis for several analytical methods in electron microscopy (Chap. 10). However, low-loss inelastic scattering is not favorable for high resolution because the inelastic-scattering process is then less localized than the elastic one. An electron can be inelastically scattered even when passing the atom at a distance of a few tenths of a nanometer. This is also illustrated by the fact that inelastic scattering is concentrated into smaller scattering angles than elastic scattering (Fig. 5.6). In order to resolve a specimen periodicity of Λ, scattering amplitudes are needed out to an angle $\theta = \lambda/\Lambda$; there are too few inelastically scattered electrons at large θ for the imaging of small spacings Λ or high spatial frequencies $q = 1/\Lambda$.

The total inelastic cross section σ_{inel} is larger than the elastic cross section σ_{el} for elements of low atomic number and smaller for high Z (Sect. 5.2.3). The energy losses in thick specimens decrease the resolution as a result of the chromatic aberration.

The largest part of the excitation energy is converted to heat (phonons) (Sect. 11.1). Excitations and ionizations in organic specimens cause bond ruptures and irreversible radiation damage (Sect. 11.2). Color centers and other point defects and clusters are generated in ionic crystals (Sect. 11.3).

Inelastic scattering can be described by a double-differential cross section $\mathrm{d}^2\sigma_{\mathrm{inel}}/\mathrm{d}(\Delta E)\mathrm{d}\Omega$, depending on the scattering angle θ and the energy loss ΔE. It becomes difficult to establish accurately such a two-dimensional cross section from theory or experiment for several reasons: the complexity of the energy-loss spectrum, its dependence on foil thickness (e.g., excitation of surface plasmon losses), and the occurrence of multiple elastic and inelastic scattering.

5.2.2 Differential Cross Section for Single-Electron Excitation

During an inelastic scattering event, an excitation energy $\Delta E = E_n - E_0$ may be transferred to an electron of the atom that is excited from the ground state (0) with energy E_0 and wave function a_{0s} ($s = 1, \ldots, Z$) to the excited state (n) with energy E_n and wave function a_{ns}. For small scattering angles, selection rules, such as $\Delta l = \pm 1$, govern the allowed excitations, similar to those for optical excitation.

The total energy and the momentum after the collision remain the same as before. We introduce the scattering vector $\boldsymbol{q}' = \boldsymbol{k}_n - \boldsymbol{k}_0$ (Fig. 5.10) with $|\boldsymbol{k}_n| < |\boldsymbol{k}_0|$ instead of $\boldsymbol{q} = \boldsymbol{k} - \boldsymbol{k}_0$ with equal magnitudes of $\boldsymbol{k}$ and $\boldsymbol{k}_0$ as used for elastic scattering. Using the relations $p = hk$ and $E = p^2/2m$ (nonrelativistic), the conservation of momentum

$$k_n^2 = k_0^2 + q'^2 - 2k_0 q' \cos\eta, \tag{5.42}$$

and the conservation of energy

$$\Delta E = \frac{h^2}{2m}(k_0^2 - k_n^2), \tag{5.43}$$

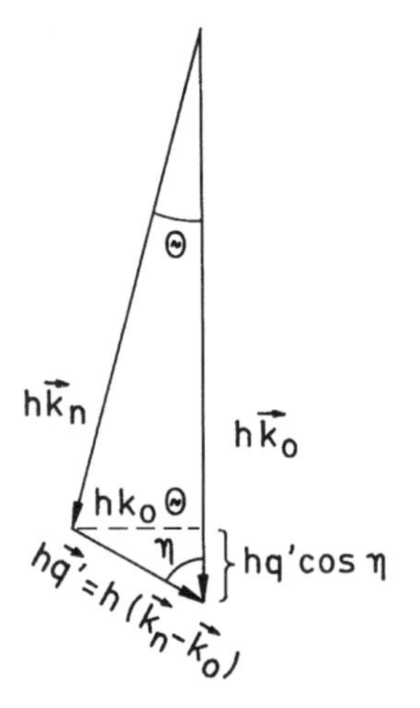

Fig. 5.10. Conservation of momentum in inelastic scattering.

we obtain the relation

$$\Delta E = \frac{h^2 k_0 q'}{m} \cos \eta. \tag{5.44}$$

Furthermore, from Fig. 5.10, the relation

$$q'^2 = (k_0\theta)^2 + (q' \cos \eta)^2 = k_0^2(\theta^2 + \theta_{\mathrm{E}}^2) \tag{5.45}$$

can be read off, in which $\theta_{\mathrm{E}} = \Delta E/2E$ or in relativistic form

$$\theta_{\mathrm{E}} = \frac{\Delta E}{E} \frac{E + E_0}{E + 2E_0} = \frac{\Delta E}{mv^2} \,. \tag{5.46}$$

Quantum-mechanical scattering theory tells us that the inelastic cross section can be calculated [5.26, 5.27] by the *golden rule*,

$$\frac{\mathrm{d}\sigma_{0n}}{\mathrm{d}\Omega} = \frac{4\pi^2 m^2}{h^4} \underbrace{\frac{k_n}{k_0}}_{\simeq 1} |\langle \psi_n | V(r) | \psi_0 \rangle|^2. \tag{5.47}$$

The wave functions $\psi_0 = a_{0s} \exp(2\pi \mathrm{i} \boldsymbol{k}_0 \cdot \boldsymbol{r}_i)$ and $\psi_n = a_{ns} \exp(2\pi \mathrm{i} \boldsymbol{k}_n \cdot \boldsymbol{r}_i)$ are products of the plane incident and scattered waves with wave vectors $\boldsymbol{k}_0$ and $\boldsymbol{k}_n$, respectively, and the atomic wave functions a_{0s} and a_{ns}; the bracketed expression in (5.47) thus becomes ($\boldsymbol{r}_j$: coordinate of atomic electrons)

$$|\langle \psi_n | V(r) | \psi_0 \rangle|^2 =$$
$$\left| \int\!\!\int \exp(-2\pi \mathrm{i} \boldsymbol{k}_n \cdot \boldsymbol{r}_i) a_{ns}^*(\boldsymbol{r}_j) V(\boldsymbol{r}_i, \boldsymbol{r}_j) a_{0s}(\boldsymbol{r}_j) \exp(+2\pi \mathrm{i} \boldsymbol{k}_0 \cdot \boldsymbol{r}_i) \mathrm{d}^3\boldsymbol{r}_i \mathrm{d}^3\boldsymbol{r}_j \right|^2 . \tag{5.48}$$

On substituting the Coulomb interaction potential

$$V(r_i) = -\frac{e^2 Z}{4\pi\epsilon_0 r_i} + \sum_{j=1}^{Z} \frac{e^2}{4\pi\epsilon_0 |\boldsymbol{r}_i - \boldsymbol{r}_j|} \tag{5.49}$$

into (5.48), the first term representing the Coulomb potential of the nucleus cancels because of the orthogonality of the atomic wave functions: $\int a_{ns} a_{0s}^* \mathrm{d}^3\boldsymbol{r}_j = \delta_{n0}$. The two exponential functions in (5.48) representing the incident and scattered waves can be combined to form $\exp(-2\pi i \boldsymbol{q}' \cdot \boldsymbol{r}_i)$. Making use of the last integral in (5.29), we obtain [5.26]

$$\frac{\mathrm{d}\sigma_{0n}}{\mathrm{d}\Omega} = \left(\frac{me^2}{2\pi\epsilon_0 h^2} \right)^2 \frac{|\epsilon(q')|^2}{q'^4} \tag{5.50}$$

with the atomic matrix element

$$\begin{aligned} |\epsilon(q')|^2 &= \left| \int a_{ns}^* \exp(-2\pi \mathrm{i} \boldsymbol{q}' \cdot \boldsymbol{r}_j) a_{0s} \mathrm{d}^3\boldsymbol{r}_j \right|^2 \\ &= \left| \int a_{ns}^* (1 - 2\pi \mathrm{i} \boldsymbol{q}' \cdot \boldsymbol{r}_j + \ldots) a_{0s} \mathrm{d}^3\boldsymbol{r}_j \right|^2 \\ &\approx 4\pi^2 q'^2 |\langle a_{ns} | \boldsymbol{u} \cdot \boldsymbol{r}_j | a_{0s} \rangle|^2 = 4\pi^2 q'^2 |x_{0n}|^2. \end{aligned} \tag{5.51}$$

The expansion of the exponential term has been truncated because of the small scattering angles of interest. The term with unity in the round bracket cancels because of the orthogonality of the atomic wave functions. The unit vector $\boldsymbol{u}$ is parallel to $\boldsymbol{q}'$. For small q', the atomic matrix element is determined by the dipole matrix element. This dipole approximation has been introduced by Bethe [5.26]. Substitution of (5.45) and (5.51) in (5.50) results in

$$\frac{\mathrm{d}\sigma_{0n}}{\mathrm{d}\Omega} = \frac{\lambda^2}{\pi^2 a_\mathrm{H}^2}|x_{0n}|^2 \frac{1}{\theta^2+\theta_\mathrm{E}^2} \; . \tag{5.52}$$

The characteristic angle θ_E (5.46) for inelastic scattering, typically 0.1 mrad for $\Delta E = 20$ eV and $E = 100$ keV, is much smaller than the characteristic angle θ_0 (5.34) for elastic scattering. Inelastic scattering is therefore concentrated within much smaller scattering angles than elastic scattering, though $(\theta^2+\theta_\mathrm{E}^2)^{-1}$ also has a long tail for larger θ. Figure 5.6 shows the decrease as θ^{-2} resulting from (5.52) when $\theta \gg \theta_\mathrm{E}$ in the case of the $\Delta E = 11.7$ eV loss of argon. (In a gas target, the influence of multiple scattering can be kept low enough.)

When a generalized oscillator strength (GOS)

$$f_{0n}(q') = \frac{2m\Delta E}{h^2}\frac{|\epsilon(q')|^2}{q'^2} = \frac{8\pi^2 m\Delta E}{h^2}|x_{0n}|^2 \tag{5.53}$$

is introduced, (5.50) and (5.52) can be written

$$\frac{\mathrm{d}\sigma_{0n}}{\mathrm{d}\Omega} = \frac{e^4}{(4\pi\epsilon_0)^2 E\Delta E}\frac{f_{0n}(q')}{\theta^2+\theta_\mathrm{E}^2} \; . \tag{5.54}$$

Whereas (5.50), (5.52), and (5.54) describe the transition between the electronic states 0 and n with a discrete energy loss ΔE, the final states form a continuum in the case of ionization. With a GOS per unit energy loss $\mathrm{d}f_{0n}(q',\Delta E)/\mathrm{d}(\Delta E)$, (5.54) becomes the double differential cross section

$$\frac{\mathrm{d}^2\sigma}{\mathrm{d}\Omega\,\mathrm{d}\Delta E} = \frac{e^4}{(4\pi\epsilon_0)^2 E\Delta E}\frac{1}{\theta^2+\theta_\mathrm{E}^2}\frac{\mathrm{d}f_{0n}(q',\Delta E)}{\mathrm{d}\Delta E} \; . \tag{5.55}$$

The GOS is identical with the optical oscillator strength for $q' \to 0$, which means that electron energy-loss spectra and the absorption spectra of light and x-ray quanta are related. The GOS satisfies Bethe's sum rule:

$$\sum_n f_{0n} = Z \quad \text{or} \quad \int \frac{\mathrm{d}f_{0n}}{\mathrm{d}\Delta E}\mathrm{d}\Delta E = Z. \tag{5.56}$$

5.2.3 Bethe Surface and Compton Scattering

A plot of $\mathrm{d}f_{0n}/\mathrm{d}(\Delta E)$ is called the Bethe surface [5.27]. Figure 5.11 shows the Bethe surface for the ionization of hydrogen or for K-shell ionization when the atomic electrons have hydrogen-like wave functions. The calculation uses hydrogen wave functions a_{0s} and plane waves a_{ns} for the ejected electron

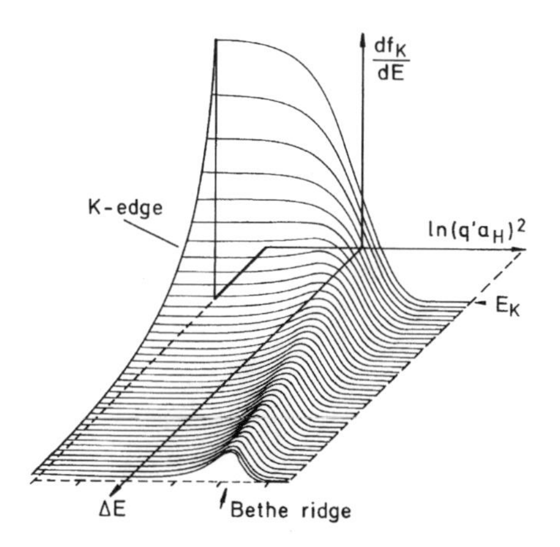

Fig. 5.11. Generalized oscillator strength (GOS) per unit energy loss as a function of scattering parameter q' and energy loss $\Delta E/E_K$ for K-shell ionization (E_K = ionization energy) with the Bethe ridge [5.27].

[5.26, 5.27]. At $\Delta E = E_K$ and $q' \to 0$, the GOS jumps to a maximum and then decreases with increasing ΔE. This can be observed experimentally either as the K-edge of the energy-loss spectrum (Fig. 5.9 and Sect. 5.3.1) or as a jump of the x-ray absorption coefficient when the quantum energy $E_x = h\nu$ exceeds E_K (Fig. 10.6). The GOS decreases with increasing q' for constant ΔE, but shows a "Bethe ridge" at larger q' (Fig. 5.11). This is caused by direct electron–electron impact. When the energy of the ejected electron is much larger than the ionization energy (weakly bound electron), the collision can be treated by classical mechanics, making use of the conservation of energy and momentum. By analogy with x-ray scattering, this process is therefore also called Compton scattering. When an electron at rest is hit by an electron of energy E, the scattering angle θ_C is strongly related to the energy loss by

$$\sin^2\theta_C = \frac{\Delta E}{E}\left[1 + \frac{E - \Delta E}{E_0}\right]^{-1} \simeq \frac{\Delta E}{E} . \tag{5.57}$$

This angle corresponds to the maximum of the Bethe ridge in Fig. 5.11 and increases with the square root of ΔE. The width of the Bethe ridge is caused by the momentum distribution of the atomic electrons on their orbits or of the valence electrons at the Fermi level [5.28]. The GOS of Fig. 5.11 has to be divided by $\Delta E(\theta^2 + \theta_E^2)$ (5.55) to become proportional to $d^2\sigma/d\Omega d(\Delta E)$.

The Bethe surface can be imaged directly with an electron spectroscopic diffraction (ESD) pattern. For example carbon film shows a diffuse ring at the Compton angle θ_C at high energy losses shown for increasing ΔE in Fig. 5.12a–d. Angle-resolved EELS (Sect. 4.6.4) is another mode that provides an image of the Bethe surface. With the diffraction pattern of an amorphous carbon film at the filter entrance plane (Fig. 4.26) and a narrow slit in this plane, the energy-dispersive plane (EDP) contains a superposition of lines across the patterns (variable θ) at different energy losses ΔE perpendicular to the θ-axis. The recorded pattern is digitized and the

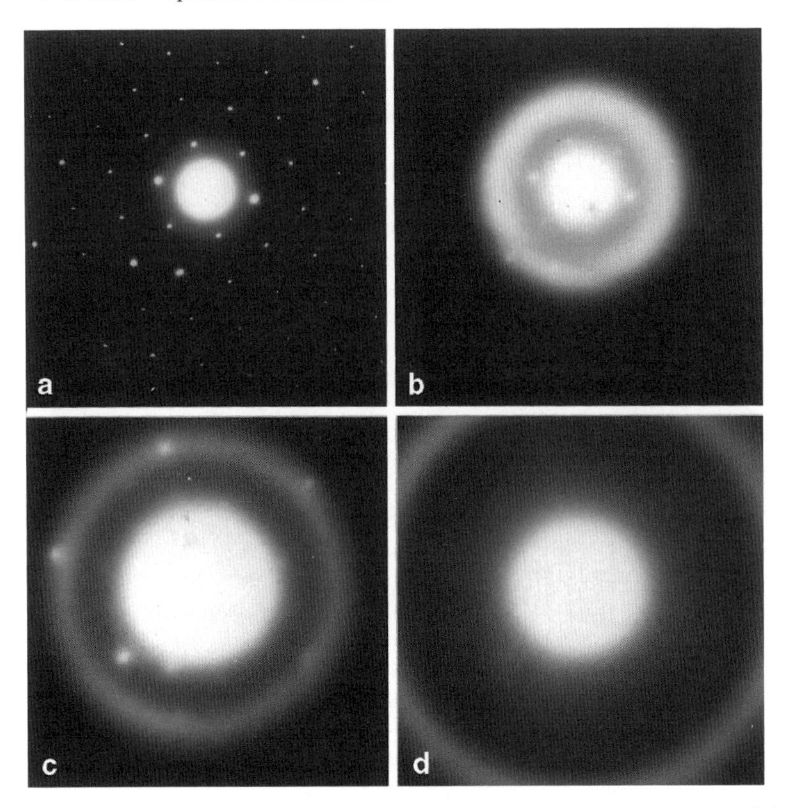

Fig. 5.12. Electron spectroscopic diffraction pattern of a graphite foil at (**a**) $\Delta E = 0$ eV, (**b**) 200 eV, (**c**) 400 eV, and (**d**) 800 eV showing a diffuse ring (Bethe ridge) caused by Compton scattering, which increases in diameter proportional to the square root of the energy loss ΔE.

intensity distribution emphasized by isodensities, which clearly show the plasmon losses, the parabolic Bethe ridge, and the carbon K edge (Fig. 5.13) [5.29].

The intensity profile of the Bethe ridge is proportional to the momentum distribution of atomic electrons in the scattering direction (z),

$$J(p_z) = \int\int n(p_x, p_y, p_z)\mathrm{d}p_x\mathrm{d}p_y, \tag{5.58}$$

where $n(\boldsymbol{p})$ is the momentum probability distribution. The Fourier transform

$$B(z) = \int J(p_z)\exp(-\mathrm{i}\,p_z z/\hbar)\mathrm{d}p_z \tag{5.59}$$

of a recorded intensity profile (e.g., across the diffuse rings in Fig. 5.12) is the autocorrelation function of the ground-state wave function. The analogous Compton scattering for x-rays [5.30, 5.31] is in common use for testing calculations of atomic orbitals in solids and this method has also been tried for electron diffraction [5.28, 5.32, 5.33, 5.34]. Advantages compared with x-ray Compton scattering are that the exposure time is much shorter and

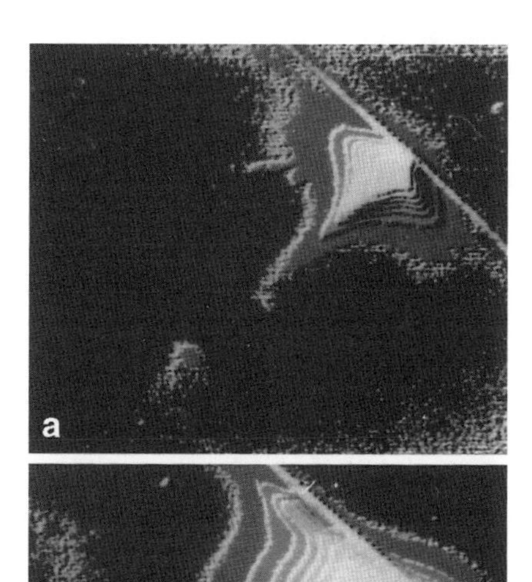

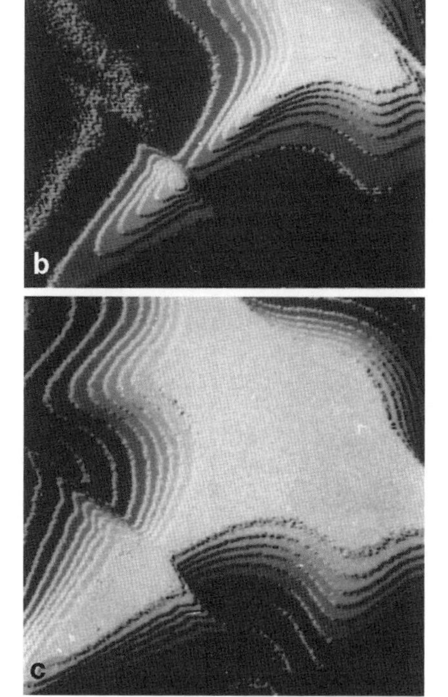

Fig. 5.13. Angle-resolved EELS of a 2.8 μg cm^{-2} carbon film in a θ–ΔE plane with exposures that increase from (**a**) to (**c**) by a factor of 16 at each step. The isodensities clearly show (**a**) the plasmon loss, (**b**) and (**c**) the K-shell ionization edge and the parabolic Bethe ridge.

nanometer-sized specimens can be studied. A disadvantage is that the information is contained in small deviations of the Compton profile from an approximately Gaussian shape. Electron diffraction has a stronger background, and, in crystalline specimens, Bragg diffraction and Kikuchi lines can disturb the profile.

5.2.4 Approximation for the Total Inelastic Cross Section

In Sect. 5.2.2, we discussed inelastic scattering in terms of a single transition $0 \rightarrow n$. In order to calculate the inelastic scattering from a complete atom, the cross sections of all allowed transitions $0 \rightarrow n$ for the $s = 1, \ldots, Z$ atomic electrons have to be summed. The following treatment of inelastic scattering does not consider the details of the energy-loss spectrum discussed in Sects. 5.2.5 and 5.3 and yields only a mean energy-loss value (mean ionization energy J). Nevertheless, this treatment gives a correct description of some important aspects of inelastic scattering, such as the concentration of inelastic scattering within smaller angles than for elastic scattering and the dependence of the ratio of the total inelastic and elastic cross sections on atomic number.

Using (5.50), this summation over n and s results in

$$\frac{\mathrm{d}\sigma_{\mathrm{inel}}}{\mathrm{d}\Omega} = \frac{(1+E/E_0)^2}{4\pi^4 a_{\mathrm{H}}^2 q'^4} \sum_{n\neq 0} \left| \sum_{s=1}^{Z} \int a_{0s} a_{ns}^* \exp(-2\pi \mathrm{i} \boldsymbol{q}' \cdot \boldsymbol{r}_j) \mathrm{d}^3\boldsymbol{r}_j \right|^2 . \tag{5.60}$$

In the inner summation, products of terms containing different values of the suffix $s = 1, \ldots, Z$ (exchange terms) can be neglected. The summations over n and s can thus be interchanged, giving

$$\begin{aligned} &\sum_{s=1}^{Z} \sum_{n\neq 0} \left| \int a_{0s} a_{ns}^* \exp(-2\pi \mathrm{i} \boldsymbol{q}' \cdot \boldsymbol{r}_j) \mathrm{d}^3\boldsymbol{r}_j \right|^2 \\ &= \sum_{s=1}^{Z} [\underbrace{\sum_{n} \left| \int a_{0s} a_{ns}^* \exp(-2\pi \mathrm{i} \boldsymbol{q}' \cdot \boldsymbol{r}_j) \mathrm{d}^3\boldsymbol{r}_j \right|^2}_{+1} - \underbrace{\left| \int a_{0s} a_{0s}^* \exp(-2\pi \mathrm{i} \boldsymbol{q}' \cdot \boldsymbol{r}_j) \mathrm{d}^3\boldsymbol{r}_j \right|^2}_{f_x^2/Z^2}] \\ &= Z - f_x^2/Z. \end{aligned} \tag{5.61}$$

The first term in the brackets is equal to unity, as can be shown using the completeness relation for the wave functions a_{ns}. The last term is the contribution of *one* electron to the x-ray scattering amplitude f_x. The approximation in which this is set equal to f_x/Z is strictly valid only if the electron-density distributions of the atomic electrons are equal (H and He atoms only). When (5.61) is substituted into (5.60) for other atoms, the resulting approximation permits us to obtain analytical formulas using the Wentzel model (5.22). The quantity q' in (5.60) contains the energy loss ΔE. Koppe [5.35] suggested substituting the mean value $\Delta E = J/2$, where J is the mean ionization energy of the atom $\simeq 13.5\, Z$ in eV. Taking f_x from (5.33), this gives [5.11]

$$\frac{\mathrm{d}\sigma_{\mathrm{inel}}}{\mathrm{d}\Omega} = \frac{(1+E/E_0)^2}{4\pi^4 a_{\mathrm{H}}^2 q'^4} Z \left[1 - \frac{1}{(1+4\pi^2 q'^2 R^2)^2} \right], \tag{5.62}$$

or with $q'^2 = (\theta^2 + \theta_{\mathrm{E}}^2)/\lambda^2$,

$$\frac{\mathrm{d}\sigma_{\mathrm{inel}}}{\mathrm{d}\Omega} = \frac{\lambda^4 (1+E/E_0)^2}{4\pi^4 a_{\mathrm{H}}^2} Z \frac{\left\{ 1 - \frac{1}{[1+(\theta^2+\theta_{\mathrm{E}}^2)/\theta_0^2]^2} \right\}}{(\theta^2+\theta_{\mathrm{E}}^2)^2} . \tag{5.63}$$

This formula for the inelastic differential cross section may be compared with its elastic counterpart (5.34). The characteristic angle θ_0, which is responsible for the decrease of the elastic differential cross section $\mathrm{d}\sigma_{\mathrm{el}}/\mathrm{d}\Omega$, is of the order of 10 mrad and the angle θ_{E}, responsible for the decrease of $\mathrm{d}\sigma_{\mathrm{inel}}/\mathrm{d}\Omega$ with increasing θ, of the order of 0.1 mrad. This confirms that inelastic scattering is concentrated within much smaller angles than elastic scattering (Fig. 5.6), though with a long tail for very large scattering angles, $\theta \gg \theta_0$ and $\theta \gg \theta_{\mathrm{E}} = J/4E$. For such large scattering angles, the ratio

$$\frac{\mathrm{d}\sigma_{\mathrm{inel}}/\mathrm{d}\Omega}{\mathrm{d}\sigma_{\mathrm{el}}/\mathrm{d}\Omega} = \frac{1}{Z} \tag{5.64}$$

depends only on the atomic number, whereas for small θ, $\mathrm{d}\sigma_{\mathrm{inel}}/\mathrm{d}\Omega > \mathrm{d}\sigma_{\mathrm{el}}/\mathrm{d}\Omega$ for all elements (Fig. 5.6).

A total inelastic cross section σ_{inel} can be defined in the same way as the elastic one σ_{el}, by using (5.5). Integration of (5.63) gives [5.11]

$$\nu = \frac{\sigma_{\mathrm{inel}}}{\sigma_{\mathrm{el}}} = \frac{4}{Z}\ln\left(\frac{h^2}{\pi m_0 J R\lambda}\right) \simeq \frac{26}{Z}, \tag{5.65}$$

and experimentally [5.36, 5.37] it is found that

$$\nu \simeq \frac{20}{Z}. \tag{5.66}$$

5.2.5 Dielectric Theory and Plasmon Losses in Solids

Only the most important theoretical and experimental results concerning energy losses in solids will be discussed. Extensive reviews have been published [5.23, 5.38, 5.39, 5.40, 5.41, 5.42].

A number of interaction processes have to be considered to explain the characteristic energy losses of a material. An atomic electron can be excited to a higher energy state by an electron–electron collision. Indeed, energy losses are found in scattering experiments on gases that can be explained as the energy differences between spectroscopic terms. Thus, a 7.6 eV loss in Hg vapor corresponds to the optical resonance line [5.44]. The electrons in the outer atomic shells of a solid occupy broad energy bands. Excitations from one band to another (interband excitation) must be distinguished from those inside one band (intraband excitation). Nonvertical interband and intraband transitions can also be observed [5.45, 5.46]. Exact information about the band structure above the Fermi level is not available for most materials, and the energy-loss spectrum is related to the light-optical constants in the visible and ultraviolet spectra by means of the so-called *dielectric theory*. In this theory, plasma oscillations are considered as longitudinal density oscillations of the electron gas [5.47].

The underlying idea of the dielectric theory can be understood in the following manner [5.38, 5.43]. The optical constants of a solid can be described either by a complex refractive index $n + \mathrm{i}\kappa$, where κ is the absorption coefficient, or by a complex permittivity $\epsilon = \epsilon_1 + \mathrm{i}\epsilon_2 = (n + \mathrm{i}\kappa)^2$. In general, the complex permittivity $\epsilon(\boldsymbol{k},\omega)$ is a function of the wave vector $\boldsymbol{k}$ and the frequency ω. Electrons that penetrate into the crystal with velocity $\boldsymbol{v}$ represent a moving point charge and can be described by a δ-function $\rho(\boldsymbol{r},t) = -e\delta(\boldsymbol{r}-\boldsymbol{v}t)$. Fourier-transforming this relation with respect to space and time, we obtain $\rho(\boldsymbol{k},\omega) = -e\delta(\boldsymbol{kv}-\nu)$, where $\omega = 2\pi\nu$. To determine the energy loss of the electron, we have to calculate the electric field acting on it. This is most easily done using Poisson's equation $\epsilon\Delta\Phi(\boldsymbol{r},t) = -\rho(\boldsymbol{r},t)$ in Fourier space $4\pi^2k^2\epsilon(\boldsymbol{k},\omega)\Phi(\boldsymbol{k},\omega) = \rho(\boldsymbol{k},\omega)$. Here we have neglected the relativistic retardation effects. The electric field is given by $\boldsymbol{E}(\boldsymbol{r},t) = -\nabla\Phi(\boldsymbol{r},t)$ or, equivalently, by $\boldsymbol{E}(\boldsymbol{k},\omega) = -2\pi i\boldsymbol{k}\Phi(\boldsymbol{k},\omega)$. The energy loss per unit length is given by

$$-\frac{\mathrm{d}W}{\mathrm{d}z} = -\frac{\mathrm{e}\boldsymbol{Ev}}{v} = e^2\frac{\boldsymbol{v}}{v}\int\frac{i\boldsymbol{k}\delta(\boldsymbol{kv}-\nu)}{2\pi k^2\varepsilon(\boldsymbol{k},\omega)}\exp[-2\pi i(\boldsymbol{kv}-\nu)t]d^3\boldsymbol{k}d\nu. \tag{5.67}$$

Assuming that the electron moves in the z-direction, we integrate over k_z and obtain

$$-\frac{dW}{dz} = \frac{+ie^2}{2\pi v^2}\int\frac{\nu d\nu d^2\boldsymbol{k}_\perp}{k^2\varepsilon(\boldsymbol{k}_\perp,\omega)}, \tag{5.68}$$

where $dk_\perp$ denotes an integration over the coordinates perpendicular to the incident beam direction. As the energy loss is a real quantity, we have to take the real part of (5.68),

$$-\frac{dW}{dz} = \frac{-e^2}{2\pi v^2}\int\mathrm{Im}\left(\frac{1}{\varepsilon(\boldsymbol{k},\omega)}\right)\frac{\nu d\nu d^2\boldsymbol{k}_\perp}{k^2}, \tag{5.69}$$

where we have used

$$\mathrm{Re}\left(\frac{i}{\varepsilon(\boldsymbol{k},\omega)}\right) = -\mathrm{Im}\left(\frac{1}{\varepsilon(\boldsymbol{k},\omega)}\right). \tag{5.70}$$

Equation (5.69) shows that the inelastic scattering is related to the imaginary part of the inverse of the dielectric function. The differential cross section is obtained from (5.55) using the relation [5.48]

$$\frac{df(\Delta E)}{dE} = \frac{-4m\varepsilon_0\Delta E}{e^2h}\mathrm{Im}\left(\frac{1}{\varepsilon(\boldsymbol{k},\omega)}\right), \tag{5.71}$$

$$\begin{aligned}\frac{\mathrm{d}^2\sigma}{\mathrm{d}\Delta E\mathrm{d}\Omega} &= \frac{1}{\pi^2 a_\mathrm{H}E}\frac{E+E_0}{E+2E_0}\frac{\mathrm{Im}\{-1/\epsilon(\Delta E,\theta)\}}{\theta^2+\theta_\mathrm{E}^2} \\ &= \frac{1}{\pi^2 a_\mathrm{H}mv^2}\frac{\mathrm{Im}\{-1/\epsilon\}}{\theta^2+\theta_\mathrm{E}^2}.\end{aligned} \tag{5.72}$$

For a free (f) electron gas with N_e electrons per unit volume, the conduction electrons of a metal or the valence electrons of a semiconductor, for example, the dependence of $\epsilon_\mathrm{f}(\omega)$ on frequency or the energy loss $\Delta E = \hbar\omega$ may be calculated using the Drude model. The alternating electric field $\boldsymbol{E} = \boldsymbol{E}_0\exp(-i\omega t)$ exerts a force on an electron given by Newton's law,

$$m^* \frac{\mathrm{d}^2 \boldsymbol{x}}{\mathrm{d}t^2} + m^* \gamma \frac{\mathrm{d}\boldsymbol{x}}{\mathrm{d}t} = -e\boldsymbol{E}. \tag{5.73}$$

This expression contains a friction term proportional to $v = \mathrm{d}x/\mathrm{d}t$, which represents the deceleration due to energy dissipation; m^* denotes the effective mass of the conduction electrons. The solution of this equation has the form

$$\boldsymbol{x} = \frac{e}{m^*\omega^2} \frac{\omega^2 - \mathrm{i}\omega\gamma}{\omega^2 + \gamma^2} \boldsymbol{E}. \tag{5.74}$$

The displacement $\boldsymbol{x}$ of the charge $-e$ causes a polarization $\boldsymbol{P} = -eN_e\boldsymbol{x} = \epsilon_0 \chi_e \boldsymbol{E}$, where χ_e is the dielectric susceptibility, $\epsilon = \epsilon_0(1 + \chi_e)$. Substituting for $\boldsymbol{x}$ from (5.74), we obtain

$$\epsilon(\omega) = \epsilon_{1,\mathrm{f}} + \mathrm{i}\,\epsilon_{2,\mathrm{f}} = \epsilon_0 \left(1 - \frac{N_\mathrm{e} e^2}{m^* \epsilon_0} \frac{1}{\omega^2 + \mathrm{i}\omega\gamma}\right) \quad \text{with} \tag{5.75}$$

$$\epsilon_{1,\mathrm{f}} = \epsilon_0 \left(1 - \frac{\omega_\mathrm{pl}^2}{\omega^2} \frac{1}{1 + (\gamma/\omega)^2}\right), \quad \epsilon_{2,\mathrm{f}} = \epsilon_0 \frac{\gamma}{\omega} \frac{\omega_\mathrm{pl}^2}{\omega^2} \frac{1}{1 + (\gamma/\omega)^2}, \tag{5.76}$$

in which ω_pl is the *plasmon frequency*

$$\omega_\mathrm{pl} = \sqrt{\frac{N_e e^2}{\epsilon_0 m^*}}\,. \tag{5.77}$$

The dependence of $\epsilon_{1,\mathrm{f}}$ and $\epsilon_{2,\mathrm{f}}$ on frequency is shown in Fig. 5.14. The factor $\mathrm{Im}\{-1/\epsilon\} = \epsilon_2/|\epsilon|^2$ that appears in (5.72) passes through a sharp maximum when the denominator reaches a minimum, which means that $\epsilon_{1,\mathrm{f}} = 0$ at $\omega = \omega_\mathrm{pl}$ for small values of the damping constant γ. This plasmon loss $\Delta E_\mathrm{pl} = \hbar\omega_\mathrm{pl}$ excites longitudinal charge-density oscillations in the electron gas, which are quantized and are known as *plasmons* [5.49]. For a large number of materials, the observed energy loss ΔE_pl agrees with that predicted by (5.77) (see Table 5.2), in which n is the number of valence or conduction electrons per atom.

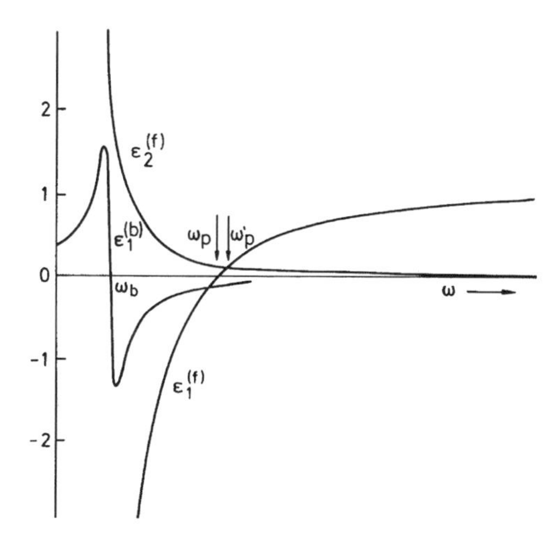

Fig. 5.14. Components of the complex dielectric permittivity $\epsilon_1 + \mathrm{i}\epsilon_2$ for the free electron gas (f) and for a bound state (b). Shift of the plasmon frequency ω_pl at $\epsilon_{1,\mathrm{f}} = 0$ to ω_pl'.

Table 5.2. Comparison of experimental and theoretical values of plasmon energy ΔE_{pl} and the half-width $(\Delta E)_{1/2}$ in eV, the constant a of the dispersion law (5.80), and the cutoff angle θ_c in mrad at 40 keV (n: number of electrons per atom) (see [5.41] for further tabulated values).

		ΔE_{pl}		$(\Delta E)_{1/2}$	a		θ_{c}	
	n	exp.	theor.		exp.	theor.	exp.	theor.
Al	3	15.0	15.8	0.6	0.40 ± 0.01	0.44	15	13
Be	2	18.9	18.4	5.0	0.42 ± 0.04	0.42	–	–
Mg	2	10.5	10.9	0.7	0.39 ± 0.01	0.37	12	11
Si	4	16.9	16.6	3.2	0.41	0.45	–	–
Ge	4	16.0	15.6	3.3	0.38	0.44	–	–
Sb	5	15.3	15.1	3.3	0.37 ± 0.03	0.38	–	–
Na	1	3.7	5.9	0.4	0.29 ± 0.02	0.25	10	9

However, the position of the plasmon losses can be influenced considerably by interband excitations (bound electrons: b). Extending (5.73) to include bound electrons, we obtain the Lorentz model

$$m^* \left(\frac{\mathrm{d}^2 \boldsymbol{x}}{\mathrm{d}t^2} + \gamma \frac{\mathrm{d}\boldsymbol{x}}{\mathrm{d}t} + \omega_{\mathrm{b}}^2 \boldsymbol{x} \right) = -e\boldsymbol{E}, \tag{5.78}$$

where ω_{b} is the resonance frequency, resulting in

$$\epsilon(\omega) = \epsilon_0 \left(1 + \frac{N_{\mathrm{e}} e^2}{m^* \epsilon_0} \frac{1}{\omega_{\mathrm{b}}^2 - \omega^2 - \mathrm{i}\omega\gamma} \right). \tag{5.79}$$

In the $\epsilon_{1,\mathrm{b}}$-curve, the bound states show a typical anomalous dispersion near the resonance frequency. In the special case of Fig. 5.14 with $\omega_{\mathrm{b}} < \omega_{\mathrm{pl}}$, the superposition of $\epsilon_{1,\mathrm{b}}$ and $\epsilon_{1,\mathrm{f}}$ (Drude–Lorentz model) causes a shift of the plasmon loss to a higher frequency ω'_{pl}, for which $\epsilon_{1,\mathrm{b}} + \epsilon_{1,\mathrm{f}} = 0$. When $\omega_{\mathrm{b}} > \omega_{\mathrm{pl}}$, the plasmon loss may be shifted to a lower frequency. Thus, the agreement between the calculated and measured values of the 15 eV plasmon loss in Al (Fig. 5.20) in Table 5.2 is accidental. The optical constants indicate that $\epsilon_{1,\mathrm{f}} = 0$ for $\Delta E = 12.7$ eV. An oscillator contribution at 1.5 eV (interband transition) shifts the loss to 15.2 eV. For silver, there is a transition of $4f$ electrons to the Fermi level at 3.9 eV, and further interband excitations occur at $\Delta E > 9$ eV. These shift the energy at which $\epsilon_1 = 0$ to 3.75 eV. This sharp energy loss (Fig. 5.8) is therefore a plasmon loss, which cannot be separated from the 3.9 eV interband loss. Figure 5.8 shows, as an example, a comparison of a measured energy-loss spectrum (full curve) with one calculated using the dielectric theory with optical values of ϵ (dotted curve). Two further maxima are also in agreement with the optical data.

These classical models can describe the most important features of plasmon excitation. A quantum-mechanical model should predict the dielectric behavior from first principles. The dielectric function of the free electron gas was calculated in the random-phase approximation by Lindhard [5.53]. This approach is not capable of calculating half-widths and results only in a δ-peak for the plasmon loss. In improved theoretical models, correlation, the periodic lattice potential, and the core polarizability must be considered (see [5.42] for further details).

One application of EELS for low energy losses is the measurement of band gaps in semiconductors and insulators [5.50, 5.52]. Furthermore, the optical constants in the ultraviolet can be determined [5.39]. The dielectric theory showed that the intensity in energy-loss spectra is proportional to $\mathrm{Im}\{-1/\epsilon\}$. The Kramers–Kronig relation (Sect. 10.2.2) allows us to calculate $\mathrm{Re}\{1/\epsilon\}$, and hence ϵ_1 and ϵ_2 can be obtained. These optical constants for 1–50 eV photons are mainly measured with synchrotron radiation. Electron energy loss spectroscopy has the additional advantage that measurements in the nanometer region are possible [5.51].

The loss spectra of different substances show characteristic differences. However, the spectra are not so specific that they can be used for elemental analysis. Furthermore, the spectra contain multiple losses and surface-plasmon losses, which depend on the foil thickness. Nevertheless, EELS can be used in many cases as an analytical tool to distinguish different phases (SiC [5.54], glass [5.55], or organic molecules [5.25, 5.56], for example).

The dependence of $\Delta E_{\mathrm{pl}} = \hbar\omega_{\mathrm{pl}} \propto \sqrt{N_e}$ on the electron density N_e (5.77) results in a weak decrease of ΔE_{pl} with increasing temperature because of the thermal expansion [5.57, 5.58]. Changes of ΔE_{pl} in alloys, due to the change of electron concentration, are of special interest because these shifts can be used for the local analysis of the composition of an alloy by EELS. Figure 5.15 shows, as an example, the shift of the plasmon energy in an Al-Mg alloy [5.59], the different phases of which can be identified by their energy losses. The plasmon energy ΔE_{pl} depends linearly on the concentration in the α, γ, and δ phases, which allows the local concentration to be measured with a spatial resolution of the order of 10 nm. In this way, the variation of Mg concentration was measured near large-angle boundaries after quenching of an Al-7wt% Mg alloy [5.60], and the variation of Cu in the Al-rich phase

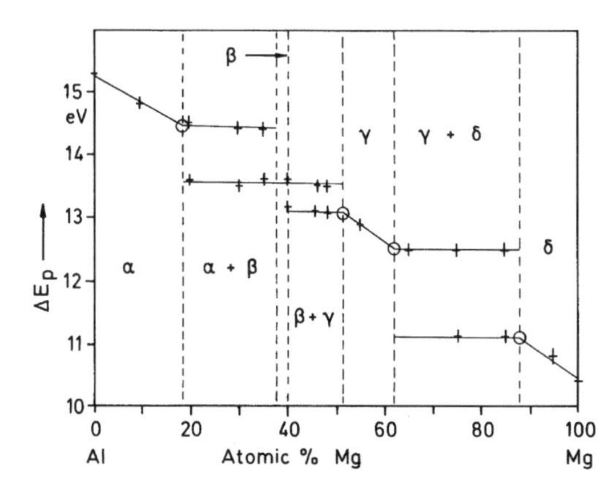

Fig. 5.15. Variation of the plasmon-loss energy ΔE_{pl} with composition in the Al-Mg system [5.59].

near $CuAl_2$ precipitates in a eutectic $CuAl_2$ alloy [5.61]. The plasmon losses of Al-Zn alloys have also been investigated [5.62]. The position of a plasmon loss can be determined with an accuracy of 0.1 eV, even if the full-width of half-maximum (FWHM) is about 1–4 eV [5.63]. However, the background of the energy loss spectrum caused by contamination can produce spurious shifts of the same order [5.64], and strains in inhomogeneous alloys can cause shifts relative to calibrations made with a homogeneous alloy [5.65]. Shifts of ± 0.1 eV have been observed at distances ± 20 nm from a dislocation [5.66].

Plasmon losses show a dispersion in the sense that the magnitude of $\Delta E_{\rm pl}$ depends on the momentum transferred and therefore on the scattering angle

$$\Delta E_{\rm pl}(\theta) = \Delta E_{\rm pl}(0) + 2Ea\theta^2 \quad \text{with} \quad a = \frac{3}{5}\frac{E_{\rm F}}{\Delta E_{\rm pl}(0)}\,, \tag{5.80}$$

where $E_{\rm F}$ is the Fermi energy. This dispersion can be directly imaged by angle-resolved EELS as a parabolic extended plasmon loss (Fig. 5.13a). The plasmon dispersion can also be seen in a series of electron spectroscopic diffraction patterns as small diffuse rings surrounding the primary beam. For single-crystal Sn films with a plasmon loss at $\Delta E = 21$ eV, these rings increase in diameter with increasing ΔE and disappear beyond $\Delta E = 28$ eV because of the existence of a cutoff angle $\theta_{\rm c}$ (see below) [5.67]. In addition, the half-width $(\Delta E)_{1/2}$ of the plasmon-loss maxima increases with increasing θ (Fig. 5.16b)

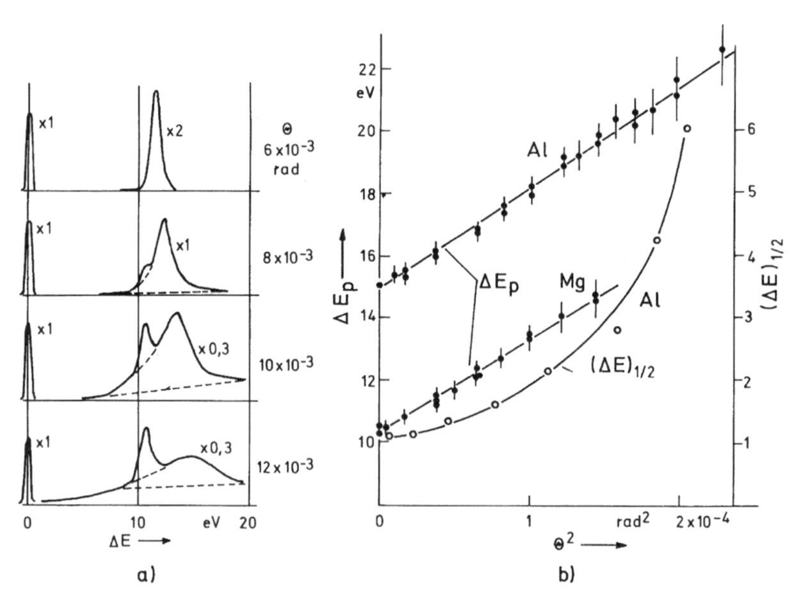

Fig. 5.16. (**a**) Dispersion of the plasmon loss of Mg with increasing scattering angle θ and the unshifted line due to elastic large-angle scattering and inelastic small-angle scattering. (**b**) Verification of the dispersion relation (5.80) by plotting $\Delta E_{\rm pl}$ versus θ^2 for the Al and Mg plasmon losses [5.73] and the broadening $(\Delta E)_{1/2}$ of the plasmon loss of Al with increasing θ [5.69, 5.70].

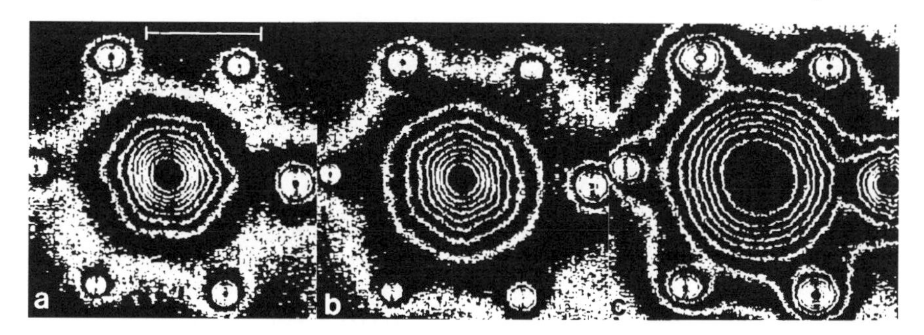

Fig. 5.17. Intensity contours (digital isodensities) in ESD patterns of a graphite foil showing the azimuthal anisotropies for the interband transitions at (**a**) $\Delta E = 7$ eV and (**b**) $\Delta E = 13$ eV and no anisotropy for (**c**) the plasmon loss at $\Delta E = 31$ eV.

[5.68, 5.69, 5.70, 5.71, 5.72, 5.73, 5.74, 5.75]. The plot of $\Delta E_{\rm pl}$ versus θ^2 in Fig. 5.16b demonstrates the validity of this dispersion law. The constant $a_{\rm exp}$ in (5.80) (Table 5.2) can be obtained from the slope of the curve. Interband excitations show no dispersion and normally have a larger half-width. However, a dispersion with $a = 0.15$ has been observed for the 13.6 eV loss of LiF, which is an exciton excitation [5.76]. If the energy loss is observed at a scattering angle θ, not only is the shifted value $\Delta E_{\rm pl}(\theta)$ observed but also the unshifted $\Delta E_{\rm pl}(0)$, which results either from primary small-angle plasmon scattering and secondary elastic large-angle scattering or vice versa (Fig. 5.16a). The values of the plasmon losses and their dispersion are anisotropic in anisotropic crystals such as graphite [5.77, 5.78] and also in cubic crystals (Al, for example) for large scattering angles [5.79]. This anisotropy can be imaged by electron spectroscopic diffraction as shown for a graphite foil in Figs. 5.17a–c by drawing isodensities around the primary beam. The anisotropy of the interband transition at $\Delta E = 7$ eV in Fig. 5.17a can be seen as a hexagon with corners directed towards the surrounding Bragg spots, whereas at $\Delta E = 13$ eV (Fig. 5.17b), the corners are directed between the Bragg spots. The plasmon loss at $\Delta E = 31$ eV (Fig. 5.17c) shows no anisotropy (circular isodensities).

Integration of (5.72) over ΔE gives the contribution of a plasmon loss to the differential cross section

$$\frac{{\rm d}\sigma_{\rm pl}}{{\rm d}\Omega} = \frac{\Delta E_{\rm pl}}{2\pi a_{\rm H} N_e E} \frac{E + E_0}{E + 2E_0} \frac{1}{\theta^2 + \theta_{\rm E}^2} G(\theta, \theta_c), \tag{5.81}$$

in which we have used

$$\int_0^\infty {\rm Im}\{-1/\epsilon(\omega)\}\hbar{\rm d}\omega = \pi \Delta E_{\rm pl}/2. \tag{5.82}$$

The cross section decreases as θ^{-2} for medium scattering angles $\theta_{\rm E} < \theta < \theta_{\rm c}$ (Fig. 5.18). The correction function $G(\theta, \theta_{\rm c})$ introduced by Ferrell [5.80] takes into account the fact that ${\rm d}\sigma_{\rm pl}/{\rm d}\Omega$ has to become zero at a cut-off angle $\theta_{\rm c}$, which implies that plasmon wavelengths shorter than the mean distance

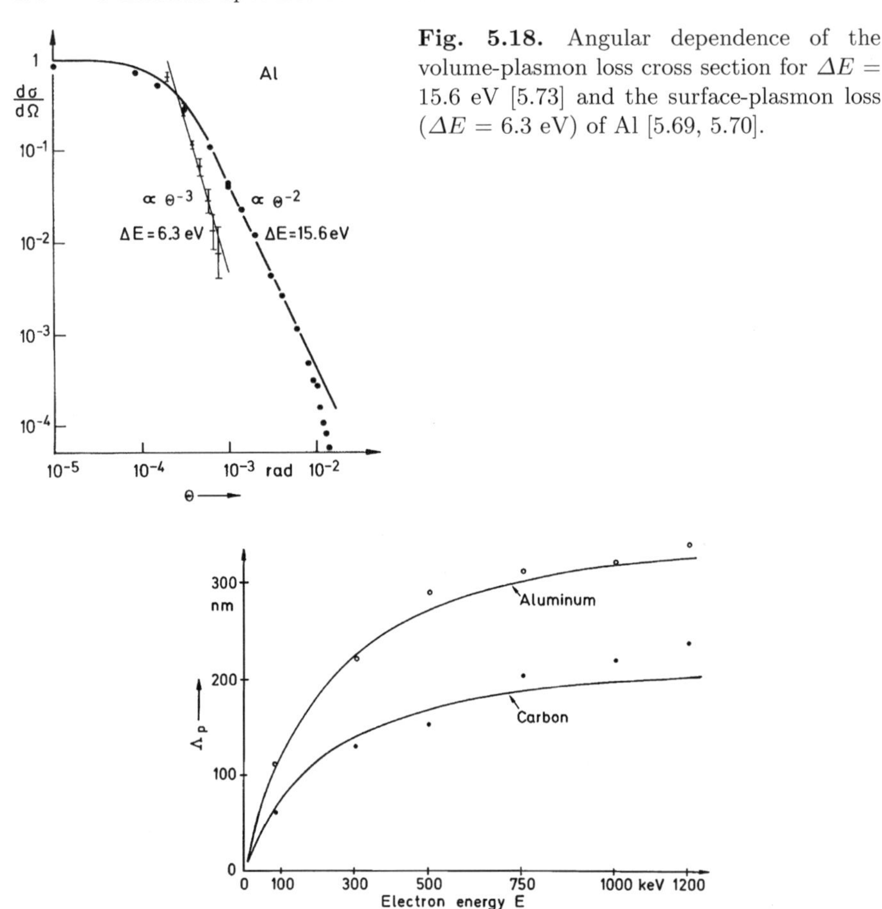

Fig. 5.18. Angular dependence of the volume-plasmon loss cross section for $\Delta E = 15.6$ eV [5.73] and the surface-plasmon loss ($\Delta E = 6.3$ eV) of Al [5.69, 5.70].

Fig. 5.19. Mean-free-path length Λ_{pl} for Al and C plasmon losses as a function of electron energy [5.81].

between valence electrons are damped more strongly. There is thus a maximum momentum that can be transferred in the inelastic collision, and scattering angles greater than θ_c are not possible.

Integration over the solid angle Ω in (5.81), using the approximation $G(\theta, \theta_c) = 1$ for $\theta < \theta_c$ and vanishing for $\theta > \theta_c$, yields the total cross section for plasmon excitation

$$\sigma_{\text{pl}} \simeq \frac{\theta_{\text{E}}}{N_e a_{\text{H}}} \ln(\theta_c/\theta_{\text{E}}) \tag{5.83}$$

assuming that $\theta_{\text{E}} \ll \theta_c$, and the corresponding mean free path becomes $\Lambda_{\text{pl}} = 1/N_e\sigma_{\text{pl}}$. Figure 5.19 shows calculated and measured values of Λ_{pl} in the range $E = 100$–1000 keV.

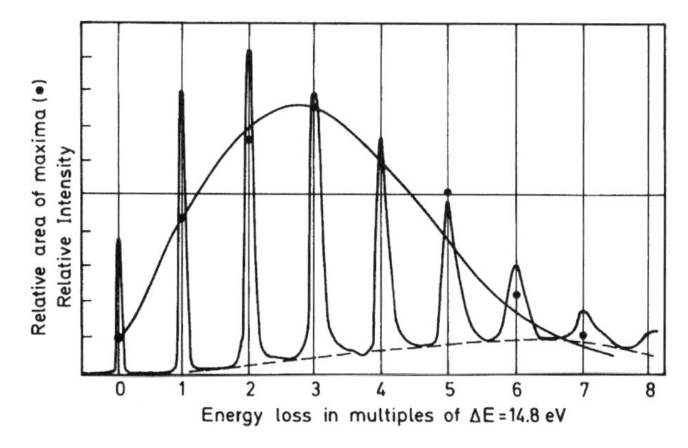

Fig. 5.20. Multiple characteristic plasmon losses of 20 keV electrons passing through a 208 nm Al film and comparison of the areas of the loss maxima (•) with a theoretical Poisson distribution [5.82].

Multiple inelastic scattering is observed in thick specimens (Sect. 5.4.2), which means that multiples of the plasmon losses appear in the loss spectrum. This can be seen, in particular, in the loss spectrum of Al, which shows one sharp plasmon loss at 15.2 eV (Fig. 5.20). The probability $P_n(t)$ for the appearance of an energy loss $\Delta E = n\Delta E_{\mathrm{pl}}$ in a specimen layer of thickness t can be described by a Poisson distribution ($n = 0$ corresponds to an elastic scattering event with no energy loss)

$$P_n(t) = \left(\frac{t}{\Lambda_{\mathrm{pl}}}\right)^n \frac{\exp(-t/\Lambda_{\mathrm{pl}})}{n!} \,. \tag{5.84}$$

The integrated intensities of the loss maxima agree well with this distribution (see, e.g., Fig. 5.20 for Al [5.82] and [5.83] for the 16.9 eV loss of Si). However, deviations from Poisson statistics can occur when a fraction of the multiple plasmon loss is scattered through angles larger than the aperture used. The convolution with the background intensity in the loss spectra of carbon and aluminum films is considered in [5.84, 5.85]. The multiple-loss spectrum can be calculated for increasing thicknesses by a double convolution over energy loss ΔE and scattering angle θ by a Fourier method using theoretical formulas for the inelastic single scattering with plasmon losses, Compton scattering, and inner-shell ionizations [5.86].

5.2.6 Surface-Plasmon Losses

The plasmon losses discussed above are so-called volume losses. Surface-plasmon losses, with lower energy-loss values, are also observed; these can be explained by the generation of surface-charge waves [5.87]. Figure 5.21 shows the distribution of the electric field for a symmetric ω^- (Fig. 5.21a) and an antisymmetric (Fig. 5.21b) ω^+ surface oscillation mode and for a single boundary in a thick layer (Fig. 5.21c). Both modes of oscillation show

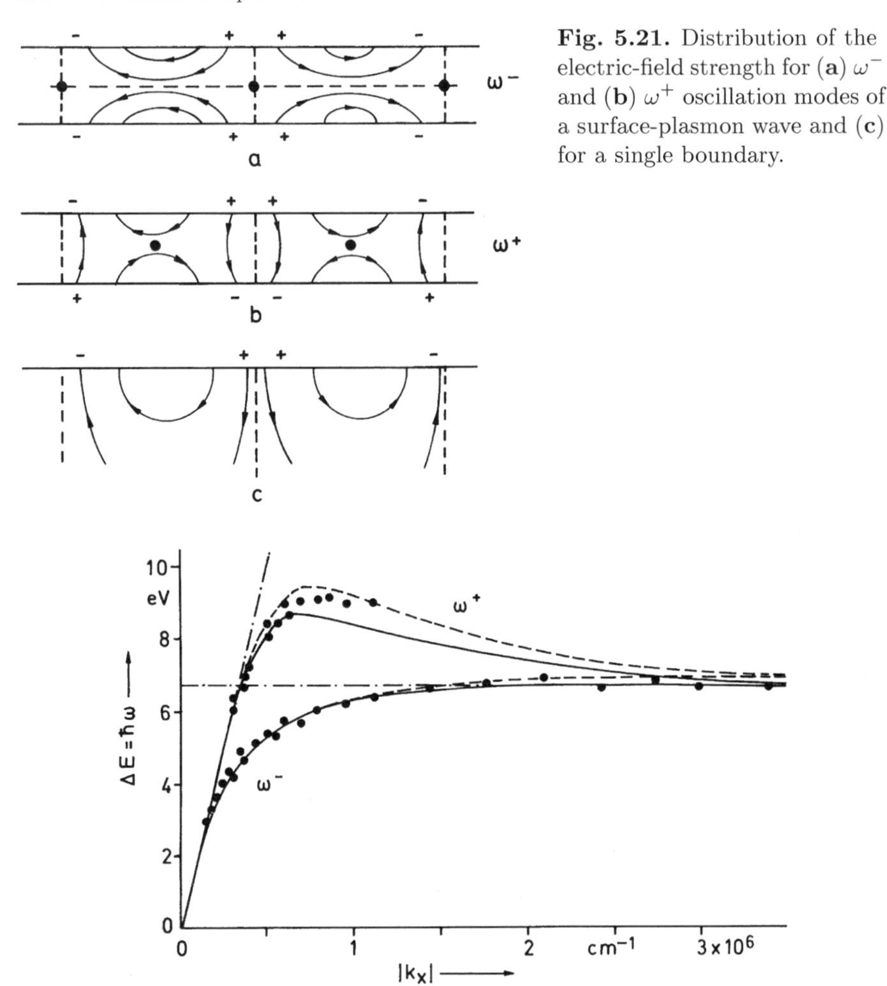

Fig. 5.21. Distribution of the electric-field strength for (**a**) ω^- and (**b**) ω^+ oscillation modes of a surface-plasmon wave and (**c**) for a single boundary.

Fig. 5.22. Dispersion of the surface-plasmon modes ω^- and ω^+ versus $k_x = \theta/\lambda$ for a 16 nm Al film covered on each side with a 4 nm oxide film. The dashed curve corresponds to ϵ for amorphous Al_2O_3 and the full curve for α-Al_2O_3 [5.91].

strong dispersions [5.73, 5.90, 5.91, 5.92] that depend not only on the wave number k_x of the surface waves but also on the specimen thickness t (see the example in Fig. 5.22).

The mode ω^+ is excited with a lower probability than ω^-. Figure 5.23 shows how the ω^+ and ω^- losses move together with increasing film thickness to a saturation value

$$\omega^{\pm} = \frac{\omega_{\mathrm{pl}}}{\sqrt{1+\epsilon}}\,, \tag{5.85}$$

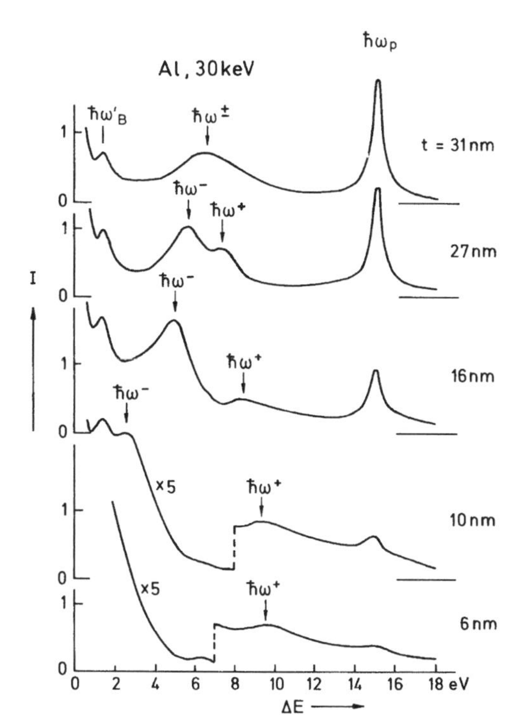

Fig. 5.23. Dispersion of the two surface plasmon losses $\hbar\omega^-$ and $\hbar\omega^+$ as a function of film thickness t and increase of the plasmon-loss intensity of $\hbar\omega_{\text{pl}}$ with increasing t [5.92].

where ϵ is the relative permittivity of the neighboring medium, typically vacuum, oxide, or supporting film. If both boundaries of the layer are limited by vacuum ($\epsilon = 1$), equation (5.85) gives $\omega^{\pm} = \omega_{\text{pl}}/\sqrt{2}$ for large thicknesses; for an oxide-coated aluminum layer, we have $\Delta E = \hbar\omega_{\pm} = 6.25$ eV with $\epsilon_{\text{oxide}} = 4.7$, for example (Fig. 5.22).

The differential cross section of surface-plasmon losses decreases as $\theta\theta_{\text{E}}/(\theta^2 + \theta_{\text{E}}^2)^2$ for increasing θ and hence as θ^{-3} for $\theta \gg \theta_{\text{E}}$ [5.88, 5.89] (Fig. 5.18). At nonnormal incidences of the primary electrons, the excitation of surface-plasmon losses has an asymmetric angular distribution [5.87, 5.91, 5.92, 5.93, 5.94].

If the electron velocity v is greater than the velocity of light in the specimen layer (e.g., for Si), energy losses $\Delta E = 3.4$ eV are observed due to the generation of Čerenkov radiation [5.95, 5.96]. Guided-light modes can be excited in thin dielectric films, such as graphite [5.96].

Surface-plasmon losses can also be excited by the polarization caused by the Coulomb field of electrons when the electron probe of a scanning transmission electron microscope passes close to a crystal without striking it [5.97, 5.98]. Correspondingly, an electron spectroscopic image with surface-plasmon loss shows a bright rim extending exponentially to about 10 nm outside a cubic MgO crystal [4.111]. The local excitation of surface plasmons can be used to determine the dielectric properties of nanotubes or nanoparticles [5.99, 5.100, 5.101].

5.3 Energy Losses by Inner-Shell Ionization

5.3.1 Position and Shape of Ionization Edges

The shells K, L, M, N, and O correspond to the main quantum numbers $n =$ 1–5, respectively. Electrons on these atomic levels have energies of the order of

$$E_n = -R(Z - \sigma_n)^2/n^2, \tag{5.86}$$

where $R = 13.6$ eV is the ionization energy of hydrogen and $Z - \sigma_n$ denotes an effective atomic number, decreased by screening.

There $l = 0, 1, \ldots, n-1$ are possible values of the azimuthal quantum number, resulting in angular momenta $L = \sqrt{l(l+1)}\hbar$ and denoted by the symbols s, p, d, f, and g for $l =$ 0, 1, 2, 3, and 4, respectively. The electron spin is described by the quantum number $s = \pm 1/2$ with an angular momentum $S = \sqrt{s(s+1)}\hbar$. The corresponding vectors $\boldsymbol{L}$ and $\boldsymbol{S}$ couple by spin-orbit interaction to form the total angular momentum $\boldsymbol{J} = \boldsymbol{L} + \boldsymbol{S}$ with $J = \sqrt{j(j+1)}\hbar$. Thus, for a $2s$ electron with $l = 0$, only $j = 1/2$ occurs, while for a $2p$ electron with $l = 1$, j can take the values 1/2 and 3/2. This results in the splitting of the L shell into three sublevels, L_1, L_2, and L_3 (Fig. 10.3). The magnetic quantum number $m = -j, \ldots, +j$ describes the $2j + 1$ possible z-components of the angular momentum $L_z = m\hbar$ and tells us that $2j + 1$ electrons can be accommodated in the corresponding subshell of quantum number j.

Following Pauli's exclusion principle, these configurations of quantum numbers are filled consecutively as the atomic number increases, though the sequence is often interrupted; in the transition metals, for example, the $3d$ shell is being filled, as is the $4f$ shell of lanthanides.

The ionization energy that is observed as an edge in electron energy-loss spectroscopy (EELS) is the energy difference between the first unoccupied energy state beyond the Fermi level and the ionized subshell. The edges observed in EELS are therefore labeled according to the ionized subshell, L_2 or L_3, for example, and L_{23} if the corresponding edges cannot be resolved.

Figure 5.24 shows the edge energy losses $\Delta E = E_I$ (I = K, L, M,...) versus the atomic number. Because E_I increases as Z^2 (5.86), the K-edge can be observed within the useful interval $\Delta E = 0 - 2$ keV only up to Si ($Z = 14$). However, Fig. 5.24 shows that higher shell ionizations occur in each element within this range of ΔE. Collections of the electron energy-loss spectra of all elements have been published [5.102, 5.103]; these also contain information about the profile of the edges discussed below.

The differential cross section of inner-shell ionizations decreases over orders of magnitude with increasing energy loss (Fig. 5.9). Energy losses beyond 2000 eV are therefore rarely used in EELS microanalysis. Beyond the edges follows a long tail of energy losses that result from the excitation of core electrons to unoccupied states of the continuum beyond the Fermi level. The plasmon and

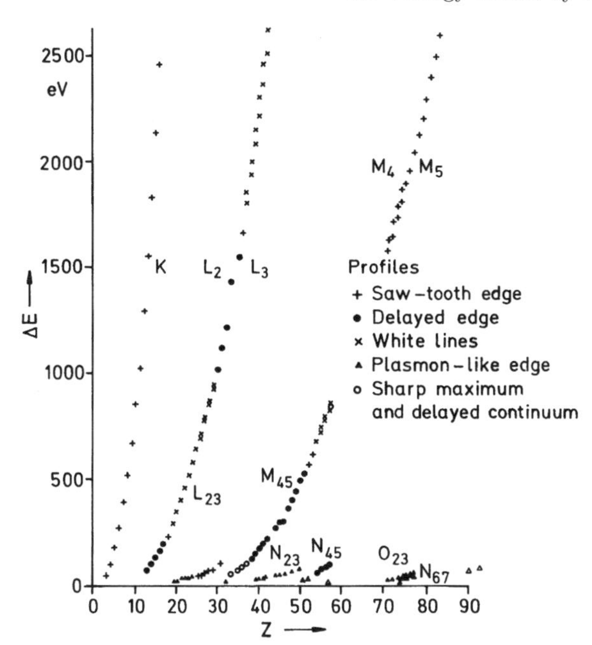

Fig. 5.24. Position of the ionization edges within the energy-loss interval $\Delta E =$ 0–2.5 keV as a function of the atomic number

interband transitions also show a background that extends to a few hundred electron volts. The decrease beyond an edge can be approximated by

$$\frac{\mathrm{d}\sigma(\alpha)}{\mathrm{d}\Delta E} \propto \Delta E^{-s}. \tag{5.87}$$

The exponent s depending on α can be determined from the slope in a double-logarithmic plot of the number of counts versus ΔE (Fig. 5.9), and s is found to be of the order of 3.5–4.5. The law (5.87) can also be applied to the background in front of the edge to permit extrapolation of the background beyond the edge (Sect. 10.3.3).

The shapes of ionization edges can be classified into groups [5.104] listed and indicated in Fig. 5.24 and discussed below with examples of recorded spectra (see also the review in [5.105]).

K Ionizations. K edges can be used from Li to Si and show a typical sawtooth shape, as shown for carbon and aluminum in Figs. 5.25a,b. Hydrogen has been investigated in metal hydrides and can be detected as a shift of the plasmon loss [5.106, 5.107]. Helium can be analyzed as condensed gas bubbles after He implantation in solids and shows a weak peak at 21–23 eV from atomic-like $1s \rightarrow 2p$ transitions [5.108, 5.109, 5.110].

L_{23} Ionizations. Whereas the profiles of K edges are nearly independent of atomic number, the L_{23} edges of elements of the third group of the periodic table (Si–Cl) show a delayed maximum 10–15 eV above the threshold as shown for Si in Fig. 5.25c. This is a consequence of the centrifugal potential

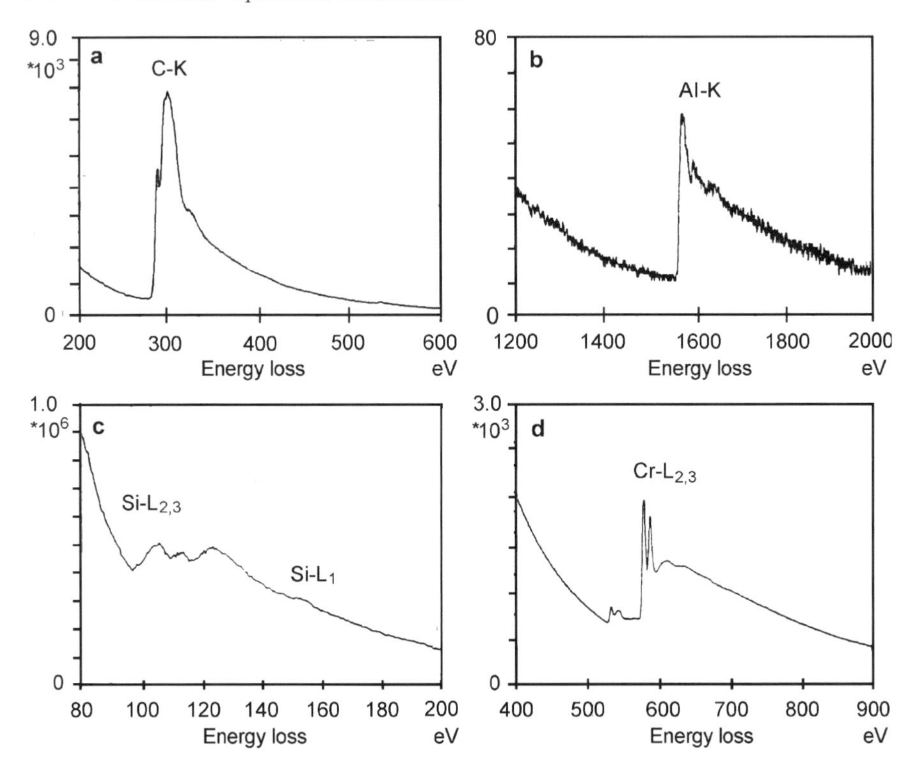

Fig. 5.25. Examples of recorded energy-loss spectra [5.103] demonstrating the different shapes of ionization edges: (**a**, **b**) K edges of C and Al; (**c**, **d**) L_{23} edges of Si and Cr [5.103].

barrier in (5.89) when $2p$ electrons are excited to final states with $l' \geq 2$. The probability of exciting the $2s$ electron to the L_1 subshell is much lower, and the corresponding maximum is often buried within the energy-loss near-edge structure (Sect. 5.3.3). In the elements of the fourth group (K–Cu), on the other hand, $2p$ electrons can be excited not only to the continuum but also to unoccupied bound d states; "white lines" are then seen at the threshold, as shown for Cr in Fig. 5.25d, where the two narrow white lines are caused by spin-orbit splitting of the $2d$ subshell. In the elements Cu–Br, no white lines occur because the d shell is filled and only rounded delayed maxima caused by transitions to the continuum are observed. Whereas Cu alone shows no white lines, the electron transfer from Cu to O in copper oxide produces unfilled d levels, and white lines appear (Fig. 5.27, Sect. 5.3.3).

M_{45} Ionizations. Elements Rb–I show edges similar to the $3d$-filled L_{23} edges. The delay of the maximum can reach 60–80 eV, as shown for a background-subtracted (stripped) M_{45} edge of Mo in Fig. 5.26a. For Cs to Yb (including the rare earths), the f shell contains bound states and white

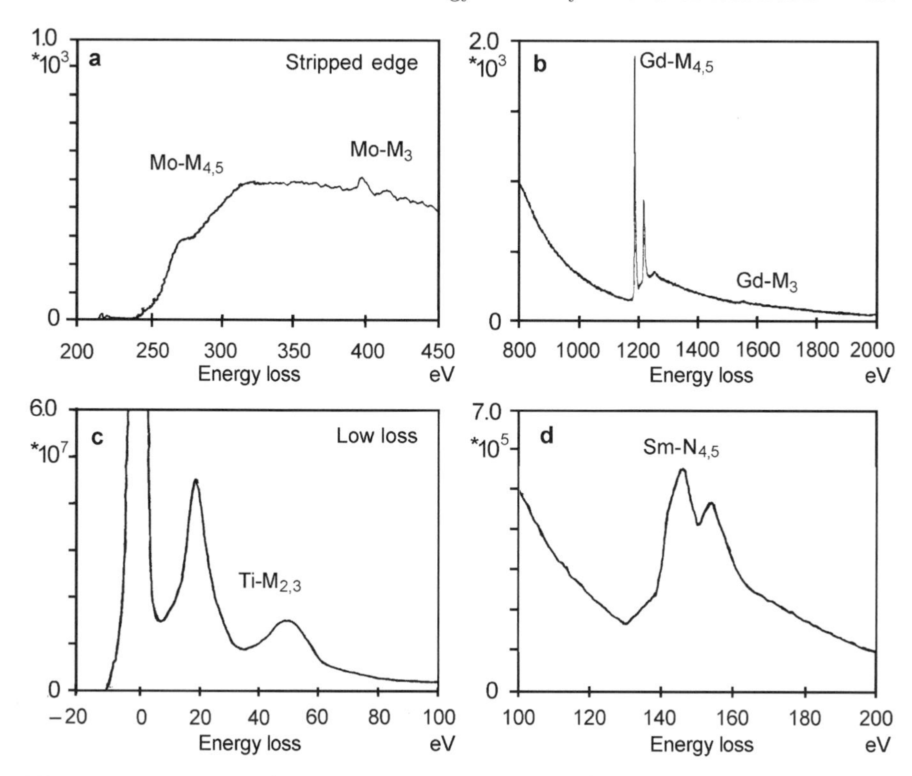

Fig. 5.26. Examples of recorded energy-loss spectra [5.103] demonstrating the different shapes of ionization edges: (**a**, **b**) M_{45} edge of Mo and Gd, (**c**) plasmon-like M_{23} edge of Ti, and (**d**) N_{45} edge of Sm [5.103].

lines again appear as shown for Gd in Fig. 5.26b. From Lu to Au, the f shell is filled and again strongly delayed maxima and no white lines are observed.

M_{23} Ionizations. The edges of K–Zn lie between 30 and 100 eV and show plasmon-like peaks superposed on the background of the valence electrons as shown for Ti in Fig. 5.26c.

N_{45} Ionizations. The elements Cs, Ba, and the lanthanides show characteristic profiles in the range 80–120 eV resulting from the excitation of $4d_{3/2}$ and $4d_{5/2}$ electrons to f states as shown for Sm in Fig. 5.26d.

5.3.2 Inner-Shell Ionization Cross Sections

The calculation of the GOS and $\mathrm{d}^2\sigma/\mathrm{d}\Omega\mathrm{d}(\Delta E)$ for inner-shell ionization has been discussed in Sect. 5.2.2. The wave functions a_0 and a_n of the initial and final states of the atomic electrons are solutions of the Schrödinger equation

$$\left[\frac{\hbar^2}{2m}\nabla^2 + E_n + V(r)\right] a_n = 0. \tag{5.88}$$

After substituting $a_{nlm} = R_{nl}(r)Y_{lm}(\Omega)/r$, where Y_{lm} is the spherical function, and separating the variables, the following equation for the radial part is obtained:

$$\left\{\frac{\hbar^2}{2m}\nabla^2 + E_n + \left[V(r) + \frac{l(l+1)\hbar^2}{2mr^2}\right]\right\} R_{nl}(r) = 0. \tag{5.89}$$

The quantum number l modifies the Coulomb potential $V(r)$ by a term that can be interpreted as a centrifugal barrier. The transitions from an initial state with quantum number l to the final state with l' obey the selection rule $\Delta l = \pm 1$ for $q' \to 0$ (optically allowed transitions), whereas, for large q', all transitions can contribute to the energy-loss spectrum.

For the calculation of K shell ionization cross sections, hydrogenic atomic functions are widely used, and they can also be used for higher Z by considering screening of the nuclear charge $+Ze$ by Z–1 electrons [5.111]. These calculations are included in the SIGMAK program for elements Li to Si [5.112].

For L shells, the centrifugal barrier (5.89) has to be considered. In the SIGMAL program, photoabsorption data [5.113] and EELS measurements [5.114] have been used to correct hydrogenic calculations [5.115, 5.116]. Hydrogenic model calculations of the M_{45} and M_{23} shells based on photoabsorption data have also been published [5.117]. These calculations are all based on the Schrödinger equation for the incident electron and thus confined to the non-relativistic case. The validity of the resulting formulas can be extended up to about 120 keV electrons by replacing the expressions for the energy and momentum transfer by their relativistically correct counterparts. For higher electron energies, a fully relativistic treatment should be used [5.119].

The use of atomic Hartree–Slater wave functions for the initial and final states [5.118, 5.120] allows us to predict the characteristic shapes of the edges, such as the sawtooth shape of the K edge and the delayed edge for L shell ionizations, but not the white lines that occur in M and L edges by transitions to unoccupied bound states. The generalized oscillator strength for the white-line components can be calculated separately [5.121].

When the objective diaphragm or the entrance slit or diaphragm of the spectrometer acts as a limiting aperture, the double differential cross section has to be integrated between 0 and α as well as between E_{I} (I = K, L, M, ...) and $E_{\mathrm{I}} + \Delta$ to get a partial cross section $\sigma(\alpha, \Delta)$. The width Δ of the energy window beyond the ionization edge at E_{I} has to be limited to 50–100 eV because of the long tail of the edges and uncertainties when subtracting the extrapolated background in front of the edge in order to extract the cross section $\mathrm{d}\sigma(\alpha)/\mathrm{d}\Delta E$, which can be compared with experiments.

The difficulties of the absolute measurement of ionization cross sections can be avoided by measuring ratios of partial cross sections of elements a and a standard element b = O, B, or C, which is present in an oxide, boride, or carbide of a,

$$k_{ab} = \frac{\sigma_b(\alpha, \Delta)}{\sigma_a(\alpha, \Delta)} = \frac{I_b(\alpha, \Delta)}{I_a(\alpha, \Delta)} \frac{N_a}{N_b}, \tag{5.90}$$

where N_a and N_b are the numbers of atoms per unit area. This ratio is analogous to the Cliff–Lorimer ratio of x-ray microanalysis (Sect. 10.2.4). For these light standard elements, absolute partial cross sections can be calculated accurately.

Experimental k-factors have been determined for K, L_{23}, M_{45}, M_{23}, and N_{45} shells [5.114, 5.122, 5.123, 5.124, 5.125]; see also [5.126, 5.127]). These data can also be represented by integrated oscillator strengths $f(\Delta)$ (parametrization), which are independent of α and E [5.128, 5.129, 5.105].

5.3.3 Energy-Loss Near-Edge Structure (ELNES)

The energy-loss near-edge structure is concentrated within a region of about 50 eV and results from the unoccupied density of states (DOS) beyond the Fermi energy, which may be regarded as a multiplicative envelope applied to the corresponding edge. Energy resolutions of less than 1 eV are necessary to resolve details.

One effect in ELNES is the so-called chemical shift of an edge, which can be observed when elements occur in different crystal structures or compounds; for example, the Al L_{23} edge shifts from 73 eV in the metal to 77 eV in Al_2O_3 and the Si L_{23} edge from 99.5 eV in Si to 103 eV in SiO and 106 eV in SiO_2. The shift of the K edge from 284 eV in graphite to 289 eV in diamond can be attributed to the 4 eV band gap. The chemical shift of amorphous silicon alloys is linearly related to the electronegativity of the ligand, which is a measure of the charge transfer from Si to the ligand [5.130].

The following examples demonstrate how the fine structure can be interpreted (see [5.105] for further details). The L_{23} edge in Cu does not show white lines, unlike the spectrum of CuO (Fig. 5.27). In the metal, the $3d$ band is filled and lies just below the Fermi level, whereas in CuO an electron exchange between Cu and O atoms produces vacant states in the $3d$ band; the Fermi level shifts into the $3d$ band with the result that pronounced white lines are seen in the CuO specimen [5.131]. Figures 5.28a,b show the C K edges

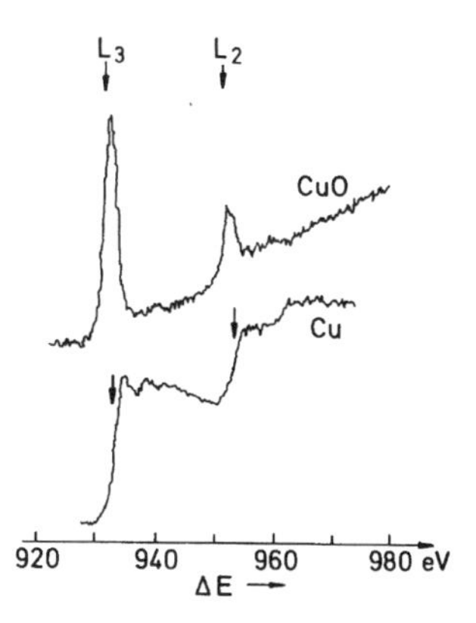

Fig. 5.27. Comparison of the L_{23} edges of Cu and CuO [5.131].

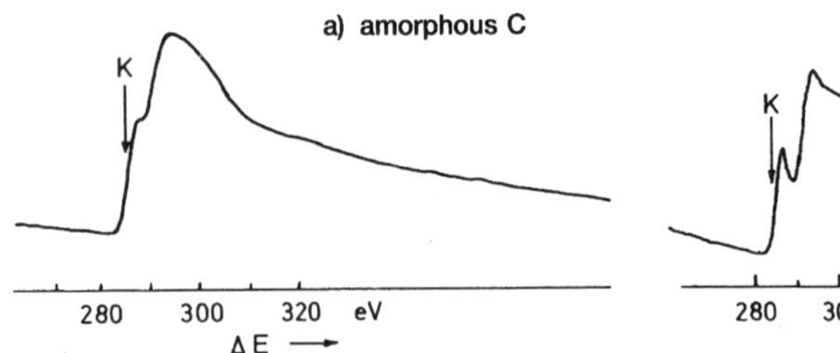

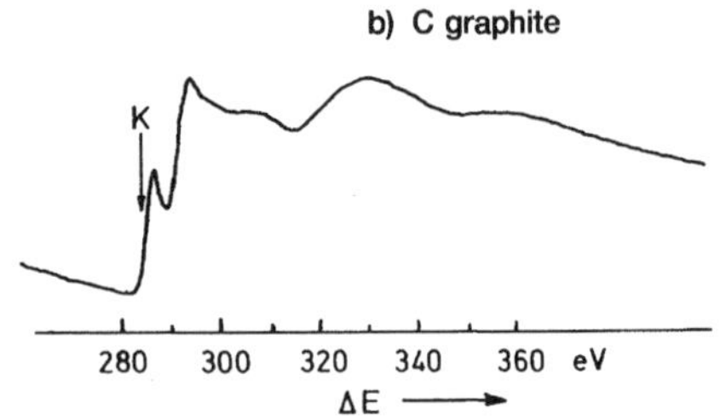

Fig. 5.28. Differences in the energy-loss spectrum at the K ionization edge for carbon in (**a**) amorphous carbon films and (**b**) graphite films [5.132].

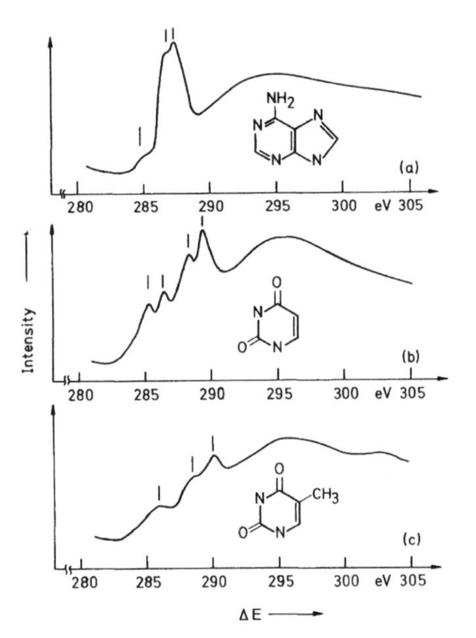

Fig. 5.29. Fine structure in the energy-loss spectrum near the carbon K ionization edge for (**a**) adenine, (**b**) uracil, and (**c**) thymine [5.25].

for amorphous carbon and graphite [5.132]. The weak preionization peaks are attributed to π^* bound states below the ionization threshold. The number of peaks and their energies observed in the C K edges of different organic compounds (Fig. 5.29) can be explained in terms of a chemical shift because the carbon atoms present at different sites within the molecule carry different charges [5.25, 5.133]. As another example, Fig. 5.30 shows the B K edge, the N K edge, and the density of free π^* and σ^* states beyond the Fermi level in hexagonal boron nitride [5.134]. The dotted areas are caused by convolution with the plasmon-loss spectrum. Such transitions to π^* and σ^* states also form the loss spectrum of graphite in Fig. 5.28b. They show different angular distributions in the EELS, and their intensity changes when different scattering angles are selected. Tilting of anisotropic crystals also alters the ELNES because π^* and σ^* orbitals are parallel and perpendicular to the c-axis, respectively [5.135, 5.136].

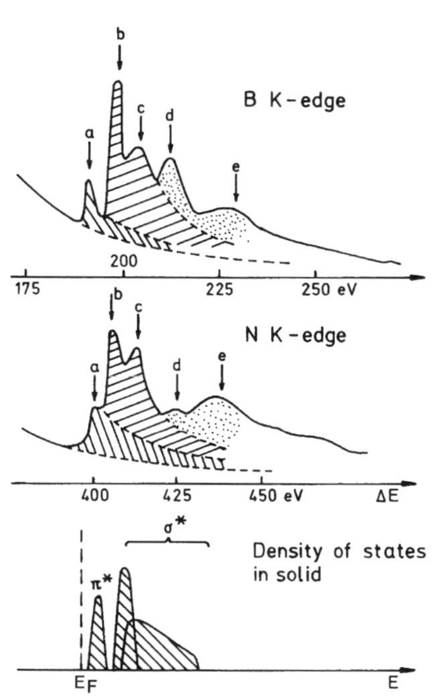

Fig. 5.30. Boron and nitrogen K-edge profiles of boron nitride with transitions to the $\pi^\star$ and $\sigma^\star$ bands (differently hatched areas) and convolution due to plasmon losses (dotted areas) [5.135].

Energy-loss near-edge structures can be used to identify elements in similar compounds or environments when reference spectra are available ("fingerprint" method) [5.137, 5.138]. As examples, we mention the use of the difference between graphite and diamond (Fig. 5.28) to identify interstellar diamond [5.139] and the use of the intensity ratio of the L3 and L2 white lines to determine the Fe^{2+}/Fe^{3+} ratio [5.140].

For more detailed studies, it is of interest to compare ELNES with theoretical calculations [5.141] using the band structure and the unoccupied DOS. The *augmented plane-wave method* [5.142] has been used to interpret ELNES for transition metals and their carbides, oxides, and nitrides [5.143, 5.144]. A *pseudopotential band theory* [5.145] is suitable for semiconductors and ceramic materials [5.146, 5.147]. A *multiple-scattering calculation* [5.148] considers the influence of nearest-neighbor shells. The excited electron wave is backscattered and interferes with itself, thus influencing the excitation probability. This is analogous to the model (Fig. 5.32) also used for EXELFS. This method is particularly suitable for large unit cells [5.149, 5.150] or oxygen compounds [5.151]. An alternative method is a *molecular orbital calculation* applied to transition metal ions in solids, for example [5.150, 5.152]. The details in the region of the white lines of the transition metals can be described using an atomic model and entering the crystal field as a perturbation [5.153].

5.3.4 Extended Energy-Loss Fine Structure (EXELFS)

In contrast to the strong effects in ELNES, EXELFS can be observed only as weak oscillations in the tail of an edge up to energy losses of about 100–200 eV beyond it (Fig. 5.31) [5.134, 5.154, 5.155, 5.156, 5.157, 5.158, 5.159, 5.160]. This fine structure is also observed in x-ray absorption spectra [5.161], where it is called EXAFS (extended x-ray-absorption fine structure). Both are generated by interference between the outgoing spherical wave of the excited electron with an excess energy $\Delta E - E_\mathrm{I}$ beyond the Fermi level and the waves backscattered at the nearest-neighbor atoms (Fig. 5.32). The variation of the cross section can be described by

$$\Delta\sigma(k) = \sum_j f_j(k) \frac{n_j \exp(-2r_j/\Lambda)}{2\pi k r_j^2} \exp(-8\pi^2\sigma_j^2 k_j^2) \sin[4\pi k r_j + \eta_j(k)], \quad (5.91)$$

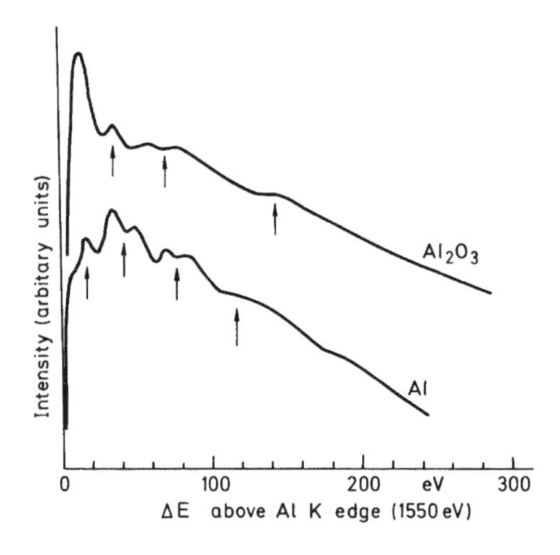

Fig. 5.31. Energy-loss maxima (extended fine structure) of Al and Al_2O_3 films above the K ionization edge of Al due to interactions with the neighboring atoms [5.134].

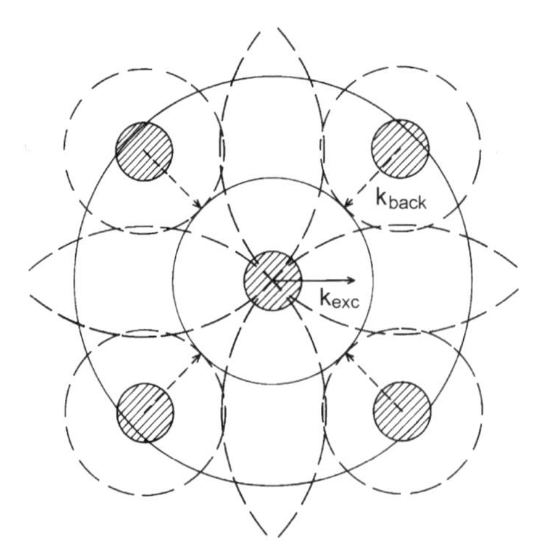

Fig. 5.32. Excitation and backscattering of secondary waves at neighboring atoms and interference with the primary wave excited by inner shell excitation to explain EXELFS. In the example, the excited and backscattered waves interfere constructively at the central atom, resulting in a maximum of $\Delta\sigma(k)$.

where

k $= [2m(\Delta E - E_{\rm I})/h^2]^{1/2}$ is the wave number of the excited electron;
r_j = distance of the n_j neighboring atoms in the jth coordination shell;
$f_j(k)$ = elastic scattering amplitude of neighboring atoms for deflection through 180°;
Λ = mean free path for inelastic scattering of the excited electron as measured by Auger electron spectroscopy, for example, with a minimum of 1 nm at 100 eV;
σ_j = Debye–Waller factor arising from thermal vibrations and/or statistical disorder of the neighboring atoms.

The first term, $4\pi k r_j$, in the sine function is the geometric phase shift of the backscattered wave corresponding to the distance $2r_j$, and $\eta_j(k)$ is a phase shift associated with backscattering.

Equation (5.91) shows that the cross section is the Fourier transform of the radial density distribution n_j/r_j^2 of the nearest-neighbor atoms. After subtracting the continuous tail beyond the edge, the density distribution can be derived by an inverse Fourier transform of the oscillatory part. Thus, the main maxima indicated by arrows in Fig. 5.31 and split by the influence of second-nearest neighbors give the value $r_1 = 0.28 \pm 0.01$ nm for Al and $r_1 = 0.20 \pm 0.01$ nm for the oxide, in agreement with crystallographic x-ray data. The EXELFS spectrum of crystalline specimens also depends on the direction of $\boldsymbol{q}'$ of the momentum transfer [5.165].

5.3.5 Linear and Circular Dichroism

On several occasions, we have discussed the dependence of the energy-loss spectrum on the orientation of the scattering vector $\boldsymbol{q}'$. For small scattering angles (dipole approximation), the double differential cross section (5.51) is related to the photoabsorption cross section for a photon energy $\Delta E = h\nu$,

$$\sigma_{Ph} = 2\pi h\alpha |\langle a_{ns}|\boldsymbol{e}\cdot\boldsymbol{r}_j|a_{0s}\rangle|^2. \tag{5.92}$$

Here $\alpha \approx \frac{1}{137}$ is the fine-structure constant and $\boldsymbol{e}$ the polarization vector of the photon. A comparison of (5.51) with (5.92) shows that the direction of the scattering vector $\boldsymbol{q}'$ corresponds to the polarization vector $\boldsymbol{e}$ of the photon. In optics, the dependence of the photoabsorption on the direction of $\boldsymbol{e}$ is called *dichroism.* This term is also used in energy-loss spectroscopy. Instead of talking about the dependence of the spectra on the orientation of the scattering vector, we can state that we can detect the linear dichroism in our specimen.

In optics, one can also measure the absorption of circularly polarized light. In some cases, particularly for magnetic materials, photoabsorption then depends on the helicity (right or left) of the incident photon. As the polarization vector $\boldsymbol{e}$ of a circularly polarized photon traveling along the z-direction is given by $\boldsymbol{e} = \boldsymbol{e_x} \pm \mathrm{i}\boldsymbol{e_y}$, equation (5.92) for the photoabsorption cross section has to be generalized correspondingly [5.163].

Recently Hébert and Schattschneider have shown that this effect can also be measured in EELS if a coherent superposition of two plane waves is used as an incident wave instead of a single plane wave [5.162]. The scattered intensity then contains an interference term [6.118] $\langle a_{ns}|\exp(-2\pi\mathrm{i}\boldsymbol{q}' \cdot \boldsymbol{r}_j)|a_{0s}\rangle\langle a_{0s}|\exp(2\pi\mathrm{i}\boldsymbol{q} \cdot \boldsymbol{r}_j)|a_{ns}\rangle$. If the two scattering vectors are small and perpendicular to each other, these matrix elements correspond to equivalent terms in the photoabsorption cross section. First experiments on Fe demonstrate that this effect can indeed be detected in energy-loss spectra [5.164].

5.4 Multiple-Scattering Effects

5.4.1 Angular Distribution of Scattered Electrons

The angular distribution of transmitted electrons consists of the peak of unscattered primary electrons of intensity I and illumination aperture α_i together with the angular distribution of scattered electrons, which can be measured by recording the current $\Delta I(\theta)$ with a detector or Faraday cage having a solid angle $\Delta\Omega$ of collection; the result is normalized by dividing by the incident current I_0. The relation

$$\frac{1}{I_0}\frac{\Delta I(\theta)}{\Delta\Omega} = \frac{N_\mathrm{A}\rho t}{A}\frac{\mathrm{d}\sigma}{\mathrm{d}\Omega} = \frac{x}{x_t}s_1(\theta), \tag{5.93}$$

where $x = \rho t$ is the mass thickness, is valid only for very small values of x; x_t denotes the mean-free-path length of (5.8), and $s_1(\theta)$ is the normalized single-scattering distribution ($\int 2\pi s_1(\theta)\theta\mathrm{d}\theta = 1$).

The intensity I of the unscattered primary beam decreases exponentially with increasing mass thickness x according to (5.8), i.e., $I/I_0 = \exp(-x/x_\mathrm{t})$ with

$$\frac{1}{x_\mathrm{t}} = \frac{1}{x_\mathrm{el}} + \frac{1}{x_\mathrm{inel}} = \frac{N_\mathrm{A}}{A}(\sigma_\mathrm{el} + \sigma_\mathrm{inel}) = \frac{N_\mathrm{A}\sigma_\mathrm{el}}{A}(1+\nu) = \frac{1+\nu}{x_\mathrm{el}}, \tag{5.94}$$

where the ratio $\nu = \sigma_\mathrm{inel}/\sigma_\mathrm{el}$ is defined in (5.66). Values of x_el are listed in Table 6.1.

For E = 100 keV, a value x_t = 12 μg/cm^{-2} or t = 120 nm is found for organic material of density ρ = 1 g/cm^3, but for evaporated Ni and Fe films the same mass thickness is obtained for $t \simeq 15$ nm [5.166]. The corresponding decrease of primary-beam intensity is important for the visibility of phase-contrast effects, which are generated by interference between the primary and scattered electron waves. In Lorentz microscopy (Sect. 6.8), the domain contrast is created by using a primary beam of very small illumination aperture, $\alpha_\mathrm{i} \leq 10^{-2}$ mrad; elastically and inelastically scattered electrons cause a blurring of the domain contrast in the Fresnel mode.

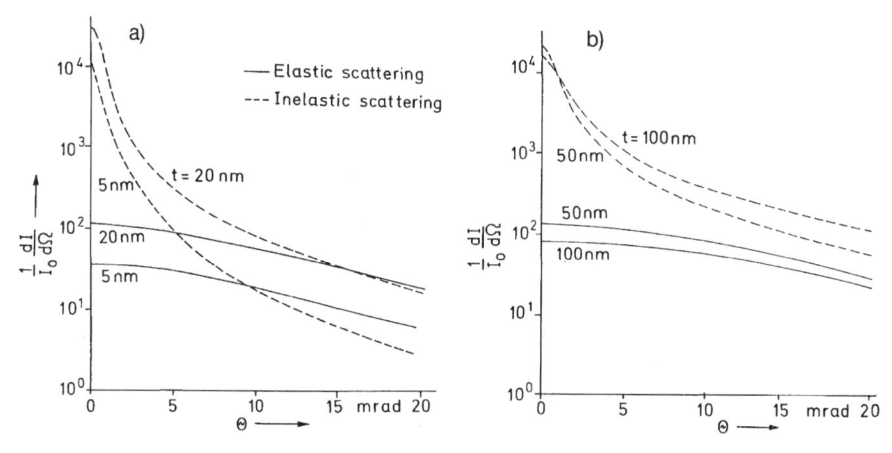

Fig. 5.33. (**a, b**) Angular distribution of the elastic and inelastic scattering intensities of 100 keV electrons in carbon films of increasing thickness $t = 5$–100 nm [5.169].

The angular distribution for multiple scattering can be obtained by evaluating a multiple-scattering integral [5.11] or by superposition of multiple-scattering distributions $s_n(\theta)$, which are calculated by an n-fold convolution of $s_1(\theta)$ defined in (5.93),

$$s_n(\theta) = s_{n-1}(\theta) \otimes s_1(\theta). \tag{5.95}$$

These are then weighted with the coefficients of a Poisson distribution [5.167]

$$\frac{1}{I_0}\frac{\Delta I(\theta)}{\Delta \Omega} = \exp\left(-\frac{x}{x_t}\right)\sum_{n=1}^{\infty}\left(\frac{x}{x_t}\right)^n \frac{s_n(\theta)}{n!} \ . \tag{5.96}$$

A procedure is available whereby the two-dimensional integration necessary for the convolution in (5.95) is reduced to a one-dimensional integration by using projected distributions [5.168].

Figure 5.33 shows the contributions of elastic and inelastic scattering calculated from (5.96) for different carbon-film thicknesses [5.169]. For thin films (5 nm), the angular distribution can be assumed to be approximately proportional to the differential cross section $d\sigma/d\Omega$ for single atoms. The intensity distribution is modified by multiple scattering as the thickness is increased. The elastic contribution at small scattering angles θ increases up to 50 nm but decreases for greater thicknesses due to elastic multiple scattering into larger angles and to inelastic scattering, which dominates for greater thicknesses.

These calculations neglect all interference effects. In crystalline specimens, destructive interference decreases the scattered intensity between the primary beam and the Bragg-diffraction spots; the scattered intensity is caused by thermal diffuse scattering (electron–phonon scattering) and inelastic scattering. In amorphous specimens, the short-range order corresponds to a radial distribution function of neighboring atoms that causes diffuse maxima and minima in the scattered intensity distribution (Sect. 7.5.1). However, this distribution oscillates around the distributions calculated here, in which interference effects were neglected.

5.4.2 Energy Distribution of Transmitted Electrons

Figure 5.34 shows the variation of the energy-loss spectrum with increasing thickness for 1.2 MeV electrons [5.81]. Analogous results are obtained with 100 keV electrons only for correspondingly thinner films because the mean free path is shorter (Fig. 5.19). In a very thin film (Fig. 5.34a), a large fraction of the electrons pass through the film without energy loss. The three multiples of the Al plasmon loss at $\Delta E = 15.2$ eV follow a Poisson distribution (5.84). The intensity of higher energy losses is very low, and an increase caused by L-shell ionization appears at $\Delta E = 80$ eV. At medium thicknesses (Fig. 5.34b), the zero-loss peak is strongly reduced, and seven plasmon losses can be detected. The plasmon losses are superposed on a broad maximum due to overlapping of the L-ionization edge and the multiple plasmon losses. The plasmon losses disappear in very thick specimens (Fig. 5.34c), and only a broad energy distribution with a most probable energy loss $\Delta E_{\rm p}$ and a full-width at half-maximum $\Delta E_{\rm H}$ is observed.

For the value of the most probable energy loss, a theory of Landau [5.170] can be used that considers the atomic structure only in terms of a mean ionization energy $J = 13.5Z$ (see also modifications in [5.171, 5.172]):

$$\Delta E_{\rm p} = \frac{N_A e^4 Z x}{8\pi\epsilon_0^2 A E_0 \beta^2}\left[\ln\left(\frac{N_A e^4 Z x}{4\pi\epsilon_0^2 J^2 A(1-\beta^2)}\right) - \beta^2 + 0.198\right] \tag{5.97}$$

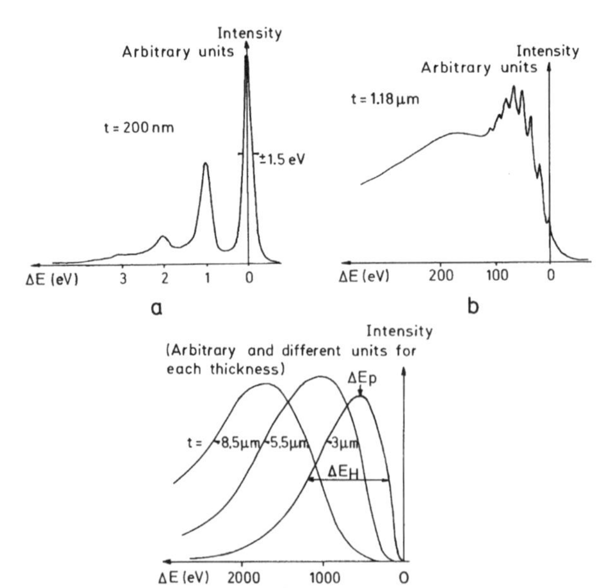

Fig. 5.34. (a-c) Energy-loss spectra of 1200 keV electrons in Al foils of increasing thickness t [5.81].

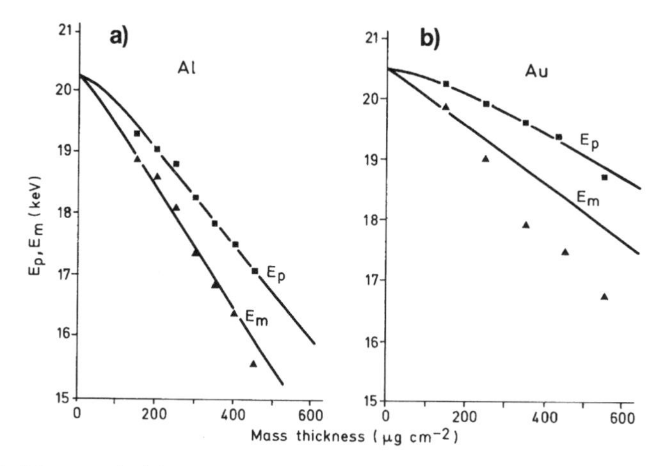

Fig. 5.35. Most probable energy $E_{\mathrm{p}} = E - \Delta E_{\mathrm{p}}$ and mean energy E_{m} of 20 keV electrons in (**a**) Al and (**b**) Au films of increasing mass thickness $x = \rho t$ calculated with (5.97) and (5.100) and comparison with measured values [5.173].

($E_0 = m_0c^2$, $\beta = v/c$, and $x = \rho t$ is the mass thickness). The validity of this formula has been confirmed at 20 keV for Ag and Al (Fig. 5.35) [5.173] and recently at 80 keV using an energy-filtering electron microscope [4.124]. However, the observed values of the full-width at half-maximum ΔE_{H} of the energy distribution are greater than the value

$$\Delta E_{\mathrm{H}} = 4.02 \frac{N_A e^4 Z X}{8\pi\epsilon_0^2 A E_0 \beta^2} \tag{5.98}$$

derived from the Landau theory. The Fourier algorithm of the Landau theory can be extended to the calculation of both angular and energy distributions simultaneously when single-scattering cross sections containing plasmon losses, their dispersion and cutoff, the Compton scattering, and ionization cross sections are used. This allows the influence of multiple scattering on energy-loss spectra to be calculated for different apertures and compared with experiment [5.86].

For many applications, it is not sufficient to characterize the energy distribution by the most probable energy loss E_{p} and the half-width ΔE_{H}; it is also of interest to know the mean energy, which can be calculated from

$$\text{(a)} \quad E_{\mathrm{m}} = \frac{\int_0^E EN(E)E\mathrm{d}E}{\int_0^E N(E)\mathrm{d}E} \quad \text{or} \quad \text{(b)} \quad E_{\mathrm{m}} = E - \int_0^x \left|\frac{\mathrm{d}E_{\mathrm{m}}}{\mathrm{d}x}\right|_{\mathrm{B}} \mathrm{d}x \tag{5.99}$$

by using a measured energy-loss spectrum or from the theoretical Bethe formula for the mean loss per unit path length measured in terms of mass thickness [5.26],

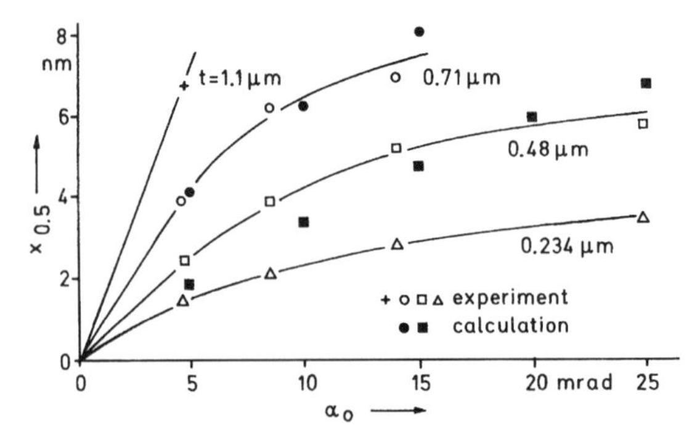

Fig. 5.36. Experimental value of the half-width $x_{0.5}$ of an edge (see Fig. 5.39a for definition) caused by the chromatic aberration as a function of objective aperture α_o for different thicknesses of polystyrene spheres. Solid points: Monte Carlo simulations [5.174].

$$\left|\frac{\mathrm{d}E_\mathrm{m}}{\mathrm{d}x}\right|_\mathrm{B} = \frac{e^4 N_A Z}{4\pi\epsilon_0^2 A E_0 \beta^2} \ln\left(\frac{E_0\beta^2}{2J}\right). \tag{5.100}$$

This formula can also be used to calculate the specimen heating (Sect. 11.1) and the radiation damage caused by ionization (Sect. 11.2). Values of the mean energy E_m obtained from measured energy distributions $N(E)$ by applying (5.99a) agree with calculations using (5.99b) and (5.100) (Fig. 5.35). The stronger decrease of the experimental values for large mass thicknesses of gold can be attributed to an increase of the effective path length caused by multiple scattering.

The energy losses impair the resolution as a result of chromatic aberration (2.63). Measurements of the width $x_{0.5}$ of the blurred intensity spread (Figs. 5.38c and d) at the edges of indium crystals placed below polystyrene spheres of different thicknesses (see Fig. 5.39a for a definition of $x_{0.5}$) are plotted in Fig. 5.36 as a function of objective aperture. The reason why the measured values of $x_{0.5}$ do not increase in proportion to α_o (2.63) is that the step intensity distribution consists of a steep central and a flat outer part. Monte Carlo simulations that take into account the plural scattering of the carbon plasmon-loss spectrum, from which $x_{0.5}$ can be obtained by the same method, predict the same dependence on aperture and agree with experimental results. Figures 5.38c and d at $E = 100$ and 200 keV, respectively, show that the effect of chromatic aberration decreases with increasing E.

5.4.3 Electron-Probe Broadening by Multiple Scattering

The angular distribution of scattered electrons in thick films (0.1–1 μm) produces a spatial distribution that in turn broadens the incident electron probe normal to the beam direction. This effect limits the resolution of the scanning transmission mode, although the chromatic aberration of the conventional

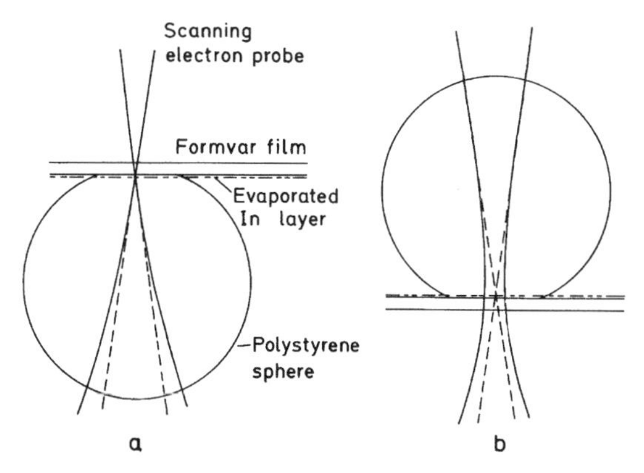

Fig. 5.37. Specimen structure and electron-beam broadening in the two cases in which the polystyrene spheres are (**a**) below and (**b**) above the evaporated indium layer.

transmission mode shown in Fig. 5.36 is avoided. It likewise limits the lateral resolution of x-ray microanalysis of thick specimens.

The multiple scattering can be observed as a *top–bottom effect* in the scanning transmission mode and is illustrated in Figs. 5.37 and 5.38. The specimen consists of a thin formvar supporting film onto which indium, which condenses as small flat crystals on the substrate, has been evaporated. This specimen is coated with polystyrene spheres of 1 μm diameter to simulate a thick specimen of known thickness. The indium layer is scanned by an unbroadened electron probe with the polystyrene sphere below the layer (Figs. 5.37a and 5.38a). The image of the indium crystals is sharp, and the subsequent scattering in the polystyrene sphere and the broadening of the beam merely decrease the intensity recorded with the STEM detector without affecting the resolution. With the polystyrene spheres uppermost, the indium layer is scanned by a broadened probe, the edges of the indium crystals are blurred, and the resolution is reduced (Figs. 5.37b and 5.38b).

A resolution parameter can be obtained by measuring the intensity distribution across the edge of the indium crystals and the width $x_{0.5}$ between the points at which the step reaches 0.25 and 0.75 of its total intensity (Fig. 5.39). Measured values of $x_{0.5}$ behind polystyrene spheres of thickness t are plotted in Fig. 5.39 for different electron energies.

A value of $x_{0.5} \simeq 10$ nm is found for $E = 100$ keV and $t = 1$ μm. The order of magnitude is the same for the chromatic aberration using objective apertures $\alpha_o \geq 10$ mrad (Fig. 5.36). Whereas the blurring of specimen structures by chromatic aberration is approximately the same over the whole specimen thickness, structures at the top of a 1.1 μm layer are imaged in the scanning mode with a better resolution. It has therefore been suggested that the chromatic aberration of a conventional TEM mode should be avoided in this way [4.67]. However, the top–bottom effect sets a limit on the improvement that

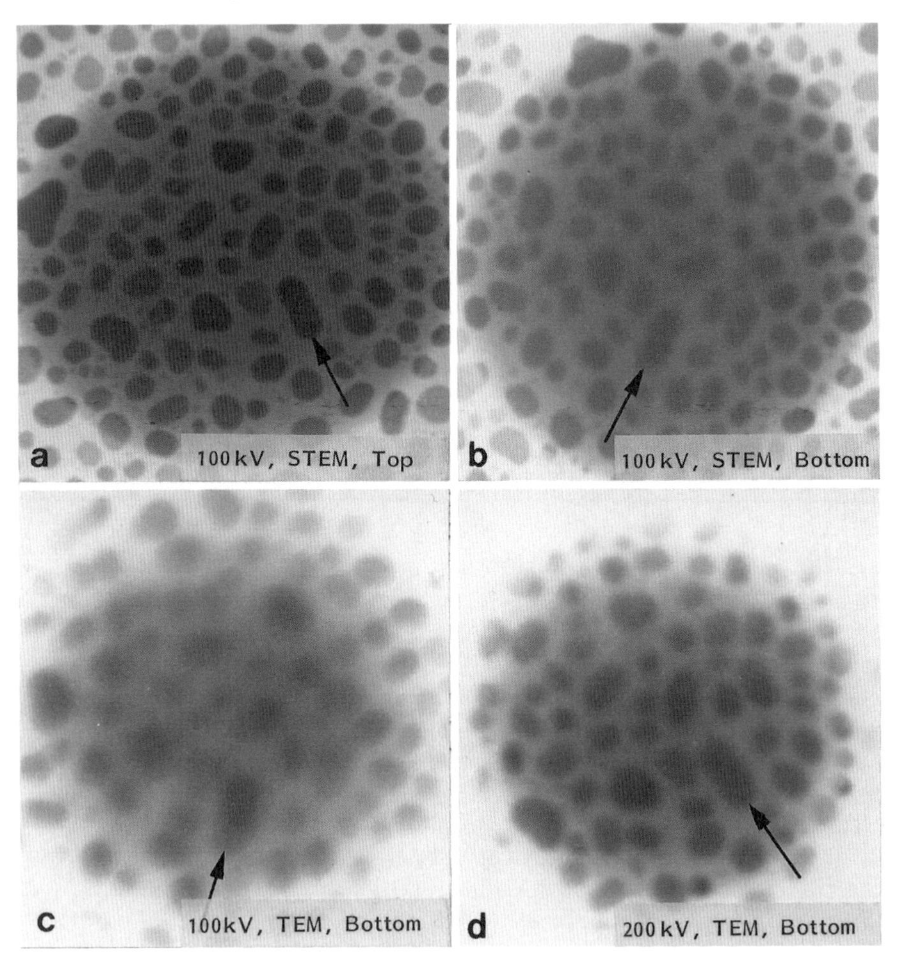

Fig. 5.38. Images of indium crystals and the same polystyrene sphere of 1.1 μm diameter in the 100 keV STEM mode with the polystyrene sphere (**a**) below and (**b**) above the indium layer to demonstrate the top–bottom effect. (**c**) and (**d**) are normal TEM images of the same area at $E = 100$ and 200 keV, respectively, blurred by the chromatic aberration and the energy losses in the polystyrene spheres [5.175].

can be achieved [5.175]. The advantages of the scanning mode can be seen in other applications (Sect. 4.5.1).

The effect of chromatic aberration decreases with increasing energy, and in high-voltage electron microscopy, multiple scattering can also cause a top–bottom effect in the conventional bright-field TEM mode. Here, however, structures at the bottom of the specimen are imaged with a better resolution [5.174]. Such a top–bottom effect is also experimentally proved at 100 keV when the chromatic aberration is avoided by zero-loss filtering with an energy-filtering transmission electron microscope [5.176].

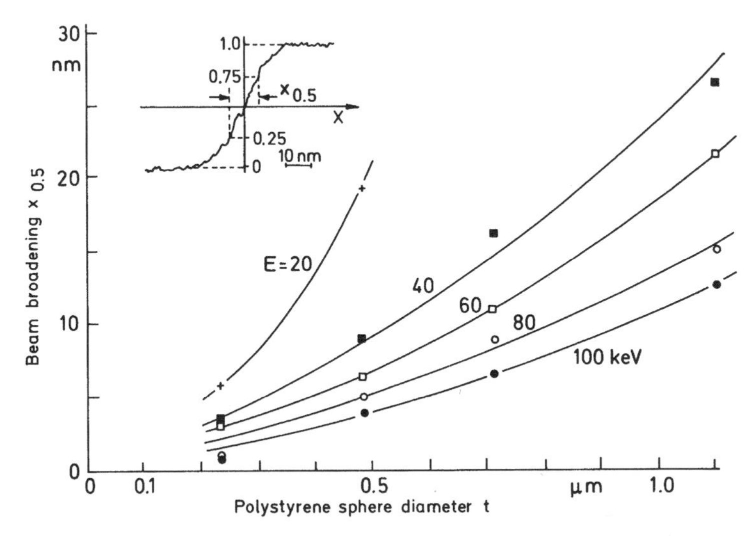

Fig. 5.39. Measurements of $x_{0.5}$ in the STEM mode caused by beam broadening for specimens below polystyrene spheres of thickness t [5.175]. (Inset: Densitometer recording across an edge of an indium crystal and definition of $x_{0.5}$).

This spatial broadening of the electron probe can be calculated from the differential cross section $d\sigma/d\Omega$ by evaluating a multiple-scattering integral [5.177], by solving the Boltzmann transport equation [5.178, 5.179], or by Monte Carlo simulations [5.180]. A disadvantage of all of these methods is that they do not lead to analytical formulas.

An approximate formula for estimating the beam broadening can be obtained if we return to a multiple-scattering theory proposed by Bothe [5.181] (see also [5.182]). The differential cross section (5.34) can be approximated by a two-dimensional Gaussian function of the form $\exp[-(\theta_x^2+\theta_y^2)/\theta_0^2]$, where $\overline{\theta^2} = \theta_0^2$. This function has the advantage that convolutions can be evaluated straightforwardly thanks to the following property of Gaussians:

$$\exp(-x^2/a^2) \otimes \exp(-x^2/b^2) \propto \exp[-x^2/(a^2+b^2)]. \tag{5.101}$$

Bothe obtained the projected probability function that expresses the likelihood of finding an electron at a depth z, a distance x from the axis, and with a projected scattering angle θ_x (Λ: mean-free-path length):

$$P(z,x,\theta_x) = \exp\left[-\frac{4\Lambda}{\theta_0^2}\left(\frac{\theta_x^2}{z} - \frac{3x\theta_x}{z^2} + \frac{3x^2}{z^3}\right)\right]. \tag{5.102}$$

Integration over x gives the projected angular distribution for a film of thickness $z = t$

$$f(t,\theta_x) \propto \exp(-\theta_x^2/\overline{\theta^2}) \quad \text{with} \quad \overline{\theta^2} = \theta_0^2 t/\Lambda. \tag{5.103}$$

This means that the width of the angular distribution increases as $t^{1/2}$. Integration over θ_x results in the projected lateral distribution

$$I(t,x) = \exp(-x^2/x_0^2) \quad \text{with} \quad x_0^2 = \frac{\theta_0^2}{3\Lambda} t^3. \tag{5.104}$$

Substitution of θ_0 from (5.34) and $\Lambda = A/N_A \rho \sigma_{\text{el}}$ from (5.40) yields

$$\begin{aligned} x_0 &= \frac{\lambda^2}{2\pi a_{\text{H}}} \left(\frac{N_A \rho}{3\pi A} \right)^{1/2} Z(1 + E/E_0)\, t^{3/2} \\ &= 1.05 \times 10^5 \left(\frac{\rho}{A} \right)^{1/2} \frac{Z}{E} \frac{1 + E/E_0}{1 + E/2E_0} t^{3/2} \end{aligned} \tag{5.105}$$

with x_0 and t in cm, and E in eV.

With the exception of the numerical factor and the relativistic correction, this formula is identical with one derived in [5.183]. For polystyrene spheres ($\rho = 1.05$ g cm^{-3}), $t = 1$ μm, and $E = 100$ keV, equation (5.105) gives $x_{0.5} = 0.96 x_0 = 20$ nm, which is larger than the measured value of 10 nm (Fig. 5.39). However, the blurred image of an edge in the scanning mode is produced only by small-angle scattering $\theta \leq \alpha_{\text{d}}$, which reduces the beam broadening actually observed.

5.4.4 Electron Diffusion, Backscattering, and Secondary-Electron Emission

Electron diffusion in bulk material is more important for scanning electron microscopy. In TEM, the specimens normally have to be thin enough to avoid the multiple-scattering effects that occur in thick films. However, a knowledge of electron interactions with solids is necessary for recording by photographic emulsions, scintillators, and semiconductors. Walls and diaphragms in the microscope are struck by electrons, and backscattered electrons (BSE) and secondary electrons (SE) from the specimen can be used as signals in the STEM mode. The most important facts about electron diffusion and BSE and SE emission from thin films will therefore be summarized here (see [1.122] for details).

Electron trajectories in a solid are curved by large-angle elastic-scattering processes. The mean electron energy decreases along the trajectory as a result of energy losses. This decrease can be described by the Bethe stopping power (5.100). Integration of (5.100) using (5.99b) yields $E_{\text{m}}(x)$ (Fig. 5.40). Setting $E_{\text{m}} = 0$ gives the *Bethe range* R_{B}, which increases with increasing atomic number Z. However, the trajectories become more strongly curved with increasing Z owing to the presence of the factor Z^2 in the Rutherford cross section (5.31). In practice, the range R is approximately independent of Z if it is measured in units of mass thickness (e.g., μg cm^{-2}) and is of the order of R_{B} only for small Z. The range can be estimated from the empirical formula

$$R \simeq \frac{20}{3} E^{5/3} \tag{5.106}$$

in the range $10 \leq E \leq 100$ keV with R in μg cm^{-2} and E in keV.

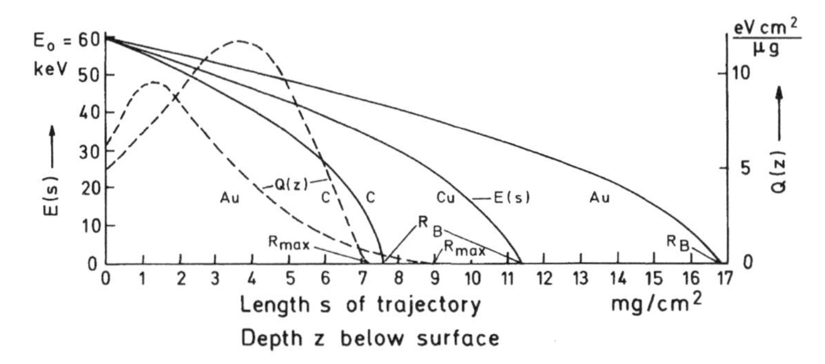

Fig. 5.40. Decrease of mean energy E_{m} along the electron trajectories for different elements and definition of the Bethe range R_{B} is shown along with the depth distribution of ionization density $Q(z)$ (dashed curves) and practical range $R_{\max}$.

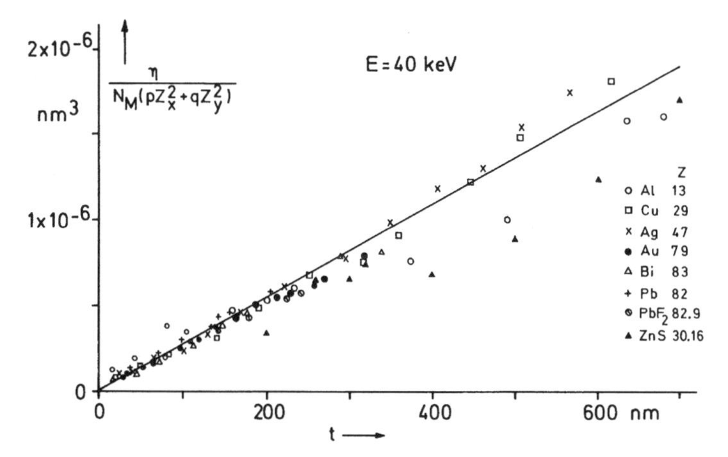

Fig. 5.41. Increase of the backscattering coefficient η with increasing film thickness t plotted as η/NZ^2 versus thickness ($N = N_A\rho/A$: number of atoms per unit volume) [5.185].

The depth distribution $Q(z)$ of energy dissipation by ionization describes the probability of producing electron-hole pairs in semiconductors, or photons in scintillators, and the generation of heat. In Fig. 5.40, $Q(z)$ curves are plotted for C and Au. These have a maximum below the surface and also demonstrate the existence of a range $R_{\max}$ approximately independent of Z.

A fraction η of the incident electrons can leave the specimen as backscattered electrons (BSE) with energies reduced by inelastic scattering; η is known as the backscattering coefficient. Integration of the Rutherford cross section (5.31) from $\theta = \pi/2$ to $\theta = \pi$ (backscattering) gives

$$\eta = \frac{e^4 Z^2}{16\pi\epsilon_0^2 E^2}\left(\frac{E+E_0}{E+2E_0}\right)^2 Nt, \tag{5.107}$$

where $N = N_A\rho/A$ is the number of atoms per unit volume. Plots of η/NZ^2 against the film thickness t are indeed approximately independent of the material (Fig. 5.41). The linear increase of η with increasing film thickness can be used to measure the latter by placing a small Faraday cage in front of the specimen [5.184, 5.185, 5.186]. For a more accurate comparison of theory and experiment, it is necessary to consider Mott instead of Rutherford cross sections [5.187]. Monte Carlo simulations are suitable for these calculations [5.188, 5.189]. The backscattering coefficient may be influenced by channeling effects and also depends on the orientation of the crystal foil relative to the electron beam. These effects can be used to record channeling patterns with BSE by rocking the incident electron beam (Sect. 8.1.2).

Electrons excited by inelastic collisions with an energy sufficiently far above the Fermi level to overcome the work function can leave the specimen as secondary electrons (SE); by convention, these have an energy $E_{\rm SE} \leq$ 50 eV and emerge from a small exit depth of the order of $t_{\rm SE} = 0.5$–10 nm [5.190, 5.191]. The secondary-electron yield δ is proportional to the Bethe loss $|\mathrm{d}E_{\rm m}/\mathrm{d}x|$ (5.100) in the surface layer and to the path length $t_{\rm SE}$ sec ϕ inside the exit depth; ϕ is the angle between the incident direction and the surface normal. The total SE yield is the sum of the SE generated by the primary beam ($\delta_{\rm PE}$) and by the backscattered electrons or the transmitted electrons on the bottom surface ($\delta_{\rm BSE}$),

$$\delta = \delta_{\rm PE} + \delta_{\rm BSE} = \delta_{\rm PE}(1 + \beta\eta). \tag{5.108}$$

The fraction β is greater than unity and can increase to values of 2–3 for compact material [5.192]. This indicates that the number of SE per BSE is greater than $\delta_{\rm SE}$ owing to the decreased BSE energy and the increased path length of BSE in the exit depth. The SE yield at the top and bottom of a thin foil can also be explained in these terms [5.193].

6

Scattering and Phase Contrast for Amorphous Specimens

Elastic scattering through angles larger than the objective aperture causes absorption of the electron at the objective diaphragm and a decrease of transmitted intensity. This *scattering contrast* can be explained by particle optics. The exponential decrease of transmission with increasing specimen thickness can be used for quantitative determination of mass thickness or the total mass of an amorphous particle, for example. The zero-loss mode of electron spectroscopic imaging allows us to increase the contrast by removing inelastically scattered electrons; alternatively, the contrast can be increased by energy filtering at higher energy losses.

The superposition of the electron waves at the image plane results in interference effects and causes *phase contrast*, which depends on defocusing and spherical aberration, on the objective aperture, and also on the particular illumination conditions.

It is possible to characterize the imaging process independently of the specimen structure by introducing the contrast-transfer function, which describes how individual spatial frequencies of the Fourier spectrum are modified by the imaging process. The contrast-transfer function of the normal bright-field mode alternates in sign and decreases at high spatial frequencies owing to the partial temporal and spatial coherence. Methods of suppressing the change in sign, by hollow-cone illumination, for example, have been proposed.

The idea of holography as an image-restoration method, originally proposed by Gabor for electron microscopy, was for many years impeded by the imperfect coherence of the electron beam. With the introduction of field-emission guns, holography can now be employed in electron microscopy for the quantitative measurement of phase shifts at the atomic scale.

Many different methods can be employed for analog or digital image restoration and for the alignment of image structures in a series of micrographs. A tilt series can be used for tomography, especially for low-dose exposures of biomacromolecules.

The phase shift caused by magnetic fields inside ferromagnetic domains can be exploited to image the magnetic structure of thin films or small particles; this is known as Lorentz microscopy.

6.1 Scattering Contrast

6.1.1 Transmission in the Bright-Field Mode

We assume in this section that the electrons move as particles through the imaging system. All electrons that do not pass through the objective diaphragm are stopped by it; this results in *scattering contrast.* This means that we shall be considering the intensity and not the wave amplitude, even in the focal plane of the objective lens, and that we shall sum intensities and not wave amplitudes in the image plane. In the purely wave-optical theory of imaging (Sect. 3.3.2), we always sum over wave amplitudes and obtain the image intensity by squaring the wave amplitude in the final image plane. The resulting *phase contrast* will be considered in Sect. 6.2. (It will be shown in Sect. 6.2.6 in (6.25)–(6.28) that the scattering contrast can be incorporated in the more general phase-contrast theory if complex scattering amplitudes are used.) The scattering contrast therefore describes the image intensity at low and medium magnifications, where phase-contrast effects do not normally have to be considered unless a highly coherent electron beam and large defocusing are employed.

In the bright-field mode, the diaphragm in the focal plane of the objective lens acts as a stop (Fig. 4.15) that absorbs all electrons scattered through angles $\theta \geq \alpha_o$ (objective aperture). Only electrons scattered through $\theta < \alpha_o$ can pass through the diaphragm. We can thus define a transmission $T(\alpha_o)$ that depends on the objective aperture α_o and also on the electron energy E, the mass thickness $x = \rho t$ (ρ: density, t: thickness) and the material composition (atomic weight A and atomic number Z). We assume that the illumination aperture α_i of the incident beam is appreciably smaller than α_o ($\alpha_i \ll \alpha_o$), which is usually the case in normal TEM work, whereas in STEM the two apertures are normally comparable (Sect. 6.1.5).

Scattering contrast is typically observed with amorphous specimens, surface replicas, or biological sections (see quantitative examples in Sect. 6.1.3). Even for amorphous specimens, the assumption that the waves scattered at single atoms add incoherently is not fully justified because the angular distribution of the scattered intensity (diffraction pattern) shows diffuse maxima (Sect. 7.5.1). However, the total number of electrons scattered into a cone of half-angle α_o is very insensitive to such diffuse maxima in the diffraction pattern because the angular distribution oscillates about that corresponding to completely independent scattering (Fig. 7.22). Polycrystalline films with very small crystals, platinum for example, can also be treated by the theory of scattering contrast [6.1]. Evaporated films that contain larger crystals, Ag or Au for example, may deviate from this simple theory owing to dynamical diffraction effects, and their mean transmission averaged over a larger film area cannot be described exactly by the formulas of scattering contrast.

Equation (5.5) can be used to calculate the number of electrons scattered through angles $\theta \geq \alpha_o$ and intercepted by the diaphragm in the focal plane of the objective lens. Substituting the small-angle approximation (5.34) for elastic scattering gives the partial cross section [5.11]

$$\sigma_{el}(\alpha_o) = 2\pi \int_{\alpha_o}^{\infty} \frac{d\sigma_{el}}{d\Omega}\theta d\theta = 2\pi \frac{4Z^2R^4(1+E/E_0)^2}{a_H^2} \int_{\alpha_o}^{\infty} \frac{\theta}{[1+(\theta/\theta_0)^2]^2} d\theta$$

$$= \frac{Z^2R^2\lambda^2(1+E/E_0)^2}{\pi a_H^2} \frac{1}{1+(\alpha_o/\theta_0)^2} . \tag{6.1}$$

The total cross section σ_{el} (5.40) is obtained by setting $\alpha_o = 0$. According to (5.9), the mean free path x_{el} between elastic scattering events becomes

$$x_{el} = \frac{1}{N\sigma_{el}} = \frac{\pi A a_H^2}{N_A Z^2 R^2 \lambda^2 (1+E/E_0)^2} , \tag{6.2}$$

where $N = N_A/A$ denotes the number of atoms per gram. If complex scattering amplitudes $f(\theta)$ are available, say from WKB or partial wave calculations, σ_{el} can be obtained from the optical theorem (5.39); subtracting the number of electrons passing through the diaphragm then gives

$$\sigma_{el}(\alpha_o) = \underbrace{2\lambda \mathrm{Im}\{f(0)\}}_{\sigma_{el}} - 2\pi \int_0^{\alpha_0} |f(\theta)|^2 \theta d\theta. \tag{6.3}$$

This means that the complex scattering amplitude needs to be known only in the region $0 \leq \theta \leq \alpha_o$.

A formula for the cross section $\sigma_{inel}(\alpha_o)$, analogous to (6.1), can be calculated by using the differential inelastic cross section (5.63). The term θ_E that contains the mean ionization energy J can be neglected in q' because it will be important only for very small scattering angles. We find

$$\sigma_{inel}(\alpha_o) = 2\pi \int_{\alpha_o}^{\infty} \frac{d\sigma_{inel}}{d\Omega}\theta d\theta \tag{6.4}$$

$$= 2\pi \frac{\lambda^4 Z(1+E/E_0)^2}{4\pi^4 a_H^2} \int_{\alpha_o}^{\infty} \frac{1}{\theta^4}\left(1 - \frac{1}{[1+(\theta/\theta_0)^2]^2}\right)\theta d\theta$$

$$= \frac{4ZR^2\lambda^2(1+E/E_0)^2}{\pi a_H^2}\left[-\frac{1}{4[1+(\alpha_o/\theta_0)^2]} + \ln\sqrt{1+(\theta_0/\alpha_o)^2}\right] .$$

The decrease of transmission $T(\alpha_o)$ through the aperture α_o with increasing mass thickness $x = \rho t$ can be obtained as in (5.6),

$$\frac{dn}{n} = -\frac{N_A}{A}[\sigma_{el}(\alpha_o) + \sigma_{inel}(\alpha_o)]\, dx. \tag{6.5}$$

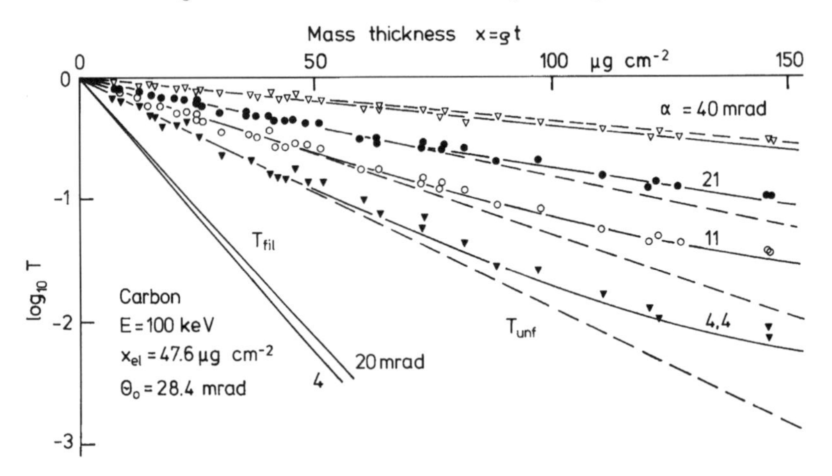

Fig. 6.1. Semilogarithmic plot of the transmission T of carbon films as a function of mass thickness $x = \rho t$ for different objective apertures α_o ($E = 100$ keV). The full curves were calculated using a multiple-scattering integral [5.11] with the constants $x_{el} = 47.6\ \mu$g cm^{-2} and $\theta_0 = 28.4$ mrad, obtained from a best fit of the initial slopes at small x. The straight lines T_{fil} correspond to measurements of zero-loss-filtered transmission. The straight lines T_{unf} correspond to unfiltered transmission.

Integration gives

$$T(\alpha_o) = n/n_0 = \exp[-x/x_k(\alpha_o)], \tag{6.6}$$

where the *contrast thickness* $x_k(\alpha_o)$ is given by

$$\frac{1}{x_k(\alpha_o)} = \frac{4}{Zx_{el}}\left[\frac{Z-1}{4[1+(\alpha_o/\theta_0)^2]} + \ln\sqrt{1+(\theta_0/\alpha_o)^2}\right] \tag{6.7}$$

and x_{el} is defined in (6.2).

The exponential decrease (6.6) of transmission with increasing mass thickness x can be checked by a semilogarithmic plot. The expected linear decrease of $\log_{10} T$ is observed for small mass thicknesses (Fig. 6.1) [6.2, 6.3, 6.4, 6.5]. The agreement is less good for larger mass thicknesses, owing to multiple scattering. A higher transmission is observed than that predicted by (6.6) because electrons first scattered through large angles can be scattered back toward the incident direction and can hence pass through the objective diaphragm. For high energies and large apertures, the situation can be reversed; T then shows a lower increase than expected from the value of x_k because electrons are scattered out of the cone with aperture α_o by multiple scattering. The full curves in Fig. 6.1 were calculated on the basis of a multiple-scattering integral [5.11] and show good agreement with the experimental results. The curves were calculated with the values x_{el} and θ_0 of Table 6.1, which were obtained by fitting the initial slopes of $\log_{10} T$ versus x curves. The limits of linearity of these curves are discussed in [6.5, 6.6, 6.7]. For very large mass thicknesses ($x \geq 100\ \mu$g cm^{-2} in Fig. 6.1), the transmission T is proportional to the solid

Table 6.1. Experimental values [6.5] of mean free path x_{el} and characteristic angle θ_0. Mean-free-path length $\Lambda_{\mathrm{el}} = 10x_{\mathrm{el}}/\rho$ with Λ (nm), x_{el} (μg cm^{-2}) and ρ (g cm^{-3}).

	C		Ge		Pt	
E (keV)	x_{el} [μg cm^{-2}]	θ_0 (mrad)	x_{el} (μg cm^{-2})	θ_0 (mrad)	x_{el} (μg cm^{-2})	θ_0 (mrad)
17.3	10.1	92.4	–	–	6.5	53.8
25.2	14.4	69.9	6.8	50.6	8.1	52.4
41.5	22.4	46.6	10.6	42.6	11.65	50.8
62.1	31.8	37.8	14.4	38.2	14.1	43.2
81.8	39.7	32.4	17.8	34.4	16.8	40.2
102.2	47.6	28.4	21.0	30.8	19.2	38.4
150	70.6	21.6	28.0	23.4	23.4	25.8
300	114.0	17.8	42.0	19.0	31.6	16.2
750	139.2	10.2	58.7	11.5	50.7	13.2
1200	168.0	6.5	62.1	6.8	46.8	8.0

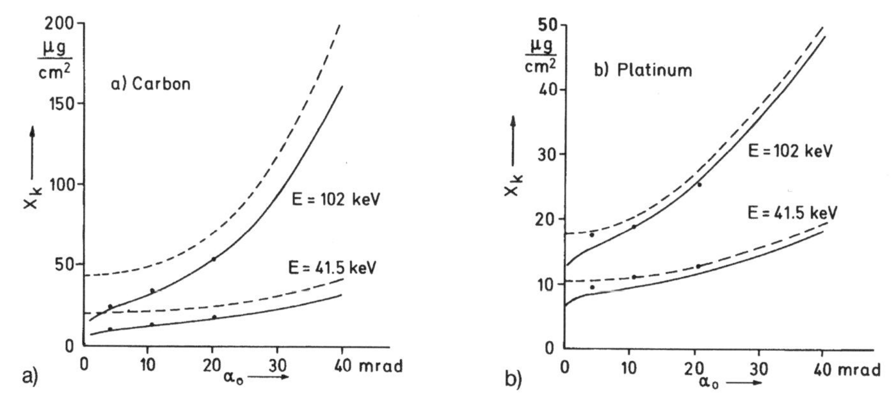

Fig. 6.2. Contrast thickness x_{k} of (**a**) carbon and (**b**) platinum for $E = 40$ and 100 keV, respectively [6.5]. (- - -) Theoretical values considering elastic scattering only and (—) considering both elastic and inelastic scattering [6.8].

angle $\pi\alpha_{\mathrm{o}}^2$ of electrons passing through the objective aperture α_{o}. This is a consequence of the broadened angular distribution of the scattered electrons, which decreases slowly with increasing θ for the range of apertures used. However, this thickness range is of no interest for conventional TEM because of the large energy losses and probe broadening due to multiple scattering. The transmission for zero-loss filtering is discussed in Section 6.1.4.

Values of contrast thickness x_{k} obtained from the initial slope of $\log_{10}T(x)$ in Fig. 6.1 are plotted in Fig. 6.2 for different apertures α_{o} and electron energies; for comparison, calculated values using (6.3) and complex scattering amplitudes $f(\theta)$ given by the WKB method (Sect. 5.1.3) (pure elastic scattering), modified to take account of the inelastic contribution [6.8], are also plotted. The calculation of $f(\theta)$ assumed dense atomic packing, represented by the muffin-tin model.

The mean free path x_{el} for elastic scattering (6.2) and the characteristic angle $\theta_0 = \lambda/2\pi R$ (5.34) should depend on only one parameter, the screening radius R, when the Wentzel potential model (5.22) is used. However, this is a consequence of the Born approximation, which fails for high Z. Nevertheless, (6.7) can still be used when the parameters x_{el} and θ_0, which appear in (6.7), are fitted to the measured values of $x_{\mathrm{k}}(\alpha_{\mathrm{o}})$. Values of these quantities are tabulated in Table 6.1, and the dependence of x_{el} on electron energy is shown in Fig. 6.3. The values for carbon differ from those given in (6.2) by only a constant vertical shift in the logarithmic scale of Fig. 6.3, which means a constant factor. The theory is thus confirmed, so far as the dependence on electron energy is considered, apart from this constant factor, which is determined by the scattering potential $V(r)$ of the atoms.

For all elements, x_{el} attains a saturation value at high electron energies (Fig. 6.3), whereas the contrast thickness $x_{\mathrm{k}}(\alpha_{\mathrm{o}})$ continues to increase for a fixed value of α_{o} (Fig. 6.4). The increase can be understood from the fact that, with increasing energy, the electrons are scattered through smaller angles (Fig. 5.5). For this reason, smaller apertures are normally used in high-voltage electron microscopy.

An empirical law [6.4, 6.5]

$$\log_{10} T = -b \frac{Z^a}{A} x \tag{6.8}$$

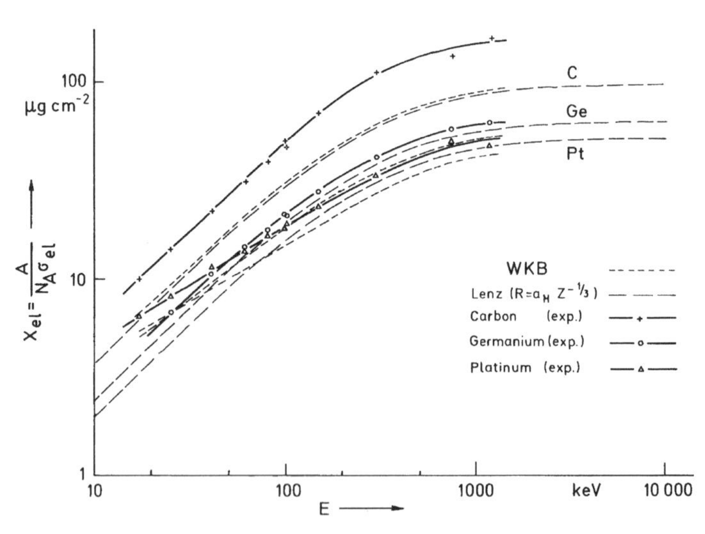

Fig. 6.3. Variation of electron mean free path x_{el} for C, Ge, and Pt films with electron energy E. (- - -) Calculations based on the Lenz theory [5.11] (Born approximation) and (—) calculations by the WKB method [6.5].

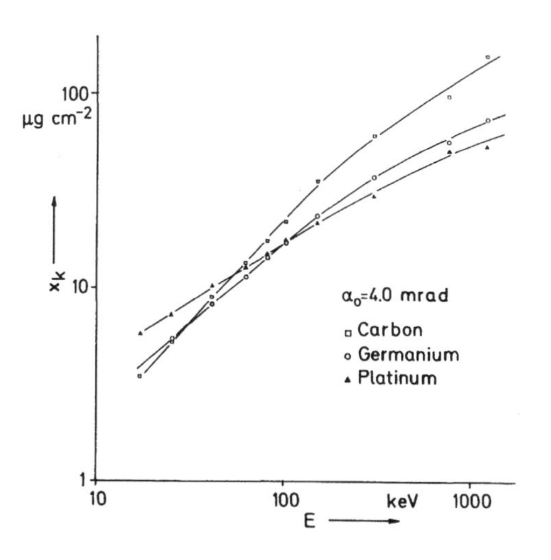

Fig. 6.4. Dependence of the contrast thickness x_{k} of carbon films on the electron energy for different objective apertures α_{o}.

can be used to describe the dependence of T on the atomic number Z for a constant α_{o} and electron energy E; a and b are aperture- and energy-dependent constants. Measurements on gases, which are ideal examples of amorphous specimens, can also be approximated by the same power law as for amorphous and polycrystalline films. The Wentzel atomic model (5.22) with $R = a_{\mathrm{H}} Z^{-1/3}$ leads to $a = 4/3$ for purely elastic scattering if (6.2) is used. Rutherford scattering would give $a = 2$. In reality, none of these exponents of Z is valid. The case in which $E = 60$ keV and $\alpha_{\mathrm{o}} = 4$ mrad is of special interest because $a = 1.1$ for these values and the slow decrease of Z/A with increasing Z is thus compensated. In consequence, the value of T is nearly constant for equal mass thicknesses x of different elements; this is of interest for the determination of mass thickness from measurements of the transmission (Sect. 6.1.6).

6.1.2 Dark-Field Mode

The bright-field mode is not convenient for specimens with very small mass thicknesses such as DNA molecules or virus particles because a decrease of transmission of at least 5% is needed for visual detection. Better contrast can be expected in the dark-field mode if a thin supporting film is used (see the example in Sect. 6.1.3). However, the requisite electron charge density in C cm^{-2} and the exposure time are greater for the dark-field mode. Dark-field imaging is also advantageous if structures with high and low mass thicknesses are to be imaged simultaneously; bacteria with cilia provide a striking example [6.9].

Dark-field images can be formed in the various ways described in Fig. 4.17. To decrease the effect of lens aberrations, the tilt method (Fig. 4.17c) is widely used, and the transition from the bright- to the dark-field mode can be effected by switching on the current in the tilt coils [6.10]. Another way of distributing the intensity of the primary beam around the circular diaphragm is to work

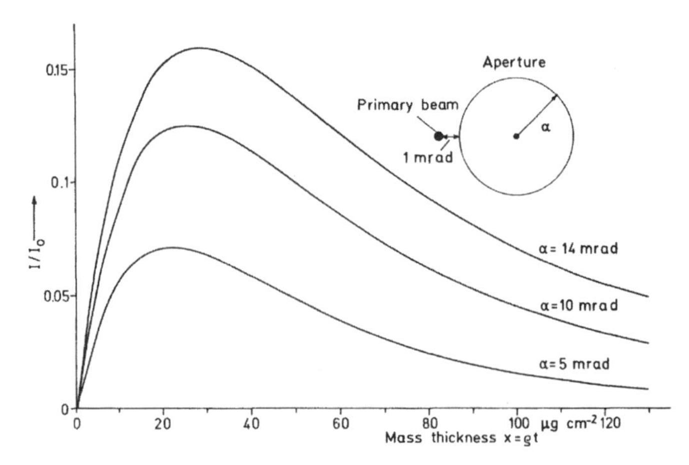

Fig. 6.5. Example of dark-field intensity I/I_0 [I_0: intensity of the incident electron beam] as a function of carbon mass thickness $x = \rho t$ for different objective apertures α_o in the tilted-beam mode; the distance of the primary beam from the periphery of the centered objective diaphragm is 1 mrad (E = 100 keV).

with an annular diaphragm in the condenser lens [6.11, 6.12] or to deflect the electron beam electronically on a cone by means of the tilt coils between condenser and objective lens (hollow-cone illumination) [6.13].

The dark-field intensity I/I_0 is plotted against mass thickness x in Fig. 6.5 for the tilted-beam mode and various centered apertures; the primary beam is at a distance of 1 mrad from the periphery of the centered diaphragm. The intensity passes through a maximum because the number of electrons scattered through the dark-field aperture first increases with mass thickness and subsequently decreases with increasing mass thickness as a result of multiple scattering to larger angles.

6.1.3 Examples of Scattering Contrast

The following quantitative examples of scattering contrast (Fig. 6.6) illustrate how the scattering contrast affects different imaging problems and how this contrast can be calculated with the aid of experimental data; they also indicate how the measured transmission can be quantitatively evaluated. The x_k values used have been calculated from (6.7) using the experimental x_{el} and θ_0 values of Table 6.1.

(a) Shadow-Casting Film (Fig. 6.6a). Shadowing surface replicas with evaporated films of heavy metals increases the contrast and resolution (Fig. 9.37). A shadow such as that shown in Fig. 6.6a is clearly recognizable. Denoting the intensity without a specimen by I_0, the intensity with the carbon supporting film by I_C, and the intensity with the evaporated Pt film by I_{Pt}, the following relations are found:

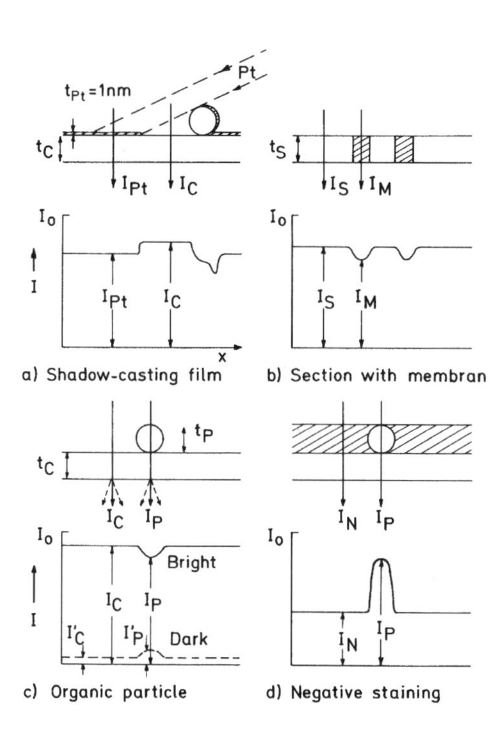

Fig. 6.6. Examples of scattering-contrast calculations.

$$I_C = I_0 \exp\left(-\frac{\rho_C t_C}{x_{k,C}}\right); \quad I_{Pt} = I_0 \exp\left[-\left(\frac{\rho_C t_C}{x_{k,C}} + \frac{\rho_{Pt} t_{Pt}}{x_{k,Pt}}\right)\right]. \tag{6.9}$$

The ratio of the platinum and carbon intensities

$$\frac{I_{Pt}}{I_C} = \exp\left(-\frac{\rho_{Pt} t_{Pt}}{x_{x,Pt}}\right) \tag{6.10}$$

is observed in the image. If t_{Pt} is small, so that the exponential law of transmission (6.6) is obeyed, the thickness of the carbon supporting film has no influence on the ratio I_{Pt}/I_C in (6.10). For a given value of I_{Pt}/I_C, the thickness of the shadowing film must be at least

$$t_{Pt} = \frac{x_{x,Pt}}{\rho_{Pt}} \ln \frac{I_C}{I_{Pt}} . \tag{6.11}$$

As a numerical example, for $E = 80$ keV, $\alpha_o = 4$ mrad, $x_{k,Pt} = 17.5\ \mu\text{g cm}^{-2}$, and $\rho_{Pt} = 21$ g cm^{-3}, we find $t_{Pt} = 0.9$ nm for $I_{Pt}/I_C = 0.9$.

(b) Stained Membrane in a Biological Section (Fig. 6.6b). Measurements at $E = 60$ keV and $\alpha_o = 5$ mrad of the transmission of a thin section of an OsO_4-stained mitochondrial membrane embedded in Vestopal result in mean values of $T_S = I_S/I_0 = 0.765$ for the embedding medium and $T_M = I_M/I_0 = 0.67$ at a membrane. The thickness of the section can be calculated from the first value by assuming that the main contribution to the contrast comes from carbon ($x_{k,C} = 14.6\ \mu\text{g cm}^{-2}$); this gives $x_S = x_{k,C} \ln(1/T_S) = 3.9$

μg cm^{-2}, so that with $\rho_S = 1.1$ g cm^{-3}, the section thickness $t_S = 35.5$ nm. For more accurate quantitative measurements, the mass loss by radiation damage (Sect. 11.2) has to be considered.

Assuming $x_{k,Os} \simeq x_{Pt} = 13.0$ μg cm^{-2}, the second value T_M implies $x_{Os} = x_{k,Pt} \ln(T_S/T_M) = 1.7$ μg cm^{-2} for the equivalent mass thickness of the incorporated osmium. The relative fraction of Os atoms becomes

$$\frac{\text{number of C atoms}}{\text{number of Os atoms}} = \frac{x_S}{x_{Pt}} \frac{A_{Os}}{A_C} = 36.$$

The same ratio $T_M/T_S = 0.88$ would be observed at $E = 1$ MeV and $\alpha_o = 1.5$ mrad for a membrane in a section of thickness $t_S = 120$ nm. For a section as thick as this, the resolution is already reduced at $E = 60$ keV by the effect of chromatic aberration.

(c) Organic Particle on a Supporting Film (Fig. 6.6c). This case is described by a formula similar to (6.10),

$$\frac{I_P}{I_C} = \exp\left(-\frac{\rho_P t_P}{x_{k,C}}\right). \tag{6.12}$$

In bright-field mode, an unstained particle with $t_P = 10$ nm and $\rho_P = 1$ g cm^{-3} generates an intensity ratio $I_P/I_C = 0.97$ for $E = 100$ keV, $\alpha_o = 10$ mrad, and $x_{k,C} = 32$ μg cm^{-2}, which is beyond the limit of visibility. However, such a particle can be seen in phase contrast at optimum defocusing (Sect. 6.2).

If the same particle of 10 nm diameter ($x_P = 1$ μg cm^{-2}) on a carbon support film of $x_C = 1$ μg cm^{-2} ($t_C = 5$ nm) is observed in the dark-field mode, the ratio I'_P/I'_S increases to 2 because the dark-field intensities are proportional to x for small thicknesses. From Fig. 6.5, the ratio I'_S/I_0 can be seen to be 0.01. A 30–50-fold longer exposure time than for a bright-field mode is therefore needed.

(d) Negatively Stained Particle (Fig. 6.6d). The same particle, 10 nm in diameter, is now negatively stained by embedding it in a thin layer of phosphotungstic acid, PWO_4 ($\rho_N = 4$ g cm^{-3}). For the same imaging conditions as in c), the contrast thickness for PWO_4 will be approximately the same as that for Pt: $x_{k,N} \simeq x_{k,Pt} = 19$ μg cm^{-2}. Where the particle is situated, an increase of the transmitted intensity ratio

$$\frac{I_P}{I_N} = \exp\left[\left(\frac{\rho_N}{x_{k,N}} - \frac{\rho_P}{x_{k,C}}\right) t_P\right] = 1.19 \tag{6.13}$$

can be expected, which means a considerable gain of contrast in comparison with the decrease $I_P/I_S = 0.97$ for an unstained particle in the bright-field mode.

6.1.4 Improvement of Scattering Contrast by Energy Filtering

Zero-Loss Filtering. The exponential decrease of transmission (6.6) in the conventional bright-field mode depends on the contrast thickness $x_k(\alpha_o)$ and is a sum $T = I_{un} + I_{el} + I_{in}$ of unscattered, elastically scattered, and inelastically scattered electrons, respectively, that pass the objective diaphragm. By zero-loss filtering, the part I_{in} can be removed and the transmission becomes [6.14]

$$T_{fil} = I_{un} + I_{el} = \exp\left[-\frac{x}{x_{el}}\left(\frac{1}{1+(\alpha_o/\theta_0)^2} + \nu\right)\right], \tag{6.14}$$

where ν (5.66) is the ratio of inelastic-to-elastic total cross sections. In the semilogarithmic plots of $T(x)$ in Fig. 6.1 measurements of the zero-loss transmission T_{fil} are compared with unfiltered values (T_{unf}) for carbon at $E = 80$ keV. Whereas carbon shows a much stronger decrease of the transmission T_{fil} and a weak dependence on aperture α_o, the differences are much less for evaporated platinum films [6.14]. This is a consequence of the difference between the values $\nu \simeq 3$ for carbon and $\nu \simeq 0.25$ for platinum. The gain of contrast for zero-loss filtering by the higher sensitivity to small variations in mass thickness and by the avoidance of chromatic aberration is therefore largest for carbon-containing specimens. The resolution and contrast of membrane structures in biological sections, for example, are much better with zero-loss filtering, and it is possible to investigate section thicknesses up to 0.5 μm where the zero-loss transmission T_{fil} falls below 10^{-3} (Fig. 6.1), which is a criterion for the practical limit of observation (see also Sect. 9.1.6). An example of the improvement of contrast and resolution by zero-loss filtering is shown in Fig. 6.7 for a section of a copolymer of polyethylene and polypropylene stained with ruthenium oxide.

Although the chromatic aberration can be avoided by zero-loss filtering, the resolution of 0.5–1 μm thick organic specimens can be limited by the

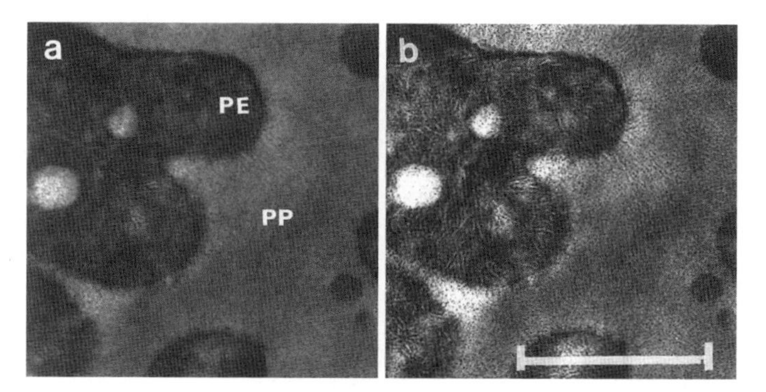

Fig. 6.7. Comparison of (**a**) an unfiltered and (**b**) zero-loss filtered image of a thin section of a copolymer of polyethylene (PE) and polypropylene (PP) stained with ruthenium oxide ($E = 80$ keV, bar $= 0.5\ \mu$m).

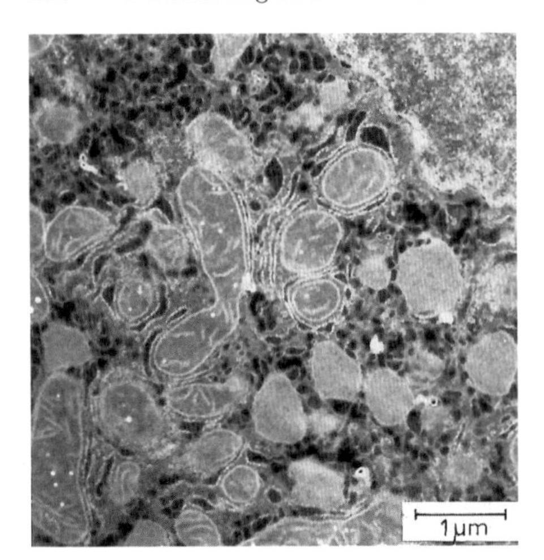

Fig. 6.8. Demonstration of structure-sensitive contrast in an electron spectroscopic image (ESI) at $\Delta E = 250$ eV of a 60 nm liver section (OsO_4-gluaraldehyde fixed, uranyl-acetate stained, and epon embedded).

top–bottom effect, as described for the STEM mode in Sect. 5.4.3 with the difference that structures at the bottom show a better resolution than those at the top [5.176].

Structure-Sensitive Contrast. With an energy-loss window just below the carbon K edge at $\Delta E = 285$ eV (Fig. 4.30), the contribution of carbon to an electron spectroscopic image will be at a minimum. Superposed contributions to the EELS intensity from the tail of plasmon losses and ionization edges below the carbon K edge of other elements give a brighter image of these components, as in a dark-field image, with a better contrast than in the conventional dark-field mode (Sect. 6.1.2) [5.37, 6.15]. This structure-sensitive contrast can be seen in an ESI image of a 60 nm liver section recorded at $\Delta E = 250$ eV (Fig. 6.8). At $\Delta E \simeq 50$ eV, the EELS of the stained part intersects that of the unstained material and the contrast changes from bright to dark field.

Contrast Tuning. The EELS from different parts of a specimen can intersect several times owing to differences in the decrease of the background intensity with increasing energy loss and overlapping of the ionization edges of different elements. This causes contrast reversals when the selected energy is tuned over a larger range of selected energy-loss windows. This technique of contrast tuning [6.16, 6.17] can be applied to thicker biological sections when the stained areas become very dark and cannot be recorded together with much brighter areas. Contrast tuning can reveal an optimum energy window in which both parts are imaged with comparable intensities. Another example is shown in Fig. 6.9 for the same copolymer as in Fig. 6.7, but the specimen is now thicker and more lightly stained. Whereas the unfiltered image and an ESI at $\Delta E = 50$ eV show no strong difference in contrast and the boundaries between PE and PP cannot be clearly distinguished, maximum contrast is

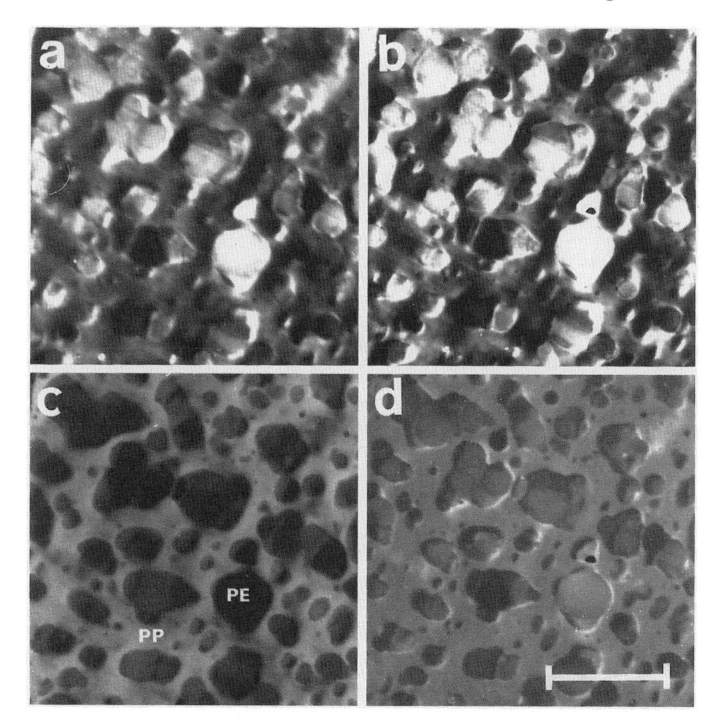

Fig. 6.9. Demonstration of contrast tuning for the example of a section of a copolymer of polyethylene (PE) and polypropylene (PP) stained with ruthenium oxide; the section is thicker than that of Fig. 6.9 and more lightly stained. (**a**) Unfiltered, (**b**) $\Delta E = 50$ eV, (**c**) $\Delta E = 200$ eV, and (**d**) $\Delta E = 350$ eV ($E = 80$ keV, bar = 5 μm).

observed at $\Delta E = 200$ eV, which can be used for a stereological measurement of relative fractions. At $\Delta E = 300$ eV beyond the carbon K edge, the contrast decreases again.

Most-Probable-Loss Imaging. When the intensity of the zero-loss transmission falls below 10^{-3} (e.g., $T \simeq 10^{-4}$ for a 1 μm thick biological section at 80 keV), the EELS shows a broad maximum between 100 and 300 eV (e.g., $\Delta E = 270$ eV for $t = 1$ μm). This most probable energy loss can be calculated by the Landau formula (5.97), which is in agreement with experiments [4.124, 5.173]. The intensity at the most probable loss is large enough to record an image either in a dedicated scanning transmission electron microscope [6.18] or by EFTEM with an imaging energy filter [6.16, 6.19]. Whereas zero-loss filtering is limited to mass thicknesses $x \leq 70$ μg/cm^2, most-probable-loss imaging allows us to investigate organic films up to $x \simeq 150$ μg/cm^2. The resolution will be limited by the broad energy window of 10–20 eV and the large aperture necessary to obtain sufficient intensity. Electron spectroscopic images of 0.7 μm sections at 80 keV are comparable with micrographs in a conventional transmission electron microscope at 200 keV, though there are differences in contrast: The ESI image shows more details [6.16].

6.1.5 Scattering Contrast in the STEM Mode

It is a characteristic of the bright-field transmission mode in TEM that the illumination aperture α_i is much smaller than the objective aperture α_o (Fig. 6.10a). In consequence, the transmission $T = I/I_0$ depends only on the objective aperture. Small shifts of the objective aperture or small inclinations of the incident beam hardly alter the intensity I that goes through the diaphragm. It was shown in Sect. 4.2.2 that the smallest possible spot size of an electron probe for STEM can be obtained only with a relatively large probe aperture $\alpha_\mathrm{p} \simeq 10$ mrad. The theorem of reciprocity (Sect. 4.5.3) indicates that the same transmission can be expected if the electron-probe aperture α_p is approximately equal to α_o, whereas the detector aperture is small: $\alpha_\mathrm{d} \simeq \alpha_\mathrm{i}$ (Fig. 6.10b). In fact, a lower intensity I_0 is recorded in the absence of a specimen because I_0 is a fraction $\alpha_\mathrm{d}^2/\alpha_\mathrm{p}^2$ of the intensity of the incident electron probe with aperture α_p. The intensity I with a specimen present is determined by the decrease of I_0 due to scattering through larger angles together with the increase due to scattering from the other directions of incidence back into the detector aperture. The same ratio $T = I/I_0$ can therefore be expected as in the TEM mode if we normalize with respect to the intensity I_0 that actually passes through the detector aperture. In practice, however, the electron irradiation must be minimized and the signal-to-noise ratio must be made as high as possible; it thus becomes more convenient to work with $\alpha_\mathrm{d} \simeq \alpha_\mathrm{p}$ in the STEM mode (Fig. 6.10c), so as to collect all of the electrons of the incident beam when no specimen is present.

Figure 6.11a shows calculated lines of equal transmission $T = I/I_0$ for a relatively thin carbon film ($t = 320$ nm) in an $\alpha_\mathrm{p} - \alpha_\mathrm{d}$ diagram. The full curves are those for which I_0 represents the intensity going through the detector aperture. The dashed lines for $\alpha_\mathrm{p} < \alpha_\mathrm{d}$ are those for which I_0 is the total

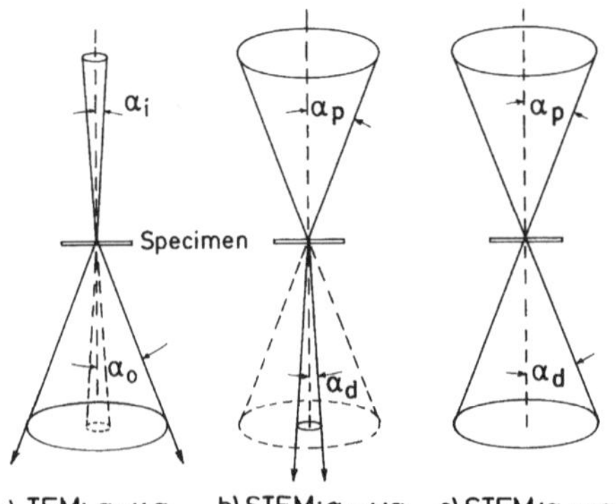

Fig. 6.10. (**a**) Apertures in the TEM mode, (**b**) reciprocal apertures in the STEM mode, and (**c**) optimum STEM mode with $\alpha_\mathrm{p} \simeq \alpha_\mathrm{d}$.

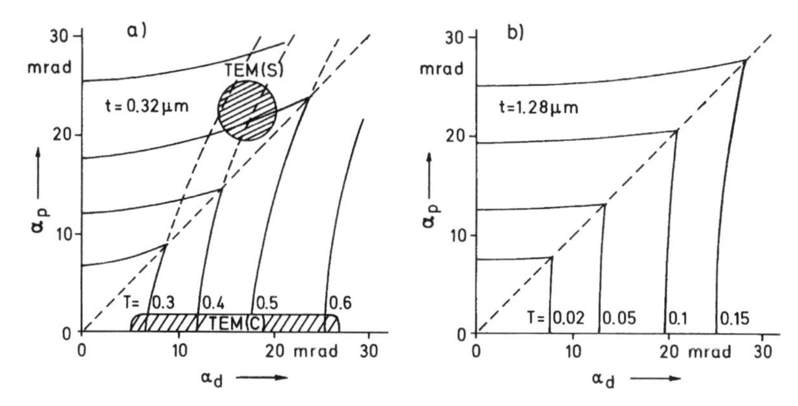

Fig. 6.11. Lines of equal transmission T in an $\alpha_p - \alpha_d$ diagram (α_p: electron-probe aperture, α_d: detector aperture) for (**a**) $t = 320$ nm and (**b**) $t = 1.28$ μm carbon films ($E = 100$ keV). The hatched areas indicate the ranges of the conventional (C) and scanning transmission (S) modes of TEM.

current in the electron probe. When thicker films are used ($t = 1.28$ μm in Fig. 6.11b), the angular width of the electron-scattering distribution becomes broader than the apertures used. The transmission then becomes less dependent on the aperture and is determined by the larger of the two apertures α_p and α_d [6.20, 6.21].

If α_d is increased while α_p is kept constant (corresponding to motion along a line parallel to the abscissa of Fig. 6.11a), the transmission decreases for small mass thicknesses x. The decrease of T with increasing mass thickness is still exponential, but the value of x_k for $\alpha_d \simeq \alpha_p$ is larger in the STEM mode than it is for $\alpha_i \ll \alpha_o = \alpha_d$ in the TEM mode.

As shown in Fig. 1.2, the signals I_{el} for electrons scattered elastically through large angles, I_{un} for unscattered electrons, and I_{in} for inelastically scattered electrons with energy losses can be recorded simultaneously. A display of the ratio I_{el}/I_{in} provides a Z-contrast image with enhanced contrast of stained and unstained biological sections [6.22, 6.23], though the ratio signal is only thickness-independent as long as both signals are proportional to the mass thickness [6.24]. Another way of discriminating between different elements is to use two annular detectors or a set of ring detectors to collect elastically scattered electrons at different scattering angles [6.25, 6.26].

6.1.6 Measurement of Mass Thickness and Total Mass

The exponential law of transmission (6.6) in the conventional TEM bright-field mode and the STEM mode can be used for a quantitative determination of the mass thickness of amorphous specimens, such as supporting films, biological sections, and microorganisms (see the examples in Sect. 6.1.3) [6.27, 6.28, 6.29]. The method can also be used to measure the loss of mass by radiation damage (Sect. 11.2). It is only necessary to know the contrast thickness x_k for the operating conditions in question (electron energy, objective aperture, material). Calibration of this value with films of known mass-thickness, established by microbalance or interferometric measurements, is preferable to theoretical calculations. If t is measured by an interferometric method (two-beam or Tolansky multiple-beam interferometry), care must be taken to ensure that the film has the same density ρ as the bulk material for the calculation of the mass thickness $x = \rho t$. The mass thickness will be directly proportional to $\log_{10}(1/T) = \log_{10}(I_0/I_s)$. The intensities I_s and I_0 with and without the specimen, respectively, can be obtained by placing a Faraday cage in the image plane, by measuring the photographic density D of a developed emulsion with a densitometer, or by using the signal from a CCD camera.

In the STEM mode, the signal provided by a scintillator–photomultiplier combination is directly proportional to the intensity. A signal proportional to the mass thickness can be obtained online by means of a logarithmic amplifier [6.30] and can be displayed as a Y-modulation trace on the cathode-ray tube (CRT). This method can also be used to plot lines of equal transmission (mass thickness) directly, and these can be superposed on the CRT image. Isodensity curves can be produced from photographic records by special reproduction techniques [6.31]. The proportionality of the dark-field signal to very small local mass thicknesses (Fig. 6.5) can also be used to provide a digital record of mass thickness in a dedicated STEM [6.32, 6.33, 6.34]. The backscattering coefficient of thin films is proportional to the thickness (Fig. 5.37). A backscattered electron signal can be recorded by placing a semiconductor or scintillation detector in front of the specimen (Fig. 4.19) and can be used in the STEM mode for the determination of the local mass thickness of biological sections [6.35].

These methods yield the local mass thickness of a specimen. The total mass of a particular particle can be evaluated by numerical integration over the projected area, which is straightforward with digital integration of a logarithmic STEM signal. A special photometric method has been employed for the bright-field [6.36, 6.37] and dark-field modes [6.12], but the methods discussed above are preferable when the microscope is linked to a computer.

These methods for the quantitative measurement of mass thickness are applicable only to amorphous specimens; in the crystalline state, a film of the same mass thickness will show a decrease of the diffuse scattering depending on specimen temperature (thermal diffuse scattering, Sect. 7.5.3), and the intensities of the Bragg reflections depend strongly on the specimen thickness and orientation. Polycrystalline films with large crystals (Cu, Ag, and Au evaporated films, for example) show an averaged transmission that can be twice the value found for an amorphous film. For films with very small crystals (such as Al, Ni, Pt), however, the transmission is of the same order as that of amorphous films of equal mass thickness, provided that the crystals are so small that their diffraction intensity is within the limits of the kinematical diffraction theory [6.1].

6.2 Phase Contrast

6.2.1 The Origin of Phase Contrast

We have shown that there is a phase shift of 90° ($\pi/2$ radians) between the unscattered and scattered waves (Sect. 5.1.3). The complex scattering amplitude of the atom creates an additional phase shift $\eta(\theta)$, which can be neglected for low-Z material. If ψ_i is the amplitude of the incident wave in the final image ($I_0 = \psi_i \psi_i^* = |\psi_i|^2$) and ψ_{sc} that of the scattered spherical wave that passes through the objective diaphragm, there will be a phase shift of $\pi/2$ if we assume that the imaging lens introduces no additional phase shift. We examine the 90° phase shift by plotting $\psi_i + i\,\psi_{sc}$ as a complex amplitude. Figure 6.12a shows that, for $\psi_{sc} \ll \psi_i$, the resulting amplitude has approximately the same absolute value as ψ_i, so that $I = |\psi_i + i\psi_{sc}|^2$ does not differ significantly from $I_0 = |\psi_i|^2$; this means that the phase object is invisible. If the phase of the scattered wave could be shifted by a further 90° (Fig. 6.12b), the superposition would become $\psi_i - \psi_{sc}$ and hence $I = |\psi_i - \psi_{sc}|^2 = \psi_i^2 - 2\psi_i\psi_{sc} + \ldots < I_0$. This is called *positive phase contrast*. If ψ_{sc} were shifted by $3\pi/2$ or $-\pi/2$, the superposition would be $\psi_i + \psi_{sc}$ (Fig. 6.12c) so that $I > I_0$; this is called *negative phase contrast*. In light microscopy, these phase shifts can be produced by inserting a Zernike phase plate in the focal plane of the objective lens; such a plate shifts the scattered wave by an optical-path-length difference of $\lambda/4$ and has a central hole through which the primary beam passes unmodified. In electron microscopy, a path difference $\lambda/4$, which corresponds to a phase shift of $\pi/2$, can be produced by passing 100 keV electrons through a 23 nm thick carbon foil with inner potential $U_i = 8$ V (Sect. 3.1.3). This possibility has been investigated in attempts to create the desired phase shift by means of a carbon foil with a central perforation (Sect. 6.4.6). However, practical difficulties arise with such phase plates because, in continuous operation, the foil becomes charged and contaminated by electron irradiation. Recently, microscopic electrostatic elements have been built, that allow application of a well-defined phase-shift to the primary beam. These will be discussed in Sect. 6.4.6.

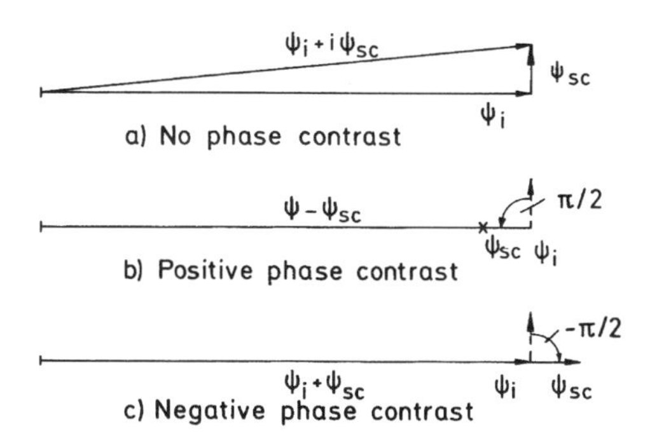

Fig. 6.12. (**a**) Vector addition of the image amplitude ψ_i and the scattered amplitude ψ_{sc} phase shifted by $\pi/2$ or 90°, (**b**) positive phase contrast produced by an additional phase shift of $+\pi/2$; (**c**) negative phase contrast produced by an additional phase shift of $-\pi/2$ or $+3\pi/2$.

The effect of spherical aberration and defocusing may be expressed in terms of the wave aberration $W(\theta)$ (Fig. 3.15). The shape of the wave-aberration curve for different values of defocusing shows that a phase shift $\varphi = -W(q) = \pi/2$ cannot be obtained simultaneously for all scattering angles; only for a limited range of scattering angles or their corresponding spatial frequencies q will $W(\theta)$ produce the desired phase shift. Defocus values for which $W(\theta)$ takes a minimum value of $-\pi/2$ are particularly favorable ($\Delta z^* = 1$ in Fig. 3.15).

6.2.2 Defocusing Phase Contrast of Supporting Films

Supporting films (especially carbon) show a characteristic granular structure at high resolution, the appearance of which changes with the defocus (Figs. 6.14); this granularity was first reported by Sjöstrand [6.38] and discussed as a phase-contrast effect by von Borries and Lenz [6.39]. Carbon films show statistical fluctuations of local mass thickness and therefore of the electron-optical phase shift. The two-dimensional Fourier transform of the phase shift contains a wide range of spatial frequencies. For this reason, carbon (or better, amorphous germanium) films are ideal test specimens for investigating the transfer characteristics of an electron-optical imaging system for different spatial frequencies.

A single spatial frequency q that corresponds to spacing or periodicity $\Lambda = 1/q$ creates a diffraction maximum at a scattering angle $\theta = \lambda/\Lambda = \lambda q$. Those spatial frequencies for which the wave aberration is an odd multiple of $\pi/2$, and thus

$$W(\theta) = (2m-1)\frac{\pi}{2}\begin{cases} m = \text{even: maximum negative phase contrast} \\ m = \text{odd: maximum positive phase contrast,} \end{cases} \tag{6.15}$$

will be imaged with maximum phase contrast. Wave aberrations (phase shifts) for which $W(\theta) = m\pi$, where m is an integer, generate no phase contrast and thus leave gaps in the spatial-frequency spectrum observed at the image. Substituting for $W(\theta)$ from (3.65) in (6.15) and writing $\theta = \lambda/\Lambda$, we obtain an equation for those values of Λ for which maximum positive or negative phase contrast is to be expected. Solving for Λ gives

$$\Lambda = \lambda\left[\frac{\Delta z}{C_s} \pm \left(\frac{\Delta z^2}{C_s^2} + \frac{(2m-1)\lambda}{C_s}\right)^{1/2}\right]^{-1/2} . \tag{6.16}$$

This formula of Thon [6.40] is an extension of the earlier expression of Lenz and Scheffels [6.41]. In the latter, only those terms of $W(\theta)$ caused by defocusing were considered, $W(\theta) = \pi\Delta z\theta^2/\lambda = \pi/2$. This led to

$$\Lambda = \sqrt{2\Delta z\lambda} . \tag{6.17}$$

This is valid for large Λ and defocusing Δz. In these conditions, it relates a specimen periodicity Λ to the optimum defocusing Δz at which the periodicity will be imaged with optimum phase contrast.

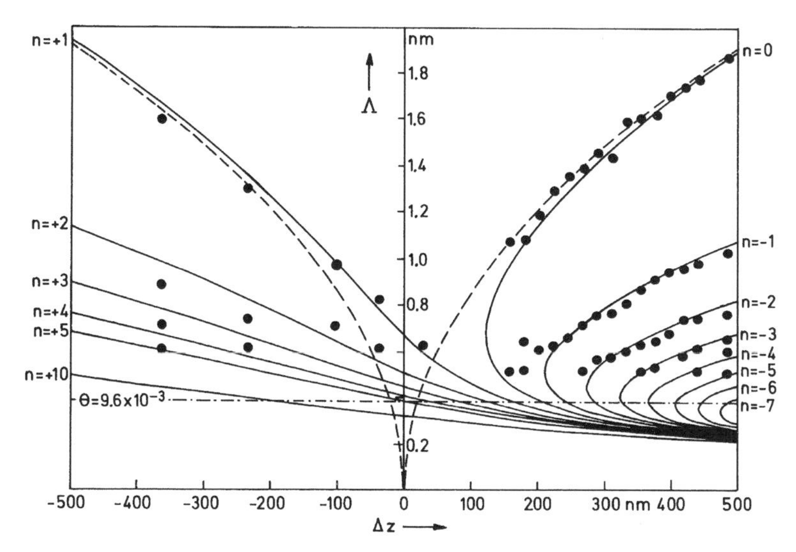

Fig. 6.13. Comparison of measured spatial frequencies q with maximum positive and negative phase contrast obtained by laser diffraction on micrographs of carbon films and theoretical curves based on (6.16) [6.40].

The periodicities Λ that are imaged with maximum positive or negative phase contrast can be measured in light-optical Fraunhofer diffraction patterns of the developed photographic emulsion or by digital two-dimensional Fourier transform (Sect. 6.4.7); typical curves are plotted in Fig. 6.13 as functions of defocusing Δz. The full curves were calculated from (6.16) and show excellent agreement. Even in focus, the granularity of the carbon film does not disappear owing to the term in $W(\theta)$ that contains the spherical aberration. The resolution is limited in this experiment (horizontal dashed line in Fig. 6.13) by the attenuation of contrast transfer caused by chromatic aberration and the finite illumination aperture (Sect. 6.4.2).

The transfer of spatial frequencies as a function of the defocus can be illustrated by calculating a linear Fourier transform of the phase-contrast image of a tilted specimen [6.42]. An example is shown in Fig. 6.14.

Crystalline areas with periodic structures have been observed in carbon foils [6.43] by using a tilted primary beam, as used for the imaging of lattice planes (Sect. 9.6.1). However, when such structures are seen in amorphous specimens with this mode of imaging, they may equally well be caused by selective filtering of spatial frequencies. This selective filtering results from modification of the contrast-transfer function caused by a tilt of the illuminating beam (Sect. 6.4.3). Bright spots of 0.2–0.5 nm diameter have been observed in dark-field imaging with an annular aperture [6.44] that were most intense in overfocus. These spots were attributed to Bragg reflections on small crystallites (see also [6.45]).

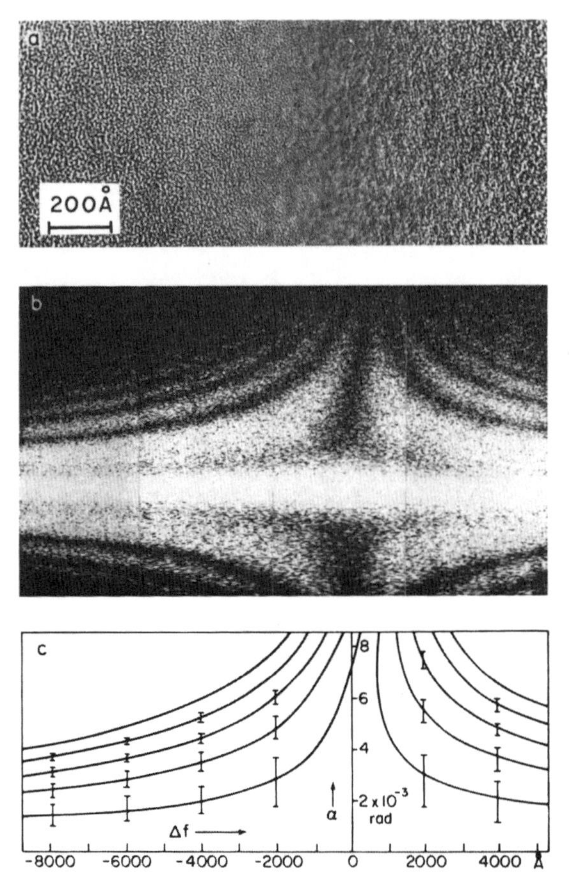

Fig. 6.14. (**a**) Bright-field electron micrograph of tilted carbon film [6.42]. *A* 1000 Å distance in the horizontal direction corresponds to a defocus range of 2145 Å. ($C_s = 1.35\,mm, \lambda = 0.037Å$). (**b**) "One-dimensional" light optical Fourier transform of a tilted film. (**c**) Match between the theoretical defocusing dependence (solid lines) and experimental transform (vertcal bars). The vertical bars indicate the approximate width of the bright bands from the experimental "one-dimensional" transform. (**b**) and (**c**) can be directly compared since their horizontal and vertical scales are identical and a vertical line will pass through the same defocus.

As we have seen, the granularity of carbon foils is very useful for investigating the contrast transfer of TEM but degrades the image of small particles, macromolecules, and single atoms by adding a noisy background. Numerous attempts have therefore been made to prepare supporting films with less granularity in phase contrast (Sect. 4.3.2).

An electron-optical method of decreasing the phase contrast of the supporting film relative to the contrast of single atoms and structures with stronger phase contrast involves using hollow-cone illumination (Sect. 6.4.3).

6.2.3 Examples of Phase Contrast

Figure 6.14 shows a through-focus series of ferritin molecules on a carbon supporting film 5 nm thick. In focus (third image in second row), the molecules show weak scattering contrast due to the iron-rich core of the molecules ($\simeq$5 nm in diameter). This contrast is caused by the loss of electrons that have been scattered through large angles and intercepted by the objective diaphragm. The part of the electron wave that passes through the objective aperture is phase shifted by 90°. An increase of contrast caused by phase contrast can be observed for underfocus ($\Delta z > 0$). The image of the molecules becomes darker at the center. Normally, the operator instinctively focuses for maximum contrast, which means underfocusing. In overfocus ($\Delta z < 0$), the phase shift $W(\theta)$ becomes positive and the molecules appear bright in the center. For a quantitative interpretation of the dependence of image intensity in the center on defocusing, see [6.46].

Reversed phase contrast may occur in some specimens, for example molecules of o-phenanthroline incorporated in electrodeposited nickel films [6.47]. The molecules are imaged as bright spots in underfocus and as dark spots in overfocus. This confirms that there really are vacancies in the nickel film ($\simeq$1 nm in diameter) that contain the organic molecules. Because of the lower inner potential U_{i} of the vacancies, the wavefront behind the inclusions will exhibit an opposite phase shift. Phase contrast can also be observed in defocused images of crystal foils with vacancy clusters [6.48].

In phase contrast, the number of electrons that pass through the objective diaphragm will be constant and all will reach the image. This means that if the intensity at some points of the specimen is increased by summing the amplitudes with favorable phase shifts, the intensity at neighboring image points will be decreased so that the mean value of the intensity is reduced only by scattering contrast. If the image of a particle is darker in the center as a result of positive phase contrast, it will be surrounded by a bright rim and vice versa (Fig. 6.15). Beyond this bright ring, further rings follow with decreasing amplitudes. In complex structures, and especially in periodic structures, these bright and dark fringes can interfere and cause artifacts. Figure 6.16 demonstrates such an effect for myelin lamellae. The contrast of the membranes can be reversed by overfocusing ($\Delta z < 0$). The width of the dark stripes increases with increasing overfocus, and at $\Delta z = -4.8$ μm, twice the number of dark lines can be seen. In underfocus, an increase of the dark contrast of the membranes can again be observed.

The two examples of Figs. 6.15 and 6.16 show that the phase-contrast effects in a defocus series can be interpreted when the specimen structure is known from a focused image or from the method of preparation. For structures smaller than 1 nm, however, this becomes difficult because the spherical aberration term of $W(\theta)$ also has to be considered. In this case, more complicated image-reconstruction methods have to be used (Sect. 6.6) to extract information about the specimen from a single micrograph or a series.

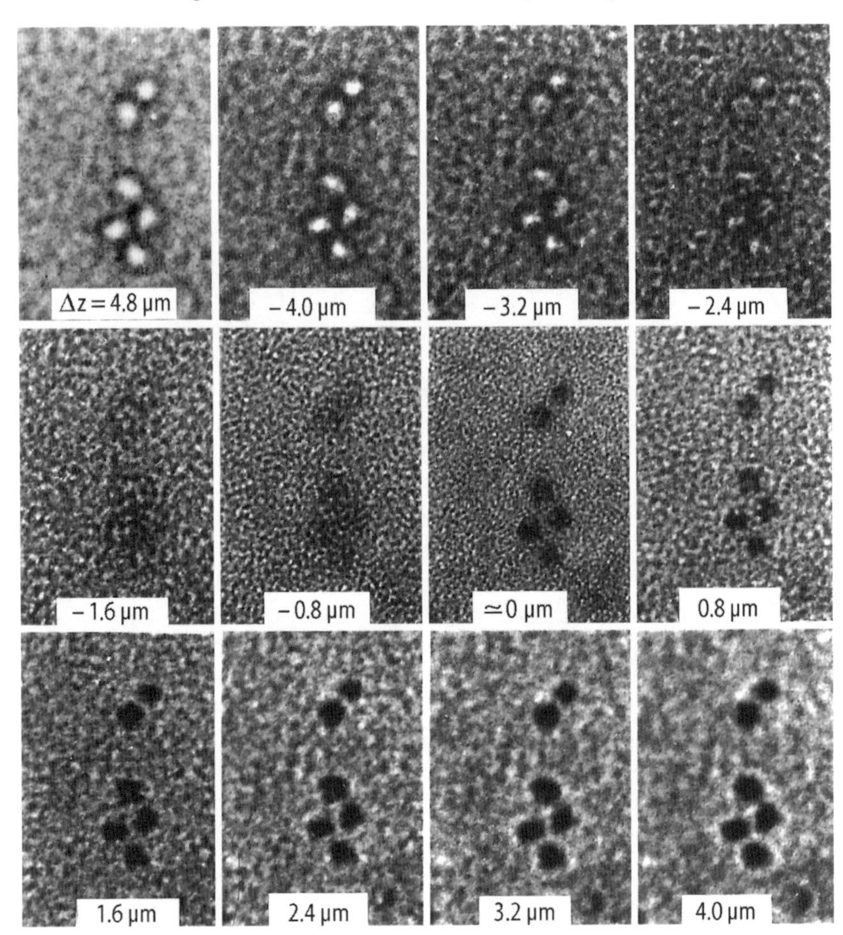

Fig. 6.15. Defocus series of ferritin molecules on a 5 nm carbon supporting film and changes in the granulation of the carbon film ($E = 100$ keV).

6.2.4 Theoretical Methods for Calculating Phase Contrast

The wave-optical theory of imaging has already been described in Sect. 3.3.2. We set out from formula (3.36) for the modified plane wave behind the specimen. The amplitude ψ_0 will be normalized to unity. The local amplitude modulation $a_s(\boldsymbol{r})$ is assumed to differ little from one: $a_s(\boldsymbol{r}) = 1 - \epsilon_s(\boldsymbol{r})$, where $\epsilon_s(\boldsymbol{r})$ is small. If the phase shift $\varphi_s(\boldsymbol{r})$ is also much less than one, then the exponential term in (3.36) can be expanded in a Taylor series

$$\psi(\boldsymbol{r}) = 1 - \epsilon_s(\boldsymbol{r}) + i\varphi_s(\boldsymbol{r}) + \ldots \, . \tag{6.18}$$

With this approximation, the specimen is said to be a weak-amplitude, weak-phase object. In practice, electron-microscope specimens thinner than 10 nm and of low atomic number do behave as weak-phase objects. The amplitude

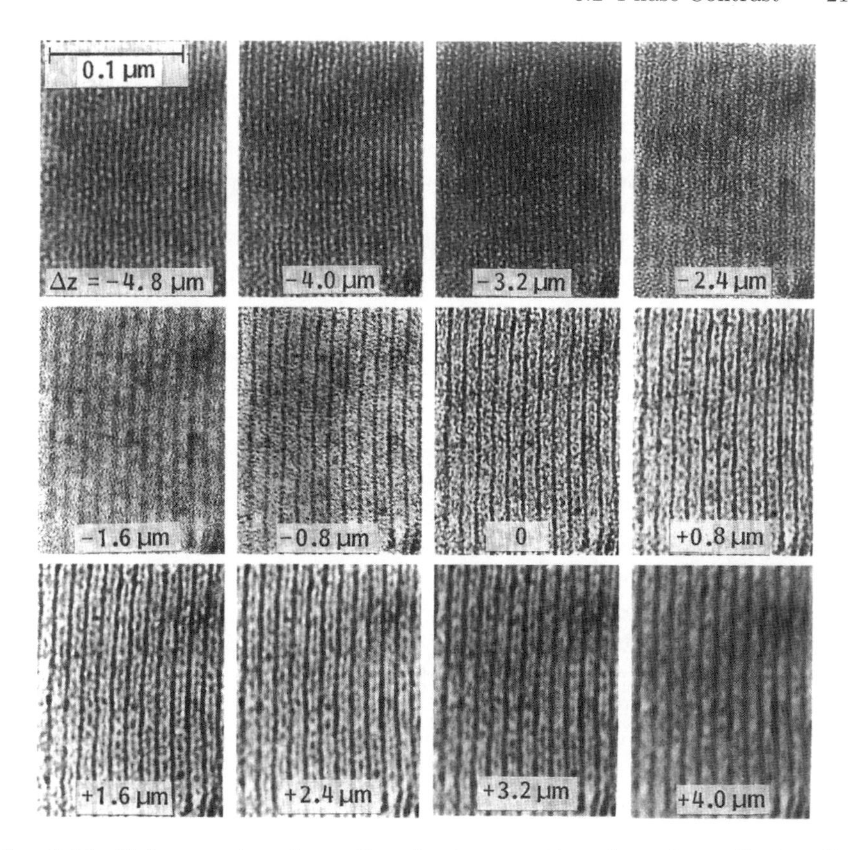

Fig. 6.16. Defocus series of an ultramicrotome section through myelin lamellae (stained with OsO_4, embedded in Vestopal).

modulation $\epsilon_s(\boldsymbol{r})$ can then be neglected. When the phase contrast of particles with high atomic number, such as colloidal gold particles, is calculated, the decrease $\epsilon_s(\boldsymbol{r})$ of amplitude, however, must be considered [6.46, 6.49, 6.50]. For low spatial frequencies the influence of the amplitude leads to an extended peak in a regime, where the phase-contrast transfer function is almost zero [6.51].

Equations (3.72) and (3.73) contain the complete mathematical treatment of phase contrast. Depending on the information required and the nature of the phase contrast, the following procedures can be used:

1. If the scattering amplitude $F(\theta)$ of a specimen is known, the image amplitude $\psi'_m(\boldsymbol{r})$ is given approximately by (3.72). For high resolution, $F(\boldsymbol{q})$ is related to the scattering amplitude $f(\theta)$ of a single atom (Sect. 5.1.3) by $F(q) = \lambda f(\theta)$. Examples are discussed in Sect. 6.3.1.
2. For constant conditions and variations of the specimen structure, it can be advantageous to use the convolution (3.73) of the object function $\psi_s(\boldsymbol{r})$

with the Fourier transform $h(\boldsymbol{r})$ of the pupil function $H(\boldsymbol{q})$ because in this case, intermediate calculation of $F(\boldsymbol{q})$ would be a waste of computation time.

3. If more general information is wanted about the contrast transfer – which spatial frequencies q are imaged with positive or negative phase contrast for a given electron lens or how the contrast transfer is influenced by a finite illumination aperture or by the energy spread of the incident electron beam – then the pupil function or the contrast-transfer function can be used (Sect. 6.4).

6.2.5 Imaging of a Scattering Point Object

For a further discussion of phase contrast, we consider an idealized point specimen that scatters isotropically into all scattering angles. It is the source of a spherical wave of amplitude $f(\theta)$, independent of the scattering angle θ. As shown in Sect. 3.3.2, the amplitude-blurring or point-spread function $h(r)$ is obtained as the image. The scattering amplitude of a single atom decreases with increasing θ. Nevertheless, to a first-order approximation, this point specimen can be pictured as a single atom, though in most cases the scattering amplitude $f(\theta)$ of single atoms already begins to decrease within $\theta \leq \alpha_o$. The resulting phase contrast of single atoms will be discussed in Sect. 6.3.

We introduce polar coordinates r' and χ in the image plane and normalize the magnification to unity ($M = 1$). The scalar product in (3.72) becomes $\boldsymbol{q} \cdot \boldsymbol{r} = qr' \cos\chi = \theta r' \cos\chi/\lambda$; we have $\mathrm{d}^2\boldsymbol{q} = \theta \mathrm{d}\theta/\lambda^2$ and $F(q) = \lambda f(\theta)$. For the bright- and dark-field modes (1 and 0, respectively, for the first term), we obtain

$$\psi_\mathrm{m}(\boldsymbol{r}') = \left.\begin{matrix}1\\0\end{matrix}\right\} + \frac{\mathrm{i}}{\lambda}\int\limits_0^{\alpha_o}\int\limits_0^{2\pi} f(\theta)\mathrm{e}^{-\mathrm{i}W(\theta)} \exp\left(\frac{2\pi\mathrm{i}}{\lambda}\theta r' \cos\chi\right)\theta\mathrm{d}\theta. \tag{6.19}$$

The difference between the bright- and dark-field modes is that, in the former, the primary incident wave (normalized in amplitude to unity) contributes to the image amplitude, whereas in the dark-field mode, it will be absorbed by a central beam stop or by a diaphragm. The factor i indicates that there is a phase shift of 90° between the primary and scattered waves.

If the specimen and the scattering amplitudes are assumed to be rotationally symmetric, the integration over χ in (6.19) gives the Bessel function J_0. The term involving $W(\theta)$ can be rewritten, using the Euler formula, as follows:

$$\begin{aligned}\psi_\mathrm{m}(r') &= \left.\begin{matrix}1\\0\end{matrix}\right\} + \frac{2\pi\mathrm{i}}{\lambda}\int\limits_0^{\alpha_o} f(\theta)[\cos W(\theta) - \mathrm{i}\sin W(\theta)]\mathrm{J}_0\left(\frac{2\pi}{\lambda}\theta r'\right)\theta\mathrm{d}\theta \\ &= \left.\begin{matrix}1\\0\end{matrix}\right\} + \epsilon_\mathrm{m}(r') + \mathrm{i}\,\varphi_\mathrm{m}(r').\end{aligned} \tag{6.20}$$

In the absence of the wave aberration $[W(\theta) = 0]$, the real part $\epsilon_{\rm m}(r')$ of (6.20) becomes zero and the same result is obtained as in Fig. 6.12a, namely that the 90° phase-shifted imaginary part $\varphi_{\rm m}(r')$ makes no contribution to bright-field image contrast because $\varphi_{\rm m}(r') \ll 1$. With nonvanishing wave aberration $W(\theta)$, the real part of (6.20), which contains $\sin W(\theta)$, is non-zero: $\epsilon_{\rm m}(r') \neq 0$. The image intensity is obtained by squaring the absolute amplitude; i.e.,

$$I(r') = \psi_{\rm m}(r')\psi^*_{\rm m}(r') = \left[\begin{matrix}1\\0\end{matrix}\right\} + \epsilon_{\rm m}(r')\Big]^2 + \varphi^2_{\rm m}(r')$$
$$= \begin{cases} 1 + 2\epsilon_{\rm m} + \epsilon^2_{\rm m} + \varphi^2_{\rm m} \simeq 1 + 2\epsilon_{\rm m}(r') + \ldots & \text{bright field} \\ \epsilon^2_{\rm m} + \varphi^2_{\rm m} & \text{dark field.} \end{cases} \tag{6.21}$$

Because both $\epsilon_{\rm m}$ and $\varphi_{\rm m}$ are very much smaller than unity, the quadratic terms can be neglected in the bright-field mode. If we consider the intensity variation

$$\Delta I(r') = I(r') - I_0 = 2\epsilon_{\rm m}(r')$$
$$= \frac{4\pi}{\lambda}\int_0^{\alpha_{\rm o}} f(\theta) \sin W(\theta)\, {\rm J}_0\left(\frac{2\pi}{\lambda}\theta r'\right) \theta {\rm d}\theta \tag{6.22}$$

for the bright-field mode relative to the background $I_0 = 1$, the integrand in (6.22) can be split into three factors, which are plotted in Fig. 6.17. The factor $\theta f(\theta)$ expresses the fact that the area of an annular element $2\pi\theta\ {\rm d}\theta$ increases as θ. The factor $\sin W(\theta)$ passes through a broad maximum when the minimum of the wave aberration in Fig. 3.15 takes the value $-\pi/2$; this

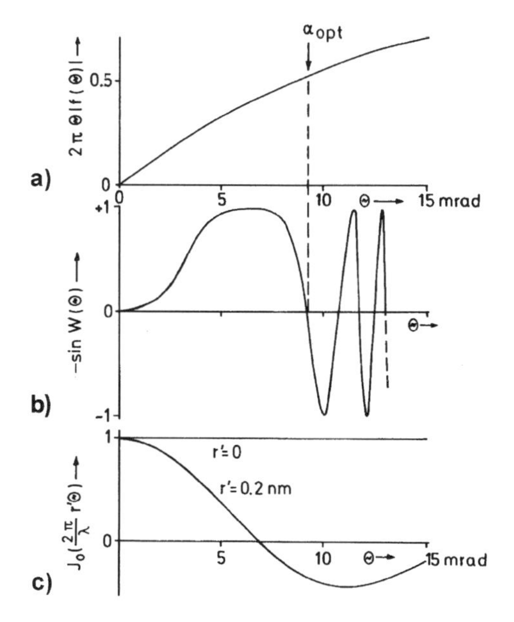

Fig. 6.17. Plot of the three factors (**a**) $\theta f(\theta)$, (**b**) – sin $W(q)$ and (**c**) $\mathrm{J}_0(2\pi\theta r'/\lambda)$ in the integrand of (6.22).

occurs for a reduced focusing $\Delta z^* = 1$. The Bessel function J_0 is unity at the center ($r' = 0$) of the atom. The reduced optimum aperture will be that for which $W(\theta)$ is again zero, that is, $\alpha^*_{opt} = \sqrt{2}$ for $\Delta z^* = 1$. The subsequent rapid oscillations of $W(\theta)$ with increasing θ will give no further contribution to the integral in (6.22). From (3.68), the values

$$\alpha_{opt} = 1.41(\lambda/C_s)^{1/4}, \quad \Delta z_{opt} = (C_s\lambda)^{1/2}, \tag{6.23}$$

are obtained for the so-called Scherzer focus with maximum positive phase contrast [3.33] and the corresponding optimum aperture.

As the distance r' from the center of the point source is increased, the oscillations of the Bessel function J_0 in (6.22) are shifted to smaller values of θ (Fig. 6.17c), which decreases the value of the integral and can even change its sign. The image amplitude (3.74) is obtained for $F(\theta) = \text{const}$ and for $\sin W(\theta) = -1$ at all scattering angles. At the Scherzer focus, the half-width of the image-intensity distribution passes through a minimum:

$$\delta_{min} = 0.67(C_s\lambda^3)^{1/4}. \tag{6.24}$$

This quantity δ_{min} is often used to define the *resolution* of TEM. However, a single number proves to be insufficient to characterize the resolution. Thus specimen details closer together than δ_{min} can be imaged by shifting the minimum of the wave aberration toward higher spatial frequencies by defocusing. However, this better resolution will be obtained only for a limited range of spatial frequencies. Furthermore, the influences of the chromatic aberration and of the finite illumination aperture have to be considered. It is therefore more informative to characterize the objective lens of a transmission electron microscope by its contrast-transfer function (Sect. 6.4).

6.2.6 Relation between Phase and Scattering Contrast

We now demonstrate that the phase and scattering contrast will both emerge from the wave-optical theory of image formation if complex scattering amplitudes (5.17) are substituted in (6.20) [6.46, 6.54, 6.55]. The phase shift $\eta(\theta)$ has to be added to the existing phase shift of 90° between primary and scattered wave; it causes a decrease of the amplitude in Fig. 6.12a even if the lens introduces no additional phase shift. To demonstrate this, we assume that $W(\theta) = 0$ and replace $f(\theta)$ by $|f(\theta)| \exp[i\eta(\theta)]$ in (6.20) for the bright-field mode. When the Euler formula is applied to $\exp[i\eta(\theta)]$, equation (6.20) becomes

$$\psi_m(r') = 1 + \frac{2\pi i}{\lambda} \int_0^{\alpha_o} |f(\theta)|[\cos\eta(\theta) + i\sin\eta(\theta)] J_0\left(\frac{2\pi}{\lambda}\theta r'\right) \theta d\theta. \tag{6.25}$$

If we assume that $|f(\theta)| \sin\eta(\theta) \simeq |f(0)| \sin\eta(0) = \text{const}$ for all scattering angles $\theta \leq \alpha_o$, the relation $\int_0^x yJ_0(y)dy = xJ_1(x)$ can be used and (6.25) becomes

$$\psi_{\mathrm{m}}(r') = 1 - |f(0)| \sin\eta(0) \frac{\alpha_{\mathrm{o}}}{r'} \mathrm{J}_1\left(\frac{2\pi}{\lambda}\alpha_{\mathrm{o}} r'\right)$$
$$+ \mathrm{i}\,|f(0)| \cos\eta(0) \frac{\alpha_{\mathrm{o}}}{r'} \mathrm{J}_1\left(\frac{2\pi}{\lambda}\alpha_{\mathrm{o}} r'\right)$$
$$= 1 + \epsilon_{\mathrm{m}}(r') + \mathrm{i}\,\varphi_{\mathrm{m}}(r'). \tag{6.26}$$

The radial variation $\Delta I(r')$ of the intensity distribution is obtained as in (6.22), but all the terms in (6.21) are now retained:

$$\Delta I(r') = 2\epsilon_{\mathrm{m}} + \epsilon_{\mathrm{m}}^2 + \varphi_{\mathrm{m}}^2. \tag{6.27}$$

The dominant first term is negative, which means that a decrease of intensity is observed in the bright field. Integrating the intensity variation $\Delta I(r')$ over the whole image disc, we obtain

$$2\pi \int_0^\infty \Delta I(r') r' \mathrm{d}r' = -4\pi |f(0)| \sin\eta(0) \alpha_{\mathrm{o}} \int_0^\infty \mathrm{J}_1\left(\frac{2\pi}{\lambda}\alpha_{\mathrm{o}} r'\right) \mathrm{d}r'$$
$$+ 2\pi |f(0)|^2 [\sin^2\eta(0) + \cos^2\eta(0)] \alpha_{\mathrm{o}}^2 \int_0^\infty \frac{\mathrm{J}_1^2\left(\frac{2\pi}{\lambda}\alpha_{\mathrm{o}} r'\right)}{r'} \mathrm{d}r'$$
$$= -2\lambda |f(0)| \sin\eta(0) + \pi\alpha_{\mathrm{o}}^2 |f(0)|^2. \tag{6.28}$$

The first term is identical with σ_{el}, as the optical theorem (5.39) shows. The last term is the elastically scattered intensity that goes through the objective diaphragm; the last integral takes the value 1/2. The whole integral is equal to $-\sigma_{\mathrm{el}}(\alpha_{\mathrm{o}})$; see (6.3). This is none other than the contribution of one atom to the decrease of intensity caused by scattering contrast. Formula (6.5) for the decrease of intensity caused by a layer of atoms is obtained by multiplying $\sigma_{\mathrm{el}}(\alpha_{\mathrm{o}})$ by the number of atoms $N_A \mathrm{d}x/A$ per unit area of a film of mass thickness $\mathrm{d}x$. If the individual atoms cannot be resolved, an average over the intensity decrease of all atoms is observed, as in (6.28).

6.3 Imaging of Single Atoms

6.3.1 Imaging of Single Atoms in TEM

One of the reasons for calculating atomic images is to study the behavior of the radial intensity distribution when different parameters are varied. Most of the calculations of the image contrast of single atoms have used real values of $f(\theta)$ in (6.19) [6.52, 6.53, 6.54, 6.55, 6.56, 6.57].

Figure 6.18 shows the calculated decrease of intensity $\Delta I/I_0$ at the center of a platinum atom ($E = 100$ keV, $C_{\mathrm{s}} = 1$ mm) in the form of lines of equal $\Delta I/I_0$ with defocus Δz and objective aperture α_{o} as coordinate axes. If the objective aperture is varied at the Scherzer optimum defocus, along the line BB′, the upper curve of Fig. 6.18 shows that an increase of α_{o} beyond the optimum aperture does not improve the image because of the rapid oscillations of

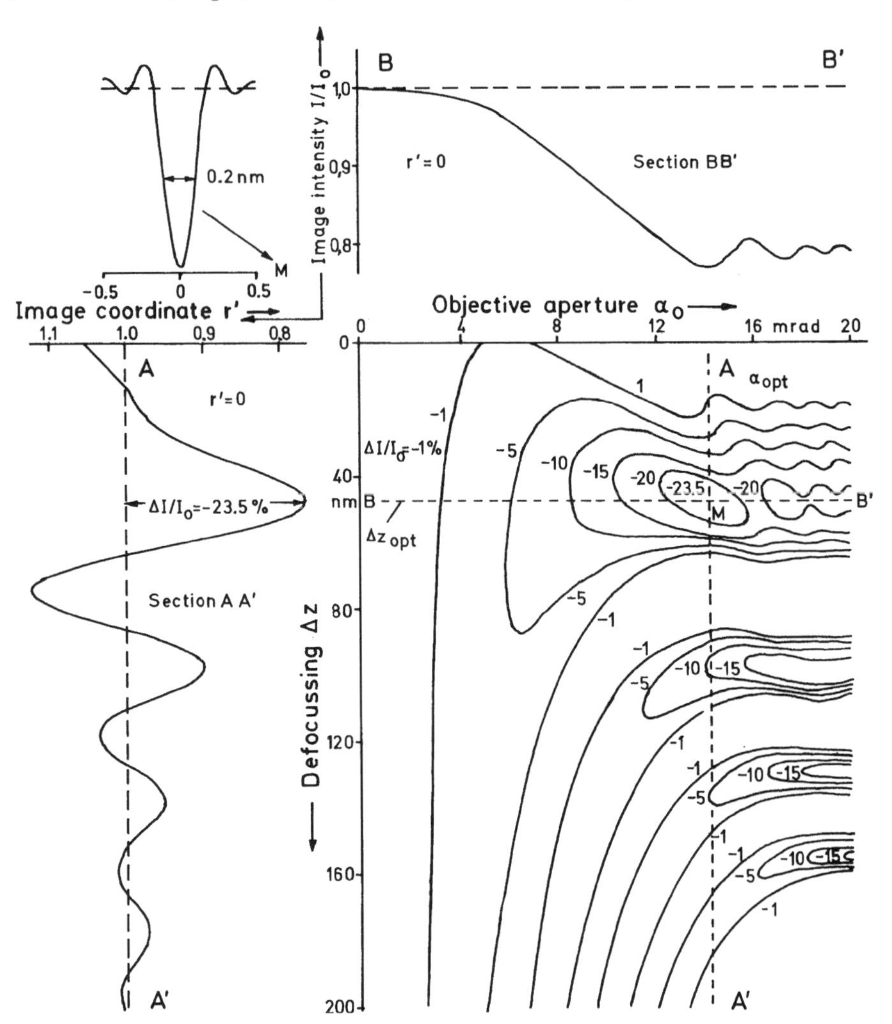

Fig. 6.18. Calculated decrease of intensity $\Delta I/I_0$ in the center of platinum atoms as a function of defocus Δz and objective aperture α_o for $E = 100$ keV and $C_s = 0.5$ mm. Sections along the lines AA′ and BB′ are shown at the bottom left and top right, respectively. The radial intensity distribution of a platinum atom at the Scherzer focus M for Δz_{opt} and α_{opt} is seen at the top left corner.

$W(\theta)$ (see also the contrast-transfer function in Sect. 6.4.1). In practice, large diaphragms should be used because a smaller diaphragm that corresponds to the optimum aperture at the Scherzer focus can become charged around the periphery, thereby causing additional phase shifts. If the defocus is varied through the Scherzer focus at constant aperture α_o along the line AA′, the left curve in Fig. 6.18 shows that the phase contrast oscillates with increasing defocus Δz. The atom is alternately imaged in positive and negative phase contrast. Positive phase contrast is observed not only at the Scherzer focus, at $\Delta z^* = 1$ ($\Delta z = 43$ nm), but again at $\Delta z^* = \sqrt{5}$ ($\Delta z = 96$ nm), where a broad interval of spatial frequencies is transferred with the phase shift $W(\theta) = -5\pi/2$ (Figs. 3.15 and 6.21c). The inset in the top left corner of Fig. 6.18 contains the radial distribution of $I(r')$ for a platinum atom at the Scherzer focus M. Once again, a bright annular ring is observed around the central darker region (Sect. 6.2.2), which reconciles the larger decrease of the intensity in the central region with the fact that the number of electrons transmitted is constant.

Figure 6.19 shows the influence of various parameters on the radial intensity distribution $I(r')$ of a single bromine atom [6.58]. A decrease of the spherical-aberration constant C_s (full curves in Fig. 6.19a) increases the

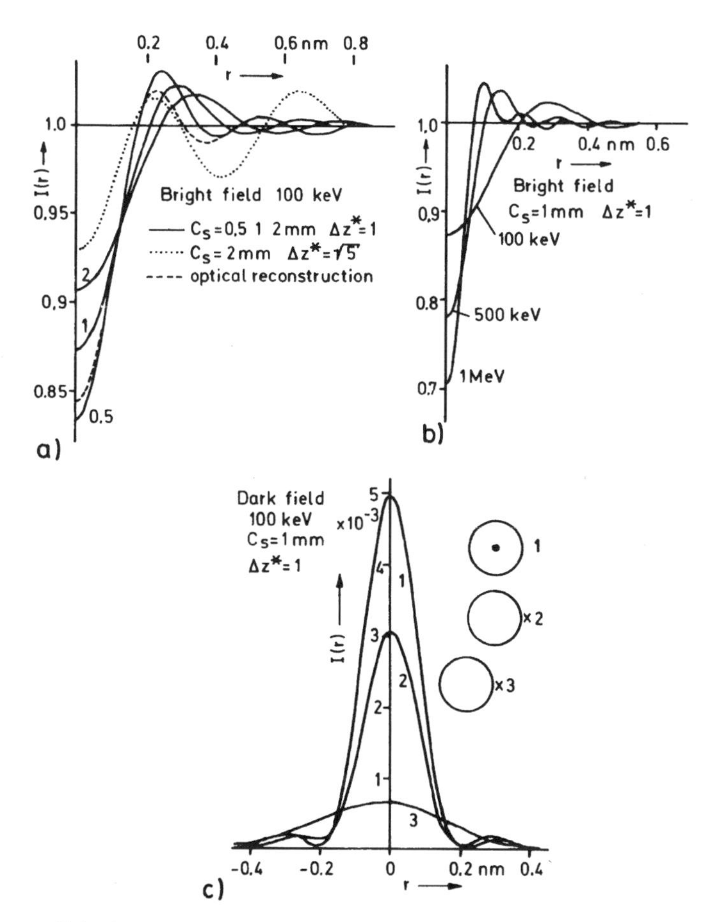

Fig. 6.19. Calculated radial intensity distribution of Br atoms (**a**) for different values of the spherical-aberration constant C_s and the reduced defocus Δz^* at $E =$ 100 keV, (**b**) for a range of electron energies, and (**c**) for three modes of dark-field imaging.

positive phase contrast at the center ($r' = 0$), strengthens the bright annular ring, and reduces the half-width. For $\Delta z^* = \sqrt{5}$ (dotted line), stronger oscillations are observed at large distances r'. If the sign is changed (phase shift of 180° in the region in which $-\sin W(\theta)$ is negative), as in an optical reconstruction scheme proposed by Maréchal and Hahn (Sect. 6.5.4), a result is obtained for $C_s = 2$ mm that is comparable with that for $C_s = 0.5$ mm at the Scherzer optimum defocus (dashed curve in Fig. 6.19a).

Figure 6.19c contains calculated dark-field intensity distributions for three different modes of dark-field imaging (Sect. 6.1.2). The advantage of using a central beam stop (1) or tilted illumination with a centered diaphragm (2) rather than a shifted diaphragm (3) can be seen clearly (see also [6.59]). In modes (2) and (3), the radial intensity distributions are somewhat asymmetrical. Illumination with a hollow cone (Sect. 6.4.3) corresponds to an incoherent superposition of images obtained with mode (2); this averages the weak asymmetry of the image discs (see also [6.60, 6.61]).

The intensity of dark-field images is much lower and is smaller than the decrease of intensity in the bright-field mode. Nevertheless, the dark-field mode has the advantage of higher contrast. Single atoms appear as relatively bright spots against the weak background intensity of the supporting film [6.63, 6.64, 6.65, 6.66]. However, longer exposure times are needed than in the bright-field mode. Furthermore, the contrast transfer of the dark-field mode is nonlinear [6.60, 6.61, 6.67]. In (6.21), $\epsilon_m(r')$ appears as a linear term in the bright-field mode but as a quadratic term in the dark field mode. The Fourier spectrum of a specimen periodicity Λ with spatial frequency $q = 1/\Lambda$ consists of the central beam and two diffracted beams of order ± 1. Removal of the central beam in the dark field mode will result in twice the spatial frequency between the two diffracted waves so that a periodicity $\Lambda/2$ will be observed in the dark field.

The presence of neighboring atoms leads to a superposition of the image amplitudes of the individual atoms that can produce parasitic structures in bright- and dark-field imaging. Consider, for example, the dotted curve in Fig. 6.19a, which corresponds to the image of a Br atom at a defocus $\Delta z^* = \sqrt{5}$. If two neighboring atoms are separated by a distance of 0.4 nm, the second minimum of $I(r')$ for one atom will coincide with the central decrease of the other, thus causing an increase in the contrast of both. If, however, they are separated by 0.8 nm, the secondary minimum at $r' = 0.4$ nm will increase; a third atom will apparently be seen, though this in fact will be an image artifact.

Hitherto we have discussed only the contribution of elastic scattering to phase contrast. The image amplitudes of the inelastically scattered electrons also have to be considered. However, elastic scattering is more concentrated within smaller scattering angles than elastic scattering and already decreases strongly with increasing θ within the objective aperture. We know that the image amplitude is the Fourier transform of the scattering amplitude $f(\theta)$. A narrow scattering distribution results in a broader image disc [6.59, 6.68].

It can also be argued that there are many fewer inelastically scattered electrons at high spatial frequencies, where they would be needed for high resolution. In a classical model of scattering, we can say that inelastic scattering is less localized than elastic scattering. An electron that passes far from an atom can nevertheless excite an atomic electron by Coulomb interaction. Inelastic scattering at low energy losses is therefore useless for obtaining high-resolution information.

Images of single heavy atoms have been observed in molecules of known structure: triangles of mercury atoms separated by distances of the order of 1 nm in triacetomercuryaurin [6.69], uranium-stained mellitic acid [6.70], monolayers of thorium-hexafluoracetylacetonate [6.71], and single W atoms and clusters [6.72]. These confirm that the calculated contrast and resolution in the bright-field mode are of the right order of magnitude. In the dark-field mode, single-atom images have been obtained for U, Os, Ir, Pd [6.10], Th [6.63], Rh [6.73], and, at high voltages (200 and 3000 keV), for U, Ba, Sr, and Fe ($Z = 26$) [6.74, 6.75]. The dark-field mode with hollow-cone illumination was employed to observe Hg [6.66] and U and Ba atoms [6.74].

These experiments merely show that single atoms can indeed be imaged in principle; they also clearly demonstrate that high resolution is limited not by the lack of contrast but by the background noise of the supporting film or organic matrix and by radiation damage.

6.3.2 Imaging of Single Atoms in the STEM Mode

It was shown in Sect. 4.2.2 that small electron probes can be obtained only with large probe apertures α_p. The theorem of reciprocity (Sect. 4.5.3) indicates that phase-contrast effects can be observed also with $\alpha_\mathrm{p} \gg \alpha_\mathrm{d}$. In the normal STEM mode with $\alpha_\mathrm{p} \simeq \alpha_\mathrm{d} \simeq 10$ mrad, however, the illumination is incoherent, which blurs phase-contrast effects (see also Sect. 6.4.4). Nevertheless, the contrast of atoms can be increased by using the following three signals, all of which can be obtained with a dedicated STEM equipped with an electron spectrometer (Fig. 4.25):

(1) The signal I_el generated by the elastically scattered electrons. All electrons scattered through angles larger than the detector aperture α_d, which is of the same order as the electron-probe aperture α_p, are collected by an annular scintillator or semiconductor detector. This signal contains a few inelastically scattered electrons, but these can be neglected because inelastic scattering is concentrated at small scattering angles. At large angles, the ratio of the elastic and inelastic differential cross sections is proportional to Z^{-1} (5.64). Similarly, some of the elastically scattered electrons remain inside the detector cone and pass into the spectrometer, where they contribute to the signal I_un of unscattered electrons. From (6.1), which is strictly valid only for parallel illumination, we can assume that about 50% of the electrons are scattered inside a cone of aperture θ_0. This characteristic angle of elastic scattering is tabulated in Table 6.1 and α_d should be appreciably smaller than θ_0.

For calculating I_{el}, the total elastic cross section σ_{el} (e.g., the approximation (5.41)) can be used.

(2) The signal $I_{un} = I_p - (I_{el} + I_{in})$ that corresponds to the unscattered electrons, which pass through the specimen and spectrometer with no energy loss (I_p: probe current).

(3) The signal I_{in} is generated by all of the inelastically scattered electrons with the exception of those scattered through angles larger than α_d; the approximation (5.52) can be used for the total inelastic cross section σ_{in}.

A homogeneous supporting film (suffix s in 6.29) containing $N = N_A\rho/A$ atoms per unit volume and of thickness t will produce the signals

$$I_{el,s} = \sigma_{el,s} N t I_p, \quad I_{in,s} = \sigma_{in,s} N t I_p. \tag{6.29}$$

The probe current I_p is concentrated within the probe diameter d_p. The current density is therefore $j_p \simeq I_p/d_p^2$. The image of a single heavy atom (suffix a in 6.30) will also take the form of an error disc of diameter d_p. At its center, a signal contribution

$$I_{el,a} = \sigma_{el,a}\, j_p \simeq \sigma_{el,a}\, I_p/d_p^2 \tag{6.30}$$

will be observed. This relation was verified experimentally for single U, Hg, and Ag atoms [6.76]. The contrast of single atoms can be increased by exploiting the fact that only this signal contributes to the high-resolution information. The inelastic scattering of a heavy atom is distributed over a larger area ([6.77] and Sect. 10.5.4) owing to the delocalization of inelastic scattering. The signal I_{el} (Fig. 6.20b) from a supporting film with varying mass thickness together with isolated individual heavy atoms (Fig. 6.20a) contains a long-range

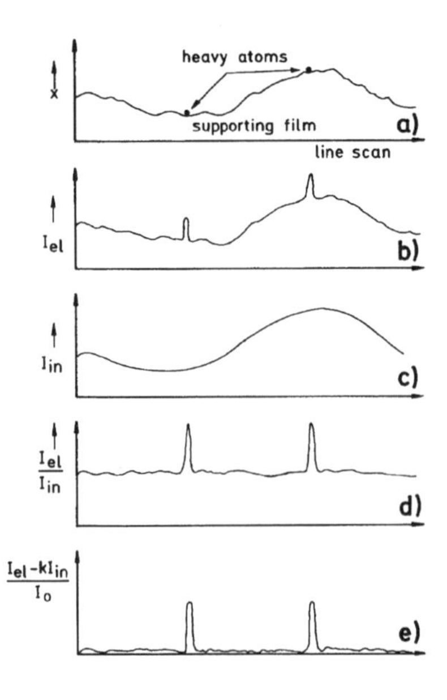

Fig. 6.20. Schematic variation of STEM signals of single heavy atoms on a supporting film (**a**), (**b**) elastic signal I_{el}, (**c**) inelastic signal I_{in}, (**d**) ratio I_{el}/I_{in}, and (**e**) difference signal $(I_{el} - kI_{in})/I_0$.

contribution that depends on the local mass thickness, in which the contributions of the single atoms of the supporting film overlap and their images are not resolved. In addition, there exists a short-range fluctuation associated with the higher spatial frequencies of the supporting film and with the local increase of elastic scattering at the individual heavy atoms. The inelastic signal I_{in} (Fig. 6.20c) also contains the long-range variation of mass thickness, but the image of the higher spatial frequencies of the supporting film and that of the single atoms are blurred on account of delocalization. The contrast of single atoms can be increased and filtered by combining the various signals online by means of analog techniques [6.78, 6.79] as follows:

1. The ratio $I_{\text{el}}/I_{\text{in}}$ renders the contrast due to the long-range variations of mass thickness of the supporting film uniform (Fig. 6.20d).
2. The difference signal $(I_{\text{el}} - kI_{\text{in}})/I_0$ can also be used to suppress the long-range variations of mass thickness (Fig. 6.20e). Division by the emission current I_0 eliminates effects due to fluctuations of this current.
3. If two annular detectors are used, the scattering angle between the two detectors can be chosen in such a way that heavy atoms scatter mainly on the outer annular detector. It is now possible to eliminate the short-range fluctuations of mass thickness from the supporting film.

The following quantitative values for Hg atoms ($Z = 80$) on a carbon ($Z = 6$) substrate [6.78] give an idea of the number of electrons per unit area needed to record a high-resolution STEM micrograph at $E = 40$ keV. The supporting film ($t = 2$ nm, $\rho = 2$ g cm^{-3}) contains $N_A\rho t/A = 200$ nm^{-2} carbon atoms, and the electron-probe area is taken to be $d_{\text{p}}^2 = 0.05$ nm^2. This gives

$$I_{\text{el,s}}/I_{\text{p}} = 2.9 \times 10^{-2}, \qquad I_{\text{el,a}}/I_{\text{p}} = 0.13,$$
$$I_{\text{in,s}}/I_{\text{p}} = 4.4 \times 10^{-2}, \qquad I_{\text{in,a}}/I_{\text{p}} \simeq 0,$$

and the ratio signals become

$$I_{\text{el,s}}/I_{\text{in,s}} = 0.65, \qquad I_{\text{el,a}}/I_{\text{in,s}} = 3.3.$$

The first ratio will be observed for the pure supporting film and the second when an Hg atom is present. The ratio of these two ratios is the increase of the signal inside the image disc of an Hg atom relative to the background of the supporting film: $I_{\text{el,a}}/I_{\text{el,s}} = 4.6$. It will be necessary to record about ten electrons per atom in order to form an image disc that can be separated from the background, for which about two electrons are needed per the same area. This implies that $n = 10I_{\text{p}}/I_{\text{el}}d_{\text{p}}^2 = 1.5 \times 10^3$ electrons nm^{-2} or a charge density of $J = j\tau = ne = 2.5 \times 10^2$ C m^{-2}. This charge density is already high enough to cause severe damage to organic material; most organic molecules will be destroyed at such large densities by irreversible radiation damage (Sect. 11.2).

The positions of atoms on carbon substrates are seen to change in a sequence of micrographs [6.80, 6.81, 6.82], whereas clusters of two or more atoms remain stationary. Examination of biological molecules stained with heavy atoms will be possible in STEM only if the atoms stay at their reaction sites.

6.4 Contrast-Transfer Function (CTF)

6.4.1 The CTF for Amplitude and Phase Specimens

The method whereby the imaging properties of an objective lens are described by a contrast-transfer function, independent of any particular specimen structure, was first developed in light microscopy and subsequently applied to electron microscopy by Hanszen and coworkers [6.62, 6.83, 6.84, 3.34].

For a specimen with a single spatial frequency q, $\epsilon_s(\boldsymbol{r})$ and $\varphi_s(\boldsymbol{r})$ in (6.18) can be replaced by $\epsilon_q \cos(2\pi qx)$ and $\varphi_q \cos(2\pi qx)$, respectively, giving

$$\psi_s(x) = 1 - \epsilon_q \cos(2\pi qx) + \mathrm{i}\varphi_q \cos(2\pi qx) + \dots \,. \tag{6.31}$$

Apart from the central peak, the Fourier transform $F(q)$ of $\psi_s(x)$ consists of two diffraction maxima of order ± 1:

$$F(\pm q) = \frac{1}{2}(-\epsilon_q + \mathrm{i}\varphi_q). \tag{6.32}$$

Equation (6.19) simplifies to a sum over the amplitudes of the primary beam and the two diffracted beams:

$$\begin{aligned} \psi_m(x') &= 1 + \sum_{\pm q} \frac{1}{2}(-\epsilon_q + \mathrm{i}\varphi_q)\mathrm{e}^{-\mathrm{i}W(q)}\mathrm{e}^{2\pi \mathrm{i} q x'} \\ &= 1 + (-\epsilon_q + \mathrm{i}\varphi_q)\mathrm{e}^{-\mathrm{i}W(q)} \cos(2\pi q x') \,. \end{aligned} \tag{6.33}$$

The image intensity becomes

$$\begin{aligned} I(x') &= |\psi_m(x')|^2 \\ &= 1 - 2\cos W(q)\, \epsilon_q \cos(2\pi q x') \quad + 2\sin W(q)\, \varphi_q \cos(2\pi q x') + \dots \\ &= 1 - \qquad D(q)\epsilon_q \cos(2\pi q x') \quad - \qquad B(q)\varphi_q \cos(2\pi q x'). \end{aligned} \tag{6.34}$$

The factor of the term ϵ_q is the CTF of the amplitude structure of the specimen:

$$D(q) = 2\cos W(q). \tag{6.35}$$

Similarly, the factor of the term containing φ_q is the CTF of the phase structure:

$$B(q) = -2\sin W(q) = -2\sin\left[\frac{\pi}{2}(C_s\lambda^3 q^4 - 2\Delta z \lambda q^2)\right]. \tag{6.36}$$

The sign of $B(q)$ is chosen so that $B(q) > 0$ for positive phase contrast. Equation (6.36) can be written in terms of the reduced coordinates (3.68) and (3.69):

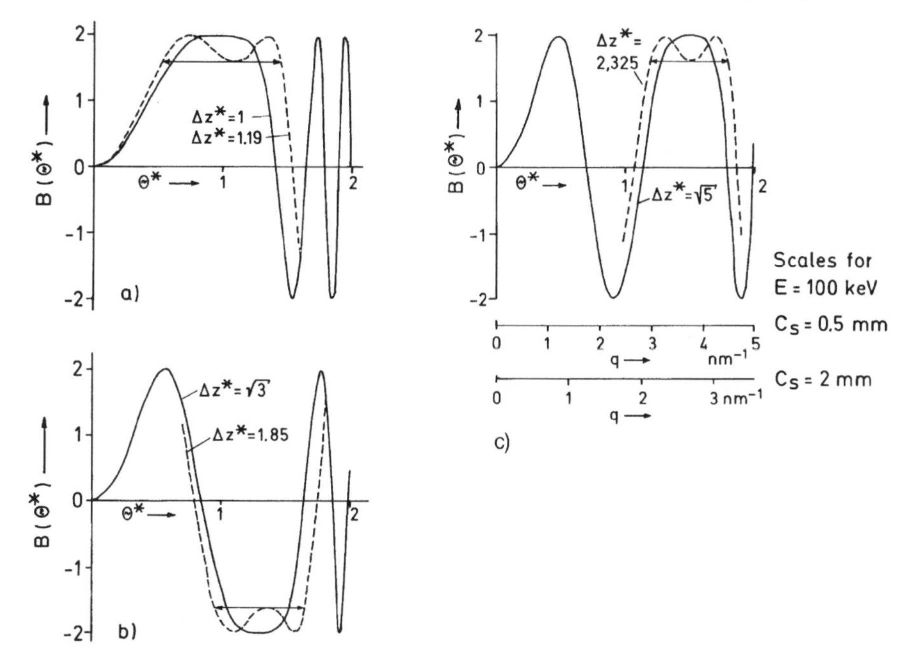

Fig. 6.21. (**a-c**) Phase-contrast-transfer functions $B(\theta^*) = -2\sin W(\theta^*)$ for weak-phase specimens in reduced coordinates $\theta^* = (C_s/\lambda)^{1/4}\theta$ for various values of reduced defocus $\Delta z^* = (C_s\lambda)^{-1/2}\Delta z$. The arrows indicate the main transfer intervals.

$$B(\theta^*) = -2\sin W(\theta^*) = -2\sin\left[\frac{\pi}{2}\left(\theta^{*4} - 2\theta^{*2}\Delta z^*\right)\right]. \tag{6.37}$$

We discuss only the more important case of the CTF for phase structures. Figure 6.21 shows the CTF $B(\theta^*)$ for three values of the reduced defocus $\Delta z^* = 1, \sqrt{3}, \sqrt{5}$, and for the neighboring values of $\Delta z^* = (C_s\lambda)^{-1/2}\Delta z$ indicated in the figure as a function of the reduced angular coordinate $\theta^* = (C_s/\lambda)^{1/4}\theta$ (3.68). The ideal CTF would take the value $B(q) = 2$ for all q. The CTFs shown in Fig. 6.21 pass through zero at certain points, around which there are transfer gaps; for the corresponding values of θ^* or q, no specimen information reaches the image. Other spatial-frequency transfer intervals are seen with negative values of $B(q)$, which means imaging with negative phase contrast for the corresponding range of q. With negative phase contrast, the maxima and minima in the image of a periodic structure are interchanged relative to those seen with positive phase contrast. Broad bands of spatial frequencies (main transfer bands) with the same sign of the CTF are expected when the minima of $W(q)$ in Fig. 3.15 are odd multiples of $-\pi/2$. The main transfer bands are indicated in Fig. 6.21 by arrows. The transfer bands become somewhat broader if the underfocus is increased slightly beyond the values $\Delta z^* = \sqrt{n}$ (see dashed CTFs with a central dip in Fig. 6.21).

In focus ($\Delta z^* = 0$), there is no main transfer band. This is also the case for overfocus ($\Delta z^* < 0$), for which the oscillations of the CTF are more frequent.

For a corrected electron microscope, the spherical aberration C_s can be adjusted at will. The wave aberration is then given by

$$W(q) = 2\pi \left(\frac{C_5}{6} \lambda^5 q^6 + \frac{C_3}{4} \lambda^3 q^4 + \frac{C_1}{2} \lambda q^2 \right) . \tag{6.38}$$

Here we have used C_5 for the constant of the fifth-order spherical aberration, $C_3 = C_s$ for the third-order spherical aberration, and $C_1 = -\Delta z$ for the defocus. With the additional flexibility to adjust C_3, one can extend the transfer band up to spatial frequencies beyond 10 nm^{-1} for phase as well as for amplitude contrast [1.74, 6.86].

6.4.2 Influence of Energy Spread and Illumination Aperture

We assumed in Sect. 6.4.1 that the electron beam is monochromatic (temporal coherence) and the incident wave is plane or spherical (point source = spatial coherence). In reality, the electron-emission process gives a beam with an energy width of $\Delta E = 1$–2 eV for thermionic guns and 0.3–0.5 eV for Schottky and field-emission guns (Sect. 4.1.2), and the electron source (crossover) has a finite size corresponding to an illumination aperture α_i. So long as $\alpha_i \ll \alpha_o$, the illumination is said to be partially spatially coherent; when α_i and α_o are of the same order, the illumination becomes incoherent. The variations of electron energy ΔE as well as those of the accelerating voltage and the lens currents ΔU and ΔI, respectively, result in variations Δf of the defocusing (2.62). The influence on the CTF has been investigated in [6.87]. The energy spread can be approximated by a Gaussian distribution

$$j(\Delta f) = \frac{2\sqrt{\ln 2}}{\sqrt{\pi} H} \exp \left[-\ln 2 \left(\frac{\Delta f}{H/2} \right)^2 \right] , \tag{6.39}$$

which is normalized so that $\int_{-\infty}^{+\infty} j(\Delta f) \mathrm{d}(\Delta f)$ is equal to unity and has the full-widths at half maxima

$$H = C_c \frac{\Delta E}{E} f_r , \quad H = C_c \frac{\Delta U}{U} f_r , \quad \text{or} \quad H = 2 C_c \frac{\Delta I}{I} f_r ,$$
$$\text{where} \quad f_r = \frac{1 + E/E_0}{1 + E/2E_0} . \tag{6.40}$$

The contributions from electrons with different values of Δf are superposed incoherently at the image. We thus have to average over the image intensities. By using (6.34), the contribution from a phase object becomes

$$\overline{I(x')} = \int_{-\infty}^{+\infty} I(x') j(\Delta f) \mathrm{d}(\Delta f) = 1 - \varphi_q \cos(2\pi q x') \int_{-\infty}^{+\infty} B(q) j(\Delta f) \mathrm{d}(\Delta f)$$
$$= 1 - B(q) K_c(q) \varphi_q \cos(2\pi q x') . \tag{6.41}$$

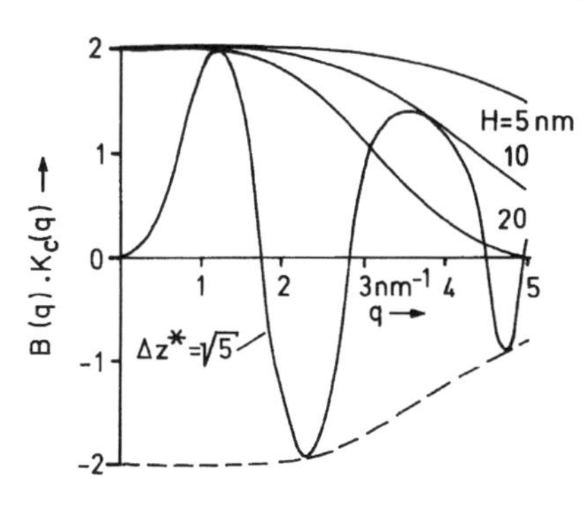

Fig. 6.22. Envelope $K_c(\theta^*)$ of the contrast-transfer function $B(\theta^*)$ for different values of the parameter H. Values of $H = 5$, 10, 20 nm correspond to $\Delta E = 1$, 2, and 4 eV, respectively, for $E = 100$ keV, $C_s = C_c = 0.5$ mm.

The value of Δz in $B(q) = -2\sin W(q)$ in (6.36) is the mean defocus $\Delta z = \Delta f$, and the result of the integration (averaging) in (6.41) is to multiply $B(q)$ by the function

$$K_c(q) = \exp\left[-\left(\frac{\pi\lambda q^2 H}{4\sqrt{\ln 2}}\right)^2\right], \tag{6.42}$$

which depends only on q and not on $B(q)$. The function $K_c(q)$ therefore acts as an *envelope function*; it damps the CTF oscillations for increasing q (Fig. 6.22). The contrast transfer of low spatial frequencies will not be affected because the spatial frequency appears in the exponent of (6.42) to the power of 4. We can define a limiting spatial frequency $q_{max} = 1/\Lambda_{min}$ for which $K_c(q) = 1/\mathrm{e} = 37\%$. The exponent in (6.42) then becomes unity. Solving for Λ_{min} gives

$$\Lambda_{min} = \left(\frac{\pi\lambda H}{4\sqrt{\ln 2}}\right)^{1/2}. \tag{6.43}$$

As a numerical example, for $E = 100$ keV, $C_c = 1$ mm, $\Delta E = 1$ eV, or $\Delta I/I = 5 \times 10^{-6}$, we find $\Lambda_{min} = 0.2$ nm ($q_{max} = 5$ nm^{-1}). To obtain this resolution, the half-widths of the energy spread, the accelerating voltage or the objective lens current must not exceed these values. A Gaussian distribution is only an approximation to the true energy-spread distribution. In reality, an asymmetric distribution similar to a Maxwellian distribution should be used. Calculations show that this asymmetry has little effect [6.87].

If a finite electron-source size and hence a finite illumination aperture is used, many of the electrons in a supposedly parallel beam in fact travel at an oblique angle to the optic axis; this angle is characterized by an angular coordinate $\boldsymbol{s} = \boldsymbol{\theta}/\lambda$. The action on the CTF is discussed in [6.88, 6.89, 6.90]. Each spatial frequency q produces diffraction maxima of order ± 1 on either side of the primary beam. The diffraction maxima with angular coordinates $\boldsymbol{q}+\boldsymbol{s}$ will pass through the objective lens with phase shifts different from those of the central beam, for which $s = 0$. For small values of s, the phase shift can be described by the first term of a Taylor series,

$$W(\boldsymbol{q} \pm \boldsymbol{s}) = W(\boldsymbol{q}) \pm \nabla W(\boldsymbol{q}) \cdot \boldsymbol{s} + \ldots, \tag{6.44}$$

where ∇ is the gradient. Equation (6.34) now becomes

$$\begin{aligned} I(x') &= 1 - \varphi_q \cos(2\pi q x')[\sin W(\boldsymbol{q}+\boldsymbol{s}) + \sin W(\boldsymbol{q}-\boldsymbol{s})] \\ &= 1 - \varphi_q \cos(2\pi q x')\{2 \sin W(\boldsymbol{q}) \cos[\nabla W(\boldsymbol{q}) \cdot \boldsymbol{s}]\}. \end{aligned} \tag{6.45}$$

If a two-dimensional Gaussian distribution is assumed for the s values, so that

$$j(s) = \frac{\ln 2}{\pi H^2} \exp\left[-\left(\frac{s}{H}\right)^2 \ln 2\right] : \quad H = \frac{\alpha_\mathrm{i}}{\lambda}; \quad 2\pi \int_0^\infty j(s) s \mathrm{d}s = 1, \tag{6.46}$$

then averaging over all s as in (6.41) again results in an envelope function

$$K_\mathrm{s}(q) = \exp\left(-\frac{[\nabla W(q)]^2 H^2}{4 \ln 2}\right) = \exp\left[-\frac{(\pi C_\mathrm{s} \lambda^2 q^3 - \pi \Delta z q)^2 \alpha_\mathrm{i}^2}{\ln 2}\right]. \tag{6.47}$$

Unlike the envelope $K_\mathrm{c}(q)$ (6.42), which depends only on q, $K_\mathrm{s}(q)$ depends also on the illumination aperture and defocusing. After first decreasing, $K_\mathrm{s}(q)$ passes through a minimum and rises again to unity where $W(q)$ reaches a minimum and hence $\nabla W(q) = 0$. The main transfer bands, in which $\nabla W(q)$ is small over a wide range of spatial frequencies, will therefore be influenced least (Fig. 6.23, curves with $\alpha_\mathrm{i}^* = 0.09$). The attenuation of the CTF by the envelope $K_\mathrm{s}(q)$ can be confirmed by laser diffraction or digital Fourier transform of the micrograph (Sect. 6.4.7) [6.91, 6.92].

Under usual conditions of TEM in the presence of both energy spread and finite source size ($\Delta E \leq 2$ eV, $\alpha_\mathrm{i} \leq 1$ mrad), the effective envelope can be approximately written as a product of the envelope functions $K_\mathrm{c}(q)$ and $K_\mathrm{s}(q)$, which describe the effects of energy spread and illumination spread separately [6.90]. For larger values of α_i or the reduced aperture $\alpha_\mathrm{i}^* = \alpha_\mathrm{i}(C_\mathrm{s}/\lambda)^{1/4}$, Fig. 6.23 shows the numerical results for $\Delta z^* = 1$ and $\sqrt{2}$ [6.87]. An envelope

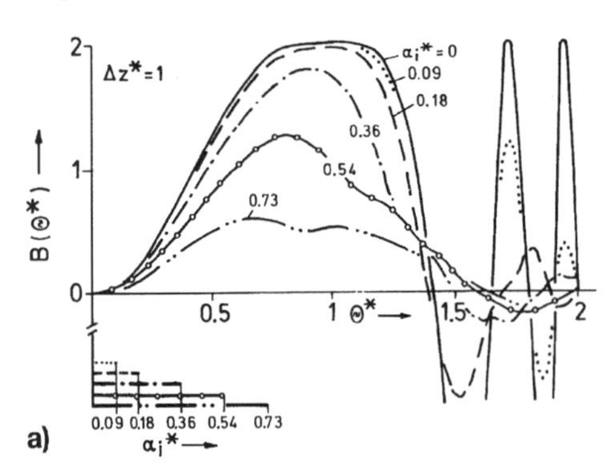

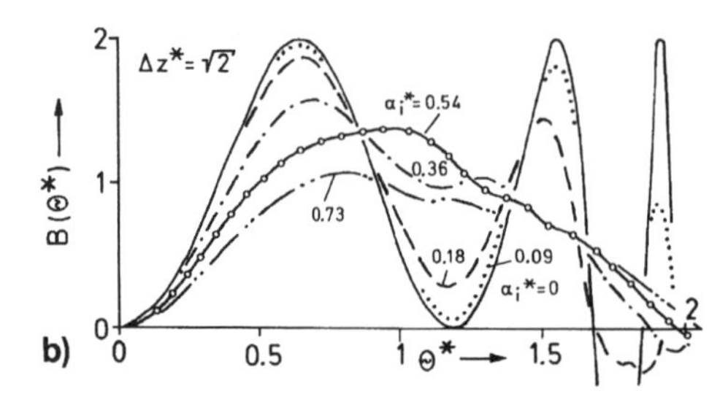

Fig. 6.23. Contrast-transfer functions for phase contrast at the defocus values (**a**) $\Delta z^* = 1$ and (**b**) $\Delta z^* = \sqrt{2}$ and increasingly large reduced illumination apertures $\alpha_\mathrm{i}^* = (C_\mathrm{s}/\lambda)^{1/4} \alpha_\mathrm{i}$ [6.87].

representation is no longer possible; $B(\theta^*)$ is now damped inside the main transfer intervals as well. For $\Delta z^* = \sqrt{2}$, for example, and a large illumination aperture, $B(\theta^*)$ has a broad interval of equal sign but with reduced amplitude, of the order of 1 instead of 2 for the maxima. A full analysis of these effects is given in Vol. 3 of [2.6].

6.4.3 The CTF for Tilted-Beam and Hollow-Cone Illumination

In the axial illumination mode, each spatial frequency contributes to the diffraction maxima of order ± 1 (double-sideband transfer). The superposition of the primary beam and the two sidebands is responsible for the gaps in the CTF. Tilted-beam illumination with the primary beam near the centered objective diaphragm cuts off one sideband (single-sideband transfer); in one direction (across the aperture), twice the maximum spatial frequency for axial illumination can be transferred. If this extended transfer is to be achieved in more than one direction, several micrographs must be recorded with different azimuths of the tilted beam. The superposition of several exposures with a range of tilted-beam-illumination azimuths is of special interest because this is equivalent to hollow-cone illumination. Single-sideband transfer can also be achieved with axial illumination by using a shifted circular diaphragm or a specially designed half-plane diaphragm (see single-sideband holography in Sect. 6.5.2). A disadvantage of all these modes is that the primary-beam spot passes near the diaphragm, which can introduce unreproducible phase shifts due to local charging. Tilted-beam or hollow-cone methods that do not require a physical diaphragm or can function with one of a larger diameter will therefore be of interest. Some important properties of these nonstandard modes will now be discussed in detail.

In the tilted-beam illumination mode, an extended range of spatial frequencies is transferred without a transfer gap but with a variable phase difference between the primary beam and the diffracted beam caused by the difference between $W(|\boldsymbol{\theta}|)$ for $\boldsymbol{\theta} = \boldsymbol{\alpha}$ corresponding to the direction (tilt) of the primary beam and for $\boldsymbol{\theta} = \boldsymbol{\alpha} + \boldsymbol{q}/\lambda$ corresponding to the diffracted beam. In axial illumination, the images of single atoms and small particles are surrounded by concentric Fresnel fringes of Airy-disc-like intensity distributions (Fig. 3.16 and 6.19). With tilted-beam illumination, the fringe system is asymmetric, with bright and dark central intensities depending on defocus and aberrations. The different phase shifts $W(\boldsymbol{\theta})$ create different lateral shifts of the corresponding specimen periodicities or Fourier components in the image. Thus, particles typically show an asymmetric contrast with bright and dark intensities on opposite sides, which resembles oblique illumination with light (pseudo-topographic contrast).

The contrast transfer of tilted-beam illumination is linear for weak amplitude and phase objects. The appropriate CTFs have been calculated and discussed in [6.87, 6.93, 6.94, 6.95, 6.96, 6.97, 6.98, 6.99], among others. The effect of partial spatial coherence (finite illumination aperture) can be expressed as an envelope function only to a first approximation. For partial temporal coherence (energy spread of the electron gun), an important finding with tilted-beam illumination is the existence of an achromatic circle [6.100]. Whereas the envelope $K_c(q)$ for axial illumination shows a rapid decrease at the resolution limit, the envelope function for tilted-beam illumination increases again to a maximum for q values around twice the resolution limit for axial illumination.

Hollow-cone illumination can be produced with an annular condenser diaphragm. However, a large fraction of the electron beam is absorbed, and it is better to move a beam of low aperture around a cone by exciting a two-stage deflection system [6.99]. This can, in practice, be reduced to superposition of a limited number of exposures; for example, eight different azimuths around a cone with a half-angle (tilt) of 10 mrad for an illumination aperture of $\simeq$0.1 mrad [6.101, 6.102]. The asymmetric fringe systems of the tilted-beam illumination are canceled and the granular contrast of supporting films also decreases, whereas the central contrast of stronger-phase objects will be the same for all the different azimuths.

The theory of hollow-cone illumination is discussed in more detail in [6.88, 6.103, 6.104, 6.105, 6.106, 6.107, 6.108]. Figure 6.24a shows calculated CTFs for phase structures at a reduced defocus $\Delta z^* = 1$ and different values of $\alpha_o^* = \alpha_o(C_s/\lambda)^{1/4}$ [6.88]. The dashed line B_{id} is the CTF for an ideal lens without aberrations, calculated on the assumption that the directions around the hollow cone or the discrete number of tilt angles superpose incoherently. A given spatial frequency with the scattering angle $\theta^* \leq 2\alpha_o^*$ can be transferred only by beams on the arcs ABC and A′B′C′ of the hollow cone (Fig. 6.24b), and the CTF becomes proportional to the ratio of these arcs to the total arc length 2π of the hollow beam

$$B_{id}(\theta^*) = \frac{4\arccos(\theta^*/2\alpha_o^*)}{2\pi} = \frac{2}{\pi}\arccos\left(\frac{\theta^*}{2\alpha_o^*}\right). \tag{6.48}$$

In the presence of a wave aberration, the phase shift $W(\theta^*)$ is not uniform over the arcs ADC and A′D′C′. Because $|\sin W(\theta^*)| \leq 1$, all the CTFs clearly lie below $B_{id}(\theta^*)$. The curve for $\alpha_o^* = 1.49$ in Fig. 6.24a shows that the ideal B_{id} is approached very closely for $\Delta z^* = 1$ with $q_{max}^* = 2\alpha_o^* = 2.98$. This CTF may be compared with that for axial illumination in Fig. 6.21a. The CTFs of hollow-cone illumination in Fig. 6.24a do not show any contrast reversals. The only disadvantage is that $B(\theta^*)$ in Fig. 6.24a reaches a maximum value of only 0.8 as compared with 2 for axial illumination.

Figure 6.25 shows that hollow-cone illumination can also be used without an aperture-limiting diaphragm [6.105]. The defocusing is $\Delta z^* = \sqrt{\pi}$ and the quantities α_1^* and α_2^* are the inner and outer apertures of the cone of finite width. This optimum condition differs from that of Fig. 6.24a mainly in that narrow regions of negative sign occur for small and large θ^* and by the presence of ripple oscillations on the CTF. The optimum conditions are those in which the phase contrast reinforces the scattering contrast.

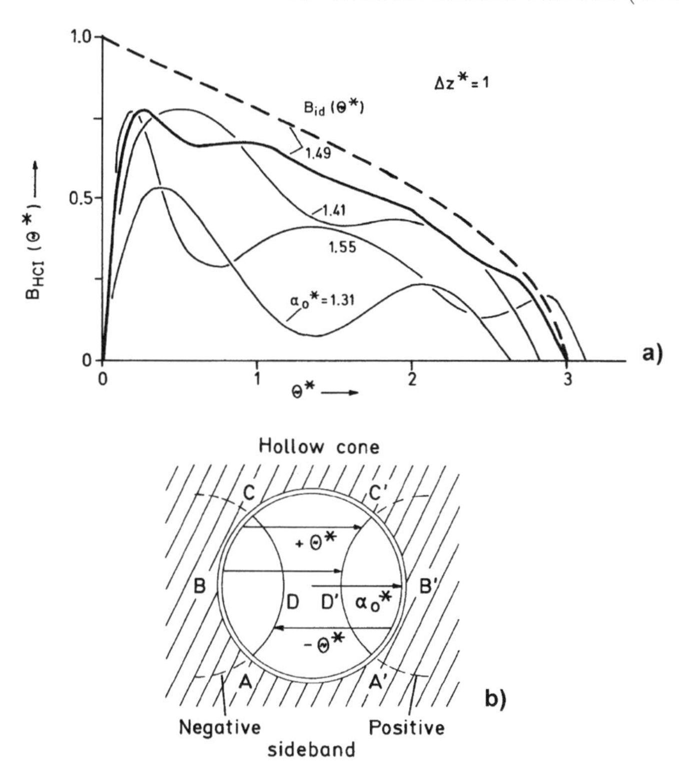

Fig. 6.24. (**a**) Phase-contrast-transfer function for hollow-cone illumination for different values of $\alpha_o^* = (C_s/\lambda)^{1/4}\alpha_o$ and a reduced defocus Δz^*=1; (– – –) ideal CTF, which is proportional to the lengths of the segments ABC and A′B′C′ in (**b**). Electrons scattered through angles θ^* can pass through the diaphragm only within the segments ABC and A′B′C′ [6.87].

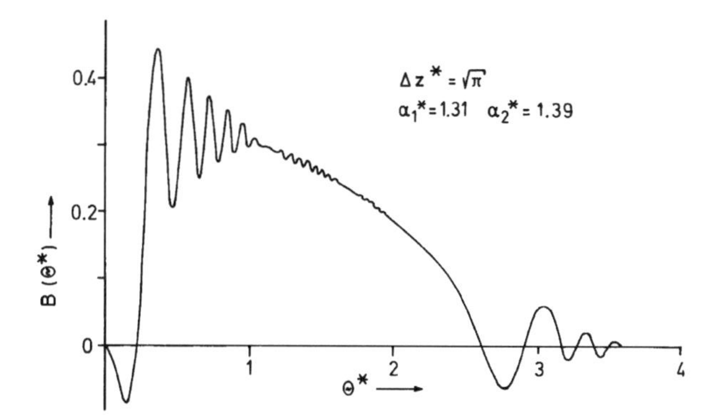

Fig. 6.25. Example of the phase CTF with hollow-cone illumination and with no objective diaphragm for a reduced defocus $\Delta z^*=\sqrt{\pi}$ and a broad cone of illumination of inner and outer diameters α_1^* and α_2^* [6.105].

6.4.4 Contrast Transfer in STEM

As discussed in Sect. 4.5.3, the phase-contrast effects in the STEM mode will be the same as in TEM if the corresponding apertures are interchanged: $\alpha_\mathrm{d} = \alpha_\mathrm{i} \simeq 0.1$ mrad and $\alpha_\mathrm{p} = \alpha_\mathrm{o} \simeq 10$ mrad (Figs. 6.10a,b). However, only a small fraction of the incident cone would be collected by the detector if these conditions were employed. There are two collection possibilities: either the electrons inside the cone of aperture $\alpha_\mathrm{d} \approx \alpha_\mathrm{p}$ are all detected or an annular detector, which collects only the mainly elastically scattered electrons, is used (Sect. 6.3.2). These modes can be described as bright and dark field, respectively. A detector that collects the whole illumination cone produces an incoherent bright-field image if interference effects between the scattered and unscattered waves need not be considered. Figure 6.26 shows the various current-density distributions in the detector plane. The dashed curve would be obtained with no specimen, whereas the full curve shows the modification caused by scattering. This implies that

$$2\pi \int_0^{\alpha_\mathrm{d}} (j_0 - j_\mathrm{DF})\theta \mathrm{d}\theta = 2\pi \int_{\alpha_\mathrm{d}}^{\infty} j_\mathrm{d}\theta \mathrm{d}\theta . \tag{6.49}$$

However, the current density for $\theta \le \alpha_\mathrm{d}$ is modulated by phase effects. Each point (direction) of the unscattered cone with direction $\boldsymbol{\theta}_\mathrm{u}(|\boldsymbol{\theta}_\mathrm{u}| < \alpha_\mathrm{d})$ corresponds to an angle of incidence in the cone of the electron probe, but the probe-forming lens shifts the phase by $W(\boldsymbol{\theta}_\mathrm{u})$. These phase shifts are responsible for the shape of the electron probe and the deviations from an Airy-disc-like probe profile. The intensity at each point of the detector plane is the result of interference between the unscattered wave of direction $\boldsymbol{\theta}_\mathrm{u}$ and a wave elastically scattered into this direction with a scattering angle $\boldsymbol{\theta}_\mathrm{s}$. The elastically scattered wave experiences a phase shift of $\pi/2$ during the scattering process (Sect. 5.1.3) and an additional phase shift relative to the unscattered wave: $W(|\boldsymbol{\theta}_\mathrm{u} - \boldsymbol{\theta}_\mathrm{s}|) - W(\boldsymbol{\theta}_\mathrm{u})$. For a fixed direction $\boldsymbol{\theta}_\mathrm{u}$, the result of superposing all possible waves with scattering angles $\boldsymbol{\theta}_\mathrm{s}$ with the condition $|\boldsymbol{\theta}_\mathrm{u} - \boldsymbol{\theta}_\mathrm{s}| < \alpha_\mathrm{d}$ has to be evaluated. This leads to a modulation of the

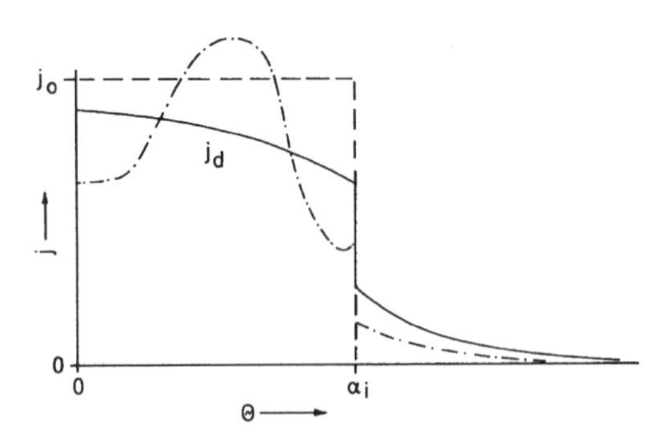

Fig. 6.26. Schematic current-density distribution in the detector plane of a scanning transmission electron microscope without a specimen (- - -), in the incoherent mode (——) and in the phase-contrast mode (- · · - · -). The electron probe is assumed to be centered on a rotationally symmetric specimen (e.g., a single atom) ($\alpha_\mathrm{i} = \alpha_\mathrm{d}$).

current-density distribution inside the illumination cone, drawn schematically as the dash-dotted line in Fig. 6.26 and consisting of zones of decreased and increased intensity.

Phase contrast can therefore be produced by dividing the detector plane into annular zones [6.109], which collect electrons with mainly constructive or mainly destructive interference. A single narrow annular detector is the counterpart of hollow-cone illumination in TEM. Hence, the CTF of this STEM mode has a triangular shape, modified by phase-contrast effects similar to those in Fig. 6.24a.

Another possible phase-sensitive detector consists of two semicircular discs, separated by a narrow gap, normal to the scan direction [6.105, 6.110, 6.111, 6.112]. This is capable of giving differential phase contrast, which represents the gradient of the object parallel to the scan direction. An obvious extension from semicircles to quadrants gives the two components of the gradient [6.113, 6.114]. To exploit these possibilities, versatile, software-configurable multichannel STEM detectors with 16 detector areas [6.115] or 30 rings split into quadrants [6.116] have been developed and tested; such detectors can also be optimally used for phase contrast in the STEM mode [6.117].

6.4.5 Phase Contrast by Inelastically Scattered Electrons

The incident and elastically scattered waves are coherent, whereas waves of inelastically scattered electrons are incoherent relative to the incident and elastically scattered waves and also to inelastic waves that differ in the final object states [6.118]. The latter are, for example, the excitation of plasmons or single electrons with different transferred momenta $h\boldsymbol{q}' = h(\boldsymbol{k}_n - \boldsymbol{k}_0)$. Plasmon excitation also shows some dependence of energy loss on the scattering angle (dispersion), and inelastic electron waves with different $\boldsymbol{k}_n$ will also be incoherent.

To a good approximation, the CTF can thus be calculated, including the partial spatial coherence of the primary beam represented by a Gaussian distribution with a half-width of about α_i, by an additional convolution with the angular distribution of inelastic scattering [5.37]. Although many inelastically scattered electrons are concentrated at low scattering angles $\theta = \lambda q$, the number $N(q)\mathrm{d}q$ with spatial frequencies between q and $q + dq$ shows a long tail because the angular distribution $\propto (\theta^2 + \theta_\mathrm{E}^2)^{-1}$ is Lorentzian. The fraction concentrated at high q totally blurs the CTF just like incoherent illumination and only the fraction with an angular width of about θ_E can be seen as a damped oscillation of CTF, with a decreased amplitude. Qualitatively, this expected decrease of phase contrast can be observed for the granularity of carbon films by comparing the zero-loss and plasmon-loss filtered images [5.37]. Phase-contrast structures with the carbon plasmon loss at lower defocusing and a width of 4 eV for the selected energy window have also been observed in the ESI mode [6.119] and in a dedicated scanning transmission electron microscope [6.120]. The counterpart of this decrease of phase contrast caused by inelastically scattered electrons is that all phase-contrast effects are enhanced by zero-loss filtering. The quantitative measurement of the phase contrast of colloidal gold particles on carbon films [6.121] likewise shows that phase contrast of the colloids can be observed with zero-loss filtering for carbon films up to a mass thickness 40 μg/cm^2 (t = 200 nm), whereas the phase contrast is invisible in unfiltered images because of the large fraction of inelastically scattered electrons.

6.4.6 Improvement of the CTF Inside the Microscope

The transfer gaps and changes of sign caused by spherical aberration render the electron-optical CTF very different from the ideal CTF with $B(q) = +2$; the latter can be attained in the light-optical phase-contrast method in which a Zernike plate shifts the phase by $+\pi/2$ in the focal plane of the objective lens. Similar methods have been proposed for the transmission electron microscope.

An objective diaphragm consisting of a plate with rings alternatively transparent and opaque to electrons could be used to suppress spatial frequencies or scattering angles that are transferred with a negative sign in $B(q)$ [6.122, 6.123]. Calculations show that no significant improvements can be expected because of the broad gaps in the CTF that correspond to spatial frequencies transferred with the wrong sign and consequently suppressed [6.53]. Two complementary zone plates have been proposed, which would cover the whole CTF without gaps for two different values of defocus [6.85].

Correcting phase shifts can be generated by means of profiled phase plates of variable thickness that can be produced by electron-beam writing or by growing a contamination layer with the required local thickness on a carbon film supported by the diaphragm [6.124]. The transfer gaps in the CTF do indeed vanish, as shown by laser diffraction [6.125]. However, no practical examples of image improvement have yet been reported.

All of these interventions in the focal plane, including single-sideband holography (Sect. 6.5.2), have the disadvantage that the diaphragm, some 100 μm in diameter, has to be adjusted precisely on-axis in the focal plane and that, whenever the electron beam strikes the transparent film or the opaque part of the diaphragm, charging can occur, which influences the phase shift unpredictably. For these reasons, none of these techniques remains in use. The present tendency is to apply a posteriori restoration methods to the final micrographs (Sects. 6.6.2 and 6.6.3).

Just recently, experiments with microfabricated electrostatic minilenses in the focal plane have demonstrated, that a $\pi/2$ phase plate can be used to obtain phase contrast for small spatial frequencies [6.126, 6.127].

6.4.7 Control of the CTF by Optical or Digital Fourier Transform

For a weak-phase specimen, a specimen periodicity Λ or spatial frequency $q = 1/\Lambda$ is linearly transferred to the image as a periodicity ΛM with an amplitude proportional to $|B(q)|$. The periodicities in the micrograph can be analyzed by light-optical Fraunhofer diffraction of the photographic record [6.129, 6.128] or by fast Fourier transform (FFT) of digitally recorded images.

The laser-diffraction technique will be advantageous for the a posteriori investigation of photographic emulsions. In practice, it is better to record a diffractogram a priori before recording and developing a photographic emulsion. This is possible by digitally recording an image with a CCD or SIT camera (Sect. 4.7.5), after which a fast Fourier transform (FFT) algorithm allows us to compute a Fraunhofer diffractogram in a few seconds or less. The digital image can alternatively be used to modulate a liquid crystal display, which allows the diffractogram to be observed online with an optical bench placed beside the microscope [6.130].

The CTF can be controlled and measured with the aid of a specimen for which the spatial frequency spectrum is like that of white noise; this implies that the amplitudes φ_q in (6.31) should be independent of q. If the spatial frequency is not too small, this is nearly true for thin carbon supporting films, as already shown in Sect. 6.2.2. A stronger granularity caused by phase contrast can be obtained with thin amorphous germanium films evaporated from heated tungsten on a glass slide and floated on water as for carbon films. The variation of the image intensity $I(x')$ is then proportional to $|B(q)|\varphi_q$ (6.34). A typical diffractogram (Fig. 6.27a) shows the transfer gaps (zero points) of $|B(q)|^2$. However, diffraction maxima that belong to regions of $B(q)$ of different sign cannot be distinguished. This requires comparison with formulas such as (6.36).

The following information can be obtained from a diffractogram:

(1) The q values that correspond to the minima in the diffractogram can be measured and plotted for a defocus series, as in Fig. 6.13. A diffractogram thus contains information about the defocusing Δz and the spherical-aberration constant C_s. The diffractogram shows maxima of $|B(q)|^2$ when $W(q) = n\pi/2$ and n is odd. The gaps of contrast transfer correspond to even values of n. By using (3.66) for $W(q)$, the relation $W(q) = n\pi/2$ can be transformed to

$$C_s\lambda^3q^2 - 2\Delta z\lambda = n/q^2. \tag{6.50}$$

Plotting n/q^2 versus q^2 results in a straight line if the numbering of n is correct. C_s can be read from the slope and Δz from the intercept of the straight line [6.131]. A tilted carbon film contains a whole range of defocus values, and it is possible to deduce the dependence on Δz in Fig. 6.13 from a single micrograph [6.42]. It is also possible to determine defocusing distances Δz as large as a few millimeters if a small illumination aperture $\alpha_i \leq 10^{-2}$ mrad is used, as in Lorentz microscopy (Sect. 6.5) [6.132, 6.133]. In this case, the term that contains the spherical aberration in (6.36) can be neglected and the relation $\Delta z = |2m - 1|/(2\lambda q^2)$, where $m = 1, 2, \ldots$, can be used to determine Δz from the q-values of the maxima in the diffractogram.

(2) Astigmatism can be included in the wave aberration $W(q)$ or CTF $B(q) = -2\sin W(q)$ by adding a term (3.67) to (3.65). This results in an

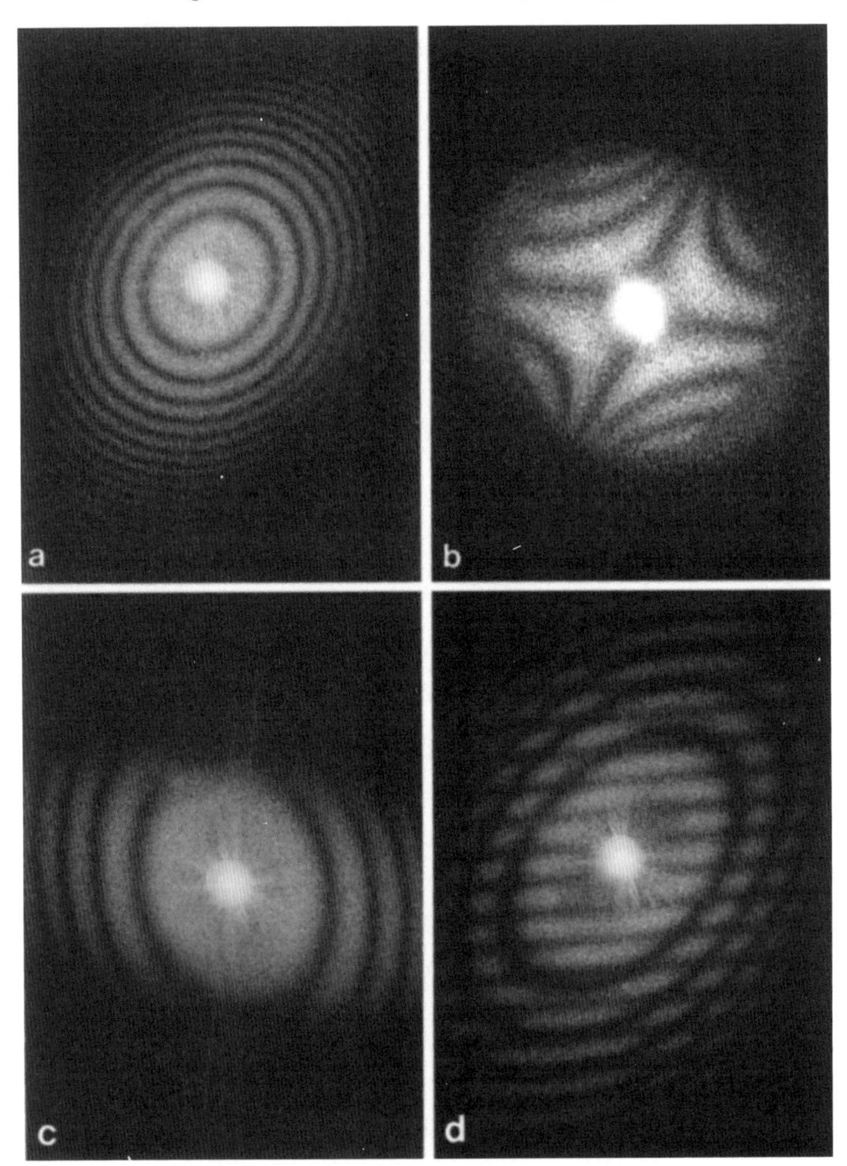

Fig. 6.27. Diffractograms of micrographs of carbon foils showing (**a**) the gaps in the contrast and a weak astigmatism (elliptical diffraction rings), (**b**) stronger astigmatism, which results in hyperbolic diffraction fringes, (**c**) detection of a continuous drift, and (**d**) a sudden jump in the specimen position during the exposure. The fringe pattern is caused by the doubling of the image structure (overlapped Young fringes).

elliptical distortion of the diffraction rings for small astigmatism (Fig. 6.27a) and a hyperbolic distortion for stronger astigmatism (Fig. 6.27b).

(3) A continuous drift of the image during exposure results in a blurring of the diffractogram parallel to the direction of drift (Fig. 6.27c). A sudden jump of specimen position during exposure duplicates the entire structure, so that a pattern of Young interference fringes is superimposed on the main pattern (Fig. 6.27d) (see also item 4 below).

(4) The envelopes $K_c(q)$ (6.42) and $K_s(q)$ (6.47) can be determined from the decrease of the diffraction-maxima amplitude for large q. The largest spatial frequency transferred is inversely proportional to the resolution limit and can be read from a diffractogram of two superposed micrographs by the following procedure [6.134, 6.135]. When the two micrographs are shifted through a small distance d, every resolved structure appears twice in the transmitted light amplitude, which means that each structure is convolved with a double source that consists of two points a distance d apart. Using the Fourier convolution theorem (3.49), we see that the diffractogram of a single micrograph will be multiplied by the Fourier transform of a double-point source. The intensity in the diffractogram is hence multiplied by $\cos^2(\pi q d)$. The diffractogram of the superposed micrographs is therefore overprinted with a pattern of Young's interference fringes with a spacing $\Delta q = 1/d$ as in Fig. 6.27d. The limit of contrast transfer can be seen from the limit of recognizable fringes. It is important to use two successive micrographs and not two copies of one micrograph. In the latter case, fringes can also be produced by clusters of silver grains generated by a single electron and reproduced in both copies. In SEM and STEM, it is necessary to shift the image on the cathode-ray tube between the two exposures by about 1 cm. Otherwise, the fringe pattern may be caused by the granularity of the CRT screen, continuing out to larger spatial frequencies [6.136].

5) The correction methods of the CTF discussed in Sect. 6.4.6 can also be controlled by studying diffractograms.

These examples of the application of laser diffraction or digital Fourier transforms show the importance of this technique for the control of the imaging process; such as the correction of astigmatism and also an exact coma-free alignment (Sect. 2.4.3).

6.5 Electron Holography

6.5.1 Fresnel and Fraunhofer In-Line Holography

The idea of holography was first introduced by Gabor [6.137] to improve the resolution of the electron microscope by (a posteriori) light-optical processing of micrographs to cancel the effect of spherical aberration. The recording of a hologram and the reconstruction of the wavefront will be described for the case of in-line holography, proposed originally by Gabor. The unscattered part

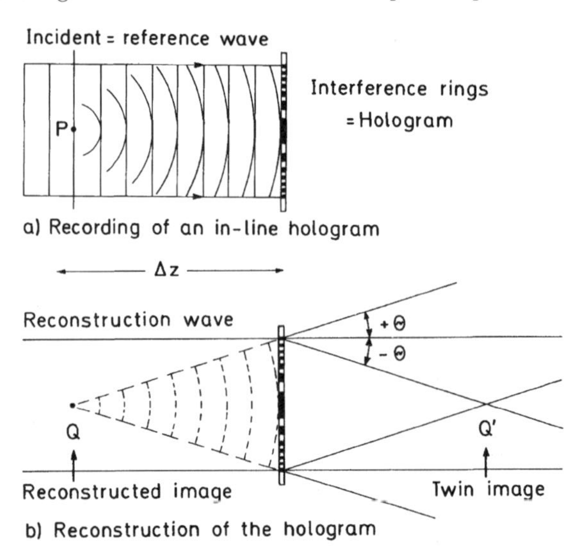

Fig. 6.28. (**a**) Recording and (**b**) reconstruction of an in-line hologram.

of an incident plane wave acts as a reference wave (Fig. 6.28a). A specimen (object) point P produces a spherical, scattered wave. The superposition of the two waves on a photographic emulsion at a distance Δz is an interference pattern, which consists of concentric fringes. The same pattern will be seen if we insert a magnifying-lens system between the object and the micrograph and record with a defocusing Δz. If the lens is not free of aberrations, the interference pattern will be modified by the additional phase shifts.

In the reconstruction (Fig. 6.28b), the micrograph is illuminated with a plane wave and acts as a diffraction grating or Fresnel-zone lens. The fringe distance decreases with increasing distance from the center, and the corresponding diffraction angle ($\pm\theta$) increases. The two diffracted waves (sidebands) form spherical waves centered at Q and Q′. We see that we reconstruct the spherical wave from P at Q behind the hologram. However, we also see that in-line holography has the disadvantage of producing a twin image at Q′. If we are looking from the right, the distance between Q and Q′ has to be so large that one of the twin images is in focus while the other is blurred. The latter is a Fresnel diffraction pattern with a defocusing $2\Delta z$ relative to the focused twin image, and a single object point is imaged as a weak concentric ring system with a large inner radius $r_1 \simeq \sqrt{\Delta z \lambda}$. A specimen structure smaller than this radius and situated in a larger structure-free area can be reconstructed without any disturbance from the twin image.

Off-axis points will create an asymmetric fringe system, and points in front of or behind the object plane behave like Fresnel-zone lenses with smaller or larger fringe diameters, respectively. Each of these fringe systems reconstructs a point source at the correct position relative to P. A hologram can therefore store and reproduce a full three-dimensional image of the specimen.

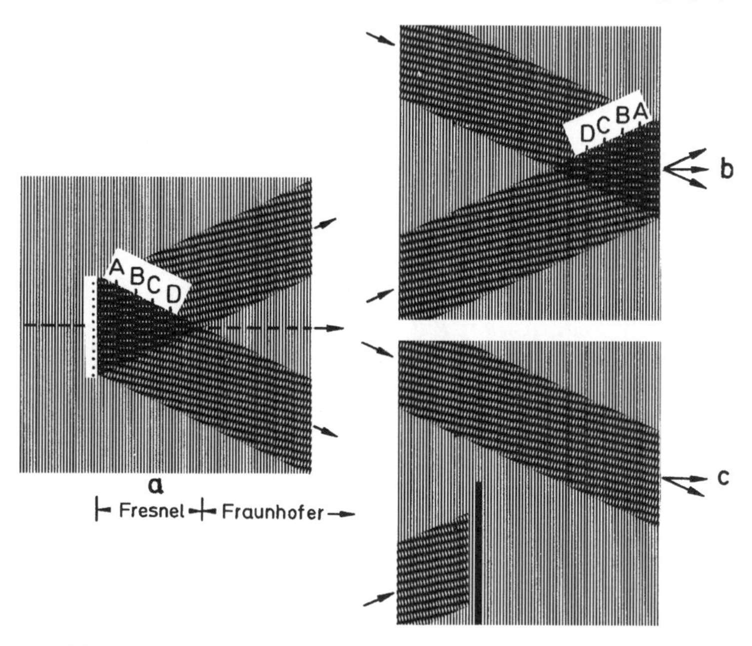

Fig. 6.29. (**a**) Plane incident wave and diffracted waves of order ± 1 in the Fresnel and Fraunhofer regions. (**b**) Overlap of the waves in the image plane. (**c**) Avoidance of the transfer gaps at A, B, C, ... by absorbing one sideband (single-sideband holography).

In-line holograms can be further classified into Fresnel, Fraunhofer, and single-sideband holograms (Sect. 6.5.2). The difference is explained in Fig. 6.29, in which a periodic object of lattice constant or spatial frequency $q = 1/\Lambda$ is used as an example. Figure 6.29a shows the superposition of the diffracted waves of order ± 1 (sidebands) and the incident plane wave. In Fresnel holography (small defocus Δz), the three waves overlap. In the planes A, B, C,..., the three-wave field results in zero intensities for defocus values $\Delta z = n\Lambda^2/\lambda$ (n integer). These defocus values correspond exactly to the zeros of the CTF $B(q) = -2\sin W(q)$ if only the defocus term of $W(q)$ is considered. Near the Gaussian image plane, the three waves overlap to form a magnified image (Fig. 6.29b). A magnification M decreases the scattering of the sidebands to $\pm\theta/M$ and increases the fringe (lattice) spacing to ΛM. Spherical aberration in the imaging system creates an additional phase shift between the plane incident wave and the two sidebands. The photographic emulsion is placed in the Fresnel region for small defocusing. Such Fresnel in-line holograms were originally proposed by Gabor.

For the light-optical reconstruction of an in-line hologram, an objective lens is needed with an appropiately scaled spherical aberration to allow for the difference between the wavelengths of the electrons used for recording and that of the light employed for reconstruction [6.138]. The reconstructed image shows a contrast-transfer function $B^2(q)$ because CTFs have to be multiplied in such a twofold imaging process. This means that the intervals corresponding to spatial frequencies with a negative sign of $B(q)$ in the original electron-optical image are reproduced with the correct sign. However, the information gaps in the CTF cannot be avoided, so that if the photographic emulsion is placed at A, B, ... in Fig. 6.29b, the Fresnel in-line hologram cannot contain any information about the corresponding spatial frequency.

In Fraunhofer in-line holograms, a larger defocusing is used so that the sidebands do not overlap and no transfer gaps occur (Fig. 6.29). This method has been tested in electron microscopy [6.139, 6.140]. It was found that gold particles smaller than 1 nm can be reconstructed, but these are also visible in the normal bright-field mode.

If a specimen area of diameter d_0 is to be recorded and reconstructed, the spatial frequencies present in the spectrum will lie between $q_{\mathrm{min}} = 1/d_0$ and q_{max} corresponding to the resolution limit; q_{min} corresponds to a diffraction angle $\theta_{\mathrm{min}} = \lambda/d_0$ so that a defocus value

$$\Delta z = \frac{d_0}{\theta_{\mathrm{min}}} = \frac{d_0^2}{\lambda} \tag{6.51}$$

will be necessary to separate the primary beam and the sidebands (Fig. 6.29). This defocusing is large even if the diameter of the specimen area d_0 is quite small. A large defocusing causes a blurring of the hologram due to the finite illumination aperture α_{i}, which sets a limit on the minimum periodicity Λ_{min} or maximum spatial frequency q_{max}:

$$\frac{1}{q_{\mathrm{max}}} = \Lambda_{\mathrm{min}} = \frac{\alpha_{\mathrm{i}} d_0^2}{\lambda} \,. \tag{6.52}$$

The following numerical example, $\Lambda_{\mathrm{min}} = 0.2$ nm, $\lambda = 3.7$ pm (100 keV), $d_0 = 100$ nm, $\alpha_{\mathrm{i}} = 7.5 \times 10^{-8}$ rad, shows the limitation of in-line holography. Only a small area of diameter d_0 can be imaged, and an extremely low aperture α_{i} is necessary, which can be obtained only with a field-emission gun.

The influence of nonaxial aberrations (coma, Seidel astigmatism, field curvature, and distortion) renders the image formation anisoplanatic [6.141, 6.142, 6.143]. These aberrations cause a shift and rotation of a transferred specimen sine wave of period Λ. If shifts and rotations of $\Lambda/8$ can be tolerated, the radius of the isoplanatic patch can be estimated to be about 100 nm.

6.5.2 Single-Sideband Holography

Figure 6.29c shows that the large defocusing needed to separate the sidebands and hence to avoid the contrast-transfer gaps (transition from Fresnel to Fraunhofer in-line holography) is not required if one of the sidebands is suppressed by a diaphragm [6.144, 6.145, 6.146, 6.147]. The best way of doing this is to insert a half-plane diaphragm in the focal plane of the objective lens with a small opening for the primary beam. Charging effects can disturb the image if the primary beam passes too close to the diaphragm; these can be reduced either by preparing the diaphragm in a special way [6.148] or by heating it [6.149].

The influence on contrast transfer can be seen from (6.33), retaining only one diffraction order (sideband) instead of both (of order ± 1). The alternative signs in the following formulas correspond to the two possible single-sideband images with complementary half-plane diaphragms; we find

$$\psi_{\mathrm{b}}(x') = 1 - \frac{1}{2}\epsilon_q \mathrm{e}^{-\mathrm{i}W(q)}\mathrm{e}^{\pm 2\pi \mathrm{i} q x'} + \frac{1}{2}\mathrm{i}\varphi_q \mathrm{e}^{-\mathrm{i}W(q)}\mathrm{e}^{\pm 2\pi \mathrm{i} q x'} \tag{6.53}$$

and hence

$$\begin{aligned} I_{\mathrm{b}}(x') &= \psi_{\mathrm{b}}\psi_{\mathrm{b}}^* \\ &= 1 - \epsilon_q \cos[2\pi q x' \mp W(q)] \mp \varphi_q \sin[2\pi q x' \mp W(q)] + \dots \; . \end{aligned} \tag{6.54}$$

This means that there are no transfer gaps, a result that can also be inferred from Fig. 6.29c. The wave aberration produces only a lateral shift of the lattice image. The image of the phase component is shifted by $\pi/2$ or $\lambda/4$ even when $W(q) = 0$. The resulting contrast is asymmetric. This is typical of single-sideband imaging. Thus edges appear bright on one side and dark on the other. This asymmetry is reversed when the complementary half-plane diaphragm is used [reversal of the sign of the last term in (6.54)].

The sign of the amplitude component in (6.54) remains unchanged. This can be used to separate the amplitude and phase components. The sum of two single-sideband holograms recorded with complementary half-plane diaphragms increases the amplitude component and cancels the phase component, whereas the difference between them cancels the amplitude and increases the phase component. The sum or difference of the holograms can be computed digitally or obtained by superposition of the micrographs; a positive and a negative copy are used for the difference [6.150]. The wave aberration $W(q)$ can be corrected in the light-optical reconstruction by using an objective lens with the appropiate spherical aberration and defocusing (Sect. 6.4.4). The focal plane of this lens again contains two sidebands. With a half-plane diaphragm complementary to that used earlier in the electron-optical imaging, the corrected intensity distributions (6.54) can be obtained free of the lateral shift caused by $W(q)$.

6.5.3 Off-Axis Holography

Off-axis or out-of-line holography, proposed by Leith and Upatnieks [6.151], uses a separate reference beam for recording the phase in the interference pattern. The superposition of a reference wave (wave vector $\boldsymbol{k}_1$), which may for example pass through a hole in the support film, and a wave ($\boldsymbol{k}_2$), which transmits the specimen structure off-axis and is modified in amplitude and phase (3.36), can be performed by means of an electrostatic biprism (Sect. 3.1.4) ($\boldsymbol{k}_1$ and $\boldsymbol{k}_2$ include an angle $\pm\beta$ with the axis) [6.140, 6.152, 6.153, 6.154].

In the narrow region of overlap of the two beams, the wave amplitude is (Fig. 6.28a)

$$\psi = \psi_0 \left[a_s(\boldsymbol{r}) e^{i\varphi_s(\boldsymbol{r})} e^{2\pi i \boldsymbol{k}_2 \cdot \boldsymbol{r}} + e^{2\pi i \boldsymbol{k}_1 \cdot \boldsymbol{r}} \right] . \tag{6.55}$$

The hologram is a record of the resulting biprism interference fringes, which have the intensity distribution

$$\begin{aligned} I(r) &= I_0\{1 + a_s^2(\boldsymbol{r}) + 2a_s(\boldsymbol{r})\cos[2\pi x/d + \varphi_s(\boldsymbol{r})]\} \\ &= I_0 \left\{ 1 + a_s^2(\boldsymbol{r}) + a_s(\boldsymbol{r}) \left[e^{i\varphi_s(\boldsymbol{r})} e^{2\pi i x/d} + e^{-i\varphi_s(\boldsymbol{r})} e^{-2\pi i x/d} \right] \right\} , \end{aligned} \tag{6.56}$$

with $x \parallel \boldsymbol{k}_1 - \boldsymbol{k}_2$, and $d = \lambda/2\beta$ is the fringe spacing. The fringes contain the amplitude modulation $a_s(\boldsymbol{r})$ of the specimen wave and the phase shift $\varphi_s(\boldsymbol{r})$ as a local shift of the interference fringes. This allows the phase and amplitude components of an image to be separated.

The technique can be put into practice in a conventional transmission electron microscope with a field-emission or Schottky-emission source. The biprism filament of $\simeq$350 nm in diameter is mounted perpendicular to the rod axis of the holder for the selected-area diaphragms in front of the intermediate lens. The holder can still contain conventional diaphragms for the routine operating modes of the transmission electron microscope. A voltage up to 300 V can be applied to the filament [1.72]. The intermediate lens has to be focused a few millimeters below the biprism in the hologram plane and a few hundred fringes with a spacing of 0.02 nm and an overlap of 30 nm can be recorded. This type of holography can also be employed in the STEM mode [6.155] and in reflection electron microscopy (Sect. 9.7.2) where surface steps result in a phase shift [6.156, 6.157].

By means of reconstruction methods described in the next section, specimens with amplitude and phase structures can be resolved with a lateral resolution of three times the fringe spacing, or twice for weak amplitude structures, which means 0.15 nm and 0.1 nm, respectively, for a fringe spacing of 0.05 nm [1.72].

Applications of off-axis holography are shown for the imaging of magnetic structures (superconducting vortices, Sect. 6.8.2f, Fig. 6.43), the imaging of electric fields (Sect. 6.8.3), and the reconstruction of phase and amplitude from holograms of the crystal structure (Sect. 9.6.4, Fig. 9.34).

6.5.4 Reconstruction of Off-Axis Holograms

Although digital reconstruction methods are often advantageous, various light-optical reconstruction methods will be discussed here because they bring out the principles in terms of optical hardware, which may be easier to appreciate than computer software.

The hologram is reconstructed by illuminating it with a coherent light wave (Fig. 6.28b). The light amplitude behind the hologram can also be represented by (6.56), but the amplitude factors are now modified by the γ-value (Sect. 4.7.2) of the emulsion. Thus, the factor $\exp(2\pi i x/d)$ in the third term of (6.56) is equivalent to a phase shift by a prism with a deflection angle $\theta = \lambda_L/d$ (λ_L denotes the wavelength of the reconstruction light wave); for the last term, the angle would be $-\theta$. This means that the specimen wave $a_s(\boldsymbol{r})\exp[i\varphi_s(\boldsymbol{r})]$ is completely reconstructed in the deflected wave (first diffraction order). This first order can be selected by placing a diaphragm in the focal plane of the lens in Fig. 6.30b, thereby suppressing the twin image. Although the wave amplitude is fully restored with the correct phase and its amplitude only modified by the γ of the recording process, the phase will again be lost in the recorded reconstruction.

However, the phase can be recovered by splitting the reconstruction wave with a Mach–Zehnder interferometer (see also Fig. 6.32) placed in front of the hologram in Fig. 6.30b [6.158]. The resulting two reconstruction waves are inclined at $\pm\theta$ to the axis, and the central beam selected by the diaphragm (aperture stop) contains the superposition of diffracted waves of orders ± 1 from the two reconstruction waves, while their primary spot is absorbed at the diaphragm. The amplitude in the reconstructed image is thus

$$\psi \propto a_s(\boldsymbol{r})\{\exp[i\varphi_s(\boldsymbol{r})] + \exp[-i\varphi_s(\boldsymbol{r})]\} \propto a_s(\boldsymbol{r})\cos[\varphi_s(\boldsymbol{r})]. \tag{6.57}$$

This merely represents an intensity distribution proportional to $\cos^2[\varphi_s(\boldsymbol{r})]$ or fringes of equal phase (Fig. 6.31c). These fringes may be lines of equal specimen thickness produced by the phase shift corresponding to the inner

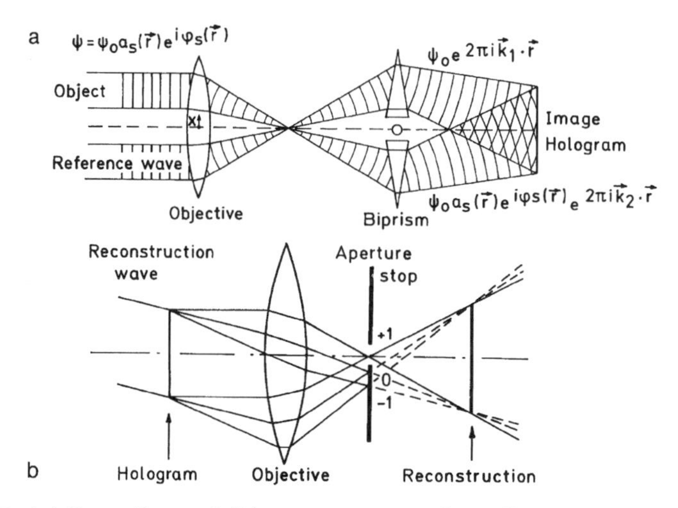

Fig. 6.30. (**a**) Recording and (**b**) reconstruction of an off-axis hologram.

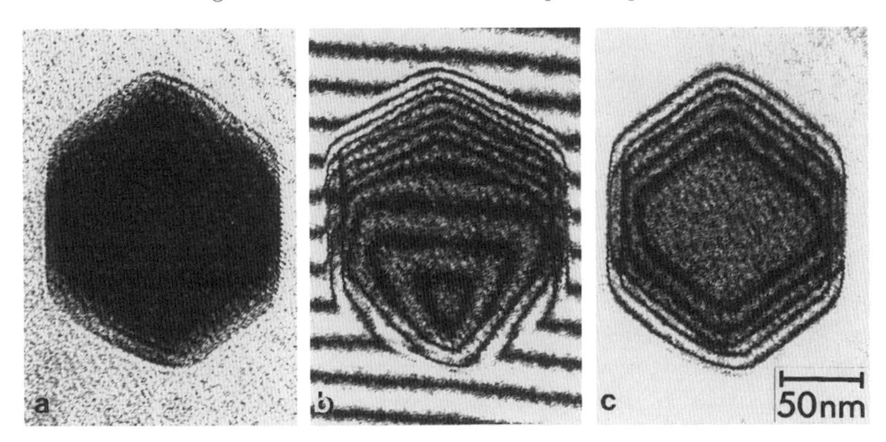

Fig. 6.31. Interference image of a decahedral Be particle: (**a**) reconstructed image, (**b**) hologram, (**c**) contour map of lines of equal phase shift (thickness) [6.158].

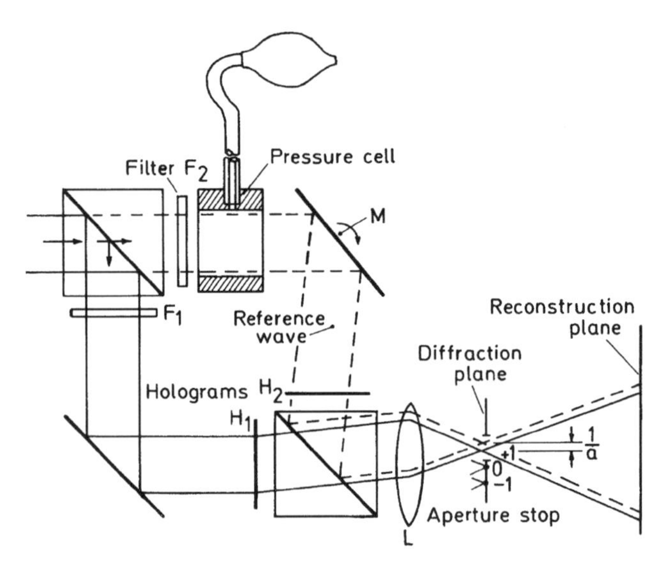

Fig. 6.32. Mach–Zehnder interferometer for recording the phase structure of holograms.

potential U_i (Sect. 3.1.3 and Table 3.2), for example, or magnetic field lines parallel to $\boldsymbol{B}$ caused by the phase shift of the magnetic vector potential (Sect. 3.1.1 and 6.8.2f) [6.160, 6.161, 1.73].

A Mach–Zehnder interferometer can also be used in another arrangement, shown in Fig. 6.32 [6.162], and there are other more compact and disturbance-free interferometers [6.163]. If only the hologram H_1 is used, the wave behind H_1 is superposed on a parallel reference wave. The tilt of this wave can be changed by means of the mirror M, its amplitude by a filter F_2, and the phase by a pressure cell. In the reconstruction, this superposition results in a new interference fringe system in which the direction, spacing, and phase can be changed. Using a copy H_2 of a hologram H_1 in the second beam, a small shift in the holograms produces a fringe shape that contains information about the gradient of the phase distribution; alternatively, the phase value can be doubled if diffraction orders of opposite sign are used for the reconstruction. A further interesting application is the study of phase shifts in crystals depending on the tilt angle relative to the exact Bragg position [6.163]. For a detailed study of electron holography, see [1.73, 6.164, 6.165].

6.6 Image Restoration and Specimen Reconstruction

6.6.1 General Aspects

We have seen in the last sections that the exit wave function $\psi_s(\boldsymbol{r})$ just behind the specimen is modified in the imaging process mainly by spherical aberration and defocusing and shows contrast-transfer gaps. The aim of image restoration is to recreate the exit wave function. This involves changing the sign of the contrast transfer, filling the transfer gaps, and extracting the amplitude and phase of $\psi_s(\boldsymbol{r})$. The best way of doing this is electron holography; otherwise, a series of micrographs at different defocuses may be used, perhaps with the additional use of a diffraction pattern.

In the case of organic specimens, low-dose exposures (noisy and highly underexposed with $<10^3$ e/nm^2) have to be used to avoid as far as possible the loss of resolution by radiation damage. A large number of specimens (separate identical copies) should be imaged with the same orientation on one or more micrographs to allow 2D averaging; this is particularly suitable for biomacromolecules and macromolecular structures. This is easier when the molecules form a 2D array (crystalline or quasi-crystalline), but more effort is necessary in the case of random orientations. In the latter situation, pattern-detection procedures are needed for the alignment of the molecular images before averaging, and misalignments in 2D arrays also have to be taken into account. The alignment is achieved by cross-correlation methods. Inorganic crystals are usually so stable that micrographs can be recorded with normal exposures. The imaging of crystal lattices is discussed in Sect. 9.6.

A 3D reconstruction of a macromolecular structure can be made by tomographic methods; views of the structure in different directions are obtained from a tilt series or from a single tilt if the molecules are randomly oriented. The methods can be applied to native or positively or negatively stained macromolecules on supporting films or cryosections of ice-embedded specimens. This kind of 3D tomography is thus on the way to becoming superior to x-ray crystallography.

Many of the methods for restoration and reconstruction were developed more or less successfully in earlier times, where the exposure of photographic emulsions was the usual method of image recording. Recording by CCD arrays now allows a series of micrographs to be recorded within a few seconds. Not only is it advantageous to obtain the image in digital form directly, but changes in specimen structure between exposures are also decreased. Autofocusing and autotuning likewise become realistic (Sect. 2.4.3). With modern microscopes connected directly to a workstation, it is also possible to display restored images online on the TV monitor.

6.6.2 Methods of Optical Analog Filtering

Although digital image processing is widely used now that micrographs can be digitized directly by means of image plates or CCD arrays, it is still useful to have a look at optical analog methods. Figure 6.33 shows the principal ray path of an optical filtering process. As for optical diffraction (Sect. 6.4.7), the micrographs can be immersed in a fluid with the same refractive index as the gelatin of the emulsion to compensate for any phase shift caused by thickness variations of the gelatin layer. Only the optical density of the silver grains will then influence the incident wave. A 1 : 1 imaging with the lenses L_2 and L_3 can be used to obtain a diffraction pattern in the focal plane of L_2, where the optical filter is situated. The diameter of the diffraction pattern can be increased to a few millimeters by making the focal length of L_2 large (30–50 cm). Filtering processes are easier to implement in the focal plane than in the electron microscope, where the diameter of the diffraction pattern in the back focal plane of the objective lens is only a few tenths of a millimeter and any filter in the form of zone plates (Sect. 6.4.6) disturbs the imaging process.

If the sole aim is optical filtering in the focal plane of L_2, then L_2 and L_3 must be free of spherical aberration within the aperture used. If the Gabor reconstruction method is to be used or if the electron-optical transfer is being simulated by light optics [6.166], the following relations between the spatial frequencies q, the objective apertures α, and the reduced defocusings Δz^* (3.68) have to be respected (subscript L: light-optical, E: electron-optical quantities):

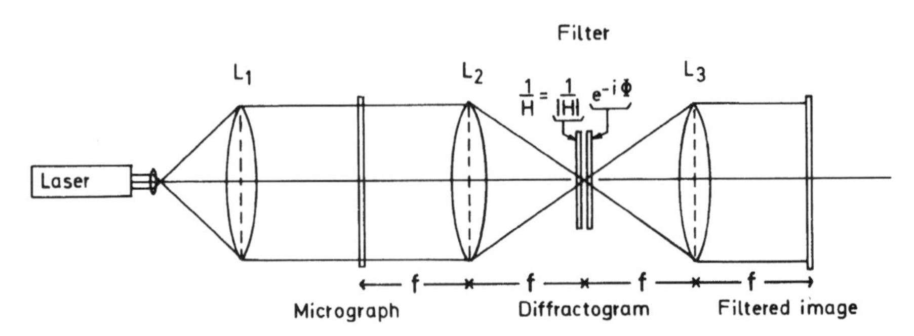

Fig. 6.33. Ray path of a light-optical image-processing system using a Fourier filter.

$$\frac{q_{\rm E}}{q_{\rm L}} = \left(\frac{\lambda_{\rm L}^3 C_{\rm s,L}}{\lambda_{\rm E}^3 C_{\rm s,E}}\right)^{1/4} \quad ; \quad \frac{\alpha_{\rm L}}{\alpha_{\rm E}} = \left(\frac{\lambda_{\rm L} C_{\rm s,E}}{\lambda_{\rm E} C_{\rm s,L}}\right)^{1/4} \quad ;$$

$$\frac{\Delta z_{\rm L}^*}{\Delta z_{\rm E}^*} = \left(\frac{\lambda_{\rm L} C_{\rm s,L}}{\lambda_{\rm E} C_{\rm s,E}}\right)^{1/2} . \tag{6.58}$$

Various imaging modes (e.g., bright-field, dark-field, and hollow-cone illumination) and their contrast-transfer characteristics can be simulated with the aid of these relations [3.35, 6.166, 6.167].

The filtering that will now be discussed applies to weak-phase objects, for which we can set $\psi(\boldsymbol{r}) = 1 - {\rm i}\varphi_{\rm s}(\boldsymbol{r})$ in (6.18); this will give an indication of the optimum design of a Fourier filter in the focal plane of L_2. The image intensity $I(\boldsymbol{r}')$ can be expressed in terms of the blurring or point-spread function $h(\boldsymbol{r})$ introduced in (3.73), which is essentially the inverse Fourier transform of the pupil function $H(q)$,

$$I(\boldsymbol{r}') = 1 + 2\varphi_{\rm s}(\boldsymbol{r}') \otimes h(\boldsymbol{r}') = 1 + 2\int\int \varphi_{\rm s}(\boldsymbol{r}) h(\boldsymbol{r}' - \boldsymbol{r}) {\rm d}^2\boldsymbol{r}. \tag{6.59}$$

Behind the micrograph (negative), the light amplitude for $\gamma = 1$ will be

$$A(\boldsymbol{r}) = 1 - 2\varphi_{\rm s}(\boldsymbol{r}) \otimes h(\boldsymbol{r}). \tag{6.60}$$

The diffracted light amplitude in the focal plane of L_2 can be represented by

$$F(\boldsymbol{q}) = \mathbf{F}\{A(\boldsymbol{r})\} = \delta(\boldsymbol{q}) - 2F_s(\boldsymbol{q}) \cdot H(\boldsymbol{q}), \tag{6.61}$$

where $\delta(\boldsymbol{q})$ represents the primary beam at $\boldsymbol{q} = 0$ and $F_s(\boldsymbol{q})$ is the Fourier transform of $\varphi_{\rm s}(\boldsymbol{q})$. The convolution theorem (3.48) maps the convolution in (6.60) into a multiplication in (6.61). If an optical filter $1/H(q)$ is present in the focal plane of L_2, the amplitude in the reconstructed image will be

$$\begin{aligned} A(\boldsymbol{r}') &= 1 - 2\varphi_{\rm s}(\boldsymbol{r}') \otimes \mathbf{F}^{-1}\left\{H(\boldsymbol{q})\frac{1}{H(\boldsymbol{q})}\right\} \\ &= 1 - 2\varphi_{\rm s}(\boldsymbol{r}') \otimes \delta(\boldsymbol{r}') = 1 - \phi_s(\boldsymbol{r}'), \end{aligned} \tag{6.62}$$

and the light intensity hence becomes $L(\boldsymbol{r}') = A(\boldsymbol{r}') \cdot A^*(\boldsymbol{r}') = 1 - 2\varphi_{\rm s}(\boldsymbol{r}')$. The originally blurred image [convolved with $h(\boldsymbol{r})$] is deblurred by the optical filter with a transmission $1/H(\boldsymbol{q})$. However, this deblurring cannot restore information lost at gaps in the contrast transfer; spatial frequencies are reconstructed only if they are present in the micrographs and are larger than the noise.

The filter can be divided into an amplitude and a phase part:

$$\frac{1}{H(\boldsymbol{q})} = \frac{1}{|H(\boldsymbol{q})|} \exp(-{\rm i}\varphi_q). \tag{6.63}$$

A negative sign of $H(\boldsymbol{q})$ can be included in the phase factor by recalling that $\exp({\rm i}\pi) = \exp(-{\rm i}\pi) = -1$.

The use of a light-optical filter with $|H(\boldsymbol{q})| = 1$ and $\exp(-{\rm i}\varphi_q) = -1$ was proposed to correct those spatial-frequency intervals for which the sign of the CTF is negative. This method generates the image that would have been produced by a system with $|B(\boldsymbol{q})|$ as CTF [6.168, 6.169]. If a more complicated amplitude filter of the form $1/|H(\boldsymbol{q})| \propto 1/|\sin W(q)|$ is used, the CTF in the transfer bands can be equalized in magnitude. However, small gaps around the zeros of sin $W(q)$ have to be tolerated [6.170].

A phase filter can also be created by means of an amplitude grating. The distance between the slits of the grating must be so small that the twin images in the first-order diffracted beams are separated from those produced by the transmitted und undiffracted wave with no overlap. In the transfer intervals with a negative sign of $B(q)$, the grating is shifted by half of the slit separation. This causes a phase shift π of the twin images, whereas the phase of the transmitted wave remains unchanged. This phase shift is a direct consequence of the translation theorem (3.44) for Fourier transforms. It is also possible to construct a combined amplitude and phase filter from a two-dimensional grating of transparent rectangles that vary in size (amplitude) and position (phase). Such binary filters can be calculated and plotted by a computer [6.171].

The noise amplitude in periodic structures can be decreased by introducing in the Fourier plane a mask that contains holes at the diffraction maxima [6.172, 6.173, 6.174]. Thus, if structures from the back and front of negatively stained particles, for example, are superimposed, these can be separated by selecting the corresponding maxima [6.175]. Care will be needed to avoid introducing artificial periodicities by this filtering method [6.176].

Another simpler method for decreasing the noise in periodic images is to produce a suitable multiple exposure of a photographic copy of the image by moving the negative or the copy by multiples of the specimen periodicity [6.177, 6.178] or by an n-fold rotation of the micrograph by multiples of $2\pi/n$ if the structure shows n-fold rotational symmetry [6.179]. This method is sometimes known as stroboscopy because the same effect can be obtained by mounting the micrograph on a turntable that rotates at a frequency f and illuminating it with a source that flashes at a frequency nf. An n-fold rotational symmetry will then be detected by the eye. Artificial structure may be produced if the correct n is established by varying n in the stroboscopic superposition. It is often more useful to superimpose different micrographs, which can be done more accurately digitally because objective criteria for alignment can be applied (Sect. 6.6.4).

6.6.3 Digital Image Restoration

All of the optical analog techniques described in Sect. 6.6.2 can equally well be exploited on a digital computer if the image intensity is first stored in a matrix array. Image plates (Sect. 4.7.3) and CCD cameras (Sect. 4.7.5) allow electron micrographs to be recorded directly in digital form. It is not possible to describe all of the various digital procedures in detail here. Our aim in this section is to give some idea of what is, in principle, possible (see [6.180] for a review). The basic routines of digital image processing are incorporated in the diverse commercial programs (see the survey in [6.181]). Digital image simulation for crystal-structure imaging is discussed in Sect. 9.6.

Digital processing becomes of special interest if two or more micrographs of a series are used for restoration of the exit wave function. The amplitude and phase distribution of the specimen can then be separated (see below). Near the resolution limit, each micrograph requires a two-dimensional restoration procedure. For this, the methods of Sect. 6.6.2 can be recast in digital form. A two-dimensional fast Fourier transform using the Cooley–Tukey algorithm provides the diffraction pattern of the micrograph, which contains information about the contrast-transfer gaps, the defocus, spherical aberration, paraxial astigmatism, and other parasitic aberrations. Filtering in Fourier space can be applied, followed by an inverse Fourier transform, to improve the image. The resulting image amplitude is complex; phase information will not be lost, as it is when the Fourier transform is performed by optical means. The ultimate aim of restoration is to acquire knowledge about the specimen amplitude and phase, $\psi_s(\boldsymbol{r}) = a_s(\boldsymbol{r})\exp[\mathrm{i}\phi_s(\boldsymbol{r})]$, without transfer gaps. Except in the case of weak-phase, weak-amplitude objects, this problem is nonlinear; it is reviewed in detail in [6.182, 6.183, 6.184]. Procedures that set out from various sets of initial data have been investigated:

1. Use of the diffraction pattern $\propto |F(q)|^2$ and a bright (or dark) field image $\propto |\psi|^2$ (Gerchberg–Saxton algorithm) [6.185, 6.186]. This method requires a periodic specimen [6.187] and has been applied to negatively stained catalase [6.188] and periodic magnetic structures [6.189], for example.
2. Two (or N) micrographs recorded at different values of defocus Δz_n [6.190, 6.191]. Schiske's original description of this procedure for restoring the exit wave function $\psi_s(\boldsymbol{r})$ of the specimen with its Fourier transform $\Psi_s(\boldsymbol{q})$ can be written [6.192]

$$
\begin{aligned}
\Psi_s(\boldsymbol{q}) &= \frac{1}{N}\sum_{n=1}^{N} \mathbf{F}\{I_n(\boldsymbol{q})\}\exp[\mathrm{i}W(q)] \\
&= \frac{1}{N}\exp(\mathrm{i}\pi C_s\lambda^3 q^4/2)\sum_{n=1}^{N} \mathbf{F}\{I_n(\boldsymbol{q})\}\exp(-\mathrm{i}\pi\lambda\Delta z q^2), \qquad (6.64)
\end{aligned}
$$

   where $\mathbf{F}\{I_n(\boldsymbol{q})\}$ is the two-dimensional Fourier transform of the intensity distribution in the nth micrograph and $W(q)$ is substituted from (3.66). Inside the sum, the phase-correction factors only depend quadratically on q. With the exception of the phase factor containing C_s, this method has been rediscovered as the focus variation or paraboloid method [6.193]. Practical difficulties arise from the need to align the micrographs to within about one-half of the desired resolution (see below) and from the contribution of inelastic scattering.
3. Bright- and dark-field micrographs taken under identical electron-optical conditions [6.187, 6.194].

4. Two micrographs with complementary half-plane diaphragms. The basic idea of this method has already been discussed as single-sideband holography in Sect. 6.5.2.

Methods 1–3 are iterative methods in which an initial approximation for amplitude and phase is guessed. Considerable thought has been given to the problem of achieving rapid convergence and a unique solution for the phase, especially in the presence of unavoidable noise; see [6.195] and Vol. 3 of [2.6].

6.6.4 Alignment by Cross-Correlation

The first step in any digital computation involving a series of micrographs with the same or different defocus is to align the individual micrographs in orientation and position. A preliminary adjustment can be made by using characteristic image details. For exact alignment, cross-correlation is needed [6.196]. The cross-correlation of two functions $f_1(\boldsymbol{r})$ and $f_2(\boldsymbol{r})$ is the integral

$$\mathrm{CCF}(\boldsymbol{r}) = f_1(\boldsymbol{r}) \star f_2(\boldsymbol{r}) = \int\int f_1(\boldsymbol{r}') f_2(\boldsymbol{r}' + \boldsymbol{r}) \mathrm{d}^2\boldsymbol{r}'$$
$$= \mathbf{F}^{-1}\{F_1(\boldsymbol{q}) \cdot F_2^*(\boldsymbol{q})\}. \tag{6.65}$$

Setting $f_1 = f_2$ gives the autocorrelation function. This integral will have a maximum at $r = 0$ for two similar, exactly aligned images because the integrand is then positive-definite over the whole area. The integral will show regularly spaced maxima for periodic structures. If two otherwise similar micrographs are not exactly aligned, the position of the maximum indicates the necessary shift (see the example in Fig. 6.34). Two micrographs taken at different values of defocus may give a very broad correlation maximum,

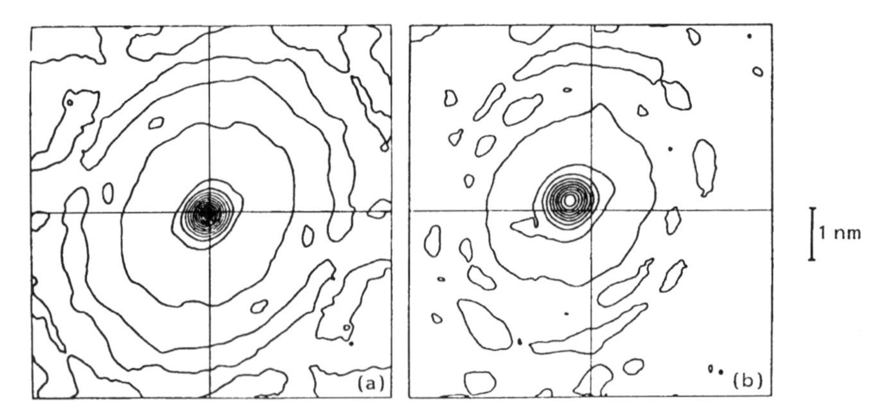

Fig. 6.34. Example of the cross-correlation of electron micrographs of carbon foils: (**a**) autocorrelation of one micrograph with a correlation peak at $x = 0$, $y = 0$, (**b**) cross-correlation of two successive micrographs, indicating an image shift between the two exposures of $x = -0.35$ nm, $y = 0.17$ nm [6.196].

which makes determination of the shift vector $\boldsymbol{r}$ more difficult [6.197]. Sharper maxima can be obtained by calculating the mutual-correlation function MCF [6.198]:

$$\mathbf{F}\{f_j(\boldsymbol{r})\} = A_j(\boldsymbol{q})\exp[\mathrm{i}\Theta_j(\boldsymbol{q})], \quad \phi_j = \mathbf{F}^{-1}\{A_j^{1/2}(\boldsymbol{q})\exp[\mathrm{i}\Theta_j(\boldsymbol{q})]\},$$
$$\mathrm{MCF}(\boldsymbol{r}) = \phi_1(\boldsymbol{r}) \star \phi_2(\boldsymbol{r}), \tag{6.66}$$

where $\phi_j(\boldsymbol{r})$ ($j = 1, 2$) are versions of the input images $f_1(\boldsymbol{r})$ and $f_2(\boldsymbol{r})$ for which the transform amplitudes have been replaced by their square roots, thus attenuating the strongest Fourier components.

For determination of the defocus, spherical aberration, and astigmatism, it is necessary to calculate the Fourier transforms $F_1(\boldsymbol{q})$ and $F_2(\boldsymbol{q})$ of $f_1(\boldsymbol{r})$ and $f_2(\boldsymbol{r})$, respectively. It is therefore of interest to note that the cross-correlation is the inverse Fourier transform of the *Wiener spectrum* $W_{12}(\boldsymbol{q}) = F_1(\boldsymbol{q})\cdot F_2^*(\boldsymbol{q})$ [see the end of (6.65)].

This method of aligning two micrographs with equal defocus can also be used in an image-difference method designed to provide information about radiation damage in the specimen between two exposures or to subtract from a macromolecular image the image of a clean supporting film obtained beforehand [6.197, 6.199]. However, successive micrographs of a clean carbon film can show variations in structure caused by contamination and radiation damage.

6.6.5 Averaging of Periodic and Aperiodic Structures

Averaging by Fourier Filtering. The signal-to-noise ratio can be improved by the following scheme [6.200]. If we consider the amplitude distribution in the image to be the specimen function $\psi_s(\boldsymbol{r})$ convolved with the blurring (point-spread) function $h(\boldsymbol{r})$ (3.53) superimposed on an additive noise distribution $n(\boldsymbol{r})$,

$$a(\boldsymbol{r}) = \psi_s(\boldsymbol{r}) \otimes h(\boldsymbol{r}) + n(\boldsymbol{r}), \tag{6.67}$$

the Fourier transform becomes

$$A(\boldsymbol{q}) = F(\boldsymbol{q}) \cdot H(\boldsymbol{q}) + N(\boldsymbol{q}) \tag{6.68}$$

with $H(\boldsymbol{q}) = -M(\boldsymbol{q})\sin W(\boldsymbol{q})$ (3.72) and $F(\boldsymbol{q}) = \mathbf{F}\{\psi_\mathrm{s}(\boldsymbol{r})\}$. Instead of applying only the filter function $H^{-1}(\boldsymbol{q})$ as in (6.62), the filter is multiplied by a further weighting function $H_\mathrm{W}(\boldsymbol{q})$; after inverse Fourier transformation, we obtain

$$a'(\boldsymbol{r}) = \psi_\mathrm{s}(\boldsymbol{r}) \otimes h_\mathrm{W}(\boldsymbol{r}) + n(\boldsymbol{r}) \otimes h_\mathrm{W}(\boldsymbol{r}) \otimes \overline{h}(\boldsymbol{r}) \tag{6.69}$$

with $h_\mathrm{W} = \mathbf{F}^{-1}\{H_\mathrm{W}\}$ and $\overline{h} = \mathbf{F}^{-1}\{H^{-1}\}$.

The convolution of $\psi_s(\boldsymbol{r})$ with $h_W(\boldsymbol{r})$ in the first term inevitably decreases the resolution. Resolution will not be lost only if H_W = const and $h_W = \delta(0)$. We therefore conclude that each noise-filtering operation will be a compromise between a loss of resolution (blurring of the image points) and a reduction of the noise amplitude.

The signal-to-noise amplitude can be calculated from the Wiener spectra $W_0 = |F(\boldsymbol{q})|^2$ and $W_n = |N(\boldsymbol{q})|^2$ and takes the following values:

$$\begin{array}{cc} \text{before filtering} & \text{after filtering} \\ \dfrac{S}{N} = \dfrac{\int\int W_0(\boldsymbol{q})|H(\boldsymbol{q})|^2 \mathrm{d}^2\boldsymbol{q}}{\int\int W_n(\boldsymbol{q})\mathrm{d}^2\boldsymbol{q}}; & \dfrac{S}{N} = \dfrac{\int\int W_0(\boldsymbol{q})|H_W(\boldsymbol{q})|^2 \mathrm{d}^2\boldsymbol{q}}{\int\int W_n(\boldsymbol{q})|H(\boldsymbol{q})|^{-2}|H_W(\boldsymbol{q})|^2\mathrm{d}^2\boldsymbol{q}} . \end{array} \tag{6.70}$$

Weighting functions that have been tested [6.200, 6.201] include:

$$H_W(\boldsymbol{q}) = \exp\left[\alpha\left(1 - \frac{1}{|\sin W(\boldsymbol{q})|}\right)\right],$$

$$H_W(\boldsymbol{q}) = \frac{H^*(\boldsymbol{q})}{|H(\boldsymbol{q})|^2 + W_n(q)/W_0(\boldsymbol{q})} \quad \text{(Wiener's optimum filter).} \tag{6.71}$$

Rotational Symmetry. For the detection of n-fold rotational symmetry the following method can be used [6.202]. The image intensity $I(r, \varphi)$ in polar coordinates is expanded in a Fourier series

$$I(r, \varphi) = \sum_{-\infty}^{+\infty} g_n(r)\mathrm{e}^{\mathrm{i}n\varphi}, \tag{6.72}$$

and the strength of an n-fold component can be calulated from

$$P_n = \int |g_n(r)|^2 r\mathrm{d}r. \tag{6.73}$$

The presence of unique n-fold symmetry will be indicated by pronounced maxima of P_n for one value of n and its multiples.

Periodic Structures. Electron microscopy of biomacromolecules needs low-exposure techniques and averaging of as many individual molecules as possible. Averaging is best performed with two-dimensional crystalline arrays, and the techniques for the 2D crystallization of membrane and water-soluble proteins have therefore been extensively developed [6.203]. For 2D arrays that are imperfectly ordered, displacement vectors can be calculated to prevent the averages from being degraded [6.204, 6.205].

The Fourier coefficients $F(\theta, \varphi)$ of a two-dimensional periodic specimen with a unit cell characterized by two translation vectors vary in a defocus series, the Fourier coefficients of a micrograph being proportional to $|F(\theta, \varphi) \sin W(\theta)|$. The value of $|F(\theta, \varphi)|$ and the amplitude component transferred, if any, can be evaluated from the series by the method of least squares. These corrected Fourier coefficients can be used to calculate a periodic image that represents an average over all unit cells and micrographs. There is an optimum increase of the signal-to-noise ratio because noise due to the electron statistics, the grain of the photographic emulsion, and the inhomogeneities of the specimen is scattered diffusely over the whole Fourier plane (diffraction pattern). This method has been applied to catalase crystals, for example [6.206]. It can also be employed to sharpen micrographs taken at very low electron exposures to reduce radiation damage [6.207, 6.208].

Images of periodic arrays of macromolecules can be obtained by using various heavy-metal stains, negative staining, freeze-drying, freeze-fracturing, and thin shadow-casting films of Ta/W, for example. They provide information about the internal structure, the external shape, and the surface relief [6.209]. The optical diffraction method (Sect. 6.4.7) can also be employed to investigate differences between shadowed periodic arrays of macromolecules using different types of thin shadowing films [6.210].

Aperiodic Structures. Noise-reduced images of aperiodic structures similar in appearance but randomly distributed over the micrograph (e.g., macromolecules or virus particles with a site of preferential attachment to the supporting film) can be obtained by averaging over a sufficiently large number of particles after alignment in position and orientation. For this, the computer must be furnished with a motif-detection capability [6.211]. The method becomes reasonably practicable when applied interactively on an image-analyzing computer [6.212, 6.213, 6.214]. The particles are selected by eye and centered by the cross-correlation methods described above. This means that the cross-correlation maxima for different shifts and rotations have to be calculated. If a low-dose exposure is employed to reduce radiation damage, a subsequent high-dose picture can be used for prealignment and selection of particles that have the most satisfactory appearance.

Figure 6.35 shows an application of noise reduction to ribosomes [6.215]. The latter are randomly distributed and can be separated into left- and right-oriented particles (Fig. 6.35a). Figure 6.35b shows a series of left-oriented images after alignment: 77 particles were used for averaging. Figures 6.35e and f show the averages of 38 and 39 arbitrarily selected particles, respectively, and Fig. 6.35c the average of all 77 particles at a resolution of 1.4 nm. Figure 6.35d shows the result of further averaging over neighboring image points with a resolution of 3.3 nm. Van Heel [6.216] has introduced a method that allows single particles to be detected automatically against an extremely noisy background. Each image element is replaced by the image variance in its environment. The method based on correspondence analysis also permits particles to be classified into groups, such that the members of each group bear a close resemblance to one another (multivariate statistical analysis). An objective and critical selection of particles can then be made before averaging by superposition [6.217, 6.218, 6.219].

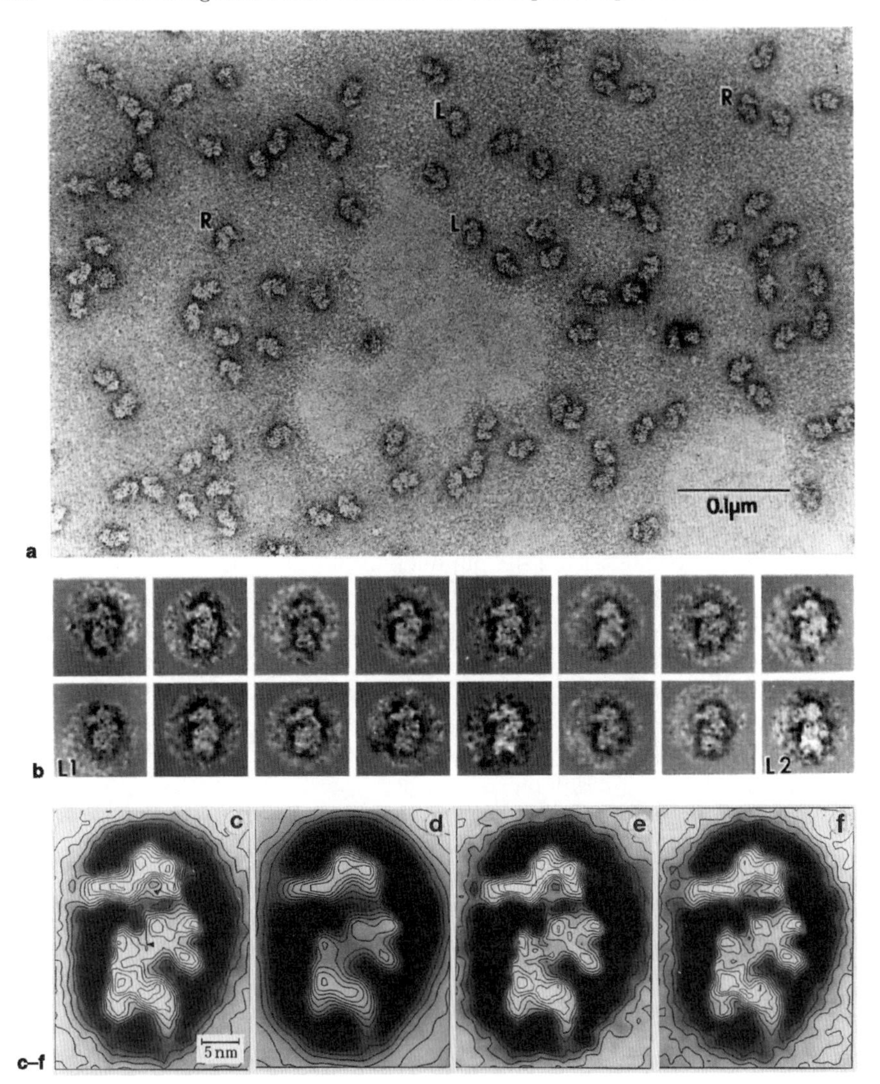

Fig. 6.35. (**a**) Micrograph showing left-oriented (L) and right-oriented (R) 40 S ribosomal subunits of HeLa cells. (**b**) Gallery of 16 L particles after alignment. (**c**) Average obtained from all 77 particles displayed at $1/1.4\ \mathrm{nm}^{-1}$ resolution and (**d**) at $1/3.2\ \mathrm{nm}^{-1}$ resolution. (**e**) and (**f**) Averages from independent sets of 38 and 39 particles, respectively [6.215].

6.7 Three-Dimensional Reconstruction

6.7.1 Stereometry

This technique is based on two tilted micrographs. Two specimen points A and B with a height difference $\Delta z = z_\mathrm{B} - z_\mathrm{A}$ are imaged with different separations

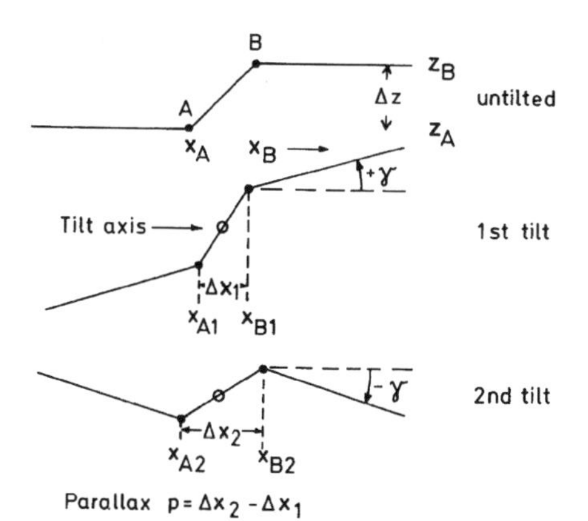

Fig. 6.36. Basis relation for stereometric reconstruction.

Δx_1 and Δx_2 at tilt angles $\pm\gamma$ (Fig. 6.36). In this simple case of parallel projection, the parallax

$$p = (x_{B2} - x_{A2}) - (x_{B1} - x_{A1}) = \Delta x_2 - \Delta x_1 = 2M\Delta z \sin\gamma \tag{6.74}$$

is directly proportional to the height difference Δz if the tilt axis passes through the center of the image area observed. For further details of stereometry, see [6.220, 6.221].

The method can be applied to surface replicas, thick biological sections, aggregates of small particles, and lattice defects in crystal foils. It is essential that the two micrographs contain sharp image details, recognizable in both micrographs; otherwise the parallax cannot be determined accurately. The accuracy for tilt angles $\gamma = \pm 10°$ is of the order of $\Delta z = \pm 3$ nm for $p/M = \pm 1$ nm.

6.7.2 Electron Tomography

Unlike stereometry, 3D reconstruction by tomography does not necessarily need sharp image details. The aim is to reconstruct the specimen density distribution $\rho(x, y, z)$ from a series of projections. Two types of methods are employed, one of which operates in Fourier space [6.222, 6.223] and the other in real space [6.224, 6.225, 6.226, 6.227, 6.228, 6.229]. The formal equivalence and the differences are explained in [6.230]. The Fourier method will be discussed in more detail, because the information content and the information gaps can be evaluated more satisfactorily.

The specimen is represented by its mass-density distribution $\rho(x, y, z)$, the three-dimensional Fourier transform of which is

$$F(q_x, q_y, q_z) = \int\int\int \rho(x, y, z) \exp[-2\pi i(q_x x + q_y y + q_z z)] dx\, dy\, dz. \tag{6.75}$$

For specimens with cylindrical symmetry, it is better to use cylindrical polar coordinates. The Fourier transform then becomes a Fourier–Bessel transform.

A micrograph with the electrons incident parallel to z corresponds to a central section through the Fourier space in the q_x, q_y plane, for example, and the transform reduces to

$$F(q_x, q_y, 0) = \int\int \underbrace{\left[\int \rho(x, y, z)\mathrm{d}z\right]}_{\sigma(x,y)} \exp[-2\pi \mathrm{i}(q_x x + q_y y)]\mathrm{d}x\,\mathrm{d}y. \tag{6.76}$$

The integral $\sigma(x, y)$ is a projection in the z direction (mass thickness distribution) of the mass-density distribution ρ; this produces an image intensity $I(x, y) = I_0 \exp[-\sigma(x, y)/x_\mathrm{k}]$, applying (6.6) for scattering contrast. Equation (6.76) tells us that a two-dimensional Fourier transform of $\sigma(x, y)$ yields a central section through Fourier space. If $F(q_x, q_y, q_z)$ is known from a tilt series, which is equivalent to a bundle of central sections through the Fourier space, $\rho(x, y, z)$ can be calculated by an inverse Fourier transform. One of the difficulties of 3D reconstruction is immediately obvious: the maximum tilt angle that can be attained using a specimen goniometer is $\pm 40° - \pm 70°$ so that a double cone of the Fourier space remains vacant ("missing cone"). This can result in a deterioration of resolution or the creation of artifacts; elongation of the particle shape normal to the specimen plane, for example. Recovery routines for filling this vacant Fourier space are introduced in [6.231, 6.232, 6.233]. Single-axis rotation in which the specimen is mounted at the tip of a microneedle or micropipette has been proposed for HVEM [6.234].

For many specimens, symmetry relations can be used to reduce the necessary number n of micrographs. Thus, for a specimen with helical symmetry, a single micrograph is sufficient (T4 phage tail); for icosahedral symmetry (e.g., tomato bushy stunt virus), we find $n = 2$. In the absence of symmetry, $n = 30$ (e.g., ribosomes). The rule of thumb $n = \pi D/d$ has been proposed, where D is the diameter and d the resolution [6.222].

In principle, any specimen can be 3D reconstructed from a tilt series (single-axis tomography), where, however, the necessary dose and radiation damage must be kept low to preserve the biomolecular structure. The radiation damage can be reduced when macromolecules are embedded in ice and being observed in cryosections on holey carbon films. Figure 6.37 shows the 3D reconstruction of thermosomes (16-meric complexes of thermosomal α-subunits from *Thermoplasma acidophilum*, expressed in *Escherichia coli*) from a tilt series in a range $\pm 54°$ with $6°$ increments. The total magnification was $40\,000\times$ at the CCD camera; thus the pixel size was 0.48 nm at the specimen. The total dose used for recording a tilt series containing 19 projections was kept as low as $2000\ \mathrm{e}^-/\mathrm{nm}^2 \simeq 300\ \mathrm{C/m}^2$. A 2D image of the particle in top-view orientation is obtained by 2D alignment and averaging 1292 individual particle images (Fig. 6.37a) and in side-view orientation obtained from 450 particles (Fig. 6.37b). The image contrast has been reversed so that the particles (protein) appear in positive contrast relative to the surrounding ice film. After 3D reconstruction of the tilt series by means of weighted backprojection, volume data of 307 individual particles could be selected. These were

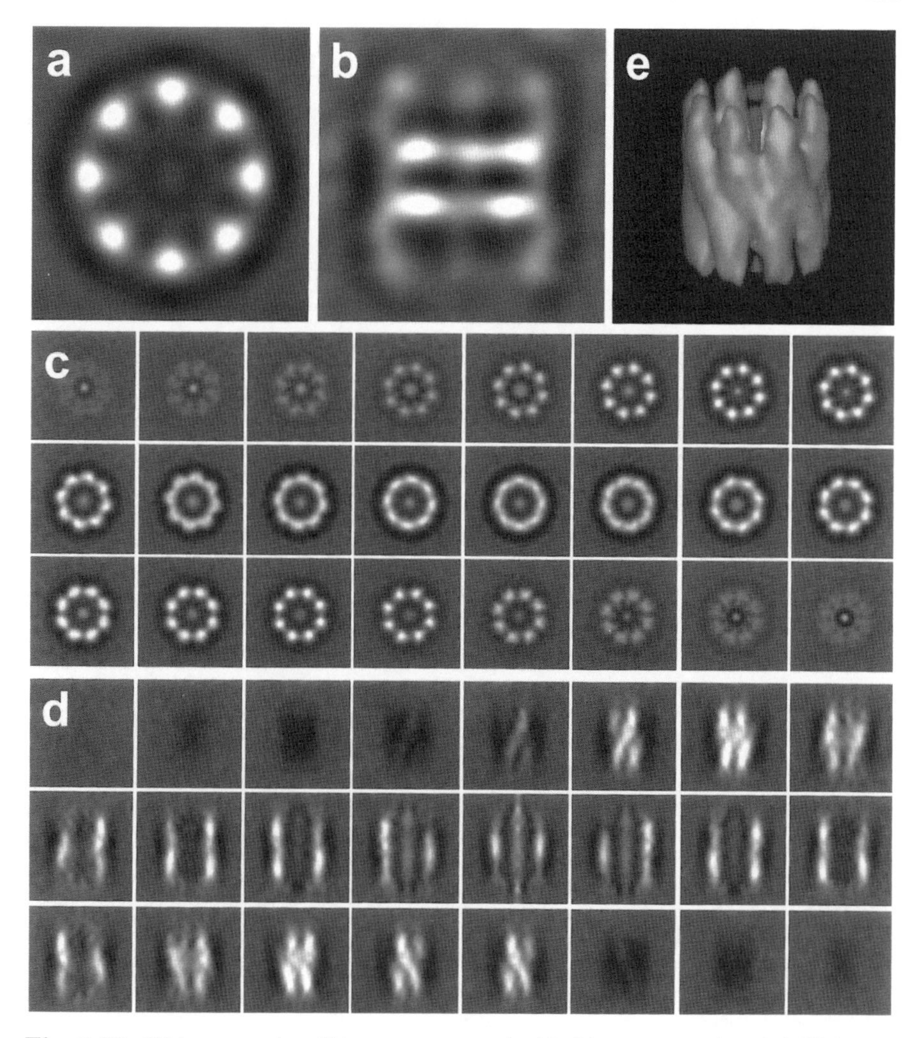

Fig. 6.37. 3D tomography of thermosomes embedded in a cryosection: (**a**) 2D image of the averaged particle in top-view and (**b**) side-view orientations, (**c**) set of xy slices and (**d**) xz slices 1 nm apart, (**e**) surface-view representation of the reconstructed particle; particle diameter 16 nm, length 17 nm (courtesy of D. Typke).

aligned in several cycles with respect to three positional and three angular parameters. Finally, a 3D reconstruction was calculated by weighted backprojection using an appropiate weighting function for the full data set. Figure 6.37c shows a set of xy slices 1 nm apart through the 3D reconstructed and averaged particle and Fig. 6.37d the same for xz slices; Fig. 6.37e shows a surface-view representation of the reconstructed particle.

When averaging over randomly oriented biomolecules in a single micrograph (Sect. 6.6.5), we again assume that identical particles lie in preferred orientations parallel to the specimen plane but in random orientations (azimuthal angles) in that plane. A micrograph at normal incidence can be used for averaging but contains no tomographic information because the images are identical. However, when the specimen is tilted through $\gamma = 45° - 60°$, images of particles with different azimuths are not identical and their directions of view lie on a cone of semiangle γ. This is the idea of random conical tilting or the single-exposure conical reconstruction technique (SECReT) [6.235, 6.236, 6.237]. An additional untilted micrograph is used to determine the azimuthal angle and the position on the cone. Cryosections contain molecules in random orientations, and more sophisticated recognition methods have to be applied to determine their orientations [6.238]. A further step toward automatic electron tomography is to refine the low-dose technique by using three to five different specimen areas for compensating the specimen displacement during tilt, autofocusing, and refocusing before recording the tilt series [6.239, 6.240].

More recently, electron tomography has also been applied to inorganic specimens using either dark-field images to determine the structure or elemental maps to quantify the 3D composition of the specimen [6.241, 6.242].

6.8 Lorentz Microscopy

6.8.1 Lorentz Microscopy and Fresnel Diffraction

It was shown in Sects. 2.1.2 and 3.1.1 that the angular deflection ϵ (2.17) of an electron beam by a transverse magnetic field can be calculated either by evaluating the Lorentz force on the electron or by introducing the phase shift (3.6) caused by the magnetic vector potential or the enclosed magnetic flux Φ_{m}. With an arbitrary origin ($x = 0$) in the specimen plane, the phase shift caused by a magnetic field parallel to the specimen plane can be written

$$\varphi_{\mathrm{m}}(x) = -\frac{2\pi e}{h}\Phi_{\mathrm{m}} = -\frac{2\pi e}{h}B_{\mathrm{s}}\,t\,x, \tag{6.77}$$

where t is the film thickness and x the coordinate in the object plane normal to the magnetic induction $\boldsymbol{B}_{\mathrm{s}}$. Thus, for a ferromagnetic film of iron with a thickness $t = 50$ nm and a spontaneous magnetization $B_{\mathrm{s}} = 2.1$ T, (2.17) gives a deflection angle $\epsilon = 0.1$ mrad for 100 keV electrons. A phase difference $\varphi_{\mathrm{m}} = \pi$ corresponding to a path difference $\lambda/2$ is found for $x = 20$ nm. A ferromagnetic film is therefore a pure but not necessarily weak phase object and can be studied with the theory of phase contrast.

As in (3.36), the wave amplitude behind the specimen is modified from $\psi_0 \exp(2\pi\mathrm{i}kz)$ to

$$\psi(x) = \psi_0 \exp[\mathrm{i}\varphi_{\mathrm{m}}(x)] \exp(2\pi\mathrm{i}kz). \tag{6.78}$$

In the focal plane of the objective lens, the amplitude distribution is given by the Fourier transform

$$F(q_x) = \psi_0 \int_{-\infty}^{+\infty} \exp[\mathrm{i}\varphi_\mathrm{m}(x)] \exp(-2\pi \mathrm{i} q_x x) \mathrm{d}x. \tag{6.79}$$

From the translation theorem of Fourier transforms (3.44), this has the form of a δ-function at $q_x = \varphi_\mathrm{m}/2\pi x = eBt/h$. Substituting $mv = h/\lambda$ and recalling that $q = \theta/\lambda$, we obtain the same angular deflection $\epsilon = \theta$ for $t = L$ as in (2.17).

Because $F(q_x)$ is concentrated within a very small range of values of q_x, only the defocusing term of the wave aberration $W(q)$ in (3.66) need be considered, and (3.72) gives

$$\psi_\mathrm{m}(x') = \frac{\psi_0}{M} \int F(q_x) \exp(\mathrm{i}\pi\lambda\Delta z q_x^2) \exp(2\pi \mathrm{i} q_x x') \mathrm{d}q_x. \tag{6.80}$$

Substituting for $F(q_x)$ from (6.79), we find

$$\begin{aligned}\psi_\mathrm{m}(x') &= \frac{\psi_0}{M} \int \{\int \exp[\mathrm{i}\varphi_\mathrm{m}(x)] \exp(-2\pi \mathrm{i} q_x x) \mathrm{d}x\} \\ &\quad \times \exp(\mathrm{i}\pi\lambda\Delta z q_x^2) \exp(2\pi \mathrm{i} q_x x') \mathrm{d}q_x \\ &= \frac{\psi_0}{M} \int (\int \exp\{\mathrm{i}\pi[2q_x(x'-x) + \lambda\Delta z q_x^2]\} \mathrm{d}q_x) \exp[\mathrm{i}\varphi_\mathrm{m}(x)] \mathrm{d}x. \end{aligned} \tag{6.81}$$

Introducing $q'^2 = 2\lambda\Delta z[q_x + (x'-x)/(\lambda\Delta z)]^2$, we can rewrite (6.81)

$$\begin{aligned}\psi_\mathrm{m}(x') &= \frac{\psi_0}{M} \frac{1}{\sqrt{2\lambda\Delta z}} \int_{-\infty}^{+\infty} \underbrace{\left[\int_{-\infty}^{+\infty} \exp(\mathrm{i}\pi q'^2/2) \mathrm{d}q'\right]}_{1+\mathrm{i}} \\ &\quad \times \exp\left\{\mathrm{i}\left[\varphi_\mathrm{m}(x') - \pi\frac{(x'-x)^2}{\lambda\Delta z}\right]\right\} \mathrm{d}x. \end{aligned} \tag{6.82}$$

The inner integral is one of the Fresnel integrals of (3.34). A further substitution $u = \sqrt{2/\lambda\Delta z}(x'-x)$ gives

$$\psi_\mathrm{m}(x') = \frac{\psi_0}{M} \frac{1+\mathrm{i}}{2} \int_{-\infty}^{+\infty} \exp(\mathrm{i}\varphi_\mathrm{m}) \exp(-\mathrm{i}\pi u^2/2) \mathrm{d}u. \tag{6.83}$$

This is none other than Fresnel diffraction from the phase distribution caused by the magnetization. This can also be seen at $\varphi_\mathrm{m} = 0$; the integral contributes a further factor $1 + \mathrm{i}$, and we have $|\psi_\mathrm{m}|^2 = |\psi_0|^2/M^2$. For uniform magnetization (B_s = const), φ_m is a linear function of x (6.77). This again results in a uniform intensity distribution in the image, but the specimen coordinates are shifted by $\epsilon M \Delta z$, which can also be deduced from particle optics by considering a plane at a distance Δz behind the specimen (Fig. 6.39). Contrast effects will be seen in the image only for nonuniform distributions of $\boldsymbol{B}_\mathrm{s}$, such as magnetic domain walls, across which the direction of $\boldsymbol{B}_\mathrm{s}$ changes, or magnetization ripple, in which the magnetic field exhibits periodic or aperiodic small-angle deviations from the mean value of $\boldsymbol{B}_\mathrm{s}$.

6.8.2 Imaging Modes of Lorentz Microscopy

All modes of Lorentz microscopy (for reviews, see [6.243, 6.244, 6.245, 6.246, 6.247]) require the illumination aperture α_i to be smaller than the deflection angle ϵ; otherwise, the illumination would become incoherent. Apertures $\alpha_\mathrm{i} \simeq 10^{-2}$ mrad can be obtained by forming a small image of the crossover at a large distance in front of the specimen by strongly exciting the first condenser lens, for example. Because the gun brightness is conserved (4.12), the current density at the specimen plane is reduced; hence long exposure times are needed for large magnifications. However, in most applications of Lorentz microscopy (the imaging of domains, for example) a magnification of 10 000 is sufficient. The holographic method (Sect. 6.5.3) requires the higher brightness of a field-emission gun. Furthermore, the magnetic distribution must not be disturbed by the magnetic field of the objective lens.

The original magnetization of the specimen and the components B_x and B_y parallel to the film or foil can only be analyzed if the z component of the magnetic lens field is small enough. The magnetic field of the objective lens at the specimen can be reduced by switching off this lens and using the intermediate lens, by lifting the specimen some millimeters and reducing the lens excitation, or by using a specially designed objective lens of long focal length and small bore to ensure that B falls off rapidly [6.248].

Coils can be used to produce a magnetic field parallel to the film and to observe the movement of ferromagnetic domain walls [6.249]. The presence of such a field normal to the electron beam also causes a deflection. Two further coils are therefore inserted above and below the specimen with opposite excitation to compensate for the deflection and maintain the beam on-axis.

(a) Small-Angle Electron Diffraction. In small-angle electron diffraction (Sect. 8.1.5a), a diffraction pattern is recorded with $\alpha_\mathrm{i} \simeq 10^{-2}$ mrad and a large camera length. A ferromagnetic layer with uniaxial anisotropy consists of domains with antiparallel directions of $\boldsymbol{B}_\mathrm{s}$ separated by 180° walls. The primary beam splits into two spots with an angular separation $\pm\epsilon$ [6.250]. Four different distributions of $\boldsymbol{B}_\mathrm{s}$ parallel to $\langle 100 \rangle$ are possible in a [100] epitaxial iron film, causing splitting into four spots (Fig. 6.38); the central spot is the primary beam passing through holes in the film. In polycrystalline films with varying directions of magnetization, the splitting of the primary beam results in a circular or sickle-shaped diffraction pattern. The splitting 2ϵ can be used to determine the specimen thickness by means of (2.17) [6.251]. However, the angle of divergence may be decreased since the value of $\boldsymbol{B}_\mathrm{s}$ can be lower in thin films.

Periodic domain-wall spacings (25 μm in a cobalt film, for example) can create diffraction maxima, which can be interpreted quantitatively in terms of diffraction at a phase grating [6.252, 6.253].

(b) Foucault Mode. In the case of a 180° domain wall, the antiparallel magnetization directions produce two spots in the focal plane of the objective lens; these are separated by a distance $\simeq 2\epsilon f$. If the objective diaphragm is

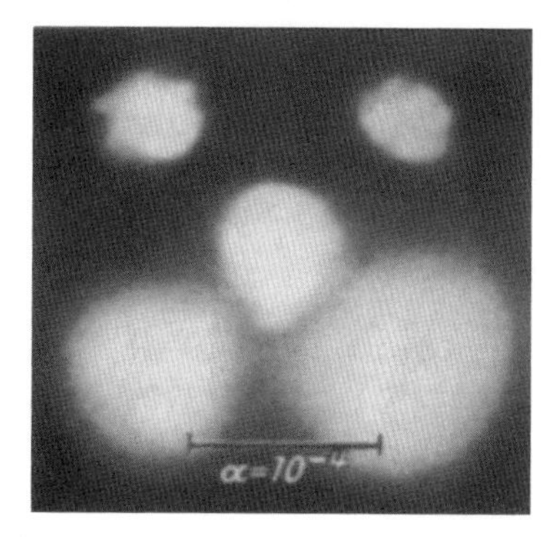

Fig. 6.38. Small-angle electron diffraction pattern of a (100) oriented epitaxially grown iron film on NaCl. The primary beam splits into four beams, corresponding to the ⟨100⟩ directions of spontaneous magnetization.

moved, one of the spots can be suppressed, and the corresponding domain becomes dark in the image [6.254, 6.255]. In this mode, the domain wall is imaged as a boundary between dark and bright areas (Fig. 6.40c,d). The objective lens then operates in focus, so that the specimen is imaged on the final screen without any defocusing [6.256]. A special lens with a long focus has to be used so that the specimen is in a nearly field-free region.

Some means of ensuring that the objective diaphragm is situated in the focal plane should be provided. Any displacement decreases the area in which the Foucault contrast can be observed. A thin-foil diaphragm (Sect. 4.4.1) is preferable; this also reduces charging of the diaphragm.

The Foucault mode can also be used to study much larger extended magnetic stray fields around thin wires or small, compact specimens [6.257, 6.258].

In another version of the Foucault mode [6.259], the small-angle deflection caused by the inner potential (refractive index) is exploited. Small crystals act like prisms, resulting in a splitting of the electron-diffraction spots and the primary beam if the illumination aperture is very small. One or more of these deflected beams can be halted by the diaphragm and can cause a contrast difference that depends on the inclination of the crystal faces. For more complex specimens, the contrast is caused by the local gradient of the optical path length; thus a dark contour for a positive gradient and a bright contour for a negative gradient lead to pseudo-topographic contrast similar to that observed in single-sideband holography (Sect. 6.5.2).

(c) Fresnel Mode. The intensity distribution at a distance Δz below or above the specimen is imaged by under- or overfocusing, respectively [6.260]. The principle of this mode will first be explained in terms of the particle model (geometric theory), in which the electron trajectories are deflected by an angle ϵ proportional to the local value of $\int B \mathrm{d}z$. Figure 6.39 shows that the electron trajectories either converge (left) or diverge (right) at a distance Δz below a specimen with antiparallel domain walls. The left domain wall will appear as a bright line and the right one as a dark line in the defocused image. Defocusing in the opposite direction reverses the contrast (Fig. 6.40a, b). The width of the gap or the overlap $b \simeq 2\epsilon\Delta z$ of the divergent and convergent

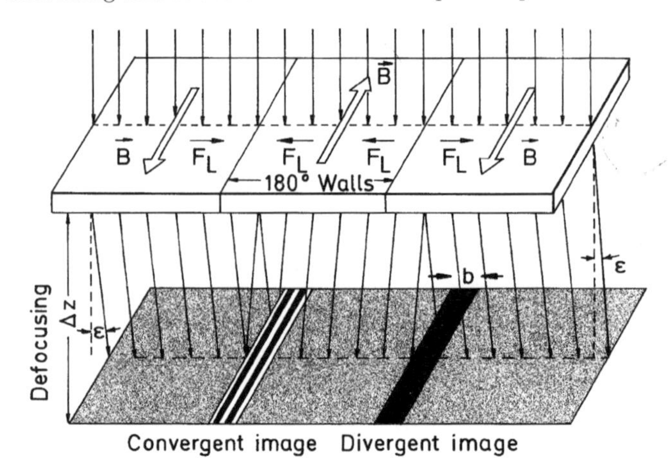

Fig. 6.39. Deflections of electron trajectories in a magnetic film with 180° domain walls that form a convergent (*left*) and divergent (*right*) wall image at a defocus Δz (Fresnel mode).

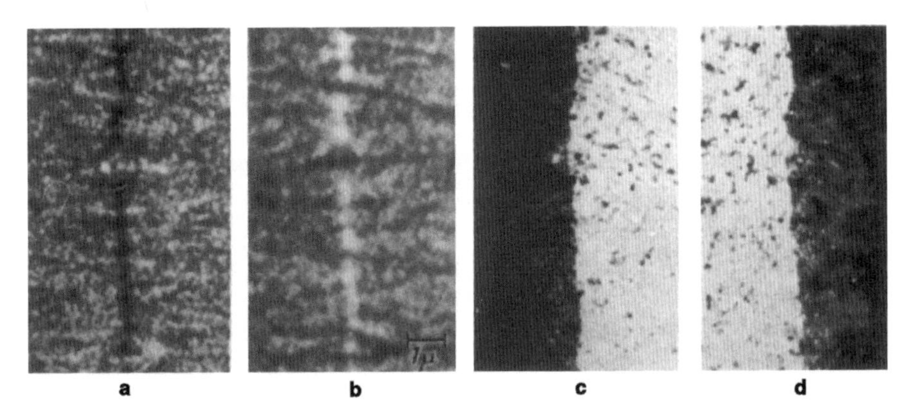

Fig. 6.40. Micrographs that show a 180° domain wall of a polycrystalline iron film in the Fresnel mode with (**a**) under- and (**b**) overfocusing and in the Foucault mode absorbing the left (**c**) and the right (**d**) deflected beams.

wall images, respectively, should be large enough to be detectable at medium magnifications, so that with $b = 0.1$ μm and $\epsilon = 0.1$ mrad, for example, a defocusing of 0.5 mm is needed.

The geometric theory cannot be used for detailed image analysis; the wave-optical theory has to be invoked. The most striking effect is the appearance of *biprism fringes* in the convergent image (Fig. 6.41), in which two coherent waves overlap with a convergence angle 2ϵ [6.261]. The fringe spacing can be calculated from (3.25) with $\beta = \epsilon$. Using (2.17), $mv = h/\lambda$, and $L = t$ gives

$$\Delta x = \frac{h}{2eBt} \,. \tag{6.84}$$

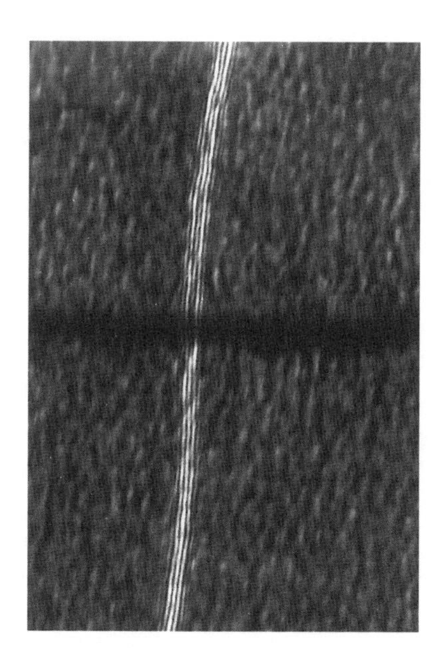

Fig. 6.41. Fresnel mode, showing biprism interference fringes in the convergent domain wall image (*top to bottom*) and a divergent wall image (*horizontal*) of a (100) oriented single-crystal iron film.

This means that the fringe spacing remains constant for constant film thickness, and the number of fringes can be increased only by increasing the overlap (defocusing Δz). Inside the zone of width Δx, the magnetic flux

$$\Phi_{\rm m} = \Delta x B t = \frac{h}{2e} \tag{6.85}$$

is enclosed between two interference fringes. This quantity is just the magnetic-flux quantum (fluxon, Sect. 3.1.1). This fluxon criterion [6.262] can be used to estimate the spacing and number of fringes observable.

Figure 6.42 shows a comparison of the intensity profile across a 180° domain wall of width $w = 80$ nm in a $t = 20$ nm Fe foil at a defocusing $\Delta z = 4$ mm calculated by using the geometric theory (left) and wave optics (right). The differences for a very coherent beam (zero illumination aperture $\alpha_{\rm i}$) are obvious. An aperture of 10^{-2} mrad already blurs the biprism fringes of the wave-optical theory, and geometric theory and wave optics result in similar intensity profiles.

The intensity profile of the divergent wall image is scarcely affected by the use of the wave-optical theory (Fig. 6.42, bottom) [6.245, 6.263]. In the geometric theory, the width b of the divergent wall image is enlarged, in the first-order approximation, by the wall width w, whereas in the convergent image it is decreased by w ($b \simeq 2\epsilon\Delta z \pm w$). Differences between the widths of divergent and convergent wall images can be used to estimate the wall thickness [6.249, 6.264, 6.265, 6.266, 6.267]. This method also requires a knowledge of the influence of elastic and inelastic small-angle scattering on image contrast [6.263]. Thicker films are better studied in a high-voltage electron microscope, in which small-angle scattering has less effect [6.265, 6.274]. The contrast can also be increased by zero-loss filtering [6.275].

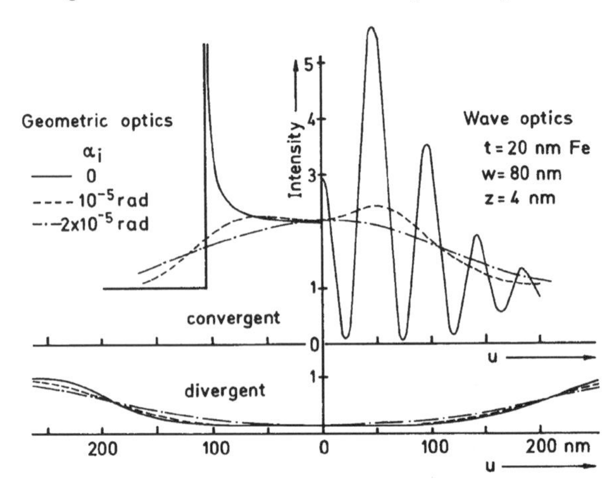

Fig. 6.42. Calculated intensity profiles of convergent and divergent domain-wall images across a 180° wall of width $w = 80$ nm in a 20 nm Fe foil at a defocus $\Delta z = 4$ mm (α_i: illumination aperture).

Comparison of convergent wall images that contain biprism fringes with wave-optical calculations based on models of the magnetization distribution inside the domain wall can be used to test the model and to distinguish between Néel walls and Bloch walls [6.268, 6.269, 6.270, 6.271]. A tilt of the specimen by ±45° allows the B_z component to be determined as well [6.272].

Besides domain walls, periodic fluctuations in the magnetization (ripple) can be observed. A ripple structure is not seen in single-crystal films or in electrolytically polished foils but is mostly observed in evaporated, polycrystalline films. The contrast of the ripple depends strongly on the length of the periodicities and on the defocusing. For quantitative determination of the ripple spectrum, therefore, the contrast-transfer characteristic has to be considered [5.166, 6.273, 6.274].

The sensitivity of the Fresnel mode increases with increasing defocus. When planes a few centimeters below the specimen are imaged (by switching off the objective lens and imaging with the intermediate lens, for example), the relation between object and image can be regarded as a projection, with the demagnified crossover below the first condenser lens as the projection center. The projected shadows are shifted by the Lorentz force in the specimen plane. This has been used to image magnetic stray fields at the surface of superconductors during the transition from the normal to the superconducting state [6.276, 6.277].

(d) Diffraction Contrast. The deflection by the Lorentz force also changes slightly the excitation error of Bragg reflections. This can result in a lateral shift of any bend contours that cross a domain wall in a single-crystal film [6.278].

(e) STEM Modes. The theorem of reciprocity (Sect. 4.5.3) tells us that all modes of Lorentz microscopy can also be used in the STEM mode. For the Fresnel mode, it will be necessary to use a very small detector aperture [6.279, 6.280]. The advantage of STEM is that a direct record of the intensity profiles across domain-wall images is available even if the image intensity is too low for direct viewing. Contamination marks can be printed on the specimen, and the deflection by the Lorentz force can be read directly in terms of the change of the spacing of these marks. An additional mode, applicable in STEM, involves the use of two half-plane detectors or a quadrant detector; the difference signal produces differential contrast similar to that of the Foucault mode [6.281, 6.282].

(f) Reconstruction and Holographic Methods. Because each defocused image may also be regarded as a hologram, the phase and the magnetization distribution can be reconstructed [6.283, 6.284]. For example, an inversion method can be used to obtain information about the magnetization in a domain wall from a divergent-wall image [6.285], or the Gerchberg–Saxton algorithm (Sect. 6.6.3) may be employed to reconstruct the distribution in stripe domains [6.189]. Another holographic recording and reconstruction method is off-axis holography [6.160, 6.161, 1.73]. Figure 6.43 shows a micrograph of a vortex lattice (quantized magnetic flux lines of magnetic flux $h/2e$) in a 70 nm superconducting niobium foil. The specimen on one side of the biprism was inclined at 45° in a 300 kV transmission electron microscope to record the normal component of $\boldsymbol{B}$. Using a Mach–Zehnder interferometer (Fig. 6.32) for the reconstruction and a 16× amplification of phase, the projected magnetic

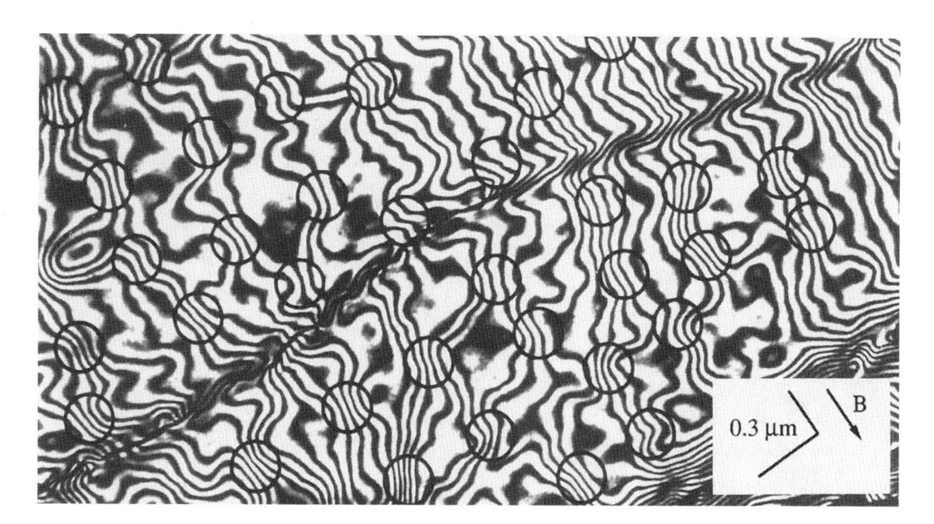

Fig. 6.43. Reconstructed hologram of vortices (magnetic flux lines with $\Phi_{\rm m} = h/2e$) in a superconducting 70 nm niobium foil inclined by 45° in a 300 kV transmission electron microscope [6.286].

lines of force are directly observed as contour fringes. The lines are concentrated locally within the circled regions, becoming narrowly spaced. A bend contour runs diagonally through the micrograph.

Stroboscopic Methods. The dynamic properties of domain walls in high-frequency magnetic fields (1–30 MHz) can be investigated by stroboscopy. The short strobe pulses of the stroboscopic illumination must be synchronized with the a.c. magnetic field applied to the specimen but with a variable time shift (phase angle). This is achieved by chopping the electron beam; the electron beam is deflected by the static electric field of a parallel-plate condenser and returned on-axis by applying a voltage pulse of a few nanoseconds duration [6.287, 6.288]. The technique can be used to investigate the forced and free oscillations of domain walls and Bloch lines, with a time resolution of the order of nanoseconds. The method allows the "mass" and relaxation times of domain walls and Bloch lines to be determined quantitatively.

6.8.3 Imaging of Electrostatic Specimen Fields

The Fresnel mode of Lorentz microscopy can also be used for the investigation of electrostatic fields caused by charging of the specimen, by ferroelectric domains, or by the electric field in the depletion layer of p-n junctions.

Electrostatic fields are generated by electron bombardment in nonconducting specimens. In a shadow projection (Fresnel mode with a very large defocusing) of collodion, formvar, and SiO supporting films, a fluctuating granulation can be observed in TEM by strongly exciting the first condenser lens and switching off the second condenser and the objective lens [6.289, 6.290]. This fluctuating charging occurs only if the beam also hits the specimen grid; otherwise a stronger charging of uniform magnitude causes a larger deflection [6.291]. From the deflection of the electron beam, local field strengths of the order of 10^8 V/m can be estimated. This fluctuation disappears when a conducting film of metal or carbon has been deposited by evaporation or when the specimen is simultaneously bombarded with low-energy electrons of a few hundred eV [6.290]. The fluctuations can be explained in terms of a statistical charge-compensation mechanism due to the secondary electrons produced at the specimen grid.

Small particles, such as MgO, NaCl, or polystyrene spheres, on a carbon or metal film become charged relative to the supporting film [6.292, 6.293]. This charging acts like a lens, focusing the electron rays some 3–6 cm below the specimen. The charge on NaCl crystals can be estimated to be equivalent to a potential of +2 V, corresponding to a field strength at the surface of the order of $10^6 - 10^7$ V/m. A positive charging by secondary-electron emission can also be observed from insulating layers on a conductive support [6.294].

Ferroelectric polarization is associated with a larger lattice deformation than that caused by magnetostriction in ferromagnetics. Ferroelectric domains in ferroelectrics can therefore be distinguished by diffraction contrast and edge fringes on oblique domain boundaries [6.295, 6.296, 6.297, 6.298, 6.299].

The action of the internal electric field in boundaries with a head-to-head polarization can be demonstrated by defocusing (Fresnel mode); however, the deflection angle is smaller than 10^{-2} mrad [6.300].

The electric field strength inside the depletion layer of a p-n junction has been imaged with the Foucault mode [6.301], the Fresnel mode [6.302, 6.303], and by holography [6.304, 6.305]. The latter records the phase shift

$$\varphi(x_0, y_0) = \frac{\pi e}{\lambda E} \int_{-\infty}^{+\infty} \Phi(x_0, y_0, z) \mathrm{d}z \tag{6.86}$$

caused by the local potential Φ.

7

Theory of Electron Diffraction

The theoretical treatment of electron diffraction on crystals needs the concepts of lattice planes and the reciprocal lattice, as in x-ray diffraction. More detailed descriptions of these matters can be found in standard textbooks on solid-state physics or crystallography [4.1, 7.1, 7.2, 7.3]. Kinematical theory leads to the Bragg condition and to a description of the influence of the structure of a unit cell and the external size of a crystal on the diffracted amplitude in terms of structure and lattice amplitudes, respectively. The observed diffraction pattern is equivalent to the points of intersection of the Ewald sphere of radius $1/\lambda$ with the reciprocal-lattice nodes.

The dynamical theory considers the interaction between the primary and reflected waves. For example, when the Bragg condition is satisfied, examination of the two-beam case reveals a complementary oscillation of the primary and reflected intensities with increasing thickness. On taking into account the boundary condition at the surface and the crystal periodicity of the wave field inside the crystal, the solution of the Schrödinger equation takes the form of a Bloch-wave field. An example of the effect of inelastic scattering is the difference between the interaction probability for Bloch waves with nodes and antinodes at the nuclei. This results in the effect known as anomalous absorption. The critical-voltage phenomenon is a typical dynamical effect that can cancel the intensity of Bragg reflections at a voltage that depends sensitively on the structure amplitude.

Inelastic scattering between the Bragg reflections is also influenced by the crystal periodicity and results in Kikuchi lines and bands. Diffraction by amorphous specimens produces diffuse diffraction maxima, which depend on the density distribution of atoms. Polycrystalline specimens can generate Debye–Scherrer rings. Energy filtering of diffraction patterns (electron spectroscopic diffraction) makes it possible to reduce the inelastic background between diffraction spots and to investigate the contribution of electrons with different energy losses to the diffraction pattern.

7.1 Fundamentals of Crystallography

7.1.1 Bravais Lattice and Lattice Planes

A crystal lattice consists of a regular array of *unit cells*, which are the smallest building blocks of the lattice. Each unit cell is a parallelepiped, built up from three noncoplanar, fundamental translation vectors $\boldsymbol{a}_1, \boldsymbol{a}_2, \boldsymbol{a}_3$. The whole crystal lattice can be generated by translation of the unit cell through multiples of the $\boldsymbol{a}_i$ (Fig. 7.1). The origins of the unit cells therefore can be described by a translation vector

$$\boldsymbol{r}_g = m\boldsymbol{a}_1 + n\boldsymbol{a}_2 + o\boldsymbol{a}_3 \qquad (m, n, o \text{ integers}). \tag{7.1}$$

The end points of these vectors form the Bravais lattice. This Bravais lattice may also be characterized by the values of $|\boldsymbol{a}_i| = a, b, c$ and the angles α, β, γ between the axes (Table 7.1). The unit cell is said to be *primitive* if one single atom in the unit cell is sufficient to describe the positions of all other atoms by translations $\boldsymbol{r}_g$. The unit cell normally contains more than one ($k = 1,...,n$) atom at the positions

$$\boldsymbol{r}_k = u_k\boldsymbol{a}_1 + v_k\boldsymbol{a}_2 + w_k\boldsymbol{a}_3, \tag{7.2}$$

$\boldsymbol{r}_1 = (0,0,0)$, and $\boldsymbol{r}_2 = (\frac{1}{2}, \frac{1}{2}, \frac{1}{2})$ in a body-centered cubic lattice, for example (Table 7.1). All other lattice points (open circles) belong to neighboring unit cells. Table 7.1 lists the unit cells of the most important crystal structures and the coordinates (u_k, v_k, w_k). The position of an atom in a Bravais translation lattice is defined by the vector sum $\boldsymbol{r}_g + \boldsymbol{r}_k$.

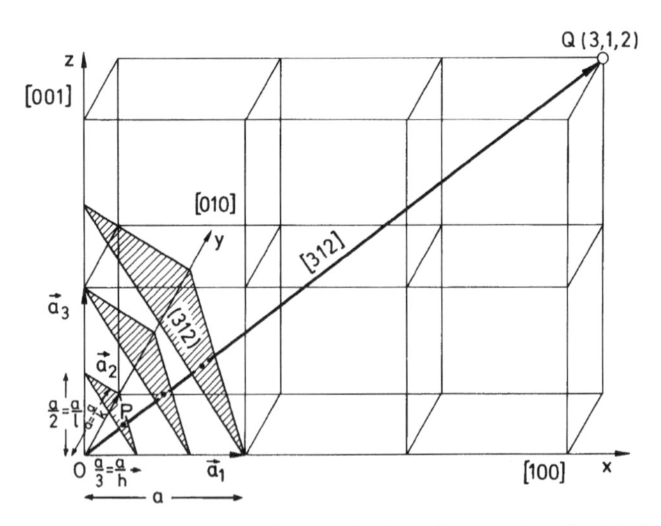

Fig. 7.1. Construction of a crystal by translation of the unit cell with fundamental vectors $\boldsymbol{a}_1, \boldsymbol{a}_2, \boldsymbol{a}_3$. Example of lattice directions [100], [010], [001], and [312] and lattice planes with Miller indices (312).

Table 7.1. List of the most common crystal types (Bravais translation lattices). Structures of the unit cell, lattice-plane spacings d_{hkl} and structure factors F_{cell}

A) Lattices

1. Cubic lattices $a = b = c$; $\alpha = \beta = \gamma = 90°$; $d = \dfrac{a}{\sqrt{h^2 + k^2 + l^2}}$

a) Simple cubic lattice (sc, e.g. Po)

b) Body-centered cubic lattice (bcc, e.g. Cr, Fe, Mo, W)
The unit cell consists of atom at
(0,0,0) and $\left(\frac{1}{2}, \frac{1}{2}, \frac{1}{2}\right)$

$F_{\text{cell}} = 0$ if $(h + k + l)$ odd
$F_{\text{cell}} = 2f$ if $(h + k + l)$ even

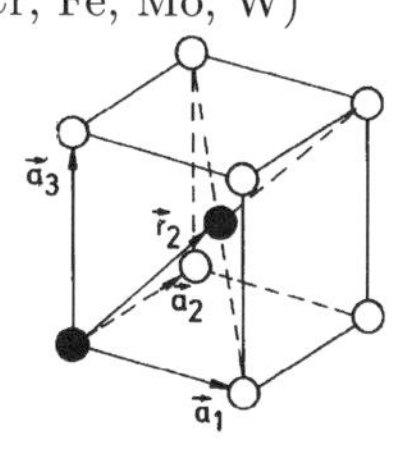

Cubic unit cell

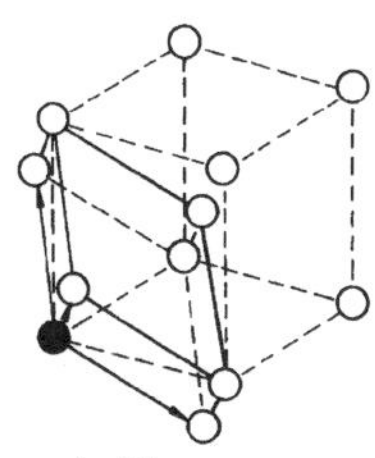

primitive unit cell

c) Face-centered cubic lattice (fcc, e.g. Al, Ni, Cu, Ag, Au, Pt)
The unit cell consists of atom at
(0,0,0), $\left(\frac{1}{2}, \frac{1}{2}, 0\right)$, $\left(\frac{1}{2}, 0, \frac{1}{2}\right)$, $\left(0, \frac{1}{2}, \frac{1}{2}\right)$

$F_{\text{cell}} = 0$ if h, k, l mixed (even and odd)
$F_{\text{cell}} = 4f$ if h, k, l all even or all odd

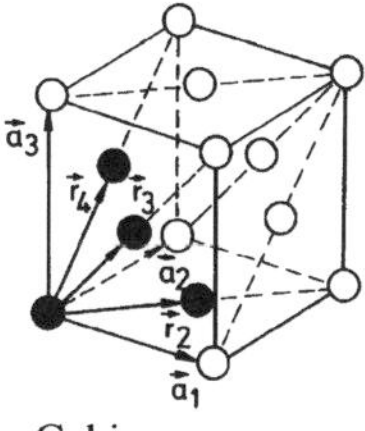

Cubic unit cell

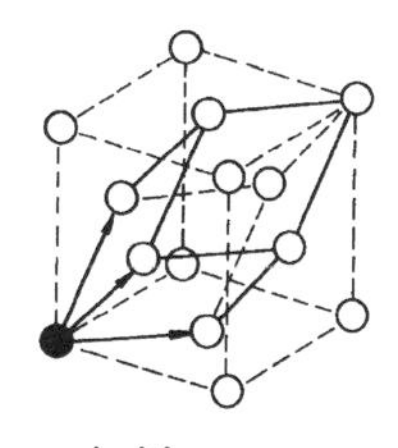

primitive unit cell

---

2. Hexagonal Lattices, $a = b \neq c$, $\alpha = \beta = 90°$, $\gamma = 120°$

$$d = \frac{a}{\sqrt{\frac{4}{3}(h^2 + k^2 + hk) + (a/c)^2 l^2}}$$

---

3. Tetragonal lattices, $a = b \neq c$, $\alpha = \beta = \gamma = 90°$

$$d = \frac{a}{\sqrt{h^2 + k^2 + (a/c)^2 l^2}}$$

---

4. Orthorhombic lattices, $a \neq b \neq c$, $\alpha = \beta = \gamma = 90°$

$$d = \frac{1}{\sqrt{(h/a)^2 + (k/b)^2 + (l/c)^2}}$$

---

5. Trigonal lattices, $a = b = c$, $\alpha = \beta = \gamma = 120°$

$$d = a\sqrt{\frac{1 - 3\cos^2\alpha + 2\cos^3\alpha}{B\sin^2\alpha + 2C(\cos^2\alpha - \cos\alpha)}}; \quad B = h^2 + k^2 + l^2, \quad C = hk + kl + hl$$

Table 7.1 (continued)

6. Monoclinic lattices, $a \neq b \neq c, \quad \alpha = \gamma = 90°, \beta \neq 90°$

$$d = \frac{1}{\sqrt{A/\sin^2\beta + k^2/b^2}}; \quad A = \frac{h^2}{a^2} + \frac{l^2}{c^2} - \frac{2hl}{ac}\cos\beta$$

7. Triclinic lattice, $a \neq b \neq c, \quad \alpha \neq \beta \neq \gamma \neq 90°$

$$d = abc\sqrt{\frac{1 - \cos^2\alpha - \cos^2\beta - \cos^2\gamma + 2\cos\alpha\cos\beta\cos\gamma}{q_{11}h^2 + q_{22}k^2 + q_{33}l^2 + q_{12}hk + q_{13}hl + q_{23}kl}}$$

$q_{11} = b^2c^2\sin^2\alpha$; $q_{22} = a^2c^2\sin^2\beta$; $q_{33} = a^2b^2\sin^2\gamma$
$q_{12} = 2abc^2(\cos\alpha\cos\beta - \cos\gamma)$
$q_{13} = 2ab^2c(\cos\alpha\cos\gamma - \cos\beta)$
$q_{23} = 2a^2bc(\cos\beta\cos\gamma - \cos\alpha)$

B) Structures

1. Cubic structures

a) Diamond structure (e.g. C, Si, Ge)
The unit cell consist of two fcc lattices shifted by $\left(\frac{1}{4}, \frac{1}{4}, \frac{1}{4}\right)$

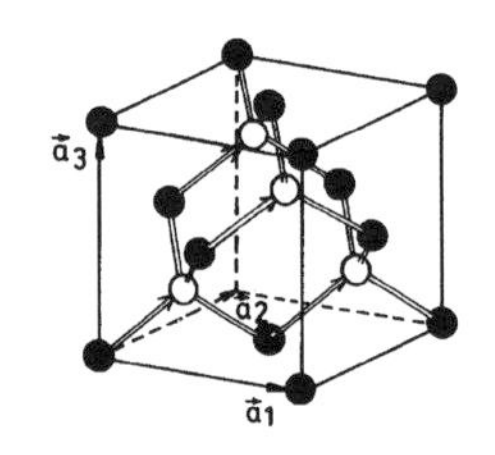

$|F_{\text{cell}}|^2 = 0$ if h,k,l mixed
$|F_{\text{cell}}|^2 = 64\ f_{\text{Ge}}^2$ if h,k,l all even and $(h+k+l) = 4n$
$|F_{\text{cell}}|^2 = 32\ f_{\text{Ge}}^2$ if h,k,l all odd
$|F_{\text{cell}}|^2 = 0$ if h,k,l all even and $(h+k+l) = 4n+2$

b) Caesium chloride structure (e.g. CsCl, TlCl)
The unit cell consist of two primitive cubic lattices shifted by $\left(\frac{1}{2}, \frac{1}{2}, \frac{1}{2}\right)$ Cs: (0,0,0); Cl: $\left(\frac{1}{2}, \frac{1}{2}, \frac{1}{2}\right)$

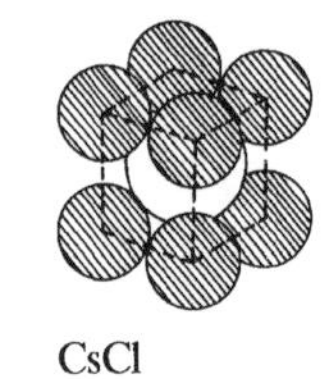
CsCl

$|F_{\text{cell}}|^2 = (f_{\text{Cs}} + f_{\text{Cl}})^2$ if $h+k+l$ even
$|F_{\text{cell}}|^2 = (f_{\text{Cs}} - f_{\text{Cl}})^2$ if $h+k+l$ odd

c) Sodium chloride structure (e.g. NaCl, LiF, MgO)
Unit cell consits of two fcc Na and Cl sublattices shifted by $\left(\frac{1}{2}, \frac{1}{2}, \frac{1}{2}\right)$

$|F_{\text{cell}}|^2 = 0$ if h,k,l mixed
$|F_{\text{cell}}|^2 = 16\ (f_{\text{Na}} - f_{\text{Cl}})^2$ if h,k,l all odd
$|F_{\text{cell}}|^2 = 16\ (f_{\text{Na}} + f_{\text{Cl}})^2$ if h,k,l all even

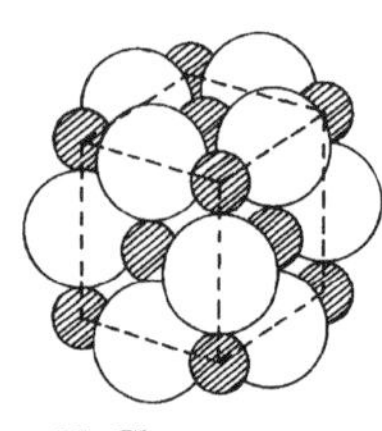
NaCl

d) Zincblende structure (e.g. ZnS, CdS, InSb, GaAs)
Unit cell consists of two fcc Zn and S sublattices shifted by $(\frac{1}{4}, \frac{1}{4}, \frac{1}{4})$

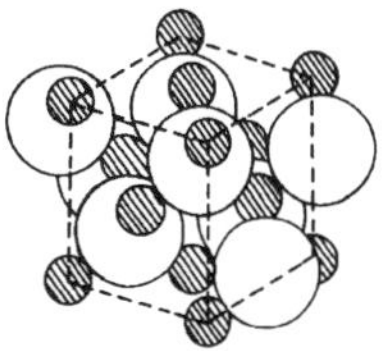
Zincblende

$|F_{\text{cell}}|^2 = 0$ if h,k,l mixed
$|F_{\text{cell}}|^2 = 16\,(f_{\text{Zn}}^2 + f_{\text{S}}^2)$ if h,k,l all odd
$|F_{\text{cell}}|^2 = 16\,(f_{\text{Zn}} + f_{\text{S}})^2$ if h,k,l all even and $(h+k+l) = 4n$
$|F_{\text{cell}}|^2 = 16\,(f_{\text{Zn}} - f_{\text{S}})^2$ if h,k,l all even and $(h+k+l) = 4n+2$

2. Hexagonal structures

a) Hexagonal close-packed structure (hcp, e.g. Mg, Cd, Co, Zn)
Unit cell consists of atoms at (0,0,0), $(\frac{1}{3}, \frac{2}{3}, \frac{1}{2})$

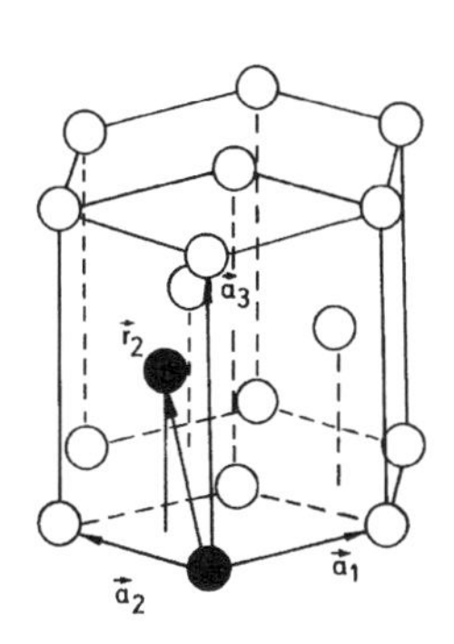

$|F_{\text{cell}}|^2 = 0$ if l odd, $(h+2k)=3n$
$|F_{\text{cell}}|^2 = 4f^2$ if l even, $(h+2k)=3n$
$|F_{\text{cell}}|^2 = 3f^2$ if l odd, $(h+2k)=3n+1$ or $3n+2$
$|F_{\text{cell}}|^2 = f^2$ if l even, $(h+2k)=3n+1$ or $3n+2$

b) Wurtzite structure (e.g. ZnS, ZnO)
Unit cell consists of two hcp Zn and S sublattices shifted by $(\frac{1}{3}, \frac{1}{3}, \frac{1}{8})$

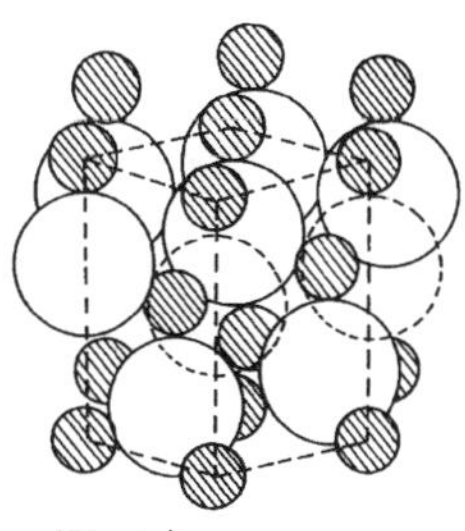
Wurtzite

The cubic lattices have the advantage that their structure can be described by a Cartesian coordinate system. However, it is worth mentioning that the face- and body-centered cubic lattices can also be described by a primitive unit cell that is, however, trigonal in shape (Table 7.1).

Direction in crystals is expressed as a vector that connects the origin O to the origin Q of another unit cell; the components of the vector are scaled so that all are integers, as small as possible (e.g., [312], [100], etc., in Fig. 7.1).

Lattice planes are parallel, equidistant planes through the crystal with the same periodicity as the unit cells. Examples of three equidistant lattice planes are shown in Fig. 7.1; one further plane goes through the origin O. Such a set of lattice planes can be characterized by *Miller indices*. The plane closest to the one that passes through the origin intercepts the fundamental translation vectors $\boldsymbol{a}_i$ at points that may be written $\boldsymbol{a}_1/h$, $\boldsymbol{a}_2/k$, $\boldsymbol{a}_3/l$ (h, k, l integers); otherwise, the system of parallel planes could not have the same periodicity as the lattice because this requirement implies that there must be an integral number h, k, l of interceptions of parallel planes that divide the translation vectors $\boldsymbol{a}_i$ of the unit cell into equal parts. The triplet (hkl) is the set of Miller indices that are the reciprocal intercepts in units of $|\boldsymbol{a}_i|$. The intercepts in Fig. 7.1 are $a_1/h = a_1/3, a_2/k = a_2/1, a_3/l = a_3/2$, and so the Miller indices are (312). Miller indices are always enclosed in parentheses to distinguish them from directions, which are always denoted by square brackets. Only in cubic lattices is the $[hkl]$ direction normal to the (hkl) lattice planes.

If a lattice plane intersects one or two axes at infinity, which means that the plane is parallel to one or two of the $\boldsymbol{a}_i$, then the corresponding Miller indices are zero (Fig. 7.2). If the plane cuts one of the axes on the negative side of the origin, the corresponding Miller indices are negative. This is indicated by placing a minus sign above the index; for example, $(1\bar{1}1)$ in Figs. 7.2, which shows further examples of indices in a cubic lattice.

For hexagonal lattices, four indices $(hkil)$ are often used; these are obtained from intercepts with the c axis and the three binary axes inclined at 120° to one another. The indices h, k, and i satisfy the relation $i = -(h+k)$.

If we wish to refer to a full set of equivalent lattice planes, such as all six cubic faces of a cubic crystal, (100), (010), (001), $(\bar{1}00)$, $(0\bar{1}0)$, $(00\bar{1})$, we enclose the Miller indices in braces (curly brackets): $\{100\}$. Thus we might say that the $\{111\}$ planes in a face-centered cubic lattice are close-packed planes. A full set of crystallographically equivalent directions or axes with all directions parallel to one of the fundamental vectors $\boldsymbol{a}_i$, for example, is denoted by angle brackets: $\langle 100 \rangle$.

Close-packed structures such as the face-centered cubic and the hexagonal close-packed structures are of special interest. Figure 7.3 shows that there are two possible sets of positions, B and C, at which a second close-packed plane can be stacked above the plane with atoms at positions A. The face-centered cubic lattice can be characterized by the sequence ABCABC..., and the $\{111\}$

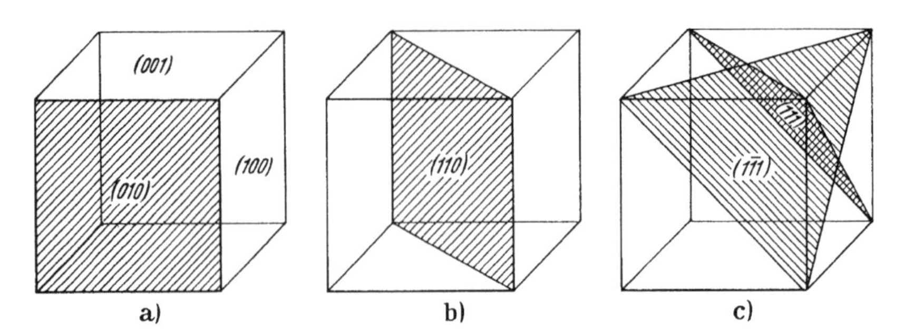

Fig. 7.2. (**a-c**) Examples of lattice planes in a cubic crystal: (**a**) cubic $\{100\}$, (**b**) dodecahedral $\{110\}$, and (**c**) octahedral $\{111\}$.

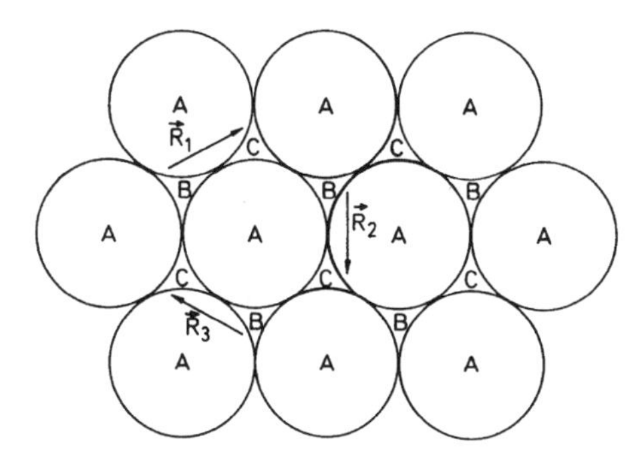

Fig. 7.3. Close-packed atoms in a layer A and positions of the atoms in neighboring layers B or C. $\boldsymbol{R}_i$ ($i = 1, 2, 3$) are the displacement vectors of the layers in a close-packed lattice.

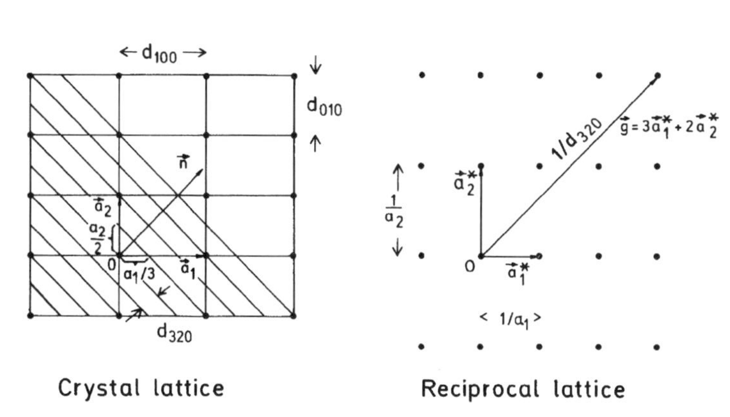

Fig. 7.4. Construction of reciprocal-lattice vectors $\boldsymbol{g}$ parallel to the normal of the crystal-lattice planes with indices (hkl) and length $1/d_{hkl}$.

planes are then close-packed, whereas the closed-packed hexagonal structure follows the sequence ABAB... and the close-packed planes are now (0001). This corresponds to a ratio $c/a = \sqrt{8/3} = 1.63$. However, the measured value of this ratio for hexagonal crystals is slightly different as a result of binding forces depending on the crystallographic directions parallel and normal to the close-packed planes.

7.1.2 The Reciprocal Lattice

The reciprocal-lattice concept is important for the understanding and interpretation of electron-diffraction patterns. There are different ways of introducing the reciprocal lattice, which will be shown to be equivalent.

We start with an intuitive, graphical construction. Each point of the reciprocal lattice will be related to a set of lattice planes of the crystal lattice with Miller indices (hkl). Such a point can be constructed by plotting a vector $\boldsymbol{n}$ normal to the (hkl) planes and of length $1/d_{hkl}$ from the origin O of the reciprocal lattice. The procedure is illustrated in Fig. 7.4 for a two-dimensional projection of a lattice (built up from the vectors $\boldsymbol{a}_1$ and $\boldsymbol{a}_2$ with $\boldsymbol{a}_3$ normal to the plane). The lattice planes $(hk0) = (320)$ intercept this plane in parallel straight lines d_{hk0} apart. Figure 7.4 shows that all points of a reciprocal lattice can be described by the reciprocal translation vectors $\boldsymbol{a}_1^*$ and $\boldsymbol{a}_2^*$ with $|\boldsymbol{a}_1^*| = 1/d_{100} = 1/a_1$ and $|\boldsymbol{a}_2^*| = 1/d_{010} = 1/a_1$ and that $\boldsymbol{g} = h\boldsymbol{a}_1^* + k\boldsymbol{a}_2^*$ is a reciprocal-lattice vector.

The next method is a more abstract mathematical construction. If the $\boldsymbol{a}_i$ are the fundamental translation vectors of a primitive unit cell, the lattice vectors $\boldsymbol{a}_i$ and the translation vectors $\boldsymbol{a}_i^*$ of the reciprocal lattice are related by

$$\boldsymbol{a}_i \cdot \boldsymbol{a}_j^* = \delta_{ij} = \begin{cases} 0 & \text{if} \quad i \neq j \\ 1 & \text{if} \quad i = j \end{cases} \quad (i, j = 1, 2, 3). \tag{7.3}$$

This system of nine equations has the solution

$$\boldsymbol{a}_1^* = \frac{\boldsymbol{a}_2 \times \boldsymbol{a}_3}{V_e}, \quad \boldsymbol{a}_2^* = \frac{\boldsymbol{a}_3 \times \boldsymbol{a}_1}{V_e}, \quad \boldsymbol{a}_3^* = \frac{\boldsymbol{a}_1 \times \boldsymbol{a}_2}{V_e}. \tag{7.4}$$

This shows immediately that the vector $\boldsymbol{a}_1^*$ is normal to $\boldsymbol{a}_2$ and $\boldsymbol{a}_3$. ($V_e = \boldsymbol{a}_1 \cdot (\boldsymbol{a}_2 \times \boldsymbol{a}_3)$ is the volume of the unit cell.)

The reciprocal lattice of a primitive cubic cell with lattice constant a is again a primitive cubic; the lattice constant of the reciprocal unit cell is $1/a$. The reciprocal lattice of a face-centered cubic (fcc) lattice can be deduced by considering the primitive trigonal cell of Table 7.1; we see that any fundamental vector of the primitive trigonal unit cell of the body-centered cubic (bcc) lattice is normal to two fundamental vectors of the primitive trigonal unit cell of the fcc lattice. The condition (7.4) for the fcc lattice therefore is satisfied by a reciprocal bcc lattice. Conversely, the reciprocal lattice of a bcc lattice is fcc (Fig. 7.5).

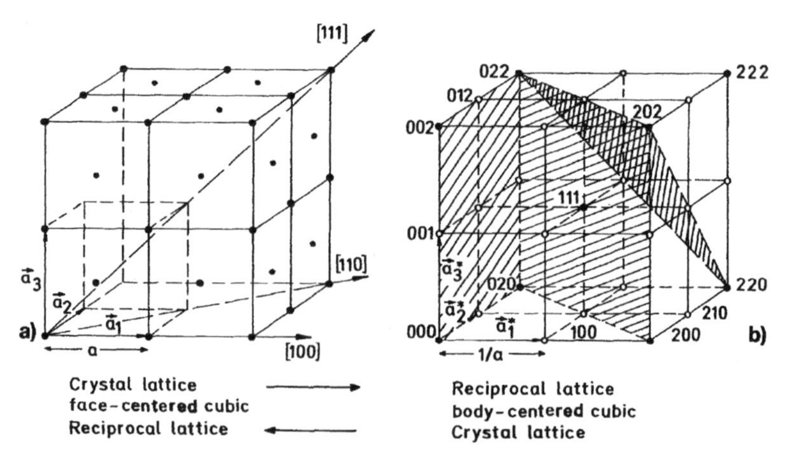

Fig. 7.5. Body-centered cubic crystal (**b**) as the reciprocal lattice of a face-centered cubic lattice (**a**) and vice versa. Only the full circles in (**b**) are reciprocal-lattice points. The open circles are forbidden by the extinction rules for the structure amplitude F. The shaded planes are used in Sect. 7.1.3 to construct the Laue zones (Fig. 7.7).

Not all of the reciprocal lattice points predicted by (7.4) can in fact be observed when the fundamental vectors of a nonprimitive unit cell are used. Some points will disappear. In the reciprocal lattice of the nonprimitive fcc structure (Fig. 7.5b), for example, the reciprocal lattice points 200, 220 etc. are allowed, but 100, 210, etc., are "forbidden". The reasons for this will become clearer when we meet the zero rules ($F = 0$) of the structure amplitude F in Sect. 7.2.2 or the interpretation of the reciprocal lattice as the three-dimensional Fourier transform of the crystal lattice (see the end of this section).

Let us now consider some important laws that can be derived from the definitions (7.3) and (7.4) of the reciprocal lattice:

(1) The reciprocal-lattice vector $\boldsymbol{g} = h\boldsymbol{a}_1^* + k\boldsymbol{a}_2^* + l\boldsymbol{a}_3^*$ (h, k, l integers) is normal to the (hkl) planes.

Proof: Figure 7.1 shows that two nonparallel vectors on the (hkl) plane can be obtained as differences between the points at which the fundamental lattice vectors intersect this plane: $\boldsymbol{r}_1 = \boldsymbol{a}_1/h - \boldsymbol{a}_2/k$, $\boldsymbol{r}_2 = \boldsymbol{a}_1/h - \boldsymbol{a}_3/l$. The scalar products of these vectors with $\boldsymbol{g}$ are zero, which means that $\boldsymbol{g} \perp \boldsymbol{r}_1, \boldsymbol{r}_2$; $\boldsymbol{g}$ is therefore also normal to all other vectors that lie in the (hkl) plane.

(2) The length of the reciprocal-lattice vector $\boldsymbol{g}$ is equal to the reciprocal lattice-plane distance $1/d_{hkl}$.

Proof: Let $\boldsymbol{u}_n$ be the unit vector normal to the (hkl) plane and hence parallel to $\boldsymbol{g}$, which means that we can write $\boldsymbol{u}_n = \boldsymbol{g}/|\boldsymbol{g}|$. From Fig. 7.1, we see that d_{hkl} is equal to the projection of $\boldsymbol{a}_1/h$, $\boldsymbol{a}_2/k$, or $\boldsymbol{a}_3/l$ on the unit vector $\boldsymbol{u}_n$:

$$d_{hkl} = \boldsymbol{u}_n \cdot \boldsymbol{a}_1/h = \frac{\boldsymbol{g}}{|\boldsymbol{g}|} \cdot \frac{\boldsymbol{a}_1}{h} = \frac{1}{|\boldsymbol{g}|} \, . \tag{7.5}$$

(3) The solution of the system of (Laue) equations

$$\boldsymbol{a}_i \cdot \boldsymbol{g} = h_i \quad (i = 1, 2, 3;\ h_{1,2,3} = h, k, l) \tag{7.6}$$

is

$$\boldsymbol{g} = h\boldsymbol{a}_1^* + k\boldsymbol{a}_2^* + l\boldsymbol{a}_3^*. \tag{7.7}$$

Proof: Substitute (7.7) into (7.6) and use (7.3).

The third way to introduce the reciprocal lattice is to define it as the Fourier transform of the crystal lattice. The Fourier integral (3.40) of a three-dimensional crystal lattice with δ-functions at the origins of the unit cells becomes a sum over the discrete lattice points $\boldsymbol{r}_g$ (7.1),

$$\begin{aligned} G(\boldsymbol{q}) &= \sum_g \exp(-2\pi \mathrm{i}\, \boldsymbol{q} \cdot \boldsymbol{r}_g) \\ &= \sum_{m,n,o} \exp[-2\pi \mathrm{i}\, \boldsymbol{q} \cdot (m\boldsymbol{a}_1 + n\boldsymbol{a}_2 + o\boldsymbol{a}_3)], \end{aligned} \tag{7.8}$$

where m, n, o are integers. This sum will be nonzero only if the products $\boldsymbol{q} \cdot \boldsymbol{a}_i$ in the exponent are all integers. If we call these integers h_i, we recover the system of equations (7.6) and have nonvanishing values of $G(\boldsymbol{q})$ for $\boldsymbol{q} = \boldsymbol{g}$.

7.1.3 Construction of Laue Zones

The product of a translation vector $\boldsymbol{r}_g$ (7.1) of the crystal lattice and a reciprocal-lattice vector $\boldsymbol{g}$ (7.7),

$$\boldsymbol{g} \cdot \boldsymbol{r}_g = mh + nk + ol = N, \tag{7.9}$$

is an integer. If $N = 0$, all the $\boldsymbol{g}$ for a given value of $\boldsymbol{r}_g$ lie in a plane through the origin of the reciprocal lattice and are normal to the zone axis $\boldsymbol{r}_g$. The system of lattice planes that belongs to these values of $\boldsymbol{g}$ forms a bundle of planes that have the zone axis as a common line of intersection (Fig. 7.6a). The reciprocal-lattice plane that contains the corresponding $\boldsymbol{g}$ is called the zero-order *Laue zone*. For $N = 1, 2, \ldots$ the first- (FOLZ), second-, and higher-order (HOLZ) Laue zones, respectively, are obtained, which are parallel to the zero-order Laue zone (Fig. 7.6b). This means that the Laue zones are parallel sections through the reciprocal lattice.

The construction of Laue zones is very useful for the indexing and computation of electron-diffraction patterns. Either triplets of integers hkl are sought that fulfill the condition (7.9) and have nonzero structure amplitude (Sect. 7.2.2) or a model of the reciprocal lattice like that of Fig. 7.5b can be used.

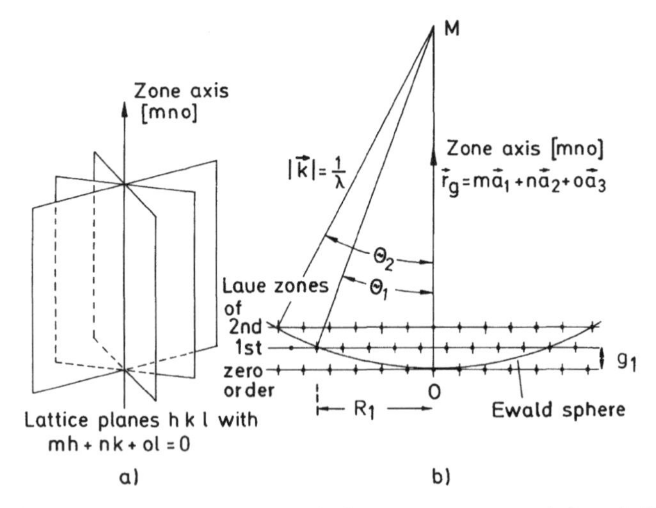

Fig. 7.6. (**a**) Bundle of lattice planes with the common zone axis $[mno]$. (**b**) Position of the zero- and higher-order Laue zones in the reciprocal lattice. The angles θ_1 and θ_2 at which the Ewald sphere cuts the higher-order zones are discussed in Sect. 8.3.4.

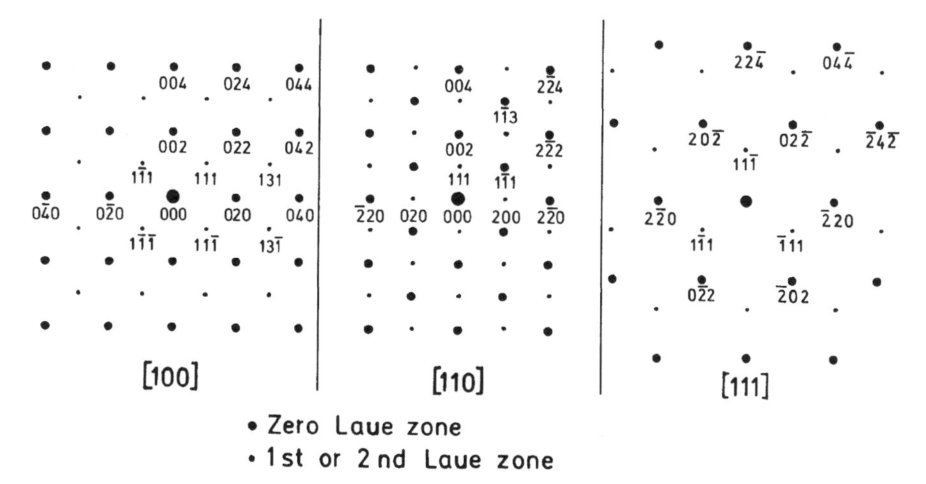

Fig. 7.7. Examples of the construction of zero- and first- or second-order Laue zones for the [100], [110], and [111] zone axes of a face-centered cubic lattice.

From Fig. 7.5b, for example, the reciprocal-lattice points for an fcc lattice can be read off; these are situated on the Laue zones for the zone axes $[mno]$ = [100], [110], and [111]. Figure 7.7 shows indexed zone patterns for the zero-, first-, and second-order Laue zones.

7.2 Kinematical Theory of Electron Diffraction

7.2.1 Bragg Condition and Ewald Sphere

The Laue conditions $\boldsymbol{q} \cdot \boldsymbol{a}_i = h_i$ (integers), which result from the Fourier transform (7.9) of the crystal lattice, guarantee that the scattered plane waves with wave vectors $\boldsymbol{k}$ do indeed overlap and interfere constructively, so that their amplitudes sum. With $\boldsymbol{q} = \boldsymbol{k} - \boldsymbol{k}_0$ (3.39) and $\boldsymbol{q} = \boldsymbol{g}$ from (7.9), the Laue conditions can be solved for $\boldsymbol{q} = \boldsymbol{k} - \boldsymbol{k}_0$ by using (7.6) and (7.7), which gives

$$\boldsymbol{k} - \boldsymbol{k}_0 = \boldsymbol{g} = h\boldsymbol{a}_1^* + k\boldsymbol{a}_2^* + l\boldsymbol{a}_3^* . \tag{7.10}$$

This is the Bragg condition in vector notation. The vector $\boldsymbol{g} = \boldsymbol{k} - \boldsymbol{k}_0$ is normal to the bisector of the angle between $\boldsymbol{k}_0$ and $\boldsymbol{k}$ (Fig. 7.8). On the right-hand side of (7.10), we have the reciprocal-lattice vector $\boldsymbol{g}$, which is normal to the lattice planes (hkl). It follows that $\boldsymbol{k} - \boldsymbol{k}_0$ is parallel to this normal. The angles of incidence and scattering θ_{B} relative to the lattice planes (Fig. 7.8) must be equal. Although this is strictly an interference phenomenon, the result can be interpreted as a reflection at the lattice planes. It differs from light-optical reflection in that only a fixed angle θ_{B} is allowed. This Bragg angle

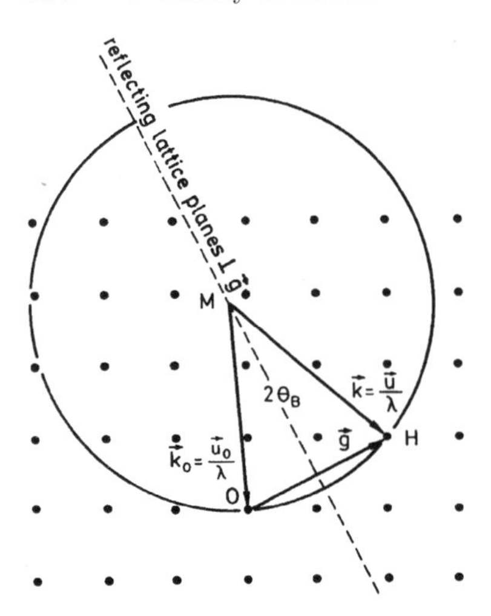

Fig. 7.8. Ewald sphere of radius $k = 1/\lambda$ in a reciprocal lattice. A Bragg reflection is excited if the sphere intersects a reciprocal-lattice point, such as H.

θ_B can be calculated by examining the magnitude of (7.10). Figure 7.8 shows that $|\boldsymbol{k} - \boldsymbol{k}_0| = 2\sin\theta_\mathrm{B}/\lambda$ and $|\boldsymbol{g}| = 1/d_{hkl}$ (7.5). Thus (7.10) results in the well-known *Bragg condition*

$$2d_{hkl}\sin\theta_\mathrm{B} = \lambda. \tag{7.11}$$

The Bragg condition is valid for x-rays and electrons. Typical back reflections with $2\theta_\mathrm{B}$ close to 180° can be obtained with x-rays, whereas the same lattice planes typically give forward reflection for electrons since the wavelength is so much smaller. The scattering amplitude for x-rays is approximately isotropic for all scattering angles, and back reflection can be observed, whereas the scattering amplitude for electrons decreases with increasing angle and the Bragg reflections are limited to a cone with an aperture of the order of 50 mrad.

Equation (7.10) can be used to generate a construction first employed by Ewald. A vector $\boldsymbol{k}_0 = \overline{MO}$ is drawn with one end at the origin O of the reciprocal lattice and with a length $|\boldsymbol{k}_0| = 1/\lambda$ (Figs. 7.6 and 7.8). The other end M (excitation point) of $\boldsymbol{k}_0$ is taken as the center of a sphere of radius $1/\lambda$. Diffraction will be observed only if this *Ewald sphere* intersects one or more points $\boldsymbol{g}$ of the reciprocal lattice (e.g., H in Fig. 7.8). The direction $\boldsymbol{k} = \overline{\mathrm{MH}}$ will be the direction of the scattered wave, and $\boldsymbol{k} - \boldsymbol{k}_0 = \boldsymbol{g}$ is the vector that connects the end points of $\boldsymbol{k}$ and $\boldsymbol{k}_0$.

The Ewald sphere in Fig. 7.8 has been drawn with a small radius, as for x-rays. In electron diffraction, the radius of the Ewald sphere, $1/\lambda = 240$ nm^{-1} for 80 keV electrons, is much larger than the distances between the reciprocal-lattice points; e.g. $1/a = 2.8\ \mathrm{nm}^{-1}$ for copper (Fig. 7.6b).

If the incident beam is parallel to a zone axis, the diffraction pattern (e.g., Fig. 7.26a) contains Bragg reflections near the primary beam from the zero-order Laue zone; at larger Bragg angles, circles of reflections occur where the Ewald sphere cuts the first- and higher-order Laue zones (see also the discussion of HOLZ patterns in Sect. 8.3.4).

7.2.2 Structure Amplitude and Lattice Amplitude

The amplitude of the scattered wave in the direction $\boldsymbol{k}$ can be obtained from the Fourier transform of the crystal lattice. Consider the kth atom ($k = 1, \ldots, n$) inside one unit cell; this atom scatters with an amplitude $f_k(\theta)$, which can be calculated by one of the methods described in Sect. 5.1.3 if a screened Coulomb potential modified by close packing of atoms in a solid (e.g., the muffin-tin model) is used. Furthermore, the crystal is assumed to be parallelepipedal in shape, with edge lengths $L_i = M_i a_i$ ($i = 1, 2, 3$) parallel to the fundamental vectors $\boldsymbol{a}_i$. The Fourier sum (7.9) becomes

$$F(\boldsymbol{q}) = \sum_{m=1}^{M_1} \sum_{n=1}^{M_2} \sum_{o=1}^{M_3} \sum_{k=1}^{n} f_k \exp[-2\pi\mathrm{i}(\boldsymbol{k} - \boldsymbol{k}_0) \cdot (\boldsymbol{r}_g + \boldsymbol{r}_k)]. \tag{7.12}$$

The summation over k, which corresponds to the different atoms of the unit cell, can be extracted, and (7.12) becomes

$$F(\boldsymbol{q}) = \underbrace{\sum_{k=1}^{n} f_k \exp[-2\pi\mathrm{i}(\boldsymbol{k} - \boldsymbol{k}_0) \cdot \boldsymbol{r}_k]}_{F_{cell}} \cdot \underbrace{\sum_m \sum_n \sum_o \exp[-2\pi\mathrm{i}(\boldsymbol{k} - \boldsymbol{k}_0) \cdot \boldsymbol{r}_g]}_{G}. \tag{7.13}$$

The first factor, F_{cell}, is called the *structure amplitude* and depends only on the positions and type of atoms inside the unit cell. The second factor, G, is called the *lattice amplitude* and depends only on the external shape of the crystal.

The structure amplitude will be of interest only for the Bragg condition. It will not be altered by small deviations from the geometry of the Bragg condition, unlike G, as our later calculations will show. Substituting for $\boldsymbol{r}_k$ from (7.2), we find

$$F_{cell} = \sum_{k=1}^{n} f_k \exp(-2\pi\mathrm{i}\boldsymbol{g} \cdot \boldsymbol{r}_k) = \sum_{k=1}^{n} f_k \exp[-2\pi\mathrm{i}(u_k h + v_k k + w_k l)]. \tag{7.14}$$

The value of F_{cell} will now be calculated for some typical examples.

(a) Body-Centered Cubic Lattice

Even though the body-centered cubic lattice is a Bravais lattice, it can be described as a simple cubic lattice with two atoms in the unit cell (Table 7.1) at $\boldsymbol{r}_1 = (0, 0, 0)$ and $\boldsymbol{r}_2 = (\frac{1}{2}, \frac{1}{2}, \frac{1}{2})$. Substitution in (7.14) gives

$$F_{cell} = f\{1 + \exp[-\pi\mathrm{i}(h + k + l)]\}.$$

Using the relation

$$\exp(-\mathrm{i}\pi n) = \begin{cases} 1 & \text{if } n \text{ is an even integer} \\ -1 & \text{if } n \text{ is an odd integer,} \end{cases}$$

we find

$$F_{cell} = \begin{cases} 2f & \text{if } h + k + l \text{ is even} \\ 0 & \text{if } h + k + l \text{ is odd.} \end{cases}$$

(b) Face-Centered Cubic Lattice

The face-centered cubic lattice can be described as a simple cubic lattice with four atoms in the unit cell at the positions
$\boldsymbol{r}_1 = (0,0,0)$, $\boldsymbol{r}_2 = (\frac{1}{2}, \frac{1}{2}, 0)$, $\boldsymbol{r}_3 = (\frac{1}{2}, 0, \frac{1}{2})$, $\boldsymbol{r}_4 = (0, \frac{1}{2}, \frac{1}{2})$.

Hence

$$F_{cell} = f\{1 + \exp[-\pi \mathrm{i}(h+k)] + \exp[-\pi \mathrm{i}(h+l)] + \exp[-\pi \mathrm{i}(k+l)]\}.$$

The rules odd + odd = even, etc., show that

$$F_{cell} = \begin{cases} 4f & \text{if } h,\ k,\ l, \text{ are either all even or all odd} \\ 0 & \text{if } h,\ k,\ l, \text{ are mixed (odd and even).} \end{cases}$$

(c) NaCl Structure

The unit cell consists of two sodium and chlorine face-centered sublattices that are shifted by one half of the body diagonal $(\frac{1}{2}, \frac{1}{2}, \frac{1}{2})$ of the unit cell. This shift can be considered by introducing a common phase factor for the chlorine sublattice,

$$\begin{aligned} F_{cell} &= \{f_{\mathrm{Na}} + f_{\mathrm{Cl}} \exp[-\pi \mathrm{i}(h+k+l)]\} \\ &\quad \times \{1 + \exp[-\pi \mathrm{i}(h+k)] + \exp[-\pi \mathrm{i}(h+l)] + \exp[-\pi \mathrm{i}(k+l)]\}, \end{aligned}$$

which results in

$$F_{cell} = \begin{cases} 4(f_{\mathrm{Na}} + f_{\mathrm{Cl}}) & \text{if } h,\ k,\ l, \text{ are all even} \\ 4(f_{\mathrm{Na}} - f_{\mathrm{Cl}}) & \text{if } h,\ k,\ l, \text{ are all odd} \\ 0 & \text{if } h,\ k,\ l, \text{ are mixed.} \end{cases}$$

Similar calculations can be made for other crystal structures (Table 7.1). The three types of cubic lattices exhibit different zero rules; i.e., different sets of (hkl) for which $F = 0$. These exclude some of the $\boldsymbol{g}$ values of the reciprocal lattice that would be found if a primitive unit cell with only one atom at the origin of each unit cell were used. The two atoms in the body-centered cell and the four atoms in the face-centered cell all scatter either in phase (constructive interference) or in antiphase, leading to $F = 0$ (destructive interference).

For more complicated structures (e.g., NaCl), the nonzero reflections can have different structure amplitudes. In the case of KCl, the difference $(f_{\mathrm{K}} - f_{\mathrm{Cl}})$ for h, k, l odd becomes very small. It is zero for x-rays because both the K^+

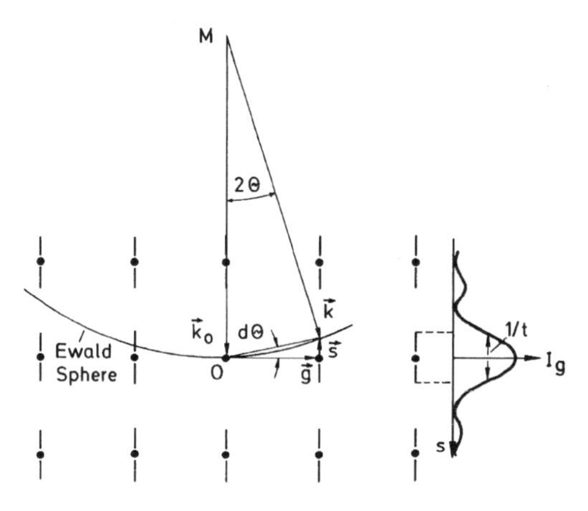

Fig. 7.9. Introduction of the excitation error s and a convolution of the reciprocal-lattice points with the needle-shaped square of the lattice amplitude $|G|^2$ for a thin foil of thickness t.

and the Cl^- ions have the same electron configuration as argon. In electron diffraction, there is a small residual difference in the nuclear charges Ze.

We now consider the triple sum of the lattice amplitude G in (7.13) and allow small deviations from the exact Bragg condition $\boldsymbol{k} - \boldsymbol{k}_0 = \boldsymbol{g}$. This deviation is described by the *excitation error* $\boldsymbol{s} = (s_x, s_y, s_z)$; this vector connects the lattice point $\boldsymbol{g}$ in the reciprocal lattice to the Ewald sphere in the direction parallel to the incident beam (Fig. 7.9). The magnitude and the tilt angle $\Delta\theta$ out of the Bragg condition are related by

$$s = g\Delta\theta = \frac{\Delta\theta}{d_{hkl}} = \frac{2\sin\theta_B}{\lambda}\Delta\theta. \tag{7.15}$$

Substituting $\boldsymbol{k} - \boldsymbol{k}_0 = \boldsymbol{g} + \boldsymbol{s}$ in (7.13) and recalling that $\boldsymbol{g} \cdot \boldsymbol{a}_i = n$ (integer) and $\exp(-2\pi i n) = 1$, we obtain an expression for the lattice amplitude

$$G = \sum_{m=1}^{M_1} \sum_{n=1}^{M_2} \sum_{o=1}^{M_3} \exp[-2\pi i(\boldsymbol{g} + \boldsymbol{s}) \cdot \boldsymbol{r}_g] = \sum_{m,n,o} \exp(-2\pi i \boldsymbol{s} \cdot \boldsymbol{r}_g). \tag{7.16}$$

The phase $2\pi i \boldsymbol{s} \cdot \boldsymbol{r}_g$ varies very slowly as we move through the crystal from one unit cell to another. The triple sum can therefore be replaced by an integral over the crystal volume $V = L_1 L_2 L_3$ (V_e: volume of the unit cell),

$$G = \frac{1}{V_e} \int_{-L_1/2}^{+L_1/2} \int_{-L_2/2}^{+L_2/2} \int_{-L_3/2}^{+L_3/2} \exp[-2\pi i(s_x x + s_y y + s_z z)] \mathrm{d}x \mathrm{d}y \mathrm{d}z. \tag{7.17}$$

Setting $V_e = a_1 a_2 a_3$ (cubic lattice) and integrating with respect to x, we find

$$\begin{aligned} G_x &= \frac{1}{a_1} \int_{-L_1/2}^{+L_1/2} \exp(-2\pi i s_x x) \mathrm{d}x \\ &= \frac{1}{\pi s_x a_1} \frac{\exp(\pi i s_x L_1) - \exp(-\pi i s_x L_1)}{2i} = \frac{\sin(\pi s_x M_1 a_1)}{\pi s_x a_1} \end{aligned} \tag{7.18}$$

and correspondingly for the y and z directions. This is the typical formula, well known in light optics, for the diffraction at a grating with M_1 slits with spacing a_1. The total diffracted intensity becomes

$$I_g \propto |F_{cell}|^2|G|^2 = \\ |F_{cell}|^2 \frac{\sin^2(\pi s_x M_1 a_1)}{(\pi s_x a_1)^2} \frac{\sin^2(\pi s_y M_2 a_2)}{(\pi s_y a_2)^2} \frac{\sin^2(\pi s_z M_3 a_3)}{(\pi s_z a_3)^2}. \tag{7.19}$$

The form of $|G|^2$ will now be discussed for some simple crystal shapes.

(1) *Thin Crystal Foils (Discs)* with the z direction of electron incidence normal to the surface (Fig. 7.10a). The last factor in (7.19) reaches a maximum value of M_3^2 for $s_z = 0$; it first falls to zero when the numerator becomes zero, which occurs when $\pi s_z M_3 a_3 = \pi$ or $s_z = 1/M_3 a_3 = 1/L_3 = 1/t$. Corresponding values are found for the x and y directions. However, the intensity first becomes zero at much lower excitation errors, $s_x = 1/L_1$ and $s_y = 1/L_2$ ($L_1, L_2 = 1/D$). The function $|G(s_x, s_y, s_z)|^2$ therefore has a needle-like shape in the z direction (Fig. 7.10a). The length of the needle in the reciprocal lattice is inversely proportional to the foil thickness $L_3 = t$. Each reciprocal-lattice point will be convolved with this $|G|^2$ function (Fig. 7.9). The needle-like extension of the lattice points provokes simultaneous excitation of a large number of Bragg reflections because the Ewald sphere can intersect

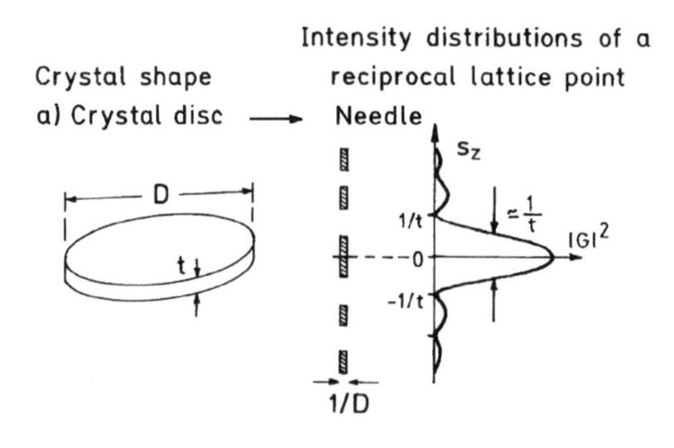

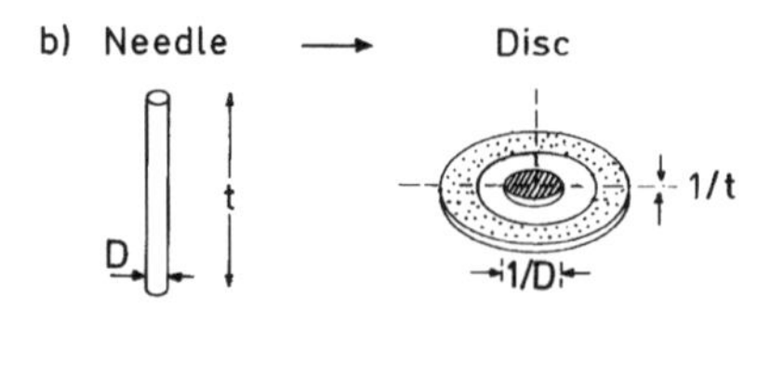

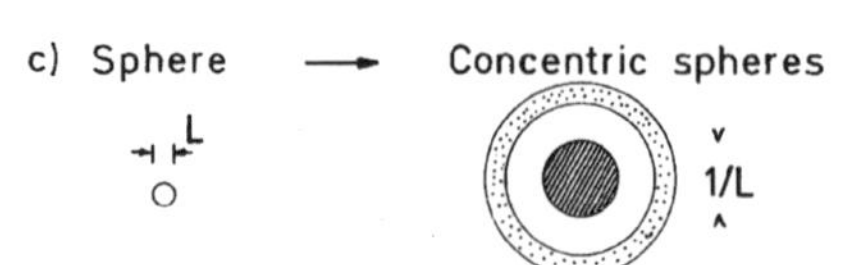

Fig. 7.10. Shape of the square of the lattice amplitude $|G|^2$ with which reciprocal-lattice points have to be convolved for (**a**) a crystal disc (thin foil of thickness t), (**b**) a needle of length t, and (**c**) a sphere of diameter L.

more needles than points (Figs. 7.6b and 7.9). Such needles in the reciprocal lattice can also be observed for plate-like precipitates in alloys. If the plates are inclined to the foil, this can cause elongations of the Bragg spots or the latter may appear to be shifted (Fig. 8.12).

(2) *Needle-like Crystals* with the long axis in the z direction (Fig. 7.10b). The first zero for $|G|^2$ is reached for small s_z and larger s_x and s_y. $|G|^2$ is thus a disc normal to the z axis.

(3) *Small Cubes (Spheres)* with edge (diameter) L. The extension of $|G|^2$ is the same in all directions: $s_x = s_y = s_z = 1/L$ (Fig. 7.10c). If the Ewald sphere intersects a reciprocal-lattice point, a broadened diffraction spot is observed. Debye–Scherrer rings will be broadened by

$$\frac{\Delta r}{r} = \frac{1/L}{1/d} = \frac{d}{L} \,. \tag{7.20}$$

This relation can be used to estimate particle dimensions in the range $L =$ 0.3–5 nm from the broadening Δr of the rings.

Finally, we can arrive at the concept of convolution of each reciprocal-lattice point with $|G|^2$ by reasoning based on the Fourier transform and in particular on the convolution theorem (3.48), which transforms a convolution of two functions into a product of their Fourier transforms; likewise a product is transformed into a convolution.

A crystal can be described by the expression $[p(\boldsymbol{r}) \otimes f(\boldsymbol{r})] \cdot g(\boldsymbol{r})$, which involves the three functions $p(\boldsymbol{r})$, $f(\boldsymbol{r})$, and $g(\boldsymbol{r})$; $p(\boldsymbol{r})$ denotes a set of δ-functions at the origin of the unit cells, while $f(\boldsymbol{r})$ describes the potential within a single unit cell. The convolution $[p(\boldsymbol{r}) \otimes f(\boldsymbol{r})]$ represents an infinite lattice in which each origin is convolved with the potential of a unit cell. Finally, $g(\boldsymbol{r}) = 1$ inside and 0 outside the crystal defines the finite crystal volume.

The Fourier transform yields

$$\mathbf{F}\{[p(\boldsymbol{r}) \otimes f(\boldsymbol{r})] \cdot g(\boldsymbol{r})\} = [P(\boldsymbol{q}) \cdot F_{cell}(\boldsymbol{q})] \otimes G(\boldsymbol{q}), \tag{7.21}$$

in which $P(\boldsymbol{q})$ denotes a set of δ-functions at the reciprocal-lattice points; $F_{cell}(\boldsymbol{q})$ is the structure amplitude. The presence of points at which $F_{cell} = 0$ reduces the number of reciprocal-lattice points if the unit cell contains more than one atom. $G(\boldsymbol{q})$ is the lattice amplitude. Each reciprocal-lattice point is convolved with this function.

7.2.3 Column Approximation

In discussions of the contrast of defects in crystalline specimens, it is useful to consider not only the amplitude $F(\boldsymbol{q})$ in the Fraunhofer diffraction plane but also the intensity at a point P just below the specimen (Fresnel diffraction).

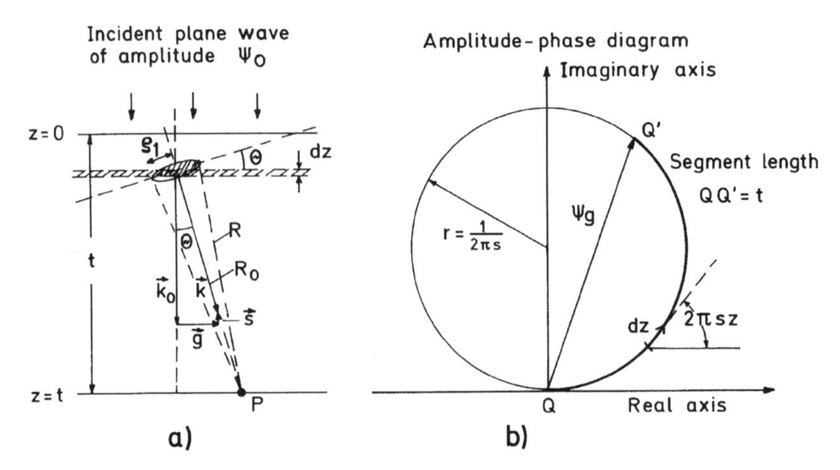

Fig. 7.11. (**a**) Column approximation for calculating the amplitude ψ_g of a diffracted wave at a point P on the bottom of a crystal foil of thickness $z = t$ ($2\rho_1$ = diameter of the first Fresnel zone). (**b**) Amplitude-phase diagram for calculating $I_g = \psi_g \psi_g^*$.

We assume $r \to \infty$ in the formula for Fresnel diffraction (Sect. 3.2.1), which means that the point source is at infinity and hence that we have a plane incident wave of amplitude ψ_0. There are $\mathrm{d}z/V_\mathrm{e}$ unit cells per unit area in an element of thickness $\mathrm{d}z$, each of which scatters with the structure amplitude $F(\theta)$. We use the scattered wavefront with a scattering angle $\theta = 2\theta_\mathrm{B}$ to calculate the contribution $\mathrm{d}\psi_g$ of the layer $\mathrm{d}z$ to the diffracted amplitude at the point P (Fig. 7.11a). Strictly, $\psi_0/\cos\theta$ should be used instead of ψ_0 to correct for the cross section of the wavefront. This correction can, however, be neglected because $\cos\theta \simeq 1$ for the small values of θ in question. Equation (3.28) becomes

$$\begin{aligned}\mathrm{d}\psi_g &= \psi_0 \frac{\mathrm{d}z}{V_\mathrm{e}} \int_S F(\theta) \frac{\mathrm{e}^{2\pi \mathrm{i} kR}}{R} \mathrm{d}S = \psi_0 \frac{2\pi \mathrm{d}z}{V_\mathrm{e}} \int_{R_0}^{R} F(\theta) \mathrm{e}^{2\pi \mathrm{i} kR} \mathrm{d}R \\ &= \mathrm{i}\psi_0 \frac{\lambda F(\theta)}{V_\mathrm{e}} \mathrm{e}^{2\pi \mathrm{i} kR_0} \mathrm{d}z = \frac{\mathrm{i}\pi}{\xi_g} \psi_0 \mathrm{e}^{2\pi \mathrm{i} kR_0} \mathrm{d}z \end{aligned} \tag{7.22}$$

with $\mathrm{d}S = 2\pi R\, \mathrm{d}R$. We have introduced the *extinction distance* ξ_g (Table 7.2), defined by

$$\xi_g = \frac{\pi V_\mathrm{e}}{\lambda F(\theta)} . \tag{7.23}$$

It was shown in Sect. 3.2.1 that the main contribution to the integral in (7.22) comes from the first Fresnel zone of radius $\rho_1 = \sqrt{\lambda R_0}$ (Fig. 7.11a). For a distance (foil thickness) $R_0 = 100$ nm and $\lambda = 3.7$ pm (100 keV electrons), we find $\rho_1 = 0.6$ nm. This means that only a column with a diameter of 1–2 nm is contributing to the amplitude at the point P, and the method is therefore called the *column approximation.*

Table 7.2. Extinction distances ξ_g [nm] for $E = 100$ keV. (Th: Thomas et al. [7.16], H: Hirsch et al. [7.17]

Face-centered cubic lattice and NaCl structure								
hkl	111	200	220	311	222	400	331	
Al	56.3	68.5	114.4	147.6	158.6	202.4	235.7	Th
Cu	28.6	32.6	47.3	57.9	61.5	76.4	88.1	Th
Ni	26.8	30.6	44.6	54.7	58.1	72.0	82.9	Th
Ag	24.2	27.2	38.6	47.4	50.4	63.0	73.0	Th
Pt	14.7	16.6	23.2	27.4	28.8	34.3	38.5	H
Au	18.3	20.2	27.8	33.6	35.6	43.5	49.5	Th
Pb	24.0	26.6	35.9	41.8	43.6	50.5	55.5	H
LiF	117.7	64.5	94.2	219.9	121.0	146.3	335.2	H
MgO	272.6	46.1	66.2	1180	85.2	103.3	1075	H
Body-centered cubic lattice								
hkl	110	200	211	220	310	222	400	
Cr	28.8	42.3	55.5	68.6	81.6	94.7	121.9	Th
Fe	28.6	41.2	53.5	65.8	78.0	90.4	116.2	Th
Nb	26.1	38.3	49.9	61.4	72.9	84.6	108.5	Th
Mo	22.9	33.6	43.2	52.7	62.0	72.3	89.7	Th
Ta	20.2	27.5	33.9	40.0	45.9	51.8	63.8	Th
W	18.0	24.5	30.2	35.5	41.0	46.2	55.6	Th
Diamond structure								
hkl	111	220	311	400	331	511 333	400	
C	47.6	66.5	124.5	121.5	197.2	261.3	215.1	H
Si	60.2	75.7	134.9	126.8	204.6	264.5	209.3	H
Ge	43.0	45.2	75.7	65.9	102.8	127.3	100.8	H
Hexagonal lattice								
hkl	1110	1120	2200	1101	2201	0002	1102	
Mg	150.9	140.5	334.8	100.1	201.8	81.1	231.0	H
Co	46.9	42.9	102.7	30.6	62.0	21.8	70.2	H
Zn	55.3	49.7	118.0	35.1	70.4	26.0	76.2	H
Zr	59.4	49.3	115.1	37.9	69.1	51.7	83.7	H
Cd	51.9	43.8	102.3	32.4	60.8	24.4	68.3	H

The amplitude ψ_g of a Bragg reflection is obtained by integrating (7.22) over the thickness. Using $\boldsymbol{k} = \boldsymbol{k}_0 + \boldsymbol{g} + \boldsymbol{s}$, $R_0 = t - z$, and $|\psi_0| = 1$ [1.26],

$$\begin{aligned}\psi_g &= \mathrm{i}\frac{\pi}{\xi_g}\exp(2\pi \mathrm{i} k_0 t)\int_0^t \exp[-2\pi\mathrm{i}(\boldsymbol{g}+\boldsymbol{s})\cdot\boldsymbol{z}]\mathrm{d}z \\ &= \mathrm{i}\frac{\pi}{\xi_g}\exp(2\pi\mathrm{i}k_0 t)\int_0^t \exp(-2\pi\mathrm{i}sz)\mathrm{d}z. \end{aligned} \tag{7.24}$$

The integral may be evaluated as for (7.18), and the diffraction intensity at the point P ($I_0 = 1$) becomes

$$I_g = \psi_g \psi_g^* = \frac{\pi^2}{\xi_g^2} \frac{\sin^2(\pi t s)}{(\pi s)^2} . \tag{7.25}$$

Substituting for ξ_g from (7.23), we obtain the same dependence on the excitation error s as in (7.19).

The last integral in (7.24) can also be solved graphically by using an amplitude-phase diagram (Sect. 3.2.1) in which lengths dz are added vectorially with slope $2\pi s z$ in the complex-number plane (Fig. 7.11b); the result is a circle of radius r. The length of the circular segment QQ′ is t. The circle is closed when the phase factor $\exp(2\pi \mathrm{i} s z)$ reaches unity, which occurs for $sz = 1$ and so the perimeter $2\pi r$ of the circle is equal to $1/s$. It follows that $r = 1/2\pi s$. The amplitude ψ_g is proportional to the square of the length of the chord $\overline{\mathrm{QQ}'}$. For foil thicknesses $t = n/s$ (n integer), Q and Q′ coincide and the diffraction intensity I_g becomes zero; when the thickness is further increased, $I_g = R$ again increases and subsequently oscillates as shown in Fig. 7.14b.

7.3 Dynamical Theory of Electron Diffraction

7.3.1 Limitations of the Kinematical Theory

The kinematical theory is valid only for very thin films for which the reflection intensity I_g is small and the decrease of the primary-beam intensity I_0 can be neglected. If the Bragg condition is exactly satisfied ($s = 0$), we obtain from (7.25) $I_g = \pi^2 t^2/\xi_g^2$ and $I_0 = 1$. The intensity I_g increases as t^2, and the condition $I_g \ll 1$ will be satisfied only for $t < \xi_g/10$. If $s \neq 0$, the intensity I_g oscillates with increasing t and reaches maximum values of $1/\xi_g^2 s^2$ (Fig. 7.14b). The condition $I_g \ll 1$ will be satisfied when $s \gg 1/\xi_g$. In Sect. 7.3.4, we shall see that, in this case, the kinematical and dynamical theories lead to identical results.

Furthermore, it must not be forgotten that the case in which only one Bragg reflection is excited, which is called the two-beam case (including the primary beam with $\boldsymbol{g} = 0$), is unusual; normally, a larger number $n > 2$ of reflections must be considered (n-beam case). Numerous small reflection intensities I_g can reduce the intensity of the primary beam more strongly than in the two-beam case. The n-beam case therefore restricts the validity of the kinematical theory to even smaller thicknesses. In the Bragg condition ($s = 0$), the dynamical theory predicts an oscillation of the intensities I_0 and I_g. A strong Bragg reflection will excite neighboring reflections with a larger amplitude than the primary beam. Therefore, in many practical situations the interaction of 30–100 Bragg reflections has to be considered. Furthermore, the intensity does not remain localized in the Bragg reflection. The diffuse electron scattering between the Bragg diffraction spots by inelastic and thermal diffuse scattering causes a decrease of intensity in the Bragg spots themselves (Sect. 7.4.2).

7.3.2 Formulation of the Dynamical Theory as a System of Differential Equations

This formulation of dynamical theory was first used by Darwin [7.4, 7.5] for x-ray diffraction and transferred by Howie and Whelan [7.6] to electron diffraction.

We first discuss the two-beam case [1.26]. An incident wave of amplitude ψ_0 and a diffracted wave of amplitude ψ_g fall on a layer of thickness $\mathrm{d}z$ inside the crystal foil. After passing through this layer, the amplitude ψ_0 will be changed by $\mathrm{d}\psi_0$ and ψ_g by $\mathrm{d}\psi_g$. These changes can be calculated from Fresnel diffraction theory using the column approximation. The contributions of ψ_0 and ψ_g to $\mathrm{d}\psi_0$ and $\mathrm{d}\psi_g$ can be obtained by using (7.22)–(7.24) with the extinction distances (7.23) $\xi_0 = \pi V_\mathrm{e}/\lambda F(0)$ and $\xi_g = \pi V_\mathrm{e}/\lambda F(2\theta_\mathrm{B})$. The result is a linear system of differential equations (Howie–Whelan equations):

$$\begin{aligned}\frac{\mathrm{d}\psi_0}{\mathrm{d}z} &= \frac{\mathrm{i}\pi}{\xi_0}\psi_0 + \frac{\mathrm{i}\pi}{\xi_g}\psi_g \mathrm{e}^{2\pi \mathrm{i} sz},\\ \frac{\mathrm{d}\psi_g}{\mathrm{d}z} &= \frac{\mathrm{i}\pi}{\xi_g}\psi_0 \mathrm{e}^{-2\pi \mathrm{i} sz} + \frac{\mathrm{i}\pi}{\xi_0}\psi_g. \end{aligned} \tag{7.26}$$

The second term of the first equation results from the scattering of the diffracted wave back into the primary beam; the sign of the excitation error s is the reverse of that for scattering in the opposite direction (first term in the second equation). This system of equations can be extended to the n-beam case by introducing the relative excitation errors s_{g-h} and extinction distances ξ_{g-h}:

$$\frac{\mathrm{d}\psi_g}{\mathrm{d}z} = \sum_{\boldsymbol{h}=\boldsymbol{g}_1}^{\boldsymbol{g}_n} \frac{\mathrm{i}\pi}{\xi_{g-h}}\psi_h \exp(2\pi \mathrm{i} s_{g-h} z) \quad \text{for} \quad \boldsymbol{g} = \boldsymbol{g}_1, ..., \boldsymbol{g}_n;\ \boldsymbol{g}_1 = 0. \tag{7.27}$$

In the final result, we are interested only in the reflection intensity I_g and we can therefore use the transformation

$$\psi_0' = \psi_0 \exp(-\mathrm{i}\pi z/\xi_0); \quad \psi_g' = \psi_g \exp(2\pi \mathrm{i} sz - \mathrm{i}\pi z/\xi_0). \tag{7.28}$$

These new quantities contain only an additional phase factor, which cancels out when we multiply by the complex conjugate. Substituting (7.28) into (7.26) yields the simpler formulas

$$\begin{aligned}\frac{\mathrm{d}\psi_0'}{\mathrm{d}z} &= \frac{\mathrm{i}\pi}{\xi_g}\psi_g',\\ \frac{\mathrm{d}\psi_g'}{\mathrm{d}z} &= \frac{\mathrm{i}\pi}{\xi_g}\psi_0' + 2\pi \mathrm{i} s\psi_g'. \end{aligned} \tag{7.29}$$

The boundary conditions for these differential equations at the entrance surface of the foil ($z = 0$) are $|\psi_0| = 1$ and $|\psi_g| = 0$. We discuss a solution of (7.29) together with the solution of the eigenvalue problem in Sect. 7.3.4.

The system of differential equations for the n-beam case can be solved by the Runge–Kutta method or a similar numerical method. The multislice method (Sect. 9.6.3) uses elements of finite thickness Δz and projects the potential inside the layer Δz onto the lower boundary of the layer element. The space between the layers is treated as a vacuum and the wave propagates by Fresnel diffraction.

7.3.3 Formulation of the Dynamical Theory as an Eigenvalue Problem

This formulation was first used by Bethe [7.7] for electron diffraction. The Schrödinger equation (3.21) is solved with a potential $V(r)$ that is the superposition of all the atomic potentials (Fig. 3.3) and therefore has the same periodicity as the lattice. This means that $V(r)$ can be expanded as a Fourier sum:

$$V(r) = -\sum_g V_g \exp(2\pi i \boldsymbol{g} \cdot \boldsymbol{r}) = -\frac{h^2}{2m} \sum_g U_g \exp(2\pi i \boldsymbol{g} \cdot \boldsymbol{r}). \tag{7.30}$$

A value V_g can be attributed to each point $\boldsymbol{g}$ of the reciprocal lattice. The V_g (eV) and U_g (cm^{-2}) are related to the structure amplitude $F(\theta)(\theta = 2\theta_B)$ of the kinematical theory because, in the Born approximation, $F(\theta)$ is also the Fourier transform of the scattering potential $V(r)$ (Sect. 5.1.3):

$$V_g = \frac{\lambda^2 E}{2\pi V_e} \frac{2E_0 + E}{E_0 + E} F(\theta) = \frac{h^2}{2\pi m V_e} F(\theta); \quad U_g = \frac{F(\theta)}{\pi V_e}. \tag{7.31}$$

The extinction distance ξ_g introduced by (7.23) can be written as follows, where $\xi_{g,100}$ denotes the extinction distance at $E = 100$ keV:

$$\xi_g = \frac{\pi V_e}{\lambda F(\theta)} = \frac{\lambda E}{2V_g} \frac{2E_0 + E}{E_0 + E} = \frac{h^2}{2m\lambda V_g} = \frac{1}{\lambda U_g}, \tag{7.32}$$

$$\xi_g = \xi_{g,100} \frac{m_{100}\lambda_{100}}{m\lambda} = \xi_{g,100} \frac{v}{v_{100}}. \tag{7.33}$$

Equation (7.33) allows us to transfer tabulated values of ξ_g for $E = 100$ keV (Table 7.2) to other electron energies. The influence of lattice vibrations (see the Debye–Waller factor $\exp(-2M)$ in Sect. 7.5.3) and the thermal expansion of the lattice cause a slow increase of ξ_g with temperature [7.8, 7.9].

If (7.30) for $V(r)$ is substituted in (3.21), the solutions will also reflect the lattice periodicity. Such solutions of the Schrödinger equation are called *Bloch waves*,

$$b^{(j)}(\boldsymbol{k}, \boldsymbol{r}) = \sum_g C_g^{(j)} \exp[2\pi i(\boldsymbol{k}_0^{(j)} + \boldsymbol{g}) \cdot \boldsymbol{r}]. \tag{7.34}$$

The summation runs over the infinite number of reciprocal-lattice vectors $\boldsymbol{g}$. As an approximation, we confine the sum over n excited points $\boldsymbol{g} = \boldsymbol{g}_1, ..., \boldsymbol{g}_n$ of the reciprocal lattice, including the incident direction ($\boldsymbol{g}_1 = 0$). The number $j = 1,\ldots,n$ of different Bloch waves is needed to represent the propagation of electron waves in a crystal and to satisfy the boundary condition at the vacuum–crystal interface. This requires a superposition of n^2 different waves with wave vectors $\boldsymbol{k}_0^{(j)} + \boldsymbol{g}$ and amplitude factors $C_g^{(j)}$.

We substitute (7.30) and (7.34) into (3.21) and introduce the abbreviation

$$K = [2m_0E(1 + E/2E_0) + 2m_0V_0(1 + E/E_0)]^{1/2}/h \tag{7.35}$$

for the wave vector inside the crystal, which is obtained from the sum of the kinetic energy and the coefficient $V_0 = eU_i$ (inner potential) (Sect. 3.1.3) of the Fourier expansion (7.30). This gives

$$4\pi^2 \sum_{\boldsymbol{g}} \left[K^2 - (\boldsymbol{k}_0^{(j)} + \boldsymbol{g})^2 + \sum_{h\neq 0} U_h \exp(2\pi \mathrm{i} \boldsymbol{h}\cdot\boldsymbol{r}) \right]$$
$$\cdot\, C_g^{(j)} \exp[2\pi\mathrm{i}(\boldsymbol{k}_0^{(j)} + \boldsymbol{g})\cdot\boldsymbol{r}] = 0 \tag{7.36}$$

for all $\boldsymbol{g}$. This system of equations can be satisfied if the coefficients of identical exponential terms simultaneously become zero. After collecting up terms containing the factor $\exp[2\pi\mathrm{i}(\boldsymbol{k}_0^{(j)} + \boldsymbol{g})\cdot\boldsymbol{r}]$, we obtain the *fundamental equations* of dynamical theory,

$$[K^2 - (\boldsymbol{k}_0^{(j)} + \boldsymbol{g})^2]C_g^{(j)} + \sum_{h\neq 0} U_h C_{g-h}^{(j)} = 0; \quad \boldsymbol{g} = \boldsymbol{g}_1, ..., \boldsymbol{g}_n. \tag{7.37}$$

The $\boldsymbol{k}_g^{(j)} = \boldsymbol{k}_0^{(j)} + \boldsymbol{g}$ are the wave vectors of the Bloch waves, the magnitudes of which are not identical with K. As in kinematical theory (Fig. 7.8), we obtain the excitation points M_j as the starting points of the vectors $\boldsymbol{k}_0^{(j)}$, which end at the origin O of the reciprocal lattice. For calculation of the position of M_j, we recall that $K \gg g$ and

$$K + |\boldsymbol{k}_0^{(j)} + \boldsymbol{g}| \simeq K + k_z^{(j)} \simeq 2K. \tag{7.38}$$

Introducing the difference (Fig. 7.12)

$$K - |\boldsymbol{k}_0^{(j)} + \boldsymbol{g}| \simeq s_g - (k_z^{(j)} - K) = s_g - \gamma^{(j)}, \tag{7.39}$$

we find that the first factor of (7.37) becomes

$$[K^2 - (\boldsymbol{k}_0^{(j)} + \boldsymbol{g})^2] = (K + |\boldsymbol{k}_0^{(j)} + \boldsymbol{g}|)(K - |\boldsymbol{k}_0^{(j)} + \boldsymbol{g}|) \simeq 2K(s_g - \gamma^{(j)}). \tag{7.40}$$

s_g is negative when the reciprocal-lattice point $\boldsymbol{g}$ is outside the Ewald sphere, as in Fig. 7.12. By using (7.40), the system of equations (7.37) can be written in matrix form after dividing by $2K$, and we have [7.10, 7.11, 7.12, 7.13]

$$\begin{pmatrix} A_{11} & A_{12} & \ldots & A_{1n} \\ \ldots & \ldots & \ldots & \ldots \\ A_{21} & A_{22} & \ldots & A_{2n} \\ A_{n1} & A_{n2} & \ldots & A_{nn} \end{pmatrix} \begin{pmatrix} C_1^{(j)} \\ C_2^{(j)} \\ \ldots \\ C_3^{(j)} \end{pmatrix} = \gamma^{(j)} \begin{pmatrix} C_1^{(j)} \\ C_2^{(j)} \\ \ldots \\ C_3^{(j)} \end{pmatrix} \quad \text{for} \quad j = 1, ..., n \tag{7.41}$$

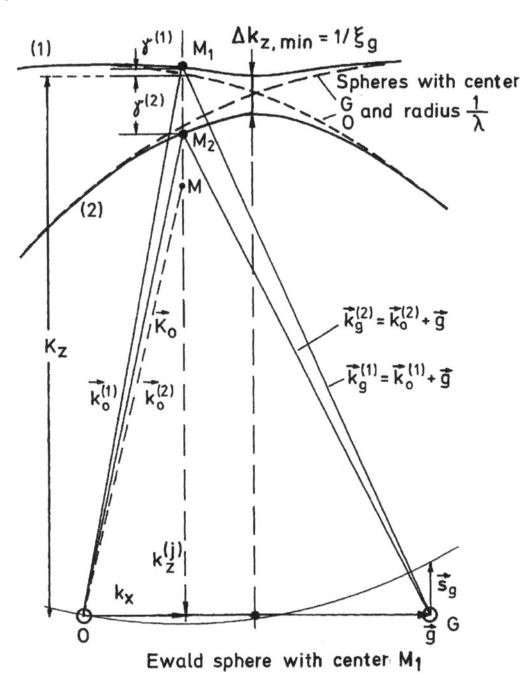

Fig. 7.12. Branches $j = 1$ and 2 of the dispersion surface for the two-beam case with a least distance $\Delta k_{z,\mathrm{min}} = 1/\xi_g$ in the Bragg condition ($k_x = 0$). Construction of the excitation points M_1 and M_2 on the dispersion surface for the tilt parameter k_x and the four wave vectors $\boldsymbol{k}_0^{(j)}$ and $\boldsymbol{k}_g^{(j)} = \boldsymbol{k}_0^{(j)} + \boldsymbol{g}$ ($j = 1, 2$) to the reciprocal lattice points O and $\boldsymbol{g}$. $\boldsymbol{K}_0$ is the wave vector of the incident wave and M its excitation point.

with the matrix elements

$$A_{11} = 0,\ A_{gg} = s_g,\ A_{hg}^* = A_{gh} = U_{g-h}/2K = \frac{1}{2\xi_{g-h}}\ .$$

This is the equation for an eigenvalue problem. A given matrix $[A]$ has n different eigenvalues $\gamma^{(j)}$ ($j = 1,\ldots,n$) with the accompanying eigenvectors $C_g^{(j)}(\boldsymbol{g} = \boldsymbol{g}_1,\ldots,\boldsymbol{g}_n)$. If we introduce the matrix $[C]$, the columns of which are the eigenvectors, so that $C_{gj} = C_g^{(j)}$ and the diagonal matrix $\{\gamma\}$ with the eigenvalues $\gamma^{(j)}$ as diagonal elements, (7.41) can be written

$$[A][C] = [C]\{\gamma\}. \tag{7.42}$$

A matrix $[A]$ is thus diagonalized by a linear transformation of the form $[C^{-1}][A][C]$.

In general, the matrix $[A]$ is Hermitian. For centrosymmetric crystals it is symmetric. Programs exist for defining the matrix and calculating the eigenvalues and eigenvectors. It is suggested that the Bloch waves should be numbered in order of decreasing $k_z^{(j)}$ [7.14]. The Bloch wave with the largest $\gamma^{(j)}$ has the index $j = 1$, etc.

The eigenvectors are orthogonal and satisfy the orthogonality relations

$$\sum_g C_g^{(i)*} C_g^{(j)} = \delta_{ij}; \quad \sum_j C_g^{(j)*} C_h^{(j)} = \delta_{gh}. \tag{7.43}$$

Changing the direction of the incident wave from $\boldsymbol{k}_0^{(j)}$ to $\boldsymbol{k}_0^{(j)} - \boldsymbol{h}$ alters the sequence of the column vectors in the matrix $[C]$, which imposes a periodicity condition on the $C_g^{(j)}$,

$$C_g(\boldsymbol{k}_0^{(j)}) = C_{g+h}(\boldsymbol{k}_0^{(j)} - \boldsymbol{h}). \tag{7.44}$$

The n eigenvalues $\gamma^{(j)}$ correspond to n Bloch waves (7.34) with wave vectors $\boldsymbol{k}_0^{(j)} + \boldsymbol{g}$. Their starting points do not lie on a sphere of radius K around O but at modified points M_j, given by (7.40). For different tilts of the specimen – equivalent to varying values of s_g in (7.40) or k_z in Fig. 7.12 – the points $\mathrm{M}_j(k_x)$ lie on a *dispersion surface.* The starting points of the wave vectors $\boldsymbol{k}_0^{(j)} + \boldsymbol{g}$ on this dispersion surface can be obtained by the following construction. The $\boldsymbol{K}_0$ vector parallel to the incident direction determines the point M in Fig. 7.12. Through M a straight line is drawn parallel to the crystal normal. The points of intersection with the n-fold dispersion surface are the excitation points M_j, which lie above one another in the case of normal incidence. This geometrical construction results from the boundary condition that the tangential components of the waves have to be continuous at the crystal boundary. For nonnormal incidence, the excitation points M_j are no longer above one another (see, e.g., [7.15]).

The total wave function (Bloch-wave field), the solution of (7.41), will be a linear combination of the Bloch waves $b^{(j)}(\boldsymbol{k}, \boldsymbol{r})$ (7.34) with the Bloch-wave excitation amplitudes $\epsilon^{(j)}$; i.e.,

$$\psi_{\mathrm{tot}} = \sum_j \epsilon^{(j)} b^{(j)}(\boldsymbol{k}, \boldsymbol{r}) = \sum_j \epsilon^{(j)} \sum_g C_g^{(j)} \exp[2\pi \mathrm{i}(\boldsymbol{k}_0^{(j)} + \boldsymbol{g}) \cdot \boldsymbol{r}]. \tag{7.45}$$

The amplitude ψ_g of a particular reflected wave can be obtained by summing over all $j = 1,\ldots,n$ waves from the excitation points M_j to the corresponding reciprocal-lattice point $\boldsymbol{g}$,

$$\psi_g = \sum_j \epsilon^{(j)} C_g^{(j)} \exp[2\pi \mathrm{i}(\boldsymbol{k}_0^{(j)} + \boldsymbol{g}) \cdot \boldsymbol{r}], \tag{7.46}$$

or

$$\psi_g = \sum_j \epsilon^{(j)} C_g^{(j)} \exp(2\pi \mathrm{i} \gamma^{(j)} z), \tag{7.47}$$

if a phase factor is omitted. The excitation amplitudes $\epsilon^{(j)}$ of the Bloch waves can be obtained from the boundary condition at the entrance of the incident plane wave into the crystal. The phase factors in (7.47) are all equal to unity for $z = 0$, and a plane wave in a vacuum and the Bloch-wave field in the crystal must be continuous. This requires

$$\psi_0(0) = \sum_j \epsilon^{(j)} C_0^{(j)} = 1,$$

$$\psi_g(0) = \sum_j \epsilon^{(j)} C_g^{(j)} = 0 \quad \text{for all } \boldsymbol{g} \neq 0, \tag{7.48}$$

or in a matrix formulation for the two-beam case

$$\begin{pmatrix} C_0^{(1)} & C_0^{(2)} \\ C_g^{(1)} & C_g^{(2)} \end{pmatrix} \begin{pmatrix} \epsilon^{(1)} \\ \epsilon^{(2)} \end{pmatrix} = \begin{pmatrix} \psi_0(0) \\ \psi_g(0) \end{pmatrix} = \begin{pmatrix} 1 \\ 0 \end{pmatrix}, \tag{7.49}$$

which can be readily extended to the n-beam case

$$[C]\boldsymbol{\epsilon} = \boldsymbol{\psi}(0), \tag{7.50}$$

where $\boldsymbol{\epsilon}$ and $\boldsymbol{\psi}(0)$ are column vectors of n components.

Comparison with the first of the orthogonality relations (7.43) shows that the boundary conditions (7.49)–(7.50) can be satisfied by writing $\epsilon^{(j)} = C_0^{(j)*}$ for normal incidence. In a more general formulation, the $\epsilon^{(j)}$ can be calculated from (7.50) by multiplying with the inverse matrix $[C^{-1}]$, which is identical with the adjoint matrix $[\widetilde{C}]$ because of the unitarity of the $C_g^{(j)}$:

$$\boldsymbol{\epsilon} = [C^{-1}]\boldsymbol{\psi}(0). \tag{7.51}$$

7.3.4 Discussion of the Two-Beam Case

In order to bring out the most important results of the dynamical theory, we now solve and discuss the two-beam case in detail, though it will be a poor approximation in practice. For high electron energies, the curvature of the Ewald sphere is so small that a large number of reflections (30–100) are excited simultaneously.

In kinematical theory, the centers M of the various Ewald spheres (Fig. 7.8) lie on a sphere of radius $k = 1/\lambda$ around the origin O of the reciprocal lattice if the direction of the incident wave is varied. When the intensity of the diffracted beam is increased by increasing the thickness and becomes larger than the intensity of the primary beam, the former can be treated as the primary wave and a sphere of radius k can also be drawn around the reciprocal-lattice point $\boldsymbol{g}$ as the geometrical surface that describes all possible values of the excitation points M (Fig. 7.12). As will be shown below, the two spheres do not intersect each other but withdraw from one another in a characteristic manner.

For the two-beam case, the fundamental equations of the dynamical theory, (7.37) and (7.41), are

$$\begin{aligned} -\gamma^{(j)}\, C_0^{(j)} + \qquad\quad \frac{U_g}{2K}\, C_g^{(j)} &= 0, \\ \frac{U_g}{2K}\, C_0^{(j)} + (-\gamma^{(j)} + s)\, C_g^{(j)} &= 0. \end{aligned} \tag{7.52}$$

Such a homogeneous linear system of equations for the $C_g^{(j)}$ has a nonzero solution if and only if the determinant of the coefficients is zero:

$$\begin{vmatrix} -\gamma^{(j)} & U_g/2K \\ U_g/2K & (-\gamma^{(j)} + s) \end{vmatrix} = \gamma^{(j)2} - s\gamma^{(j)} - U_g^2/4K^2 = 0. \tag{7.53}$$

This is a quadratic equation for the eigenvalues $\gamma^{(j)}$. In the n-beam case this characteristic equation is of order n. Before discussing the solution, it will be shown that the Howie–Whelan equations (Sect. 7.3.2) lead to the same characteristic equation. If we substitute for ψ'_g and $\mathrm{d}\psi'_g$ from the first equation of (7.29) into the second, we obtain

$$\frac{\mathrm{d}^2\psi'_0}{\mathrm{d}z^2} - 2\pi\mathrm{i}s\frac{\mathrm{d}\psi'_0}{\mathrm{d}z} + (\pi/\xi_g)^2\psi'_0 = 0. \tag{7.54}$$

A similar equation is obtained for ψ'_g. If we look for a solution of the form $\psi' = A\exp(2\pi\mathrm{i}\gamma^{(j)}z)$, the same equation (7.53) is found for the $\gamma^{(j)}$ because $\xi_g = K/U_g$ (7.32). This shows that the two different ways of treating the dynamical theory lead to the same solution.

Solving the quadratic equation (7.53) gives

$$\begin{aligned}\gamma^{(j)} &= \frac{1}{2}\left[s - (-1)^j\sqrt{(U_g/K)^2 + s^2}\right] = \frac{1}{2}\left[s - (-1)^j\sqrt{1/\xi_g^2 + s^2}\,\right] \\ &= \frac{1}{2\xi_g}\left[w - (-1)^j\sqrt{1 + w^2}\right],\end{aligned} \tag{7.55}$$

in which the parameter $w = s\xi_g$ characterizes the tilt out of the Bragg condition ($w = 0$). This solution is plotted in Fig. 7.13a as a function of w and in Fig. 7.12 for a Ewald sphere of a relatively small radius. The two circles around O and G in Fig. 7.12 correspond to the straight lines (asymptotes of the hyperbola) in Fig. 7.13a. The two Ewald spheres (asymptotes) do not intersect but approach most closely for the Bragg condition $w = 0$; their separation is then

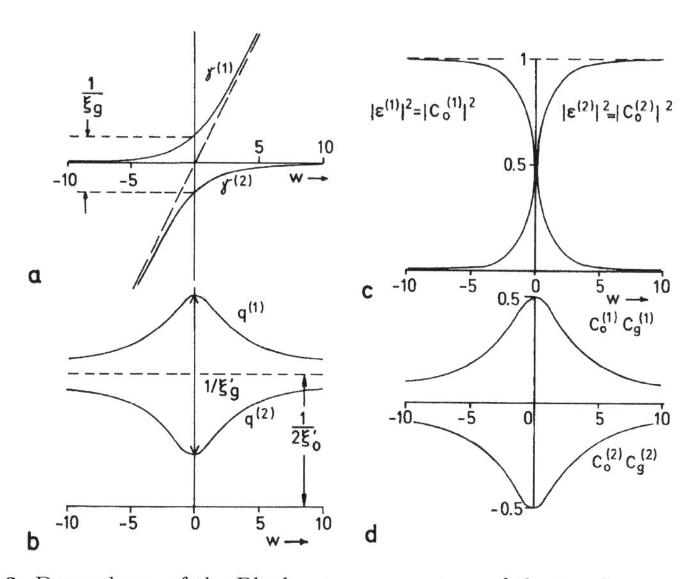

Fig. 7.13. Dependence of the Bloch-wave parameters of the two-beam case on the tilt parameter $w = s\xi_g$ out of the Bragg condition ($w = 0$). (**a**) $\gamma^{(j)}$; $\gamma^{(1)} - \gamma^{(2)}$ is the distance between the two branches of the dispersion surface (Fig. 7.12). (**b**) Absorption parameters $q^{(j)}$ and (**c**), (**d**) wave amplitudes of the four excited waves.

$$\Delta k_{z,\min} = \gamma^{(1)} - \gamma^{(2)} = U_g/K = 1/\xi_g. \tag{7.56}$$

By using the eigenvalues $\gamma^{(j)}$, the linear system of equations (7.52) can be solved for the $C_g^{(j)}$. For the amplitudes $\epsilon^{(j)}C_g^{(j)} = C_0^{(j)}C_g^{(j)}$ of the four Bloch waves with wave vectors $\boldsymbol{k}_0^{(j)} + \boldsymbol{g}$ for normal incidence, we obtain

$$C_0^{(j)}C_0^{(j)} = \frac{1}{2}\left[1 + (-1)^j \frac{w}{\sqrt{1+w^2}}\right]; \quad C_0^{(j)}C_g^{(j)} = -\frac{1}{2}\frac{(-1)^j}{\sqrt{1+w^2}}. \tag{7.57}$$

In the Bragg condition $w = 0$, all four waves have the amplitude 1/2 (Fig. 7.13c,d).

Sometimes the substitution $w = \cot\beta$ is used for the two-beam case. The matrix $[C]$ of the eigenvectors then becomes

$$[C] = \begin{pmatrix} C_0^{(1)} & C_0^{(2)} \\ C_g^{(1)} & C_g^{(2)} \end{pmatrix} = \begin{pmatrix} \sin(\beta/2) & \cos(\beta/2) \\ \cos(\beta/2) & -\sin(\beta/2) \end{pmatrix}. \tag{7.58}$$

In order to calculate the intensity $I_0 = \psi_0\psi_0^*$ of the primary beam, which we call the *transmission* T, and the intensity of the reflected beam $I_g = \psi_g\psi_g^*$ or *reflection* R, we use (7.47) and substitute the specimen thickness t for the z component of the vector $\boldsymbol{r}$:

$$\psi_0(t) = \sum_{j=1}^{2} C_0^{(j)}C_0^{(j)} \exp(2\pi i k_z^{(j)} t),$$

$$\psi_g(t) = \sum_{j=1}^{2} C_0^{(j)}C_g^{(j)} \exp(2\pi i k_z^{(j)} t)\exp(2\pi i g x). \tag{7.59}$$

Substituting the values given in (7.55) and (7.57) and omitting the common phase factor $\exp(2\pi i K_z t)\exp(\pi i w t/\xi_g)$, we find

$$\psi_0(t) = \cos\left(\pi\sqrt{1+w^2}\frac{t}{\xi_g}\right) - \frac{iw}{\sqrt{1+w^2}}\sin\left(\pi\sqrt{1+w^2}\frac{t}{\xi_g}\right),$$

$$\psi_g(t) = \frac{i}{\sqrt{1+w^2}}\sin\left(\pi\sqrt{1+w^2}\frac{t}{\xi_g}\right)\exp(2\pi i g x). \tag{7.60}$$

The intensities (transmission T and reflection R) become

$$\underbrace{\psi_g\psi_g^*}_{R} = \underbrace{1 - \psi_0\psi_0^*}_{1-T} = \frac{1}{1+w^2}\sin^2\left(\pi\sqrt{1+w^2}\frac{t}{\xi_g}\right). \tag{7.61}$$

Recalling that $w = s\xi_g$, we see that for $w \gg 0$ (large tilt out of the Bragg condition) (7.61) is identical with the solution (7.25) of the kinematical theory. Otherwise, however, the kinematical theory predicts that for $w = 0$, R increases as t^2 and becomes larger than one, which is in contradiction with the conservation of intensity $T + R = 1$. The formula (7.61) given by the dynamical theory results in $R = 1 - T = \sin^2(\pi t/\xi_g)$ for $w = 0$. This means that, even in the Bragg condition, the electron intensity oscillates between the primary and the Bragg-reflected beam with increasing film thickness (Fig. 7.14a)

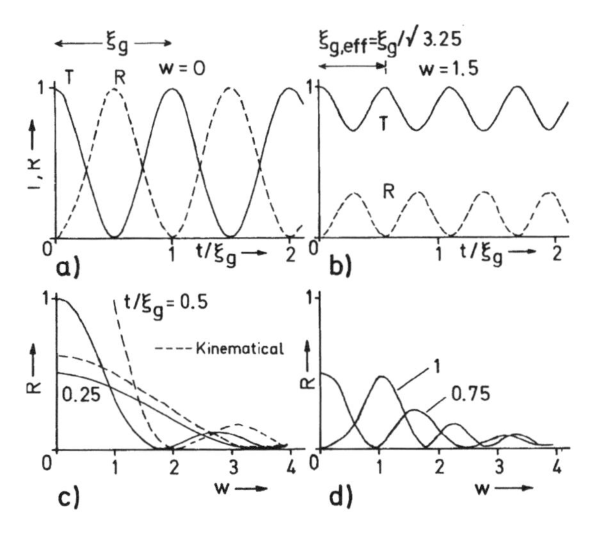

Fig. 7.14. Dynamical two-beam case without absorption. Thickness dependence of the transmitted (T) and Bragg-reflected intensity (R) (**a**) in the Bragg condition $w = 0$ and (**b**) for a tilt parameter $w = s\xi_g = 1.5$; (**c**) and (**d**) tilt dependence (rocking curve) of R for the different relative thicknesses t/ξ_g (between 0.25 and 1). (- - -) comparison with the kinematical theory.

("pendellösung" of the dynamical theory). We now clearly see the meaning of the extinction distance ξ_g (Table 7.2); it is the periodicity in depth of this oscillation. There are thicknesses $t = (n + 1/2)\xi_g$ for which the intensity is completely concentrated in the Bragg reflection and others, $t = n\xi_g$, for which the whole intensity returns to the direction of incidence. These oscillations result from the superposition of the two waves with wave vectors $\boldsymbol{k}_0^{(1)} + \boldsymbol{g}$ and $\boldsymbol{k}_0^{(2)} + \boldsymbol{g}$, which are somewhat different in magnitude: $|\boldsymbol{k}_z^{(1)} - \boldsymbol{k}_z^{(2)}| = 1/\xi_g$ (7.56).

For $w \neq 0$, the amplitude (7.61) of the oscillation decreases as $(1 + w^2)^{-1}$ and the depth of the oscillations can be described by a reduced effective extinction distance (Fig. 7.14b)

$$\xi_{g,\text{eff}} = \xi_g/\sqrt{1 + w^2}\,. \tag{7.62}$$

The dependence of T and R on the tilt angle $\Delta\theta$ of the specimen or the excitation error s or tilt parameter w for a fixed thickness t is called a *rocking curve* (Fig. 7.14c,d). In the absence of absorption, the condition $T + R = 1$ is everywhere satisfied, and T and R are, as can be seen from (7.61), symmetric in w. (This will cease to be the case for T when we consider absorption in the next section.) Figures 7.14c and d show R for $t/\xi_g = 0.25$–1. We observe that $R = 0$ for $w = 0$ and $t/\xi_g = 1$ (Fig. 7.14a). If the specimen is tilted ($w \neq 0$), R increases again (Fig. 7.14d), reaching a maximum at $w \simeq 1$. The distances Δw between the minima ($R = 0$) of the rocking curve become narrower with increasing t. Figure 7.14c also contains the results of the kinematical theory (dashed lines). Deviations from the kinematical theory are observed for larger values of t/ξ_g, especially when w is small.

The relation $E = h^2k^2/2m$ between energy and momentum $p = hk$ can be used to reveal an analogy between the dispersion surface as a function of k_x (Fig. 7.12) and the Fermi surface of low-energy conduction electrons. If there is no interaction between the electrons and the lattice (no excitation of a low-order reflection), the dispersion surface degenerates to a sphere around the origin of the reciprocal lattice (Fermi surface of free electrons). In the theory of conduction electrons, the Fermi surface also splits into energy bands with forbidden gaps if there exists an interaction with the lattice potential, and $\mathrm{d}E/\mathrm{d}k$ becomes zero at the boundary of the Brillouin zone for which the Bragg condition is satisfied. The same behavior can be seen in Fig. 7.12; the Brillouin zone is the midplane between O and G. The splitting $\Delta k_{z,\min}$ of the energy gap is directly proportional to V_g and therefore to the interaction with the crystal lattice.

7.4 Dynamical Theory Including Absorption

7.4.1 Inelastic-Scattering Processes in Crystals

If the energy of the electron falls from the initial value E_m to the final value E_n during a scattering process with an energy loss $\Delta E = E_m - E_n$, the dispersion surfaces for these two energies are different (Fig. 7.15). The surfaces have the

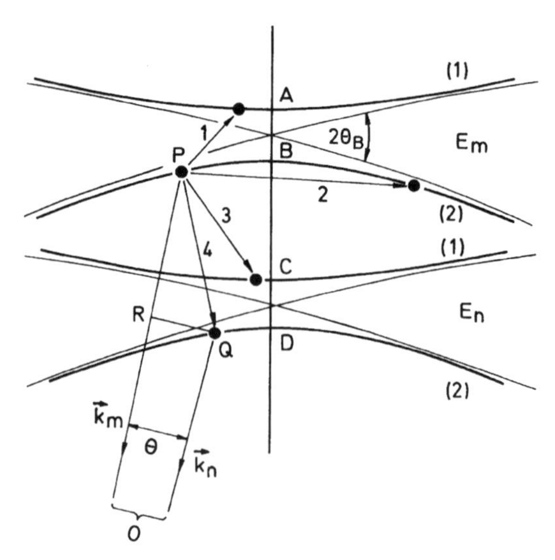

Fig. 7.15. Shift of the dispersion surface caused by electron excitation $m \to n$ with energies E_m and E_n, respectively, and interband (1, 3) and intraband (2, 4) transitions.

same shape because $\Delta E \ll E_m$ but are shifted by $\Delta k_z = |\boldsymbol{k}_m - \boldsymbol{k}_n|_z = \Delta E/hv$ [7.18]. The excitation point P corresponds to the excitation of a Bloch wave of type $j = 2$. The transitions 1–4 are called

$$\left.\begin{array}{l}\text{1. Elastic}\\ \text{3. Inelastic}\end{array}\right\}\text{ interband transition}\qquad \left.\begin{array}{l}\text{2. Elastic}\\ \text{4. Inelastic}\end{array}\right\}\text{ intraband transition.}$$

The symmetry (type) of Bloch waves is changed in an interband transition and preserved in an intraband transition. The vector $\overline{PQ}$ corresponds to $\boldsymbol{q}'$ in Fig. 5.10 and, according to (5.45),

$$q'^2 = \overline{PQ}^2 = \overline{RQ}^2 + \overline{PR}^2 = K^2(\theta^2 + \theta_{\mathrm{E}}^2). \tag{7.63}$$

The shift Δk_z of the dispersion surface for E_n and E_m depends on the scattering process. For thermal diffuse scattering (electron–phonon scattering), the difference can be neglected. For the Al 15 eV plasmon loss, the shift becomes $\Delta k_z = \Delta E/hv = 2 \times 10^{-2}$ nm^{-1} at $E = 100$ keV, whereas the distance $\overline{AB} = \overline{CD}$ between the branches of the dispersion surface is $1/\xi_g = 1/56.3 = 1.8{\times}10^{-2}$ nm^{-1}. The distance and the shift are thus of the same order of magnitude.

Inelastic scattering by a single atom was treated in Sect. 5.2.2 in terms of a model in which the incident wave is plane and the scattered wave is also plane far from the scattering atom. In a crystal, the primary wave function ψ_m ($m = 0$) as well as the scattered wave function ψ_n are Bloch waves that are solutions of the Schrödinger equation (3.22) for the complete system of the incident electron (coordinate $\boldsymbol{r}$) and atomic electrons (coordinates $\boldsymbol{r}_j$) and nuclei ($\boldsymbol{R}_k$) [7.18, 7.19, 7.20]:

$$\left[-\frac{h^2}{2m}\nabla^2 + H_{\mathrm{c}} + H'\right]\Psi = E\Psi. \tag{7.64}$$

The first term of the Hamiltonian represents the propagation of free electrons, the term H_{c} the interaction of the bound electron and ions, and

$$H' = \frac{1}{4\pi\epsilon_0}\left(\sum_j \frac{e^2}{|\boldsymbol{r}-\boldsymbol{r}_j|} - \sum_k \frac{e^2 Z_k}{|\boldsymbol{r}-\boldsymbol{R}_k|}\right) \tag{7.65}$$

the interaction energy between the incident electron and the crystal.

The total wave function Ψ can be expanded as a series

$$\Psi(\boldsymbol{r},\boldsymbol{r}_j,\boldsymbol{R}_k) = \sum_n a_n(\boldsymbol{r}_j,\boldsymbol{R}_k)\psi_n(\boldsymbol{r}), \tag{7.66}$$

where the a_n are the wave functions of the crystal electrons in the nth excited energy state ϵ_n determined by

$$H_{\mathrm{c}}a_n = \epsilon_n a_n. \tag{7.67}$$

$\psi_0(\boldsymbol{r})$ is the wave function of the incident and elastically scattered electron and $\psi_n(\boldsymbol{r})$ that of the inelastically scattered electron of energy $E_n = E - \epsilon_n$ for an energy loss $\Delta E = E - E_n = \epsilon_n$.

Substitution of (7.66) into (7.64), multiplication by a_n^*, and integration over the coordinates $\boldsymbol{r}_j$ and $\boldsymbol{R}_k$ (crystal volume) lead to a set of equations for the ψ_n,

$$\left[-\frac{h^2}{2m}\nabla^2 - E_n + H_{nn}\right]\psi_n = -\sum_{m\neq n} H_{nm}\psi_m, \quad n = 0, 1, ..., \tag{7.68}$$

with the matrix elements

$$H_{nm}(\boldsymbol{r}) = \int_V a_n^*(\boldsymbol{r}_j, \boldsymbol{R}_k)H'(\boldsymbol{r}, \boldsymbol{r}_j, \boldsymbol{R}_k)a_m(\boldsymbol{r}_j, \boldsymbol{R}_k)\mathrm{d}^3\boldsymbol{r}_j\mathrm{d}^3\boldsymbol{R}_k$$
$$= \langle a_n|H'|a_m\rangle. \tag{7.69}$$

The diagonal elements

$$H_{nn}(\boldsymbol{r}) = H_{00}(\boldsymbol{r}) = -\sum_g V_g \mathrm{e}^{-2\pi \mathrm{i}\boldsymbol{g}\cdot\boldsymbol{r}} \tag{7.70}$$

represent the usual potential $V(r)$ in (7.30). The off-diagonal elements, which appear on the right-hand side of (7.68), characterize the probability of an inelastic transition from $a_m \cdot \psi_m$ to $a_n \cdot \psi_n$ caused by the Coulomb interaction H' and are small compared with H_{nn}. It can be seen that the normal case of elastic electron scattering with the Bloch-wave solution is obtained if all of the off-diagonal elements H_{nm} are zero and no inelastic scattering occurs.

The H_{nm} have the same periodicity as the lattice and can be expanded in a Fourier series,

$$H_{nm}(\boldsymbol{r}) = \exp(-2\pi \mathrm{i}\boldsymbol{q}_{nm}\cdot\boldsymbol{r})\sum_g H_g^{nm}\exp(2\pi\mathrm{i}\boldsymbol{g}\cdot\boldsymbol{r}), \tag{7.71}$$

where q_{nm} is the wave vector of the crystal excitation created in the transition $m \rightarrow n$.

If all of the $H_{nm}(\boldsymbol{r})$ with $n \neq m$ are small compared with $H_{00}(\boldsymbol{r})$ and the amplitudes ψ_n of the inelastically scattered waves are small compared with ψ_0, the set of equations (7.68) can be written [7.19]

$$\left[-\frac{h^2}{2m}\nabla^2 - E_0 + H_{00}\right]\psi_0 = -\sum_{n\neq 0} H_{0n}\psi_n,$$
$$\left[-\frac{h^2}{2m}\nabla^2 - E_n + H_{nn}\right]\psi_n = -H_{n0}\psi_0. \tag{7.72}$$

Yoshioka [7.19] also omitted H_{nn} in (7.72) and solved the system with the aid of the Green's function for scattered spherical waves. However, in a crystal, both the incident and scattered waves can propagate only as Bloch waves. The solution with a Green's function constructed of Bloch waves is discussed in [7.21, 7.22]. As the crystal potential is hardly influenced by the transition, H_{nn} can be approximated by H_{00}. Howie [7.18] considered a long-range interaction potential H' for the excitation of plasmons. Single-electron excitation is discussed in [7.23, 7.24, 7.25] and electron–phonon scattering in [7.26, 7.27]. Solutions ψ_n of (7.72) in the form of a series of Bloch waves $b^{(i)}$

(7.34) can be sought. These Bloch waves are solutions of $[-(h^2/2m)\nabla^2 - E_n + H_{nn}]\psi_n = 0$ but have z-dependent amplitudes $\epsilon_n^{(i)}(z)$

$$\psi_n(\mathbf{r}) = \sum_i \epsilon_n^{(i)}(z) b^{(i)}(\mathbf{k}_n^{(i)}, \mathbf{r})$$
$$= \sum_i \epsilon_n^{(i)}(z) \sum_g C_g^{(i)}(\mathbf{k}_n^{(i)}) \exp[2\pi \mathrm{i}(\mathbf{k}_n^{(i)} + \mathbf{g}) \cdot \mathbf{r}]. \quad (7.73)$$

Substituting (7.73) into (7.72), neglecting the small terms $\mathrm{d}^2\psi_n/\mathrm{d}z^2$, multiplying both sides by $b^{*(j)}$, and integrating over the x, y plane containing the reciprocal-lattice vectors $\mathbf{g}$, we obtain the following relation between the Bloch-wave amplitudes $\epsilon^{(j)}$ of the incident wave and $\epsilon^{(i')}$ of the scattered wave (the latter indicated by a dash):

$$\frac{\mathrm{d}\epsilon^{(i')}}{\mathrm{d}z} = \sum_{m \neq n} \sum_j c_{mn}^{i'j} \epsilon_m^{(j)}. \quad (7.74)$$

The matrix elements

$$c_{mn}^{i'j} = -\frac{\mathrm{i}m}{h^2 [k_n^{(i')}]_z} \exp[2\pi \mathrm{i}(k_m^{(j)} - k_n^{(i')} - q_{nm}^{i'j})z] \sum_{h,g} C_g^{(i')} H_{g-h}^{mn} C_h^{(j)} \quad (7.75)$$

describe the transition probabilities between branches j and i' of the dispersion surfaces ($i' = j$: intraband transition; $i' \neq j$: interband transition).

It can be seen from (7.75) that the scattering in a crystal depends on both eigenvector components $C_g^{(i')}$ and $C_h^{(j)}$. This is none other than a reciprocity theorem [7.28], which means that if a primary wave travels in the reverse direction along the path of the scattered wave, it will be scattered with the same probability in the former primary direction. In a crystal, the interaction is inevitably a scattering from one Bloch-wave field into another because incident and scattered waves have to exhibit the lattice periodicity. The scattered intensity therefore depends not only on the scattering angle θ, as for a single atom or for amorphous material, but also on the excitation probabilities of the Bloch waves in the incident and scattered directions. This observation will be used in the discussion of the intensities of Kikuchi lines and bands (Sect. 7.5.4).

Plasmon scattering is not concentrated at the nucleus but within a relatively large volume of 1–10 nm diameter. This process is therefore limited to small scattering angles, and therefore the term H_0^{mn} dominates in (7.75). The sum over the matrix elements in (7.75) can thus be approximated by

$$\sum_g C_g^{(i')} H_0^{mn} C_g^{(j)} = H_0^{mn} \delta_{i'j}, \quad (7.76)$$

which is nonzero only for $i' = j$. This means that plasmon scattering causes predominantly intraband scattering. Image contrast by Bragg reflection is consequently preserved [7.18]. The same is true for the ionization processes in inner shells, so long as $H_0^{mn} \gg H_g^{mn}$ [7.23].

Interband scattering can be observed for large scattering angles, which is equivalent to a narrower localization of the scattering process near the nucleus [7.23]. On the other hand, electron–phonon scattering (thermal diffuse scattering) is predominately interband scattering for small scattering angles and intraband scattering for large scattering angles between the Bragg reflections [7.27]. It therefore contributes mainly to the absorption parameters of the Bloch-wave field.

7.4.2 Absorption of the Bloch-Wave Field

The transition from the initial state $m = 0$ to any $n \neq m$ and i' in (7.74) results in an exponential decrease of $\epsilon_0^{(j)}$, with the value $\epsilon_0^{(j)}(0) = C_0^{(j)}$ at the entrance surface of the crystal ($z = 0$) being determined by the boundary conditions discussed at the end of Sect. 7.3.3.

This exponential decrease can also be incorporated in the Schrödinger equation (3.21) and in the fundamental equations of dynamical theory (7.37) by introducing an additive imaginary lattice potential V_g': $V_g \rightarrow V_g + \mathrm{i}V_g'$ or by replacing U_g by $U_g + \mathrm{i}U_g'$ [7.19]. The V_g' values can be converted to U_g' as in (7.30). Returning to (7.32), $V_g + \mathrm{i}V_g'$ obliges us to replace $1/\xi_g$ by $1/\xi_g + \mathrm{i}/\xi_g'$, where ξ_0' is the mean absorption distance and the ξ_g' are anomalous absorption distances. Values of the imaginary Fourier coefficients V_g' are listed in Table 7.3; see also [7.29, 7.30] for relative values of ξ_g/ξ_g' and [7.31, 7.32, 7.33, 7.34] and [7.35] for contributions to V_g' by thermal diffuse scattering and inner-shell ionization, respectively.

The V_g are assumed to be independent of electron energy because they are defined as Fourier coefficients of the lattice potential $V(r)$. The V_g' are proportional to v^{-1}. In view of (7.33), this implies that $\xi_g \propto v$ and $\xi_0', \xi_g' \propto v^2$, which can be confirmed experimentally [7.36, 7.37, 7.38].

Replacing $1/\xi_g$ by $1/\xi_g + \mathrm{i}/\xi_g'$ in (7.26), we obtain the form of the Howie–Whelan equations in which these absorption effects are considered. In the formulation of the dynamical theory as an eigenvalue problem (Sect. 7.3.3), the matrix $[A]$ in (7.41) now contains the elements

$$A_{11} = \mathrm{i}\,U_0'/2K, \quad A_{gg} = s_g + \mathrm{i}\,U_0'/2K, \quad A_{gh} = (U_{g-h} + \mathrm{i}\,U_{g-h}')/2K,$$

with the result that the eigenvalues become complex: Instead of $\gamma^{(j)}$, we write $\gamma^{(j)} + \mathrm{i}\,q^{(j)}$. The characteristic equation (7.53) for the complex eigenvalues becomes more complicated. Assuming that $\xi_0', \xi_g' \gg \xi_g$, approximately the same values are obtained for the real part $\gamma^{(j)}$ (7.55), and for the two-beam case the imaginary absorption parameters become

$$q^{(j)} = \frac{1}{2}\left[\frac{1}{\xi_0'} - \frac{(-1)^j}{\xi_g'\sqrt{1+w^2}}\right], \tag{7.77}$$

which are plotted in Fig. 7.13b.

Table 7.3. Imaginary Fourier coefficients V_g' (eV) for different substances at $E = 100$ keV and room temperature (and in a few cases at 150 K). The absorption distance ξ_0' is equal to $340/V_0'$ nm and $\mu_0 = V_0'/54$ nm^{-1}.

Substance	V_0'	hkl V_g'				Ref.
Al	–	111	220	311		
theor.	0.85	0.18	0.14	0.13	–	[7.39]
	0.58	0.16	–	–	–	[7.40]
exp.	0.37	0.23	–	–	–	[7.36]
	0.6	–	0.11	0.13	–	[7.41]
	0.54	0.17	–	–	–	[7.40]
Si		220	331	422	–	
theor.	0.70	0.11	0.07	0.08	–	[7.39]
exp.	0.68	0.11	0.08	0.08	–	[7.42]
	0.62	0.14	0.08	–	–	[7.41]
Cu		111	200	220	311	
theor.	3.48	0.83	0.79	0.68	0.63	[7.39]
exp.	1.48	0.81	0.92	–	–	[7.43, 7.44]
	1.35	–	–	0.49	0.45	[7.41]
Ge		220	400	422		
theor.	1.56	0.54	0.48	0.43		[7.39]
exp.	1.25	0.52	–	0.36		[7.42]
	1.35	–	0.32	–		
Au		220	331	440		
theor.	7.57	2.8	2.3	1.87		[7.39]
exp.	2.64	2.0	–	1.5		[7.45]
	–	–	1.62	–		[7.41]
150 K	6.71	–	1.81	–		[7.39]
	2.5	–	1.12	–		[7.46]
MgO		200				
theor.	1.8	0.16				[7.39]
exp.	1.5	0.13				[7.47]
NaCl		220	420			
theor.	1.63	0.20	0.14			[7.39]
exp.	–	0.21	0.15			[7.48]
PbTe(150 K)		422				
theor.	4.7	0.98				[7.39]
exp.	1.8	0.67				[7.46]

The $q^{(j)}$ can also be calculated for the n-beam case. If, in the additional elements of the matrix $[A]$, the U'_g are smaller than 0.1 U_g, then the familiar first-order perturbation method of quantum mechanics can be applied, giving

$$q^{(j)} = \left\langle b^{(j)}(\boldsymbol{k},\boldsymbol{r}) \left| \frac{U'}{2K} \right| b^{(j)}(\boldsymbol{k},\boldsymbol{r}) \right\rangle = \frac{1}{2K} \sum_g \sum_h U'_{g-h} C_h^{(j)} C_g^{(j)} , \tag{7.78}$$

in which the $C_g^{(j)}$ are the components of the eigenvectors of the unperturbed matrix without the imaginary components, corresponding to the situation in which absorption is disregarded. The Bloch-wave formula (7.34) has to be modified to

$$b^{(j)}(\boldsymbol{k},\boldsymbol{r}) = \exp(-2\pi q^{(j)} z) \sum_g C_g^{(j)} \exp[2\pi \mathrm{i}(\boldsymbol{k}_0^{(j)} + \boldsymbol{g}) \cdot \boldsymbol{r}]. \tag{7.79}$$

The first factor with $q^{(j)}$ in the real exponent describes an exponential decrease of the Bloch-wave amplitude with increasing depth z below the surface. From the first term of (7.77), we know that all Bloch-wave amplitudes decrease as $\exp(-\pi z/\xi'_0)$. Differences of the $q^{(j)}$ due to the second term of (7.77) can be understood from the following Bloch-wave model. Let us combine the four possible waves of the two-beam case (Fig. 7.12), but not as we did in (7.47), when we calculated the amplitudes of the primary and Bragg-reflected beams. Now, the waves with wave vectors $\boldsymbol{k}_0^{(1)}$ and $\boldsymbol{k}_g^{(1)}$ form "Bloch wave 1" and $\boldsymbol{k}_0^{(2)}$ and $\boldsymbol{k}_g^{(2)}$ "Bloch wave 2", where $\boldsymbol{k}_g^{(1,2)} = \boldsymbol{k}_0^{(1,2)} + \boldsymbol{g}$. The superposition of two inclined waves propagates in the direction of the angle bisector, which is parallel to the reflecting lattice planes hkl. The superposition results in interference fringes in the x direction, perpendicular to the direction of propagation, with a periodicity equal to the lattice-plane spacing d_{hkl}. From (7.57), we see that all of the $C_0^{(j)} C_h^{(j)}$ (j =1, 2; $\boldsymbol{h}$ = 0, $\boldsymbol{g}$) take the values $\pm 1/2$ in the Bragg condition ($w = 0$). The $C_h^{(j)}$ are symmetric for $j = 1$ (equal signs) and antisymmetric for $j = 2$ (opposite signs). Substitution in (7.34) and (7.79) results in

$$\begin{aligned} |b^{(1)}| &\propto \cos(\pi \boldsymbol{g} \cdot \boldsymbol{r}) = \cos(\pi x/d_{hkl}), \\ |b^{(2)}| &\propto \sin(\pi \boldsymbol{g} \cdot \boldsymbol{r}) = \sin(\pi x/d_{hkl}), \end{aligned} \tag{7.80}$$

because the $q^{(j)}$ are equal for the same branch (j) of the dispersion surface.

The probability densities $|b^{(j)}|^2$ of the Bloch waves 1 and 2 are therefore proportional to $\cos^2(\pi x/d_{hkl})$ and $\sin^2(\pi x/d_{hkl})$, respectively. This results in minima (nodes) at the lattice plane for the antisymmetric wave (j =2) and in maxima (antinodes) for the symmetric wave ($j = 1$) (Fig. 7.16). This is important for the absorption of these Bloch waves. In particular, thermal diffuse scattering is caused by the deviations from the ideal lattice structure due to thermal vibrations of the lattice. Because the amplitude of lattice vibrations is small, the symmetric Bloch wave 1 with maxima at the nuclei will be scattered more strongly than the antisymmetric wave, the values of the absorption parameters $q^{(j)}$ will be larger, and the Bloch-wave amplitude will

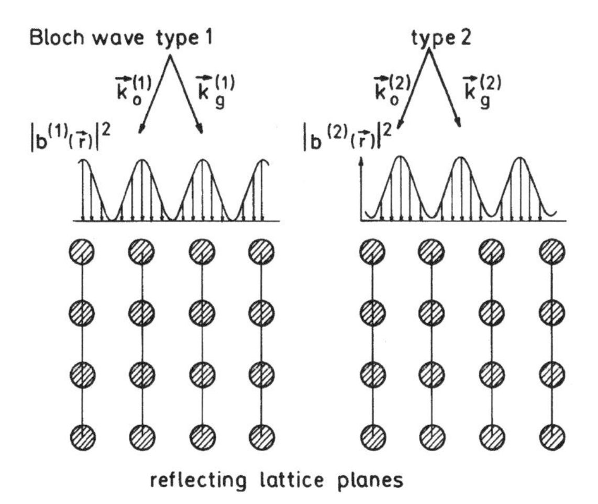

Fig. 7.16. Squared amplitudes (interaction probabilities) $\psi\psi^* \propto |b^{(j)}|^2$ of Bloch waves of type 1 and type 2 in the two-beam case (independent Bloch-wave model) with antinodes and nodes at the nuclei and lattice planes, respectively.

decrease more rapidly with increasing z; the antisymmetric Bloch wave 2 with nodes at the nuclei interacts less strongly ($q^{(2)} < q^{(1)}$, see also Fig. 7.13b). A consequence of thermal diffuse scattering is that ξ_0' and ξ_g' depend strongly on temperature.

There are tilt angles out of the Bragg position for which the antisymmetric Bloch waves show a large transmission (low $q^{(2)}$) and a large excitation amplitude $\epsilon^{(2)}$ (anomalous transmission). As shown in Fig. 7.13, this is the case for positive excitation parameters w for which the squared amplitude $|\epsilon^{(2)}|^2 = |C_0^{(2)}|^2$ is large and $q^{(2)} < q^{(1)}$.

The following analytical formula for T and R can be derived for the two-beam case [7.49]. Using the abbreviations

$$\mu_g = 2\pi U_g'/K = 2\pi/\xi_g' \quad \text{and} \quad \mu_0 = 2\pi/\xi_0',$$

we find

$$T = \frac{e^{-\mu_0 z}}{2(1+w^2)} \left[(1+2w^2)\cosh\frac{\mu_g z}{\sqrt{1+w^2}} + 2w\sqrt{1+w^2}\sinh\frac{\mu_g z}{\sqrt{1+w^2}} + \cos\left(2\pi\frac{\sqrt{1+w^2}}{\xi_g}z\right)\right], \tag{7.81}$$

$$R = \frac{e^{-\mu_0 z}}{2(1+w^2)} \left[\cosh\frac{\mu_g z}{\sqrt{1+w^2}} - \cos\left(2\pi\frac{\sqrt{1+w^2}}{\xi_g}z\right)\right]. \tag{7.82}$$

Certain characteristic differences are found when the expressions above are compared with the two-beam case without absorption (7.61). The rocking curve of the transmission is not symmetric about the Bragg position (Fig. 7.18). This asymmetry is a consequence of the second term in the square

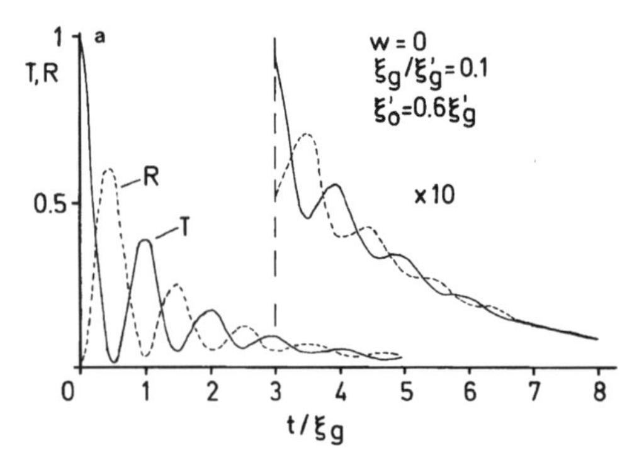

Fig. 7.17. Thickness dependence of the transmitted (T) and Bragg-reflected (R) intensities in the Bragg condition ($w = 0$) with absorption.

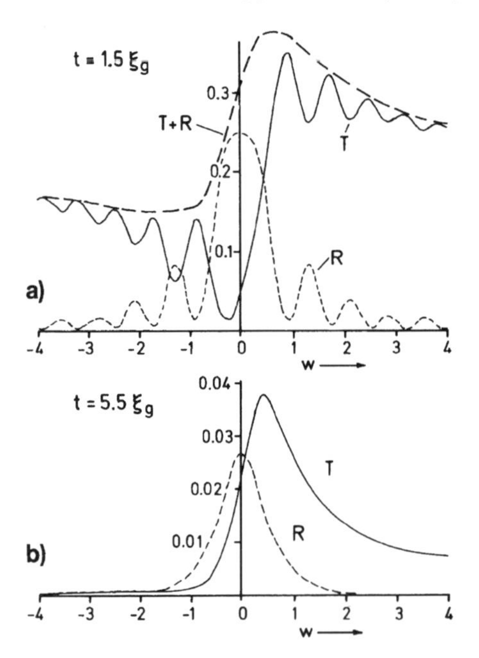

Fig. 7.18. Tilt dependence (rocking curve) of the transmitted (T) and Bragg-reflected (R) intensities for the dynamical two-beam case with absorption for foil thicknesses (**a**) $t = 1.5\xi_g$ and (**b**) $t = 5.5\xi_g$.

brackets of (7.81). The reflected intensity R still remains symmetric [there are no terms in odd power of w in (7.82)]. The relation $T + R = 1$ is no longer valid. The amplitude of the pendellösung fringes decreases with increasing thickness (Fig. 7.17). Only a broad transmission band (anomalous transmission) with extremely weak pendellösung fringes is observed for very thick specimens (Fig. 7.18b).

In the case of anomalous transmission, only the antisymmetric Bloch wave 2 with low $q^{(2)}$ remains after passage through a thick specimen layer. This observation might seem to imply that the Bloch waves can be regarded as approximately independent. However, this independent Bloch-wave model cannot satisfy the boundary condition at the entrance plane of the crystal and experiments show that there are situations in which the whole Bloch-wave field (dependent Bloch-wave model) has to be considered – in the study of the probability of large-angle scattering, for example [7.50].

7.4.3 Dynamical n-Beam Theory

The n-beam case must normally be treated numerically by solving the eigenvalue problem (7.41) for a large number of excited reciprocal-lattice points $\boldsymbol{g}$ near the Ewald sphere or by applying the Howie–Whelan equations. The dispersion surface splits into n branches, and $j = 1,\ldots, n$ absorption parameters $q^{(j)}$ of the n Bloch waves have to be considered.

A special case of the n-beam theory is the systematic row, where the reflections $-n\boldsymbol{g},\ldots,-2\boldsymbol{g},-\boldsymbol{g},0,+\boldsymbol{g},+2\boldsymbol{g},\ldots,+n\boldsymbol{g}$ are excited. The overlap of $-\boldsymbol{g}$ and $+\boldsymbol{g}$ reflections in the rocking curve forms reflection bands, which can be seen in Kikuchi diagrams, channeling patterns, and images of bent crystal foils (bend contours, Sect. 9.1.1). As an example, Fig. 7.19a shows the absorption parameters $q^{(j)}$ of a three-beam case for Cu at $E = 100$ keV with $\boldsymbol{g} = \overline{2}\overline{2}0$, 0, 220 excited. The tilt parameter k_x/g is now zero for the symmetric incidence and +0.5 for the excitation of $\boldsymbol{g} = 220$. For a thickness $t = 40$ nm, Fig. 7.20a shows the rocking curve for the intensity I_0 of the primary beam and I_{220} of one Bragg reflection. $I_{\overline{2}\overline{2}0}$ will show a similar curve with its center

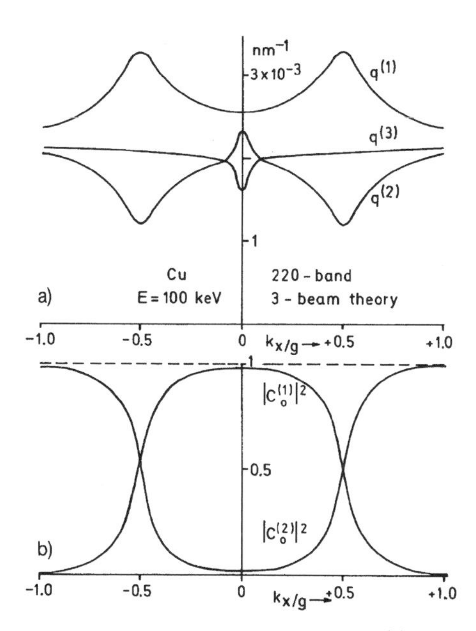

Fig. 7.19. Dependence of (**a**) absorption parameters $q^{(j)}$ and (**b**) the squared amplitudes $|C_0^{(j)}|^2$ of the Bloch waves on the tilt parameter k_x/g for a 220 band in Cu at $E = 100$ keV (three-beam case with excitation of $\overline{2}\overline{2}0, 0, 220$).

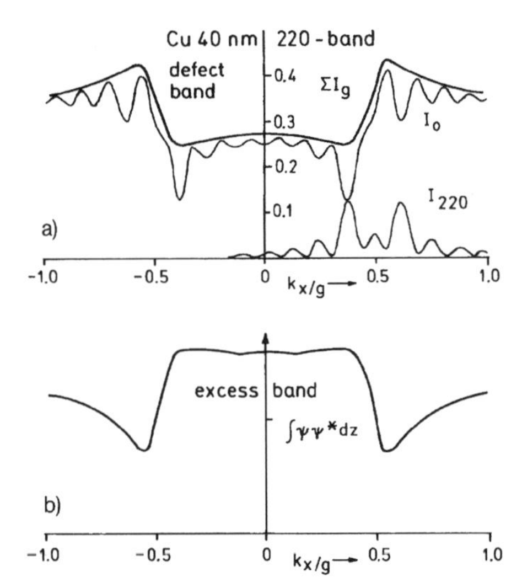

Fig. 7.20. Dependence of (**a**) the primary-beam intensity I_0 and Bragg-reflected intensity I_{200} (rocking curve) and ΣI_g (defect band) and (**b**) the large-angle scattering probability $\int \psi\psi^* \mathrm{d}z$ (excess band) as a function of the tilt parameter k_x/g for the three-beam case in Cu at $E = 100$ keV.

at $k_x/g = -0.5$. The rocking curve for I_{220} is symmetric about the Bragg condition (see also the discussion of the two-beam case in Sect. 7.4.2), whereas I_0 exhibits the asymmetry associated with anomalous absorption. This can be seen from Fig. 7.19. For $k_x/g < 0.5$, the Bloch-wave intensity $|C_0^{(1)}|^2$ with a large value of $q^{(1)}$ is larger than $|C_0^{(2)}|^2$, which results in stronger absorption. For $k_x/g > 0.5$, the intensity $|C_0^{(2)}|^2$ is larger for the lower parameter $q^{(2)}$, leading to a higher value of the transmission. (The intensity $|C_0^{(3)}|^2$ is so low in this special case that it is not included in Fig. 7.19b.). Tilting to negative values of k_x/g results in anomalous absorption for $k_x/g > -0.5$ and anomalous transmission for $k_x/g < -0.5$.

In the sum of intensities $\Sigma_g I_g = I_{-g} + I_0 + I_{+g}$, the pendellösung fringes cancel. However, the sum is not constant, as it is in the dynamical theory without absorption, but varies with k_x/g and a defect band of reduced intensity is created (Fig. 7.20a). This cancellation of the pendellösung fringes can be shown analytically by using (7.47) and (7.79) for the ψ_g, changing the order of summation and employing the orthogonality relations (7.43) for the eigenvector components

$$\begin{aligned}\sum_g I_g &= \sum_g \psi_g \psi_g^* \\ &= \sum_g |\sum_j C_0^{(j)*} C_g^{(j)} \exp(-2\pi q^{(j)} z) \exp[2\pi \mathrm{i}(\boldsymbol{k}^{(j)} + \boldsymbol{g}) \cdot \boldsymbol{r}]|^2 \\ &= \sum_g \sum_{i,j} C_0^{(i)} C_0^{(j)*} C_g^{(i)*} C_g^{(j)} \exp[-2\pi(q^{(i)} + q^{(j)})z] \exp[2\pi \mathrm{i}(\boldsymbol{k}^{(j)} - \boldsymbol{k}^{(i)}) \cdot \boldsymbol{r}]\end{aligned}$$

$$= \sum_{i,j} \underbrace{[\sum_{g} C_g^{(i)*} C_g^{(j)}]}_{\delta_{ij}} C_0^{(i)} C_0^{(j)*} \exp[-2\pi(q^{(i)} + q^{(j)})z] \exp[2\pi \mathrm{i}(\boldsymbol{k}_0^{(j)} - \boldsymbol{k}_0^{(i)}) \cdot \boldsymbol{r}]$$

and hence

$$\sum_g I_g = \sum_i |C_0^{(i)}|^2 \exp(-4\pi q^{(i)} z). \tag{7.83}$$

Inside the defect band, only the Bloch-wave intensity $|C_0^{(1)}|^2$ with a large value of $q^{(1)}$ is excited, and the intensity therefore decreases.

Scattering effects that are strongly localized at the nuclei (e.g., scattering through large angles and backscattering or excitation of an inner shell or x-ray emission) are proportional to the probability density $\psi\psi^*$ at the atom positions. For a finite thickness t, the scattering probability is therefore proportional to $\int \psi\psi^* \mathrm{d}z$. Inside the defect band, the intensity $|C_0^{(1)}|^2$ of the Bloch wave with its antinodes at the nuclei is large, which causes increased large-angle scattering or an increased probability of inner-shell ionization, resulting in an excess band (Fig. 7.20b).

For high electron energies, a semiclassical method can be used to calculate the Bloch-wave amplitude distribution in the lattice and the corresponding $q^{(j)}$. If the electrons travel along an atomic row, the interaction can be expressed approximately in terms of a projected potential valley. Quantum mechanics tells us that the electrons occupy quantized states in such a valley, e.g. $1s$, $2s$, $2p$, ... Branch 1 of the dispersion surface corresponds to the most strongly bound state $1s$ with the largest probability density at the nuclei. The $1s$ state therefore shows a strong absorption (blocking) and a $2p$ state (branch 2) with a low probability density a weak absorption (channeling). Calculations can be based on a semiclassical approach, particularly for shorter values of electron wavelength, and there is an analogy with the classical theory of channeling for ion beams with the difference that, owing to the opposite sign of the charge (the same is also true for positrons), ion channeling is observed when electrons are blocked and vice versa. This model is especially suitable for calculating the parameters of the dispersion surface and the amplitude distribution of Bloch waves at high electron energies [7.51, 7.52, 7.53].

7.4.4 The Bethe Dynamical Potential and the Critical Voltage Effect

In the following discussion, an n-beam case with one strongly excited low-order reflection ($\boldsymbol{g}_1 = 0$ and $\boldsymbol{g}_2 = \boldsymbol{g}$) and a number of weakly excited reflections $\boldsymbol{g}_n$ ($n > 2$) is considered (Bethe's approximation [7.7]). In the fundamental equations (7.37), the first equations are the same, and in those for $\boldsymbol{g}_n$ ($n > 2$), only terms with the largest amplitude factors $C_0^{(j)}$ and $C_g^{(j)}$ need be considered in a first-order approximation:

$$[K^2 - (\boldsymbol{k}_0^{(j)} + \boldsymbol{g})^2]C_g^{(j)} + \sum_{h \neq 0} U_h C_{g-h}^{(j)} = 0 \quad \text{for} \quad \boldsymbol{g}_1 = 0 \text{ and } \boldsymbol{g}_2 = \boldsymbol{g}$$

$$[K^2 - (\boldsymbol{k}_0^{(j)} + \boldsymbol{h})^2]C_h^{(j)} + U_h C_0^{(j)} + U_{h-g} C_g^{(j)} = 0 \text{ for } \boldsymbol{h} = \boldsymbol{g}_3, \ldots, \boldsymbol{g}_n. \quad (7.84)$$

The second set of equations is not coupled and can be solved for the $C_h^{(j)}$; these can then be substituted in the first equations for $\boldsymbol{g}_1 = 0$ and $\boldsymbol{g}_2 = \boldsymbol{g}$. This yields a two-beam case analogous to (7.52) but with a corrected potential coefficient $U_{g,\mathrm{dyn}}$ or *dynamic potential*

$$V_{g,\mathrm{dyn}} = V_g - \frac{2m_0}{h^2}(1 + E/E_0) \sum_{h \neq 0} \frac{V_g V_{g-h}}{K^2 - (\boldsymbol{k} + \boldsymbol{h})^2} = V_g - V_{g,\mathrm{corr}}. \quad (7.85)$$

Depending on the sign of the denominator in (7.85), $V_{g,\mathrm{dyn}}$ may be larger or smaller than V_g. The latter will be decreased when the reciprocal-lattice point is inside the Ewald sphere. From the relation $\xi_g = (\lambda U_g)^{-1}$ (7.32), the extinction distances will also be changed by this dynamical interaction.

The value of $V_{g,\mathrm{dyn}}$ in (7.85) decreases with increasing electron energy and can vanish for a certain *critical voltage* V_c or energy $E_c = eV_c$ if the excitation error is positive (reciprocal-lattice point inside the Ewald sphere). At this voltage, $\xi_g \to \infty$, which means that the corresponding reflection will not be excited. This effect was first observed in electron-diffraction patterns [7.54, 7.55]. The minimum distance $\Delta k_{z,\mathrm{min}}$ between the branches of the dispersion surface falls to zero and the branches intersect.

The use of (7.85) alone can explain the existence of a critical voltage but is not sufficient for an accurate calculation. With increasing energy, first the dynamical potential $V_{g,\mathrm{dyn}}$ of a systematic row $(0, \pm g, \pm 2g, \ldots)$ vanishes because the two terms in (7.85) cancel for the second-order reflection (e.g., 400 in a 200 row) when the $2g$ reflection is fully excited. We therefore assume a three-beam case $0, g, 2g$ [7.56] with $s_{2g} = 0$ and $s_g = K - \sqrt{K^2 - g^2} \simeq g^2/2K$. The eigenvalue equation (7.41) can be rewritten

$$\begin{pmatrix} -2K\gamma & U_g & U_{2g} \\ U_g & (g^2 - 2K\gamma) & U_g \\ U_{2g} & U_g & -2K\gamma \end{pmatrix} \begin{pmatrix} C_0 \\ C_g \\ C_{2g} \end{pmatrix} = 0. \quad (7.86)$$

The solution can be simplified by considering the symmetry of the matrix, which leads us to distinguish the two cases

$(a)\ C_0 = C_{2g},\quad C_g \neq 0$ and
$(b)\ C_0 = -C_{2g},\ C_g = 0.$

On substituting in (7.86), we obtain the reduced system of equations

$$(a) \quad (U_{2g} - 2K\gamma)C_0 + U_g C_g = 0,$$
$$2U_g C_0 + (g^2 - 2K\gamma)C_g = 0,$$

$$\text{giving} \quad K\gamma^{(1,2)} = \frac{1}{4}\left[U_{2g} + g^2 \pm \sqrt{8U_g^2 + (U_{2g} - g^2)^2}\right],$$

$$(b) - 2K\Delta k_z C_0 - U_{2g} C_0 = 0, \quad \text{giving} \quad K\gamma^{(3)} = -\frac{1}{2}U_{2g}.$$

The values of γ for case (a) are obtained by setting the determinant equal to zero. The difference

$$\gamma^{(2)} - \gamma^{(3)} = \frac{1}{4K}\left[3U_{2g} + g^2 - \sqrt{8U_g^2 + (U_{2g} - g^2)^2}\right] \tag{7.87}$$

vanishes when the quantity in the square brackets is zero. This yields

$$U_g^2 = U_{2g}^2 + U_{2g}g^2 . \tag{7.88}$$

When U_g is replaced by using (7.30), the critical energy $E = E_c$ appears in the factor $m = m_0(1 + E/E_0)$ and solving for E results in

$$E_c = eV_c = \left[\frac{h^2 g^2 V_{2g}}{2m_0(V_g^2 - V_{2g}^2)} - 1\right] E_0 . \tag{7.89}$$

The critical voltage can also be obtained by calculating the Bloch-wave amplitudes $C_0^{(j)}$ because the symmetry of Bloch waves 2 and 3 changes for the second-order reflection when the electron energy exceeds the critical energy [7.16]. We find

	$V < V_c$	$V > V_c$
$j = 2$	symmetric	antisymmetric
$j = 3$	antisymmetric	symmetric

and

1. $V < V_c$ when $|C_0^{(2)}| > |C_0^{(3)}|$ or $V > V_c$ when $|C_0^{(2)}| < |C_0^{(3)}|$ for the second-order reflection at $k_x = 0$ (symmetrical incidence of the electron beam)
2. $V < V_c$ when $|C_0^{(3)}| > |C_0^{(4)}|$ or $V > V_c$ when $|C_0^{(3)}| < |C_0^{(4)}|$ for the third-order reflection at $k_x = 0.5g$ (Bragg condition for the third order).

In Table 7.4, some experimental values of the critical voltage are listed. Most of the values of V_c are much greater than 100 keV and hence in the HVEM range. Vanishing of the reflection $2g$ in the Bragg condition can be observed with Kikuchi lines in electron-diffraction patterns [7.57], with bend contours in electron micrographs or in convergent-beam diffraction patterns [7.58, 7.59]. Figure 7.21 shows the rocking curve near the 220 Bragg condition for a 300 nm Cu foil. The critical voltage V_c can be determined with an accuracy of a few kilovolts.

The following applications of the critical voltage effect are of interest:

(1) Accurate measurement of V_c and study of the excitation of other reflections predicted by the dynamical theory allow us to measure the coefficients V_g of the lattice potential and to calculate the structure amplitude $F(\theta)$ at the corresponding scattering angle $\theta = 2\theta_B$ by using (7.31) [7.56, 7.59, 7.60, 7.61, 7.62, 8.9, 7.64, 7.65]. The differences between $F(\theta)$ and the calculated values of $f(\theta)$ for free atoms can be used to obtain information about variations in the electron-density distribution caused by the packing of atoms in a solid (e.g., for Ge and Si [7.66]). Another possibility is to evaluate $F(\theta)$ from convergent-beam electron diffraction using energy-filtering microscopy (Sect. 8.3.3).

Table 7.4. Examples of experimental critical voltages V_c [kV] at room temperature [7.16].

Face-centered cubic metals				
hkl	111	200	220	331
Al	425, 430	895, 918	–	–
Co	276	555	1745	2686
Ni	295, 298	588, 587	1794	2730
Cu	310, 325	600, 605	1750	2700
Ag	55	225	919	1498
Au	(<0)	108	726	1266
$AuCu_3$	175	425	–	–
Body-centered cubic metals				
hkl	110	200	211	
V	230, 238	–	–	
Cr	259, 265	1238	–	
Fe	305	1249	–	
Nb	35	749	1595	
Mo	35	789	1729	
Ta	–	6651	–	
W	–	660	>1100	
Diamond cubic crystals				
hkl	111	220	400	
Si	1113	>1150	–	
Ge	925	1028	>1100	
GaP	1026	1098	>1100	

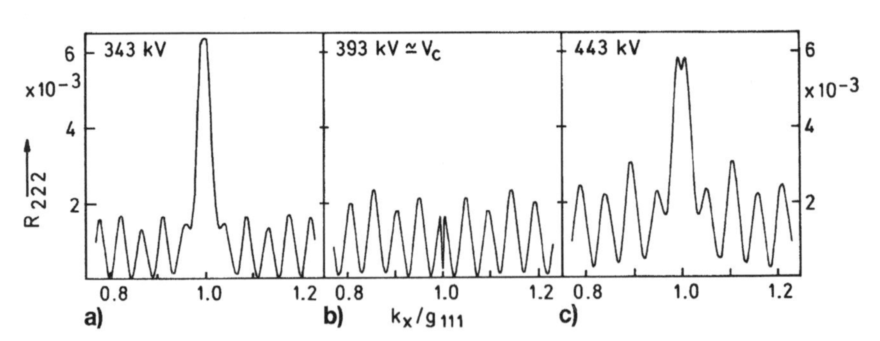

Fig. 7.21. Calculated rocking curves near the 222 Bragg condition (**a**) 50 kV below the critical voltage, (**b**) at $V_c = 393$ kV with the critical-voltage effect at $k_x/g_{111} = 1$, and (**c**) 50 kV beyond the critical voltage of copper ($t = 300$ nm) [7.58]. The sharp dip at the center of (**b**) is a computation artifact.

(2) The critical voltage depends on the composition of an alloy. It decreases from $V_c = 590$ kV for pure Ni to $V_c \simeq 450$ kV for a Ni-10 mol% Au alloy, for example [7.67]; this can be used for local measurements of concentration.

(3) Measurements of V_c at different temperatures give $F(\theta)\exp(-M_g)$ (Sect. 7.5.3) and can be used to obtain an accurate value of the Debye temperature θ_D [7.56] and its dependence on orientation in noncubic crystals (e.g., Zn, Cd [7.68]) or alloys (e.g, Cu-Al, Cu-Au [7.69]).

(4) The study of V_c can shed light on ordering in solids. Thus, an increase of V_c from 166 kV to 175 kV is observed during the transition from a disordered to an ordered state in $AuCu_3$ [7.16]. Ordered states of short range in Fe-Cr and Au-Ni alloys act like frozen-in lattice vibrations and make a temperature-independent contribution to the Debye temperature. By measuring the temperature dependence of V_c, the contributions from distortion of the long-range order and thermal vibrations can be separated if it is assumed that the corresponding $\langle u^2 \rangle$ can be added [7.67, 7.70].

These few examples show how the critical-voltage effect offers interesting possibilities for the quantitative analysis of metals and alloys. The critical-voltage effect and the intersecting Kikuchi-line technique [7.71, 7.72] are closely related and both need at least a three-beam diffraction condition. In the second technique, distances between intensity anomalies in intersections in either Kikuchi or convergent-beam diffraction patterns are measured. Such intersections can be found for high-order reflections at any voltage and also allow the potential coefficients V_g to be determined.

7.5 Intensity Distribution in Diffraction Patterns

7.5.1 Diffraction at Amorphous Specimens

The diffraction pattern of amorphous films – carbon supporting films, polymers, silicon and aluminum oxides, glass, and ceramics – consist of diffuse rings (Fig. 7.22a). Each amorphous structure contains a nearest-neighbor ordering, which can be described by a radial distribution function $\rho(r)$. The probability of finding the centers of neighboring atoms inside a spherical shell between r and $r + \mathrm{d}r$ is $4\pi r^2 \rho(r)\mathrm{d}r$ (Fig. 7.22b).

The observed intensity distribution I_{exp} in the diffraction pattern oscillates about the intensity distribution $N|f(\theta)|^2$ that would be seen if all of the N atoms scattered independently, without interference [7.73, 7.74],

$$\frac{1}{I_0}\frac{\mathrm{d}I_{\mathrm{exp}}}{\mathrm{d}\Omega} = N|f(q)|^2 \left[1 + \int_0^\infty 4\pi r^2 \rho(r) \frac{\sin(2\pi q r)}{2\pi q r}\mathrm{d}r\right] + \frac{1}{I_0}\frac{\mathrm{d}I_{\mathrm{inc}}}{\mathrm{d}\Omega}, \tag{7.90}$$

with the incident electron current I_0, the spatial frequency $q = \theta/\lambda$, and an incoherent background $\mathrm{d}I_{\mathrm{inc}}/\mathrm{d}\Omega$ caused by inelastic scattering. After producing a normalized function

$$i(q) = \frac{1}{I_0 N|f(q)|^2}\left[\left(\frac{\mathrm{d}I_{\mathrm{exp}}}{\mathrm{d}\Omega} - \frac{\mathrm{d}I_{\mathrm{inc}}}{\mathrm{d}\Omega}\right) - I_0 N|f(q)|^2\right] \tag{7.91}$$

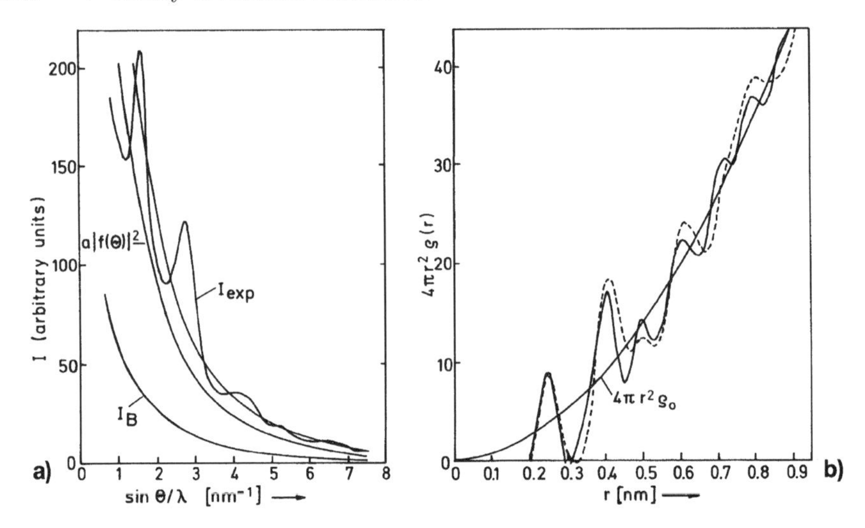

Fig. 7.22. (**a**) Measured intensity distribution of the diffuse diffraction rings of an amorphous Ge film ($E = 60$ keV) and fits of the background. (**b**) Calculation of the radial density distribution $4\pi r^2 \rho(r)$ oscillating about a mean value $4\pi r^2 \rho_0$. Values obtained by x-ray diffraction (– – –) for comparison [7.73].

from the experimentally observed distribution, the transformation (7.90) can be reversed, giving the reduced density function

$$4\pi r^2[\rho(r) - \rho_0] = 8\pi \int_0^\infty i(q) \sin(2\pi q r) q \mathrm{d}q. \qquad (7.92)$$

Zero-loss filtering of an amorphous diffraction pattern can remove the inelastic and incoherent background [7.74, 7.75], whereupon the maxima and minima become more pronounced and it is easier to fit the elastic contribution $N|f(q)|^2$; this will not, however, be exactly proportional to the differential elastic cross section because of multiple elastic scattering, which also occurs in thin films. This action of zero-loss filtering is demonstrated in Fig. 7.23, which shows the recorded radial intensity distribution of a 27 nm amorphous germanium film, whereas the oscillations between maxima and minima are much lower for inelastically scattered electrons with an energy loss $\Delta E = 17$ eV.

7.5.2 Intensity of Debye–Scherrer Rings

Polycrystalline specimens with random crystal orientations produce diffraction spots distributed randomly in azimuth. If the irradiated area is large and/or the crystal size is small, the high density of diffraction spots forms a continuous Debye–Scherrer ring for each allowed set of hkl values with non-vanishing structure amplitude F (Fig. 8.11).

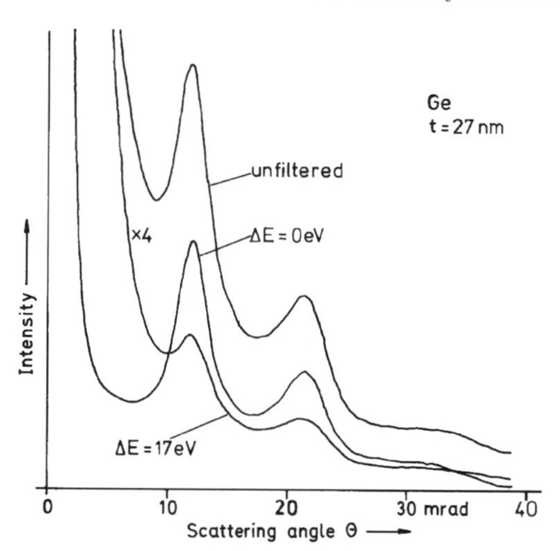

Fig. 7.23. Radial intensity distribution of the diffraction pattern of an evaporated amorphous 27 nm germanium film: unfiltered, zero-loss, and plasmon-loss filtered at $\Delta E = 17$ eV.

The intensity of the rings is obtained by averaging over all crystal orientations, that is, by integrating over s in (7.19) or (7.61). The kinematical theory gives the following expression for the integrated intensity I_{hkl} of the total ring with Miller indices hkl [7.76]:

$$I_{hkl} = j_{\mathrm{s}} \frac{2\pi^2 m^2 e^2}{h^4} K N V_{\mathrm{e}} p_{hkl} |V_g|^2 \exp(-2M_g) \lambda^2 d_{hkl}, \tag{7.93}$$

where j_{s} denotes the current density in the specimen plane, K the number of crystals with on average N unit cells, V_{e} the volume of the unit cell, p_{hkl} the multiplicity of the hkl planes (e.g., $p_{100} = 6$ since there are six possible cubic planes, $p_{110} = 12$, $p_{111} = 8$, etc.), and V_g the lattice potential with the Debye–Waller factor $\exp(-M_g)$ (Sect. 7.5.3).

This implies that the ring intensity depends only on the total number KN of unit cells in the electron beam and not on the shape and dimensions of the crystals. The intensity ratios of various rings are independent of wavelength and crystal dimensions and depend on p_{hkl}, V_g, and d_{hkl}.

If dynamical two-beam theory is employed, the following differences relative to kinematical theory are found [7.77]:

$$\frac{I_{\mathrm{dyn}}}{I_{\mathrm{kin}}} = \frac{1}{A_{hkl}} \int_0^{A_{hkl}} \mathrm{J}_0(2x)\mathrm{d}x, \tag{7.94}$$

where J_0 is the Bessel function and $A_{hkl} = 2\pi e m_0 \, t \, V_g / h^2$.

The dynamical theory predicts the same results as the kinematic theory ($I_{\mathrm{dyn}}/I_{\mathrm{kin}} \simeq 1$) if A_{hkl} is small either because the crystal diameter t is small or the wavelength λ is short. The intensity ratios of different rings do not remain independent because they also depend on A_{hkl}. The curves in Fig. 7.24 show the decrease of $I_{\mathrm{dyn}}/I_{\mathrm{kin}}$ with increasing A_{hkl} and measurements of the ratio for evaporated Al films with small and large crystals [7.76]. The ratio

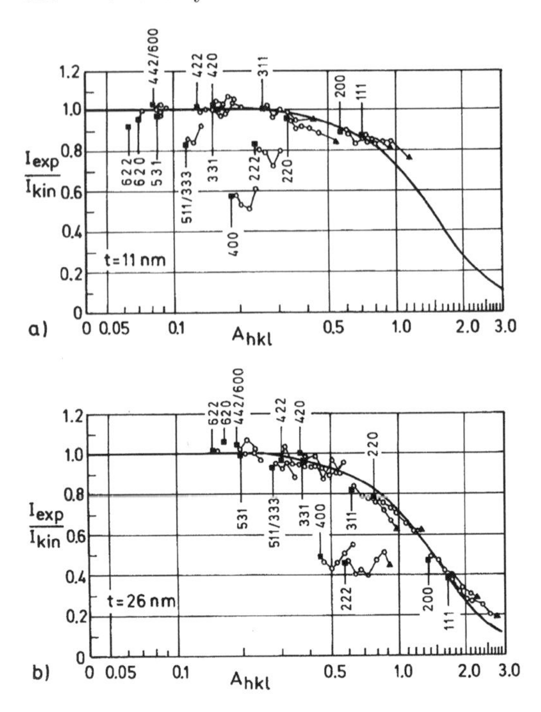

Fig. 7.24. Comparison of the diffracted intensity I_{exp} of Debye–Scherrer rings of evaporated Al films of thickness (**a**) $t = 11$ nm and (**b**) $t = 26$ nm with the value I_{kin} calculated by the kinematical theory. The decrease of $I_{\text{exp}}/I_{\text{kin}}$ for large A_{hkl} and high-order systematic reflections (e.g., 222, 400) indicates deviations from the kinematical theory. (——) Calculations of the intensity ratio with the dynamical theory (7.94).

$I_{\text{dyn}}/I_{\text{kin}}$ is close to unity for crystals of the order of 10 nm (Fig. 7.24a). The intensities of Bragg reflections with low indices are the first to decrease for larger crystals (Fig. 7.24b), whereas the kinematical theory still holds for high indices. Reflections of higher order (e.g., 400 or 222) do not lie on the curves because reflections of low order are also excited. The two-beam theory is no longer valid, and many-beam theory has to be used.

The laws governing Debye–Scherrer ring intensities assume that the orientation of the crystals is random, with no preferential orientation (texture). A texture causes strong changes of the ring intensities and also an azimuthal variation of ring intensity for oblique electron incidence to the fiber axis of the texture (Sect. 8.2.2).

Zero-loss filtering of diffraction patterns with Debye–Scherrer rings considerably increases the contrast [7.78, 7.79, 7.80, 7.81, 7.82], as shown for an evaporated aluminum layer ($t = 230$ nm) in Fig. 7.25. The background between the rings can be attributed to thermal diffuse scattering. This is important for the investigation of thick films and the detection of weak ring intensities. The gain of contrast (e.g., of the ratio of a (111) ring-to-background ratio) is largest for low atomic number ($\simeq 25\times$), as for the example in Fig. 7.25, and low for high atomic number (e.g., platinum films).

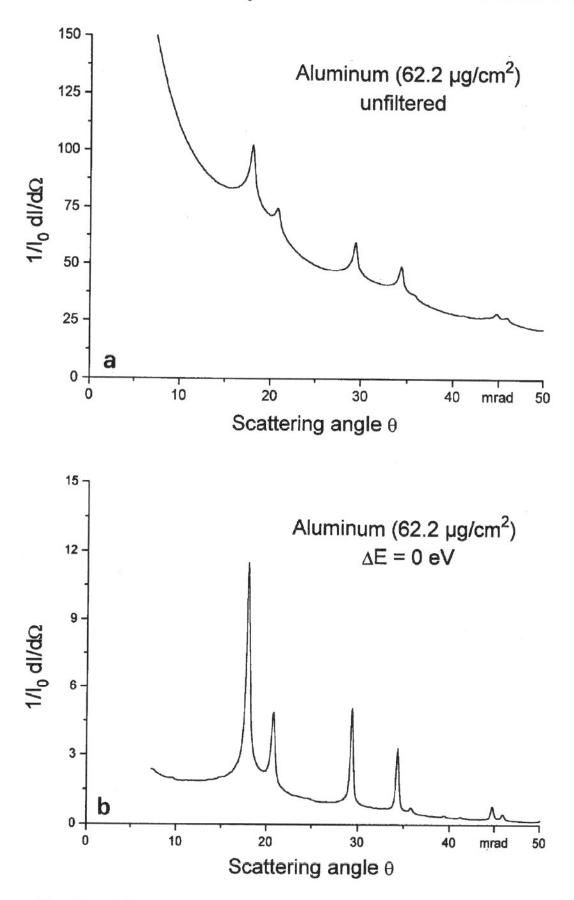

Fig. 7.25. Normalized radial intensity distributions (note the different scales) of (**a**) unfiltered and (**b**) zero-loss filtered diffraction patterns of an evaporated aluminum film ($x = 62\ \mu\text{g/cm}^2$, $t = 230$ nm, $E = 80$ keV) [7.82].

7.5.3 Influence of Thermal Diffuse Scattering

Thermal vibrations of the atoms (nuclei) cause a distortion of the lattice periodicity and produce the following effects:

1. decrease of the effective potential V_g by the Debye–Waller factor with increasing temperature, thus influencing the extinction distances ξ_g and the critical voltage V_c,
2. increase of the absorption parameters $q^{(j)}$ of the dynamical theory and decrease of the absorption distances ξ'_0 and ξ'_g with increasing temperature,
3. thermal diffuse scattering in the background between and near the Bragg spots and formation of Kikuchi lines and bands.

These interactions can be treated as electron–phonon scattering because the lattice vibrations are quantized (phonons of momentum $p = hk$ and energy $E = \hbar\omega$).

The influence of lattice vibrations on the potential coefficients V_g can be understood from the following simplified model. If the atoms are shifted through a distance $\boldsymbol{u}(\boldsymbol{r})$ from their equilibrium positions $\boldsymbol{r}$, the potential is changed in first order only by a translation $V(\boldsymbol{r}) \to V(\boldsymbol{r} + \boldsymbol{u}(\boldsymbol{r}))$. Using the Fourier expansion (7.30) of the lattice potential, each Fourier coefficient V_g contains the following factor ($\langle\ \rangle$: mean value):

$$\begin{aligned}\langle \mathrm{e}^{2\pi \mathrm{i}\boldsymbol{u}\cdot\boldsymbol{g}} \rangle &= \langle 1 + 2\pi\mathrm{i}\boldsymbol{u}\cdot\boldsymbol{g} - 2\pi^2(\boldsymbol{u}\cdot\boldsymbol{g})^2 + \ldots \rangle \\ &= 1 - 2\pi^2\langle(\boldsymbol{u}\cdot\boldsymbol{g})^2\rangle + \ldots \simeq \mathrm{e}^{-2\pi^2\langle u^2\rangle g^2} = \mathrm{e}^{-M_g} . \end{aligned} \tag{7.95}$$

The mean value $\langle \boldsymbol{u} \rangle$ in the Taylor series becomes zero and only the quadratic term in $\langle(\boldsymbol{u}\cdot\boldsymbol{g})^2\rangle$ has a nonzero value. It can be demonstrated that this can be written as an exponential *Debye–Waller factor* $\exp(-M_g)$ [7.83]. The Debye–Waller factor results formally in a reduction of the Fourier coefficient V_g to $V_g \exp(-M_g)$, and the reflection amplitudes of the kinematical theory will be decreased by the same factor; the diffraction intensities are thus attenuated by $\exp(-2M_g)$. An increase of ξ_g is also expected due to (7.23) and is observed experimentally [7.9].

The mean-square value $\langle u^2 \rangle$ of the lattice vibrations depends on the phonon spectrum of the crystal. The Debye model for a monoatomic cubic crystal gives

$$\langle u^2 \rangle = \frac{3h^2}{4\pi^2 M k \theta_\mathrm{D}} \left(\frac{1}{4} + \frac{T^2}{\theta_\mathrm{D}^2} \int_0^{\theta_\mathrm{D}/T} \frac{x\,\mathrm{d}x}{\mathrm{e}^x - 1} \right) , \tag{7.96}$$

where M is the atomic mass and θ_D the Debye temperature. The term in brackets is tabulated (*International Tables for X-Ray Crystallography*, Vol. II). The term 1/4 results from the zero-point vibrations, which are present even at $T = 0$, as the quantum-mechanical treatment of a harmonic oscillator shows. The quantity $\langle u^2 \rangle$ will also depend on the direction of $\boldsymbol{g}$ for noncubic crystals. It differs for different types of atoms in the unit cell.

The absorption parameters $q^{(j)}$ or imaginary parts V_g' of the lattice potential (Sect. 7.4.2) contain a large contribution from thermal diffuse scattering and depend strongly on temperature (see, for example, calculations [7.39, 7.84, 7.85] and experiments [7.43]).

Scattering between the Bragg spots cancels the destructive interference in ideal crystals. That part of the background caused by thermal diffuse scattering between the spots increases with increasing temperature, as does the background near strongly excited Bragg spots. The scattering is inversely proportional to the square of the phonon frequency $\nu(q)$. Diffuse streaks connecting Bragg spots [7.86, 7.87, 7.88, 7.89, 7.90, 7.91] are therefore mainly generated by transverse acoustic phonons of low frequency with wave vectors $\boldsymbol{k}$ perpendicular to one of the atomic-chain directions and polarization vectors parallel to it.

The influence of lattice vibrations is small at large scattering angles. An exception is the contrast of the excess Kikuchi bands in electron back-scattering patterns (EBSP, Sect. 8.1.4), which decreases with increasing temperature. Temperature-independent large-angle scattering is found as a result of summing the scattering processes at individual nuclei, which means that the scattering process is concentrated near a nucleus.

The background of electron diffraction patterns of polycrystalline films depends more strongly on temperature than the Debye–Waller factor predicts [7.92, 7.93]. The influence of thermal diffuse scattering on the background has been investigated for Al [7.94], Ag [7.95], and Au [7.96]. The intensities of the primary and reflected beams also depend more strongly on temperature than would be expected from the influence of the Debye–Waller factor (7.95) [7.97, 7.98, 7.99, 7.100]. These effects can be explained by the dependence of the absorption parameters $q^{(j)}$ on temperature.

7.5.4 Kikuchi Lines and Bands

The background between the Bragg-diffraction spots of a diffraction pattern contains a structure that can be characterized as excess and defect Kikuchi lines and bands (Fig. 7.26).

Excess and defect Kikuchi lines are formed by the following mechanism. Electrons scattered diffusely by thermal diffuse or inelastic scattering can be Bragg-reflected at lattice planes with reciprocal lattice vector $\boldsymbol{g}$ if the Bragg angle is $\pm\theta_{\mathrm{B}}$ or if the direction of incidence $\boldsymbol{k}'_0$ lies on one of the *Kossel cones*, which have an aperture of $90° - \theta_{\mathrm{B}}$ and the $\boldsymbol{g}$ direction (normal to the lattice planes) as cone axis (Fig. 7.27). The Bragg-reflected beam also lies on the opposite cone in the plane defined by $\boldsymbol{g}$ and $\boldsymbol{k}'_g$ and results in a bright excess Kikuchi line along the hyperbola in which the Kossel cone intersects the plane of observation (diffraction pattern). The lines are approximately straight owing to the low values of θ_{B}. On the other side, the Bragg reflection decreases the intensity of the incident direction $\boldsymbol{k}'_0$, and the intersection of the incident Kossel cone with the plane of observation results in a dark defect Kikuchi line. This mechanism therefore generates a set of corresponding excess and defect Kikuchi lines separated by an angular distance $2\theta_{\mathrm{B}}$. The system of Kossel cones behaves as though fixed to the crystal, which means that the Kikuchi lines move if the crystal is tilted, whereas the position of the Bragg-reflection spot is fixed at the plane of observation (angle $2\theta_{\mathrm{B}}$ with the primary beam), and the Bragg spots are visible only for a limited tilting range around the Bragg position of the primary beam. In the Bragg position (excitation error $s = 0$), the excess Kikuchi line coincides with the diffraction spot and the defect line with the primary beam. The displacement between the Bragg spot and the corresponding Kikuchi line can be used to measure the excitation error.

Excess Kikuchi bands are formed in each scattering process between the Bragg-reflection spots. As discussed in Sect. 7.4.1, the inelastically and

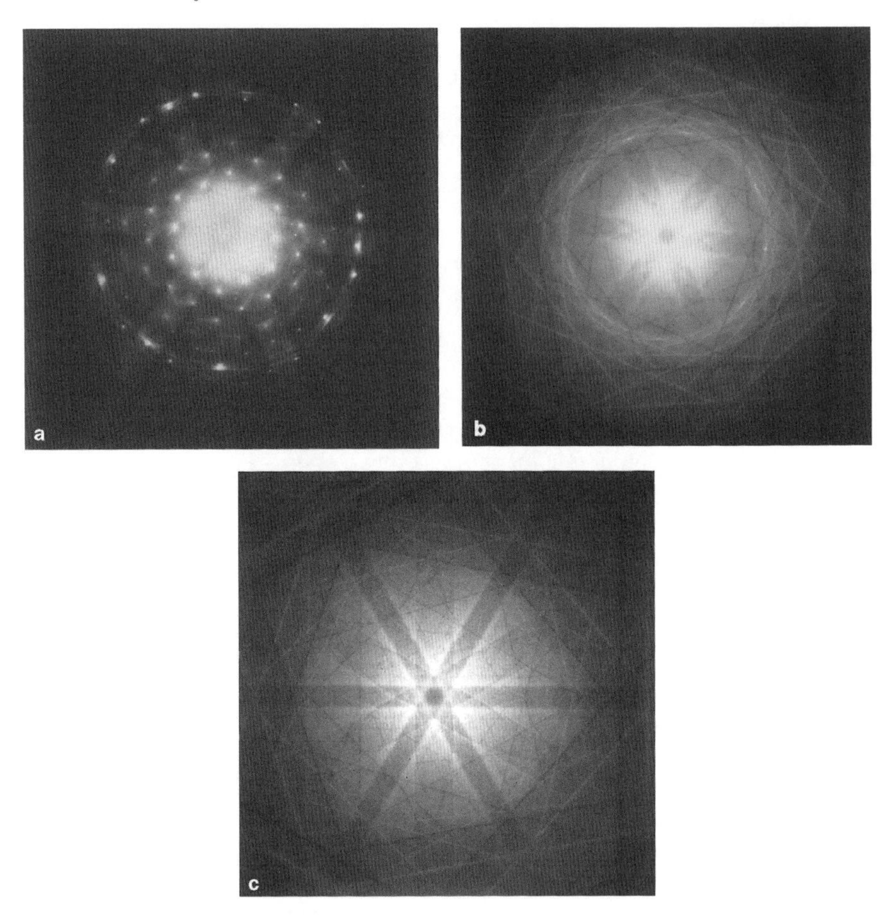

Fig. 7.26. Electron-diffraction patterns of Si foils at $E = 100$ keV with increasing thickness (**a**) $t = 80$ nm, (**b**) 800 nm, and (**c**) 1500 nm with the electron beam parallel to [111]. Pattern (**a**) shows diffraction spots of the zero- and first-order Laue zones; (**b**) shows defect and excess Kikuchi lines at medium angles and defect Kikuchi bands at low angles. In (**c**), the center shows only Kikuchi bands, and the region of excess and defect Kikuchi lines is shifted toward larger angles.

thermal diffusely scattered electrons are scattered again as Bloch waves, so that a theorem of reciprocity can be established. A primary Bloch-wave field has a larger scattering probability if there are antinodes at the nuclei. The scattered intensity becomes proportional to $\int \psi\psi^* \mathrm{d}z$ and depends on the tilt parameter k_x, which means that the whole intensity of large-angle scattering including backscattering varies, as shown in Fig. 7.20b, if the direction of incidence is changed. If the rocking beam forms a raster, the signal from a large-solid-angle detector designed to collect backscattered electrons or forward-scattered electrons with scattering angles $\theta \geq 5° - 10°$ generates

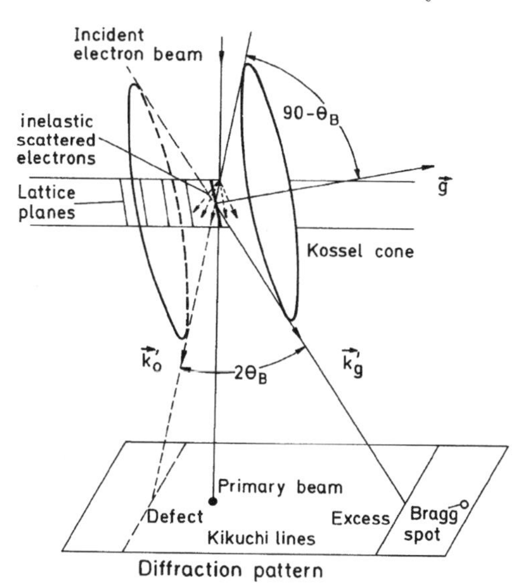

Fig. 7.27. Generation of excess and defect Kikuchi lines by Bragg scattering of diffusely scattered electrons regarded as the intersection of the Kossel cones of angular aperture $90° - \theta_B$ with the plane of observation.

an electron channeling pattern (ECP), well-known in scanning electron microscopy (SEM), with excess-band intensity distributions as in Fig. 7.20b.

When we observe a stationary electron-diffraction pattern, the direction of the primary beam is fixed. The scattering of electrons into larger angles depends on the Bloch-wave intensity at the nuclei, but the scattering probability will be large only when the Bloch-wave field of the scattered Bloch wave also shows antinodes at the nuclei. The scattered intensity depends on the observation angle and the Bloch-wave intensity that would appear if the scattered Bloch wave struck the crystal opposite the direction of observation. The whole angular distribution of scattered electrons is therefore not uniform but modulated by a system of excess Kikuchi bands. These can be seen at large scattering angles in diffraction patterns of thin single crystals and can also be observed as electron backscattering patterns (EBSP) recorded on a screen near the specimen in a scanning electron microscope (Sect. 8.1.4).

The excess Kikuchi bands can also show a contrast reversal that apparently converts them into defect bands (Fig. 7.26b,c) by the following mechanism [7.17, 7.101]. A set of excess and defect Kikuchi lines can be observed so long as more electrons hit the lattice planes from one side than from the other (Fig. 7.26b). With increasing foil thickness, this will be the case for large scattering angles, for which the scattered intensity distribution decreases more rapidly with increasing angle (Fig. 7.26c). For small scattering angles, however, the angular distribution is so diffuse and uniform that equal numbers of electrons hit the lattice planes from both Kossel cones, and cancellation of the pattern of excess and Kikuchi lines is to be expected. However, the resulting intensity distribution will be a defect Kikuchi band due to the influence of anomalous absorption. For a direction of observation near the Kossel cone, the intensity will be the sum of the intensities of the incident and Bragg-reflected beams; that is, $T+R$ for the two-beam case and ΣI_g for the n-beam case. The dependence of this sum on the tilt angle has already been discussed in Fig. 7.20a and results in a defect band. Only in kinematical theory is $T+R=$ const, which results in the cancellation of Kikuchi-line contrast without the formation of defect Kikuchi bands.

These are the basic mechanisms – in a somewhat simplified presentation – whereby excess and defect Kikuchi lines and bands are formed; their appearance in diffraction patterns depends on the distribution of the diffusely scattered electrons. With increasing foil thickness, the central region shows Bragg spots, pairs of excess and defect Kikuchi lines, and excess Kikuchi bands extending to large scattering angles. This central region expands with increasing thickness to larger scattering angles for thin films [7.102, 7.103, 7.104], and excess and defect Kikuchi lines are observed in only that angular region for which the scattered intensities decrease strongly and different intensities hit the two sides of the lattice planes. This is also the region in which the central excess bands show a contrast reversal, which converts them to defect Kikuchi bands, produced by the mechanism discussed above. More details about the attempts to describe this phenomenon quantitatively are to be found in [7.28, 7.101, 7.105, 7.106, 7.107, 7.108, 7.109, 7.110, 7.111]. The study of Kikuchi-line patterns from high-order Laue zones (HOLZ patterns) is described in Sect. 8.3.4.

7.5.5 Electron Spectroscopic Diffraction

An electron spectroscopic diffraction (ESD) pattern can be obtained with a diffraction image at the entrance plane of an imaging energy filter (Sect. 4.6.4). The advantages of energy filtering of diffraction patterns from amorphous and polycrystalline specimens have already been described in Sects. 7.5.1 and 7.5.2. At high energy losses, amorphous and crystalline specimens show a diffuse ring caused by Compton scattering (Sect. 5.2.3). The method of angle-resolved EELS allows us to image a two-dimensional map of diffraction intensity in an ΔE–θ plane by selecting a line across a diffraction pattern with a slit in the filter entrance plane (Figs. 5.12). The zero-loss filtering of CBED and LACBED patterns (Sects. 8.3 and 8.4) allows a better quantitative analysis to be made. Here, some typical effects and applications of ESD on single-crystal specimens are reported.

Zero-loss filtering can increase the contrast of weak reflections, such as superlattice reflections, by removing the inelastic background. This makes it easier to adjust dark-field images on axis. The streaks caused by electron-phonon scattering (Sect. 7.5.3) are increased in contrast. With increasing thickness, these streaks also appear in multiple plasmon losses as a result of elastic-inelastic multiple scattering; they become more diffuse because of the convolution with the angular distribution of plasmon losses. The background of plasmon and other low-energy losses near the Bragg spots can be

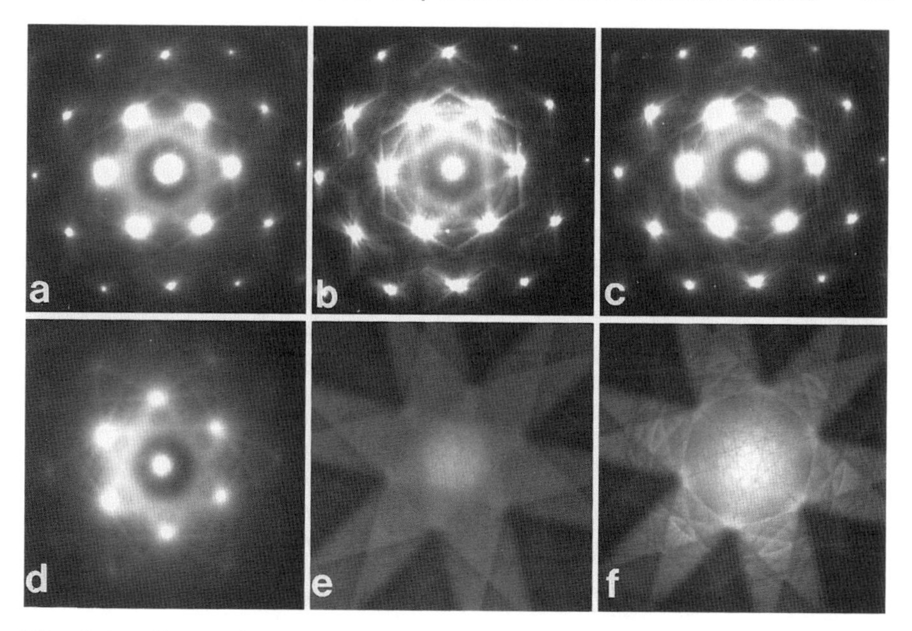

Fig. 7.28. Series of electron spectroscopic diffraction (ESD) patterns of a 111-oriented 50 nm Si foil: (**a**) unfiltered, (**b**) $\Delta E = 0$ eV, (**c**) 16 eV, (**d**) 100 eV, (**e**) 1800 eV, and (**f**) 2000 eV.

particularly harmful if the intensities and positions of the reflections are to be analyzed quantitatively or if weak reflections surround very strong reflections, as is the case in large-unit-cell crystals.

The various contributions of inelastically scattered electrons to the intensity of single-crystal diffraction patterns can be seen in a series of electron spectroscopic diffraction (ESD) patterns at increasing energy losses [4.112, 7.81, 7.112, 7.113]. Figure 7.28a shows the unfiltered diffraction pattern of a 50 nm 111-oriented Si foil. The zero-loss filtered pattern (Fig. 7.28b) sharpens the Bragg spots and enhances the excess Kikuchi lines through the 220 reflections and the thermal diffuse streaks. This shows that Kikuchi lines will also be produced by quasi-elastic thermal diffuse scattering. In the plasmon-loss filtered pattern (Fig. 7.28c) the Bragg spots become diffuse as a result of the convolution with the angular distribution of inelastic scattering. The increasing broadening with increasing energy loss (Fig. 7.28d) results in a strong decrease of high-order diffraction for $\Delta E = 100$ eV; the low-order diffraction spots are totally blurred at energy losses of a few hundred electron volts. The ESD patterns at 1800 eV below (Fig. 7.28e) and at 2000 eV beyond the K edge (Fig. 7.28f) of silicon show excess Kikuchi bands. These are generated by the same mechanism as described in Sect. 7.5.4 for large-angle elastic scattering. Also, these high-energy losses are concentrated near the nuclei. A large scattering intensity is observed only in directions where the Bloch wave of the inelastically scattered electron shows antinodes at the nuclei. Beyond the Si K edge, the ESD pattern (Fig. 7.28f) not only shows the expected jump in intensity compared with Fig. 7.28e but sharper Kikuchi lines and dark HOLZ lines are also seen in the central region. Below the edge, the background of the EELS contains a mixture of multiple high-energy losses, whereas just beyond the edge the largest fraction comes from a single ionization process. A further increase of foil thickness decreases the jump ratio of the K edge; both the background and the region beyond the edge are produced mainly by multiple scattering. Both patterns then show defect Kikuchi bands.

8

Electron-Diffraction Modes and Applications

Electron-diffraction methods are employed for the identification of substances by measuring the lattice-plane spacings and for the determination of crystal orientations in polycrystalline films (texture) or single-crystal foils. Extra spots and streaks, caused by antiphase structures or plate-like precipitates, for example, may also be observed when imaging a selected area.

The selected-area diffraction technique, in which an area of the order of 0.1–1 μm across is selected by a diaphragm in the first intermediate image, is a standard method. The introduction of additional scan and rocking coils into an instrument capable of producing an electron probe of the order of a few nanometers renders micro-area diffraction techniques feasible. In particular, convergent-beam electron diffraction (CBED) and the observation of high-order Laue zone (HOLZ) lines inside the primary-beam spot provide further information about the crystal structure. The charge-density distribution in unit cells can be obtained by a best fit of CBED intensities. Lattice defects can be analyzed by their influence on HOLZ lines. The overlap of convergent-beam Bragg spots can be avoided by the technique of large-angle CBED (LACBED).

8.1 Electron-Diffraction Modes

8.1.1 Selected-Area Electron Diffraction (SAED)

The cone of diffracted electrons with an aperture of the order of a few tens of mrad can pass through the small polepiece bores of the final lenses only if the back focal plane of the objective lens that contains the first diffraction pattern is focused on the screen. Figure 4.16 shows the ray diagram of this technique [8.1, 8.2, 8.3]. A selector diaphragm of diameter d situated in the intermediate image plane (magnification $M \simeq 20 - 50$) in front of the intermediate or diffraction lens selects an area of the specimen of diameter d/M. This area can be chosen in the normal bright-field mode (Fig. 4.16a), in which the primary beam passes through the objective diaphragm. When the excitation of the

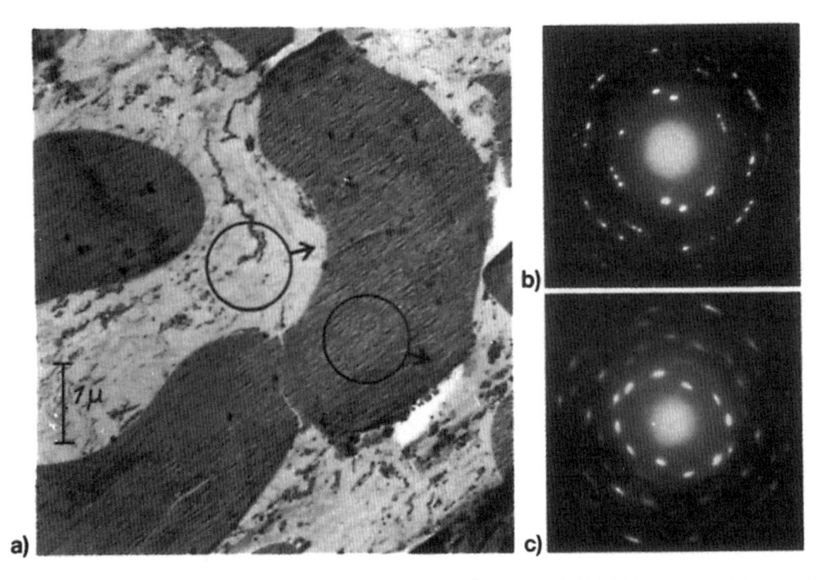

Fig. 8.1. Example of a selected-area electron diffraction (SAED) from a thin section of an Al-Cu eutectic cut with a microtome. (**b**) (Al matrix) and (**c**) ($AlCu_2$) contain the SAED pattern of the circles indicated in (**a**).

intermediate lens is decreased, its focal length is increased and the diffraction pattern in the focal plane of the objective lens can be focused on the final screen after removing the objective diaphragm (Fig. 4.16b). The excitations of the later projector lenses are unchanged. These lenses magnify either the intermediate image or the diffraction pattern behind the intermediate lens. Figure 8.1 shows an example of SAED from 1 μm diameter areas of Al and Al_2Cu in a section of an Al-Cu eutectic alloy cut with a diamond knife.

The diameter of the area selected cannot be decreased below 0.1–1 μm owing to the spherical aberration of the objective lens. The intermediate images of the Bragg reflections (dark-field images) are shifted relative to the bright-field image [8.4, 8.5] by a distance

$$\Delta s = (C_s \theta_g^3 - \Delta z \theta_g)M, \tag{8.1}$$

as can be seen from (3.63), which depends on the defocusing Δz and the constant C_s ($\theta_g = 2\theta_B$ and θ_B is the Bragg angle). It is of course possible to compensate for the shift by a suitable choice of the defocus Δz, but only for one Bragg reflection, not for the whole diffraction pattern simultaneously. The consequence is that Bragg reflections of high order with large θ_g do not come from the area that was selected in the bright-field mode. Thus, for $2\theta_B = 50$ mrad and $C_s = 1$ mm, the shift is 0.125 μm. The diffraction angle θ_g decreases linearly with λ as the electron energy is increased. A further selection error can result if the position of the intermediate image is shifted when the intermediate lens changes over from the imaging to the diffraction mode.

Diffraction patterns from smaller areas can only be obtained by using the rocking-beam technique (Sect. 8.1.2) or by producing a small electron probe (Sect. 8.1.3).

The resolution $d/\Delta d$ of an SAED pattern can be defined in terms of the smallest lattice-spacing difference Δd that can be resolved and may be estimated from the ratio $\Delta r/r$. Here r denotes the distance from a diffraction spot to the center of the diffraction pattern in the focal plane of the objective lens, $r = 2\theta_B f = \lambda f/d$ (f: focal length), and Δr is the diameter of the spot, which is equal to the diameter $2\alpha_i f$ of the primary beam (α_i: illumination aperture):

$$\frac{d}{\Delta d} = \frac{r}{\Delta r} = \frac{\lambda}{2\alpha_i d} . \tag{8.2}$$

Thus, for $\lambda = 3.7$ pm (100 keV), $d = 0.1$ nm, and $\alpha_i = 0.1$ mrad, we find $d/\Delta d = 200$. The resolution can be increased only by reducing α_i, but this reduces the pattern intensity.

The spherical aberration of the objective lens can cause barrel and spiral distortion of the SAED pattern [8.6, 8.7, 8.8] but this is, however, smaller than 1%; an elliptic distortion can arise due to astigmatism of the intermediate lens. The most severe distortion is caused by the projector lens. For the accurate determination of lattice spacings d, the diffraction (camera) length L (Sect. 8.2.1) must be calibrated by using a diffraction standard.

8.1.2 Electron Diffraction Using a Rocking Beam

A rocking beam with varying angle of incidence γ in the specimen plane can be generated by means of scanning coils situated between the specimen and the final condenser lens. The following two diffraction techniques can be employed.

In the first technique, described by Fujimoto et al. [8.9], the specimen area (0.2–4 μm) that contributes to the electron-diffraction pattern (EDP) is defined by a selector diaphragm at the first intermediate image, as in the SAED mode (Sect. 8.1.1). The lenses below the objective lens produce a magnified image of the first EDP, formed in the focal plane of the objective lens. When the incident beam is rocked, the primary beam and all of the diffraction spots shift, so that the diffraction pattern is scanned bodily across the final screen plane. A small fixed detector diaphragm selects the direction in the EDP that just coincides with the microscope axis. The detector signal measures the electron intensity and is then fed to the intensity modulation of a cathode-ray tube, which is scanned in synchrony with the rocking. The theorem of reciprocity tells us that the intensity of the on-axis beam is the same as in a stationary SAED technique [8.10].

The angular resolution of the recorded ECP can be varied by altering either the diameter of the detector diaphragm or the magnification camera length (L) of the EDP in the final screen plane, but it can never be smaller than the illumination aperture of the rocking beam.

This technique has the advantage that the EDP is recorded in the scanning mode and that the spot intensities can therefore be displayed directly, by the Y-modulation technique, for example [8.11, 8.12]. Furthermore, the selection error of SAED is avoided because all the beams recorded pass through the microscope on-axis. The diameter of the selected area is, however, limited by the diameter of the selector diaphragm, which cannot be smaller than 5 μm, owing to charging and contamination.

In a second technique, described by van Oostrum et al. [8.13], a highly magnified image ($M \geq 100\,000$) is formed on the final screen (detector plane). The diameter d of the detector diaphragm selects a small area of the image, which, back-projected into the specimen plane, can be much smaller than the selected area in SAED (e.g., 3–10 nm for $M = 100\,000$ and $d = 0.3$–1 mm as in [8.14, 8.15]).

The angular resolution is provided by the diaphragm in the focal plane of the objective lens, which again cannot be smaller than 5 μm. If the primary beam passes through this diaphragm, a bright-field image appears in the detector plane. The primary beam is intercepted by the diaphragm if the beam is tilted. A diffraction spot passes through the diaphragm if the tilt angle $\gamma = 2\theta_B$ and can then generate a dark-field image in the detector plane. This means that, during the rocking, a bright-field image is seen followed sequentially by a series of dark-field images that correspond to different diffraction spots. The dark-field images are not shifted toward the bright-field image as they are with the SAED technique because all the beams recorded are on-axis.

In conclusion, we see that the Fujimoto technique can give a better angular resolution but the area selected is limited by the diameter of the selector diaphragm, whereas the van Oostrum technique can select smaller areas but the angular resolution is limited by the size of the objective diaphragm.

8.1.3 Electron Diffraction Using a Stationary Electron Probe

The selection error of SAED can also be avoided by using small electron probes. It was shown in Sect. 4.2.3 that electron-probe diameters of 2–5 nm can be produced with a thermionic electron gun, whereas diameters less than 1 nm require a Schottky or field-emission gun. The smallest possible probe diameter will be obtained with a large probe aperture α_p of the order of about 10 mrad. Decreasing this aperture increases the probe diameter. It is therefore impossible to obtain EDPs with sharp diffraction spots from areas as small as 2–10 nm because apertures less than 1 mrad would be needed [8.16, 8.17]. However, the convergent-beam diffraction patterns described below are in many points even more informative than conventional EDPs.

Figures 8.2a–c and 8.3a–c show how the EDP changes, from a spot pattern to a convergent-beam and a Kossel pattern, as the electron-probe aperture α_p is increased. A spot pattern as obtained by SAED requires $\alpha_p \ll \theta_B$ (Figs. 8.2a and 8.3a), so that the primary beam and the Bragg-reflection spots are sharp. The system of Kikuchi lines and bands will not be affected by the illumination

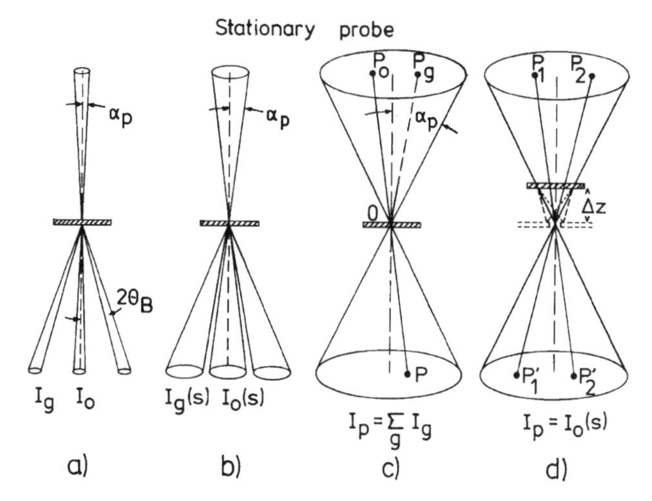

Fig. 8.2. Electron-diffraction techniques that use a small electron probe. (**a**) A Bragg-diffraction spot pattern as obtained by SAED, (**b**) a convergent-beam electron-diffraction (CBED) pattern, (**c**) a Kossel pattern in which the intensity at a point P is proportional to ΣI_g, and (**d**) large-angle convergent-beam electron diffraction (LACBED) obtained by raising the specimen a distance Δz and selecting the primary spot by a virtual diaphragm conjugate to a selected-area diaphragm.

aperture because they are normally generated by the cone of electrons diffusely scattered inside the specimen, and the necessary angular divergence can also be aided by any initial divergence (convergence) of the incident beam.

Increasing the aperture ($\alpha_p < \theta_B$, Figs. 8.2b and 8.3b) increases the Bragg-spot diameter. The primary beam and the diffraction spots are extended to circular discs with sharp edges if the cone of the illumination aperture is sharply limited by a condenser diaphragm inside which the current density is uniform. However, each point in the discs corresponds to one distinct direction of incidence in the illumination cone. The intensity within the discs of the primary beam and the diffraction spots varies owing to the variation of the excitation errors, and the intensity distribution inside the discs corresponds to a two-dimensional index rocking curve of the dynamical theory of electron diffraction. This convergent-beam electron-diffraction (CBED) technique was introduced by Kossel and Möllenstedt [8.19, 8.20] and has become a routine method for electron diffraction of small areas with an electron probe. The information obtainable from a CBED pattern will be discussed in Sect. 8.3.

An undisturbed CBED pattern can be expected only if there is no strong lattice distortion or crystal bending inside the irradiated area. For this reason, CBED patterns can normally be recorded only with a small electron probe. A limitation can be the rapid growth of a contamination needle on the irradiated area, which can be reduced either by specimen heating [8.21] or specimen cooling [8.22]; see also the discussion of contamination in Sect. 11.4.2.

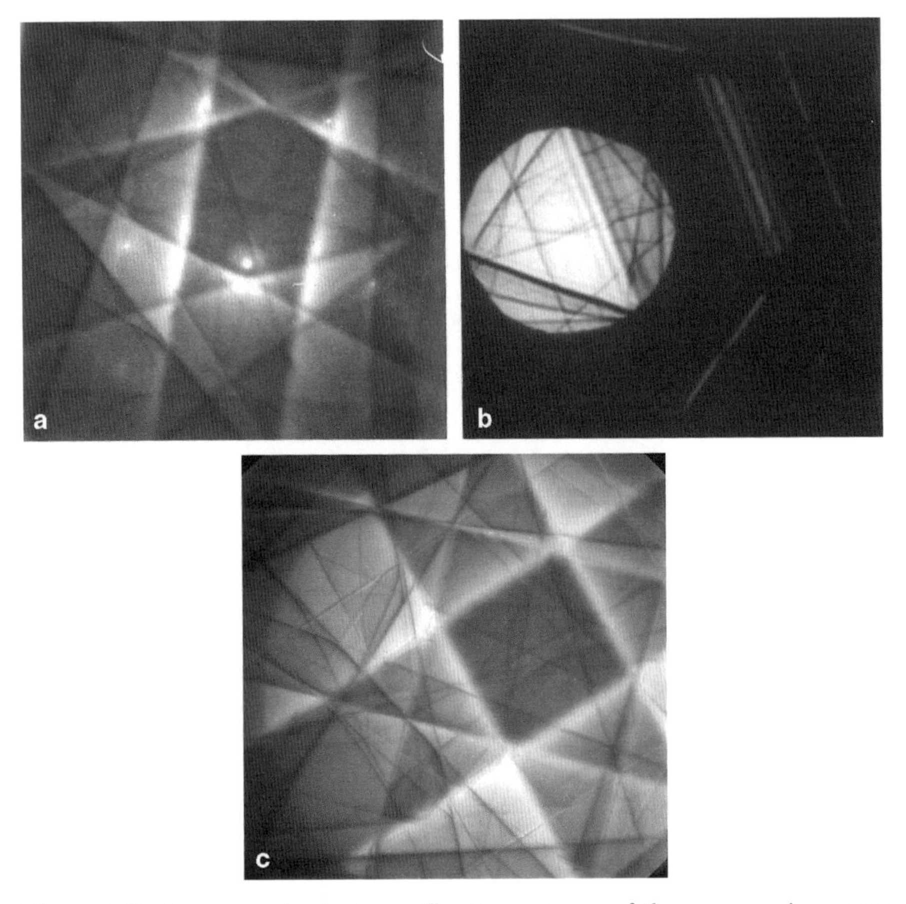

Fig. 8.3. Stationary-probe electron-diffraction patterns of the same specimen area of a 150 nm thick Cu foil with the three illumination conditions of Figs. 8.2a–c. (**a**) $\alpha_\mathrm{p} \ll \theta_\mathrm{B}$, primary beam (*left*) and 220 Bragg-diffracted spot (*right*). (**b**) $\alpha_\mathrm{p} < \theta_\mathrm{B}$, CBED pattern with the spot circles of the primary beam (*left*), showing the anomalous absorption, and that of the Bragg reflection (*right*), showing the symmetric pendellösung fringes. (**c**) $\alpha_\mathrm{p} > \theta_\mathrm{B}$, Kossel pattern with an intensity distribution proportional to ΣI_g [8.18].

When a highly coherent field-emission source is used, interference fringes can be observed where neighboring convergent discs [8.23, 8.24, 8.25, 8.26] overlap. The phase difference between diffracted waves can be read from the relative positions of the fringes. Energy filtering increases the contrast.

A further increase of probe aperture, $\alpha_\mathrm{p} > \theta_\mathrm{B}$ (Fig. 8.2c), increases the overlap of the extended diffraction spots [8.27, 8.28]. The intensity at a point P of the Kossel pattern does not consist only of the contribution from the primary-beam direction P_0. In addition, the directions P_g contribute with the corresponding Bragg-diffraction intensities. The intensity $I_\mathrm{P} = \Sigma_g I_g$, including $\boldsymbol{g} = 0$, is a multibeam rocking curve. According to Fig. 7.20a, there is cancellation of the pendellösung fringes, but an intensity distribution in the form of defect Kossel bands persists (Figs. 8.3c and 8.6a), caused by the dependence of anomalous transmission on the excitation error, see also (7.83). Residual pendellösung-fringe contrast in Fig. 8.3c can result from buckling of the Cu foil inside the irradiated area.

The following nomenclature is proposed [8.29] to distinguish between Kikuchi and Kossel bands. In EDPs with Kikuchi bands, the angular divergence is caused by scattering in directions between the diffraction spots, whereas Kossel bands are produced by the convergence of the external probe aperture. Kossel patterns without diffraction spots can be obtained for all film thicknesses, whereas Kikuchi bands appear only in thicker specimens.

In the standard CBED method, the maximum angle of the central circular disc corresponds to the value of $2\theta_B$ for the nearest low-indexed reflection so as to prevent any overlap of the discs. For large unit cells especially, this is a major handicap. To get two-dimensional rocking curves over a large angular range without overlap of the discs of Bragg reflections, the technique of large-angle convergent-beam diffraction (LACBED) introduced by Tanaka [8.30] can be employed (Fig. 8.2d). When a stationary electron probe is focused on the specimen plane with a large convergence angle, an image of this spot can be seen in the normal imaging mode. When the specimen is raised by a distance Δz, additional spots are generated by Bragg reflection and form a small spot-diffraction pattern, not as is usually the case in the focal plane of the objective lens but now in the specimen plane. One of these spots can be selected in the image mode by a diaphragm, which is best placed in the conjugate intermediate image by adjusting a 5 μm selector diaphragm normally used for SAED. On switching the intermediate (diffraction) lens to the diffraction mode, the large-angle convergent-beam (LACBED) pattern appears with the angular width of the primary cone. Figure 8.4a shows an energy-filtered LACBED pattern obtained from Si along the $\langle 331 \rangle$ zone axis and Fig. 8.4b a computer simulation of the pendellösung fringes of the FOLZ reflections and the location of the HOLZ lines matched to the experimental pattern in Fig. 8.4a for an acceleration voltage 121.7 keV.

If a small selector diaphragm and a large Δz are used, the diaphragm selects only small scattering angles (down to $\simeq 0.1$ mrad) around the selected spot. A much smaller fraction of the elastic and inelastic diffuse background then contributes to the LACBED. This can be seen as a kind of energy filtering [8.31], though real zero-loss filtering generates a further significant increase of contrast.

This geometry has the consequence that a larger area of 100–1000 nm diameter contributes to the LACBED. This can be an advantage for radiation-sensitive specimens because the current density is orders of magnitude lower than for CBED with the electron probe focused on the specimen. Otherwise, the pattern is an overlap of real and reciprocal space information. Each point in the LACBED pattern corresponds simultaneously to a distinct specimen

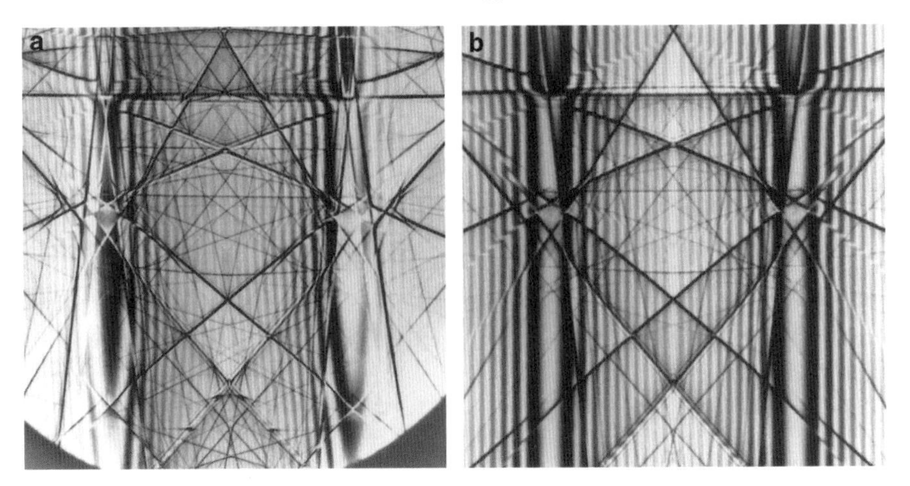

Fig. 8.4. (**a**) Energy-filtered LACBED pattern from Si along the ⟨331⟩ zone axis. (**b**) Computer simulation matched to the experimental pattern (**a**) for an acceleration voltage of 121.7 keV.

point and distinct excitation error s_g. Applications of this technique will be discussed in Sect. 8.3.

These results show that information about the crystal structure does not necessarily require the use of small electron-probe apertures α_p for a diffraction pattern with Bragg spots; small electron probes with large apertures are more suitable for CBED, LACBED, and Kossel pattern studies. A conventional spot pattern does not always provide the fullest information about crystal structure and symmetry as shown in Sect. 8.3.

8.1.4 Electron Diffraction Using a Rocking Electron Probe

Electron-probe diffraction patterns can be displayed on a CRT by post-specimen deflection of the diffraction pattern across a detector diaphragm or by rocking the electron probe by means of scan coils placed in front of the specimen (Fig. 8.5a). However, it is not easy to avoid shifting the electron probe during rocking. The spherical aberration of the probe-forming lens causes an unavoidable shift, and the electron probe moves along a caustic figure. This shift can be partially compensated for by adding a contribution to the deflection-coil current proportional to θ^3 [8.32], and the shift can be kept below 0.1 μm in TEM. The rocking causes the EDP to move across the detector plane, like the rocking beam in the Fujimoto technique (Sect. 8.1.2). If a small probe aperture $\alpha_p \ll \theta_B$ is used, a spot pattern will be recorded by a small detector aperture ($\alpha_d \ll \theta_B$) as the theorem of reciprocity shows; with $\alpha_d < \theta_B$, a convergent-beam pattern is obtained, and with $\alpha_d > \theta_B$, a Kossel pattern. However, to avoid the shift caused by spherical aberration, it is better to use a stationary probe and postspecimen deflection. The advantage of

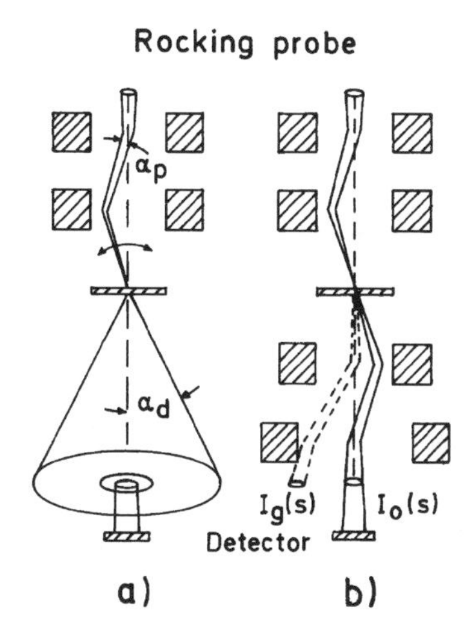

Fig. 8.5. Electron-diffraction techniques that use a rocking electron probe. (**a**) A rocking probe of small aperture produces a spot, CBED, or Kossel pattern, depending on the detector aperture α_d. (**b**) Double-rocking technique.

rocking the electron probe is that types of information not available from a conventional EDP become accessible. Figure 8.6a shows a stationary electron-probe Kossel pattern of the 111 pole of a Si foil. Using a rocking probe and a large detector aperture, a similar pattern can be recorded with the cone of electrons scattered through the detector aperture (Fig. 8.6b). These diagrams contain defects due to anomalous electron transmission. Suppose now that the annular detector is placed below the specimen [4.68]. Electrons scattered through large angles, normally not used in TEM, will be collected, and a contrast reversal is observed (Fig. 8.6c). The diagram now contains excess Kikuchi bands generated by direct scattering of electrons out of the Bloch-wave field into larger angles. The pattern is the same as that recorded with backscattered electrons (BSE) (Fig. 8.6d) known as an electron channeling pattern (ECP) in SEM. The only difference is that the noise is larger in the BSE pattern (Fig. 8.6d) than in the pattern recorded with (Fig. 8.6c) because many fewer electrons are backscattered than forward scattered through large angles; we recall that the Rutherford cross section varies with angle as $\mathrm{cosec}^4(\theta/2)$.

If two semiconductor detectors are used, one annular in shape, the other occupying the central region, the first will record a Kossel pattern and the second a spot pattern [8.33, 8.34]. The two signals can be added or subtracted to obtain a Kossel pattern on the CRT with superposed bright or dark diffraction spots. This facilitates accurate determination of orientation because the position of the Kossel bands relative to the spots and the position of the central beam in the diagram can be established with high accuracy.

Another interesting variant is the double-rocking method [8.35, 8.36, 8.37] (Fig. 8.5b). A second postspecimen scan-coil system is arranged in such a way that the primary beam falls on the detector at all rocking angles. A convergent-beam pattern can be obtained for the primary beam over a very much larger angular range undisturbed by the overlap of other Bragg reflections. In a

Fig. 8.6. Electron diffraction patterns near the 111 pole of an Si foil. (**a**) Stationary-probe Kossel pattern recorded with the transmitted electrons. (**b**) Rocking-probe diffraction pattern with the transmitted electrons, (**c**) with the transmitted electrons scattered through large angles ($\theta > 10°$), and (**d**) with the backscattered electrons ($\theta > 90°$) [8.18].

similar fashion, a convergent-beam pattern can be obtained from a Bragg reflection if the detector diaphragm is shifted and only the diffracted intensity I_g recorded (see also zone-axis pattern in Sect. 8.1.5e). This technique is the rocking variant of LACBED.

8.1.5 Further Diffraction Modes in TEM

(a) Small-Angle Electron Diffraction Small-angle x-ray diffraction is successfully used for the investigation of periodicities and particles of the order of 10–100 nm. This method can also be used with electrons [8.38]. Diffraction at spacings of $d = 100$ nm requires a primary beam with an illumination aperture

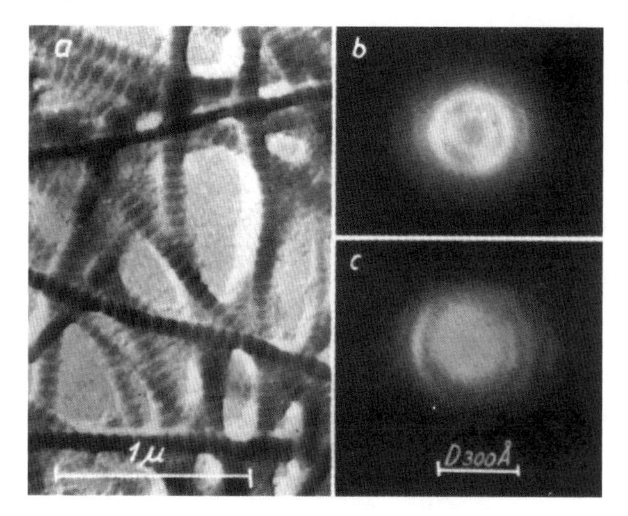

Fig. 8.7. Small-angle electron-diffraction patterns (**b** and **c**) of a shadow-cast collagen specimen (**a**) showing the periodicities of the collagen fibers [8.38].

smaller than the diffraction angle: $\alpha_i < \lambda/d = 0.03$ mrad for $E = 100$ keV. With a double-condenser system, $\alpha_i = 0.01$ mrad can be achieved (Sect. 4.2.1) and the neighborhood of the primary beam can be magnified by the projector lens (thereby increasing the effective camera length to several meters). An extensive account of small-angle electron diffraction has been given in [8.39].

A typical application of this method is the study of evaporated films with isolated crystals, showing diffuse rings with diameters inversely proportional to the mean value of the crystal separation [8.40]. Further applications involve periodicities in collagen (Fig. 8.7), conglomerates of latex spheres and virus particles [8.41, 8.42], catalase [8.43], and high polymers [8.44]. Organic specimens have to be coated with a metal conductive layer to prevent charging, which would perturb the primary beam. Periodicities can also be resolved in a micrograph, and the same information is obtainable by laser (Fraunhofer) diffraction on the developed film or plate (Sect. 6.4.7) or by Fourier transformation of digitized images. Small-angle diffraction can also be used in Lorentz microscopy (Sect. 6.8.2a) because the primary beam is split by the Lorentz force inside the magnetic domains.

(b) Scanning Electron Diffraction For the quantitative interpretation of EDPs, it can be useful to record the intensities directly by scanning the diffraction pattern across a detector diaphragm (Grigson mode). The detector may be a Faraday cage, a semiconductor, or a scintillator–photomultiplier combination [7.79, 8.45]. This method also can be used for energy filtering of an EDP [7.80] by means of a retarding-field filter [7.79, 8.47] or a magnetic prism spectrometer and is of special interest for ultrahigh-vacuum experiments [8.48]. The EDP can also be recorded digitally by means of a CCD

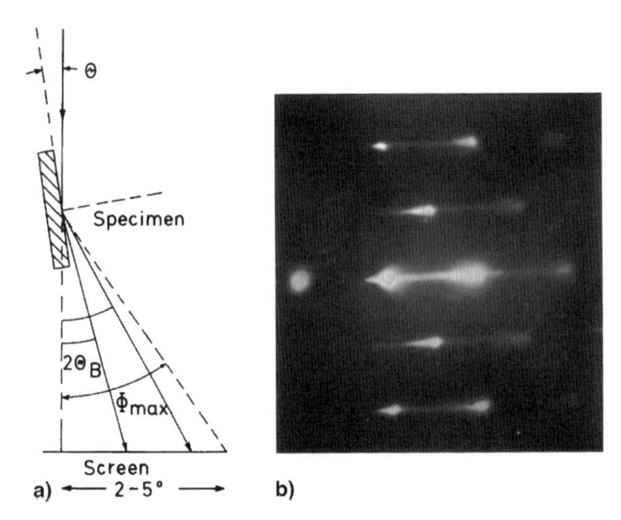

Fig. 8.8. (**a**) Schematic ray diagram for recording a high-energy reflection of an NaF film evaporated on a (100) NaF cleavage plane of 270°C (H. Raether).

camera or an image plate; the beam-rocking methods described in Sect. 8.1.2 can likewise be employed [8.49].

(c) Reflection High-Energy Electron Diffraction (RHEED) Bulk material can be investigated by allowing the electron beam to fall obliquely on the specimen at a glancing angle $\theta < \theta_B$ (Fig. 8.8). The electrons penetrate only a few atomic layers into the material. Because the interaction volume is so thin, the reciprocal-lattice points will be drawn out to needles normal to the surface, and the Bragg spots will be elongated. Plane surfaces also show Kikuchi lines and bands (Fig. 8.9). The influence of refraction, which shifts the Bragg spots to smaller Bragg angles, has to be considered. The method is as sensitive to surface layers as low-energy electron diffraction (LEED). The surface can be cleaned by heating or by electron and/or ion bombardment [8.50]. Charging effects can cause problems with insulating materials. They can be avoided by irradiating the specimen with 200–1000 eV electrons from a separate source, which increases the secondary-electron production and avoids the buildup of a large negative charge, or by heating (increasing electrical conductivity).

Imaging of the surface by selected Bragg-diffraction spots is used in reflection electron microscopy (REM, Sect. 9.7). Micro-area diffraction can be achieved by employing a nanometer probe in a STEM [8.51], and this technique can also be used at low electron energies (0.5–20 keV) [8.52].

(d) Electron-Backscattering Pattern and Electron-Channeling Pattern Increasing the angle θ between beam and specimen (Fig. 8.8a) to 5°–30° and increasing $\Phi_{\max}$ to 20°–30° yields electron-backscattering patterns (EBSP) [8.53] that contain excess Kikuchi bands (Fig. 8.9). Increasing θ still further causes contrast reversal in defect bands for small take-off angles

Fig. 8.9. Electron-backscattering pattern (EBSP) with excess Kikuchi bands of a germanium single crystal recorded with 100 keV electrons striking the crystal with an angle of 10° (see Fig. 8.8a).

[8.54, 8.55]. This is analogous to the extension of defect bands in the central area of a transmission EDP with increasing thickness (Sect. 7.5.4, Fig. 7.26).

Electron-backscattering patterns can be obtained in TEM by positioning the specimen about 10 cm above the final screen and photographic plate and deflecting the electron beam on the specimen by means of coils situated below the projector lens [8.55].

The formation of an electron-channeling pattern (ECP) [8.56, 8.57] by backscattered or forward-scattered electrons (Fig. 10.66c,d) is related to that of EBSP by the theorem of reciprocity [8.55, 8.58].

(e) Zone-Axis Pattern (ZAP) In convergent-beam electron diffraction (CBED) (Sect. 8.3), each point corresponds to a particular direction of electron incidence inside the convergent electron probe. If a crystal foil is bent two-dimensionally by distortion and the electron beam hits the foil nearly parallel to a low-indexed zone axis, then each point of the foil in a bright-field image corresponds to another direction of incidence of the parallel electron beam relative to the lattice planes (Fig. 9.3). If the two main radii of curvature of the foil have the same sign and equal order of magnitude (dome- or cup-shaped), the bend contours (Sect. 9.1.1), which form a zone-axis pattern (ZAP), exhibit the same intensity distribution as a CBED pattern. There is also a resemblance to the double-rocking technique (Sect. 8.1.3 and Fig. 8.5b) because the tilt of the foil can be larger than the Bragg angles, resulting in overlap of the reflection circles in CBED but not in the ZAP.

Unlike CBED, which produces a two-dimensional rocking curve in Fourier space (diffraction plane), the ZAP is a real-space phenomenon. Whereas a pattern is formed by an electron probe of a few nanometers in diameter, a ZAP extends over a few micrometers. Zone axis patterns contain a lot of information about the crystal symmetry [8.59, 8.60, 8.61] and can be used to determine the space group [see also discussion of the high-order Laue-zone patterns (HOLZ) in Sect. 8.3.4]. If the incident electron energy is varied, the intensity distribution near the zone axis changes and gives information about Bloch-wave channeling and the critical voltage [8.61, 8.62, 8.63].

8.2 Some Uses of Diffraction Patterns with Bragg Reflections

8.2.1 Lattice-Plane Spacings

Calculations of the lattice-plane spacing d_{hkl} from the Bragg condition (7.11) require a knowledge of the electron wavelength λ (3.1) and the Bragg angle θ_B. Since θ_B is small, the sine that occurs in the Bragg condition can be replaced by the tangent in a first-order approximation,

$$\frac{\lambda}{d} = 2\sin\theta_B \simeq \tan(2\theta_B) = \frac{r}{L}, \qquad d \simeq \frac{\lambda L}{r}, \tag{8.3}$$

where r is the distance of the diffraction spot from the primary beam or the radius of a Debye–Scherrer ring, and L is the diffraction (or camera) length. For higher accuracy, a further term of the series expansion can be included,

$$d = \frac{\lambda L}{r}\left[1 + \frac{3}{8}\left(\frac{r}{L}\right)^2 + \ldots\right]. \tag{8.4}$$

However, this formula is valid only if the diffraction pattern is magnified without any barrel or pin cushion distortion. The pin cushion distortion of the projector lens (Sects. 2.3.4 and 8.1.1) contributes a further term in r^2 so that a value larger than 3/8 may be found for the constant in (8.4) during calibration [8.8].

The diffraction (or camera) length L cannot be measured directly. The product λL must therefore be determined by calibration with a substance of known lattice constant. Only substances with the following properties should be used for calibration:

1. many sharp rings with known d_{hkl};
2. chemically stable and no change of lattice parameters under electron irradiation;
3. correspondence with x-ray lattice constant;
4. easy preparation.

Table 8.1. Calibration standards for electron diffraction.

Standard	Lattice type	Lattice constant (nm)	Ref.
LiF	NaCl	$a = 0.4020$	[8.69]
TlCl	CsCl	$a = 0.3841$	[8.64, 8.70]
MgO	NaCl	$a = 0.4202$	[8.70]
ZnO	Wurtzite	$a = 0.3243$	[8.71]
		$c = 0.5194$	
Au	fcc	$a = 0.40783$	
Si	Diamond	$a = 0.54307$	[8.68]

Table 8.1 contains some suitable calibration substances and their lattice parameters. The value for TlCl has been compared with the x-ray data in a high-precision experiment ($\Delta d/d = \pm 3 \times 10^{-5}$) [8.64].

When single-crystal standards are used, the reciprocal-lattice points have a needle-like shape parallel to the surface normal (Figs. 7.9 and 7.10a). If the crystal normal is inclined to the electron beam, the intersections of these needles with the Ewald sphere can alter the positions of Bragg-diffraction spots [8.65].

Tilt coils for the electron beam can be used for the calibration of diffraction patterns because the deflection angle θ is proportional to the coil current. The diffraction pattern is shifted by $L\theta$ [8.66, 8.67]. Double exposures with known and previously calibrated θ allow the distortion of the EDP to be determined. Alternatively, the lattice spacing can be measured directly in the microscope with an accuracy of 0.1% by recording the x and y coil currents needed to bring the diffracted beam on-axis [8.68].

8.2.2 Texture Diagrams

For many polycrystalline specimens, the distribution of crystal orientations is not random; instead, one lattice plane may lie preferentially parallel to the specimen plane. In this plane, however, the crystals are rotated randomly around a common axis F = $[mno]$, the fiber axis of the fiber texture.

The existence of this fiber axis means that the reciprocal-lattice points are distributed around concentric circles centered on the fiber axis (Fig. 8.10a). (They would lie on a sphere for a totally random distribution.)

If the electron beam is parallel to the fiber axis, the Ewald sphere intersects the circles and a Debye–Scherrer ring pattern is observed that does not show all possible hkl but only those that fulfill the condition

$$(hkl)[mno] = hm + kn + lo = 0. \tag{8.5}$$

The limitation of the number of observable rings can mimic extinction rules $|F|^2 = 0$. As a consequence, wrong conclusions may be drawn about the crystal structure. In a weak fiber texture, the fiber axis will be no more

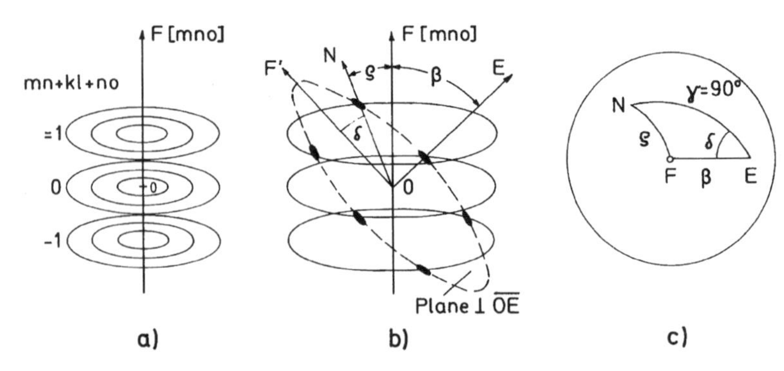

Fig. 8.10. (**a**) Distribution of the $\boldsymbol{g}$ vectors of the crystallites in reciprocal space for foils with a fiber texture with $[mno]$ as fiber axis. (**b**) Intersection of the Ewald sphere (– – –) with these rings resulting in sickle-shaped segments of the Debye–Scherrer rings. (**c**) Relations between the angles (E: electron-beam direction, N: intersection of Ewald sphere and reciprocal lattice, F: fiber axis).

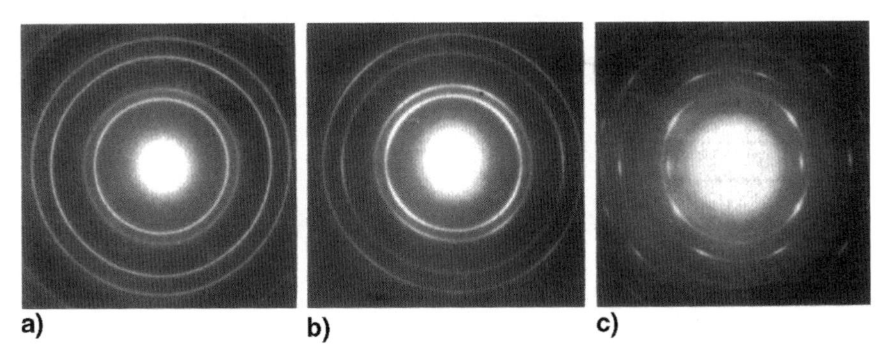

Fig. 8.11. (**a**) Electron-diffraction pattern at normal incidence of an evaporated Au film with a weak (111) fiber texture and (**b**) the same specimen tilted 45° to the electron beam showing a weak dependence of the ring intensities on the azimuth. (**c**) Strong fiber texture of an evaporated Zn film tilted at 45° to the electron beam.

than a preferential direction; all possible rings now appear but with the wrong intensity ratios. The diffraction pattern of an evaporated gold film (Fig. 8.11a), for example, shows a 220 ring that is more intense than would be expected from a random distribution.

A fiber texture can be recognized clearly if the direction of electron incidence E is tilted through an angle β relative to the fiber axis F; in practice, the specimen normal is tilted relative to the electron beam in a goniometer stage. The intersections of the Ewald sphere with the concentric circles then become sickle-shaped (Fig. 8.10b), as can be seen for the weak fiber texture of a gold film in Fig. 8.11b and for the stronger texture of a zinc film in Fig. 8.11c. The sickles become narrower for a larger tilt angle $\beta = 45° - 60°$. The texture should be most clearly detectable with $\beta = 90°$, but this is possible only in a RHEED experiment (Sect. 8.1.5).

Quantitative results about the fiber axis can be obtained by measuring the azimuths δ of the centers of the sickle-shaped ring segments [8.72, 3.17]. The intersection of the plane that contains the directions E and F with the Ewald sphere defines a direction F′ in the diffraction pattern that gives the origin of δ (Fig. 8.10b). The angle ρ between (hkl) and $[mno]$ can be calculated by evaluating their scalar product. The hkl are obtained from the ring diameter. For the spherical triangle FEN in Fig. 8.10c, with $\gamma = 90°$, we have

$$\cos\rho = \cos\beta\cos\gamma + \sin\beta\sin\gamma\cos\delta = \sin\beta\cos\delta. \tag{8.6}$$

For known values of the tilt angle β and azimuth δ, the value of cos ρ can be calculated and compared with theoretical values for different possible fiber axes. A procedure for noncubic crystals is described in [8.74, 8.75]. Measurement of the azimuthal distribution of the ring intensities can be used for the quantitative characterization of a texture [8.45].

8.2.3 Crystal Structure

The method described for measuring lattice-plane spacings d_{hkl} can be used for direct comparison of different substances. If the chemical composition is known or has been established by x-ray microanalysis or EELS, the crystal structure can be identified by comparing the spacings with tabulated x-ray values of the d_{hkl} A.S.T.M. Index or using the search/match programs developed for electron diffraction listed in [8.46]. However, a problem is that the d_{hkl} cannot be measured with an accuracy better than about 1% and the results may fail to coincide with the tabulated data. Only in the case of a standard (evaporated film or small particle, Table 8.1) at the same position in the specimen plane can accuracies of 0.1% be achieved. In single-crystal EDPs, the symmetry of the spot diagram, the extinction rules, and the angles between the diffraction spots can be used for a further identification of the structure.

The complete determination of crystal structure by Fourier synthesis using electron-diffraction patterns is superior to x-ray analysis if the material under investigation exists in only small quantities or produces diffuse x-ray diffraction because the particles are small. In x-ray crystal-powder diffraction patterns, the Debye–Scherrer rings already begin to broaden for crystals smaller than 100 nm. This broadening can be detected in EDPs only when the crystals are smaller than 5 nm.

However, Fourier synthesis requires exact values of the reflection intensities, and the following difficulties are encountered when using conventional electron-diffraction patterns with Bragg spots:

1. The transition from the kinematical to the dynamical theory is significant for thicknesses or dimensions as small as 5 nm.
2. Forbidden and weak reflections can be excited by multiple diffraction.
3. In small particles or small irradiated areas, only a few reflections, for which the reciprocal-lattice points are near the Ewald sphere, are excited.

4. Foils of large area are often bent and exhibit Bragg-spot intensities that depend randomly on the fraction of the irradiated area that contributes to the Bragg reflection.

Methods have been described for obtaining quantitative values of the structure factors $|F(\theta)|^2$ from polycrystalline ring and texture patterns for subsequent Fourier synthesis of the atomic positions in the unit cell [3.17, 8.74, 8.75]. These methods assume that the kinematical theory is completely valid and that there is no influence on the intensity of high orders of a systematic row (see, e.g., Fig. 7.24). In consequence, the results have to be interpreted with care [8.76]. A correction using the Blackman formula (7.94) has been applied to Ni_3C, for example [8.77]; see also [8.78].

The problems in determining crystal structure from electron-diffraction patterns with Bragg spots is discussed in [8.79, 8.80], and a method is described for measuring the integrated intensity of Bragg spots. Each diffraction spot is expanded to a square by a pair of deflection coils, which allows the integrated intensity to be determined more easily by photometry. It has also been suggested that the specimen should be oscillated around an axis in the specimen plane to average over the reflection range of a reciprocal-lattice point [8.81]. Vincent and Midgley have proposed to rock the beam conically around the optic axis with a double-rocking technique [8.82]. Recent progress is discussed in [8.83].

Thin crystal lamellae are often parallel to the specimen plane. The Bragg spots around the primary beam are an "image" of the zero-order Laue zone (HOLZ) of the reciprocal lattice. If the Bragg spots in high-order Laue zones (HOLZ) are regarded as a two-dimensional net, this net can be continued to the origin of the reciprocal lattice. This results in an overlap like that seen in Fig. 7.7. The diameters of the high-order Laue zones (Sect. 8.3.4) can be used to extract information about the lattice-plane spacings normal to this plane. A goniometer that provides tilt angles up to $\pm 45°$ and adjustment of the tilt axis with an accuracy of 0.1–1 μm permits us to explore the three-dimensional structure of the reciprocal lattice [8.84, 8.85, 8.86]. It will be useful to rotate the crystal around the specimen normal, so that a systematic row is parallel to the tilt axis. In most cases, the unit cell can be reconstructed with only a few tilts, for which the electron beam is again parallel to a zone axis. If necessary, a further tilt around another row can be performed [8.87, 8.88, 8.89].

For complete structure analysis, not only should a goniometer be used but also convergent-beam diffraction (Sect. 8.3) to get information about the point and space group and to measure the lattice-potential coefficients V_g and the crystal thickness from the spacing of the pendellösung fringes (Sect. 8.3). An electron micrograph contains information about the crystal size and lattice defects that can be included in the analysis of EDP. For thin crystals ($t \leq 5$ nm), it may be possible to resolve the projected lattice structure directly (Sect. 9.6).

X-ray diffraction is influenced by the electron-density distribution of the atomic shell and electron diffraction by the screened Coulomb potential of the nuclei. The scattering amplitudes of single atoms are proportional to Z for x-rays (5.33) and to $Z^{2/3}$ for electrons (5.34). The difference between light and heavy atoms is smaller for electrons than for x-rays. Light atoms can be localized better in the presence of heavy atoms [8.90, 8.91].

The different scattering mechanisms for x-rays and electrons can be observed with KCl, for example. Both elements have the electron configuration of argon, and the scattering amplitudes for x-rays are equal; the structure amplitude $4(f_\mathrm{K} - f_\mathrm{Cl})$ therefore vanishes for all hkl odd (Sect. 7.2.2). In electron diffraction, the nuclear charges are different, so that for odd values of hkl, the amplitude is weak but not zero, whereas the amplitude for hkl even, $4(f_\mathrm{K} + f_\mathrm{Cl})$, is strong [8.92].

8.2.4 Crystal Orientation

A knowledge of the exact orientation of crystals is important for investigating lattice defects and for establishing the relative orientations of different phases of the matrix and precipitates.

Many diffraction spots are observed if the foil is very thin, so that the reflection range of the reciprocal-lattice points is enlarged, or if the foil is bent. Both effects limit the accuracy of orientation determination. For this, the diffraction spots R_1, R_2, ... at the distances r_n from the central beam O are used to calculate d_n by taking the product λL from a calibration pattern (Sect. 8.2.1). When the lattice structure and lattice constant are known, the corresponding reciprocal-lattice points $\boldsymbol{g}_n = (h_n, k_n, l_n)$ can be indexed. For cubic crystals, the ratio method can be applied by using

$$\frac{r_1^2}{r_2^2} = \frac{h_1^2 + k_1^2 + l_1^2}{h_2^2 + k_2^2 + l_2^2} . \tag{8.7}$$

From a table of values of this ratio for all possible combinations of $|\boldsymbol{g}_1|$ and $|\boldsymbol{g}_2|$, the indices of both reflections can be identified without knowing λL. To be sure that spots have been indexed with the correct signs, agreement between the observed angle α_{12} between OR_1 and OR_2 and the theoretical value,

$$\cos\alpha_{12} = \frac{\boldsymbol{g}_1 \cdot \boldsymbol{g}_2}{|\boldsymbol{g}_1||\boldsymbol{g}_2|} = \frac{h_1h_2 + k_1k_2 + l_1l_2}{(h_1^2 + k_1^2 + l_1^2)^{1/2}(h_2^2 + k_2^2 + l_2^2)^{1/2}} , \tag{8.8}$$

must be checked. There may be small differences between measured and calculated α_{12} [8.93]. The direction of the electron incidence (normal to the foil) is parallel to the vector product of two reciprocal-lattice vectors

$$\boldsymbol{n} \parallel \boldsymbol{g}_1 \times \boldsymbol{g}_2 = (k_1l_2 - k_2l_1, l_1h_2 - l_2h_1, h_1k_2 - h_2k_1). \tag{8.9}$$

The needle-like extension of the reciprocal-lattice points (Sect. 7.2.2) widens the tilt range, which can be $\pm 10°$ for low-order reflections [8.94]. The orientation therefore becomes more accurate if a large number of reflections is used, especially high-order reflections at the periphery of the EDP. An accuracy of $\pm 3°$ can then be achieved [8.95, 8.96]. Further improvements can be made by considering the intensities of the diffraction spots [8.94].

The orientation determination becomes unique and more accurate if three reflections $\boldsymbol{g}_1, \boldsymbol{g}_2, \boldsymbol{g}_3$ are employed [8.97, 8.98]. A circle is drawn through the three spots to establish the correct numbering of the sequence. If the central beam O is inside the circle, the reflections are numbered counterclockwise; otherwise they are numbered clockwise. With this convention, the determinant

$$\boldsymbol{g}_1 \cdot (\boldsymbol{g}_2 \times \boldsymbol{g}_3) = \frac{1}{V} \begin{vmatrix} h_1 & k_1 & l_1 \\ h_2 & k_2 & l_2 \\ h_3 & k_3 & l_3 \end{vmatrix} \tag{8.10}$$

should be positive. If it is not, the signs of the hkl must be reversed so that for combinations of two $\boldsymbol{g}_i$, (8.8) is also obeyed. The direction of the normal antiparallel to the electron beam is then given by the mean value

$$\boldsymbol{n} \parallel |\boldsymbol{g}_1|^2(\boldsymbol{g}_2 \times \boldsymbol{g}_3) + |\boldsymbol{g}_2|^2(\boldsymbol{g}_3 \times \boldsymbol{g}_1) + |\boldsymbol{g}_3|^2(\boldsymbol{g}_1 \times \boldsymbol{g}_2). \tag{8.11}$$

However, the orientation is not unique if all of the reflections happen to belong to a single zone with odd symmetry [8.93, 8.99, 8.100]. After rotation of the crystal through 180°, the same diffraction pattern is obtained. The orientation can be established uniquely from a second EDP obtained after tilting the specimen. A goniometer should also be used if the EDP does not contain two or three diffraction spots convenient for calculation. With these precautions, an accuracy of $\pm 0.1°$ can be attained.

A high accuracy is also obtained if the Kikuchi lines are used [8.93, 8.98, 8.101, 8.102]. A reflection is in the exact Bragg condition if the excess Kikuchi line goes through the diffraction spot. If the distance from the line to the reflection g_n is a_n, the tilt out of the Bragg position or the excitation error s_g is

$$\Delta\theta = \frac{\lambda}{d}\frac{a_n}{r_n}, \qquad s_g = g\Delta\theta = \frac{\lambda}{d^2}\frac{a_n}{r_n}. \tag{8.12}$$

Equation (8.11) implies exact Bragg positions. If the three terms are multiplied by $a_n = (r_n + 2a_n)/r_n$, an accuracy of $\pm 0.1°$ is obtained [8.98]. For determination of the relative orientation of two crystals from the Kikuchi pattern, see [8.103].

The relative orientation between a matrix and coherent or partially coherent precipitates can be determined by means of a transfer matrix that relates the coordinate systems of the two phases [8.104, 8.105, 8.106].

The orientation can be checked by using specimen details that are visible in the micrograph, provided that precipitates, stacking faults, or dislocations are recognizable in different planes and show traces with measurable relative angles. The traces of structures in octahedral planes can be used, for example, to analyze the accuracy and uniqueness of the orientation determination [8.107, 8.108].

In older microscopes, the diffraction pattern of a specimen area is rotated relative to the image of the same area because of image rotation by magnetic lenses. This rotation angle can be calibrated by using MoO_3 crystal lamellae. These are prepared as smoke particles by heating an Mo wire in air with a high a.c. current in an open high-vacuum evaporator and collecting the vapor on a glass slide. After floating the crystal film on water and collecting it on a formvar-coated grid, crystal lamellae parallel to the specimen plane are obtained. The long edge of these crystals is parallel to [100]. The rotation angle can be read from a double exposure of an image and the SAED pattern [8.109].

If the orientation of the diffraction pattern relative to the specimen is of interest, the angle of rotation between specimen and micrograph also has to be measured [8.110]. Kikuchi lines can again be used for this purpose [8.111]. The system of Kossel cones is fixed to the crystal so that the Kikuchi lines move if the specimen is tilted about an axis of known orientation. This direction can be compared with the direction of shift of the Kikuchi lines.

8.2.5 Examples of Extra Spots and Streaks

Electron-diffraction patterns contain not only the Bragg-diffraction spots that are expected from the structure of the unit cell of a perfect crystal but also additional spots and streaks. Not every effect can be discussed here because they vary from one specimen to another. A few typical examples will be described that should provide some guidance in the discussion of particular diffraction patterns.

Forbidden Reflections In dynamical theory, the intensity of a beam diffracted at hkl lattice planes can be equal in magnitude to the primary beam. A second Bragg reflection at lattice planes $h'k'l'$ can therefore produce reflections with indices $h-h', k-k', l-l'$. Spots that are forbidden by the extinction rules for the structure amplitude F (Sect. 7.2.2) may therefore be seen: an $00l$ spot (l odd) in FeS_2 [8.112], a 222 spot in Ge by double excitation at the allowed $(\bar{1}11)$ and (311) lattice planes [8.113, 8.114], or a 00.1 spot in hexagonal cobalt coming from $(h0.1)$ and $(\bar{h}0.0)$ [8.81]. This explanation of forbidden reflections is restricted to particular crystal orientations with relatively low values of s_g (accidental interaction). In other situations, more Bragg reflections can be excited simultaneously in dynamical theory, resulting in more intense forbidden-reflection spots (systematic interaction).

Twins and Precipitates Twinning results from a mirror reflection of the crystal structure about special lattice planes. For example, face-centered cubic crystals show a twin formation with a mirror reflection about the $\{111\}$ planes, and Ag films evaporated on [100] cleavage planes of NaCl or Ni films electrolytically deposited on copper show an epitaxy with frequent twin lamellae. The reciprocal lattice of these twin lamellae can be obtained from the reciprocal lattice of the matrix by mirror reflection about the $\{111\}$ planes, and additional reciprocal lattice points occur on one-third of the neighboring points in the $\langle 111 \rangle$ directions. Extra spots are seen (Fig. 9.4), which are strongest if the foil is tilted through about 16° out of the [100] orientation [8.115]. Precipitates with a fixed orientation to the matrix cause similar effects. In order to identify the origin of extra spots, the specimen must be imaged in the dark-field mode, selecting these extra spots only; the parts of the image that contribute to the spot will then appear bright (Sect. 9.1.2).

Stacking Faults and Planar Precipitates The finite extension of a crystal plate results in a needle-shaped extension of the reciprocal-lattice points normal to the plate (Sect. 7.2.2). If the electron beam is parallel to this normal, the diffraction pattern is not changed and reflections will appear over a larger tilt range. However, small shifts in the position of diffraction spots can result from the intersection of the Ewald sphere with the needles (Fig. 7.9). The needles at the reciprocal-lattice points create diffuse streaks in the diffraction pattern if the angle between the normal to the crystal plates and the electron beam is near 90°. The streaks can extend from one Bragg spot to another and are normally sharper than streaks caused by electron–phonon scattering (Sect. 7.5.3). The existence of streaks indicates that the specimen contains planar faults such as high density, precipitate lamellae, or Guinier–Preston zones, for example. Figure 8.12 shows an example of {111} stacking faults in Fe_4N particles (fcc, $a = 0.378$ nm) extracted from an Fe-0.1wt.%N alloy heat-treated to 370°C. In Fig. 8.12a, the {111} planes are inclined more parallel to the foil, and in Fig. 8.12b their normals are inclined at an angle of nearly 90° to the electron beam [8.116].

Ordered Alloys with a Superlattice Structure The most important effects will be discussed for the example of Cu-Au alloys. The alloy $AuCu_3$ is not ordered at temperatures $T \geq 399$°C, where it consists of a solid solution with a random distribution of Au and Cu atoms at the sites of an fcc lattice.

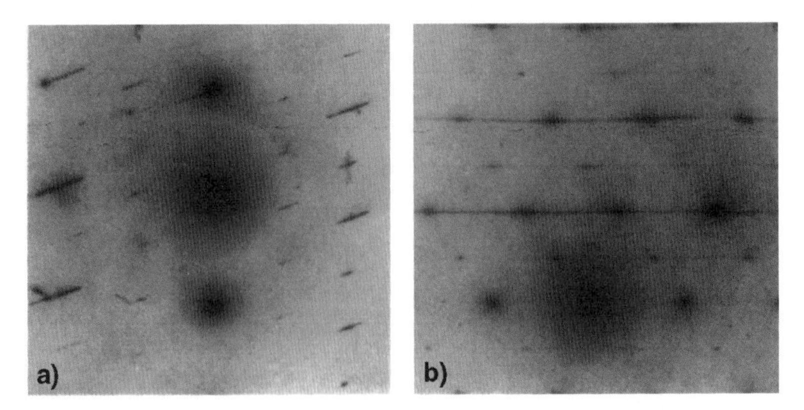

Fig. 8.12. Selected-area electron-diffraction patterns of plate-like Fe_4N precipitates with stacking faults in the {111} planes in two different orientations: (**a**) {111} oblique to the electron beam and (**b**) one of the planes parallel to the electron beam [8.116].

The diffraction pattern contains the reflections allowed by the extinction rules for this lattice type (Table 7.1). Below the transition temperature, the alloy acquires an ordered structure with the Au atoms at the corners and Cu atoms at the face centers of the cubic unit cell. In this situation, $F = f_{\rm Au} + 3f_{\rm Cu}$ for all even and all odd hkl and $F = f_{\rm Au} - f_{\rm Cu}$ for mixed hkl; for the latter, $F = 0$ in the nonordered structure. Additional Bragg spots therefore appear below 388°C, and this can be used for calibration of the specimen temperature, for example (Sect. 11.1.1).

In an Au-Cu alloy (50:50 wt%), the transition from the random phase to the ordered CuAu II phase, in which alternate (002) planes consist wholly of Cu and wholly of Au atoms, occurs at 410°C. Every 2 nm, the Cu and Au atoms change places, and the consequence is a domain structure with antiphase boundaries (Fig. 9.17). The resulting unit cell is elongated with a spacing of 4 nm, and the Bragg-diffraction spots are split in the $\langle 100 \rangle$ directions by this increased spacing (Fig. 8.13) [8.117]. The antiphase structure can be imaged in the dark-field mode in which one Bragg-reflected beam and the surrounding superlattice reflections contribute to the image intensity. Similar superlattice reflections can be observed in the EDPs of other alloys (see, for example, [8.118]).

Below 380°C, the phase CuAu I is stable; this consists of a face-centered tetragonal lattice ($c/a = 0.92$), again with alternating (002) planes of Au and Cu atoms but without the closely spaced antiphase boundaries of the CuAu II phase. These boundaries are planar faults and can be identified by their fringe pattern in electron micrographs (Sect. 9.3.3).

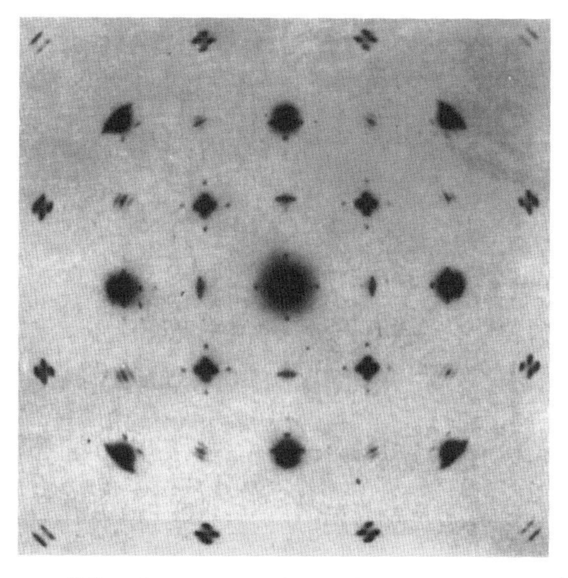

Fig. 8.13. Electron diffraction pattern of an ordered CuAu II film [8.117].

8.3 Convergent-Beam Electron Diffraction (CBED)

8.3.1 Determination of Point and Space Groups

Symmetries in CBED patterns allow the point and space groups to be determined in defect-free regions not buckled inside the contributing area. Thirty-one different diffraction groups can be distinguished by using the ten possible two-dimensional point groups for the determination of the point symmetries of the whole pattern, the dark Holz-line pattern inside the primary-beam CBED disc, or mirror lines in Bragg reflections and in $\pm \boldsymbol{g}$. Tables are published that list these symmetries for the different diffraction groups, relate the 31 diffraction groups to the 32 crystal point groups, and give the expected diffraction symmetries at any particular zone axis of each of the 32 crystal point groups [8.119, 8.120, 8.121, 8.122, 8.123].

In x-ray crystallography, the crystal space group can be obtained from forbidden reflections whenever the structure factor F (Sect. 7.2.2) vanishes for a crystal structure containing screw axes or glide planes. In electron diffraction, forbidden reflections can often appear quite strongly due to multibeam dynamical theory. However, in well-aligned zone-axis CBED, dynamic extinction conditions exist that appear as dark bars or crosses (Gjønnes–Moodie lines) [8.124] and can be used to determinate the space group [8.123, 8.125].

8.3.2 Determination of Foil Thickness

The generation of CBED patterns has been described in Sect. 8.1.3. The intensity distribution inside the circles of a convergent-beam diffraction pattern (Fig. 8.3b) is none other than a stationary, two-dimensional rocking curve (Sect. 7.3.4, Figs. 7.18 and 7.20), which contains information about the local thickness t, and the extinction and absorption distances ξ_g and ξ'_g, which are reciprocal to the lattice potentials V_g and V'_g, respectively. A necessary condition for CBED is that the circles do not overlap, which restricts the range of rocking. A larger rocking angle can be employed without overlap by the double-rocking technique (Sect. 8.1.4), by studying ZAP patterns (Sect. 8.1.5), or by large-angle CBED (LACBED, Sect. 8.3.6).

The thickness can be obtained from the positions of the subsidiary minima of the *pendellösung* fringes. The two-beam rocking curve (7.61) has minima if the argument of the sine term is an integral multiple n of π. The corresponding excitation errors s_n are then given by

$$\text{(a)}\ s_n^2 = \frac{n^2}{t^2} - \frac{1}{\xi_g^2} \quad \text{or} \quad \text{(b)}\ \left(\frac{s_n}{n}\right)^2 = \frac{1}{t^2} - \frac{1}{n^2 \xi_g^2} \,. \qquad (8.13)$$

Plotting s_n^2 against n^2 [8.126] gives us t from the slope and ξ_g from the intercept with the ordinate; alternatively, plotting $(s_n/n)^2$ against $1/n^2$ gives t from the intercept with the ordinate [8.127]. In both cases, the correct starting number n_1 of the first minimum has to be known. For a foil thickness between $m\xi_g$ and $(m+1)\xi_g$, the appropriate value is $n_1 = m+1$. If a wrong value of n_1 is used, the plot does not give a straight line. Errors arising from the two-beam case can be decreased by exciting reflections equal to or higher than 200, 220, and 311 for Al, Cu, and Au, respectively [8.128].

However, the two-beam case can never be perfectly realized, and many-beam calculations are necessary. Using rough values of t and ξ_g given by the two-beam method, many-beam fits must be computed in which t, ξ_g, and ξ'_g are allowed to vary (see also the next section).

The CBED pattern also contains a diffuse background with Kikuchi bands. This background has to be subtracted before seeking a best fit with many-beam calculations. A photometric record of the background very near the selected trace through the CBED pattern can be obtained by placing a thin wire across the circular diaphragm in the condenser lens, which casts a shadow across the CBED circles and allows the diffuse background inside the shadow to be measured [8.129]. The background intensity can be decreased by zero-loss filtering of electron spectroscopic diffraction patterns (Sect. 7.5.5).

8.3.3 Charge-Density Distributions

X-ray crystallography allows us not only to determine the position of atoms in the unit cell but also to obtain charge-density maps. Familiar examples are NaCl and the diamond structure. In NaCl the resulting charge density around the nuclei has an approximate spherical symmetry because the Na^+ and Cl^- ions have the closed shells of a noble gas. In a diamond structure, the tetrahedral bonds, regarded as overlaps of the sp^3 hybrid wave functions, result in bridges in the charge density. In x-ray diffraction, the scattering amplitude f_x (5.33) is influenced only by the charge distribution ρ_e of the jellium and tends to Z when $\theta \to 0$; for electrons, on the other hand, the structure factor may even show an enhanced sensitivity to bonding effects for low-order Bragg reflections because the structure amplitude (Fourier coefficient V_g) is proportional to $Z - f_x$ (5.30). Small changes in the scattering amplitudes f_x have a large effect on $Z - f_x$. Many difficulties with the x-ray method, such as crystal defects and dispersion correction, can be avoided with electrons, and CBED allows defect-free small areas or even nanometer-sized crystals to be investigated [8.130].

Measurements of V_g have been based on three-beam dynamical theory in the analysis of three-phase structure invariants [8.131], the critical-voltage method (Sect. 7.4.4), and the analysis of degeneracies in centrosymmetric [8.132] and noncentrosymmetric crystals [8.133, 8.134]. For accounts of structure-factor determination, see [1.81, 8.135].

Modern methods record CBED patterns by CCD arrays and apply energy filtering [7.82] to reduce the inelastic background. Line scans across the pendellösung fringes of a CBED pattern are fitted by varying the thickness and the V_g and V'_g for a large number of reflections (about 30) in the first- and high-order Laue zones. For noncentrosymmetric crystals, both the real and the imaginary (absorptive) parts of the crystal potential are complex, which requires the introduction of additional phase factors ϕ_g:

$$V(\boldsymbol{r}) = \sum_g \left[|V_g| \mathrm{e}^{\mathrm{i}\phi_g} + \mathrm{i}\, |V_g'| \mathrm{e}^{\mathrm{i}\phi_g'} \right] \exp(-2\pi \mathrm{i} \boldsymbol{g} \cdot \boldsymbol{r}). \tag{8.14}$$

The parameter space for the best fit can contain 12–16 nonseparable parameters, which can be divided into geometrical and physical parameters. The first are the starting and end points of the line scan, the incident beam direction, and the radius of the CBED discs, for example, and the last are the wanted thickness and Fourier coefficients. Different iterative procedures [8.136, 8.137, 8.138, 8.139] have been developed to find the minimum of

$$\chi^2 = \sum_{i=1}^{n} \frac{w_i (c f_i^{\mathrm{exp}} - f_i^{\mathrm{theo}})^2}{\sigma_i^2}, \tag{8.15}$$

where f_i^{exp} and f_i^{theo} are the experimental and theoretically calculated values of the intensities at the points i within the line scan. Examples of results have been published for GaAs [8.130], MgO [8.137], and BeO [8.140], for example.

8.3.4 High-Order Laue Zone (HOLZ) Patterns

A high-order Laue zone (HOLZ) diffraction pattern is obtained when the electron beam is incident on the specimen parallel to a low-index zone axis. The Ewald sphere intersects the needles of the zero-zone reciprocal-lattice points, producing the convergent-beam circles of the Bragg reflections around the primary beam; at larger angular distances, $\sin\theta_n = \lambda R_n$, the next higher Laue zones of order n are intersected in the reciprocal lattice with radii R_n (Figs. 7.6b and 8.14). These radii can be evaluated from Fig. 7.6b using the result given by elementary geometry,

$$R_n^2 = g_n \left(\frac{2}{\lambda} - g_n \right), \qquad g_n = \frac{n}{a\sqrt{m^2 + n^2 + o^2}}, \tag{8.16}$$

where $2/\lambda$ is the diameter of the Ewald sphere and g_n denotes the distance between the nth Laue zone and the zero-order Laue zone for the zone axis [mno]. Quantitative measurement of R_n gives information about the third dimension of the reciprocal lattice, especially for materials with layer structures such as $ZrSe_2$, $TaSe_2$, NbS_2, TaS_2, and MoS_2 [8.141, 8.142, 8.143]. For different crystal structures, the HOLZ rings appear with different relative intensities (Fig. 8.14). Laue zones may disappear completely if the structure amplitude F is zero for them, but these forbidden reflections can reappear with weak intensity as a result of many-beam systematic interactions. Higher-order Laue zone reflections correspond to low values of V_g and hence to large values of ξ_g, and their intensities seem to be directly related to the structure factor $|F|^2$ for these beams, even for thicker specimens, so that first-order perturbation theory is usually applicable [8.144]. However, for crystal thicknesses greater than a few tens of nanometers, dynamical interaction effects can occur, and these

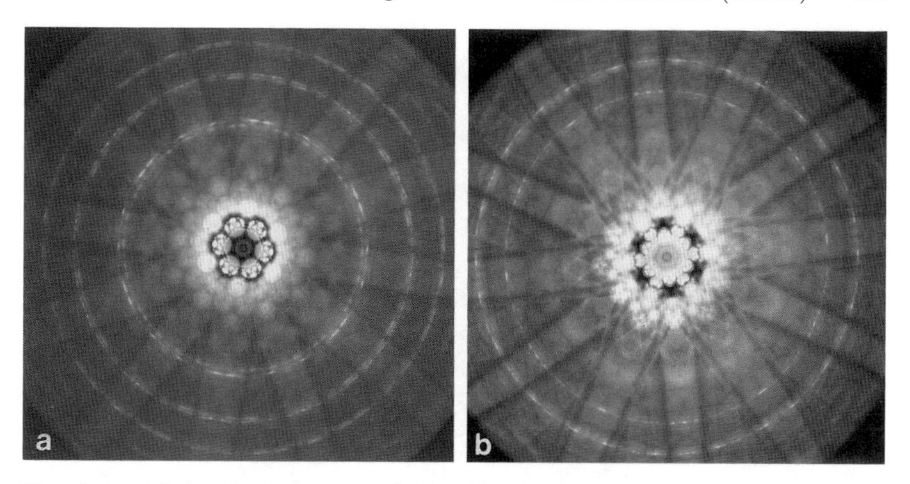

Fig. 8.14. High-order Laue zone (HOLZ) diffraction patterns of 2 H polytypes of (**a**) MoS_2 and (**b**) $MoSe_2$ with structure amplitudes $(f_{Mo} - 1.4f_C)$, f_{Mo}, and $(f_{Mo} + 1.4f_C)$ (C: chalcogen) for the first- to third-order Laue zones, respectively, showing that the first-order Laue zone (FOLZ) of $MoSe_2$ is practically invisible owing to the very small contribution from $(f_{Mo} - 1.4f_{Se})$[8.145].

furnish information about the crystal potential and the dispersion surface [8.145, 8.146, 8.147, 8.148]. The high-order reflections decrease more strongly with increasing temperature due to the Debye–Waller factor (Sect. 7.5.3), so that specimen cooling increases the intensity of HOLZ rings and HOLZ lines in the CBED of the primary beam (see below).

8.3.5 HOLZ Lines

In a CBED pattern, the HOLZ reflections become bright HOLZ lines, and each bright line in the outer HOLZ rings appears inside the primary-beam circle as a dark line because there is a relationship between excess and defect Kikuchi lines and the HOLZ lines. Figure 8.15b shows the central (000) disc with the six surrounding {220} discs for a 111-oriented Si foil. The central disc is filled with crossing dark (defect) HOLZ lines and pendellösung fringes; these are not straight as they would be for two-beam excitation but concentric, due to the interaction with the six strongly excited 220 reflections.

Whereas Kikuchi lines are hard to observe for thin specimens and become clearer as the thickness is increased, HOLZ lines are narrower ($\Delta\theta \leq 0.1$ mrad) and are most readily visible at thicknesses for which the Kikuchi lines are still weak. The angular width of a HOLZ line can be estimated from the first zero of the sine term in the two-beam approximation (7.61) to be

$$\Delta\theta = 2/g\xi_g \quad \text{for} \quad t \gg \xi_g \quad \text{and} \quad \Delta\theta = 2/gt \quad \text{for} \quad t \ll \xi_g . \tag{8.17}$$

We have the first case for zero-order (HOLZ) reflections with $\xi_g \simeq 20-50$ nm and the second case for high-order reflections with $\xi_g \geq 1000$ nm.

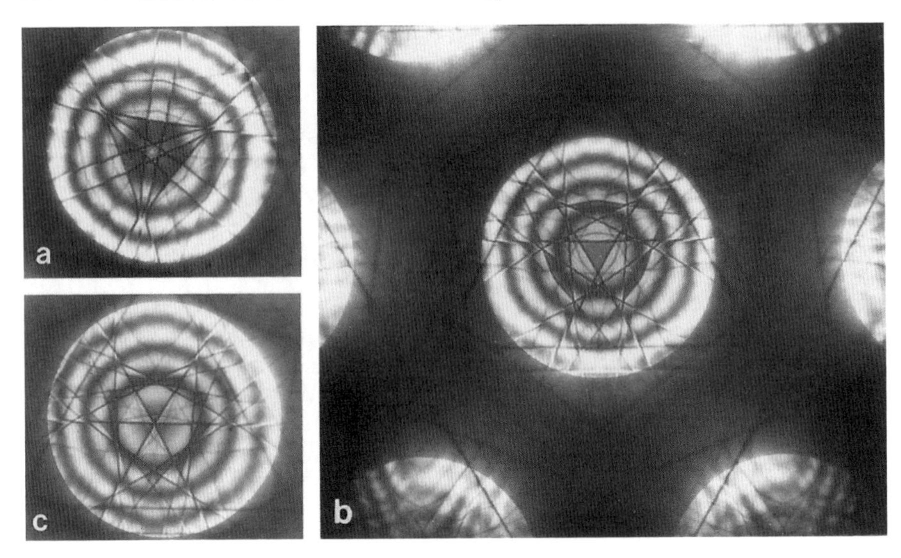

Fig. 8.15. HOLZ lines in the primary circle of a CBED pattern [six neighboring 220 reflections are visible in (**b**)] from (111) Si at three different electron energies: (**a**) $E = 96.5$ keV, (**b**) 100.5 keV, (**c**) 103.5 keV. The bright rings are pendellösung fringes that depend on film thickness and result from the interaction with the neighbored 220 reflections of the zero-order Laue zone [8.149].

Thicker specimens should therefore be used to decrease the detrimental effect of thin-film relaxations. Because of the Debye–Waller factor $\exp(-2M_g) = \exp(-4\pi^2\langle u^2\rangle g^2)$ (7.95), cooling of the specimen, which is essential to suppress contamination (Sect. 11.4.2), increases the contrast of HOLZ lines with high g; these decrease more strongly at room temperature due to g^2 in the exponent.

The pattern of overlapping HOLZ lines (Fig. 8.15) in the (000) CBED disc depends very sensitively on the electron energy E (accelerating voltage U) [8.149] and/or the lattice constant a. This is shown in Fig. 8.15a–c for small increments in electron energy. The HOLZ lines from the third Laue zone move faster than those from the second zone when the acceleration voltage U is varied. Lines from the same side of the HOLZ ring move in the same direction, and those from the diametrically opposite side move in the opposite direction.

The lattice dimension can be evaluated with high precision once the beam voltage has been calibrated with a lattice of known dimensions, such as Si; alternatively, a relative change Δa of the lattice constant due to a change of composition or to electron-beam heating [8.150] can be obtained from a shift of the HOLZ line when the pattern of crossing lines is compared with a set of computer maps.

The accuracy is $\Delta a/a = \Delta E/2E = 2 \times 10^{-4}$ at $E = 100$ keV. Thus, local concentrations of Al in Cu-Al alloys can be measured with an accuracy of 1 at% [8.142] or strains at planar interfaces can be determined in a fashion similar to the chemical changes [8.151]. The symmetry of HOLZ lines in the center of a convergent-beam pattern can be used for determination of the crystal space group. [8.152, 8.153, 8.146, 8.154]

The relative shift of HOLZ lines can be detected very sensitively at points where three HOLZ lines intersect simultaneously (see the center of Fig. 8.15c, for example). Such a situation can be created by using a suitable acceleration voltage. However, the positions of the HOLZ lines are influenced by the crystal potential for electrons because the HOLZ lines arise from intersections of their dispersion spheres with the zero-order dispersion surface. This can cause deviations of a few keV in the effective acceleration voltage or a few percent in the lattice parameter when calculating the HOLZ line positions kinematically [8.155]. A fully dynamical calculation is thus needed to reduce the errors below 10^{-4} in acceleration voltage or lattice parameters [8.156, 8.157, 8.158].

The dark high-order Kikuchi lines in the central spot of a CBED can be converted to bright lines of higher contrast by tilting the primary beam and moving it around a hollow cone with an angle θ_n equal to that of the nth Laue zone. Only the (now excess) HOLZ lines, which are not disturbed by low-order reflections of the zero-order Laue zone, appear at the cone center [8.159, 8.160].

8.3.6 Large-Angle CBED

As shown in Sect. 8.1.3, the LACBED pattern is a two-dimensional rocking curve containing FOLZ and HOLZ contour lines without any overlap of the CBED discs from neighboring reflections, each point corresponding to an image point as well as a distinct excitation error. Crystal boundaries, for example, show two LACBED patterns to the left and right of the projected boundary image, while strain fields cause a bending of the LACBED lines.

An important advantage of LACBED is that lattice defects can be investigated [8.161, 8.162]. As a typical example to illustrate the method [8.163], we discuss dislocations crossing the contributing area parallel to the x axis; the pattern of a crossing Bragg contour parallel to the y axis is then the intensity as a function of the distance y from the dislocation core and the excitation error $s_g(x)$. If we are dealing with high-order reflections, the extinction distances are much larger than 100 nm and the kinematical approach (9.4) can be applied to calculate the local intensity $1 - |\psi_g|^2$ for a black contour line of reciprocal-lattice vector $\boldsymbol{g}$ in the x direction, for example. This results in a characteristic splitting when the contour line approaches and intersects the dislocation core. In the example of Fig. 8.16, dislocation lines cross five different contours with indicated indices. For a product $\boldsymbol{g} \cdot \boldsymbol{b} = n$ ($\boldsymbol{b} = \frac{1}{3}[\bar{1}2\bar{1}0]$: Burgers vector of dislocation), we find $n - 1$ subsidiary maxima. The line will of course be unaffected by the dislocation when the invisibility criterion $\boldsymbol{g} \cdot \boldsymbol{b} = 0$ (Sect. 9.4.1) holds. Unlike the conventional Burgers vector determination from two invisibility criteria as described in Sect. 9.4.4, several $\boldsymbol{g} \cdot \boldsymbol{b} = n$ products can be read in one LACBED when several contour lines cross the core line, and the solution for $\boldsymbol{b}$ can be directly correlated to the vector parallel

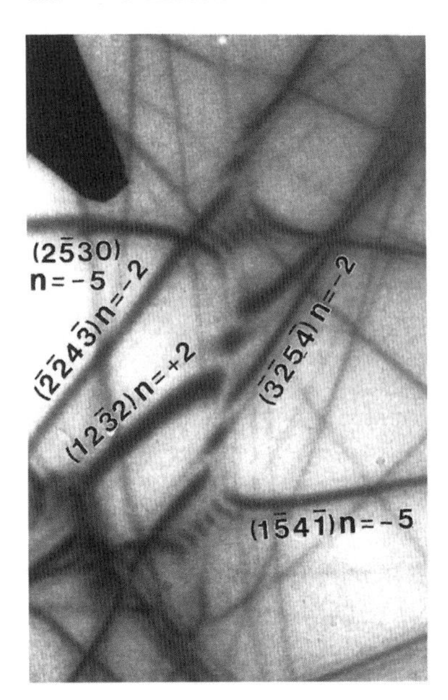

Fig. 8.16. Splitting of dark contour lines in a bright-field LACBED pattern crossed by dislocations in a quartz sample with different products $\boldsymbol{g}\cdot\boldsymbol{b}=n$ resulting in $n-1$ subsidiary maxima [8.163].

to the dislocation to distinguish edge and screw dislocations, for example. The problem of establishing the rotation of a diffraction pattern relative to the image, an inconvenience of the conventional method, does not arise. The image of the defect moves inside the LACBED pattern if the specimen is slightly shifted, and a position can be selected for which no other contour lines cross the core line nearby. In contrast to the imaging mode, where the dislocation core line can be seen over a large range of excitation errors, a dislocation can be seen in LACBED only near the intersection of a defect and a core line. In the case of low-order defect lines, dynamical calculations are necessary. Large-angle CBED patterns have the advantage that a whole pattern is visible as in Fig. 8.16, whereas a splitting of the contour lines is seen in CBED patterns only when the probe approaches a dislocation [8.164].

A further application is the determination of the displacement vector $\boldsymbol{R}$ of stacking faults [8.162]. Splitting of the contour lines allows strains and periods in bicrystals and multilayers with the boundaries parallel to the surface (plane view) to be investigated even when cross-sectional samples cannot be easily prepared [8.165, 8.166, 8.167, 8.168, 8.169]. The ability to display rocking curves over a wide angle enables us to analyze complex multilayers in plane view such as a thin GaAs quantum well between thicker layers of AlGaAs. The kinematical theory of rocking curves results in a periodic variation of the magnitude of the rocking curve maxima from which the depth and thickness of intermediate layers can be analyzed. In principle, such rocking-curve profiles can also be obtained in bend contours in imaging; these are, however, more or less accidental since the local s_g is not known exactly.

Otherwise, the structure of Bragg contours crossing multilayers in cross-section samples can also be used for strain measurements [8.170].

9

Imaging of Crystalline Specimens and Their Defects

A crystal can be imaged with the primary beam (bright field) or with a Bragg reflection (dark field). The local intensity depends on the thickness, resulting in thickness (or edge) contours, and on the tilt of the lattice planes, resulting in bend contours, which can be described by the dynamical theory of electron diffraction. In certain cases, the intensity of a Bragg reflection depends so sensitively on specimen thickness that atomic surface steps can be observed. The most important application of diffraction (Bragg) contrast is the imaging of lattice defects such as dislocations, stacking faults, phase boundaries, precipitates, and defect clusters. The contrast depends on the Bragg reflection excited and its excitation error, the type of the fault, and its depth inside the foil. The Burgers vector of a dislocation or the displacement vector of a boundary can thus be determined quantitatively. The resolution of the order of 10 nm when a strongly excited Bragg reflection is used can be reduced to the order of one nanometer by the weak-beam technique, which allows us to measure the width of dissociated dislocations, for example. Different types of contrast for precipitates are associated with coherent and incoherent precipitates, which can hence be distinguished. Electron spectroscopic imaging can remove the inelastically scattered electrons in the background of a diffraction pattern and increase the contrast and resolution of defect images.

With crystalline specimens, the interference of the primary and a Bragg-reflected wave in the final image creates images of lattice planes. When the objective aperture is large and includes a large number of Bragg reflections, the exit distribution of electrons can be imaged. Irradiation along zone axes produces a projection-like image of the crystal lattice with a resolution of 0.1–0.2 nm. For reliable interpretation, such images must be compared with a computer simulation that takes into account the thickness, the potential coefficients, and the wave aberration. The high resolution of the crystal-structure image can be exploited to investigate lattice defects and interfaces. By using electrons scattered through large angles, contrast increasing with atomic number can be superposed on the crystal-structure image. Atomic surface steps and surface-reconstruction structures can be investigated by special methods, notably by reflection electron microscopy (REM).

9.1 Diffraction Contrast of Crystals Free of Defects

9.1.1 Edge and Bend Contours

The diffraction angle $\theta_g = 2\theta_B$ between the primary beam and reflected beams is often larger than the objective aperture α_o; for Cu, for example, the 111 lattice spacing d is 0.208 nm and so at $E = 100$ keV ($\lambda = 3.7$ pm) we find $\theta_g \simeq \lambda/d = 18$ mrad. This means that then the Bragg-diffracted beams are halted by the objective diaphragm in the bright-field mode and do not contribute to the image intensity. The image intensity thus becomes equal to that of the primary beam (transmission T), which depends on the excitation error s_g or tilt parameter $w = s\xi_g$ of the Bragg reflection and on the specimen thickness t. Furthermore, contributions to the image can also come from electrons that have been scattered elastically or inelastically into the diffuse background of the diffraction pattern and pass through the objective diaphragm. Like the scattering at amorphous specimens, this contribution depends on the objective aperture α_o [9.1, 9.2]. Crystals show a higher transmission in regions without strong Bragg reflections than do amorphous films of equal mass thickness because the diffuse scattering between the Bragg reflections is reduced by destructive interference [9.3, 9.4].

We now use the results of the dynamical theory of electron diffraction (Figs. 7.17 and 7.18) to discuss diffraction contrast. The pendellösung of the dynamical theory (Fig. 7.17) can be seen as edge contours (Fig. 9.1) in the images of specimen edges of electropolished metals or of small cubic crystals (MgO), for example. A high intensity will be observed for thicknesses (Fig. 9.2a) at which the Bragg-reflected intensity is scattered back to the primary beam; maximum transmission thus occurs at a thickness equal to integral multiples of the extinction distance $\xi_{g,\mathrm{eff}}$; see (7.62) and Fig. 7.14. The spacing of the edge contours is greatest in the Bragg position ($w = 0$) and decreases with increasing positive or negative tilt, owing to the decrease of the effective extinction distance $\xi_{g,\mathrm{eff}}$. This dependence on the tilt can be seen in Fig. 9.1 on a bent crystal edge. With the aid of a tilting stage or specimen goniometer, the specimen can be brought into the Bragg position, where the largest spacing is observed. The extinction distances can be measured when the edge profile is known, or the local thickness can be estimated by using tabulated values of ξ_g (Table 7.2). However, the extinction distance for $w = 0$ can also be influenced by the excitation of other Bragg reflections (see dynamical Bethe potentials in Sect. 7.4.4). A quantitative measurement of the intensity of edge contours can be used to determine the absorption distances ξ'_0 and ξ'_g [9.1]. A large number of contours can be observed in high-voltage electron microscopy [9.5, 9.6] because ξ_g increases as v and ξ'_0, ξ'_g are proportional to v^2 (Sect. 7.4.2).

Edge contours can also be observed at crystal boundaries that are oblique to the foil surface (Fig. 9.2b). When the orientation of the second crystal is such that it does not show strong Bragg reflections, the diffraction contrast of

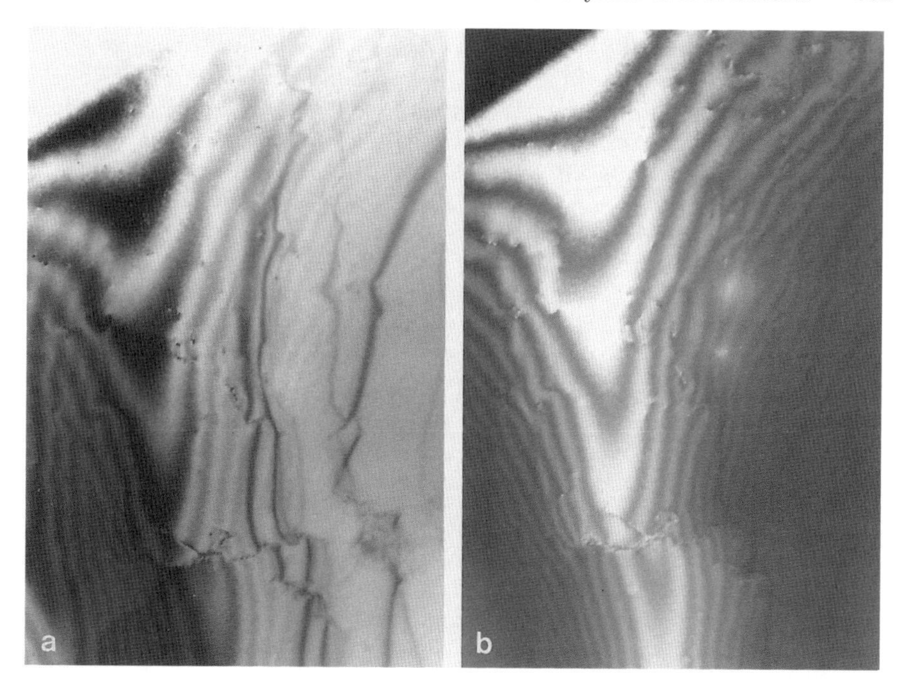

Fig. 9.1. (**a**) Bright- and (**b**) dark-field micrographs of edge contours at a bent Al foil of increasing thickness (top to bottom) with a maximum extinction distance ξ_g at the Bragg position and smaller $\xi_{g,\mathrm{eff}}$ for positive and negative tilt.

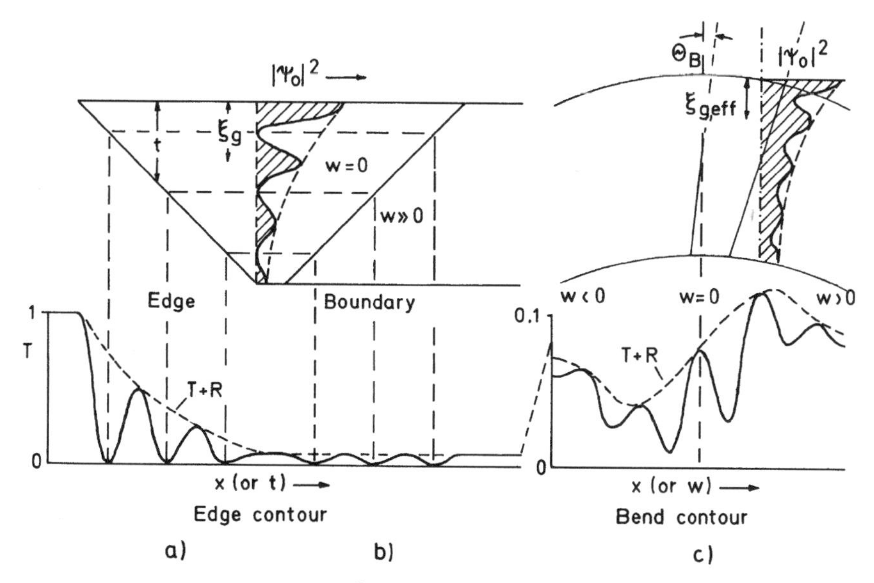

Fig. 9.2. Oscillations of $T = |\psi_0|^2$ in a crystal foil and generation of edge contours at (**a**) a crystal wedge and (**b**) a boundary. (**c**) Formation of bend contours in a bent foil.

the first crystal at the local depth of the boundary will not be changed much by the second crystal and will suffer only the exponential absorption.

Bent crystal foils of constant thickness show extended and curved bend contours (Figs. 9.2c and 9.3) caused by the local variation of the excitation errors. Lines of equal intensity represent lines of equal inclination of the lattice planes to the electron beam. In some cases, it will be possible to reconstruct the three-dimensional curvature of a crystal (e.g., for lenticular cavities in PbI_2 crystals [9.7]). When the bend contours are nearly straight, the image intensity across the contour is a direct image of the rocking curve (Figs. 7.18 and 7.20a). A bending radius R of the crystal corresponds to a tilt $\Delta\theta = x/R$ at a distance x from the exact Bragg position. The width of the bend contour is therefore proportional to R, and the contours move when the specimen is tilted in a goniometer. Bending and tilting can also be induced by specimen heating caused by intense electron irradiation or by mechanical stresses generated by the buildup of contamination layers.

Broad dark bands are normally superpositions of low-index hkl and $\bar{h}\bar{k}\bar{l}$ reflections for which the regions of diffraction contrast overlap (Figs. 7.20a and 9.3). Subsidiary maxima of the rocking curve and high-order reflections can be observed for medium foil thicknesses. For large specimen thicknesses ($t \gg \xi_0'$), only a narrow zone of anomalous transmission on each side of the extinction band shows sufficient transmission.

If the foil is bent two-dimensionally and the foil normal is near a low-index zone axis, the system of bend contours forms a two-dimensional zone-axis pattern (ZAP) (Fig. 9.3) that is comparable to the zone-axis pattern of convergent-beam electron diffraction (Sects. 8.1.4e and 8.3).

9.1.2 Dark-Field Imaging

A dark-field image can be formed by allowing one Bragg-diffracted beam to pass through the objective diaphragm. Only those specimen areas that contribute to the selected Bragg-diffraction spot appear bright (Fig. 9.1b). Either the objective diaphragm is shifted (Fig. 4.17b) or the primary beam is tilted (Fig. 4.17c), in which case the diffracted beam is on-axis. With the first method, a chromatic error streak (Sect. 2.3.6) decreases the resolution. With the second method, only the axial chromatic aberration is present. The position of the objective diaphragm and the direction of the primary and diffracted beams can be controlled with the aid of the selected-area electron-diffraction pattern (SAED, Sect. 8.1.1). For a well-adjusted microscope, the transition from the bright-field to the dark-field mode can be effected simply by switching on the currents in the deflection coils used for beam tilting. The theorem of reciprocity indicates that the intensities of the primary and diffracted beams should not be changed if the directions of the two beams are interchanged.

When the focal length of the intermediate lens is changed in order to pass from SAED to the image mode, the Bragg reflections are broadened to

Fig. 9.3. Bend contours in a NIMONIC75 thin foil, forming a [100] zone-axis pattern (courtesy of P. Tambuyser).

shadows of the selector diaphragm containing the dark-field contours. This makes it possible to determine the Miller indices of an edge or bend contour. The excitation error s_g can be measured by observing the relative positions of Kikuchi lines and Bragg-reflection spots (Sect. 8.2.4). Such information about the crystal orientation is needed to determine the Burgers vector of dislocations and other defects (Sect. 9.4.4).

Crystal phases with different crystal structures or orientations and hence with different diffraction spots can be separated in the dark-field mode by selecting the corresponding diffraction spots in the SAED pattern. If necessary, the diameter of the objective diaphragm can be decreased to 5 μm, but this will not be favorable for high resolution owing to contamination and charging of such a small diaphragm. Figure 9.4 shows an example in which microtwins in Ni layers, electrolytically deposited on a copper single crystal, are identified [9.8]. Double reflection at the matrix and the microtwins produces typical twin reflections such as a and b in Figs. 9.4c and f. Dark-field micrographs obtained with these twin reflections indicate the different orientations (Figs. 9.4a,b) of the twin lamellae that are superposed in the bright-field

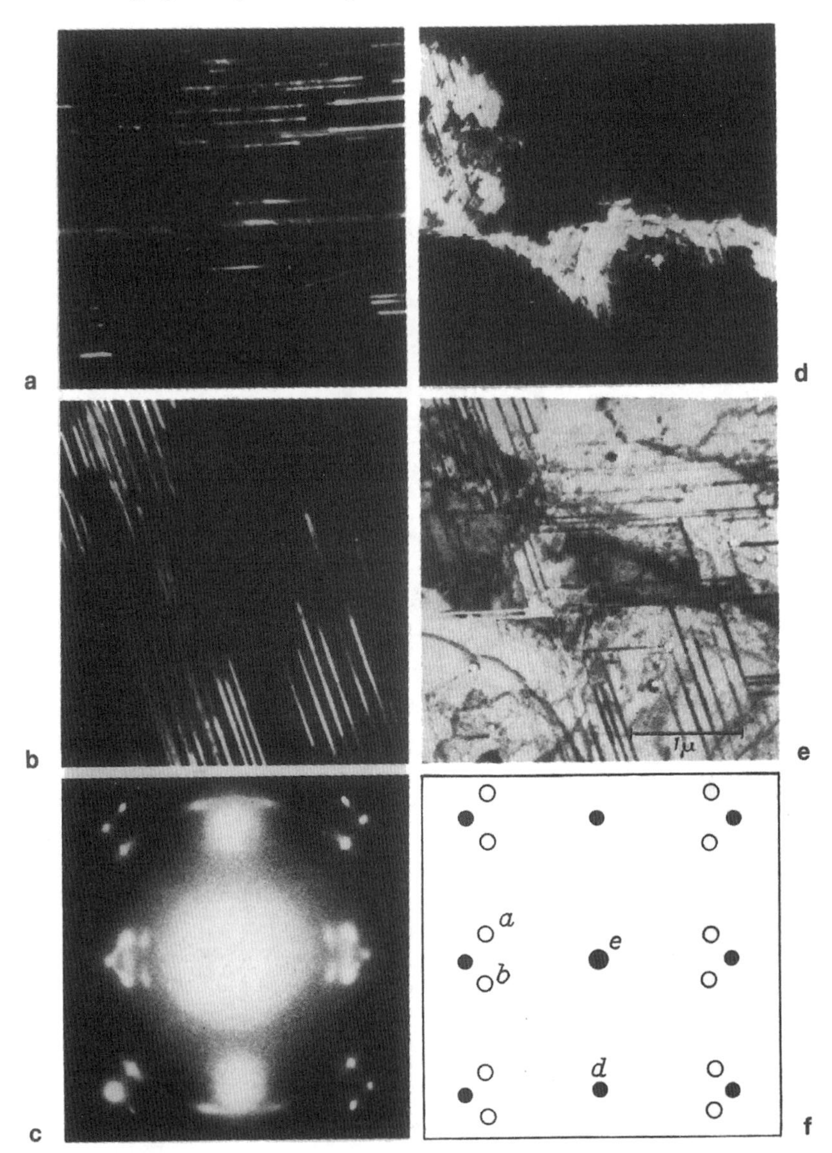

Fig. 9.4. Example of dark-field micrographs obtained by selecting diffraction spots in the diffraction patterns (**c**) and (**f**) of an electrolytically deposited nickel film on a copper substrate. (**a**) and (**b**) Selection of two sets of microtwins. (**e**) Bright-field and (**d**) dark-field micrographs with the 200 Bragg reflection.

micrograph (Fig. 9.4e). Figure 9.4d shows the contribution of the 220 Bragg reflection to the dark bend contour across the imaged area.

The dark-field images of different phases or crystal orientations with different Bragg angles are shifted parallel to the $\boldsymbol{g}$ vector by an amount proportional to $C_s\theta_g^3 M - \Delta z \theta_g M$ with defocus Δz. When two micrographs with different defocus are examined with a stereo viewer, the specimen structures appear to have different heights. This method can be useful for separating different phases or orientations [9.9, 9.10].

Dark-field imaging of lattice defects with weak beams is the most effective way of obtaining highly resolved micrographs of lattice defects (Sect. 9.4.3).

9.1.3 Moiré Fringes

When two crystal foils overlap and are rotated through a few degrees relative to one another or when their lattice constants are different, a pattern of interference fringes is observed. These are known as moiré fringes and were first observed on graphite lamellae [9.11]. The generation of these patterns will be discussed for the example of two crystal foils with different lattice constants. Such specimens can be prepared by epitaxy (e.g., palladium with $d_{220} = 0.137$ nm on a single-crystal gold film with $d_{220} = 0.144$ nm; Fig. 9.5 [9.176]). However, the two crystal foils need not be in direct contact. We expect to see a double electron diffraction pattern with Bragg reflections at different distances OP and OQ from the origin O (primary beam) (Fig. 9.6a).

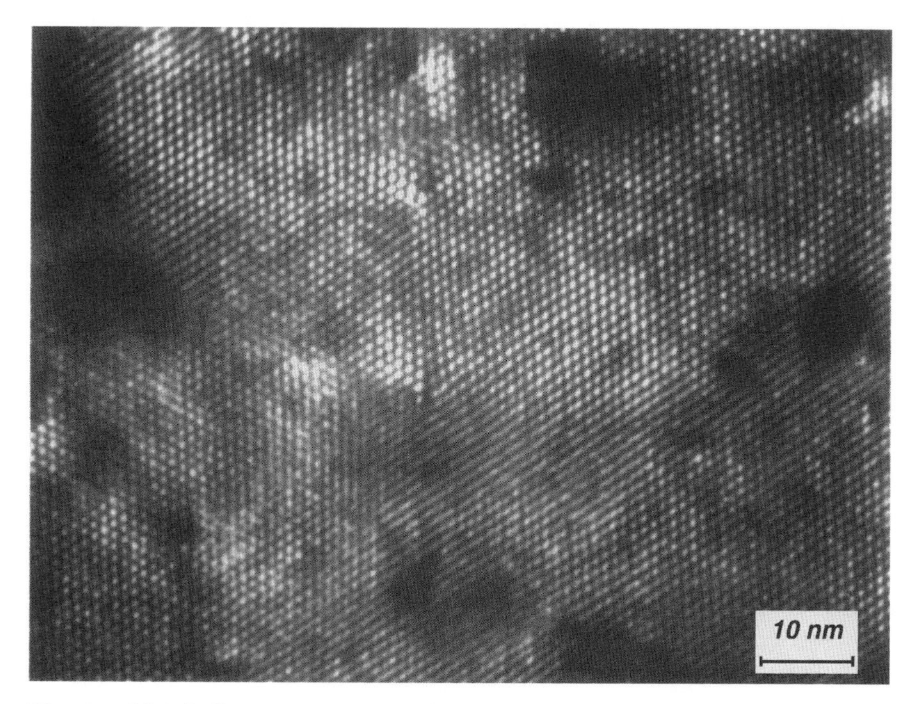

Fig. 9.5. Moiré effect in separately prepared and superposed single-crystal films of palladium and gold; parallel and rotation moiré fringes are seen (courtesy of G.A. Basset et al.).

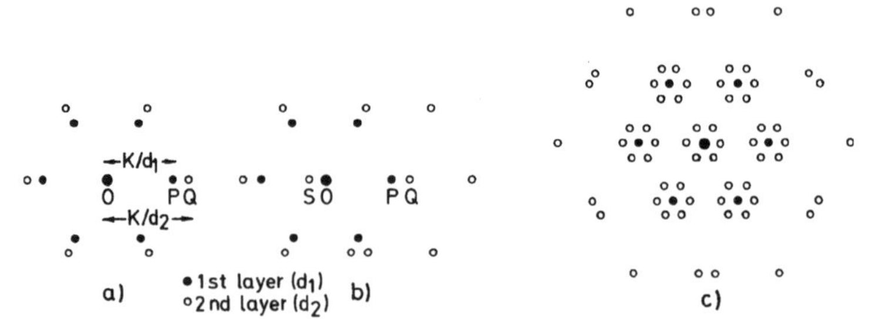

Fig. 9.6. Steps in the construction of the diffraction pattern of superposed crystals with different lattice-plane spacings d_1 and d_2. (**a**) Double diffraction pattern produced by the primary beam O only. (**b**) The beam P in the first crystal is diffracted in the second and produces a spot S near O. (**c**) All six diffraction spots equivalent to P act as a primary beam in the second crystal.

The diffracted beam in the first layer can be further diffracted in the second. This results in a shifted diffraction pattern, with the direction P as the new primary beam (Fig. 9.6b). For all strong Bragg reflections, this produces the diffraction pattern of Fig. 9.6c. The diffraction spots around the primary beam O can pass through the normal objective diaphragm and contribute to the image contrast. These diffraction spots can formally be attributed to a material with a larger lattice-plane distance d_M. If K denotes a diffraction constant proportional to the camera length, the distance OS becomes

$$\mathrm{OS} = \mathrm{PS} - \mathrm{OP} = \mathrm{OQ} - \mathrm{OP} = \frac{K}{d_2} - \frac{K}{d_1} = \frac{K}{d_\mathrm{M}} \tag{9.1}$$

and the moiré fringes have the *apparent lattice constant*

$$d_\mathrm{M} = \frac{d_1 d_2}{d_1 - d_2} \tag{9.2}$$

($d_\mathrm{M} = 2.9$ nm for the example of the Pd-Au double layer).

When one crystal is rotated relative to the other by a small angle α with the foil normal as axis, the moiré-fringe spacing becomes

$$d_\mathrm{M} = \frac{d_1 d_2}{(d_1^2 + d_2^2 - 2 d_1 d_2 \cos\alpha)^{1/2}} . \tag{9.3}$$

A two-dimensional moiré pattern as in Fig. 9.6 is formed when several Bragg reflections show double reflections. A rotation moiré can also be formed with the rotation axis parallel to the foil [9.13]. This effect can be observed in lenticular cavities of crystal lamellae. For an exact discussion of the image contrast, it is necessary to use the dynamical theory of electron diffraction [9.14, 9.15].

The generation of moiré fringes can also be understood in terms of a more intuitive model in which the imaging of lattice planes is regarded as a projection. Figures 9.7a and b contain two parallel lattices, with that in Fig. 9.7a

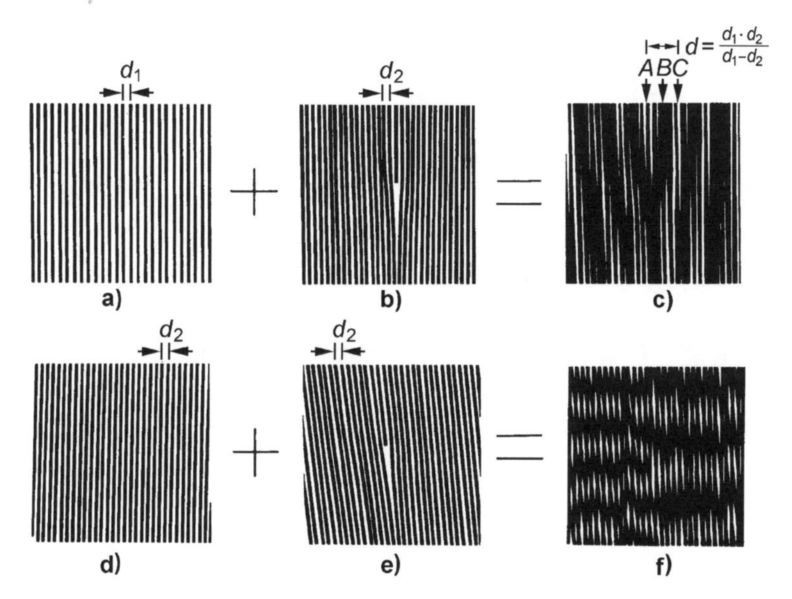

Fig. 9.7. Optical analogue that demonstrates the imaging of dislocations with the moiré effect. Lattices with lattice-plane spacings d_1 (**a**) and d_2 (**b**) superpose to form (**c**). Two lattices (**d**) and (**e**) of equal d_2 superpose to give a rotation moiré (**f**).

having a lattice constant d_1 while that in Fig. 9.7b having a lattice constant d_2 and an additional edge dislocation. The superposition of transparent foils [9.16] results in Fig. 9.7c. Dark and bright areas of width d_M alternate when the effect of the dislocation is neglected. The narrow lines with the lattice-plane spacing will be resolved only in high resolution; normally, only the fringes with the larger period d_M are resolved. In light optics, the same impression is obtained by looking at Fig. 9.7 from a larger distance. Figures 9.7d–f represent the optical analogue of a rotation moiré of two lattices with equal lattice constants ($d_1 = d_2$). In this case, the dislocation is also *imaged* but with the wrong orientation. Care is therefore necessary when interpreting the moiré patterns of lattice defects.

The moiré effect can be successfully used not only for the investigation of epitaxy [9.17, 9.18] but also for UHV in situ experiments [9.19] and measurements of the lattice displacement across boundaries [9.17]. Moiré fringes also can be observed between a matrix and precipitates (Sect. 9.5.1).

9.1.4 The STEM Mode and Multibeam Imaging

As shown in Sect. 4.5.3, the ray diagram of the STEM mode is the reciprocal of that of the TEM mode (Fig. 4.20). The illumination aperture α_i of the TEM mode corresponds to the detector aperture α_d of the STEM mode. For high-resolution STEM, the electron-probe aperture α_p must be large, of

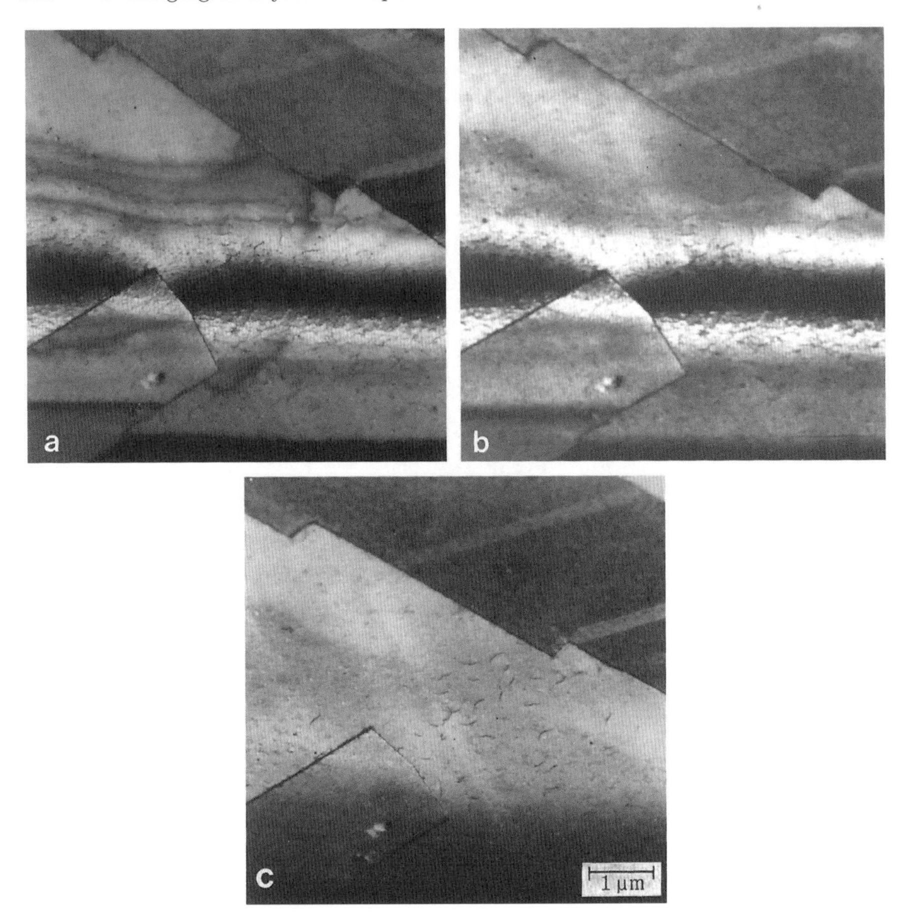

Fig. 9.8. Comparison of 100 keV TEM and STEM micrographs of a polycrystalline copper foil ($\simeq$100 nm). (**a**) TEM mode with $\alpha_i \ll \alpha_o$, (**b**) similar contrast in the STEM mode with $\alpha_d < \alpha_p \simeq \alpha_o$, and (**c**) blurring of the bend contours with $\alpha_d \simeq \alpha_p$.

the same order as the objective aperture α_o of a TEM (Sect. 4.5). The same contrast would be seen in the TEM and STEM modes if a small detector aperture α_d were used in STEM (Fig. 9.8a,b). However, a large fraction of the incident cone of aperture α_p would then not pass through the detector aperture α_d even in the absence of a specimen. It is therefore better to work with $\alpha_d \simeq \alpha_p$ (see also the discussion of the contrast of amorphous specimens in the STEM mode in Sect. 6.1.4). By collecting all the incident electrons in this way, the angle of incidence (excitation error) to the lattice planes varies widely and the image contrast is a superposition (averaging) of images with a broad spectrum of excitation errors s_g. This superposition is incoherent when interference between the different rays of the incident electron probe does not occur. The contrast of the edge contours at thicker parts of an edge is thus reduced owing to the variation of $\xi_{g,\mathrm{eff}}$ for different values of s_g. Bend contours are also blurred, and contributions from high-order reflections disappear (Fig. 9.8c) [9.20, 9.21, 9.22, 9.23, 9.24]. The diffraction contrast of a polycrystalline specimen becomes more uniform inside a single crystal, and differences in anomalous transmission can be seen mainly for different crystal orientations; the contrast of lattice defects (e.g., dislocations) is, however, preserved, though with reduced contrast (Figs. 9.8 and 9.18).

If the detector aperture α_{d} is comparable to or larger than the Bragg- diffraction angle $\theta_g = 2\theta_{\mathrm{B}}$, the primary beam and one or more Bragg-diffracted beams can pass through the detector diaphragm and the image contrast will be a superposition of these beams. In this *multibeam imaging* (MBI) mode [9.25], the subsidiary maxima of the rocking curve vanish (see ΣI_g in Fig. 7.20a) and the image intensity is influenced only by the dependence of the anomalous transmission on the excitation error. A STEM image of a crystalline specimen will normally be a mixture of MBI and the blurring caused by the presence of a broad spectrum of excitation errors [9.23]. The MBI mode can be of advantage for the imaging of lattice defects (Sects. 9.3.2 and 9.4.1). If the intensities in the bright- and dark-field images are complementary, the contrast will be canceled, whereas it is enhanced if the images are anticomplementary. This allows us to decide how an extended defect is situated relative to the top and bottom of the foil. This technique was first proposed for HVEM [9.38]. In HVEM, the diffraction angle θ_g decreases with increasing energy and, because the spherical aberration has less effect at smaller angles, a sharp image can be obtained with an objective aperture capable of transmitting the primary and the diffracted beams, whereas in a 100 keV transmission electron microscope the spherical aberration for a Bragg-reflected beam is large enough to shift the dark-field image. Although defocusing may cause the dark- and bright-field images to overlap, this will be possible for only one Bragg reflection, not for all simultaneously. Multibeam imaging in the STEM mode can also be employed at 100 keV because the spherical aberration has no influence behind the specimen.

9.1.5 Energy Filtering of Diffraction Contrast

The diffraction (Bragg) contrast of crystalline specimens is a result of the dynamical theory of electron diffraction and caused mainly by the elastically diffracted electrons in Bragg reflections; intraband inelastic scattering (Fig. 7.15) with plasmon losses or inner-shell ionizations of low ionization energy, for example, also preserve the initial elastic Bloch-wave field and consequently the diffraction contrast, whereas interband scattering (thermal diffuse electron–phonon scattering, for example) does not (Sect. 7.4.1). Zero-loss filtered images formed by thermal diffuse scattering of the electrons with the objective diaphragm between Bragg spots show no thickness and bend contours but only a structure due to anomalous absorption, and the intensity is proportional to $\Sigma_g I_g$ in (7.83). These theoretical results [7.18, 9.26] have been confirmed experimentally [9.27, 9.28, 9.29, 9.30, 9.31, 9.32]. In the case of intraband scattering, the inelastically scattered electrons around the primary beam and the Bragg spots belong to Bloch-wave fields that differ from the elastic Bloch-wave field only in their wave vectors $\boldsymbol{k}_g^{(j)}$. The Bragg contrast observed is nearly the same as in zero-loss imaging and therefore thickness fringes, bend contours, and images of lattice defects also appear in the plasmon-loss image. However, the angular distribution of inelastically scattered electrons represents a spectrum of excitation errors s_g. This causes a blurring of edge and bend contours with increasing energy loss [9.33, 9.34, 9.35]. Incoherent superposition of thickness and bend contours with the spectrum of s_g and consideration of the Poisson distribution of multiple plasmon losses can qualitatively describe the observed blurring effects [9.32]. This blurring has less influence on stacking-fault fringes, which can be observed up to energy losses of 300 eV [9.31].

The inelastically scattered electrons are not only of less value because of these contrast effects but also because of the chromatic aberration, which increases with increasing thickness. No useful diffraction contrast will be observed when the number of inelastically scattered electrons that contributes to the image becomes much larger than the number of elastically scattered electrons. Whereas the practical limit of transmission $T = 10^{-3}$ is reached for amorphous specimens at mass thicknesses $x \simeq 70$ μg/cm^2, zero-loss filtering of crystalline specimens can be applied up to 150 μg/cm^2 [9.32] with 80 keV electrons. The reason for this is that the Bloch waves with minima at the atomic positions experience a lower mean attenuation (see also differences in $T(\alpha_o)$ for amorphous and crystalline sections in [6.1]). Even thicker specimen areas can be investigated for orientations showing anomalous transmission [9.36].

Just as for amorphous specimens, the increase of contrast by zero-loss filtering is large for low atomic numbers with a large ratio of total inelastic-to-elastic cross section ν (5.66). Figure 9.9 shows a comparison of unfiltered and zero-loss filtered images of a cleaved graphite foil with moiré structures in bright-field mode and of microtwins in an epitaxially grown silver film on rock salt in dark-field mode. Zero-loss filtering can also increase the contrast of other crystal defects. Weak-beam dark-field images (Sect. 9.4.3) likewise show a gain in contrast after elimination of the incoherent inelastically scattered background [9.37].

9.1.6 Transmission of Crystalline Specimens

Crystalline specimens can be prepared as thin foils by cleavage, deposition as epitaxially grown films on single-crystal substrates, or by thinning metals, alloys, minerals, and other materials by chemical etching, electrolytic polishing, or ion-beam etching. It is necessary to prepare specimens with thicknesses in the range 0.1–0.5 μm for imaging crystal defects at 100 keV but less than

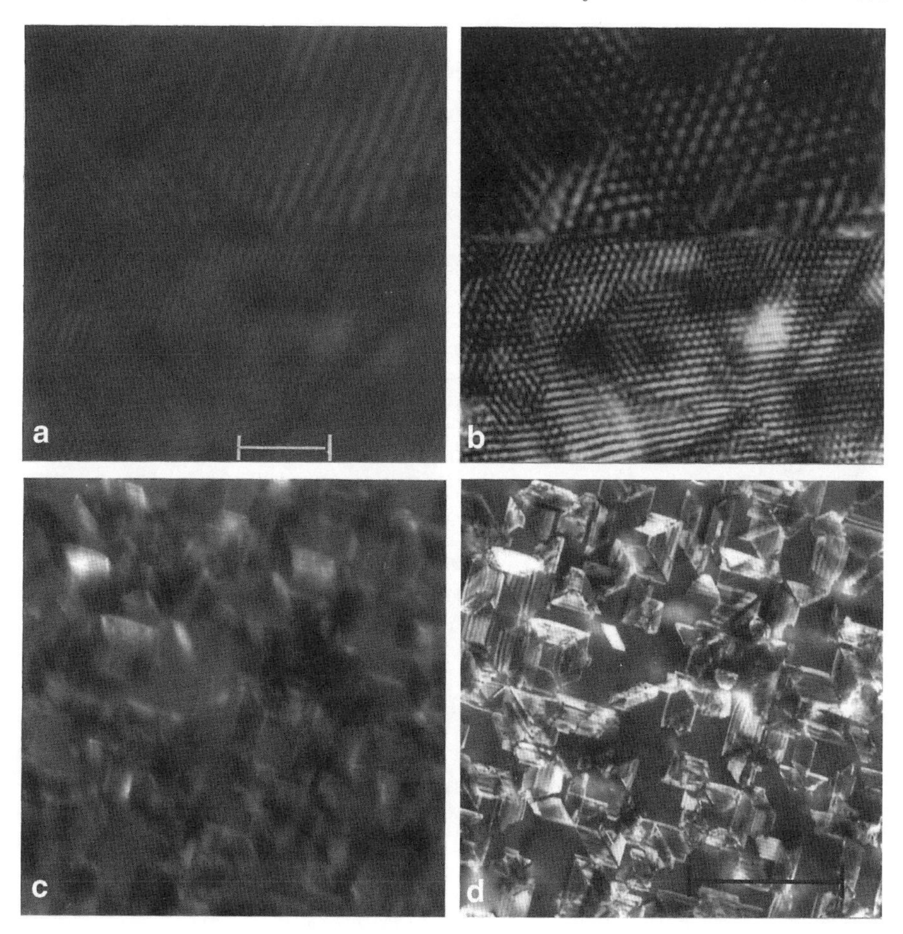

Fig. 9.9. Comparison of unfiltered (**a,c**) and zero-loss filtered (**b,d**) images of a cleaved graphite foil (top) with moiré structures in bright-field mode (bar = 1 μm) and an epitaxially grown Ag film on NaCl (bottom) with microtwins in dark-field mode (bar = 0.5 μm, E = 80 keV).

10 nm for high-resolution studies of the crystal structure (Sect. 9.6). These thinning methods yield sufficiently thin areas only in wedge-shaped foils near holes. For many applications – observation of dislocation structure and in situ experiments, for example – the question arises whether such thin foils are representative of the bulk material. Dislocations can be rearranged and/or migrate to the foil surface, and in situ precipitation and electron irradiation experiments are influenced by the surface, which plays the role of a sink for mobile point defects.

The useful specimen thickness is limited by the decrease of the Bragg-diffraction intensities and by the chromatic aberration associated with energy losses. Absorption of the Bloch-wave field and the probability of energy

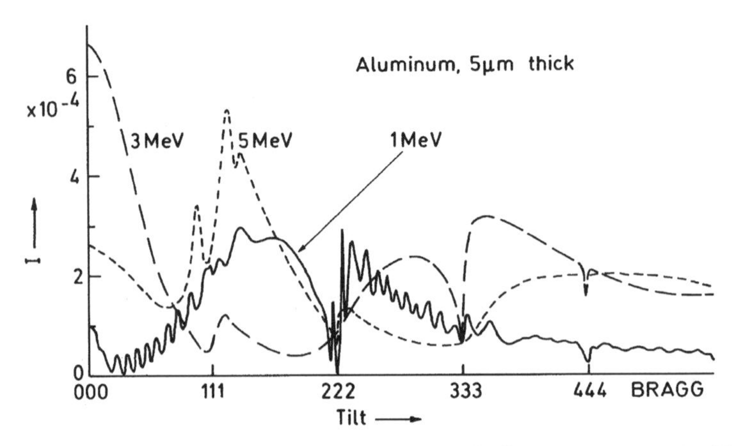

Fig. 9.10. Bright-field rocking curves for a 5 μm Al foil at 1 MeV, 3 MeV, and 5 MeV [9.41]

losses decrease with increasing electron energy; larger thicknesses can hence be used in HVEM. A very important consequence of this increase of the useful thickness is that a very much larger specimen area can be examined at high voltage than at 100 keV, for which the transmission is mostly limited to very small zones near holes.

The many-beam dynamical calculations show that the anomalous transmission depends strongly on the excitation condition and varies from element to element [9.39, 9.40]. Thus the transmission curves for a 111 row of aluminum (Fig. 9.10) reach a maximum at 1 MeV; for tilts on the positive side of the 111 Bragg position, it occurs at 5 MeV, whereas, at 3 MeV, the maximum occurs at the symmetric excitation (the pendellösung oscillations drawn for 1 MeV are smoothed out at 3 and 5 MeV). These effects are caused by differences in the Bloch-wave channeling [9.41, 7.30, 9.42]. Figure 9.11 indicates the foil thicknesses of Al, Fe, and Au for which the transmission T is 10^{-3} of the incident beam at the beam tilts of maximum transmission. This shows that optimum transmission can be obtained for Al at about 3 MeV, for Fe at 5 MeV, and for gold at 10 MeV, though the increase is modest beyond 3 MeV. It is therefore useful to be able to vary the accelerating voltage of a high-voltage electron microscope and to match the voltage to the specimen orientation or to tilt the specimen into an optimum orientation. The definition of a maximum useful thickness depends on the criterion adopted and the type of contrast. The $T = 10^{-3}$ criterion of Fig. 9.11 is arbitrary but agrees rather well with experience.

In the Kikuchi patterns of MoS_2 lamellae, the Bragg spots disappear when $t \geq 2.7\beta^2$ [9.43] ($\beta = v/c$, t measured in μm), and the anomalous transmission near low-order Bragg positions vanishes for $t \geq 5.7\beta^2$; dislocations are visible up to these thicknesses. Stacking faults in Si and Fe have been observed for thicknesses below 9 and 2 μm, respectively, at 1 MeV [9.44] (see also further experiments in [1.26, 9.45, 9.46, 9.47]).

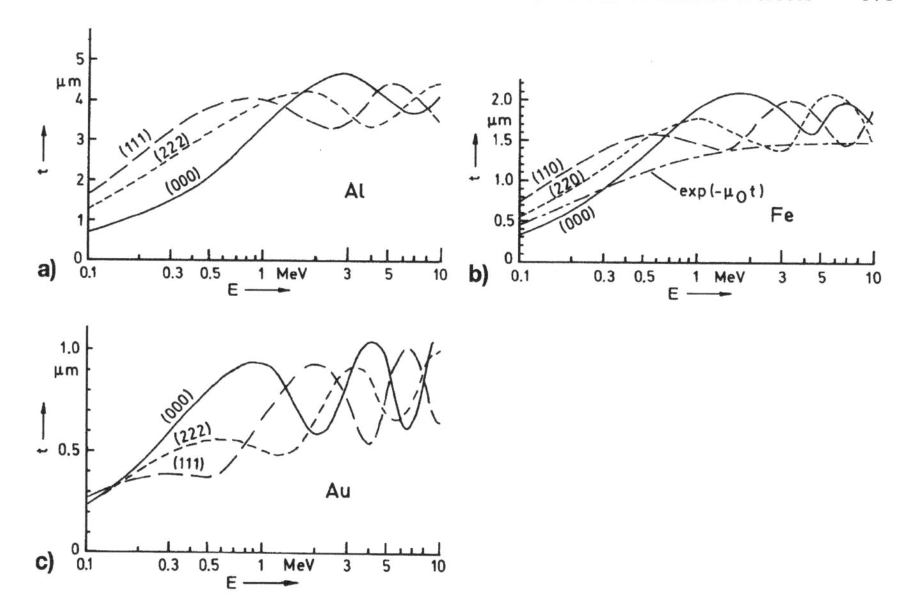

Fig. 9.11. Specimen thickness t that corresponds to a transmission $T = 10^{-3}$ for three different orientations [symmetric (000) position and on the positive side of the indicated (hkl) positions at the maximum of the rocking curve] for (**a**) Al, (**b**) Fe, and (**c**) Au [9.41].

9.2 Calculation of Diffraction Contrast of Lattice Defects

9.2.1 Kinematical Theory and the Howie–Whelan Equations

The most simple theory for the investigation of the image contrast of dislocations, stacking faults, and other defects uses the kinematical column approximation (Sect. 7.2.3) [1.26, 9.48, 9.49]. A unit cell at depth z near a lattice defect is assumed to be displaced by a vector $\boldsymbol{R}(z)$ relative to an ideal lattice without a defect (Fig. 9.12). Equation (7.24) for the reflected amplitude becomes

$$\begin{aligned}\psi_g &= \frac{\mathrm{i}\pi}{\xi_g}\int_0^t \exp\{-2\pi\mathrm{i}(\boldsymbol{g}+\boldsymbol{s})\cdot[\boldsymbol{z}+\boldsymbol{R}(z)]\}\mathrm{d}z \\ &= \frac{\mathrm{i}\pi}{\xi_g}\int_0^t \exp\{-2\pi\mathrm{i}[sz+\boldsymbol{g}\cdot\boldsymbol{R}(z)]\}\mathrm{d}z, \end{aligned} \quad (9.4)$$

in which we have written $\exp(-2\pi\mathrm{i}\boldsymbol{g}\cdot\boldsymbol{z}) = 1$ because the product of a reciprocal-lattice vector $\boldsymbol{g}$ and a translation vector $\boldsymbol{z} = \boldsymbol{r}$ is always an integer; $\boldsymbol{s}\cdot\boldsymbol{R}$ has been neglected because it is the product of two quantities that are small relative to g and z. The integral of $\exp(-2\pi\mathrm{i}sz)$ was analyzed graphically in Sect. 7.2.3 with the aid of an amplitude-phase diagram (APD) and resulted

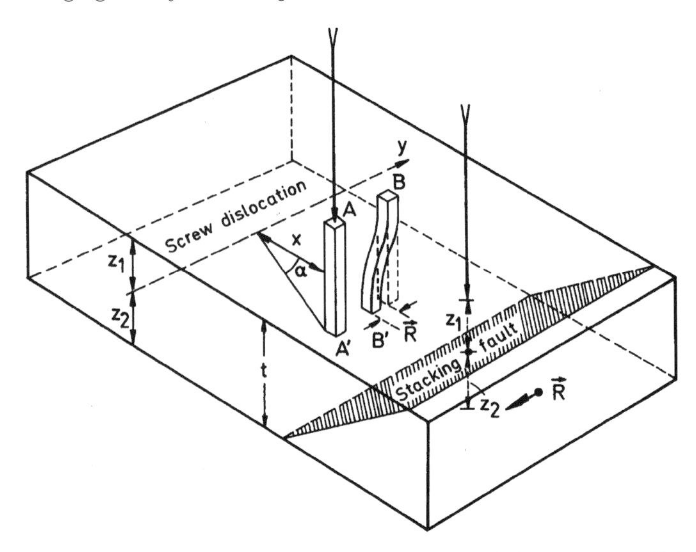

Fig. 9.12. A crystal foil of thickness t contains a screw dislocation parallel to the foil surface at a depth z_1 ($z_1 + z_2 = t$). The column AA′ of the ideal lattice is deformed to BB′ by a displacement vector $\boldsymbol{R}(z)$. A stacking fault displaces the lower part of the lattice by a constant displacement vector $\boldsymbol{R}$ relative to the upper part.

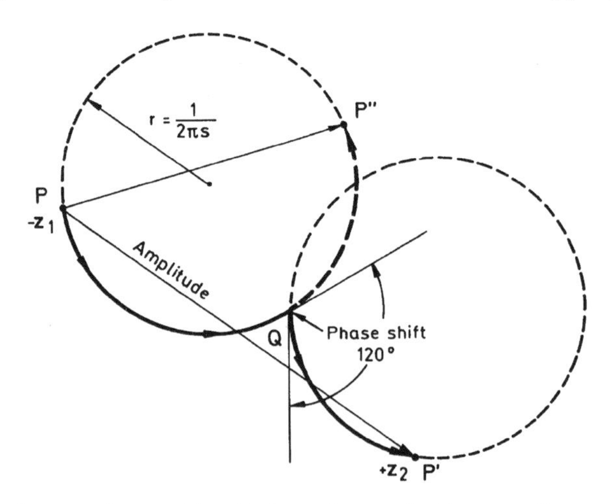

Fig. 9.13. Influence of a stacking fault at Q at a depth z_1 on the amplitude-phase diagram (phase shift by $-2\pi/3$ equivalent to a kink of 120°).

in a circle of radius $r = (2\pi s)^{-1}$. The phase and hence the radius of curvature of the APD are changed by the additive term $-2\pi i \boldsymbol{g} \cdot \boldsymbol{R}(z)$ in the exponent of (9.4). Examples of such modified APDs are shown in Fig. 9.13 for a stacking fault and in Fig. 9.19 for a dislocation. Equation (9.4) furnishes the important rule that the scalar product $\boldsymbol{g} \cdot \boldsymbol{R}$ has to be nonzero for the imaging of lattice defects. Bragg reflections $\boldsymbol{g}$ for which the contrast of lattice defects disappears because $\boldsymbol{g} \cdot \boldsymbol{R} = 0$ can be used to determine the direction and magnitude of the displacement $\boldsymbol{R}$.

A local lattice distortion can be included in the dynamical theory by adding a term $2\pi\boldsymbol{g} \cdot \boldsymbol{R}(z)$ to the phase in (7.26), and (7.27) becomes [1.26]

$$\frac{\mathrm{d}\psi_g}{\mathrm{d}z} = \sum_{h=g_1}^{g_n} \frac{\mathrm{i}\pi}{\xi_{g-h}} \psi_h \exp[2\pi\mathrm{i}s_{h-g}z + 2\pi\mathrm{i}(\boldsymbol{h} - \boldsymbol{g}) \cdot \boldsymbol{R}(z)]. \tag{9.5}$$

By using the transformations

$$\begin{aligned} \psi_0' &= \psi_0 \exp[-\mathrm{i}\pi z/\xi_0], \\ \psi_g' &= \psi_g \exp[2\pi\mathrm{i}sz - \mathrm{i}\pi z/\xi_0 + 2\pi\mathrm{i}\boldsymbol{g} \cdot \boldsymbol{R}(z)], \end{aligned} \tag{9.6}$$

analogous to those employed in the two-beam case (7.28), which do not affect the intensities because $|\psi_g'|^2 = |\psi_g|^2$, (9.5) becomes

$$\begin{aligned} \frac{\mathrm{d}\psi_0'}{\mathrm{d}z} &= \frac{\mathrm{i}\pi}{\xi_g}\psi_g', \\ \frac{\mathrm{d}\psi_g'}{\mathrm{d}z} &= \frac{\mathrm{i}\pi}{\xi_g}\psi_0' + [2\pi\mathrm{i}(s+\beta)]\psi_g' \quad \text{with} \quad \beta = \frac{\mathrm{d}}{\mathrm{d}z}[\boldsymbol{g} \cdot \boldsymbol{R}(z)], \end{aligned} \tag{9.7}$$

or when the absorption terms are considered,

$$\begin{aligned} \frac{\mathrm{d}\psi_0'}{\mathrm{d}z} &= -\frac{\pi}{\xi_0'}\psi_0' + \pi\left(\frac{\mathrm{i}}{\xi_g} - \frac{1}{\xi_g'}\right)\psi_g', \\ \frac{\mathrm{d}\psi_g'}{\mathrm{d}z} &= \pi\left(\frac{\mathrm{i}}{\xi_g} - \frac{1}{\xi_g'}\right)\psi_0' + \left[-\frac{\pi}{\xi_0'} + 2\pi\mathrm{i}(s+\beta)\right]\psi_g'. \end{aligned} \tag{9.8}$$

This linear system of differential equations is especially useful for calculating the contrast of dislocations and similar strain distributions but cannot be used for stacking faults because the constant phase shift between the two regions above and below such a fault is lost with the transform (9.6).

Equation (9.8) shows that a strain field modifies the excitation error s to $s + \beta$ and alters the deviation from the Bragg condition because the lattice planes are locally bent more into or out of the Bragg position (see also Fig. 9.21). The column approximation can be used for strongly excited Bragg reflections. For the weak-beam method in which the excitation errors are large and a resolution better than 2–5 nm is required, the Schrödinger equation has to be solved directly (Sect. 8.4.3).

9.2.2 Matrix-Multiplication Method

The amplitude ψ_g of a Bragg reflection at a depth z in a single-crystal foil can be calculated by using (7.47) when the eigenvalues $\gamma^{(j)}$, the components $C_g^{(j)}$ of the eigenvectors, and the Bloch-wave excitation amplitudes $\epsilon^{(j)}$ are known. Like (7.42) and (7.49), this equation can be written in matrix form [1.26]. Using (7.50), we have

$$\boldsymbol{\psi}(z) = [C]\{\exp(2\pi i\gamma^{(j)}z)\}\boldsymbol{\epsilon} = [C]\{\exp(2\pi i\gamma^{(j)}z)\}[C^{-1}]\boldsymbol{\psi}(0) = [S]\boldsymbol{\psi}(0)$$
$$\text{with} \quad \boldsymbol{\psi}(0) = \begin{pmatrix} 1 \\ 0 \end{pmatrix}. \tag{9.9}$$

The matrix $[S] = [C]\{\exp(2\pi i\gamma^{(j)}z)\}[C^{-1}]$ is called the scattering matrix, where $\{\exp(2\pi i\gamma^{(j)}z)\}$ is a diagonal matrix with the elements $\exp(2\pi i\gamma^{(j)}z)$. The anomalous absorption of Bloch waves can be introduced by replacing $\gamma^{(j)}$ by $\gamma^{(j)} + iq^{(j)}$. This method can be used for the calculation of lattice-defect contrast by dividing the foil into distinct layers; the kth layer is between z_{k-1} and z_k. The amplitude at the exit surface can be calculated from the amplitude at the entrance surface by using (9.9):

$$\boldsymbol{\psi}(z_k) = [S_k]\boldsymbol{\psi}(z_{k-1}) \quad \text{and} \quad \boldsymbol{\psi}(t) = [S_n][S_{n-1}]\dots[S_1]\boldsymbol{\psi}(0). \tag{9.10}$$

Lattice defects are considered by replacing U_g by $U_g \exp(2\pi i\boldsymbol{g}\cdot\boldsymbol{R})$ in the off-diagonal elements of the matrix $[A]$ (7.41) or by defining a fault matrix $[F] = \{\exp(2\pi i\boldsymbol{g}\cdot\boldsymbol{R})\}$ and replacing $[A]$ by $[F^{-1}][A][F]$. This modifies $[C]$ to $[F^{-1}][C]$ and $[S]$ to $[S_k] = [F^{-1}][S][F]$. This method is of interest for cases in which the foil contains large undisturbed regions or volumes of constant displacement vector $\boldsymbol{R}$ [9.50, 9.51]. Thus planar faults can be described by shifting the lower part of the foil, of thickness t_2, relative to the upper part, of thickness t_1, by a constant displacement vector $\boldsymbol{R}$. For the two-beam case, the fault matrix then takes the form

$$[F] = \begin{pmatrix} 1 & 0 \\ 0 & \exp(2\pi i\boldsymbol{g}\cdot\boldsymbol{R}) \end{pmatrix}; \quad [F^{-1}] = \begin{pmatrix} 1 & 0 \\ 0 & \exp(-2\pi i\boldsymbol{g}\cdot\boldsymbol{R}) \end{pmatrix}. \tag{9.11}$$

The amplitudes behind the foil are

$$\boldsymbol{\psi}(t) = [S_2][S_1]\boldsymbol{\psi}(0) = [F^{-1}][S(t_2)][F][S(t_1)]\boldsymbol{\psi}(0). \tag{9.12}$$

The matrix formulation can also be used to obtain the following symmetry rule for lattice-defect contrast [9.52]. Friedel's law for centrosymmetric crystals implies that the matrices $[A]$ and $[S]$ are symmetric. If the displacement in a column at depth z is $\boldsymbol{R}(z)$ in one foil, and if the $\boldsymbol{R}(z)$ of a second foil is obtained by inversion with respect to a point at the center of the column, equal contrast in bright-field images results when $\boldsymbol{R}(z) = -\boldsymbol{R}(t-z)$. Symmetry rules for dark-field images are discussed in [9.53].

9.2.3 Bloch-Wave Method

For some applications in which a better understanding of the alterations of the Bloch-wave field caused by lattice defects may be helpful, it is useful to rewrite the Howie–Whelan equations (9.7) for the reflection amplitude ψ_g as equations for the Bloch-wave excitation amplitudes $\epsilon^{(j)}$ (7.45) [9.54]. The ψ_g and $\epsilon^{(j)}$ are related by (7.47) or $\boldsymbol{\psi}(0) = [C]\boldsymbol{\epsilon}$ at the entrance surface and $\boldsymbol{\psi}(t_1) = [F^{-1}][C]\{\exp(2\pi i\gamma z)\}\boldsymbol{\epsilon}$ directly below a planar fault at a depth t_1. Substitution in (9.12) gives the Bloch-wave amplitude $\epsilon'^{(j)}$ at the depth t_1:

$$\boldsymbol{\epsilon}' = \{\exp(-2\pi\mathrm{i}\gamma t_1)\}[C^{-1}][F][C]\{\exp(2\pi\mathrm{i}\gamma t_1)\}\boldsymbol{\epsilon}(0). \tag{9.13}$$

The $\epsilon^{(j)}$ and $\epsilon'^{(j)}$ are constant inside an undisturbed lattice volume.

Each Bloch wave has to be multiplied by the exponential absorption term $\exp(-2\pi q^{(j)}z)$ of (7.79). The fault matrix $[F]$ is equal to the unit matrix $[E]$ in a fault-free crystal, and in this case $\epsilon^{(j)} = \epsilon'^{(j)}$; there is thus no interband scattering (Sect. 7.4.1), and the Bloch waves propagate independently. When defects are present, the fault matrix $[F] \neq [E]$ and, in the two-beam case, the excitation amplitude $\epsilon'^{(j)}$ becomes a linear combination of $\epsilon^{(1)}$ and $\epsilon^{(2)}$ so that interband scattering does occur. In Sect. 9.3.2, this reasoning will be used to discuss stacking-fault contrast.

If the defect is not a planar fault but varies continuously in depth, the following alteration of the Bloch-wave amplitude across a depth element Δz is obtained:

$$\begin{aligned}\boldsymbol{\epsilon}' &= \boldsymbol{\epsilon} + \frac{\mathrm{d}}{\mathrm{d}z}\boldsymbol{\epsilon} \\ &= \{\exp(-2\pi\mathrm{i}\gamma z)\}[C^{-1}]\left([E] + \frac{\mathrm{d}}{\mathrm{d}z}[F]\Delta z\right)\{\exp(2\pi\mathrm{i}\gamma z)\}\boldsymbol{\epsilon}. \end{aligned} \tag{9.14}$$

If the terms of $[F]$ can be expanded as $\exp(\mathrm{i}\alpha_g) = 1 + \mathrm{i}\alpha_g + \ldots$, with $\alpha_g = 2\pi\boldsymbol{g}\cdot\boldsymbol{R}$ and β_g small, (9.7) gives

$$\frac{\mathrm{d}}{\mathrm{d}z}\boldsymbol{\epsilon} = \{\exp(-2\pi\mathrm{i}\gamma z)\}[C^{-1}]\{2\pi\mathrm{i}\beta_g\}[C]\{\exp(2\pi\mathrm{i}\gamma z)\}\boldsymbol{\epsilon}. \tag{9.15}$$

This equation for the Bloch waves is equivalent to the Howie–Whelan equations (9.7). Multiplying the matrices and using the abbreviation $\Delta\gamma_{ij} = \gamma^{(j)} - \gamma^{(j)}$, we find

$$\frac{\mathrm{d}}{\mathrm{d}z}\epsilon^{(j)} = \sum_i \epsilon^{(i)}(z)\exp(2\pi\mathrm{i}\Delta\gamma_{ij}z)\sum_g C_g^{(j)}C_g^{(i)}2\pi\mathrm{i}\beta_g \quad (j = 1,\ldots,n). \tag{9.16}$$

For weak interband scattering by a lattice defect, the initial value $\epsilon^{(j)}(0) = C_0^{(j)}$ can be used at the entrance surface, and integration over a column in the z direction gives

$$\epsilon'^{(j)}(t) = \epsilon^{(j)}(0) + 2\pi\mathrm{i}\sum_i\sum_g C_0^{(i)}C_g^{(j)}\int_0^t \beta_g\exp(2\pi\mathrm{i}\Delta\gamma_{ij})\mathrm{d}z. \tag{9.17}$$

If, for small defects, β_g decreases strongly inside the foil, the limits of integration in (9.17) can be extended to $\pm\infty$, and the integral becomes a Fourier transform [9.55, 9.56, 9.57].

If $\Delta\gamma_{ij}$ is large or, in other words, the extinction distance $\xi_{ij} = 1/\Delta\gamma_{ij}$ responsible for the transition $i \to j$ is small, then only those columns for which β_g changes appreciably inside one extinction distance will contribute. This means that only the core of the defect will be imaged. This is the situation in weak-beam imaging (Sect. 9.4.3).

9.3 Planar Lattice Faults

9.3.1 Kinematical Theory of Stacking-Fault Contrast

The structure of a stacking fault will be illustrated by two examples, the face-centered cubic (fcc) lattice and the close-packed hexagonal lattice. A close-packed plane can be positioned on sites B or C of a layer A (Fig. 7.3). The fcc lattice with the {111} planes as close-packed planes can be described by the layer sequence ABCABC... and the hexagonal lattice by the sequence ABAB... . The lattice contains a stacking fault if one part of the crystal is shifted relative to the other by the displacement vector $\boldsymbol{R}_i$ (i = 1, 2, 3) of Fig. 7.3. The displacement vector $\boldsymbol{R}_1$, for example, transfers an A layer into a B layer and the new sequence is ABCABC|BCABC... . The line (|) indicates the position of the stacking fault. This fault can also be generated by removing an A layer from the crystal. It is then called an intrinsic stacking fault. An extrinsic stacking fault arises when an additional layer is introduced, a B layer in the sequence ABCABC|BABC, or when intrinsic faults occur in two neighboring planes.

In the kinematical theory of image contrast, the integral in (9.4) can be solved with the aid of the amplitude-phase diagram (APD) (Fig. 7.11), which gives a simple graphical solution of the contrast effects to be expected [9.49]. In Fig. 9.12, an inclined stacking fault crosses the foil, and the part below the fault is displaced by a constant vector $\boldsymbol{R} = a(u, v, w)$ (a: lattice constant). In (9.4), this lower part contributes with an additional phase shift

$$\alpha = 2\pi \boldsymbol{g} \cdot \boldsymbol{R} = 2\pi(hu + kv + lw) \tag{9.18}$$

because in an fcc lattice $\boldsymbol{R} = \langle 112 \rangle a/6$ and the hkl are all even or all odd (to satisfy the extinction rules of the structure amplitude F; see Table 7.1). This gives $\alpha = 2\pi n/3$ with $n = 0, \pm 1, \pm 2, \ldots$. The phase shift α is normally limited to the interval $-\pi \leq \alpha \leq \pi$ so that only the values $\alpha = 0, \pm 2\pi/3$ need be distinguished when considering the influence on image contrast. The stacking fault becomes invisible for $\alpha = 0$ because there will be no difference between (9.4) and the equation (7.24) for a perfect crystal. In some crystal structures (hexagonal AlN or rutile, for example) and in antiphase boundaries of ordered alloys (Sect. 8.3.3), stacking faults with $\alpha = \pi$ also have to be considered [9.58, 9.59].

A nonzero value of α has to be added to all phases below the stacking fault. This means that at the point Q of the APD corresponding to an $\alpha = 2\pi/3$ fault, a kink of 120° has to be introduced for a 220 reflection, for example (Fig. 9.13). The intensity of the Bragg-reflected beam is proportional to $\overline{\mathrm{PP'}}^2$ and will be greater than that of an undisturbed crystal, which is proportional to $\overline{\mathrm{PP''}}^2$, for the example shown in Fig. 9.13. The stacking fault will then be brighter in the dark-field image. The total curve length in the APD is equal to the foil thickness t and therefore the 120° kink moves for an inclined fault in such a way that $t = t_1 + t_2$ and t_1 is the depth of the fault in the column considered. The kinematic theory therefore predicts that the contrast will be symmetric about the foil center.

If $I_g = \psi_g \psi_g^*$ is calculated from (9.4), the integral can be split into two parts, with the limits 0, z_1 and z_1, t. By introducing the distance of the fault from the foil center $z' = z_1 - t/2$, formula (9.19) can be obtained [9.60],

$$I_g = \frac{1}{(\xi_g s)^2}[\sin(\pi ts + \alpha/2) + \sin^2(\alpha/2) \\ -2\sin(\alpha/2)\sin(\pi ts + \alpha/2)\cos(2\pi s z')], \tag{9.19}$$

and the symmetry relative to the foil center can be seen from the fact that, for a fixed value of foil thickness t and tilt parameter s, the position z' of the fault appears only in the last term of (9.19).

A stacking fault is imaged as a pattern of parallel equidistant fringes if the foil thickness is larger than a few extinction distances $\xi_{g,\mathrm{eff}}$ (Fig. 9.15). The number of fringes increases with increasing foil thickness, near edges for example, and the fringes split in the foil center. Intermediate thicknesses for which the term $\sin(\pi ts + \alpha/2)$ in (9.19) becomes zero show vanishing fringe contrast. If (9.18) leads to $\alpha = 0$, the stacking-fault contrast vanishes for all s and t, as mentioned above. This will be the case for $\boldsymbol{R} = [1\bar{2}1]a/6$ and $\boldsymbol{g} = 31\bar{1}$ or $1\bar{1}3$, for example. The direction of the displacement vector $\boldsymbol{R}$ can thus be determined if, by tilting the specimen, two excitations of different $\boldsymbol{g}$ can be found for which the stacking-fault contrast vanishes. Further information about the type of the fault can be obtained from the intensity laws of the dynamical theory (Sect. 9.3.2).

The number of fringes also increases with increasing tilt parameter s owing to the decrease of $\xi_{g,\mathrm{eff}}$, and the contrast difference between the dark and bright fringes is less.

For two overlapping stacking faults, the phase $\alpha = 2\pi/3$ of one fault is doubled in the region where the projected images of the faults overlap. The total phase $\alpha = 4\pi/3$ is equivalent to $-2\pi/3$ and the contrast is correspondingly modified; dark and bright fringes are interchanged. The total α becomes zero for three overlapping faults and the contrast vanishes in the overlap region for all $\boldsymbol{g}$.

9.3.2 Dynamical Theory of Stacking-Fault Contrast

The dynamical theory has to be used near a Bragg position [9.51, 9.60]. The contrast in the two-beam case can easily be calculated by the matrix method of Sect. 9.2.2 using (9.12). The matrices $[F]$ and $[F^{-1}]$ are defined in (9.11) and $[S]$ is given by (9.9) with the $[C]$ matrix of (7.58). The abbreviation $\Delta k = \gamma^{(1)} - \gamma^{(2)} = 1/\xi_{g,\mathrm{eff}}$ is used, and a common phase factor $\exp(\pi \mathrm{i} \Delta k\, t)$, unimportant for the calculation of intensity, is omitted:

$$\psi_0(t) = \cos(\pi kt) - \mathrm{i}\cos\beta\sin(\pi\Delta kt) \\ +\frac{1}{2}\sin^2\beta(\mathrm{e}^{\mathrm{i}\alpha} - 1)[\cos(\pi\Delta kt) - \cos(2\pi\Delta kz')]$$

$$\psi_g(t) = \mathrm{i}\sin\beta\sin(\pi\Delta kt) + \frac{1}{2}\sin\beta(1-\mathrm{e}^{\mathrm{i}\alpha})$$
$$\times\{\cos\beta\,[\cos(\pi\Delta kt) - \cos(2\pi\Delta kz')] - \mathrm{i}\sin(\pi\Delta kt)$$
$$+\,\mathrm{i}\sin(2\pi\Delta kz')\}. \tag{9.20}$$

β becomes equal to $\pi/2$ in the exact Bragg position $w = 0$. This results in the following values of the bright-field intensity $I_0 = \psi_0\psi_0^*$:

$$\begin{aligned}
&\text{for} \quad \alpha = \frac{2\pi}{3}: \quad I_0 = \frac{3}{4}\cos^2\left(\frac{2\pi z'}{\xi_g}\right) + \frac{1}{4}\cos^2\left(\frac{\pi t}{\xi_g}\right),\\
&\text{for} \quad \alpha = \pi \quad : \quad I_0 = \cos^2\left(\frac{2\pi z'}{\xi_g}\right).
\end{aligned} \tag{9.21}$$

The image intensity remains symmetric about the foil center because z' appears only in the cosine terms and, as absorption has been neglected, we have $I_0 + I_g = 1$. However, the cosine term containing z' is quadratic in (9.21) so that the number of fringes is double that found for large w in the kinematical theory [linear cosine term in (9.19)]. The fringe spacing now corresponds to alterations of $\xi_g/2$ with the stacking-fault depth. The transition from the pure kinematical theory ($w \gg 1$) to the dynamical theory in Bragg position ($w = 0$) proceeds by the formation of subsidiary fringes. The amplitude of these fringes increases with decreasing w, resulting in equal amplitudes and twice the number of fringes in the Bragg position.

The influence of anomalous absorption has been calculated [7.49], and the results are confirmed in experiments with larger specimen thicknesses. The following effects can be observed for an $\alpha = 2\pi/3$ stacking fault.

(1) The dark-field fringe pattern becomes asymmetric but the bright-field pattern remains symmetric about the foil center (Fig. 9.14). At the electron

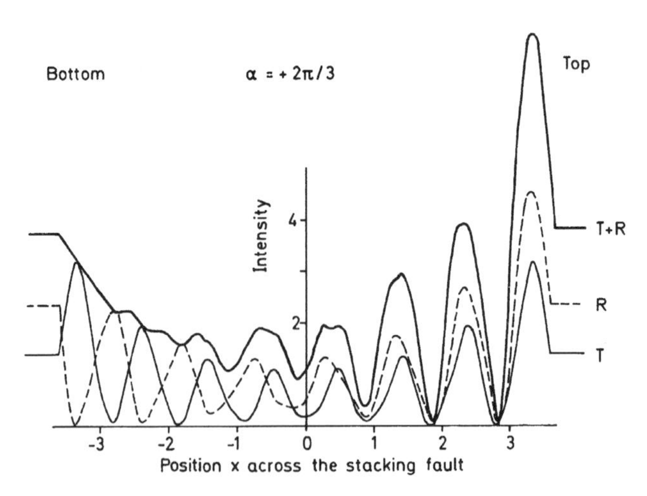

Fig. 9.14. Intensity of the primary beam T and the Bragg reflection R (bright- and dark-field intensities) across an inclined stacking fault. Note the complementarity of bright- and dark-field intensities at the bottom and anticomplementarity at the top.

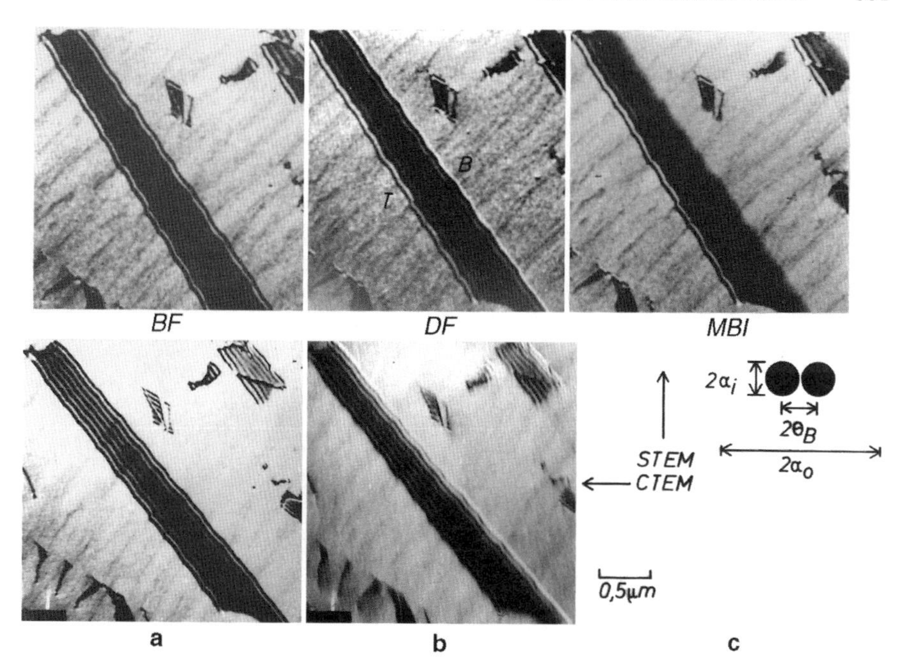

Fig. 9.15. Imaging of stacking faults in a Cu-7wt.% Al alloy in the STEM and TEM modes (top and bottom, respectively) with the indicated values of the electron probe aperture $\alpha_p = \alpha_i$, the Bragg angle θ_B, and the detector aperture $\alpha_d = \alpha_o$. (**a**) Bright field, (**b**) dark field obtained by selecting only the primary and reflected beams, respectively, and (**c**) $T + R$ multibeam image with vanishing fringe contrast at the bottom due to the complementarity of bright- and dark-field intensities.

entrance side (top), the dark- and bright-field fringes are anticomplementary; that is, maxima and minima appear at the same places for the same depths of the stacking fault. At the exit side (bottom), the bright- and dark-field fringes are complementary; maxima occur in bright-field fringes where minima are seen in the dark field and vice versa (Fig. 9.15a,b). This contrast effect can be used to decide how the stacking fault is inclined in the foil. In multibeam imaging (MBI, Sect. 9.1.3), the fringes persist at the top of the foil but are canceled at the bottom ($T + R$ in Figs. 9.14 and 9.15c).

(2) The fringe contrast decreases in the central region of the foil in the bright- and dark-field modes and completely vanishes in thick foils. Although, in the Bragg position, twice as many fringes are seen in the central part for medium thicknesses, only the normal number of fringes is observed at the top and the bottom.

(3) The first fringe in the bright field is bright for $\alpha = +2\pi/3$ (Fig. 9.15a) and dark for $\alpha = -2\pi/3$. If the direction of $\boldsymbol{g}$ is known, this can be used to establish the stacking-fault type [7.49, 9.61, 9.62].

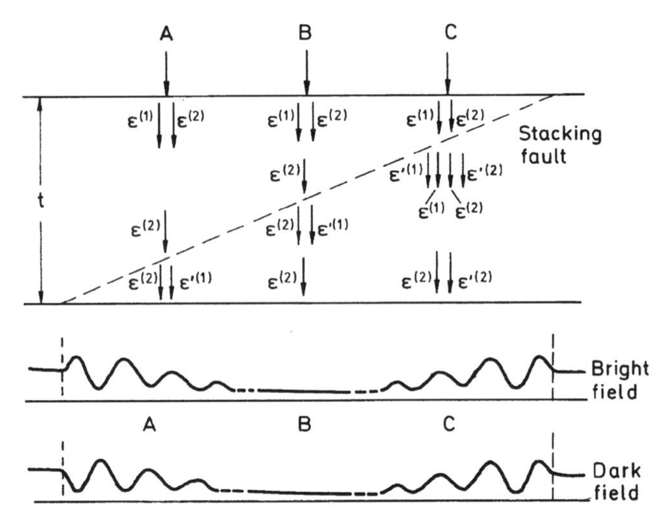

Fig. 9.16. Explanation of the complementarity of dark and bright field intensity at the bottom (A) and the anti-complementarity at the top (C) of a stacking fault and the vanishing stacking-fault contrast in the center (B) of a thick foil due to the differences in absorption of the Bloch waves.

Observations 1 and 2 can be understood in terms of the dynamical theory including absorption from the following model (Fig. 9.16). In the top part of the foil, first the Bloch waves with the excitation amplitudes $\epsilon^{(1)}$ and $\epsilon^{(2)}$ are present. In the region A, only the Bloch wave $\epsilon^{(2)}$ with decreased absorption (anomalous transmission) survives at the bottom of the foil. When the stacking fault is penetrated, a further division into $\epsilon^{(2)}$ and $\epsilon'^{(1)}$ occurs to satisfy the boundary condition at the fault plane; this can also be interpreted as interband scattering (Sect. 9.2.3). These Bloch waves create a complementary fringe pattern in dark and bright field images, analogous to the edge contours found in thin layers when absorption is neglected. In region C, the Bloch waves exist with nearly unaltered amplitudes $\epsilon^{(1)}$ and $\epsilon^{(2)}$ at the boundary. They are split and form additional waves $\epsilon'^{(1)}$ and $\epsilon'^{(2)}$. The waves $\epsilon'^{(1)}$ and $\epsilon^{(1)}$ are strongly absorbed in the lower part of the foil. The superposition of $\epsilon'^{(2)}$ and $\epsilon^{(2)}$ results in an anticomplementary fringe pattern in bright and dark field images. Correspondingly, only $\epsilon^{(2)}$ remains at the bottom in the central region B and the fringe contrast vanishes.

The following differences can be observed for π stacking faults. There is no difference of contrast for $\pm\pi$. Bright- and dark-field images are always symmetric about the foil center and anticomplementary. The central fringe is invariably bright in the bright field image and dark in the dark field image. The fringes are parallel to the central line and not to the intersection of the fault with the surface. Only two new fringes per extinction distance appear at the surface when the thickness is increased.

In contrast to Fig. 9.16, these effects can be explained by the fact that $\alpha = \pi$ corresponds to a displacement vector $\boldsymbol{R}$ of half the lattice-plane spacing. Bloch waves of type 1 with antinodes at the lattice planes become Bloch waves of type 2 with nodes below the fault plane. This means that the fault causes complete interband scattering $\epsilon^{(1)} \to \epsilon^{(2)}$ and $\epsilon^{(2)} \to \epsilon^{(1)}$ and there is no splitting into twice the number of Bloch waves, as there was for a $2\pi/3$ fault.

Stacking faults limited by partial dislocations of dissociated dislocations are discussed in Sect. 8.4.4.

9.3.3 Antiphase and Other Boundaries

Stacking-fault-like contrast with $\alpha = 0, \pm\pi$, or $\pm 2\pi/3$ can also be observed at antiphase boundaries in ordered alloys [9.63, 9.64, 9.65]. Figure 9.17 shows an example of the geometry of an antiphase boundary in an AuCu alloy. The imaging of periodic antiphase boundaries (AuCu II phase) by the extra diffraction spots has been discussed in Sect. 8.2.5 (Fig. 8.13). Here we consider only the contrast of single boundaries such as those present in the AuCu I phase generated by the primary beam and a single Bragg reflection.

The contrast is influenced by the displacement vector $\boldsymbol{R}$, and the Bragg reflection $\boldsymbol{g}$ of the ordered structure and its structure amplitude $F(\theta_g)$. The antiphase boundaries can be imaged with the superlattice reflections but not with the fundamental reflections, which are also present in the disordered state with a random distribution of the atoms on the lattice sites. This is a possible way of distinguishing antiphase boundaries from stacking faults.

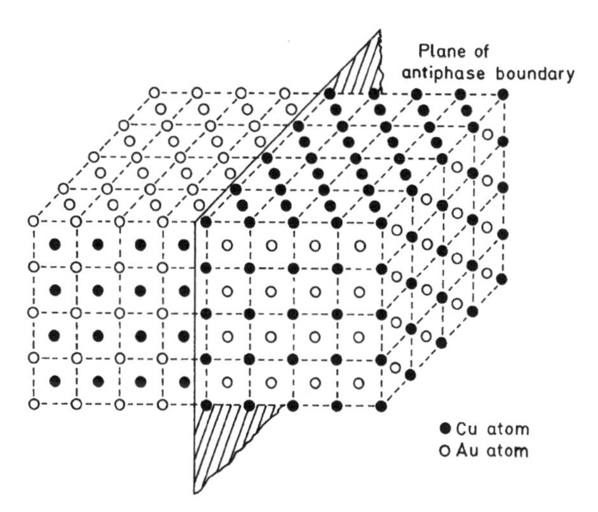

Fig. 9.17. Atomic positions at an antiphase boundary in the fcc lattice of an Au-Cu alloy.

The contrast of π and $2\pi/3$ boundaries can be analysed in the same way as that of stacking faults, and $\boldsymbol{R}$ can be derived from the condition $\boldsymbol{g} \cdot \boldsymbol{R} = 0$ of nonvisibility. Unlike those of stacking faults, the structure amplitude of the superlattice reflections are smaller owing to the differences listed in Table 9.1. The extinction distances are consequently two to three times larger than those of the fundamental reflections, which means that fewer fringes are seen.

Stacking-fault-like α-fringes can also be observed in noncentrosymmetric crystals at the boundary between enantiomorphic phases [9.66, 9.67, 9.68, 9.69]. Domain contrast is also observed for some orientations.

So-called δ-fringes are observed when the boundary cannot be described by a displacement vector $\boldsymbol{R}$ but instead either the two phases have different lattice constants or there is a tilt of the lattice with $\Delta \boldsymbol{g} = \boldsymbol{g}_1 - \boldsymbol{g}_2 \neq 0$. We then have $\delta = s_1\xi_{g1} - s_2\xi_{g2}$ [9.70, 9.71, 9.72, 9.73]. This contrast can also be interpreted as moiré fringes (Sect. 9.6.1); such effects are observed at ferroelectric domain boundaries [6.296] and antiferromagnetic domain boundaries in NiO [9.74]. δ-fringes can also be observed at precipitates when the active Bragg reflections of matrix and precipitate are almost the same [9.75]. If $\Delta \boldsymbol{g}$ is larger and a single reflection is strongly excited, only thickness (edge) contours (Sect. 9.1.1) are observed.

The following characteristic differences between δ-fringes and α-fringes can be used to distinguish the two types. The spacing of δ-fringes can be different at the top and bottom if $\xi_{g1} \neq \xi_{g2}$. The dark-field image becomes symmetric for $\xi_{g1} = \xi_{g2}$. The fringes are parallel to the surface so that, with increasing thickness, new fringes are generated in the center. The contrast of the outer fringes is dark or bright, depending on the sign of δ. Further differences can occur in the magnitude of the contrast modulation and in the background at the top and bottom of the foil.

Twin boundaries are also of interest because they are frequently encountered. Whereas single nonoverlapping boundaries show edge contours in one part of the crystal for large differences of s_1 and s_2, many-beam excitation can cause complicated fringe patterns [9.76]. In thin twin lamellae, for which no separation of the top and bottom boundary is seen in the projected image, the regions above and below the lamellae are displaced by a vector $\boldsymbol{R}$, which depends on the number of lattice planes in the lamellae; $\boldsymbol{R}$ therefore can also become zero whereupon the contrast vanishes for all Bragg reflections of the matrix.

Large cavities also cause strong diffraction contrast [9.77]; the resulting contrast may be brighter or darker than that of the matrix, depending on the excitation of the matrix reflection and the depth in the foil.

Other applications are the imaging of boundaries in minerals [1.33] and martensitic transformations [9.78].

Table 9.1. Examples of observable phase shifts $\alpha = 2\pi \boldsymbol{g} \cdot \boldsymbol{R}$ at antiphase boundaries in different crystal systems.

1) AuCu I phase (tetragonal distorted fcc lattice)

$$\text{Cu}: \boldsymbol{r}_1 = (0,0,0),\ \boldsymbol{r}_2 = \left(\tfrac{1}{2},\tfrac{1}{2},0\right); \quad \text{Au}: \boldsymbol{r}_3 = \left(\tfrac{1}{2},0,\tfrac{1}{2}\right),\ \boldsymbol{r}_4 = \left(0,\tfrac{1}{2},\tfrac{1}{2}\right)$$

Fundamental reflections: $F = 2(f_{\mathrm{Au}} + f_{\mathrm{Cu}})$ hkl mixed (odd and even)
Superlattice reflections: $F = 2(f_{\mathrm{Au}} - f_{\mathrm{Cu}})$ hkl even, even, odd or odd, odd, even

$$\boldsymbol{R} = \tfrac{a}{2}(011),\quad \tfrac{a}{2}(01\bar{1}),\quad \tfrac{a}{2}(101)\quad \tfrac{a}{2}(10\bar{1})$$

$$\alpha = \begin{cases} 0 & \text{Fundamental reflections} \\ 0, \pm\pi & \text{Superlattice reflections}\,(hkl)\,\text{mixed} \end{cases}$$

2) B2 structure (CsCl-type, FeAl, β-CuZn)

$$\text{A}: \boldsymbol{r}_1 = (0,0,0); \quad \text{B}: \boldsymbol{r}_2 = \left(\tfrac{1}{2},\tfrac{1}{2},\tfrac{1}{2}\right)$$

Fundamental reflections: $F = f_{\mathrm{A}} + f_{\mathrm{B}}$ $h + k + l$ even
Superlattice reflections: $F = f_{\mathrm{A}} - f_{\mathrm{B}}$ $h + k + l$ odd

$$\boldsymbol{R} = \tfrac{a}{2}\langle 111\rangle$$

$$\alpha = \pi(h+k+l) = \begin{cases} 0 & \text{Fundamental reflections} \\ 0, \pm\pi & \text{Superlattice reflections} \end{cases}$$

3) $\mathrm{L1_2}$ structure (Cu_3Au, Ni_3Al, Ni_3Mn)

$$\text{B}: \boldsymbol{r}_1 = (0,0,0); \quad \text{A}: \boldsymbol{r}_2 = \left(\tfrac{1}{2},\tfrac{1}{2},0\right),\ \boldsymbol{r}_3 = \left(\tfrac{1}{2},0,\tfrac{1}{2}\right),\ \boldsymbol{r}_4 = \left(0,\tfrac{1}{2},\tfrac{1}{2}\right)$$

Fundamental reflections : $F = 3f_{\mathrm{A}} + f_{\mathrm{B}}$ hkl odd or even
Superlattice reflections : $F = f_{\mathrm{A}} - f_{\mathrm{B}}$ hkl mixed

$$\boldsymbol{R} = \tfrac{a}{2}\langle 110\rangle,\ \alpha = \begin{cases} 0 & \text{Fundamental reflections} \\ 0, \pm\pi & \text{Superlattice reflections} \end{cases}$$

$$\boldsymbol{R} = \tfrac{a}{6}\langle 112\rangle\ \ \alpha = \begin{cases} 0, \pm 2\pi/3 & \text{Fundamental reflections (stacking faults)} \\ 0, \pm\pi/3, \pm 2\pi/3 & \text{Superlattice reflections} \end{cases}$$

9.4 Dislocations

9.4.1 Kinematical Theory of Dislocation Contrast

Dislocations parallel to the foil surface can be imaged as dark lines in bright-field images and bright lines in dark-field images; inclined dislocations are

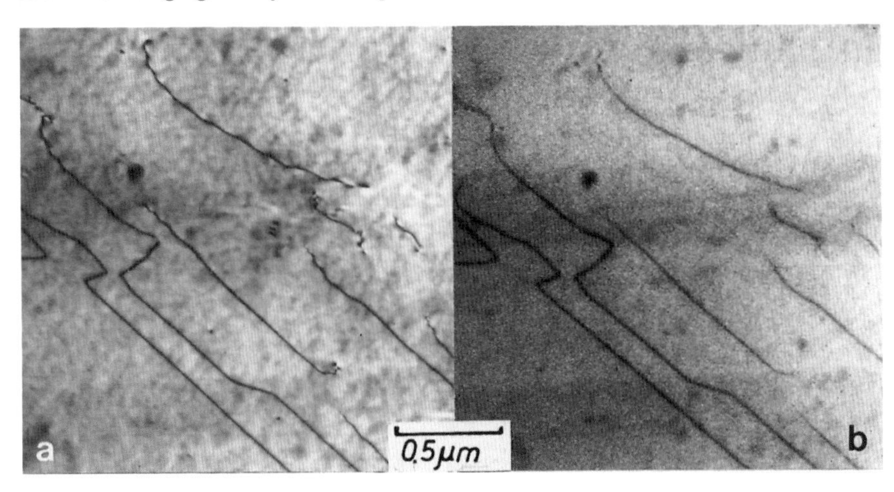

Fig. 9.18. Dislocations in an Al foil imaged in (**a**) the 100 keV TEM mode and (**b**) the STEM mode. Dislocations parallel to the surface show uniform contrast, inclined dislocations alternating contrast in the TEM mode and blurring of the alternating contrast in the STEM mode.

seen with a dotted-line or zig-zag contrast if $t \gg \xi_g$ (Fig. 9.18). Most of the contrast effects observable with large excitation errors can be explained on the basis of the kinematical theory, whereas the dynamical theory (Sect. 9.4.2) is necessary near the Bragg position and for weak-beam excitation.

The local displacement vector $\boldsymbol{R}(z)$ must be known before the contrast can be calculated. A simple analytical formula for $\boldsymbol{R}$ can be established for a screw dislocation (Fig. 9.12) with a Burgers vector $\boldsymbol{b}$ parallel to the unit vector $\boldsymbol{u}$ along the dislocation line. The column AA′ of a perfect crystal is bent to BB′ by a screw dislocation in the y direction. The unit cells are displaced in the y direction parallel to the Burgers vector $\boldsymbol{b}$. A circle around the dislocation line on a lattice plane does not close, its end being shifted by $\boldsymbol{b}$ relative to the origin. Assuming that the isotropic theory of elasticity is applicable, the displacement vector of a screw dislocation becomes

$$\boldsymbol{R} = \boldsymbol{b}\frac{\alpha}{2\pi} = \frac{\boldsymbol{b}}{2\pi}\arctan(z/x). \tag{9.22}$$

This displacement of a general dislocation, for which $\boldsymbol{b}$ is not parallel to $\boldsymbol{u}$ and the glide plane is parallel to the foil surface, becomes

$$\boldsymbol{R} = \frac{1}{2\pi}\left(\boldsymbol{b}\alpha + \frac{1}{4(1-\nu)}\{\boldsymbol{b}_\mathrm{e} + \boldsymbol{b} \times \boldsymbol{u}\,[2(1-2\nu)\ln|r| + \cos(2\alpha)]\}\right), \tag{9.23}$$

where ν is Poisson's ratio and $\boldsymbol{b}_\mathrm{e}$ is the edge component of the Burgers vector. Isotropic elasticity is again assumed to be valid, and relaxation of elastic strain at the free foil surfaces is neglected.

When (9.22) is substituted into (9.4), the scalar product $\boldsymbol{g} \cdot \boldsymbol{b}$ is an integer n because it is the product of a reciprocal-lattice vector and a crystal vector;

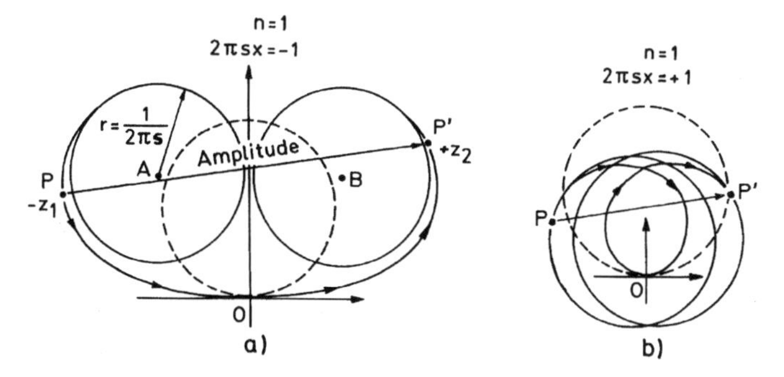

Fig. 9.19. Amplitude-phase diagram (ADP) for a column through the crystal close to a screw dislocation with $\boldsymbol{g}\cdot\boldsymbol{b}=n=1$ for (**a**) $2\pi sx=-1$ and (**b**) $2\pi sx=+1$ [9.49].

$\boldsymbol{g}\cdot\boldsymbol{b}=n=1$ for $\boldsymbol{g}=111$ and $\boldsymbol{b}=[110]a/2$ in an fcc lattice, for example. This yields

$$\psi_g=\frac{\mathrm{i}\pi}{\xi_g}\int\limits_{-z_1}^{-z_2}\exp\left[-\mathrm{i}\left(2\pi sz+n\arctan\frac{z}{x}\right)\right]\mathrm{d}z=\frac{\mathrm{i}\pi}{\xi_g}\int\limits_{-z_1}^{-z_2}\mathrm{e}^{\mathrm{i}\varphi}\mathrm{d}z. \tag{9.24}$$

The origin of the z coordinate is placed at the depth of the dislocation line ($z_1+z_2=t$). The phase φ, the slope of the element of arc length dz in an APD, becomes smaller or larger than the phase $2\pi sz$ of a perfect crystal, depending on the signs of s and x. If $sx<0$, the curvature at $z=0$ becomes smaller than the curvature $1/r=2\pi s$ for the perfect crystal (Fig. 9.19a). If $sx>0$, the curvature becomes greater (Fig. 9.19b). The value of $\arctan(z/x)$ saturates to a constant value $\pi/2$ for large z. The APD again tends asymptotically to a circle with the same radius $r=1/2\pi s$ for large z. The scattered amplitude ψ_g is proportional to the length $\overline{\mathrm{PP}'}$. The arc length between P and P′ is equal to $z_1+z_2=t$. If such ADPs are plotted for all values of $2\pi sx$, the intensity distribution $I(x)=\psi_g\psi_g^*=(\pi/\xi_g)^2\overline{PP'}^2$ normal to the dislocation line can be obtained for the dark-field mode and different values of $\boldsymbol{g}\cdot\boldsymbol{b}=n$ (Fig. 9.20). In this figure, only the distance AB between the centers of the asymmetric circles is plotted, which cancels oscillations caused by the particular depth z_1 of the dislocations and gives an average kinematical image.

Analogous calculations can be made for edge dislocations or for mixed dislocations with the glide plane parallel to the foil surface [9.79]. Apart from the width of the intensity maximum, no other important differences occur.

This simple kinematical theory predicts the following characteristics concerning the position and the width of the dislocation image. The image of the dislocation is not at the core ($x=0$), but the maximum in the dark-field mode is shifted to one side by a distance of the same order as the half-width. Figure 9.21 demonstrates schematically that the maximum will be on the side of the dislocation on which the lattice planes are bent nearer to the exact Bragg position. The position of the dislocation image therefore changes to the

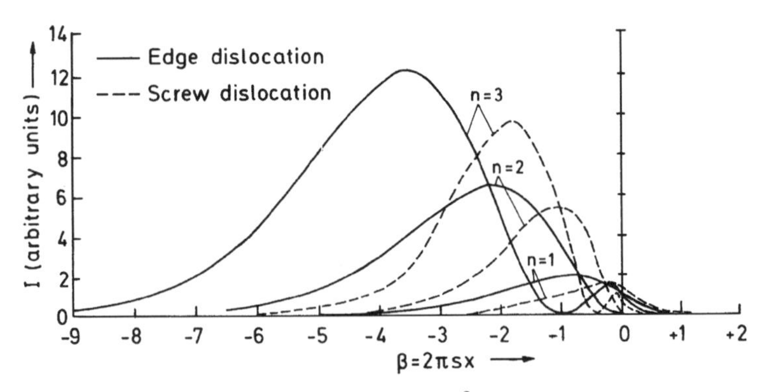

Fig. 9.20. Intensity profiles proportional to $\overline{\mathrm{AB}}^2$ in Fig. 9.19 across the dark-field images of edge and screw dislocations with different values of $\boldsymbol{g} \cdot \boldsymbol{b} = n$. The center of the dislocation is at $x = 0$ [9.49].

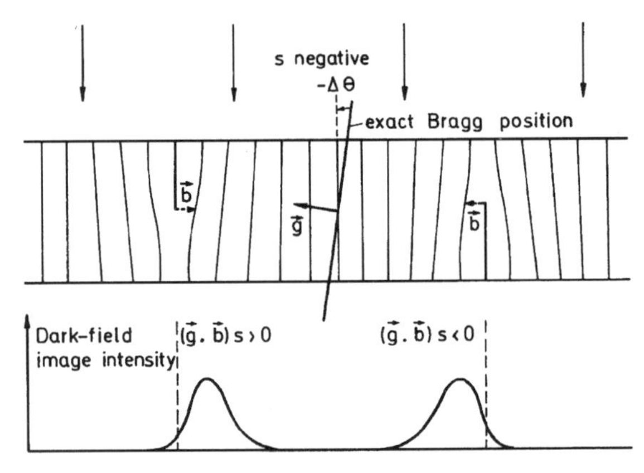

Fig. 9.21. Diagram showing on which side of its core the dislocation image is situated.

opposite side when a dislocation crosses the bend contour of the Bragg reflection (shaded in Fig. 9.22a), which changes the sign of x. For $\boldsymbol{g} \cdot \boldsymbol{b} = n = 2$, the dynamical theory indicates that a double image can arise in the Bragg position. Dislocation loops change their size for different signs of s (in-line and out-line contrast, Fig. 9.22b).

The half-width $x_{0.5}$ of the dislocation image is of the order of $1/2\pi s$ but depends also on the type of dislocation. The half-width of an edge dislocation is approximately twice that of a screw dislocation for equal values of s (Fig. 9.20). For low excitation errors, the width will be of the order of a few tens of nanometers and, in the Bragg position, of the order of $\xi_g/\pi \simeq 10 - 20$ nm (dynamical theory). Increasing s reduces $x_{0.5}$ and also the shift of the dislocation image relative to its core, but it also reduces the dark-field intensity. This is the principal reason why dislocation lines are much narrower ($x_{0.5} \simeq 1$ nm) with the weak-beam technique, which is, however, characterized not only by the

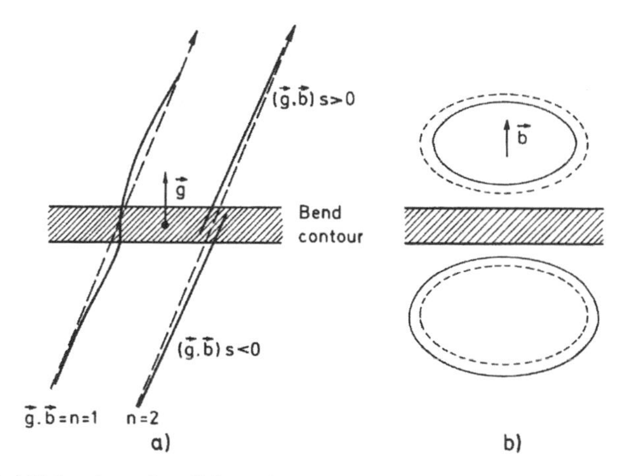

Fig. 9.22. (**a**) Behavior of a dislocation image crossing a bend contour and (**b**) of a dislocation loop image for positive and negative values of the product $(\boldsymbol{g}\cdot\boldsymbol{b})s$ as seen by an observer looking from below the foil.

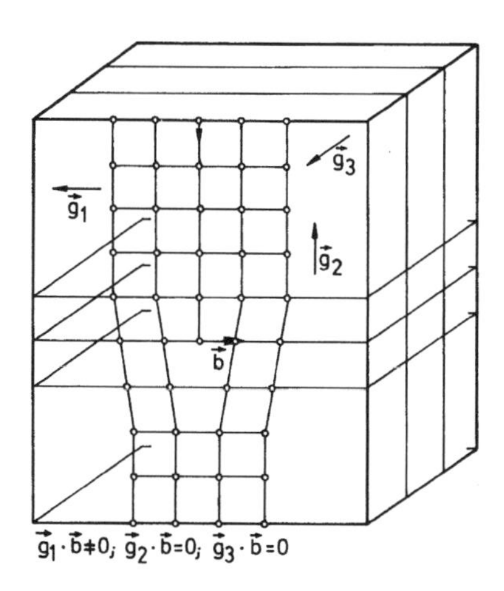

Fig. 9.23. Demonstration of the $\boldsymbol{g}\cdot\boldsymbol{b}=0$ rule for an edge dislocation. Only the lattice planes that belong to $\boldsymbol{g}_1$ are strongly bent, so that $\boldsymbol{g}_1\cdot\boldsymbol{b}\neq 0$, whereas $\boldsymbol{g}_2\cdot\boldsymbol{b}=\boldsymbol{g}_3\cdot\boldsymbol{b}=0$.

dark-field imaging with a Bragg reflection of high excitation error but also by the simultaneous strong excitation of another Bragg reflection. Weak-beam contrast (Sect. 9.4.3) thus becomes more of a dynamical contrast effect.

As shown in Sect. 9.2.1, the contrast of a lattice defect vanishes for Bragg reflections for which $\boldsymbol{g}\cdot\boldsymbol{R}=0$, which is equivalent to $\boldsymbol{g}\cdot\boldsymbol{b}=0$ for a dislocation. Figure 9.23 shows schematically that, in the presence of an edge dislocation, the lattice planes are bent for $\boldsymbol{g}_1$ but not for $\boldsymbol{g}_2$ and $\boldsymbol{g}_3$. This $\boldsymbol{g}\cdot\boldsymbol{b}=0$ rule is therefore important for the determination of the Burgers vector, which will be discussed in detail in Sect. 9.4.4.

9.4.2 Dynamical Effects in Dislocation Images

If strong low-order reflections or many beams are excited, the dynamical theory with absorption has to be used to analyze the image contrast of dislocations. The contrast and intensity profiles strongly depend on the depth of the dislocation, as can be seen in images of inclined dislocations [9.52]. Similar effects can be observed with stacking faults, as discussed in Sect. 9.3.2 and shown in Fig. 9.15. Thus the bright-field image is symmetrical about the foil center, whereas the dark-field image is asymmetrical and similar (anti-complementary) to the bright-field image near the top surface of the foil and complementary at the bottom. This asymmetry can be used to attribute the ends of an inclined dislocation to the top and bottom surfaces. In the multi-beam image (MBI) that can be formed in the STEM mode (Sect. 9.1.3) or a high-voltage electron microscope, the sum of the dark- and bright-field intensities cancels at the bottom surface [9.23]. For thicker foils, the contrast in the foil center becomes more uniform, which can be explained in terms of anomalous absorption, rather like the vanishing of the middle region of the fringes of a stacking fault (Fig. 9.16).

For $\boldsymbol{g} \cdot \boldsymbol{b} = n = 1$, the image of a screw dislocation at the foil center consists of a sharp dark peak of width $\xi_g/5$ for both DF and BF. Because the extinction distance ξ_g is between 20 and 50 nm for most metals, the width ranges from about 4 to 10 nm. The splitting of the dislocation image for $n = 2$ in the Bragg position has already been discussed in Sect. 9.4.1. Typical contrast effects for partial dislocations are also discussed in [9.52].

Although $\boldsymbol{g} \cdot \boldsymbol{b}$ vanishes for a screw dislocation normal to the foil and parallel to the electron beam, the dislocation can be seen as a black and white spot near the Bragg position. This contrast effect can be explained by considering surface relaxation and changes in the lattice parameter caused by the foil surface [9.80]. The same contrast for edge dislocations is very weak.

Dark dislocation lines often show an asymmetry in the longe-range background intensity. Thus, if this contrast alternates between nearly parallel dislocations, the signs of the $\boldsymbol{g} \cdot \boldsymbol{b}$ product of the dislocations alternate. As shown by calculations using the Bloch-wave method [9.81], this is an intrinsic-contrast phenomenon rather than a consequence of stress relaxation at the foil surfaces as was once believed [9.82].

Owing to the complexity of dynamical effects, comparison with a simulated image is indispensable; the image parameters must be known as accurately as possible, and the micrographs are compared with a set of computed fault images. Inclined defects need two-dimensional simulation. Calculations of line profiles across a dislocation, for example, is not sufficient. The computation time can be considerably reduced by using symmetries and the fact that the Howie–Whelan equations (9.7) are linear. This means that there can be only two independent solutions of these equations and that, once these are known, all other solutions (e.g., for different depths of the defect) can be obtained by forming a linear combination with the initial condition $\psi_0 = 1$ and $\psi_g = 0$ at the top surface. Such computation methods are described in [9.83, 9.84, 9.85, 9.86].

9.4.3 Weak-Beam Imaging

The Bragg contrast of dislocations using $\boldsymbol{g}$ vectors with low values of hkl and a small excitation error s_g creates broad images (Fig. 9.24a) with a half-width $x_{0.5} \simeq 5$–20 nm. Dense dislocation networks or weakly dissociated dislocations consisting of two partial dislocations on either side of a stacking fault cannot be resolved. Because the product of s and x appears in the abscissa $(2\pi sx)$ of Fig. 9.20, the width of dislocation images falls to 1.5–2 nm if large excitation errors are employed: $s_g \geq 0.2\ \mathrm{nm}^{-1}$ (Fig. 9.24b).

The basic principle of this weak-beam imaging technique [9.87, 9.88, 9.89] is demonstrated in Fig. 9.25 for a two-beam case. In Fig. 9.25a, the lattice planes are in the Bragg position ($s = 0$), giving the usual depth oscillation of I_0 and I_g with a repeat distance ξ_g. In Fig. 9.25b, a large excitation error s causes a decrease of the amplitude of the oscillation and the periodicity drops to $\xi_{g,\mathrm{eff}} \simeq 1/s \ll \xi_g$ [see also Fig. 7.14b and (7.61)]. The strain field on one side of a dislocation is idealized in Fig. 9.25c by regions (columns) AB and CD without distortion, and in the region BC the lattice planes are tilted into the Bragg position. The corresponding intensity I_g initially follows the low-intensity oscillation of periodicity $\xi_{g,\mathrm{eff}} = 1/s$, but below the depth B,

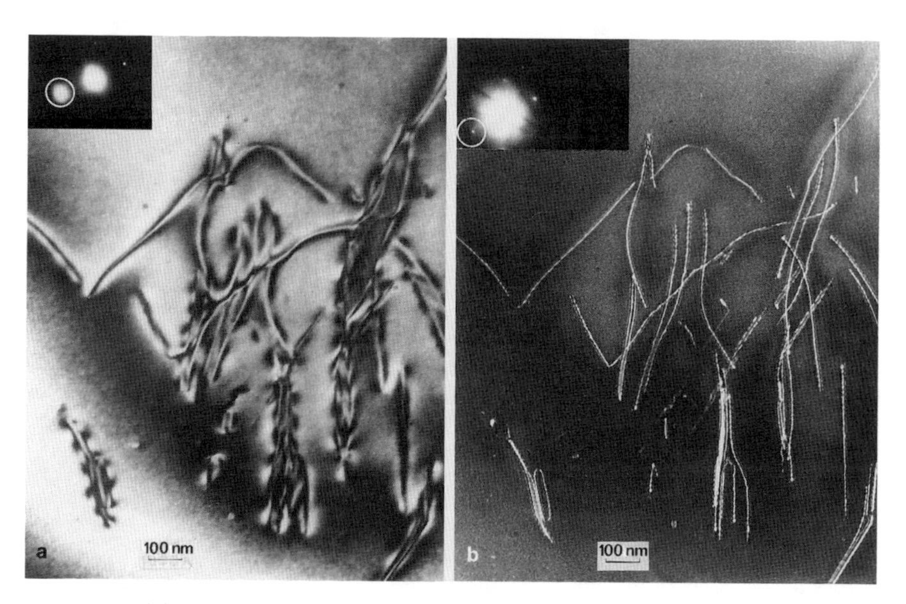

Fig. 9.24. (**a**) Dislocations in heavily deformed silicon imaged with a strong $2\bar{2}0$ diffracted beam. (**b**) Weak-beam $2\bar{2}0$ dark-field image of the same area showing the increase of the resolution of dislocation detail. The insets show the diffraction conditions used to form the images [9.90].

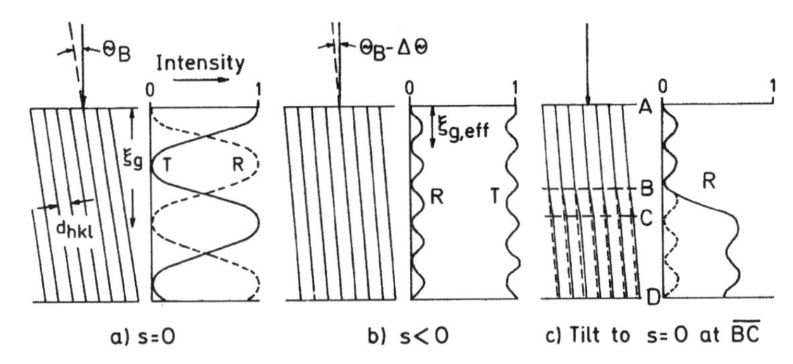

Fig. 9.25. Schematic explanation of weak-beam contrast. (**a**) Oscillation of the transmitted T and reflected-beam intensity R in the Bragg position ($s = 0$), (**b**) reduction of $\xi_{g,\text{eff}}$ and the oscillation amplitudes of T and R for $s < 0$, and (**c**) reflected intensity in the presence of a lattice-plane tilt to $s \simeq 0$ between B and C.

the intensity increases as in Fig. 9.25a. Beyond C, the decreased intensity will again show the low-intensity oscillations of Fig 9.25b. A region BC with $s \simeq 0$ is found only near the core of the dislocation. The model also explains why inclined dislocations show profile oscillations that correspond to ξ_g, which results in the alternating contrast seen in Fig. 9.24a, whereas in the weak-beam image of Fig. 9.24b, the repeat distance is much shorter because $\xi_{g,\text{eff}}$ is so much smaller.

This principle can also be applied to the two-beam Howie–Whelan equations (7.29) and (9.7) without and with a lattice defect, respectively. The Bragg position in (7.29) corresponds to $s = 0$, and the last term in the second equation becomes zero. The same condition for (9.4), which is equivalent to $s = 0$ in the region BC of Fig. 9.25, will be $s + \beta = 0$. The tilt parameter s can be compensated for by an opposite tilt of the lattice planes caused by the displacement field $\boldsymbol{R}(z)$ of the defect. The image peak occurs for those columns in which

$$s + \boldsymbol{g} \cdot \frac{\mathrm{d}\boldsymbol{R}}{\mathrm{d}z} = 0 \tag{9.25}$$

at a turning point of $\boldsymbol{g} \cdot \mathrm{d}\boldsymbol{R}/\mathrm{d}z$, where $\boldsymbol{g} \cdot \mathrm{d}^2\boldsymbol{R}/\mathrm{d}z^2 = 0$. When the displacement fields $\boldsymbol{R}(z)$ of screw and edge dislocations (9.22) and (9.23) are substituted, the image maximum is found to be at

$$x_W = -\frac{\boldsymbol{g} \cdot \boldsymbol{b}}{2\pi s}\left(1 + \frac{\epsilon}{2(1-\nu)}\right) \quad \text{with } \epsilon = \begin{cases} 1 \text{ edge} \\ 0 \text{ screw} \end{cases} \text{dislocation,} \tag{9.26}$$

whereas, for $\boldsymbol{g} \cdot \boldsymbol{b} = 2$, the average kinematical image in Fig. 9.20 (distance between the centers of the initial and final circles in the APD of Fig. 9.19) reaches a maximum at

$$x_K = -\frac{a}{2\pi s} \quad \text{with } a = \begin{cases} 2.1 \text{ edge} \\ 1.0 \text{ screw} \end{cases} \text{dislocation.} \tag{9.27}$$

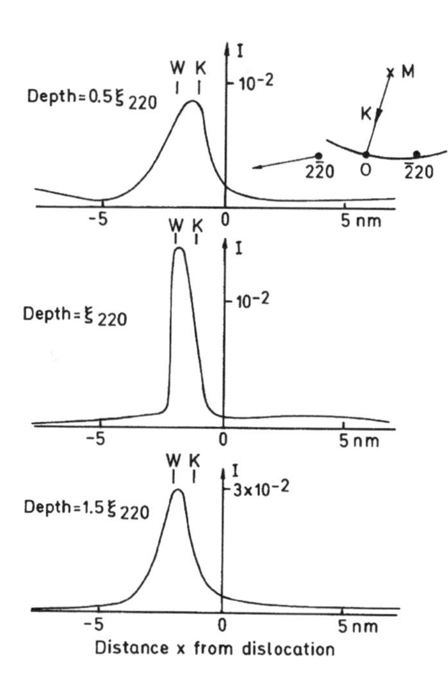

Fig. 9.26. Computed weak-beam-image profiles of an undissociated edge dislocation in copper at various depths ($\boldsymbol{g}\cdot\boldsymbol{b}=2$, $\boldsymbol{g}=220$, $t=2\xi_g$, $E=100$ keV, isotropic elasticity). W, K: image positions predicted by (9.26) and (9.27), respectively [9.89].

Values of x_W and x_K are indicated in Fig. 9.26, which shows a six-beam calculation of the image profile of an undissociated edge dislocation in copper at various depths. Dynamical effects influence the image profiles, especially if another reciprocal lattice vector $\boldsymbol{g}$ is strongly excited: A double-image line may appear, the lines may change their positions for different depths of inclined dislocations, or the contrast may disappear even if $\boldsymbol{g}\cdot\boldsymbol{b}\neq 0$. Such effects have to be considered for quantitative measurement of the size of dissociated dislocations or dislocation dipoles [9.90], and it will also be necessary to check the results against computed images based on the assumed model and the excitation parameters.

The column approximation is of limited validity for the high-resolution conditions of weak-beam imaging and can predict incorrect line profiles. The direct method of solving the Schrödinger equation assumes that the wave function $\psi_s(r)$ and the V_g vary very slowly over distances equal to the dimensions of the unit cell [9.91]. Thus, the maxima of the weak-beam dislocation profile can shift to the opposite side with increasing foil thickness [9.92].

The strong excitation of another $\boldsymbol{g}$ vector together with the positions of Kikuchi lines can be used to reach a value $s_g \geq 0.2$ nm^{-1} necessary for the weak-beam diffraction condition to be satisfied. In copper, for example, a $\bar{2}\,\bar{2}\,0$ or 600 reflection has to be strongly excited to get $s_g \simeq 0.2$ nm^{-1} for $\boldsymbol{g}=220$ [$g(\bar{g})$ or $g(3g)$ condition, respectively]. Conditions for other materials can be established using the Ewald sphere construction.

Whereas the bright-field mode is normally used when imaging with low excitation errors, it is necessary to use the dark-field mode for the weak-beam technique, which has the advantage that the dislocation image is narrow and well-contrasted but the disadvantage that a long exposure time, of the order of 10–30 s, is needed, compared with 1–2 s for a bright-field exposure.

The weak-beam technique has been applied to many problems, the quantitative measurement of stacking-fault energy by the separation of Shockley partial dislocations, for example, in Ag and Cu [9.93], Au [9.94], Ni [9.95], and stainless steel [9.96]. Dislocations in Ge and Si are generally found dissociated, and the dependence of the dissociation width on orientation is in good agreement with the anisotropic theory of elasticity [9.97]. Dissociations of the order of 0.5–1 nm can be measured in high-resolution lattice images of the dislocation core (Sect. 9.6.5).

9.4.4 Determination of the Burgers Vector

A knowledge of the direction and magnitude of the Burgers vector $\boldsymbol{b}$ is important for the interpretation of dislocation images for distinguishing between edge and screw dislocations, with Burgers vectors normal and parallel to the dislocation line, respectively. The sign of $\boldsymbol{b}$ allows us to distinguish between right-hand and left-hand screw dislocations. The image contrast depends on s, $\boldsymbol{b}$, and $\boldsymbol{g}$, and their signs and directions relative to the image must be known. Methods of recognizing the top and bottom of inclined dislocations are discussed in Sect. 9.4.2.

The first step is to align the diffraction pattern correctly relative to the image because the angle of rotation φ is not the same for both, the excitations of the intermediate lens being different in the imaging and SAED modes. In Sect. 8.2.4, methods for performing this alignment and for determining the crystal orientation are described.

The direction of the Burgers vector can be calculated if the dislocation contrast disappears for the Bragg excitations of two nonparallel vectors $\boldsymbol{g}_1$ and $\boldsymbol{g}_2$ due to the $\boldsymbol{g} \cdot \boldsymbol{b} = 0$ cancellation rule of Sect. 9.4.1. The Burgers vector will then be parallel to $\boldsymbol{g}_1 \times \boldsymbol{g}_2$. For example, $h\bar{h}l$ reflections with $h \neq 0$, l are needed to distinguish the invisibility of a $[110]a/2$ dislocation from that of other $\langle 110 \rangle a/2$ dislocations in an fcc crystal. The magnitude of $\boldsymbol{b}$ can be obtained from the behavior of the dislocation image when crossing the corresponding bend contour (e.g., $\boldsymbol{g} \cdot \boldsymbol{b} = 1$ or 2 in Fig. 9.22a). The sign of the Burgers vector is obtained from the sign of s and the position of the dislocation image relative to the dislocation center. The direction can be evaluated geometrically by using Fig. 9.21. The position of the dislocation center can be established by tilting the specimen to an opposite s and changing the dislocation image to the opposite side [9.98]. The sign of s can be determined by observing the position of the Kikuchi line relative to the correlated Bragg-diffraction spot because the system of Kikuchi lines (Kossel cones) remains fixed relative to the crystal foil during tilt, whereas the Bragg-diffraction spots do not change in position but only in intensity. The excitation error is defined as positive (see also Sect. 7.3.3) if the reciprocal lattice point is inside the

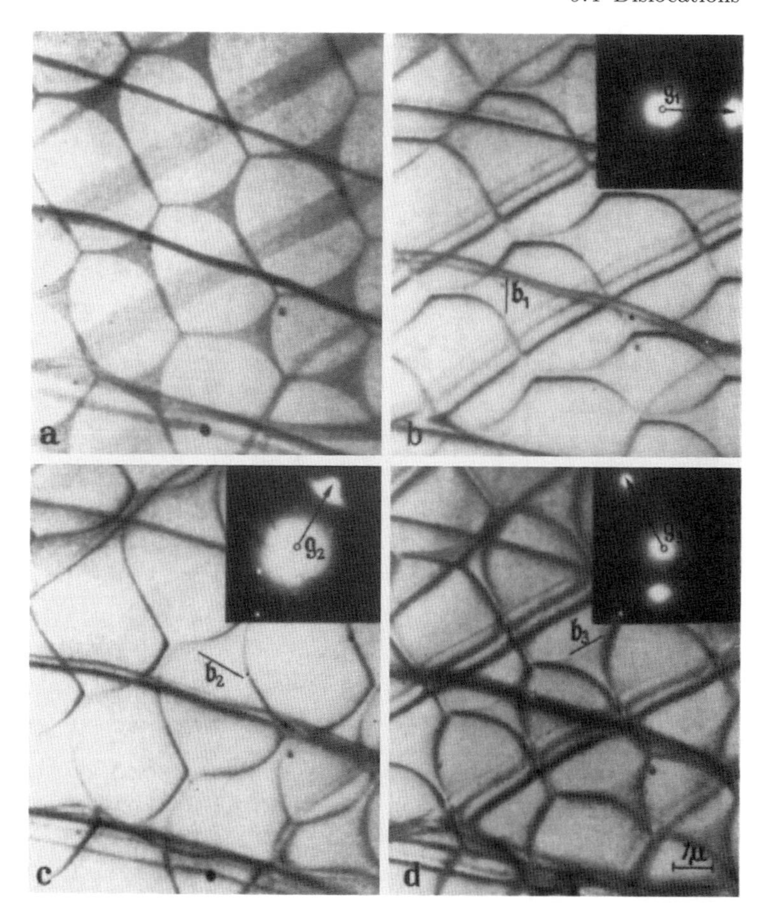

Fig. 9.27. Hexagonal network of stacking faults in graphite bounded by partial dislocations, imaged with different Bragg reflections excited (see insets). (**a**) Stacking-fault contrast only; (**b**), (**c**), and (**d**) vanishing contrast for one of the three types of partial dislocations with different Burgers vectors [9.99].

Ewald sphere and the Kikuchi line outside the diffraction spot. Kikuchi line and spot coincide for $s = 0$. The shift Δx of the Kikuchi line is directly proportional to the tilt $\Delta\theta$ of the Bragg position ($\Delta x = L\Delta\theta$, L: diffraction camera length), and s and $\Delta\theta$ are related by (7.15).

Partial dislocations that enclose a stacking fault can result from dissociation of a dislocation; e.g.,

$$\frac{a}{2}[\bar{1}\,0\,1] \rightarrow \frac{a}{6}[\bar{1}\,\bar{1}\,2] + \frac{a}{6}[\bar{2}\,1\,1]. \tag{9.28}$$

They become invisible if $\boldsymbol{g}\cdot\boldsymbol{b} = \pm\,1/3$ [9.50, 9.52]. An example of a dissociated dislocation network is shown in Fig. 9.27. In Fig. 9.27a, $\boldsymbol{g}\cdot\boldsymbol{b} = \pm 1/3$ for all partial dislocations and only the stacking-fault contrast (dark) can be seen. On tilting the specimen and imaging with the $\boldsymbol{g}$ excitations (indicated in the diffraction patterns of Figs. 9.27b–d), the stacking-fault contrast and one of the three partial dislocations vanish. The size or width of the stacking-fault area can be used to determine the stacking-fault energy [9.99]. With large stacking-fault energy, the width is of the order of a few nanometers, which can only be resolved by the weak-beam technique.

In practice, difficulties can arise in finding a second $\boldsymbol{g}$ for which the contrast disappears. Complexities in the form of residual contrast may appear if the edge component of the Burgers vector is large, and $\boldsymbol{g} \cdot (\boldsymbol{b} \times \boldsymbol{u})$ should also be zero if the displacement $\boldsymbol{R}(z)$ can be described by (9.23). Furthermore, elastic anisotropy and displacement relaxation at the foil surface have to be considered. Ambiguities can be associated with the determination of $\boldsymbol{b}$ from a disappearance condition with large $\boldsymbol{g}$, such as $\boldsymbol{g} \geq 311$ in copper [9.100]. Any detailed investigation of Burgers vectors must therefore be accompanied by computation of defect images. Asymmetries in the image of inclined dislocations related to the sign of $\boldsymbol{g} \cdot \boldsymbol{b}$ or $\boldsymbol{g} \cdot (\boldsymbol{b} \times \boldsymbol{u})$ can then be exploited [9.101].

This sometimes tedious method for the determination of $\boldsymbol{b}$ can be avoided by using LACBED patterns (Sect. 8.3.6).

9.5 Lattice Defects of Small Dimensions

9.5.1 Coherent and Incoherent Precipitates

Three types of precipitates can be distinguished, which depend on the fit between the lattice of the precipitate and the matrix: coherent, partially coherent and incoherent (Fig. 9.28). The resulting types of contrast can be separated into the following effects; these normally appear superposed but one of them often dominates.

(1) *Scattering Contrast.* For materials of large density and high atomic number, many electrons are scattered through large angles and intercepted by the objective diaphragm. A precipitate of larger density therefore appears darker in the bright-field image. On this scattering contrast, however, stronger dynamical contrast is superimposed.

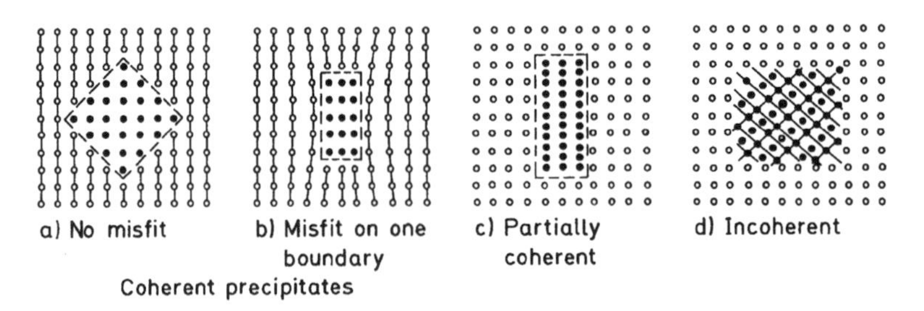

Fig. 9.28. (a-d) Schematic view of lattice distortions near coherent, partially coherent, and incoherent precipitates.

(2) *Structure-Factor Contrast* is caused by differences between the extinction distances ξ_g of the matrix and $\xi_g^\star$ of the precipitate. This contrast dominates in coherent precipitates without misfit. Coherent precipitates with misfit create a lattice strain field around the particle. This causes strain contrast, see point 6 below. For $w = s\xi_g = 0$ (Bragg position), the structure-factor contrast acts like a formal change of the specimen thickness (t: foil thickness, $\Delta t^\star$: thickness of the precipitate):

$$t_{\text{eff}} = t + \xi_g \Delta t^\star \left(\frac{1}{\xi_g^\star} - \frac{1}{\xi_g} \right) = t + \Delta t. \tag{9.29}$$

Differentiating $I_g = R = \sin^2(\pi t/\xi_g)$ given by (7.61) and multiplying by Δt, we obtain the intensity variation [9.102]

$$\Delta I_g = -\Delta I_0 = \frac{\mathrm{d}I_g}{\mathrm{d}t}\Delta t = \pi \Delta t^* \left(\frac{1}{\xi_g^\star} - \frac{1}{\xi_g} \right) \sin\left(\frac{2\pi t}{\xi_g} \right), \tag{9.30}$$

provided that $\Delta t^\star \ll \xi_g$. Maximum visibility with uniform bright contrast occurs where $t/\xi_g = 1/4$, $5/4, \ldots$ and with dark contrast where $t/\xi_g = 3/4$, $7/4, \ldots$ if $\xi_g^\star > \xi_g$. This structure-factor contrast will clearly be most effective in thin areas of the foil because the Bloch-wave absorption (not considered in (9.30)) decreases the effective contrast produced by small changes in thickness.

(3) *Orientation Contrast* can be observed for partially coherent and incoherent precipitates if they show a strong reflection and the matrix a weak reflection, or vice versa. In the dark-field image, the corresponding precipitate appears bright; this condition favors the determination of particle size and number. Large precipitates show edge fringes analogous to crystal boundaries if the Bragg reflection is excited strongly in only one part (matrix or precipitate). The spacing of edge fringes corresponds to the value of $\xi_{g,\text{eff}}$ (7.62) for the matrix or the precipitate, depending on which crystal is strongly excited.

(4) *Moiré Contrast* (Sect. 8.1.3), regarded as a special type of orientation contrast, will be seen if double reflection in the matrix and the precipitate results in extra diffraction spots near the primary beam, producing moiré fringes when superposed in the image.

(5) *Displacement-Fringe Contrast* is observed if, for example, a plate-like precipitate causes a normal displacement vector in a partially coherent precipitate,

$$\boldsymbol{R}_n = \delta \Delta t \boldsymbol{u}_n - n\boldsymbol{b}_n. \tag{9.31}$$

This displaces the matrix lattice planes in opposite directions on either side of the precipitate, where $\delta = 2(a_1 - a_2)/(a_1 + a_2)$ is the misfit parameter and n is the number of dislocations in the peripheral interface with a Burgers vector component $\boldsymbol{b}_n$ parallel to the normal. A phase shift $\alpha = 2\pi \boldsymbol{g} \cdot \boldsymbol{R}_n$ is generated as for the stacking-fault contrast, but $\boldsymbol{R}_n$ is not necessarily a lattice vector.

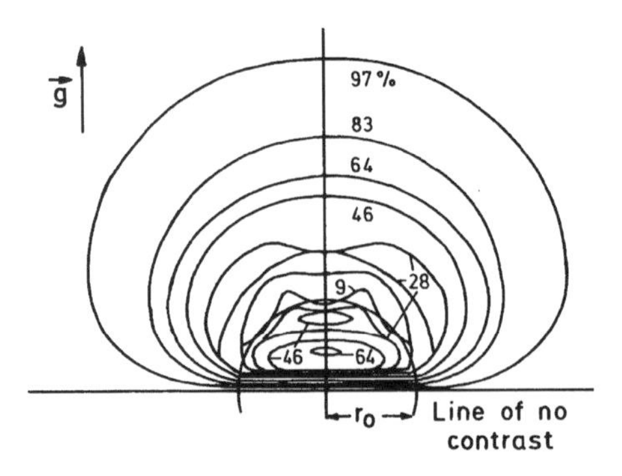

Fig. 9.29. Calculated bright-field intensity contours in the upper half-plane around a spherical, symmetrical strained precipitate of radius $r_0 = 0.25\xi_g$ in the center of a foil of thickness $t = 5\xi_g$ and a distortion $\epsilon g \xi_g = 10.2$. The figure is reflected at the line of no contrast, resulting in "butterfly" contrast.

Otherwise, many characteristics of stacking-fault contrast are observed when the plates are inclined to the foil surface. This contrast becomes most marked if the matrix is excited strongly and the precipitate weakly.

(6) *Matrix Strain-Field Contrast.* Coherent precipitates often create large strains in the matrix: Guinier–Preston zones II in Al-Cu, for example, or Co precipitates in Co-Cu alloys. Besides scattering, structure-factor, and displacement-fringe contrast, a long-range strain contrast can be observed [9.102]. The displacement vector of the strain field of a precipitate of radius r_0 is

$$\boldsymbol{R} = \epsilon \boldsymbol{r} \quad \text{for } r < r_0 \quad \text{and} \quad \boldsymbol{R} = \frac{\epsilon r_0^3}{r^3}\boldsymbol{r} \quad \text{for } r > r_0, \tag{9.32}$$

with $\epsilon = 3K\delta[3K + 2E(1+\nu)]^{-1}$, where K denotes the bulk modulus of the precipitate and E and ν are Young's modulus and Poisson's ratio for the matrix. Small cubic or tetrahedral precipitates also provoke such a symmetric strain field over a larger distance. Substitution of the radial displacement field $\boldsymbol{R}$ in (9.4) shows that a line exists along which $\boldsymbol{g} \cdot \boldsymbol{R} = 0$ and there is hence no contrast (Fig. 9.29), with the result that a butterfly or coffee-bean contrast is seen. The no-contrast line is normal to $\boldsymbol{g}$ and changes direction if other $\boldsymbol{g}$ are used by tilting the specimen. Plate-like precipitates create an anisotropic strain field, and the no-contrast line does not change much if $\boldsymbol{g}$ is varied. The width of the zone of equal contrast is proportional to $\epsilon g r_0/\xi_g$ and can be used to calculate the misfit of a precipitate [9.102, 9.103].

9.5.2 Defect Clusters

Frenkel defects (pairs of vacancies and interstitials) can be produced by bombardment with high-energy radiation. Single vacancies or interstitials cannot be resolved. The elastic strain of the lattice is too small and the phase contrast is insufficient. The defect clusters must be 1–2 nm in size for the strain field to give observable diffraction-contrast effects that can be used to analyze the type of cluster. Large clusters result in dislocation loops, stacking-fault tetrahedra, or cavities.

The symmetry of the strain field, the orientation and the Burgers vector $\boldsymbol{b}$ of the loops, and the distinction between interstitial and vacancy clusters and their depth and size distributions are of interest.

Kinematical image conditions result in black spots [9.104]. In the Bragg position, alternating black–white contrast is observed if the faults are near the top or bottom regions of the foil [9.105]. In the center of the foil, Bloch-wave absorption (see stacking-fault contrast, Sect. 8.2.2) leads to black points only [9.106].

As an example, typical contrast effects of small edge-dislocation loops of the vacancy type will now be described. A vector $\boldsymbol{l}$ from the dark to the bright spot of a bright-field image is introduced, and the following contrast effects result [9.107, 9.108, 9.109].

(1) $\boldsymbol{l}$ is parallel to the projection of $\boldsymbol{b}$ onto the foil plane and with only a few exceptions is independent of the direction of the excited $\boldsymbol{g}$. (For strain fields with spherical symmetry, $\boldsymbol{l}$ is always parallel or antiparallel to $\boldsymbol{g}$, see also Sect. 9.5.1.)

(2) The sign of $\boldsymbol{g} \cdot \boldsymbol{l}$ depends on the depth of the loop in the foil and differs for loops of vacancy and interstitial type. The diagram of Fig. 9.30 shows the typical contrast behavior. The zone L_1, for example, has a length $\xi_g/3 \simeq$ 5–10 nm. As for stacking faults and dislocations, dark- and bright-field micrographs are complementary only at the bottom surface of the foil. It is therefore necessary to know the sign of $\boldsymbol{g} \cdot \boldsymbol{l}$ and the depth before clusters of vacancies and interstitials can be distinguished. The depth can be measured with a high accuracy of $\pm$ 2 nm by recording a stereo pair [9.106, 9.107, 9.108, 9.109].

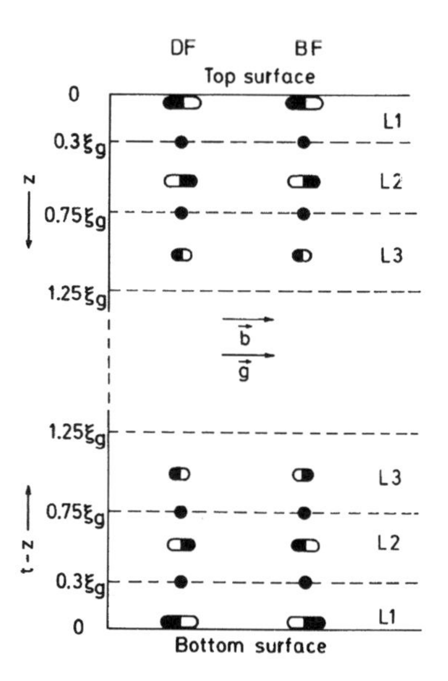

Fig. 9.30. Schematic plot of the depth oscillation of the black–white contrast from small dislocation loops of vacancy type in the dark-field (DF) and bright-field (BF) modes at the top and bottom surfaces of a thick foil. The direction of $\boldsymbol{g}$ must be reversed for loops of interstitial type [9.106].

(3) Loops with $\boldsymbol{g} \cdot \boldsymbol{b} = 0$ can exhibit weak residual contrast. The $\boldsymbol{g} \cdot \boldsymbol{b} = 0$ rule for dislocations has to be used with care for these defects, too. Computer simulation is essential for a detailed analysis [9.110, 9.111].

Disordered zones in displacement cascades in irradiated ordered alloys can be successfully investigated by examining dark-field superlattice reflections, which reveal the disordered regions as dark spots [9.112, 9.113].

Small cavities in irradiated material (diameter $\simeq$ 1–2 nm) show a dependence on depth like that discussed as structure-factor contrast in Sect. 9.5.1 but can be best observed with slight defocusing because a phase-contrast contribution has to be considered [9.114].

9.6 High-Resolution Electron Microscopy (HREM) of Crystals

9.6.1 Lattice-Plane Fringes

The crystal-lattice planes can be imaged and resolved, provided that the information about the lattice structure – that is, the primary and the Bragg-reflected beam – can pass through the objective diaphragm and that the contrast is not destroyed by insufficient spatial and temporal coherence. The primary beam and Bragg reflection are inclined at an angle $\theta_g = 2\theta_{\mathrm{B}}$ ($2d \sin\theta_{\mathrm{B}} = \lambda$). The plane waves interfere in the image plane with smaller angular separation θ_g/M (M: magnification) between the primary and reflected waves. The distance between the maxima of the resulting two-beam interference fringes is Md. In the first such observation, lattice fringes in copper phthalocyanine (d = 1.2 nm) were resolved [9.115]. Today it is possible to resolve lattice-plane fringes with $d \simeq 0.1$ nm. Figure 9.31a shows lattice-plane fringes from an evaporated gold film with a 20° crystal boundary (d_{220} = 0.204 nm) [9.116]. Diffraction contrast from the dense dislocations at the grain boundary is superposed on the lattice-fringe image.

For more extensive calculations and discussion of the contrast of lattice fringes in the two-beam case [9.14, 9.117], we use (7.60) for the amplitudes ψ_0 and ψ_g. These equations can be rewritten

$$\begin{aligned} \psi_0 &= |\psi_0|\mathrm{e}^{-\mathrm{i}\phi} \quad \text{with } |\psi_0|^2 = T \quad \text{and} \\ &\qquad \tan\phi = \frac{w}{\sqrt{1+w^2}} \tan\left[\frac{\pi\sqrt{1+w^2}}{\xi_g}\, t\right], \\ \psi_g &= \mathrm{i}|\psi_g|\mathrm{e}^{2\pi \mathrm{i} g x} \text{ with } |\psi_g|^2 = R. \end{aligned} \tag{9.33}$$

The superposition of the two waves in the image plane results in an image amplitude

$$\psi = \psi_0 + \psi_g \mathrm{e}^{-\mathrm{i}W(\theta_g)} \tag{9.34}$$

including the wave aberration $W(\theta_g)$ (3.65) in the diffracted beam. With $|\boldsymbol{g}| = 1/d$, the image intensity becomes

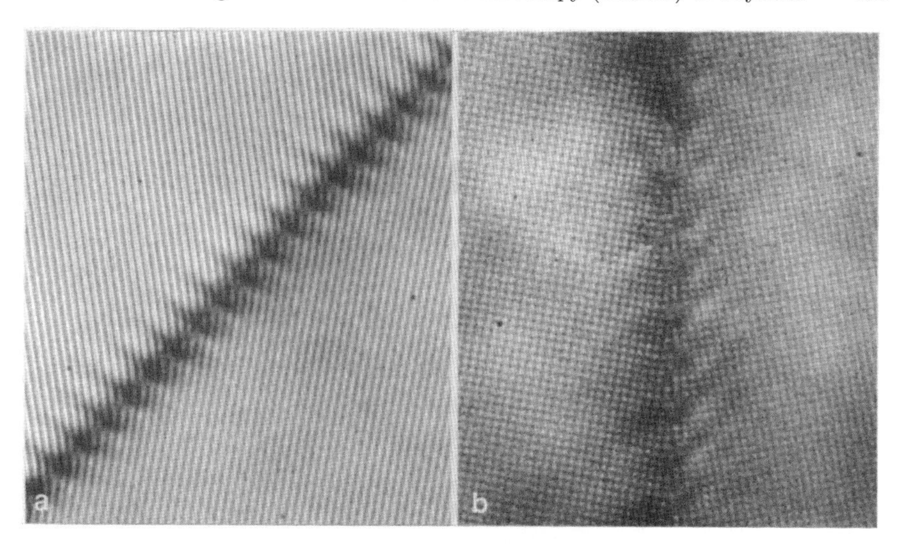

Fig. 9.31. High-resolution image of 20° [001] tilt boundaries in epitaxially grown Au films on NaCl bicrystals. (**a**) 0.204 nm lattice fringes in common [110] direction. (**b**) Crossed lattice image at a boundary in [010] orientation [9.116].

$$I(x) = \psi\psi^* = |\psi_0|^2 + |\psi_g|^2 - 2|\psi_0||\psi_g| \sin\left[\frac{2\pi x}{d} + \phi - W(\theta_g)\right] . \tag{9.35}$$

This means that the image contrast depends on the thickness t and the tilt parameter w of the crystal foil. The highest contrast will be observed for thicknesses $t = \xi_g/4$, $3\xi_g/4$, $5\xi_g/4$, etc., where $|\psi_0| = |\psi_g|$ in the Bragg position ($w = 0$); for intermediate thicknesses with $|\psi_0| = 0$ or $|\psi_g| = 0$, the contrast vanishes. It also decreases with increasing tilt parameter w because the maximum possible value of $|\psi_g|$ decreases.

The phase contribution ϕ in (9.35) can vary with x in edge contours (varying t) and/or bend contours (varying w) and causes a shift of the lattice-fringe positions. The local variation of the observed lattice-fringe spacing d' can be calculated [9.118] by writing

$$\frac{2\pi}{d'} = \frac{\mathrm{d}}{\mathrm{d}x}\left(\frac{2\pi x}{d} + \phi\right) = \frac{2\pi}{d} + \frac{\mathrm{d}\phi}{\mathrm{d}x} . \tag{9.36}$$

In edge and bend contours, $\mathrm{d}\phi/\mathrm{d}x$ is normally so small that this correction can be neglected. It must, however, be taken into account if an accurate value of d is being determined by counting a large number of fringes for the investigation of spinodal alloys, for example, which exhibit modulations in composition with a wavelength of the order of 10 nm [9.119, 9.120].

With the primary beam on-axis and the diffracted beam at an angle $\theta_g = 2\theta_\mathrm{B} \simeq \lambda/d$ to the axis, the wave aberration $W(\theta_g)$ in (9.35) causes a shift of the fringe pattern by one lattice-plane spacing if $W(\theta_g) = 2\pi$. This defocusing shift does not affect the visibility (resolution and contrast) of the fringes. However, the contrast of closely spaced lattice planes or large spatial frequencies is influenced by the attenuation characterized by the contrast-transfer envelope $K_c(q)$ in (6.42) caused by the chromatic aberration and by the energy spread of the electron gun. The resolution can therefore be increased by using tilted illumination with the primary beam tilted at an angle θ_B to the axis [9.121]. The wave aberration $W(\theta_g)$ will be the same for both waves, and the contrast-transfer envelope $K_c(q)$ decreases the fringe contrast to the same value for half the spatial frequency. In this way, lattice-plane spacings down to 0.1 nm can be resolved.

In a homogeneous material, the lattice spacing d can be determined more accurately by electron diffraction and the fringe distance can be used for calibration of magnification. As discussed in Sect. 4.4.3, this method is accurate to within $\pm 2\%$ and the limited reproducibility of the magnification contributes a further error of $\pm 1\%$.

Another application is the imaging of lattice planes in and near crystal defects, which complements the information given by the contrast effects discussed in Sects. 9.3 and 9.4. The latter are caused by the lattice displacements, which affect the amplitude of the primary and diffracted beams. Translational antiphase boundaries, for example, cause a fringe shift [9.122]; Guinier–Preston zones of Al-Cu consisting of single atomic planes enriched in Cu in the Al lattice can be imaged [9.123].

The imaging of lattice planes only requires coherent superposition of the primary beam and a reflected beam (or a systematic row) in the image plane. If several nonsystematic reflections are used, a cross-grating of lattice planes can be resolved (Fig. 9.31b); in the first version of this method, four reflections (000, 200, 020, and 220) entered the objective lens, with the optic axis in the center of the square defined by the four diffraction spots [9.124].

9.6.2 General Aspects of Crystal-Structure Imaging

For thin crystals, the amplitude of the Fourier transform of the crystal potential can be determined by electron diffraction. However, the phase information is completely lost, so that a reconstruction by a mere inverse Fourier transform is not possible. For crystal-structure imaging by high-resolution TEM (HREM), the inverse Fourier transform – from the diffraction pattern to the image – takes place directly inside the microscope without any loss of phase even though additional phase shifts are introduced by the wave aberration. Imaging of the crystal structure therefore becomes a special case of phase-contrast image formation.

One of the first examples of imaging larger unit cells in $Nb_{22}O_{54}$ (Fig. 9.32) illustrates the problems that arise on increasing the thickness (from left to right) and the defocusing (top to bottom) [9.125]. A good demonstration that the crystal images correspond to a projection of the crystal structure is the fact that tunnels in the 3×3 and 3×4 block structures of ternary oxides of Nb, W, or Ti and other materials are seen as bright spots (top left of

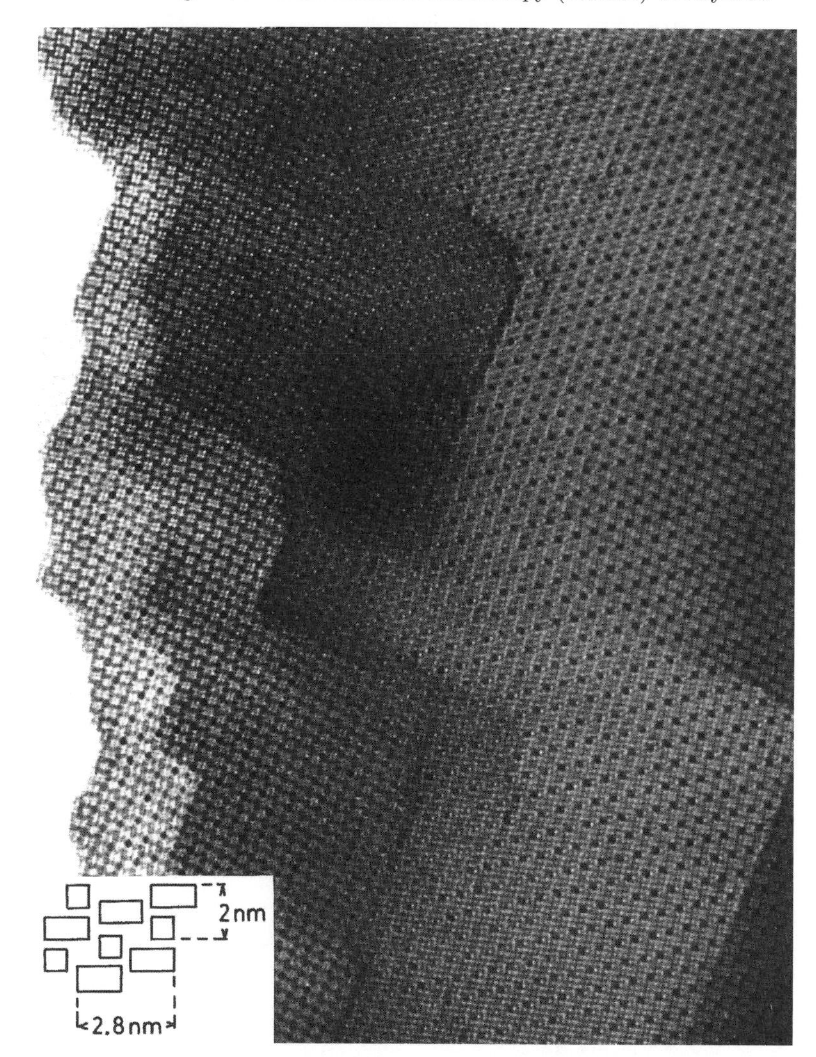

Fig. 9.32. Crystal-structure imaging of $Nb_{22}O_{54}$ containing octahedra of NbO_6, which form 3×3 and 3×4 blocks (see inset for dimensions), and its change with increasing specimen thickness (from left to right) and defocusing (from top to bottom). The crystal structure is imaged correctly at the top left. A recurrence of the same structure can be seen at the bottom right [9.125].

Fig. 9.32). The channels that contain Nb atoms are imaged as white dots, the regions where the octahedra have common edges, or shear planes, exhibit grey contrast, and the region near each tetrahedrally coordinated metal atom, surrounded by eight metal atoms in the nearest octahedra, is black.

With increasing thickness, the Bragg-diffraction intensities vary with thickness and excitation errors (dynamical theory), and the intensity distribution in the image of the lattice shows drastic changes, which can result in imaging artifacts and even contrast reversals of the main structure. If the thickness is still further increased, to $\simeq$50 nm, the structure that is observed for thicknesses below 5 nm may reappear (bottom right of Fig. 9.32) [9.126, 9.127].

The following instrumental conditions are essential for efficient HREM of crystal structures:

1. high electronic stability of acceleration voltage and lens currents,
2. vibration-free mechanical stability,
3. medium acceleration voltages of 200–400 kV (or high resolution 1 MV),
4. high gun brightness using a field-emission or Schottky electron gun,
5. low spherical-aberration constant $C_s \simeq 0.5$ mm of the objective lens (or a C_s corrector for sub Å resolution),
6. low energy spread $\Delta E \leq 0.7$ eV of the electron gun and low chromatic-aberration constant C_c for good temporal coherence,
7. "parallel" beam illumination (low illumination aperture $\alpha_i \leq 0.1$ mrad) for good spatial coherence,
8. coma-free on-axis alignment of the electron beam (Sect. 2.4.3),
9. precise alignment of crystal foils parallel to a low-index zone axis with a drift-free goniometer of high precision and an accuracy better than 0.25 mrad [2.47],
10. image control and recording by a CCD camera, and a
11. computer workstation for measurement of C_s and defocus Δz, autotuning, and online image simulation.

For the imaging of crystal structure, it is necessary to ensure that as many Bragg reflections as possible contribute to the image. The ideal image will be a projection of atomic rows, which appear as black dots when the irradiation is exactly parallel to a low-index zone axis.

The main limitations of crystal-structure imaging with a large number of reflections are the additional phase shifts $W(\theta_g)$ introduced by the wave aberration of the objective lens, which are different for the various reflections, and the attenuation of the contrast-transfer function by lack of spatial and especially temporal coherence (Sect. 6.4.2).

The present-day standard 200–400 kV HREMs allow a point-to-point resolution of about 0.15–0.2 nm to be attained, and $\simeq$0.1 nm has been reached with a 1 MV instrument [9.128]. More recently, aberration correctors have been developed (Sect. 2.4.2, [2.40, 9.129]) that allow the spherical aberration of the objective lens to be chosen at will. Then the main limitation is the partial temporal coherence caused by the energy spread of the electron gun. The latter can be reduced with a monochromator below the cathode. Zero-loss filtering of unscattered and elastically scattered electrons by imaging energy filters (Sect. 4.6) can remove the contribution from inelastically scattered electrons, and the fit with simulated images should be better. Combining an HREM with an electron monochromator unit and an imaging energy filter is not just futuristic but is now technically realizable [9.130] and allows subangstrom resolution to be obtained with 200 kV microscopes [9.131]. Using such a microscope, one can enhance the contrast of light atoms by setting the contrast to a negative value [9.132, 9.133].

9.6.3 Methods for Calculating Lattice-Image Contrast

Phase-Grating Approximation. For thin crystals, the phase-grating theory can be applied to simulate images of crystal lattices. By using (3.17) for the refractive index, the exit wave function behind the specimen is found to be

$$\begin{aligned}\psi_s(x,y) &= \psi_0 \exp\left[-\frac{2\pi\mathrm{i}}{\lambda E}\frac{E+E_0}{E+2E_0}\int_0^t V(x,y,z)\mathrm{d}z\right] \\ &= \psi_0 \exp[-\mathrm{i}\sigma(x,y)]. \end{aligned} \tag{9.37}$$

In the weak-phase-grating approximation, appropriate for very thin crystals, only the first two terms in the series expansion of the exponential are retained,

$$\psi_s(x,y) = \psi_0\left[1 - \frac{2\pi\mathrm{i}}{\lambda E}\frac{E+E_0}{E+2E_0}\int_0^t V(x,yz)\mathrm{d}z + \ldots\right], \tag{9.38}$$

and the phase-contrast theory of Sect. 6.2.4 can be applied, except that the integration over the Fourier plane now becomes a summation over the limited number of diffraction spots.

Bloch-Wave Method. For the simulation of crystal-lattice images when the unit cells are not too large, the exit wave function can also be determined by calculating the eigenvalues $\gamma^{(j)} + \mathrm{i}q^{(j)}$ and eigenvector components $C_g^{(j)}$ of Bloch waves by solving the fundamental equations of the dynamical theory (Sect. 7.3.3). The image intensity distribution is found to be

$$\begin{aligned} I\ (\boldsymbol{r}) &= \sum_{g,h}\sum_{i,j} C_0^{(i)*} C_0^{(j)} C_g^{(i)} C_h^{(j)*} \exp[-2\pi(q^{(i)} + q^{(j)})t] \\ &\times \exp\{\mathrm{i}[2\pi(\gamma^{(i)} - \gamma^{(j)})t + 2\pi(\boldsymbol{g}-\boldsymbol{h})\cdot\boldsymbol{r} - W(\theta_g,\Delta z) + W(\theta_h,\Delta z)]\}. \end{aligned} \tag{9.39}$$

The phase shift $W(\theta_g, \Delta z)$ contains the defocusing Δz and the spherical-aberration coefficient C_s; it is hence impossible to image a large number of hkl Bragg reflections around an $[mno]$ zone axis with $mh + nk + ol = 0$ with the same phase shifts, so that the superposition of the reflection amplitudes in the image becomes incorrect. These difficulties are less severe for specimens with very large unit cells $\geq$1 nm with a large number of reflections at low θ_g. The practical implementation of this technique and some typical applications are summarized in [1.69, 9.126, 9.134]. Experiments and calculations for very thin specimens confirm that, for thicknesses less than 5 nm, the crystal can be regarded as a weak-phase specimen, which is equivalent to saying that the weak-phase-grating approximation is valid. The maximum allowable thickness increases with increasing size of the unit cell. Model calculations (e.g., comparison with the multislice or Bloch-wave methods) are necessary for each particular structure to check whether the weak-phase-grating approximation is indeed valid.

For high-resolution imaging, the value of the defocusing Δz that gives a broad main transfer band of the contrast-transfer function (CTF; Sect. 6.4.1) will be most satisfactory, just as for amorphous specimens. Once again, the limiting factor will be the decrease of the CTF at large q caused by the finite illumination aperture and energy spread (spatial and temporal partial coherences, respectively). The phase shift $W(\theta_g)$ due to spherical aberration decreases as $\theta_g^4/\lambda \propto \lambda^3$ with increasing energy, and the attraction of using higher acceleration voltages for crystal-structure imaging is clear.

For a C_s-corrected microscope, the resolution is limited by envelope due to the energy spread. Therefore, to improve the resolution, these instruments are equipped with a monochromator.

Multislice Method. For large unit cells, lattice defects and interfaces, the multislice many-beam dynamical theory of Cowley and Moodie [9.135, 9.136, 9.137, 9.138, 9.139] is in widespread use. The crystal foil is cut into a larger number of thin slices of equal thickness Δz perpendicular to the incident electron beam. Just as in the phase-grating approximation (9.37), the projected potential and phase shift $\sigma(x, y)$ is determined for each slice by integrating $V(x, y, z)$ between the limits $(n-1)\Delta z$ and $n\Delta z$ for the nth slice. After applying this phase shift at the center or bottom of the nth slice to the wave function of the preceding slice by multiplying by $\exp[-i\sigma(x, y)]$, the wave propagates in a "vacuum" by Fresnel diffraction to the center or bottom of the next slice. This Fresnel propagation over a distance Δz is equivalent to convolving the wave function with

$$p(x, y) = \frac{1}{i\lambda\Delta z} \exp\left[\pi i \frac{x^2 + y^2}{\lambda\Delta z}\right], \tag{9.40}$$

which is obtained from (3.33) for $r_0 \to \infty$ (plane-wave approximation) and $R_0 = \Delta z$. It is therefore convenient to perform the calculation in Fourier space and make use of the convolution theorem of Fourier transforms. We write

$$\begin{aligned} \Psi_n(q_x, q_y) &= \Psi_n(\boldsymbol{q}) = \boldsymbol{F}\{\psi_n(x, y)\}, \\ P(q_x, q_y) &= \mathbf{F}\{p(x, y)\} = \exp[i\pi\Delta z\lambda(q_x^2 + q_y^2)], \end{aligned} \tag{9.41}$$

where $\psi_n(x, y)$ is the wave function just above the nth slice. The Fourier transform of the exit wave function (diffraction amplitude) is given by the recursion

$$\begin{aligned} \Psi'_{n-1}(\mathrm{x, y}) &= \exp[-i\sigma(x, y)]\Psi_{n-1}(x, y), \\ \Psi_n(\boldsymbol{q}) &= P(\boldsymbol{q})\Psi'_{n-1}(\boldsymbol{q}). \end{aligned} \tag{9.42}$$

This formalism can also be applied to nonperiodic and even to amorphous specimens when the positions of individual atoms are stored. In the case of periodic crystal structures, only the reciprocal-lattice points $\boldsymbol{g}$ need to be considered. For the calculation of the influence of high-order Laue zones, it is necessary to select multiples of Δz smaller than the size of the unit cell. Distribution of inelastically and thermal diffusely scattered electrons between the

Bragg reflections can also be calculated by the multislice method [7.20, 7.91]. A real-space multislice theory has been presented in [9.140, 9.141]. Various applications of the multislice method are shown in [1.19].

9.6.4 Simulation, Matching, and Reconstruction of Crystal Images

We have seen that defocus Δz and specimen thickness t influence the contrast of lattice images. Not only can contrast reversals occur as in Fig. 9.32 but the positions of black rows can shift at interfaces or boundaries and other imaging artifacts can occur due to the phase shifts caused by the wave aberration. When the imaging parameters ($C_\mathrm{s}, C_\mathrm{c}, \alpha_\mathrm{i}, \alpha_\mathrm{o}$) are known, the simplest approach is to compare visually one micrograph or a series with simulated images at different Δz and t calculated by the multislice method (see Fig. 9.35). A more objective comparison can be made by plotting in a $\Delta z - t$ plane the values of cross-correlation maxima between the experimental image and a large number of simulated images in Fourier space covering the whole defocus-thickness range near the expected value [9.142]. The method also enables the Δz and t values belonging to an experimental through-focus series to be established.

Nonlinear least-squares methods seek to minimize the residual image over all the pixels (label i) in a high-dimensional parameter space [9.143]:

$$\sum_{i=1}^{k} f_i(x)^2 = \min \quad \text{with} \quad f_i(x) = [f_i^{\mathrm{exp}} - b^{\mathrm{fit}} - f_i^{\mathrm{calc}}(x)]/W_i. \tag{9.43}$$

x refers to the s parameters $x_1, x_2, \ldots, x_s$ used in the image simulation, which can be not only the imaging parameters but also the locations of atomic rows projected along the viewing direction (e.g., a limited number of atomic columns surrounding an interface); b^{fit} takes into account any background caused by amorphous layers on the top and bottom surfaces, and $W_i = \sigma_i^{\mathrm{exp}} + 0.05 f_i^{\mathrm{exp}}$ allows for the uncertainty of the ith pixel value. In an iterative digital imaging matching program (IDIM) [9.144], parameter subspaces are separably optimized.

Another approach is to reconstruct the exit-surface wave function $\psi(\boldsymbol{r})$ from a series of micrographs. The method using a defocus series and described for amorphous specimens in Sect. 6.6.3 is the easiest way to reconstruct the amplitude and phase of the exit wave function [9.145]. In another proposal [9.146], free of assumptions about the specimen, the micrographs of a defocus series first have to be aligned by cross-correlation because image shifts of the order of a few nanometers cannot be avoided during defocusing when voltage-center alignment has been used and mechanical shifts are superposed. The intensities $I_{\mathrm{exp}}(\boldsymbol{g}, \Delta z)$ of different reflections $\boldsymbol{g}$ in the digital diffractograms of the micrographs at different defocusings Δz are fitted to theoretically calculated intensities $I_{\mathrm{theor}}(\boldsymbol{g}, \Delta z)$. All amplitudes and phases of the object wave function, as well as all the imaging parameters, contribute to the latter. A suitable measure of the goodness of fit may be the sum of the squared differences between experiment and theory, and the optimum values of the parameters are obtained by a method called "survival of the fittest" (see also "genetic algorithm" in [9.147]). The comparison of experimental and theoretical reflection intensities as a function of defocus can be used to check the result. Nonlinear relationships between the clicks of the defocus knob and the actual defocus can probably be attributed to temperature variations of the objective coil (amplitude of 0.1 K with a period of 5 min).

An important application of HREM is the imaging of heterostructures of semiconductor devices in thinned cross sections. Special procedures are proposed for extracting information about composition profiles and interfacial roughness with a resolution of 1–2 monolayers by matching routines. The contents of unit cells in the micrograph are compared either by coding them as vectors containing $m \times n$ pixel intensities [9.148, 9.149] or by using the Fourier coefficients of unit cells [9.150]. For Si/Si_xGe_{1-x} interfaces, the latter method yields local composition value

$$x = \frac{J - J_{\mathrm{Ge}}}{J_{\mathrm{Si}} - J_{\mathrm{Ge}}}, \tag{9.44}$$

where J is the measured Fourier coefficient for a reciprocal lattice vector $\boldsymbol{g}$ in each unit cell and J_{Si} and J_{Ge} are values of J from reference unit cells of pure Si and Ge, respectively. Figure 9.33a shows a 7-beam [110] cross-section lattice image of a Si/Ge strained-layer superlattice and Figure 9.33b the quantitative grey-level representation of the local Ge content.

Single-sideband holography (Sect. 6.5.2) shows no transfer gaps or sign reversals of the CTF and can also be applied to a crystalline specimen [9.151, 9.152]. The objective diaphragm is shifted off-axis so that only the primary beam and one sideband of the diffracted beams are transmitted. A new alignment of the microscope is necessary after shifting the diaphragm, which indicates that the edge of the diaphragm influences the primary beam.

Off-axis holography (Sect. 6.5.3) can also be employed to reconstruct the phase and amplitude of the exit wave function with a resolution approaching 0.1 nm [6.153, 6.154, 9.153]. Figure 9.34a shows a hologram with fringe spacings of 0.05 nm of a Si foil; the so-called dumbbell structure, which is a projection of the unit cell with diamond structure in the direction of the [110] zone axis, can be seen [9.153]. The digital diffractogram shows the central primary beam surrounded by the spatial frequencies of the contributing Bragg reflections (central band) and two sidebands [one of these is shown in Figure 9.34b], which are caused by diffraction on the fringes; both bands show reflections up to 004. Digital reconstruction of the hologram, including the elimination of the aberrations, allows us to calculate the phase (Figure 9.34c) and amplitude (Figure 9.34d) contributions to the exit wave function. The small asymmetry visible in the amplitude and phase of the corrected wave suggests that there is a residual crystal or beam tilt away from the zone axis.

A simulated thickness-defocus tableau for C_{s} is shown in Fig. 9.35 for different specimen thicknesses. It reveals that already small deviations of

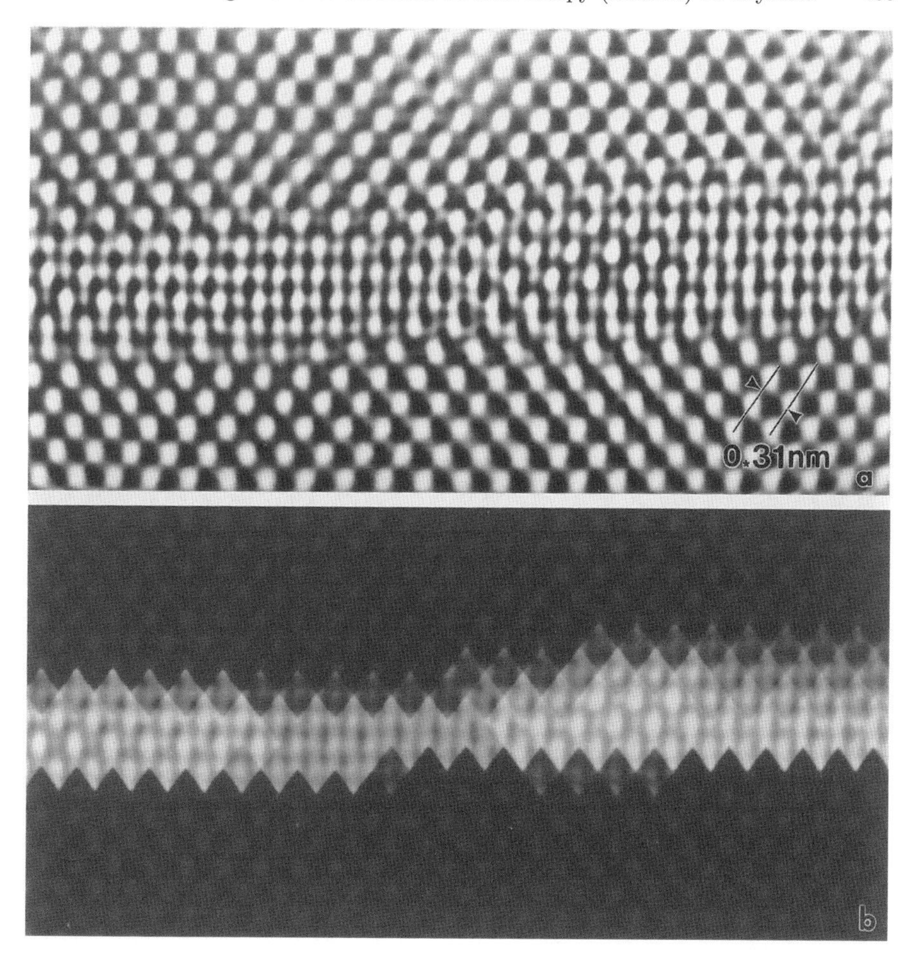

Fig. 9.33. (**a**) 7-beam [110] cross-section lattice image of an Si/Ge strained-layer superlattice and (**b**) the quantitative grey-level representation of the local Ge content [9.150].

about 5 nm in defocus drastically change the image contrast of amplitude and phase. Comparing the corrected object wave amplitude and phase from Fig. 9.34 with the defocus tableau reveals a good agreement with the simulations. With a C_s-corrected microscope, amplitude-contrast images can be obtained by choosing a small defocus and C_s to minimize the phase shift of the lenses [6.86].

9.6.5 Measurement of Atomic Displacements in HREM

In well-aligned lattice images parallel to a zone axis, the displacements of (black) lattice row images can be directly related to lattice strains. If components of a multilayer system show a misfit, strains can be measured in HREM

Fig. 9.34. Electron hologram (**a**) of an Si foil in the [110] zone axis and one sideband of the digital diffractogram (**b**). Reconstruction of the phase (**c**) and amplitude (**d**) contributions of the exit wave function [9.153].

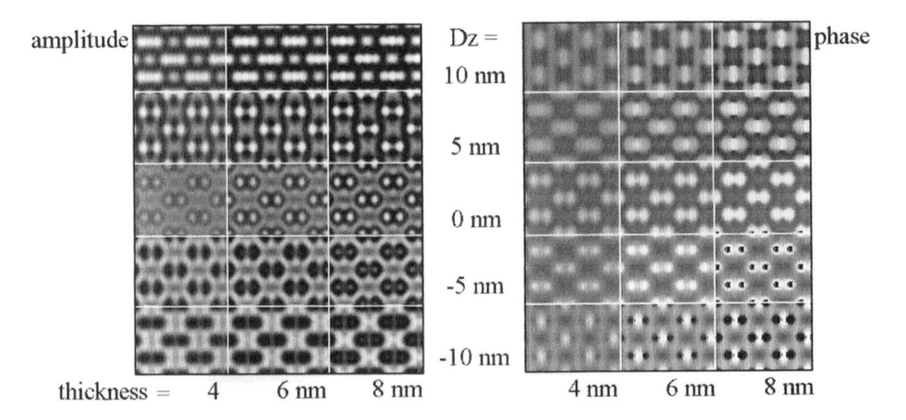

Fig. 9.35. Simulated thickness-defocus tableau of Si [110] for a correction to $C_s = 0$. Comparison with Figs. 9.34c and d reveals a good agreement for $\Delta z = 0$ [9.153].

micrographs near the boundaries, as has been shown, for example, for 1–4 monolayers of Ni embedded between 5nm Au layers [9.154]; alternatively, the displacements can be used to determine the grain boundary volume expansion [9.155]. The positions of atomic rows can be determined with an accuracy of about one-sixth of the resolved atomic row distance [9.156], better than 0.03 nm [9.157]. Nevertheless, the results must always be simulated by multislice calculations using a range of values of the imaging parameters (defocus and tilt) near the assumed values, taking into account dynamically forbidden reflections if necessary because they can also influence the image [9.158]. The core structure of grain boundaries can be compared with a molecular static computation of atomic structure [9.159].

The displacements near the core of dislocations have been measured for example in semiconductors [9.160, 9.161, 9.162, 9.163] and metals [9.164, 9.165, 9.166] and showed a remarkable agreement with the elastic theory outside the core. Dissociations of dislocations of the order of 0.5–1 nm can be measured, which allows larger stacking-fault energies to be determined than with the weak-beam method (Sect. 9.4.3).

9.6.6 Crystal-Structure Imaging with a Scanning Transmission Electron Microscope

To demonstrate the principle of crystal-structure imaging with a scanning transmission electron microscope, we again make use of the reciprocity theorem (Sect. 4.5.3) for a three-beam case (Fig. 9.36). In a conventional transmission electron microscope (bottom to top), an electron beam of aperture α_i incident on the specimen at P produces two Bragg-diffracted beams at angles $\pm 2\theta_B$ to the primary beam. These beams are focused by the objective lens

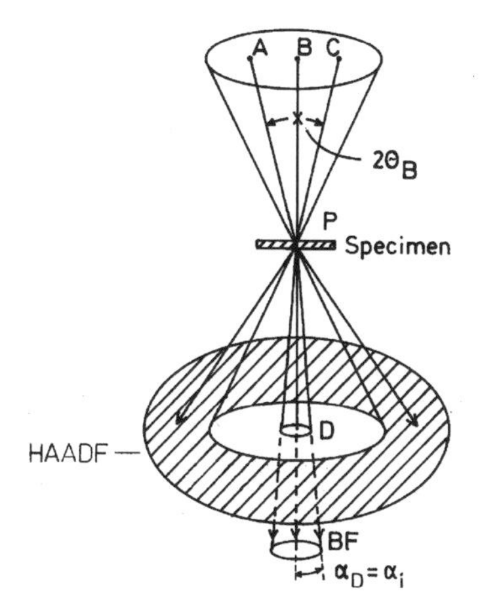

Fig. 9.36. Imaging of crystal lattices in the STEM bright-field (BF) mode reciprocal to the TEM bright-field mode and in the high-angle annular dark-field (HAADF) mode with Z-dependent contrast.

and form a lattice-fringe image in the image plane. In the STEM bright-field (BF) mode (top to bottom), the source is focused in the electron probe at P with a cone angle (probe aperture) $\geq 2\theta_B$. Directions in the cone corresponding to the primary beam BP and the diffracted beams AP and CP form three discs (convergent-beam electron-diffraction patterns), which overlap at D in the detector plane. The interference between these three discs at D produces the lattice image when the electron probe scans across the specimen [9.167].

Another way of imaging crystal lattices is the high-angle annular dark-field (HAADF) method (Fig. 9.36); the inner diameter then excludes the ZOLZ convergent-beam discs. This STEM mode with a probe diameter less than the dimension of a unit cell results in bright images of atomic rows for a zone-axis orientation of the crystal foil and shows an intensity increasing with atomic number Z [9.168, 9.169]. The contrast is caused by thermal diffuse electron scattering, and no reversals are seen when the thickness and/or defocus are varied. It can be interpreted as an incoherent contrast transfer. Simulations based on a "frozen-phonon" version of the multislice method agree with experimental results [9.170, 9.171, 9.172, 9.173].

9.7 Imaging of Atomic Surface Steps and Structures

9.7.1 Imaging of Surface Steps in Transmission

When investigating crystal growth and surface structures, it is of interest to resolve surface steps of atomic dimensions. By the replica technique (shadowing with a platinum film about 1 nm thick, using a carbon supporting film about 10 nm thick), steps of 1–2 nm can be resolved (Fig. 9.37) [9.174, 9.175]. Another possibility is the decoration technique, in which small crystals, of silver or gold for example, nucleate on alkali halides mainly at atomic surface steps (Fig. 9.38) [9.176, 9.177, 9.178]. The surface is coated with a thin layer of carbon, and the crystals and carbon layer are stripped off together. These are thus preparation methods for bulk specimens.

Normally, the contrast in bright- and dark-field images arising from the change in the transmission is too faint to reveal thickness differences d of one atomic step because $d \ll \xi_g$. However, weak-beam excitations with a large s_g [9.179] or forbidden reflections [9.180] correspond to a much smaller extinction distance, $\xi_{g,\text{eff}}$, and can be used for dark-field imaging of atomic steps. For example, surface steps on MgO become visible when the image is formed with a weakly excited 200 reflection and a strongly excited 600 reflection or with 400 and 200 reflection [9.179]. With so-called forbidden reflections, steps on Au and Si foils can be detected by contrast differences [9.180], though it will be shown below that this case can also be interpreted as weak-beam excitation of a reciprocal-lattice point in a first-order Laue zone.

The principle of the method will now be discussed by considering the example of a [111]-oriented gold film. Figure 9.39a shows the unit cell with

Fig. 9.37. Surface replica of a deformed copper single crystal (8% stretched). The mean step height is 1.8 nm. The carbon film is reinforced with a collodium backing and shadowed with a palladium film (courtesy of S. Mader).

Fig. 9.38. Gold decoration of surface steps (0.28 nm) on NaCl generated by a screw dislocation after sublimation for 6 h at 350°C (courtesy of H. Bethge).

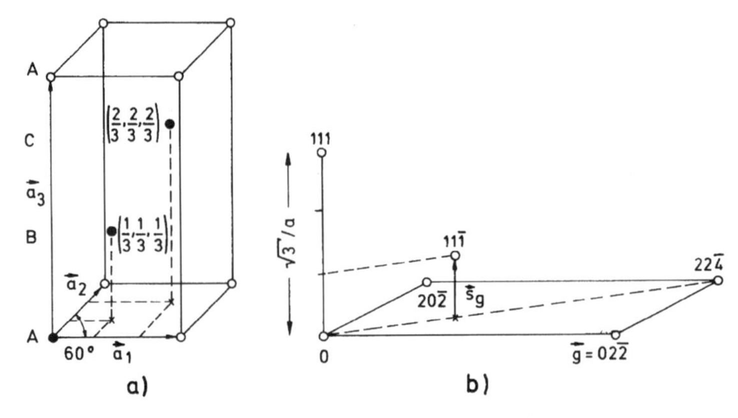

Fig. 9.39. (**a**) "Hexagonal" unit cell of an fcc lattice. (**b**) Reciprocal lattice in cubic notation showing that the excitation error s_g is $(\sqrt{3}/a)/3$ for the 111 reflection.

the ABC packing sequence (Fig. 7.3) in a hexagonal notation. The structure amplitude (7.14) becomes

$$F = f\{1 + \exp[2\pi i(h + k + l)/3] + \exp[4\pi i(h + k + l)/3]\}$$
$$= 0 \quad \text{if} \quad h + k + l = 3n + 1 \quad \text{or} \quad h + k + l = 3n + 2 \,. \tag{9.45}$$

This is the condition for a *forbidden reflection* in hexagonal notation. These reflections are situated between the primary beam and 220 Bragg spots (cubic notation). However, F is zero only if the number of close-packed layers N is a multiple of 3: $N = 3m$. This means that there must be an integral number of complete unit cells in hexagonal notation. With one layer more or less ($N = 3m{\pm}1$), F does not vanish because $F = 0$ is a consequence of destructive interference between the scattered waves for all of the atoms of a unit cell. These additional or missing layers therefore lead to nonzero values of F and are imaged as bright contrast in a dark-field micrograph if this weak Bragg spot is selected.

In another equivalent explanation, this contrast may be interpreted as a weak-beam effect with a large excitation error s_g. In the reciprocal lattice shown in Fig. 9.39b, the first-order Laue zone contains a 111 reflection, which is allowed (hkl odd in cubic notation) but normally not excited owing to the large excitation error s_g, which denotes the distance from the Ewald sphere in the zero-order Laue zone if the electron beam is parallel to the [111] zone axis. This means that s_g is equal to the distance between the first- and zero-order Laue zones, $s_g = (\sqrt{3}/a)/3$. Substituting for s_g in (7.25) or (7.61) gives the same result because $w = s_g \xi_g \gg 1$ and

$$I_g \propto \sin^2(\pi t s_g) = \sin^2(\pi \sqrt{3} t/3a) \tag{9.46}$$

so that I_g becomes zero if $t = 3m \times (a\sqrt{3}/3)$. The last factor in the bracket is, however, just the distance between the close-packed layers, namely one-third

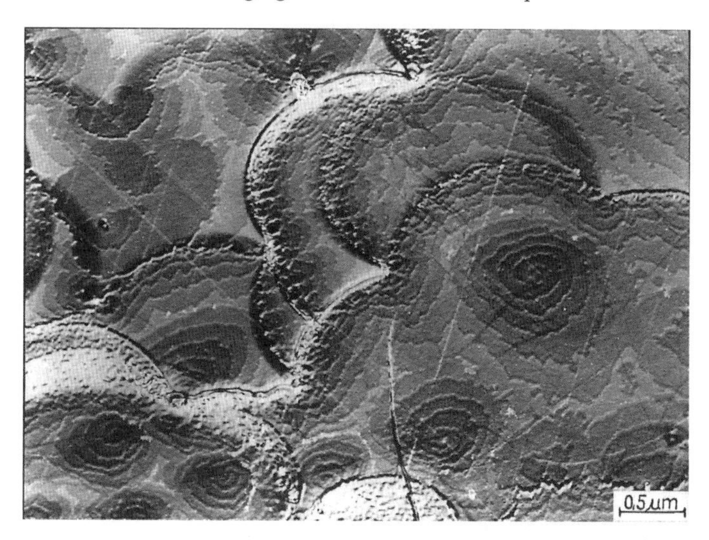

Fig. 9.40. Bright-field image of a 3.4 monolayer Au(111) film with a direction of incidence near ⟨315⟩. In addition to the contrast from different thicknesses due to the short extinction length, the steps are imaged by Fresnel-like black–white fringes due to phase contrast (courtesy of M. Klaua and H. Bethge).

of the space diagonal $a\sqrt{3}$ in the cubic unit cell, which is the same result as that given by (9.45).

Large variations in contrast are also found for strongly excited reflections. For example, a contrast maximum is found for incidence near the ⟨111⟩ zone axis of MgO with a tilt of about 0.4 of the Bragg angle at a thickness of 7 nm; under these conditions, 20% contrast is seen for an additional step of 0.12 nm [9.181]. This is confirmed by dynamical n-beam calculations. An advantage is the short exposure time of a few seconds compared with the long exposures necessary for the weak-beam method discussed above, while a disadvantage is the limited thickness range and the necessary adjustment of the orientation for optimum contrast. Up to six intensity levels for surface layers of different height have been observed (Fig. 9.40) when imaging ultrathin gold films in the exact [111] direction of incidence [9.182], and other orientations can be found that also increase the contrast for a special thickness range. In all these transmission modes for imaging surface steps, contributions from steps at the top and the bottom overlap, which is not the case in the reflection mode discussed in the next section.

Apart from these diffraction-contrast modes, which yield areas of different grey levels for surface terraces of constant thickness, the position of the step edge and its sign can be determined by phase contrast, slightly under- or overfocused [9.183, 9.184]. It is necessary to avoid low-index reflections, and the crystal has to be freely mounted on a microgrid.

The so-called reconstructed surface layers, in which the equilibrium configurations of atoms are different from the positions of atoms in the bulk

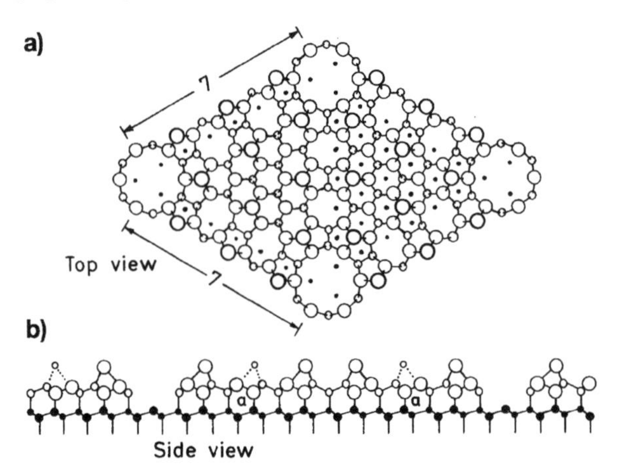

Fig. 9.41. DAS model of the Si(111) 7×7 reconstructed surface in (**a**) top and (**b**) side view containing (D) dimers (*small open circles*), and (A) adatoms (*large open circles*). Atoms in the bulk surface are shown as small full circles on the top layer and (S) a stacking-fault layer (*medium open circles*).

crystal owing to the modification of the binding forces at the surface, are of special interest. A typical example is the Si(111) 7×7 structure, top and side views of which are shown in Fig. 9.41 [9.185], containing 98 atoms in the two-dimensional unit cell in the double top layer. Such a structure and other planar superstructures different from 7×7 have been detected by LEED (low-energy electron diffraction) and result in superstructure reflections between the primary beam and the main reflections of the bulk crystal. Using specially designed TEM specimen chambers with an ultrahigh vacuum, these superstructure reflections can also be observed in the transmission diffraction mode, and the periodicity of the 7×7 structure can be imaged as 2.8 nm superlattice fringes in the dark-field mode using one main reflection and the surrounding superlattice reflections [9.185]. The intensity of these reflections can be calculated using the kinematical approach [9.186]. The surface topography of the 7×7 structure can also be recorded by scanning tunneling microscopy [9.187].

9.7.2 Reflection Electron Microscopy

In the reflection electron microscope (REM) mode, the incident electron beam is tilted by 1°–2° to strike the specimen surface at glancing incidence. Electrons scattered at low glancing angles can pass through the objective diaphragm. Attempts to use REM were made in the early years of electron microscopy [9.188, 9.189, 9.190, 9.191] to image surface topography. Owing to the strong foreshortening of the image by a factor of 20–50 and the limitation of the resolution (10–20 nm) by the chromatic error of the objective lens, this mode attracted little interest because better images of surfaces could be obtained by using surface replicas and later by scanning electron microscopy.

The revival of REM and the corresponding scanning technique (SREM) in the 1970s can be attributed to the improved vacuum and higher performance of the modern transmission electron microscope and to the possibility of using the objective diaphragm to select single Bragg-reflection spots, which consist mainly of elastically scattered electrons in the reflection high-energy electron-diffraction (RHEED) pattern (Sect. 8.1.4c) by the objective diaphragm; the resulting resolution is about 1 nm. Further progress will be made when REM is used in an energy-filtering transmission electron microscope; this will allow a further reduction of the contribution of inelastically scattered electrons and an increase in contrast [9.192].

Important applications (see also [9.193]) are the imaging of reconstructed surface structures such as the Si(111) 7×7 structure of Fig. 9.41, which can be imaged either with the (444) diffracted beam as nucleated dark regions relative to a 1×1 structure when the specimen is cooled below the transition temperature at 830°C (Fig. 9.42) or as 2.3 nm lattice fringes [9.194, 9.195, 9.196]. Surface atomic steps can be imaged as a result of both phase-contrast and strain-contrast effects. The interference between waves reflected from the top and bottom monoatomic steps produces dark or bright Fresnel-like fringes in defocus from which the sense of a step (up or down) can be determined [9.197, 9.198, 9.199, 9.200]. If there is a strain field in the crystal associated with the surface step, varying the lattice-plane orientation will give diffraction contrast. Double images of steps can be observed under surface-resonance

Fig. 9.42. Reflection electron microscope (REM) image of the (111) face of an Si crystal obtained with the (444) diffracted beam and showing the nucleation and growth of 7×7 structures (dark regions) at surface steps when the specimen is cooled from the transition temperature at 830°C, where it shows the 1×1 structure (bright regions) (courtesy of H. Yagi).

conditions [9.198]. The intensities in the RHEED pattern can be calculated by a modified multislice method, which also explains surface-resonance effects of particular diffraction spots [9.201, 9.202]. The dynamics of domains and superlattice structures of submonolayer deposits of Ag and Au on (111) silicon can be followed, for example, as the temperature is varied [9.203]. An ultrahigh vacuum is essential for quantitative work, though the typical image structures of steps and layers can also be observed in normal TEM conditions [9.204]. A scanning reflection mode [8.51] can also be employed, which has the additional advantage of providing RHEED patterns from microareas [9.205].

9.7.3 Surface-Profile Imaging

Progress in crystal-structure imaging (Sect. 9.6) and the imaging of atomic rows seen end-on as black spots allows us to observe the faceting and configuration of atomic rows about 2–5 nm long at an edge; a projected profile of the surface at the atomic scale is seen [9.206], provided the specimen is thin and that optimum defocus and crystal alignment are guaranteed. Other essentials are a good vacuum (better than 10^{-5} Pa) to avoid contamination and the possibility of insitu cleaning and annealing. Carbon contamination layers deposited outside the vacuum can be removed by electron bombardment with a high current density of about 2×10^5 A/m^2 [9.207].

As an example, we mention the 2×1 gold (110) surface for which a missing row model has been confirmed and a first-layer expansion of 20±5% has been measured [9.208, 9.209, 9.210]. Further examples are the detection of a novel 1×1 structure on a silicon (113) surface [9.211], the observation of surface dislocations [9.207, 9.212], decoration of the surface of $FeZnCrO_4$ with ZnO after a catalytic reaction [9.213], and the observation of oxidation processes. As in crystal-lattice imaging, great care is necessary because the image spots do not necessarily coincide with the atomic rows. Comparison with calculated images is therefore indispensable [9.208, 9.214]; the positions of the rows can then be determined with an accuracy of 0.01–0.02 nm.

With the aid of an image intensifier with a TV camera, the reconstruction of the surface after different treatments – e.g., removal of contamination layers, oxidation or annealing – and the dynamical internal and surface rearrangement of atoms can be observed in real time [9.206, 9.215, 9.216].

The electron-current density necessary for profile imaging is high, about $1-4 \times 10^5$ A/m^2, and so electron-beam-stimulated processes have to be taken into account. For example, growth of a metallic layer by radiolysis has been observed on some transition-metal oxides but not on rare-earth oxides [9.217].

10

Elemental Analysis by X-ray and Electron Energy-Loss Spectroscopy

The inner-shell ionization of atoms results in an emission of characteristic x-ray quanta or Auger electrons. A wavelength- or an energy-dispersive x-ray spectrometer can be coupled to a transmission electron microscope to record x-ray quanta emitted from the specimen. The quantitative methods developed for the x-ray microanalysis of bulk materials can be transferred to the investigation of thin specimens.

Electron energy-loss spectroscopy (EELS) gives information about the electronic structure and the elemental composition of the specimen. This technique is more efficient than x-ray analysis for low-Z elements because the spectrometer can collect a large fraction of the inelastically scattered electrons, which are concentrated within small scattering angles. By deconvolution and background subtraction, a net signal can be obtained from the ionization edge of an element for subsequent quantitative analysis.

An imaging electron-energy filter makes it possible to work in the electron spectroscopic imaging modes, which can be used for mapping the elemental distribution.

10.1 X-ray and Auger-Electron Emission

10.1.1 X-ray Continuum

The x-ray continuum is a result of the acceleration of electrons in the Coulomb field of the nucleus. It is well-known from electrodynamics that an accelerated charge can emit an electromagnetic wave. If the acceleration $a(\tau)$ is periodic, a monochromatic wave will be emitted (dipole radiation). Because the spectral distribution emitted is proportional to the square of the Fourier transform of $a(\tau)$ [2.8, 10.1] and the interaction time τ is very short, the spectrum resulting from the passage of an electron is very broad, and the x-ray quanta can be emitted with energies E_{x} in the range $0 \leq E_{\mathrm{x}} \leq E = eU$, in which $E_{\mathrm{x}} = eU$ is the maximum x-ray energy of the spectrum (Duane–Hunt law) (see [10.2, 10.3] for reviews).

A semiclassical formula of Kramers [10.4] gives the cross section $\mathrm{d}\sigma_\mathrm{x}/\mathrm{d}E_\mathrm{x}$ for the generation of quanta in the x-ray energy range between E_x and $E_\mathrm{x} + \mathrm{d}E_\mathrm{x}$. From this the number of continuous x-ray quanta generated in a thin layer of mass thickness $x = \rho t$ with N_A/A atoms per gram is found to be

$$N(E_\mathrm{x})\mathrm{d}E_\mathrm{x} = \frac{N_\mathrm{A}x}{A}\frac{\mathrm{d}\sigma_\mathrm{x}}{\mathrm{d}E_\mathrm{x}}\mathrm{d}E_\mathrm{x} = \frac{N_\mathrm{A}x}{A}a\frac{Z^2}{\beta^2 E_\mathrm{x}}\mathrm{d}E_\mathrm{x}, \tag{10.1}$$

where $\beta = v/c$ and $a = 5.54 \times 10^{-31}$ m^2 is the Kramers constant. More accurate quantum-mechanical calculations [10.5, 10.6] give an improved value of the Kramers constant a as a function of E, E_x, and Z, and experiments then show a better agreement with (10.1) ([10.7], for example).

The angular distribution of the x-ray continuum is very anisotropic (Fig. 10.1). The number $\mathrm{d}N_x$ of quanta emitted per unit time, with energies between E_x and $E_x + \mathrm{d}E_x$, into a solid angle $\mathrm{d}\Omega$ at an angle θ relative to the forward direction of electron incidence inside a film of thickness t and density ρ (no multiple scattering) with $N_\mathrm{A}\rho t/A$ atoms per unit area can be described by [10.6, 10.8, 10.9]

$$\mathrm{d}N_x = \frac{I(\theta)}{E_x}\frac{N_\mathrm{A}\rho t}{A}\frac{I_\mathrm{p}}{e}\mathrm{d}E_x\mathrm{d}\Omega, \tag{10.2}$$

where I_p is the probe current and

$$I(\theta) = I_x\frac{\sin^2\theta}{(1-\beta\cos\theta)^4} + I_y\left(1 + \frac{\cos^2\theta}{(1-\beta\cos\theta)^4}\right). \tag{10.3}$$

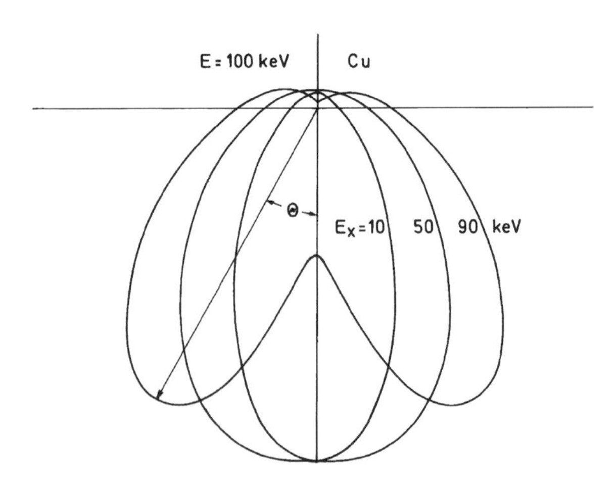

Fig. 10.1. Angular characteristics (polar diagram) of the x-ray continuum emitted by a thin Cu foil irradiated with $E = 100$ keV electrons for three quantum energies $E_\mathrm{x} = 90$, 50, and 10 keV. The curves are normalized to the same maximum value.

Analytical formulas for the coefficients I_x and I_y have been published [10.9] that approximate the exact value with an accuracy of 2%. To a first-order approximation, I_x and I_y are proportional to Z^2/E. The polar diagrams of emitted x-ray intensities in Fig. 10.1 are based on (10.3).

The x-ray continuum can be used to calibrate the film thickness in the microanalysis of biological sections, for example (Sect. 10.2.5). It also contributes to the background below the characteristic x-ray lines, thereby decreasing the peak-to-background ratio. The forward characteristics of the continuous x-ray emission illustrated in Fig. 10.1 increase with increasing electron energy. The x-ray emission in analytical TEM is observed at an angle θ between 90° and 135° relative to the incident electron beam. As a result, the line-to-continuum ratio increases with increasing energy because the emission of characteristic x-ray quanta is isotropic.

10.1.2 Characteristic X-ray and Auger-Electron Emission

Ionization of an inner shell results in an energy loss ΔE of the incident electron (Sect. 5.3.1) and a vacancy in the ionized shell (Fig. 10.2a). The electron energy E has to be greater than the ionization energy E_{nl} of a shell with quantum numbers n and l. E_{nl} is the energy difference to the first unoccupied state above the Fermi level (e.g., $E_{nl} = E_{\mathrm{K}}$ in Fig. 10.3).

The ionization cross section can be calculated from a formula given in [3.2, 5.26, 10.10],

$$\sigma_{nl} = \frac{\pi e^4 Z_{nl}}{(4\pi\epsilon_0)^2 E E_{nl}} b_{nl} \ln\left(\frac{4E}{B_{nl}}\right), \tag{10.4}$$

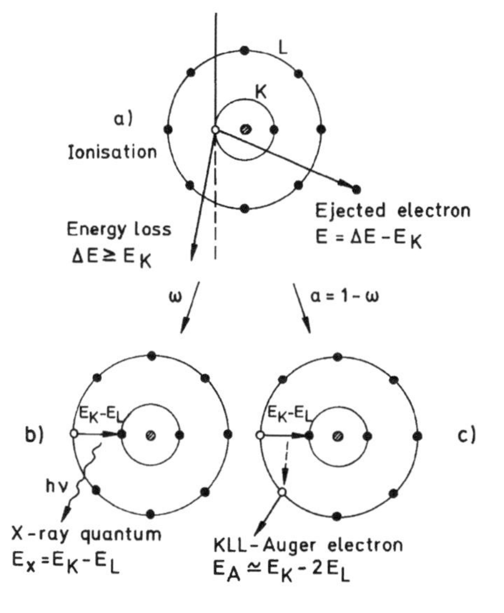

Fig. 10.2. Schematic representation of (**a**) the ionization process, (**b**) x-ray emission, and (**c**) Auger-electron emission.

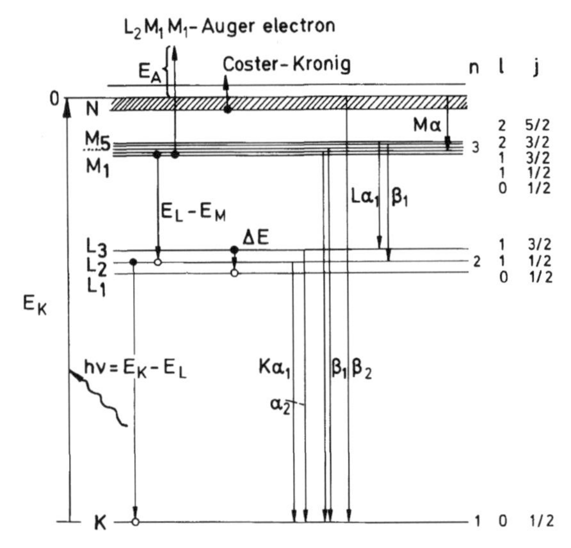

Fig. 10.3. Energy levels of the atomic subshells with quantum numbers n, l, j and ionization of the K shell; possible transitions of electrons to fill the vacancies in inner shells and nomenclature of emitted x-ray lines are shown. Example of the emission of a KLL Auger electron and a Coster–Kronig transition.

where b_{nl} and B_{nl} are numerical constants [e.g., $b_K = 0.35$, $B_K = 1.65\ E_K$ for the K shell $(n = 1)$] and Z_{nl} denotes the number of electrons with quantum numbers n, l ($Z_{nl} = 2$ for the K shell). The ratio $u = E/E_{nl} \geq 1$ is called the *overvoltage ratio*. An empirical formula

$$B_K = [1.65 + 2.35 \exp(1 - u)]E_K \tag{10.5}$$

has been proposed [10.11], which gives $B_K = 1.65E_K$ for large u but $B_K = 4E_K$ for u near unity. The use of u leads to the following formula for K-shell ionization:

$$\sigma_K = \frac{2\pi e^4}{(4\pi\epsilon_0)^2 E_K^2 u} b_K \ln \frac{4uE_K}{B_K} \ . \tag{10.6}$$

The product $\sigma_K E_K^2$ should therefore depend only on u. This is confirmed for low and medium atomic numbers, as is shown for the K-shell ionization of C, N, O, Ne, Al, Ni, and Ag in Fig. 10.4 with $E_K = 283$, 401, 532, 867, 1560, 8330, 25 500 eV, respectively, using (10.6) and a formula of Gryzinski [10.12]

$$\sigma_K = \frac{\pi e^4 Z_{nl}}{(4\pi\epsilon_0)^2 E_K^2 u} \left(\frac{u-1}{u+1}\right)^2 \times \left\{1 + \frac{2}{3}\left(1 - \frac{1}{2u}\right) \ln[2.7 + (u-1)^{0.5}]\right\} . \tag{10.7}$$

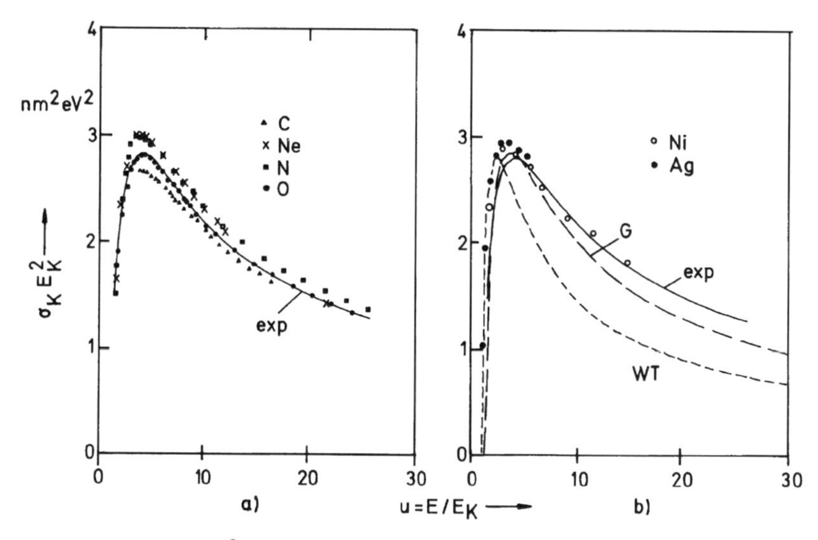

Fig. 10.4. Plot of σ_K/E_K^2 (σ_K: ionization cross section of the K shell, E_K: ionization energy) versus the overvoltage ratio $u = E/E_K$ for (**a**) C, N, O, Ne and (**b**) Ni and Ag atoms. Comparison of formulas of Worthington and Tomlin (WT) (10.6) and *Gryzinski* (G) (10.7) [10.13].

For high atomic numbers, σ_K becomes larger [10.15]. Cross sections σ_K for some elements are also plotted in Fig. 5.3 for comparison with other important cross sections of electron–specimen interactions. The theoretical and experimental cross sections of the K and L shells are reviewed in [10.13, 10.14].

The vacancy in the inner shell is filled by electrons from outer shells (Fig. 10.2b). The energy difference, $E_K - E_L$ for example, can be emitted as an x-ray quantum of discrete energy $E_x = h\nu = E_K - E_L$. Because of the quantum-mechanical selection rules, $\Delta l = \pm 1$ and $\Delta j = 0, \pm 1$, the Kα lines that result from the transition of an L ($n = 2$) electron to the K shell ($n = 1$) consist of only a doublet (Kα_1 and Kα_2 in Fig. 10.3).

To a first approximation, in which the subshells are disregarded, the quantum energies $E_{x,K}$ of the K series can be estimated from modified energy terms (5.86) of the Bohr model ($E_1 = E_K$),

$$E_{x,K} = E_n - E_1 = -R(Z-1)^2(1/n^2 - 1/1^2), n = 2 : \mathrm{K}\alpha, n = 3 : \mathrm{K}\beta, \quad (10.8)$$

where $R = 13.6$ eV denotes the ionization energy of the hydrogen atom. The reduction of Z by 1 represents the screening of the nuclear charge Ze by the remaining electron in the K shell. Likewise, for the L series ($E_2 = E_L$),

$$E_{x,L} = E_n - E_2 = -R(Z-7.4)^2(1/n^2 - 1/2^2), \quad n = 3, 4, \dots . \quad (10.9)$$

Exact values of x-ray wavelengths are tabulated [10.16, 10.17]. The x-ray wavelength λ and the quantum energy E_x are related by the formulas $h\nu = E_x$ and $\nu\lambda = c$, giving

$$\lambda = \frac{hc}{E_x} = \frac{1.24}{E_x} \tag{10.10}$$

with λ in nm and E_x in keV.

The ratio $K\alpha_1/K\alpha_2$ of the intensity of the $K\alpha_1$ line from the transition $L_2 \rightarrow K$ to that of the $K\alpha_2$ line ($L_3 \rightarrow K$) is proportional to the number of electrons in the corresponding subshells, which is $4/2 = 2$ (sum rule). The ratio $K\alpha_1/K\beta_1$ decreases from 10 for Al ($Z = 13$) to 3 for Sn ($Z = 50$). The reason for this variation of the transition probability is the gradual filling of the N and M subshells. Strong deviations from the sum rule are observed for the L series, which can be attributed to *Coster–Kronig transitions*, in which a vacancy in an L_1 or L_2 subshell is filled by an electron from another subshell (L_3). The energy is transferred to an electron near the Fermi level (Fig. 10.3). The lines with L_3 as the lowest sublevel are relatively enhanced by this effect. For experimental values of the intensity ratios, see [10.3].

Unlike the x-ray continuum, the angular emission of the characteristic quanta is isotropic. The half-widths of the emitted lines are of the order of 1–10 eV.

However, not every ionization of an inner shell results in the emission of an x-ray quantum. This process occurs with a probability ω, the *x-ray fluorescence yield.* Alternatively, the energy $E_L - E_K$, for example, may be transferred to another atomic electron without emission of an x-ray quantum. The latter electron leaves the atom as an *Auger electron* with an excess energy $E_A = (E_L - E_K) - E_I$; E_I is the ionization energy of this electron in the presence of a vacancy in one subshell, taking relaxation processes into account. An Auger electron is characterized by the three electronic subshells that are involved in the emission (e.g., KLL in Fig. 10.2c or $L_2M_1M_1$ in Fig. 10.3). The probability of this process is the *Auger-electron yield* $a = 1 - \omega$. The quantities ω and a are plotted in Fig. 10.5 as a function of atomic number Z [10.18, 10.19, 10.20].

Because the value of ω_K is very low for light elements, detection of these by x-ray microanalysis is inefficient. Electron energy-loss spectroscopy here becomes more attractive (Sect. 10.3). The other alternative, Auger electron spectroscopy, cannot be used in a conventional transmission electron microscope, though instruments have been developed using a specially designed spectrometer [10.21].

The continuous and characteristic x-ray quanta emitted interact with the solid by the following three processes.

1. *Photoionization.* The quantum is totally absorbed and the energy is used to ionize an atom in an inner shell if $E_x \geq E_{nl}$. Filling this shell results in the emission of an x-ray quantum of lower energy (x-ray fluorescence) or an Auger electron.
2. *Compton effect.* The quanta are elastically scattered by single atomic electrons (conservation of kinetic energy and momentum). The x-ray energy decreases by an amount equal to the energy of the ejected electron.

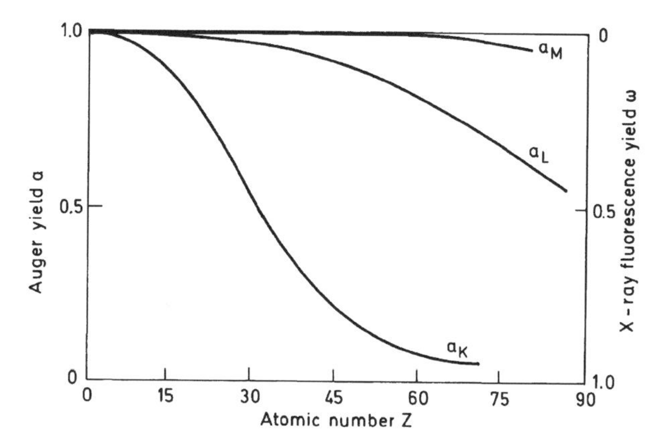

Fig. 10.5. Auger-electron yield a and fluorescence yield $\omega = 1 - a$ as a function of atomic number Z.

3. *Thomson scattering.* X-rays are scattered without energy loss at the electron shell of the atom. This effect is responsible for x-ray diffraction.

The first process dominates for $E_x < 100$ keV and results in an exponential decrease of x-ray intensity as the mass thickness $x = \rho t$ increases:

$$N = N_0 \exp(-\mu t) = N_0 \exp[-(\mu/\rho)x] \, . \tag{10.11}$$

The *mass-attenuation coefficient* $\mu/\rho = N_A \sigma_x / A$ (g^{-1} cm^2) is related to a single-atom *fluorescence cross section* σ_x and decreases as E_x^{-n} ($n = 2.5$–3.5) with increasing E_x, apart from abrupt increases when $E_x \geq E_{nl}$, the ionization energy for an atomic shell (Fig. 10.6). Numerical values and formulas for μ/ρ are to be found in *International Tables for X-Ray Crystallography* (Kynoch Press, Birmingham) and in [10.22, 10.23, 10.24, 10.25, 10.26].

10.2 X-ray Microanalysis in a Transmission Electron Microscope

10.2.1 Wavelength-Dispersive Spectrometry

A wavelength-dispersive spectrometer (WDS) makes use of the Bragg reflection of x-rays by a single crystal ($2d \sin\theta_B = n\lambda$). Better separation of narrow characteristic lines and a larger solid angle of collection, $\Delta\Omega \simeq 10^{-3}$ sr, can be obtained by focusing. The electron-irradiated spot on the specimen acts as an entrance slit, while the analyzing crystal and the exit slit are mounted on a Rowland circle of radius R (Fig. 10.7). The lattice planes of the crystal are bent so that their radius is $2R$, and the surface of the crystal is ground to a radius R. Behind the slit, the x-ray quanta are recorded by a proportional counter. The detection efficiency of the Bragg reflection and the proportional

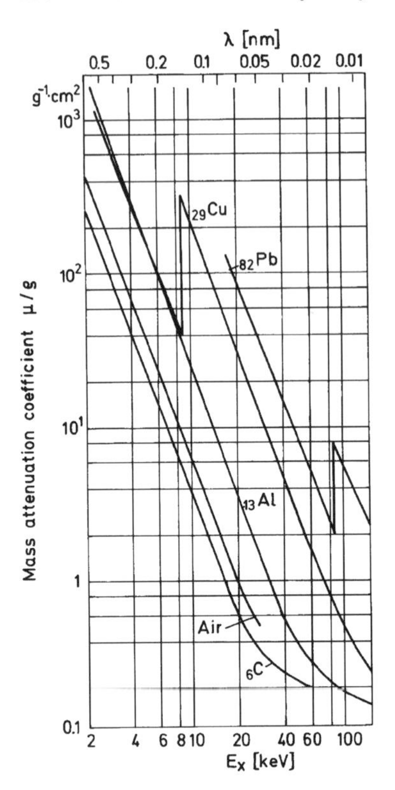

Fig. 10.6. Dependence of mass attenuation coefficient μ/ρ on quantum energy E_x (lower scale) and x-ray wavelength λ (upper scale).

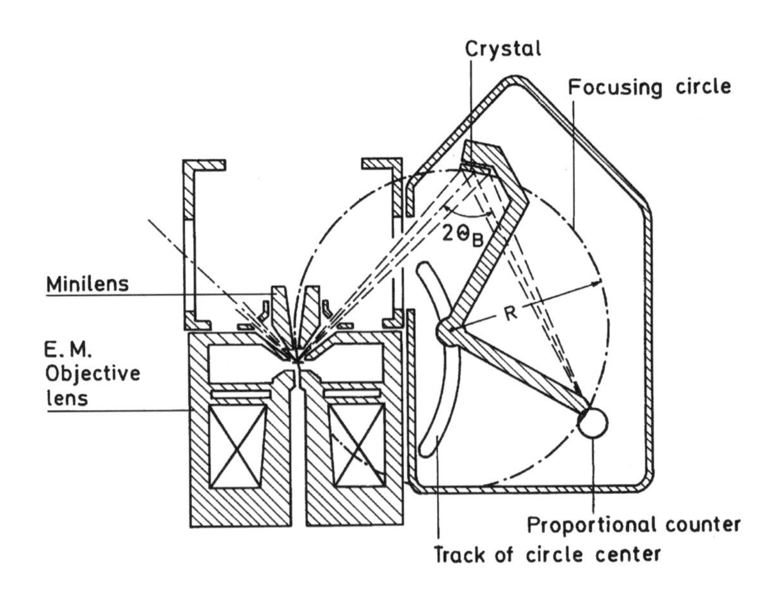

Fig. 10.7. Incorporation of a wavelength-dispersive x-ray spectrometer with large takeoff angle in a transmission electron microscope.

counter is about 10–30%. The number of electron–ion pairs generated and the resulting current pulse created by charge multiplication in the high electric field near the central wire of the counter are proportional to the quantum energy $E_x = h\nu$. Bragg reflections of higher order ($n > 1$) can be eliminated by pulse-height discrimination because such reflections are caused by quanta of lower wavelength or higher quantum energy. For the recording of a spectrum or for altering the focusing condition to another wavelength, the exit slit, the proportional counter, and the analyzing crystal can be moved by means of a pivot mechanism. X-ray microanalyzers are equipped with two or three spectrometers, which permit different wavelengths to be recorded simultaneously. Crystals with different lattice spacings have to be used so that the whole wavelength range can be analyzed. Special proportional counters with very thin mylar windows are available for analyzing the weak Kα radiation of low-Z elements; the lower limit is beryllium.

Exact adjustment of the specimen height is necessary to satisfy the focusing condition. Small variations of height or a shift of the electron beam can cause large variations of the x-ray intensity recorded. Wavelength-dispersive spectrometers are therefore used mainly in x-ray microanalyzers; they can also be mounted in transmission or scanning electron microscopes, though energy-dispersive spectrometers are more commonly used with these instruments.

Thanks to the preselection of the radiation by the analyzing crystal, WDSs can work with a high count rate for a particular characteristic line, of the order of 100 000 c.p.s. when using electron-probe currents of the order of $10^{-8} - 10^{-7}$ A in conjunction with a proportional counter. This implies lower statistical errors and more accurate determination of low elemental concentrations.

Figure 10.7 shows a cross section through the objective lens and one of the two crystal spectrometers that used to be employed in the AEI/Kratos EMMA instruments (electron-probe size: 0.1 μm, probe current: 2–5 nA at 100 keV, x-ray takeoff angle: 45°) [10.27, 10.28].

10.2.2 Energy-Dispersive Spectrometry (EDS)

Either lithium drifted silicon [Si(Li)] detectors [10.29, 10.30, 10.31] or high-purity intrinsic germanium detectors (HPGe, or IG for short) [10.32] are used. These consist of a reverse-biased p-i-n junction (p-type, intrinsic, and n-type) 2–5 mm in diameter (Fig. 10.8). In silicon, the intrinsic zone is generated by the diffusion of Li atoms. X-rays absorbed in the active detector volume create electron–hole pairs. The number of charge carriers is proportional to the x-ray quantum energy E_x and the possibility of analyzing the pulse height of the charge collected form the physical basis of the energy-dispersive spectrometer [10.30]. If the x-ray quantum is absorbed by photoionization of an inner shell, any residual x-ray energy contributes to the kinetic energy of an excited photoelectron. Any remaining energy is transferred either to an Auger electron or to an x-ray quantum of energy E'_x characteristic of the detector atoms (x-ray fluorescence). If this x-ray quantum is again absorbed in the

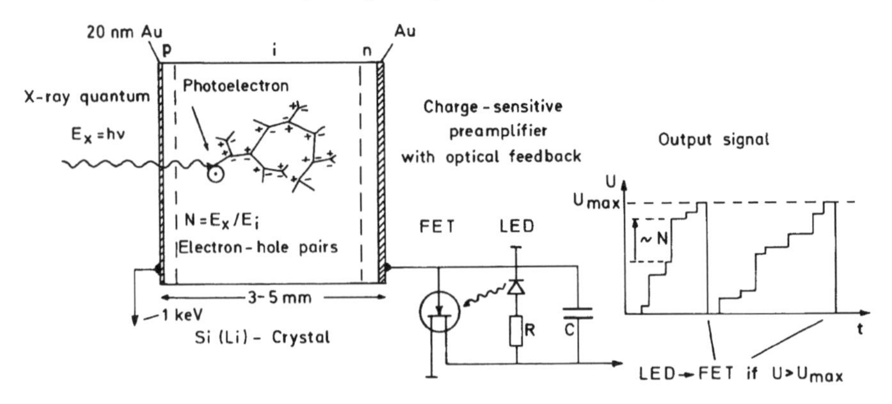

Fig. 10.8. Creation of electron–hole pairs in an Si(Li) energy-dispersive x-ray detector coupled to a charge-sensitive preamplifier with an optical feedback circuit. The latter switches the output signal to zero by means of a light-emitting diode (LED), which irradiates the field-effect transistor (FET) if $U \geq U_{\mathrm{max}}$.

detector, the whole quantum energy E_{x} serves to create charge carriers. If, on the other hand, this fluorescence x-ray quantum leaves the detector, its energy E'_{x} does not contribute to the number of charge carriers created. This can cause a weak *escape peak* of energy $E_{\mathrm{x}} - E'_{\mathrm{x}}$ in the pulse-height spectrum recorded. For an Si(Li) detector, the escape peak energies E'_{x} are 1.739 keV (Si Kα) and a weaker contribution of 1.836 keV (Si Kβ). In IG detectors, more and stronger escape peaks corresponding to Ge Kα and Kβ (9.876 and 10.983 keV) and also to Ge Lα and Lβ (1.188 and 1.219 keV) are observed, as are Kα + Lα (10.974 keV) and Kβ + Lβ (12.202 keV).

The electron–hole pairs created by x-ray photoabsorption in the active volume can be separated by applying a reverse bias of the order of 1 kV. The charge pulse collected is proportional to the quantum energy E_{x} and is converted to a voltage pulse by means of a field-effect transistor (FET), which acts as a charge-sensitive amplifier. The detector crystal and the FET are continuously cooled by liquid nitrogen, the first to avoid diffusion of Li atoms and the second to reduce noise. Further reduction of noise and an improvement of the preamplifier time constant are provided by charge accumulation. The output signal of the preamplifier consists of voltage steps proportional to E_{x}. When the voltage reaches a preset value ($\simeq$10 V), a light-emitting diode (LED) is switched on and the leakage current induced in the FET by the emitted light returns the output voltage to zero (pulsed optical feedback, Fig. 10.8). Another alternative is RC feedback with a high feedback resistance, which results in exponential decay of the pulses.

The main amplifier not only amplifies the output of the preamplifier but also shapes the pulses. The corresponding time constant is of the order of 5–10μs and is longer than the rise time of a voltage step in the preamplifier. A pulse-pileup rejector cancels the pulses when the time between two voltage steps is equal to or shorter than the processing time constant of the main amplifier. Otherwise, pulse pileup can result in artificial lines of 2 Fe Kα or Fe Kα + Fe Kβ, for example. A live-time corrector prolongs the preset counting time by one processing time constant per pulse-pileup rejection. Without control of the linearity, the number of counts per second should not exceed 2000–5000. By electron-beam blanking shortly after recording a pulse and unblanking after processing the pulse, the linearity can be extended to about 8000 cps [10.33, 10.34].

The shaped pulses from the main amplifier are collected, sorted, stored, and displayed by a multichannel analyzer (MCA). Pulse amplitudes from 0 to 10 V are linearly converted to numbers between 0 and 1024 by analog-to-digital conversion.

The collection solid angle can be calculated from the active detector area and the specimen–detector distance. It is of the order of $\Delta\Omega = 0.1$ to 0.01 sr but can be increased to 0.2–0.3 sr by special design [10.34]; this value is one to two orders of magnitude larger than for WDS. In SEM, it is necessary to place a permanent magnet in front of the detector to deflect the backscattered electrons for electron energies above 25 keV [10.35, 10.36], but in TEM the lens field acts as an electron trap.

The resolution δE_{x} can be characterized by the FWHM of a characteristic x-ray line, which will be broadened by the statistical nature of electron–hole pair creation and by the electronic noise of the detector and preamplifier. A mean ionization energy $\overline{E_{\mathrm{i}}} = 3.86$ eV (Si) and 2.98 eV (Ge) is necessary for each electron-hole pair created. The average number $N = E_{\mathrm{x}}/\overline{E_{\mathrm{i}}}$ of electron–hole pairs forming charge pulses has a standard deviation

$$\sigma_{\mathrm{x}} = \sqrt{FN} = \sqrt{FE_{\mathrm{x}}/\overline{E_{\mathrm{i}}}}. \tag{10.12}$$

This value would be $\sigma_{\mathrm{x}} = \sqrt{N}$ if the number of pairs created obeyed Poisson statistics. The Fano factor $F = 0.087$ (Si) and 0.06 (Ge) takes into account departures from the Poisson distribution caused by the gradual deceleration of the photoelectrons. Denoting the standard deviation of the electronic noise by σ_{n}, the FWHM of a line becomes

$$\delta E_{\mathrm{x}} = 2.35\sqrt{\sigma_{\mathrm{x}}^2 + \sigma_{\mathrm{n}}^2}\,. \tag{10.13}$$

Values of $\delta E_{\mathrm{x}} \simeq 140$–160 eV (Si) and ≤ 125 eV (Ge) can be obtained for $E_{\mathrm{x}} = 5.893$ keV (Mn Kα), depending weakly on the processing time of the main amplifier. Incomplete charge collection can result in a tail on the low-energy side of a characteristic peak.

The resolution of EDS is thus more than ten times worse than that of WDS and not all overlapping lines can be separated, especially those of the L series of one element and the K series of another. Furthermore, an energy window of the order of the FWHM is necessary to ensure that all the characteristic quanta of a given line are counted and more background counts are recorded than by WDS.

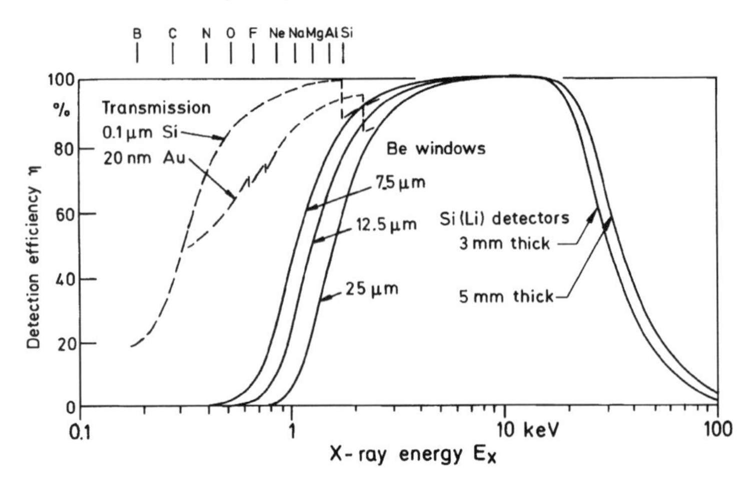

Fig. 10.9. Detection efficiency $\eta(E_x)$ of an Si(Li) energy-dispersive spectrometer for various thicknesses of the Be window and the detector; transmission (dashed lines) for a 20 nm Au coating and a 0.1 μm Si dead layer.

The detection efficiency $\eta(E_x)$ of an Si(Li) detector is nearly 100% in the range $E_x = 3$–15 keV (Fig. 10.9). The decrease at low E_x is caused by the absorption of x-rays in the window material and the conductive coating of the crystal. The old generation of Si(Li) detectors were protected by a Be window ($\geq 6 \mu$m) that strongly absorbed x-ray quanta with energies below Al K. Windowless detectors [10.37, 10.38] have the disadvantage of contamination, and nowadays ultrathin windows of $\simeq$ 500 nm of diamond-like carbon or boron compounds, or metal-coated plastic films (e.g., 300 nm pyrolene + 20–40 nm Al capable of withstanding atmospheric pressure) are in use; with these, quantum energies as low as a few hundred eV can be analyzed. Maxima produced by Kα quanta from C, N, O can be recorded, but the lines may overlap with the L lines from $Z = 16$–30 since their FWHM is large. Figure 10.9 shows as dashed lines the transmission of a 20 nm Au contact layer and a 0.1 μm Si dead layer caused by surface recombination; this is the ultimate limit of an Si(Li) detector. The decrease of efficiency at high quantum energies is caused by the increasing probability of traversing the crystal without photoionization. For example, a 3 mm thick detector crystal shows 10% transmission for 23 keV in Si and 74 keV in Ge [10.39, 10.40]. IG detectors are thus especially suitable for high-voltage microscopes. Figure 10.10 shows the x-ray spectrum of a hard metal of the type WC-TaC-TiC-Co recorded with an ultrathin-window IG detector fitted on a 200 kV microscope; x-ray lines from C K to W Kβ can be seen [10.41].

Unlike WDS, for which the irradiated point has to be adjusted on the Rowland circle, EDS does not need any mechanical adjustment and can hence be used when the specimen is large and/or rough. EDS is therefore commonly used in SEM. In TEM, an energy dispersive spectrometer has the advantage

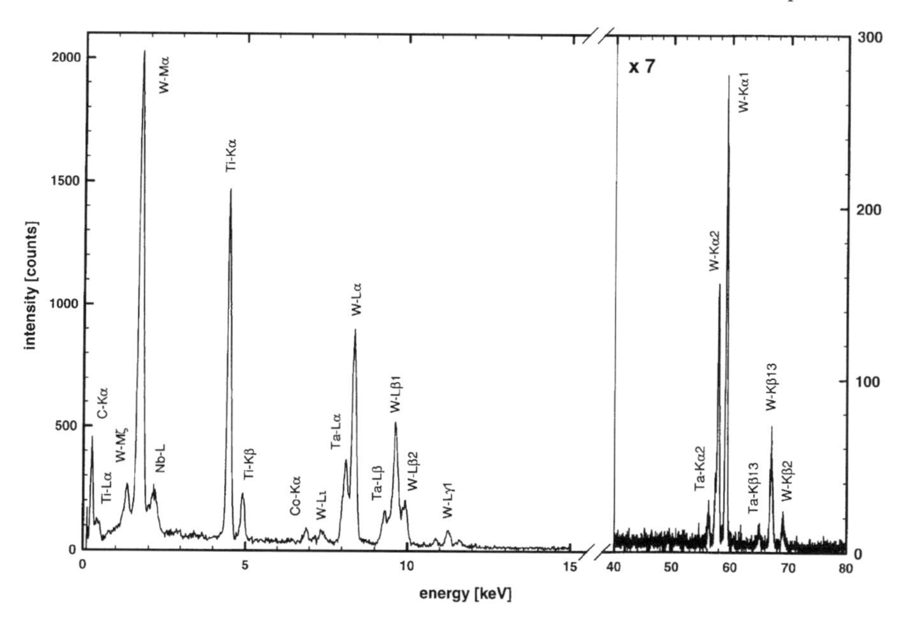

Fig. 10.10. Energy-dispersive x-ray spectrum of a hard metal (WC-TaC-TiC-Co) recorded with an ultrathin-window IG detector in a 200 kV microscope [10.41].

of occupying little space; it is merely necessary to design the cryostat tube and the objective lens in such a way that the x-ray quanta from the transparent specimen in the polepiece gap are collected with as large a solid angle of collection and takeoff angle as possible (Fig. 4.19). A further advantage of EDS is that all lines are recorded simultaneously.

Recently, x-ray microcalorimeters have been developed that use the temperature rise caused by the absorbed photon to determine its energy [10.43]. These detectors provide a high energy resolution of better than 10 eV. Unfortunately, they cannot accommodate high count rates. In order to keep the specific heat of the absorber low, they have to be cooled with liquid He. It will be interesting to see if they can be developed into a TEM attachment for routine use.

10.2.3 X-ray Emission from Bulk Specimens and ZAF Correction

In TEM, x-ray microanalysis (XRMA) is applied only to thin foils and small particles. It is therefore sufficient to discuss the count rate N of characteristic x-ray lines from thin films. Nevertheless, the so-called ZAF correction (atomic number **Z** – **A**bsorption – **F**luorescence) for the XRMA of bulk material will be described briefly (for details, see [1.117, 1.119, 1.120, 1.121, 1.122, 1.124, 1.130]) because certain corrections of the formula are important for x-ray emission from thin foils.

The XRMA of a bulk material relies on two measurements of the counts of a characteristic x-ray line for equal incident electron charge: N_a, the number of quanta emitted from the element a with a concentration c_a in the specimen under investigation, and N_s, the number of quanta emitted from a pure standard of the element a. The ratio $k = N_a/N_s$ leads only to a first-order approximation of the concentration c_a.

Atomic Number Correction. The number of x-ray quanta $\mathrm{d}N_a$ generated by $n_0 = I_\mathrm{p}\tau/e$ incident electrons along an element of path length $\mathrm{d}x = \rho \mathrm{d}s$ of the electron trajectory can be calculated using the ionization cross section σ_a, the x-ray fluorescence yield ω_a, and the ratio p_a of the observed line intensity to the intensity of all lines of the series, $p_{\mathrm{K}\alpha} = N_{\mathrm{K}\alpha}/(N_{\mathrm{K}\alpha}+N_{\mathrm{K}\beta})$, for example. With the notation $n_a = \rho c_a N_A/A_a$ for the number of atoms of element a per unit volume (ρ is the density of the material and A_a is the atomic weight of a), $\mathrm{d}N_a$ becomes

$$\mathrm{d}N_a = n_0\omega_a p_a \sigma_a n_a \mathrm{d}s = n_0 \omega_a p_a \frac{c_a N_A}{A_a} \frac{\sigma_a}{\mathrm{d}E/\mathrm{d}x}\mathrm{d}E, \tag{10.14}$$

in which $\mathrm{d}x$ denotes the mass-thickness element, $\mathrm{d}x = \rho\ \mathrm{d}s$. The stopping power is given by

$$S = \left|\frac{\mathrm{d}E}{\mathrm{d}x}\right| = \frac{N_A e^4}{8\pi\epsilon_0^2 E}\sum_i c_i \frac{Z_i}{A_i}\ln(E/E_i), \tag{10.15}$$

which is a modified form of the Bethe formula (5.100) for $i = a, b, \ldots$ elements in the specimen with mean ionization energies E_i. X-ray quanta can be generated along the whole trajectory so long as the decreasing electron energy E' is greater than the ionization energy E_I of the I = K, L, or M shell and the electron does not leave the specimen by backscattering or transmission. Integrating (10.14) then gives

$$N_a = n_0 \omega_a p_a c_a \frac{N_A}{A_a} R \int_{E_\mathrm{I}}^{E} \frac{\sigma_a}{S}\mathrm{d}E', \tag{10.16}$$

where R is the *backscattering correction factor*:

$$R = \frac{\text{total number of quanta actually generated in the specimen}}{\text{total number of quanta generated if there were no backscattering}}\ .$$

This quantity depends on the energy distribution of backscattered or transmitted electrons.

The ratio $k = N_a/N_s$ becomes

$$k = c_a \frac{R_a}{R_s} \frac{\int_{E_\mathrm{I}}^{E} \frac{\sigma_a}{S_a}\mathrm{d}E'}{\int_{E_\mathrm{I}}^{E} \frac{\sigma_a}{S_s}\mathrm{d}E'} = c_a \frac{1}{k_Z}\ . \tag{10.17}$$

The values of R for the specimen and the standard will in general be different, and k_Z also depends on differences between the stopping powers S.

For thin films of mass thickness $x = \rho t$, (10.14) simplifies to

$$N_a = n_0 \omega_a p_a \sigma_a c_a \frac{N_A}{A_a} x. \tag{10.18}$$

The atomic number correction k_Z is not very different from unity and can be neglected if the following conditions are satisfied:

1. The ionization cross section σ_a does not significantly increase because of a decrease of the mean electron energy in the specimen caused by energy losses.
2. The path lengths are not increased by multiple scattering.
3. Differences in the backscattering and transmission of specimen and standard do not influence the ionization probability.

Absorption Correction. The absorption correction is necessary because of the depth distribution $n(z)$ of x-ray emission and the absorption of x-rays inside the specimen. If the x-ray detector collects x-rays emitted with a takeoff angle ψ relative to the foil surface, x-ray quanta emitted at a depth z below the surface are attenuated along an effective length $z\,\mathrm{cosec}\psi$ and the fraction of quanta leaving the specimen is

$$f(\chi) = \frac{\int_0^x n(z) \exp(-\chi z) \mathrm{d}z}{\int_0^x n(z) \mathrm{d}z} \quad \text{with} \quad \chi = (\mu/\rho)\mathrm{cosec}\psi, \tag{10.19}$$

and (10.17) becomes

$$k = c_a \frac{1}{k_Z} \frac{f(\chi_a)}{f(\chi_s)} = c_a \frac{1}{k_Z k_A} . \tag{10.20}$$

Assuming that x-ray quanta are generated uniformly in a foil of mass thickness $x = \rho t$, (10.19) becomes

$$f(\chi) = \frac{1}{\chi x}[1 - \exp(-\chi x)] \simeq 1 - \frac{1}{2}\chi x + \ldots . \tag{10.21}$$

For thin specimens, the absorption therefore can be neglected only if $\chi x \ll 1$. The correction is essential for thick specimens and low quantum energies, especially if $E_{\mathrm{x},a}$ is just above $E_{\mathrm{I},b}$ for a matrix of atoms b; that is, just beyond the jump of μ/ρ in Fig. 10.6 [10.42, 10.44, 10.45, 10.46]. When the x-ray quanta are detected through the polepiece at an angle of 90° to the electron beam, the specimen has to be tilted to increase ψ. With a takeoff angle ψ = 20°–30°, the specimens can be investigated at normal incidence. Care must be taken with specimens that are severely bent. The influence of the absorption correction can be measured quantitatively by varying the thickness and/or the tilt of the specimen [10.47].

Fluorescence Correction. If characteristic quanta of an element b are generated with an energy $E_{\mathrm{x},b} \geq E_{\mathrm{I},a}$, these quanta and also the fraction of the continuum that satisfies this condition can be absorbed by photoionization of atoms of element a, thus causing an increase of N_a by x-ray fluorescence. Equation (10.20) becomes

$$k = \frac{N_a}{N_s} = c_a \frac{1}{k_Z k_A} \frac{1 + r_a}{1 + r_s} = c_a \frac{1}{k_Z k_A k_F} , \quad (10.22)$$

where the ratio $r = N_f/N_d$ denotes the ratio of the quanta generated by fluorescence (N_f) and directly by the electron beam (N_d). The contribution N_f from fluorescence is generated in a much larger volume than N_d and can normally be neglected in thin films [10.48]; it can, however, be significant in FeCr alloys, for example, where the Cr Kα fluorescence is excited by Fe Kα [10.49]. Formulas for parallel-sided and wedge-shaped foils are discussed in [10.50].

Influence of Crystal Orientation. Electron waves in single crystals are Bloch waves with nodes and antinodes at the nuclear sites depending on the crystal orientation (Sect. 7.4.2). Any x-ray emission is therefore anisotropic (depending on the tilt angle) because the ionization of an inner shell is localized near the nuclei. The number of x-ray quanta emitted can vary by as much as 50% when the specimen is tilted through a few degrees near a zone axis [10.51, 10.52, 10.53, 10.54]. The Cliff–Lorimer factor k_{ab} (see below) is almost independent of orientation in nonaxial orientations of the incident beam. However, even in this nonaxial case, the k-factor can decrease when the takeoff direction of the x-rays is oriented along a low-index zone axis as a result of the anomalous absorption (Bormann effect) of emitted x-rays [10.55]. Quantitative results on single-crystal foils are therefore reliable only if the measurement is insensitive to a small specimen tilt.

In compounds, the maxima of antinodes of Bloch waves can appear at different tilt angles for different atomic sites in a crystal lattice. This is the basis for ALCHEMI (Atom Location by CHanneling Enhanced MIcroanalysis). The method can be used to determine which sites are occupied by substitional impurity atoms by measuring the dependence on tilt of the characteristic x-ray emission [10.56, 10.57, 10.58, 10.59, 10.60, 10.61, 10.62]. The tilt can be determined exactly from the location of Kikuchi lines in a simultaneously recorded diffraction pattern. The method can also be applied in electron energy-loss spectroscopy by measuring the EELS signal intensity at ionization edges [10.63, 10.64].

10.2.4 X-ray Microanalysis of Thin Specimens

An important problem for XRMA of thin films in TEM is the unwanted contribution from continuous and characteristic quanta generated not in the irradiated area but anywhere in the whole specimen and specimen cartridge by electrons scattered at diaphragms above and below the specimen and from x-ray fluorescence due to x-ray quanta generated in the column. Additional diaphragms have to be inserted at suitable levels to absorb scattered electrons and x-rays, whereas the objective diaphragm should be removed during XRMA. The number of unwanted quanta can be further reduced by constructing the specimen holder from light elements such as Be, Al, or high-strength carbon [10.65].

The procedure involving the use of bulk pure-element standards for the XRMA of bulk specimens, as described in Sect. 10.2.3, can, in principle, be adapted for the XRMA of thin films [10.66, 10.67]. However, in many applications, it is the ratio c_a/c_b of the concentrations that is of interest, so that only the ratio N_a/N_b of the counts of the peaks of elements a and b is needed. Equation (10.18) has to be multiplied by the collection efficiency $\eta(E_x)\Delta\Omega/4\pi$ and the absorption correction (10.21) to correct the count ratio [10.42, 10.44, 10.68, 10.69]

$$\frac{N_a}{N_b} = \frac{\omega_a p_a \sigma_a \eta(E_a) A_a}{\omega_b p_b \sigma_b \eta(E_b) A_b} \frac{c_a}{c_b} \frac{1-\frac{1}{2}\chi_a x}{1-\frac{1}{2}\chi_b x}$$
$$= \frac{1}{k_{ab}}\left[1-\frac{1}{2}(\chi_a-\chi_b)x\right]\frac{c_a}{c_b}, \tag{10.23}$$

where k_{ab} denotes the Cliff–Lorimer ratio. This ratio method is independent of the local mass thickness x of the specimen if the absorption correction can be neglected. In work on an Al-Zn-Mg-Cu alloy using the ratio method, the $N_{\mathrm{Cu}}/N_{\mathrm{Al}}$ ratio was found to depend on the foil thickness, which could be interpreted by assuming that a Cu-rich surface layer about 15 nm thick was present, probably formed during electropolishing [10.70, 10.71]. The local thickness of a specimen can be measured by allowing contamination spots to form on the top and bottom of the foil. Tilting the specimen separates the two spots, and the thickness can be calculated from the tilt angle and the separation [10.44].

The Cliff–Lorimer factor k_{ab} in (10.23) is commonly determined experimentally by measuring the count rates of pure-element films reduced to equal thickness and incident electron charge $I_{\mathrm{p}}\tau$. Further $k_{a,\mathrm{Si}}$ values for the K, L, and M lines are reported in [10.74] and $k_{a,\mathrm{Fe}}$ ratios in [10.75]. The factor cannot be calculated very accurately (Fig. 10.11) because accurate values of the ionization cross sections σ and fluorescence yields ω are not known. Attempts to fit these quantities to polynomials are reviewed in [10.72]. A method for the parametrization of cross sections can be used to calculate k_{ab}-factors for other electron energies [10.73].

The diameter of the smallest possible area that can be analyzed is limited by the superposition of electron-probe size and the spatial broadening of the electron probe by multiple scattering (Sect. 5.4.3). Probe diameter d_{p} and probe current I_{p} are related by $I_{\mathrm{p}} \propto d_{\mathrm{p}}^{8/3}$ (4.21). The spatial broadening in foils of different composition can be calculated by Monte Carlo simulations [10.76]. Resolutions of 1–10 nm can be achieved in the STEM mode of a transmission electron microscope; this is sufficient for measuring segregation and composition profiles at grain boundaries or quantum-well layer structures,

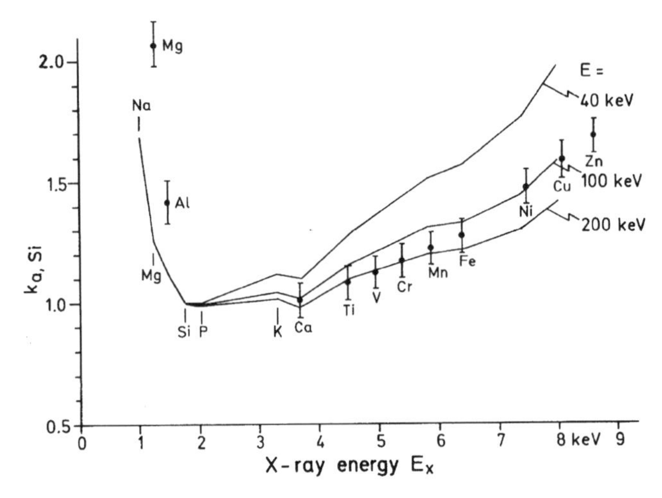

Fig. 10.11. Cliff–Lorimer factor $k_{a,\mathrm{Si}}$ relative to Si as a function of x-ray quantum energy E_x of the Kα lines, measured at 100 keV and calculated for 40, 100, and 200 keV [10.46].

for example [10.77, 10.78, 10.79]. When the counting time is long, there is a danger of specimen or electron-probe drift, and this can limit the number of points in a line scan [10.80].

10.2.5 X-ray Microanalysis of Organic Specimens

For biological specimens, either the volume concentration in mmol/l or the mass concentration in mmol/kg is of interest. The latter can be considered per total mass using hydrated cryosections or per dry mass using freeze-dried sections. The Cliff–Lorimer ratio method (Sect. 10.2.4) can rarely be applied to organic specimens because the characteristic peak of carbon shows a low fluorescence yield ω and a low detection efficiency and the absorption inside the specimen is strong.

In the Hall method [10.81, 10.82, 10.83], it is suggested that the counts N_B of the continuous background should be used to provide a signal proportional to the mass inside the irradiated volume and to eliminate the effect of local mass thickness, which can show large variations in biological specimens, especially in freeze-dried preparations. The Kramers formula (10.1) shows that the continuum background intensity is proportional to Z^2/A and so the numbers of counts for element a and the background (B) become (k_1, k_2 are constants)

$$N_a = k_1 m_a \quad \text{and} \quad N_\mathrm{B} = k_2 G m \tag{10.24}$$

with

$$G = \overline{Z^2/A} = \frac{1}{m}\left(\sum_a m_a Z_a^2/A_a + \sum_i m_i Z_i^2/A_i\right), \tag{10.25}$$

where m_i are the masses of elements not seen in the spectrum and $m = \sum m_a + \sum m_i$ the total mass. After measuring the values of N_a and N_B for a standard (s), the desired concentration will be given by

$$c_a = \frac{m_a}{m} = c_{a,s} \frac{(N_a/N_\mathrm{B})_\mathrm{Object}}{(N_a/N_\mathrm{B})_\mathrm{Standard}} \frac{G_\mathrm{Standard}}{G_\mathrm{Object}}$$

$$= k_\mathrm{H}(N_a/N_\mathrm{B})_\mathrm{Object} \frac{G_\mathrm{Standard}}{G_\mathrm{Object}} . \tag{10.26}$$

This method can be used iteratively by first neglecting the first term in (10.25) and then calculating a more accurate value of G with the approximate value of c_a obtained by (10.26). The constant k_H can be measured by using a standard of similar composition and having a known concentration $c_{a,s}$; the thickness of the specimen and the standard need not be known. The only requirements are that both be "thin" and that the analytical conditions be the same for specimen and standard. The energy window of the continuum should be near the characteristic peak to achieve similar absorption conditions.

Standards can be prepared by different methods [10.84, 10.85]. Known concentrations of elements can be embedded in resins by adding a salt solution or organometallic compounds to epoxy resins [10.86, 10.87, 10.88, 10.89] or to aminoplastic resins [10.90, 10.91]. Chelex biostandard beads with calibrated concentrations can also be embedded and sectioned [10.92]. In order to ensure that the sections have the same thickness, the tissue can be freeze-frozen together with a surrounding solution of albumin containing known concentrations of salts [10.93, 10.94, 10.95]. Similar standards can be used for freeze-dried or frozen hydrated specimens.

Any change in concentration by extraction and shrinkage during the preparation can be avoided by using only cryomethods for quantitative work. However, a severe problem is the radiation damage suffered by biological specimens, especially the loss of mass of the organic material (Sect. 11.2.2) [10.83, 10.86]. A high current density is needed for XRMA, and most of the mass is lost in a few seconds. Owing to the nonuniform composition, the mass loss can vary locally, thereby perturbing the standardization based on the Hall method.

10.3 Electron Energy-Loss Spectroscopy

10.3.1 Recording of Electron Energy-Loss Spectra

The instrumentation for recording an electron energy-loss spectrum either by a prism spectrometer or an imaging energy filter is described in Sect. 4.6. We distinguish between serial [10.96, 10.97] and parallel recording [10.98, 10.99, 10.100] of an electron energy-loss spectrum.

Serial Recording. The energy-loss spectrum in the energy-dispersive plane is scanned across a slit in front of a scintillator–photomultiplier combination or a semiconductor detector (Sect. 4.7.6). These detector systems have a low noise and a high recording speed. A scintillator or semiconductor detector can be used either in the analog-signal mode with an analog–digital (ADC) or voltage–frequency (VFC) converter or in the single-electron counting mode. The P-46 or P-47 powder scintillators or YAG (cerium-doped yttrium aluminum garnet) single crystals have a time constant of 75 ns, and pulse rates up to 10^6 cps are possible. The scintillation light pulses or the charge-collection pulses in semiconductor detectors are high enough for it to be possible to separate and discriminate pulses of 100 keV electrons from the noisy background. A disadvantage is that only one energy window is recorded at a time. This implies that the recording time is far longer in serial recording than in parallel recording, by some two orders of magnitude. This may be unacceptable for beam-sensitive organic materials and some inorganic ones. The high-intensity low-loss part of the spectrum (plasmon losses) can be scanned in a shorter time, with a dwell time per pixel of a few milliseconds, whereas 50 ms and more are necessary for higher energy losses. Photomultipliers exposed to a high light level can show a transient increase in dark current, which can in turn result in a tail of the zero-loss peak, for example. Recording should therefore start at high energy losses where the intensity is low. The intensities range over several orders of magnitude, and the gain must hence be changed when a whole spectrum is being recorded. This can be achieved by altering either the gain of the photomultiplier or the dwell time. Better quantitative fit of the gain jump can be achieved by overlapping regions with different gains.

Parallel Recording. Parallel recording can be realized by means of a linear array of photodiodes or two-dimensionally with a CCD camera (Sect. 4.7.5). The latter is coupled optically or better with a fiber-optic plate to a powder-scintillator layer or a thin YAG single-crystal disc. For low intensities, the exposure time of a CCD array can be increased to 10–1000 s when cooling the device to about −30°C to −50°C by means of a Peltier element; this reduces the background noise.

The advantages of EELS become fully apparent only when such parallel recording of the spectrum is used. The gain due to the larger probability of detecting an inner-shell ionization by EELS than by an energy-dispersive x-ray analysis with low fluorescence yield and small solid angle of collection is lost if the inner-shell loss spectrum cannot be recorded in parallel.

Whereas the strong zero-loss peak is suppressed in serial recording by the slit, for larger energy losses, the backscattering of the zero-loss beam can become a problem in parallel recording; this can be solved by incorporating a beam-stop aperture when the beam falls outside the array. The point-spread function of the CCD (Fig. 4.34c) shows a full-width at half-maximum (FWHM) of about three channels and extended tails. This can be removed by deconvolution [10.100].

10.3.2 Kramers–Kronig Relation

The electron energy-loss spectrum intensity gives information about

$$\mathrm{Im}\{-1/\epsilon(\omega)\} = \epsilon_2/(\epsilon_1^2+\epsilon_2^2), \tag{10.27}$$

and the Kramers–Kronig relation then allows us to calculate $\mathrm{Re}\{1/\epsilon\} = \epsilon_1/(\epsilon_1^2+\epsilon_2^2)$ and hence the dependence of the complex dielectric function $\epsilon = \epsilon_1 + \mathrm{i}\epsilon_2$ on $\hbar\omega$ [5.23, 5.39, 5.41]. Another application of the Kramers–Kronig relation in EELS is the determination of the foil thickness, the mean free path, the effective number of electrons per atom contributing to an energy loss and the Bethe stopping power. We now summarize the mathematics of this relation using a derivation presented in [10.101].

We consider the frequency-dependent inverse complex dielectric function $\epsilon^{-1}(\omega) - 1$, for which the temporal function

$$\epsilon^{-1}(\tau) - \delta(\tau) = p(\tau) + q(\tau) \tag{10.28}$$

is obtained by the inverse Fourier transform of $\epsilon^{-1}(\omega) - 1$, split into an even part p and odd part q. Since the dielectric function is causal,

$$\epsilon^{-1}(\tau) - \delta(\tau) = 0 \quad \text{for} \quad \tau < 0. \tag{10.29}$$

By considering positive and negative τ separately and recalling that $p(\tau)$ is finite for all τ, (10.28) and (10.29) yield

$$p(\tau) = \mathrm{sgn}(\tau)q(\tau), \tag{10.30}$$

where sgn(τ)= –1, 0, +1 for $\tau <,=,> 0$, respectively.

The temporal and frequency functions are related by the following Fourier transforms:

$$\begin{aligned}\mathbf{F}\{\epsilon^{-1}(\tau)-\delta(\tau)\} &= \epsilon^{-1}(\omega)-1 \text{ where } \epsilon^{-1}(\omega)=\mathrm{Re}[\epsilon^{-1}(\omega)]+\mathrm{iIm}[\epsilon^{-1}(\omega)],\\ \mathbf{F}\{p(\tau)\} &= \mathrm{Re}[\epsilon^{-1}(\omega)]-1 \;\text{ and }\; -\mathrm{i}\mathbf{F}\{q(\tau)\} = \mathrm{Im}[\epsilon^{-1}(\omega)],\\ \mathbf{F}\{\mathrm{sgn}(\tau)\} &= \mathrm{i}/\pi\omega.\end{aligned} \tag{10.31}$$

We now Fourier transform (10.30); the right-hand product becomes a convolution (3.47) of the Fourier transforms of the individual functions:

$$\mathrm{Re}[\epsilon^{-1}(\omega) - 1 = \frac{1}{\pi}\mathrm{Im}\left[\int\limits_{-\infty}^{0}\frac{\epsilon^{-1}(\omega')}{\omega'-\omega}\mathrm{d}\omega' + \int\limits_{0}^{+\infty}\frac{\epsilon^{-1}(\omega')}{\omega'-\omega}\mathrm{d}\omega'\right]. \tag{10.32}$$

In the first integral, we replace ω' by $-\omega'$ and make use of the assumption that $q(\omega)$ is odd, which means that $q(-\omega) = -q(\omega)$. This yields

$$\mathrm{Re}[\epsilon^{-1}(\omega)] - 1 = \frac{1}{\pi}\mathrm{Im}\left[\int\limits_{0}^{\infty}\frac{\epsilon^{-1}(\omega')}{\omega'+\omega}\mathrm{d}\omega' + \int\limits_{0}^{\infty}\frac{\epsilon^{-1}(\omega')}{\omega'-\omega}\mathrm{d}\omega'\right] \tag{10.33}$$

or

$$1 - \mathrm{Re}\{1/\epsilon(\omega)\} = \frac{2}{\pi}\int\limits_{0}^{\infty}\frac{\omega'\mathrm{Im}\{-1/\epsilon(\omega')\}}{\omega'^2-\omega^2}\mathrm{d}\omega'. \tag{10.34}$$

The real part can thus be calculated directly from the imaginary part $\mathrm{Im}\{-1/\epsilon(\omega)\}$, which is proportional to the electron energy-loss spectrum intensity at $\Delta E = \hbar\omega$. Although the integrals in (10.33) and (10.34) have a singularity at $\omega' = \omega$, the integral converges because the singularity is antisymmetric in $\omega' - \omega$. A small zone around ω' can therefore be excluded from the integral, and the result is hence the Cauchy principal part of the integral.

For $\omega = 0$, (10.34) yields the Kramers–Kronig sum rule

$$1 - \mathrm{Re}\{1/\epsilon(0)\} = \frac{2}{\pi}\int_0^\infty \mathrm{Im}\{-1/\epsilon(\omega')\}\frac{\mathrm{d}\omega'}{\omega'} \,. \tag{10.35}$$

The left-hand side of (10.35) becomes $1 - 1/n^2$ from the Maxwell relation $\epsilon = n^2$ (n = optical refractive index). For metals, n is very high and the left-hand side becomes equal to unity.

Using (5.72), integrating between $0 \le \theta \le \alpha$ results in a single-scattering electron energy-loss spectrum intensity distribution that can be defined by

$$s_1(\Delta E) = \frac{I_0 t}{\pi a_H m v^2}\,\mathrm{Im}\{-1/\epsilon(\omega)\}\ln[1 + (\alpha/\theta_\mathrm{E})^2], \tag{10.36}$$

where I_0 is the zero-loss intensity and t the foil thickness. Combining of (10.35) and (10.36) gives

$$t = \frac{2a_H m v^2}{I_0(1 - 1/n^2)}\int_0^\infty \frac{s_1(\Delta E)}{\Delta E \ln[1 + (\alpha/\theta_\mathrm{E})^2]}\,\mathrm{d}\Delta E. \tag{10.37}$$

This formula can be used to estimate the foil thickness only if $t \le \Lambda$ [10.102] or $s_1(\Delta E)$ is obtained by a deconvolution.

The multiple scattering already present in this thickness range can be approximately considered by including a correction factor $C \simeq 1 + 0.3(t/\Lambda)$ in the denominator of (10.37). The formula can also be corrected for the surface-plasmon-loss contributions to $s_1(\Delta E)$ [1.78]. The mean free path Λ can be obtained from

$$\Lambda = \frac{2a_H m v^2}{(1 - 1/n^2)}\frac{1}{\int s_1(\Delta E)\mathrm{d}\Delta E}\int_0^\infty \frac{s_1(\Delta E)}{\Delta E \ln[1 + (\alpha/\theta_\mathrm{E})^2]}\,\mathrm{d}\Delta E. \tag{10.38}$$

Bethe's sum rule (5.56) together with (5.55) and (5.72) can be used to estimate an "effective number" of electrons contributing to energy losses between zero and ΔE,

$$n_\mathrm{eff} = \frac{2\epsilon_0 m}{\pi\hbar^2 e^2 n_a}\int_0^{\Delta E} \Delta E' \mathrm{Im}\{-1/\epsilon(\Delta E')\}\,\mathrm{d}\Delta E', \tag{10.39}$$

where n_a is the number of atoms or molecules per unit volume. This integral increases with ΔE to a saturation value $n_\mathrm{eff} = 4$ for carbon, for example, indicating that all four valence electrons contribute to the plasmon losses. With the definition

$$\frac{\mathrm{d}E_\mathrm{m}}{\mathrm{d}s} = n_\mathrm{a}\int\int \frac{\mathrm{d}^2\sigma}{\mathrm{d}\Delta E \mathrm{d}\Omega}\Delta E\,\mathrm{d}\Delta E\,\mathrm{d}\Omega, \tag{10.40}$$

the mean energy dissipation per unit path length (Bethe's stopping power, Sect. 5.4.2) can also be calculated from the energy-loss spectrum by writing the differential cross section in terms of $\mathrm{Im}\{-1/\epsilon\}$.

10.3.3 Background Fitting and Subtraction

For quantitative EELS (Sect. 10.3.5) and for the determination of partial cross sections (Sect. 5.3.2), it is necessary to extrapolate the background in front of an ionization edge to energy losses beyond the edge and then subtract the extrapolated background (Fig. 10.12) to get a net signal $s_{\mathrm{I}}(\alpha, \Delta E)$, which can be attributed to the ionization of the I = K, L, M shells. The background is formed by the tails of the low-loss region and any preceding ionization edges, multiple energy losses, and any instrumental background. According to the theory, the background should satisfy a power law (5.87)

$$s(\Delta E) = A\,\Delta E^{-r} \quad \Longrightarrow \quad \ln s = \ln A - r \ln \Delta E. \tag{10.41}$$

The exponent $-r$ can vary between the extreme values -2 and -6.5 [10.103] and depends not only on the background contributions mentioned above but also on the aperture α, and the composition and thickness of the specimen.

The background is normally linearly least-squares fitted by a straight line in a double-logarithmic plot (e.g., Fig. 5.9) of the recorded electron energy-loss spectrum intensity versus the energy loss over a large energy window of width Δ below the edge of interest; the slope of the line is $-r$. Such a plot can also be used to test the validity of the power law. The logarithmic transformation makes the use of a weighted least-squares fit preferable to an unweighted one [10.104]. In order to circumvent the logarithmic transformation, a ravine-search technique [10.105], a maximum-likelihood approximation

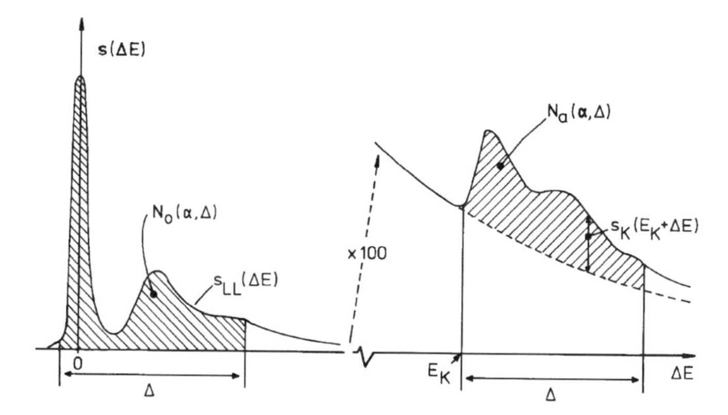

Fig. 10.12. Principle of background extrapolation and subtraction and of the Fourier-ratio method. $s(\Delta E)$: EEL spectrum; $s_{\mathrm{LL}}(\Delta E)$: low-loss part; $s_{\mathrm{K}}(E_{\mathrm{K}}+\Delta E)$: background-subtracted net signal of K ionizations; Δ: width of energy window; α: spectrometer acceptance angle.

[10.106], a simplex optimization method [10.107], or an iteration method [10.108] can be employed to search for a minimum of the error function in A–r space.

When investigating ionization edges below $\simeq$100 eV, the power law often fails. Polynomials and exponentials can be used to describe the background more accurately [10.109, 10.110], or a model for the valence-loss intensity can be included [10.111, 10.112].

Procedures for fitting the background have the disadvantage that a best fit can only be obtained over an interval Δ below the edge, and a best fit over this region is not necessarily a good fit for the region beyond the edge. The fitting can therefore be improved across the edge by fitting the spectrum with a function

$$F(\Delta E) = A\Delta E^{-r} + k\, S_{\mathrm{I}}(\Delta E), \tag{10.42}$$

where $S_{\mathrm{I}}(\Delta E)$ is a standard edge profile, which can be either a calculated cross section $\mathrm{d}\sigma/\mathrm{d}\Delta E$ [10.113] or a measured and background-subtracted reference spectrum from the element of interest [10.114]. A least-squares fit gives values for A, r, and k; the latter can be a direct measure of the concentration of the element. However, the superposed ELNES structure (Sect. 5.3.3) may be different when the atoms in the specimen and the reference differ in their coordinations.

10.3.4 Deconvolution

Even in thin specimens, multiple scattering falsifies an electron energy-loss spectrum. Multiple plasmon losses show a Poisson distribution (5.84); the mean free path Λ_{pl} between plasmon excitations is the characteristic thickness parameter. Although the mean free path between subsequent inner-shell ionizations is orders of magnitude larger, Λ_{pl} is also responsible for the convolution of the structure of an ionization edge (including ELNES and EXELFS) with the electron energy-loss spectrum in the plasmon-loss region. In addition, the spectrum is convolved with an instrumental broadening $r(\Delta E)$ caused by the energy spread of the electron gun, the aberration of the spectrometer, the width of the energy slit, and the point-spread function of the detector. These influences can be recognized in the shape and half-width of the zero-loss peak. Assuming that the instrumental function has been normalized so that $\int r(\Delta E)\mathrm{d}\Delta E = 1$, the measured spectrum $s^*(\Delta E)$ is obtained by a convolution of the ideal spectrum $s(\Delta E)$ with $r(\Delta E)$:

$$s^*(\Delta E) = \int_{-\infty}^{+\infty} r(\Delta E - \Delta E')\, s(\Delta E')\, \mathrm{d}\Delta E' = r(\Delta E) \otimes s(\Delta E). \tag{10.43}$$

For deconvolution of the instrumental broadening and the influence of multiple scattering, the Fourier-log method [1.78, 10.115, 10.116, 10.117], the Fourier-ratio method [1.78, 10.117, 10.118], and a matrix method [10.119] have all been used. The first two methods have been widely used and will be described in more detail. A Fourier convolution method can also be used to calculate the angle-resolved electron energy-loss spectrum with increasing thickness [5.86] and to deconvolve measured angle-resolved spectra [10.120].

Fourier-Log Method. We introduce a reduced thickness $p = t/\Lambda$, where the total mean free path of all inelastic scattering processes is $\Lambda < \Lambda_{\rm pl}$. The total intensities I_n of all n-fold inelastically scattered electrons obey Poisson statistics (5.84),

$$I_n = IP_n = I\frac{p^n {\rm e}^{-p}}{n!}, \tag{10.44}$$

where I is the incident electron current. The index $n = 0$ corresponds to unscattered and elastically scattered electrons with no energy loss, which are concentrated in the exponentially attenuated zero-loss peak of the spectrum with an intensity distribution

$$s_0(\Delta E) = r(\Delta E)I_0 = r(\Delta E)I\,{\rm e}^{-p}. \tag{10.45}$$

Single scattering ($n = 1$) is described by the distribution $s_1(\Delta E)$, which satisfies with (10.44) the condition

$$\textstyle\int s_1(\Delta E){\rm d}\Delta E = I_1 = I\,p\,{\rm e}^{-p} = pI_0. \tag{10.46}$$

The distribution s_2 for double scattering ($n = 2$) is proportional to $s_1 \otimes s_1$. From (10.46), we have $\int s_1 \otimes s_1 {\rm d}\Delta E = p^2 I_0^2$ and, using (10.44), we find that

$$s_2(\Delta E) = \frac{1}{2!I_0} s_1 \otimes s_1 \quad \text{and} \quad I_2 = \textstyle\int s_2 {\rm d}\Delta E = I_0 p^2/2!\,. \tag{10.47}$$

Likewise,

$$s_3(\Delta E) = \frac{1}{3!I_0^2} s_1 \otimes s_1 \otimes s_1 \tag{10.48}$$

and so on. The total measured electron energy-loss spectrum can now be described by

$$\begin{aligned} s^*(\Delta E) &= r(\Delta E) \otimes s(\Delta E) \\ &= r(\Delta E) \otimes [I_0\delta(\Delta E) + s_1 + s_2 + s_3 + \ldots] \\ &= r(\Delta E) \otimes I_0[\delta(\Delta E) + \frac{s_1}{I_0} + \frac{s_1 \otimes s_1}{2!I_0^2} + \frac{s_1 \otimes s_1 \otimes s_1}{3!I_0^3} + \ldots]\,. \end{aligned} \tag{10.49}$$

In order to extract $s_1(\Delta E)$ from $s^*(\Delta E)$ (by deconvolution), we use the convolution theorem (3.48) of Fourier transforms. Using capitals for the Fourier transforms of the functions appearing in (10.50) and expressing the "loss frequency" Ω in units of ${\rm eV}^{-1}$, we have

$$S^*(\Omega) = R(\Omega)I_0 \left[1 + \sum_{n=1}^{\infty} \frac{S_1^n(\Omega)}{n!I_0^n}\right] = S_0(\Omega)\exp[S_1(\Omega)/I_0], \tag{10.50}$$

which can be solved for $S_1(\Omega)$:

$$S_1(\Omega) = I_0 \ln[S^*(\Omega)/S_0(\Omega)]. \tag{10.51}$$

The single-scattering distribution $s_1(\Delta E)$ is then obtained by an inverse Fourier transform of (10.51). We should mention one problem, however, that is related to an ambiguity of the phase. By definition, the logarithm yields a phase between $-\pi$ and π. When an increasing phase exceeds π, it jumps back to $-\pi$. This jump is a mathematical artifact and leads to spurious oscillations in the inverse Fourier transform. Therefore these phase jumps have to be removed before using the inverse Fourier transform.

When the Fourier-log method is applied to a spectrum containing ionization edges, multiple-scattering effects are removed from the plasmon-loss region, the ionization edge, and the background. However, this Fourier-log method can create artifacts. At large energy losses, the spectrum should not be abruptly truncated but tend smoothly to zero. The experimental spectrum $s^*(\Delta E)$ is limited in the scattering angles recorded by the acceptance aperture α of the spectrometer with the result that part of the intensity is cut off, depending also on the energy loss. The plasmon losses, for example, fail to follow a Poisson distribution because the width of the angular distribution also increases with increasing energy loss. Multiple losses are more attenuated than single losses. The spectrum should therefore be recorded with as large an aperture as possible, $\alpha \simeq 10$ mrad; the resulting error should then be less than 10% for $E = 80$ keV and $\Delta E \leq 1600$ eV. Elastic scattering also attenuates the spectrum by scattering through angles larger than α. This contribution can be assumed to be approximately the same for all energy losses; for deviations, see Sect. 10.3.5. The narrow instrumental distribution $r(\Delta E)$ has a broad Fourier transform $R(\omega)$. The division by $S_0(\omega)$, the Fourier transfrom of $r(\Delta E)I_0$, in the logarithmic term of (10.51) can also increase the noise at high-loss frequencies ω. This means that we cannot expect to resolve fine structures that were not visible in the original spectrum; only maxima and minima will become more pronounced.

Fourier-Ratio Method. This method only deconvolves inner-shell ionization edges from the influence of low-loss electrons. A background-subtracted signal $s_{\mathrm{I}}(E_{\mathrm{I}} + \Delta E)$ (I = K, L, M) (Fig. 10.12) is convolved with the low-loss part $s_{\mathrm{LL}}(\Delta E)$ of the spectrum including the zero-loss peak:

$$s_{\mathrm{I}}^*(E_{\mathrm{I}} + \Delta E) = s_{\mathrm{I}}(E_{\mathrm{I}} + \Delta E) \otimes s_{\mathrm{LL}}^*(\Delta E). \tag{10.52}$$

A Fourier transform yields

$$S_{\mathrm{I}}^* = S_{\mathrm{I}} \cdot S_{\mathrm{LL}} \quad \rightarrow \quad S_{\mathrm{I}}(\Omega) = S_{\mathrm{I}}^*(\Omega)/S_{\mathrm{LL}}^*(\Omega), \tag{10.53}$$

from which the deconvolved spectrum $s_{\mathrm{I}}(E_{\mathrm{I}} + \Delta E)$ can be obtained by an inverse Fourier transform.

10.3.5 Elemental Analysis by Inner-Shell Ionizations

The basic laws of inner-shell ionization are summarized in Sect. 5.3. Electron energy-loss spectroscopy can be used for elemental analysis using the ionization edges in the range $0 < \Delta E < 3$ keV or even up to 5 keV [1.78].

The information obtainable from the energy-loss near-edge structure (ELNES) and the extended energy-loss fine structure (EXELFS) has been discussed in Sects. 5.3.2 and 5.3.3.

For quantitative EELS, the numbers of electrons $N_0(\alpha, \Delta)$ and $N_a(\alpha, \Delta)$ with an energy window of width Δ and in the ranges $0 \le \Delta E \le \Delta$ and $E_\mathrm{I} \le \Delta E \le E_\mathrm{I} + \Delta$ (I = K, L, M), respectively, are measured (shaded areas in Fig. 10.12). Most of the n_0 incident electrons are concentrated in the low-energy window, and any decrease caused by multiple scattering and by the use of a limited acceptance angle acts on both N_0 and N_I. We can therefore write

$$N_0(\alpha, \Delta) \simeq n_0 \exp[-x/x_\mathrm{k}(\alpha)], \tag{10.54}$$

where $x = \rho t$ is the mass thickness, and the exponential attenuation characterizes the loss by elastic and inelastic scattering through angles $\theta \ge \alpha$ using the contrast thickness $x_\mathrm{k}(\alpha)$ (6.7). This attenuation factor becomes important for high-Z elements in a low-Z matrix, for example. The formula (10.18) for x-ray microanalysis has to be modified to

$$N_a(\alpha, \Delta) = n_0 \sigma_a(\alpha, \Delta) c_a \frac{N_A}{A_a} x \exp[-x/x_\mathrm{k}(\alpha)]. \tag{10.55}$$

$n_a = c_a N_\mathrm{A} x / A_a$ is the number of atoms of element a per unit area and $\sigma(\alpha, \Delta)$ the partial cross section for those ionizations that result in scattering of the primary electrons within $0 \le \theta \le \alpha$ and the energy-loss region $E_\mathrm{I} \le \Delta E \le E_\mathrm{I} + \Delta$. Equations (10.54) and (10.55) can be combined to give the formula

$$n_a = \frac{1}{\sigma_a(\alpha, \Delta)} \frac{N_a(\alpha, \Delta)}{N_0(\alpha, \Delta)} \tag{10.56}$$

for quantitative analysis if the partial cross section $\sigma_a(\alpha, \Delta)$ is known.

These cross sections can be measured using oxides or nitrides of an element as a standard (see Sect. 5.3.2). For an estimation of the influence of α and Δ, we write $\sigma_a(\alpha, \Delta) = \sigma_a \eta_\alpha \eta_\Delta$, where η_α and η_Δ are corresponding collection efficiencies. Although it is possible to estimate the total number of energy losses and to measure the total cross section σ_a by using large values of Δ and α (Sect. 5.3), only a small window, of the order of $\Delta = 50$ eV, is needed for quantitative analysis; the power law (10.41) indicates that this contains a fraction

$$\eta_\Delta = 1 - \left(1 + \frac{\Delta}{E_\mathrm{I}}\right)^{1-r} \simeq (r-1)\frac{\Delta}{E_\mathrm{I}} + \ldots \quad \text{for } \Delta \ll E_\mathrm{I}. \tag{10.57}$$

The angular distribution of the scattered electrons is concentrated within a cone of semiangle $\theta_\mathrm{E} = E_\mathrm{I}/mv^2$ (5.46); a spectrometer with an acceptance angle α receives the fraction

$$\eta_\alpha = \frac{\ln(1 + \alpha^2/\theta_E^2)}{\ln(2/\theta_E)} \tag{10.58}$$

for apertures smaller than the Compton angle $\alpha < \theta_C = (E_I/E)^{1/2}$ [see (5.57) for the meaning of the Compton angle θ_C].

A fraction $\eta(\alpha, \Delta) = \eta_\alpha \eta_\Delta$ of 0.1–0.5 can readily be collected under optimum conditions. This is orders of magnitude more than the fraction of x-ray quanta collected, and moreover the low fluorescence yield ω_a for low Z does not appear in (10.55).

In these calculations, multiple elastic and inelastic scattering in the foil has not been considered. Increasing the foil thickness broadens the angular distribution because the mean-free-path lengths Λ_{el} and Λ_{in} for elastic scattering and plasmon losses, respectively, are orders of magnitude smaller than Λ_K for K-shell ionization. The influence of elastic scattering on quantitative electron energy-loss spectrum analysis using the formulation of elastic scattering proposed by Lenz ([5.11], Sect. 5.1.3) and a Poisson distribution results in a stronger decrease for high energy losses. The intensity ratio of the O K to Al K edges in Al_2O_3 decreases by about 20% for $t = 2\Lambda_{in}$, for example [10.121]. If this scattering occurs before and/or after a K-shell ionization, for example, the number of electrons with K-shell losses inside the acceptance angle α decreases; this is partly counterbalanced by the initial increase proportional to the foil mass thickness x (10.55). A maximum of K-loss electrons inside an aperture α is observed for $t \simeq \Lambda_{pl}$ because, for low-Z elements, $\Lambda_{pl} < \Lambda_{el}$ [10.122, 10.123]. This means that EELS is firmly confined to thin films, and EELS of thicker films is possible only in a high-voltage electron microscope, the thickness being proportional to the increase of Λ_{pl} with energy (Fig. 5.19). A further advantage of EELS in HVEM is that θ_E decreases with increasing E [10.124]. Observation of the ratio of the zero- and first-plasmon-loss intensities is important to ensure that the limits of the useful thickness range have not been exceeded.

As in x-ray microanalysis (Sect. 10.2.4), a ratio method can be applied [10.125] if the relative number n_a/n_b of two elements is required, and (10.55) gives

$$\frac{n_a}{n_b} = \frac{N_a}{N_b} \frac{\sigma_b(\alpha, \Delta)}{\sigma_a(\alpha, \Delta)} . \tag{10.59}$$

Just as x-ray emission depends on crystal orientation (Sect. 10.2.3), so, too, does the number of K-loss electrons depend on the orientation of the incident electron beam relative to the lattice planes. However, the inelastic interaction also has to be highly localized near the nuclei, and a small illumination aperture must be used to observe any influence of crystal orientation. This is barely detectable for the C K edge [10.125], shows a weak influence on the O K edge [10.64, 10.126], and differs by a factor of 2 for Mg and Al K edges [10.127]. A high localization requires that the average impact parameter $\langle b \rangle \simeq \lambda/4\pi\theta_E$ be smaller than the lattice-plane distance d [9.31]. When the collection aperture is decreased and Bragg spots are excluded, the

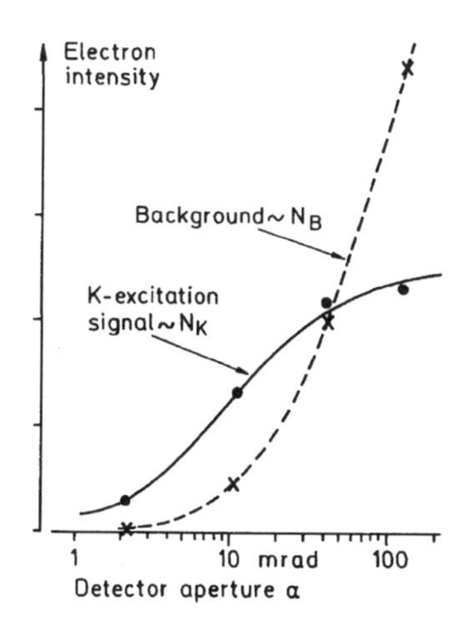

Fig. 10.13. K-loss signal and background in carbon ($E = 80$ keV) as a function of acceptance aperture α [10.123].

ratio $N_a(\alpha, \Delta)/N_0(\alpha, \Delta)$ should be constant even at the dark bend contours if Bragg contrast is preserved. However, the high energy losses show a stronger broadening by inelastic scattering than the low losses. This results in an increase of the effective aperture for high losses, and the ratio can be changed by about 10% near Bragg positions [10.125] when normal apertures are used and even more for small illumination apertures. Otherwise, a large illumination aperture as used in a focused electron probe nearly cancels all of these orientation effects.

The background below the excitation edge increases with α, Δ, and t, and the signal of the K-shell ionization decreases with decreasing concentration c_a and saturates when Δ and α are increased sufficiently (Fig. 10.13) [10.123]. The conditions for maximum signal-to-background ratio do not coincide with those for optimum signal-to-noise ratio, as shown by an analysis of oxygen and boron in B_2O_3, for example [10.128]. A small acceptance angle α is desirable for a large signal-to-background ratio (Fig. 10.14a) because the K-loss electrons are concentrated at low scattering angles, whereas the background is spread over larger angles. The use of a low value of α, however, decreases the number of electrons detected and increases the statistical noise. The signal-to-noise ratio reaches a maximum for $\alpha = 10$–20 mrad (Fig. 10.14c); this value of the spectrometer acceptance angle can just be used without too strong a decrease of the energy resolution by aberrations.

10.4 Element-Distribution Images

10.4.1 Elemental Mapping by X-Rays

Elemental analysis with x-rays can be performed with the scanning mode of a transmission electron microscope or in dedicated STEM, by using a stationary electron probe and counting times of the order of a hundred

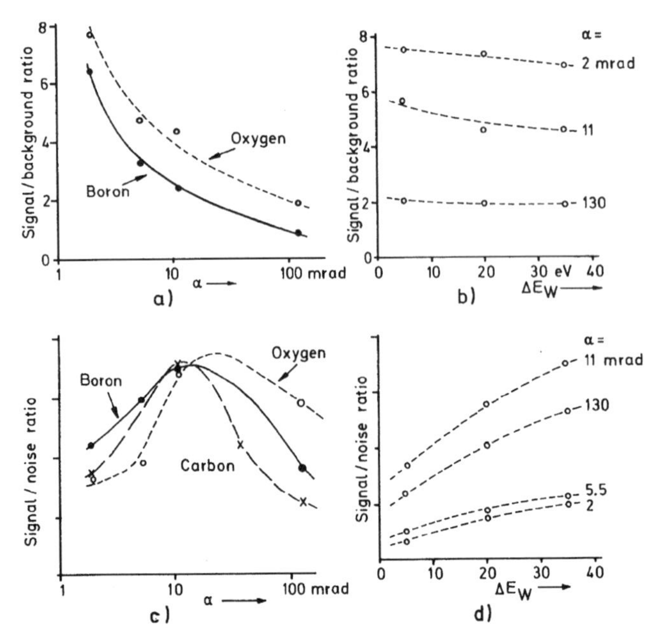

Fig. 10.14. Signal-to-background ratios (**a, b**) and signal-to-noise ratios (**c, d**) of the electron energy-loss spectrum of oxygen and boron in B_2O_3 as a function of acceptance aperture α and width Δ of the energy window [10.128].

seconds or by scanning the electron probe and generating elemental maps [10.129, 10.130, 10.131]. A simple and fast method is to record a dot on the screen when a quantum is recorded within a selected window of a characteristic line during a scan. The density of dots is a measure of the concentration. However, such images are very noisy and contain many unspecific dots from the continuum in areas not containing the element of interest. With longer dwell times of seconds, it is possible either to count in a few windows containing the continuum or to record the whole spectrum and then subtract an extrapolated background from the characteristic counts. For organic specimens, the Hall method can be used to produce elemental concentrations corrected for local thickness. However, even with a dwell time per pixel of only a few seconds, the recording time for elemental maps with 64×64 or 128×128 pixels extends over many hours. The unavoidable drift has to be compensated by digital computation of image shifts.

10.4.2 Element-Distribution Images Formed by Electron Spectroscopic Imaging

The generation of element-distribution images from electron spectroscopic images is superior to x-ray mapping because only short exposure times of the order of seconds are needed and the images can be recorded directly by a CCD camera. The rapid progress of this method is documented in the proceedings of the European Workshops on Electron Spectroscopic Imaging and Analysis Techniques (Vol. 32 of *Ultramicroscopy* and Vols. 162, 166, 174, and 182 of the *Journal of Microscopy*) and the proceedings of the International EELS workshops (Vol. 2 of *Microscopy, Microanalysis and Microstructures* and Vols. 59, 78, 96, and 106 of *Ultramicroscopy*). The most frequent application in biology concerns the element-distribution images of P, Ca, and S. It was shown in Sect. 10.3.3 how the background below an ionization edge can be fitted and extrapolated by measuring the decrease of the electron energy-loss spectrum intensity below the edge. A similar method is used to process element distribution images pixel by pixel. In a scanning transmission electron microscope, we record an energy-loss spectrum for every pixel. These spectra can be treated as discussed before. The STEM method is particularly useful for the detection of low concentrations of an element. Furthermore, it allows the local bonding states to be determined via the near-edge structure of the relevant edges.

For a series of ESI images obtained with an imaging filter, the number of images can be reduced to a minimum of three using energy windows $\Delta =$ 5–50 eV. Often a larger number of windows is used to obtain a more reliable background subtraction. The two-window method (subtracting one scaled image below from the image beyond the edge) should be avoided because large spurious signals can be generated that are not characteristic for the element of interest. If the concentration of the element is sufficiently high, one can obtain a jump-ratio image by dividing the image beyond the edge by an image below the edge. This image delivers a map of the elemental distribution but cannot be quantified to obtain the concentration of the element.

10.4.3 Three-Window Method

Two images at energy losses ΔE_1 and ΔE_2 are recorded below the edge and one at ΔE_{max} at the maximum of the electron energy-loss spectrum beyond the edge (Fig. 10.15a), each with a width $\Delta = 5$–20 eV. If the background distribution is assumed to have the form $S(\Delta E) = A\Delta E^{-r}$, the net signal becomes

$$S_{\mathrm{N}} = S(\Delta E_{\mathrm{max}}) - A\Delta E_{\mathrm{max}}^{-r} \tag{10.60}$$

with

$$r = \frac{\log[S(\Delta E_1)/S(\Delta E_2)]}{\log(\Delta E_2/\Delta E_1)} \quad \text{and} \quad A = \Delta E_1^r S(\Delta E_1). \tag{10.61}$$

Figure 10.16 shows as an example the imaging of the elemental distribution of oxygen in amorphous grain boundary films in a sintered Si_3N_4 ceramic [10.132]. The bright-field image (Fig. 10.16a) shows the grain boundaries at low magnification and the high-resolution image (Fig. 10.16b) a single grain boundary with a 0.7 nm amorphous intergranular layer. The elemental distribution image for oxygen (Fig. 10.16c) is obtained with three electron

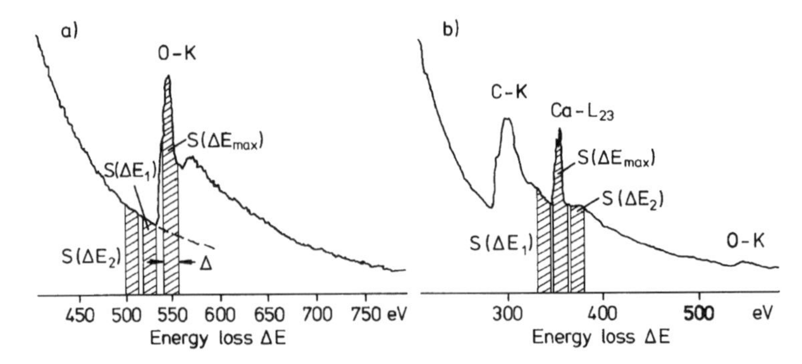

Fig. 10.15. Signals and energy windows used for (**a**) the three-window method and (**b**) the white-line method to subtract the background for element distribution images.

spectroscopic images at 484±15, 514±15, and 547±15 eV. The concentration of oxygen at the boundaries corresponding to approximately two monolayers of oxygen is clearly visible.

10.4.4 White-Line Method

This method can be used when the three-window method cannot be applied because of convolutional effects coming from low-loss electrons or edge superpositions and the element of interest shows a white line at the edge [10.133] (e.g., a Ca L white line superposed on the ELNES of C K, Fig. 10.15b). The background is then linearly extrapolated from signals at ΔE_1 below and ΔE_2 beyond the line, and the net signal becomes

$$S_{\mathrm{N}} = S(\Delta E_{\mathrm{max}}) - \left\{ S(\Delta E_1) + \frac{\Delta E_{\mathrm{max}} - \Delta E_1}{\Delta E_2 - \Delta E_1} [S(\Delta E_2) - S(\Delta E_1)] \right\} . \tag{10.62}$$

Optimal window energies ΔE_1, ΔE_2, ΔE_{max} are given in [5.103].

10.4.5 Correction of Scattering Contrast

Structures that already appear darker in the unfiltered and zero-loss filtered images due to stronger elastic scattering through angles $\theta \geq \alpha$, caused by a larger thickness or higher atomic number than the matrix, also attenuate the intensity of the ionization edges by the same amount because of elastic–inelastic multiple scattering. When, for example, epon-embedded calcium-containing apatite crystals in the early stages of bone formation appear dark in the bright-field image because of stronger elastic scattering, the white lines of the Ca L edge are also weaker. In the elemental-distribution image obtained by the white-line method, an apparently lower concentration of Ca is indicated

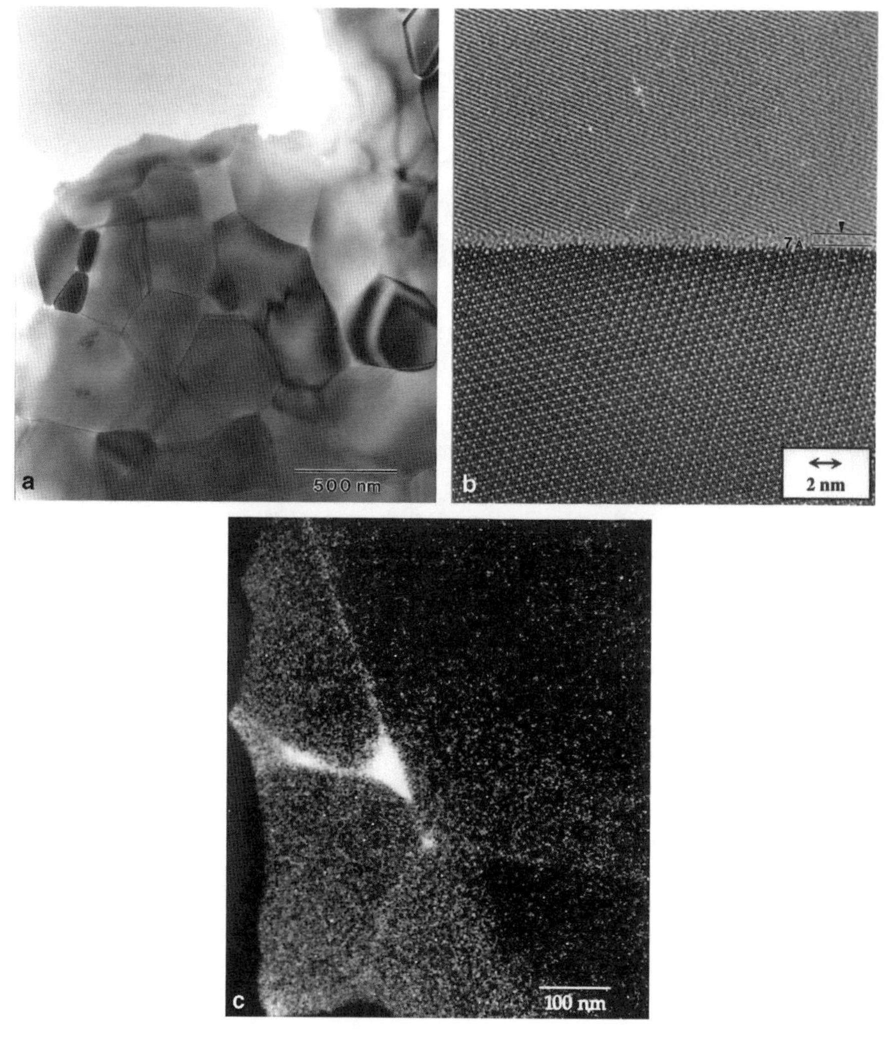

Fig. 10.16. (**a**) Bright-field image of an Si_3N_4 material containing grain boundaries. (**b**) High-resolution image of a grain boundary with a 0.7 nm layer containing oxygen. (**c**) Oxygen-distribution map of grain boundaries [10.132].

and the nonspecific noise of the matrix is relatively high (Fig. 10.17a). In quantitative EELS, this attenuation of the edge intensities can be compensated for by dividing by the low-loss intensity $N_0(\alpha, \Delta)$ in (10.56). The same result can be achieved for elemental-distribution images by recording a low-loss image of signal $S_0(\alpha, \Delta)$ for $0 \le \Delta E \le \Delta$ and performing a pixel by pixel division S_N/S_0 [10.134, 10.135, 10.136]. The value of this method is demonstrated in Fig. 10.17b. The noisy nonspecific background is decreased by the division by the bright transmission of the matrix, whereas the Ca-containing parts are

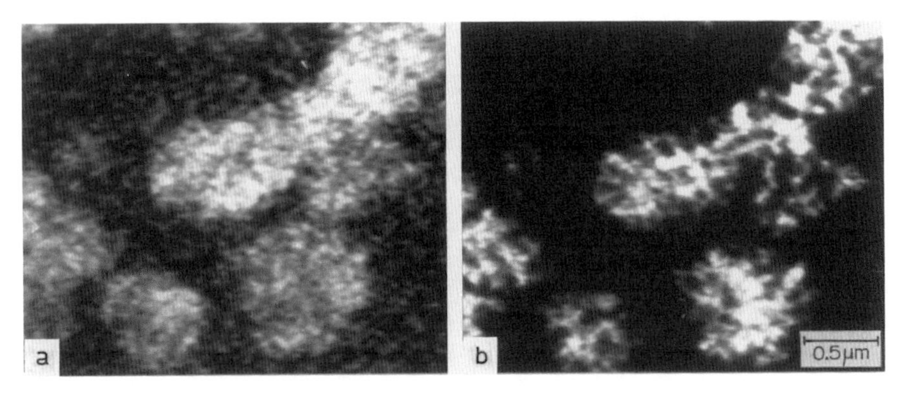

Fig. 10.17a,b. Elemental distribution image of Ca in apatite crystals formed in the early stage of bone formation and embedded in epon. (**a**) Net image with the signal S_N of the white-line method and (**b**) the ratio signal S_N/S_0 for correction of elastic large-angle scattering.

divided by the more strongly attenuated transmission. Normalizing the total image to full scale suppresses the nonspecific signals and increases the contrast of the apatite-containing parts, which are also seen with better resolution.

10.5 Limitations of Elemental Analysis

10.5.1 Specimen Thickness

An important condition for EELS and ESI is that the thickness must be smaller than the mean free path Λ_{pl} for plasmon losses, which is about 80 nm in biological sections and about 40 nm in aluminum foils for 80 keV electrons. Deconvolution methods make it possible to investigate specimens of 3–5 times this thickness by EELS but not by ESI because the whole loss spectrum is then needed. In the case of inorganic material, EELS and ESI are often possible only at the edges of thinned foils, which may not be representative of the bulk material and can have a different composition. This restriction to thin specimens is a handicap of EELS and ESI not shared by x-ray microanalysis and elemental mapping. An increase of the mean free path by using higher electron energies will therefore become important for the investigation of thicker specimen areas, though the mean free path of plasmon losses saturates at electron energies of a few hundred keV (Fig. 5.19).

10.5.2 Radiation Damage and Loss of Elements

A major problem with biological material is the accidental elimination of elements during chemical fixation, dehydration, embedding, and sectioning (contact with water during floating). Although such elimination is less serious with the newer embedding resins such as Nanoplast [10.137], only by cryofixation, cryosectioning and cryotransfer to a cryostage in the microscope can element loss be avoided completely. The zero-loss mode of electron spectroscopic imaging can be used to increase the contrast of unstained frozen-hydrated cryosections [10.138, 10.139] because the ratio of inelastic-to-elastic total cross sections (5.66) increases to $\nu \simeq 4$ for ice.

The mass loss of organic material by radiation damage (Sect. 11.2) results in a nearly complete removal of H, O, and N atoms and some volatile carbon molecules, and after about one second the specimen consists of a cross-linked carbon-rich polymer. A fraction of any phosphorus, sulfur, and sodium, for example, can also leave the specimen, though mass loss of these elements may saturate at two to three orders of magnitude higher charge densities of a few tens or hundreds of C/cm^2. Although the mass loss by radiation damage is reduced at low specimen temperatures, and embedding in ice can protect against the mass loss of organic material, the irradiation of ice can cause other artifacts by radiolysis (Sect. 11.2). An alternative is the freeze-drying of specimens followed by transfer to the microscope at room temperature. This also makes it possible to observe thicker sections, but the strong variations in the mass thickness after drying are an obstacle to analysis by EELS and ESI, whereas the Hall method of x-ray microanalysis (Sect. 10.2.5) can be applied without hesitation.

Inorganic specimens are normally more resistant because most of the damage processes caused by ionization are reversible (Sect. 11.3). For example, irradiation of alkali halides causes additional maxima to appear in the plasmon-loss region, which can be attributed to the formation of alkali colloids [10.140], while NiO_2 loses oxygen at very high doses larger than 10^8–10^9 C/m^2 [10.141].

10.5.3 Counting Statistics and Sensitivity

X-Ray Microanalysis. The sensitivity of XRMA is limited by the counting statistics. If a quantity that is subject to statistical variations is measured n times, giving a set of values $N_i (i = 1, \ldots, n)$ with mean value $\overline{N}$, the standard deviation σ will be given by

$$\sigma^2 = \frac{1}{n-1} \sum_{i=1}^{n} (N_i - \overline{N})^2. \tag{10.63}$$

For Poisson statistics and large n,

$$\sigma^2 = \overline{N}. \tag{10.64}$$

There is a probability of 68.3% that a measured value N_i will lie in the confidence interval $\overline{N} \pm \sigma$ and a probability of 95% that it lies in $\overline{N} \pm 2\sigma$. The analytical sensitivity and the minimum mass fraction will now be estimated as examples.

The *analytical sensitivity* is the smallest difference $\Delta c = c_1 - c_2$ of concentrations that can be detected. $\overline{N}_1$ and $\overline{N}_2$ denote the expected mean values of counts for the two concentrations. We assume that $\overline{N}_1 \simeq \overline{N}_2 = \overline{N} \gg \overline{N}_{\mathrm{B}}$, where $\overline{N}_{\mathrm{B}}$ corresponds to the background. The two values $\overline{N}_1$ and $\overline{N}_2$ are significantly different at the 95% confidence level if

$$\overline{N}_1 - \overline{N}_2 \geq 2\sqrt{\sigma_1^2 + \sigma_2^2} \simeq 2\sqrt{2\overline{N}}, \tag{10.65}$$

and the analytical sensitivity becomes [10.142]

$$\frac{\Delta c}{c} = \frac{\overline{N}_1 - \overline{N}_2}{\overline{N}} \simeq \frac{3}{\sqrt{\overline{N}}} \,. \tag{10.66}$$

Thus, a sensitivity $\Delta c/c$ of 1% requires $\overline{N} \geq 10^5$ counts. The mean value $\overline{N}$ is proportional to the product of electron-probe current and counting time; the counting time should not exceed about 10 min.

If the concentration of an element is as low as about 1 wt.%, $\overline{N}$ is no longer much larger than $\overline{N}_{\mathrm{B}}$ and it ceases to be clear whether an element is present in the sample. The question then arises as to what *minimum-detectable mass fraction* (MMF) will be detectable within a certain confidence interval. Equation (10.65) becomes

$$\overline{N} - \overline{N}_{\mathrm{B}} \geq 2\sqrt{\sigma^2 + \sigma_{\mathrm{B}}^2} \simeq 2\sqrt{\overline{N} + \overline{N}_{\mathrm{B}}} \,. \tag{10.67}$$

When a standard with concentration c_s of the element being investigated is used, the MMF becomes, with $\overline{N} \simeq \overline{N}_{\mathrm{B}}$,

$$\mathrm{MMF} = \frac{\overline{N} - \overline{N}_{\mathrm{B}}}{\overline{N}_s - \overline{N}_{\mathrm{B,s}}} c_s \simeq \frac{c_s}{\overline{N}_s - \overline{N}_{\mathrm{B,s}}} 3\sqrt{\overline{N}_{\mathrm{B}}} \,. \tag{10.68}$$

Estimates based on (10.68) and especially comparisons of EDS and electron energy-loss spectroscopy (EELS) have the disadvantage that certain assumptions have been made that are not valid in all practical cases (e.g., concerning the magnitude $\overline{N}_{\mathrm{B}}$ of the background). For example, a calculated value of 3–5% for the MMF is found for elements in a 100 nm thick Si foil irradiated at $E = 100$ keV with $j = 2\times10^5$ A/m^2, a spot size of 10 nm diameter, and a counting time of 100 s, whereas the minimum detectable mass is of the order of 0.5–1×10^{-19} g [10.143].

Typical values for WDS in an x-ray microanalyzer are $\simeq$ 100 ppm for MMF and $\simeq 10^{-15}$ g for the minimum detectable mass. The MMF is thus more favorable in an x-ray microanalyzer with WDS, but the minimum detectable mass is greater due to the larger probe size.

Electron Energy-Loss Spectroscopy. Estimates of EELS sensitivities can be found in [1.78, 10.122, 5.104, 10.144]. The number N_a of electrons collected as a result of an ionization of shell I = K,L,... in the energy interval $E_{\mathrm{I}} \leq \Delta E \leq E_{\mathrm{I}} + \Delta$ (10.55) and the number N_{B} of those from the background due to the other n_{t} atoms per unit area with a mean cross section σ_{B} are

$$N_a = \eta n_0 n_a \sigma_a \exp(-x/x_k), \quad N_{\mathrm{B}} = n_0 n_{\mathrm{t}} \sigma_{\mathrm{B}} \exp(-x/x_k), \tag{10.69}$$

where $n_0 = I_{\mathrm{p}}\tau/e$ (I_{p}: electron-probe current, τ: counting time, η: detector collection efficiency). The standard deviation of the recorded signal $N = N_a + N_{\mathrm{B}}$ becomes

$$\sigma = \sqrt{\mathrm{var}N} = \sqrt{N_a + hN_{\mathrm{B}}}, \tag{10.70}$$

where $h \simeq 2 - 10$ depends on the width of the energy interval used to fit the background and allows for the risk that the background noise below the edge will be "amplified" in the process of extrapolation beyond the edge.

If it is assumed that $N_a \ll N_{\mathrm{B}}$ for a small number of detectable atoms and that N_a is greater than $\kappa\sigma$ ($\kappa = 3 - 5$), the *minimum detectable atomic fraction* (MAF) and the corresponding *minimum detectable number* (MDN) of atoms inside the probe of diameter d will be

$$\mathrm{MAF} = n_a/n_t = \frac{\kappa}{\sigma_a}\left[\frac{eh\sigma_{\mathrm{B}} \exp(x/x_k)}{\eta I_{\mathrm{p}} \tau n_{\mathrm{t}}}\right]^{1/2}, \tag{10.71}$$

$$\mathrm{MDN} = \frac{\pi d^2}{4} n_{\mathrm{t}}\, \mathrm{MAF}. \tag{10.72}$$

MAF and MMF are related by MMF = $(n_{\mathrm{t}} A_a / \Sigma_i n_i A_i)$ MAF.

As a numerical example for the detection of iron atoms on (or in) a 10 nm carbon film [10.144], with $h = 8$, $\eta = 1$ (ideal parallel detection), $\kappa = 3$, $E = 100$ keV, and hydrogen-like cross sections, the detection of 100 Fe atoms within $d = 1$ nm needs 10^6 e$^-$/nm^2, corresponding to a dose of $q \simeq 1.6 \times 10^5$ C/m^2 (see Tables 11.4 and 11.6 for the terminal doses for the destruction of organic compounds).

The minimum detectable mass in EELS has been estimated to lie between 10^{-22} and 10^{-18} g and the lowest concentration to be $10^{-4} - 10^{-3}$ [10.122, 10.145]. Thus, a signal from the M-ionization loss of Fe can be easily recorded from a single ferritin molecule with an electron probe 50 nm in diameter; it indicates the presence of about 5×10^{-19} g Fe, corresponding to some 4000 iron atoms.

Leapman and Rizzo [10.146] and Suenaga [10.147] have demonstrated that under favorable conditions the detection limit is close to one atom.

Spatial Resolution. The spatial resolution is not necessarily identical with the electron-probe diameter in the STEM mode or with the instrumental resolution. The practical resolution will often be worse due to electron–specimen interactions and/or the choice of detection mode.

In thick specimens, the electron-beam broadening discussed in Sect. 5.4.3 has to be considered for x-ray microanalysis. The distribution of scattered electrons can show a broad tail; in low-Z material, few electrons scattered through $\simeq 90°$ can travel distances of the order of a micrometer in thicker sections of biological material before leaving the specimen. This can mimic the presence of a low concentration of calcium near a high concentration in a calcified tissue, for example. This effect can show a higher probability than x-ray fluorescence, but it is not observed in EELS analysis [10.151].

10.5.4 Resolution and Detection Limits for Electron Spectroscopic Imaging

For inner-shell losses in the STEM mode, the spatial resolution is almost equal to the electron-probe diameter [10.152]. In thicker specimens, it is broadened due to the electron–specimen interactions. In energy-filtering TEM, the efficiency of the detector is an important factor. The introduction of slow-scan CCD cameras (Sect. 4.7.5) has led to a great improvement compared with the early work using photographic plates for image recording [10.148, 10.149, 10.150]. Furthermore, the resolution and detection limits are strongly influenced by the imaging conditions. It is therefore important to select optimized imaging conditions to obtain a good signal-to-noise ratio. The influence of illumination angle, objective aperture angle, defocus, and energy width on the image can be described by an inelastic transfer function (ITF). These parameters can then be optimized to obtain the maximal signal-to-noise ratio [10.153]. We can only give a qualitative account of the essential results.

Illumination Angle. Contrary to the phase-contrast transfer function (Sect. 6.4.2), the inelastic transfer function depends only slightly on the illumination angle. It is therefore helpful to maximize the current density on the specimen as much as possible. This can be done by increasing the illumination angle up to about the objective aperture angle. Any further increase still increases the current and thus the beam damage. The corresponding signal, however, would not improve because most of the additionally scattered electrons are then intercepted by the objective aperture.

Magnification. As the intensity in an inner-shell loss image is very low, the influence of the noise of the detector has to be considered. To maximize the signal-to-noise ratio, we therefore choose the magnification so as to obtain a current density as high as possible in the detector plane without limiting the resolution. The magnification should therefore be set to a value that ensures that the desired value of the resolution limit corresponds to about 2–4 pixels of the camera system. Therefore the magnifications are often rather small compared with what one is used to in high-resolution electron microscopy.

Position of the Energy Windows. If we assume a power-law model AE^{-r} for the background, we can visualize the essential results of a tedious calculation [10.153] in a double-logarithmic plot (Fig. 10.18). The background yields a straight line with a negative slope. In the three-window method, we use two windows below the edge. To avoid a long range for the extrapolation, we should choose one energy window E_1 as close as possible to the edge energy. The second window E_2 should be chosen as far away as possible to obtain a good estimate for the slope $-r$ of the curve. It is obvious that a lower limit for E_2 is given by the energy of any preceding edge. The optimum position

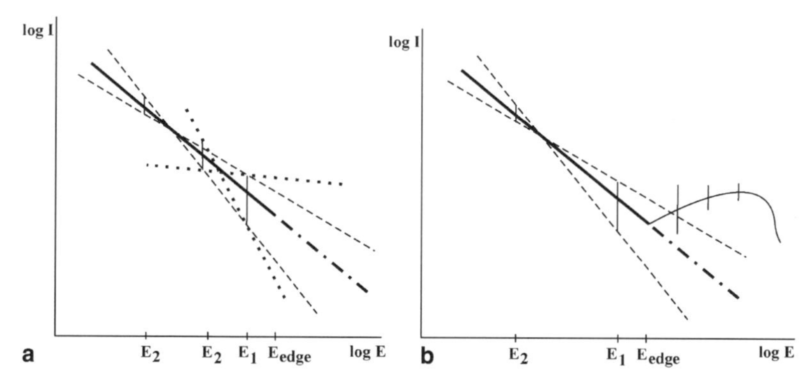

Fig. 10.18. Schematic diagram showing the influence of the choice of the energy windows on the extrapolation error **(a)** and on the signal-to-noise ratio **(b)**. **(a)** The extrapolation error is illustrated by lines passing through the extreme values of the error bars. It can be significantly reduced by choosing E_2 far away from the edge. **(b)** The choice of E_s for a delayed edge is a compromise between a small extrapolation error and a small signal (close to the edge) and a larger extrapolation error and a large signal at the maximum of the edge. In this logarithmic plot, the geometric length of the error bar is given by the *relative* error.

for the energy window E_s beyond the edge depends strongly on the shape of the edge. For a sawtooth-like shape, common for K edges, E_s should be chosen as close to the edge as possible. For a delayed edge, one has to find a compromise between the smallest extrapolation error (close to the edge) and the delayed maximum. A detailed investigation of the choice of optimum edge positions starting from measured spectra has been performed by Kothleitner et al. [10.154]. Often it is advantageous to use more than two windows for the extrapolation of the background. It has been shown that in the four-window method, using three windows below the edge, two of them should be close to the edge and one far apart [10.155].

Objective Aperture Angle, Defocus, Width of the Energy Windows and Pixel Size. Unfortunately, in the optimization procedure, the remaining imaging parameters are closely linked with one another. It is therefore difficult to give any general rules for the choice of parameters. Any increase of the objective aperture angle increases the detected signal but also increases the influence of the spherical and chromatic aberrations and thus limits the resolution. Therefore one needs to find a compromise between the resolution and the signal-to-noise ratio. The defocus is used to partly counterbalance the influence of the spherical aberration. Increasing the width of the energy window again increases the signal but also the influence of the chromatic aberration, thus deteriorating the resolution.

For a rough estimate, approximate expressions have been proposed to describe the degradation of the resolution in electron spectroscopic imaging [1.78, 10.156, 10.157, 10.158]. The principle of exact calculations is described

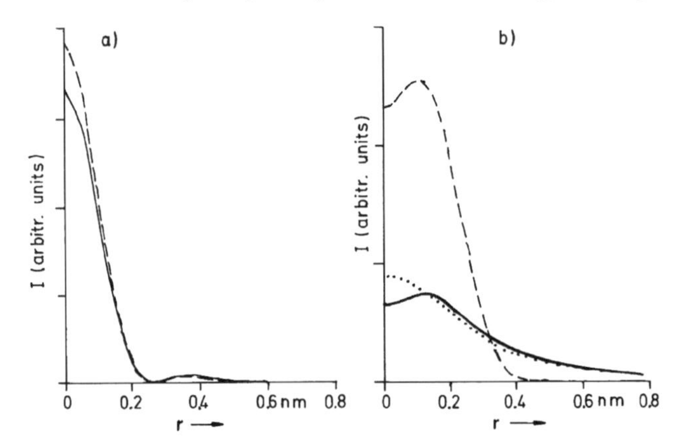

Fig. 10.19. Radial-intensity distribution of the image of an oxygen atom formed (**a**) by elastic dark-field imaging and (**b**) by electron spectroscopic imaging with the O K edge. Dashed curves: ideal lens ($C_s = C_c = 0$), solid curves: $C_s = C_c = 2.7$ mm, dotted curve in (**b**): $\alpha_i = \alpha = 7.8$ mrad [10.153].

in [6.112, 6.118, 10.159, 10.160, 10.161, 10.162, 10.163]. As an example, for 120 keV electrons, Fig. 10.19a shows the radial intensity distribution for an elastic dark-field image of an oxygen atom and Fig. 10.19b that for an image with the oxygen K loss at $\Delta E = 540$ eV. The dashed curves are for an ideal lens and the solid curves for $C_s = C_c = 2.7$ mm with a width $\Delta = 20$ eV of the energy window and an optimum defocus $\Delta z = 0.82\sqrt{C_s\lambda} = 78$ nm in Fig. 10.19a and 9.5 nm in Fig. 10.19b. The dotted curve in Fig. 10.19b results from a calculation with $\alpha_i = \alpha = 7.8$ mrad [10.153].

The definition of a figure of merit to compare different instruments with one another is very difficult because the definition of an *object resolution* includes the energy loss and a parameter defining the statistical accuracy of the background subtraction [10.162]. In practice, one has to distinguish between the attainable resolution and the detection limit. An object detail (e.g., a sphere) may be detected, but under optimal detection conditions it appears as a large spot in the image. Using these conditions, the exact size and shape of the object cannot be determined from the image.

Experimentally the object resolution limit has been determined from periodic specimens, such as oxygen planes in an AlON ceramic [4.119] or multilayers such as Cr layers in Ni_{80} Fe_{20} [10.164] or SiGe heterostructure systems with a spatial resolution of $\simeq$1 nm [10.165]. Likewise, crystals have been imaged looking down an appropriate direction [10.166]. To interpret such images, one has to take care to make sure that the contrast is due to a single inelastic scattering process and not the elastic contrast due to multiple elastic, plus one inelastic, scattering processes [10.167, 10.168], which is also true for STEM imaging [10.169].

11

Specimen Damage by Electron Irradiation

Most of the energy dissipated in energy losses is converted into heat. The rise in specimen temperature can be limited by keeping the illuminated area small.

The electron excitation of organic molecules causes bond rupture and loss of mass and crystallinity. The damage is proportional to the charge density in C m^{-2} at the specimen. This limits the high-resolution study of biological specimens.

Inorganic crystals can be damaged by the formation of point defects, such as color centers in alkali halides, and defect clusters. High-energy electrons can transfer momentum to the nuclei, which results in displacement onto an interstitial lattice site when the energy transferred exceeds a threshold value of the order of 20–50 eV.

Hydrocarbon molecules condensed from the vacuum of the microscope or deposited on the specimen during preparation and storage can form a contamination layer by radiation damage and cross-linking. When only a small specimen area is irradiated, the hydrocarbon molecules are particularly likely to diffuse on the specimen and be cracked and fixed by the electron beam.

11.1 Specimen Heating

11.1.1 Methods of Measuring Specimen Temperature

The specimen temperature can be calculated only for simple geometries such as a circular hole covered with a homogeneous foil (Sect. 11.1.3). In practice, therefore, methods of measuring the specimen temperature during electron irradiation are necessary. These can also be used for the calibration of specimen heating and cooling devices. Table 11.1 summarizes the methods that have been proposed. The continuous methods (I) can be employed in the whole temperature range indicated, while the other methods (II) use a fixed transition temperature as reference.

Table 11.1. Methods of measuring specimen temperatures (TC: thermocouple, ED: electron diffraction, DF: dark-field mode, BF: bright-field mode).

Specimen	T	Observed effect	Mode	Ref.
I. Continuous methods				
1. Cu-constantan	<600°C	Thin-film thermocouple	TC	[11.1]
2. Ag-Pd		Thin-film thermocouple	TC	[11.2]
3. Au-Ni		Thin-film thermocouple	TC	[11.3]
4. Pb	<300°C	Thermal expansion	ED	[11.4]
5. Ni	<700°C	Thermal expansion	ED	[11.4]
6. Al and others		Critical voltage	ED	[11.5]
7. Single-crystal foils		Shift of HOLZ lines	ED	[8.150]
II. Phase transitions				
a) Reversible indicators				
8. $AuCu_3$	388°C	Superstructure	ED	[11.6]
9. Ag_2S	179°C	Rhombohedral → cubic	ED	[11.7]
10. Ice	−70°C	Cubic ice	ED	[11.8]
11. Thin In layer	156°C	Melting	DF	[11.9, 11.10]
12. Thin Pb layer	327°C	Melting	DF	[11.9, 11.10]
13. Thin Ge layer	958°C	Melting	DF	[11.9, 11.10]
14. Paraffin	70°C	Melting before damage	ED	[11.11]
15. Supporting film	10–30 K	Condensation of gases	ED,BF	[11.19]
b) Irreversible indicators				
16. Thick Pb layer	327°C	Melting	BF	[11.25]
17. Al_2O_3 layer	600°C	Amorphous→crystalline	BF	[11.25]
18. Thick Sn layer	232°C	Melting	BF	[11.13]
19. Fe layer	920°C	bcc → fcc	BF	[11.13]
20. Fe_3C	720°C	Dissolving in Fe matrix	BF	[11.13]
21. PbO needles	890°C	Melting	BF	[11.14]
22. NaCl crystals	700°C	Sublimation	BF	[11.15, 11.16]

The first group of methods allows us to measure the increase of temperature continuously. Evaporated thermocouples can only be applied to special geometries owing to the difficulty of ensuring proper contact between the evaporated films. Furthermore, the electromotive force of a thermocouple depends on the thickness and structure of the evaporated layers [11.17, 11.18]. Another possibility is to use the thermal expansion of a crystal lattice since the Bragg angle in Debye–Scherrer ring patterns or single-crystal diffraction diagrams decreases with increasing temperature. (As a numerical example: for an evaporated Pb film, a temperature rise of $\Delta T = 100$°C results in $\Delta d/d = 1.8\times10^{-3}$.) The temperature of a small specimen area can be determined by selected-area electron diffraction [11.4]. The variation of critical voltage (Sect. 7.4.4) [11.5] or the shift of HOLZ lines in CBED patterns [8.150] with temperature provide other methods for small areas.

The methods (II) using phase transitions can be divided into those for which the modification is reversible and those for which it is irreversible.

Those of the first kind have the advantage that the transition temperature can be crossed repeatedly. This makes it easier to determine the current density necessary to bring the specimen to the fixed transition temperature. If the excitation of the condenser lens is not changed, the temperature rise due to electron bombardment is proportional to the incident electron current, which can be changed by means of the Wehnelt bias. This enables us to estimate the temperature for other values of the electron current. However, large variations of specimen temperature ensue when the diameter of the irradiated area is changed (Sect. 11.1.3).

A very simple and straightforward method is based on the behavior of evaporated indium layers with an island structure in the vicinity of the melting point. The melting or solidification of the small crystals is indicated by the vanishing or reappearance of the Bragg reflections in the dark-field mode. However, the presence of an evaporated film can alter the rate of heat generation and the thermal conductivity of the specimen. The method can be used for local measurement of temperature without greatly affecting the generation of heat and the thermal conductivity if a 10 μm-diameter spot is evaporated through an optically aligned 10 μm diaphragm [11.9, 11.10].

The specimen temperature in a liquid-helium-cooled stage can be estimated from the condensation of gases (Xe, Kr, O_2, A, N_2, Ne) in the range 10–70 K [11.19]. The gas-sublimation temperatures depend, however, on the gas pressure, which is not known accurately because of the cryopumping effect of the cold shield. This temperature increases from 58 K to 68 K for xenon and from 8.5 K to 10 K for neon in the pressure range $10^{-4} - 10^{-2}$ Pa, for example [11.20].

Another method of measuring the rise of temperature uses the known temperature dependence of the climb rate of Frank dislocation loops in materials with high stacking-fault energy [11.21]. An average local rise of 6°C is found when an Al-1.5wt% Mg alloy is irradiated under normal bright-field operation.

11.1.2 Generation of Heat by Electron Irradiation

A knowledge of the mean contribution $\Delta Q/\Delta x$ of one electron per unit mass thickness ($\Delta x = \rho \Delta t$) to specimen heating is required for a theoretical discussion and calculation of the specimen temperature.

If it is assumed that only plasmon energy losses ΔE_{pl} with a mean-free-path length Λ_{pl} contribute to heat generation, $\Delta Q/\Delta x$ can be calculated with the aid of (5.84):

$$\frac{\Delta Q}{\Delta x} = \frac{\sum\limits_{n=1}^{\infty} n \Delta E_{\mathrm{pl}} P_n(t)}{\rho t} = \frac{\mathrm{e}^{-t/\Lambda_{\mathrm{pl}}} \Delta E_{\mathrm{pl}}}{\rho \Lambda_{\mathrm{pl}}} \sum_{n=1}^{\infty} \frac{(t/\Lambda_{\mathrm{pl}})^{n-1}}{(n-1)!} = \frac{\Delta E_{\mathrm{pl}}}{\rho \Lambda_{\mathrm{pl}}} . \quad (11.1)$$

Thus, for Al with a plasmon loss at $\Delta E_{\mathrm{pl}} = 15.3$ eV and a mean free path $\Lambda_{\mathrm{pl}} = 70$ nm for $E = 60$ keV, we find $\Delta Q/\Delta x = 0.83$ eV μg^{-1} cm^2. This value

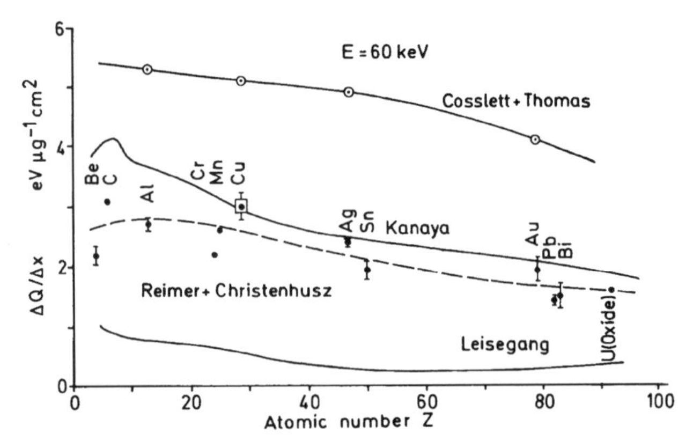

Fig. 11.1. Dependence of energy dissipated per unit mass thickness $\Delta Q/\Delta x$ (eV μg^{-1} cm^2) on the atomic number Z for 60 keV electrons. Full curve calculated with the Bethe formula (11.2) [11.24]. The calculations of Leisegang [11.12] and the measurements of Cosslett and Thomas [11.26] at E = 10–20 keV and transferred to 60 keV, respectively, under- and overestimate the value of $\Delta Q/\Delta x$.

is much smaller than the experimental value of 2.3 eV μg^{-1} cm^2 (Fig. 11.1). This means that Bethe losses, which appear in Fig. 5.20 as a continuous background to the energy-loss spectrum, also contribute strongly to specimen heating.

It therefore seems more reasonable to use the Bethe formula (5.100) to estimate the rate of heat generation. This model has been applied to the problem of specimen heating [11.22] and yields for nonrelativistic energies

$$\frac{\Delta Q}{\Delta x} = \left|\frac{\mathrm{d}E}{\mathrm{d}x}\right| = 7.8 \times 10^4 \frac{Z}{A}\frac{1}{E}\ln\frac{E}{J}, \tag{11.2}$$

where E (eV) and $\Delta Q/\Delta x$ (eV μg^{-1} cm^2), and $J \simeq 13.5Z$ denotes the mean ionization energy in eV. Figure 11.1 shows values calculated from this formula for E = 60 keV. The dependence on the atomic number Z is confirmed by experiment [11.23, 11.24]. In these experiments, the energy loss converted into heat was measured directly by irradiating a large transparent thermocouple on a supporting film of 1 cm diameter. The energy loss was calibrated by measuring the heat generated by an electric current passing through the film and producing the same temperature. The values calculated by Leisegang [11.12] underestimate $\Delta Q/\Delta x$; the experimental values of Cosslett and Thomas [11.26] obtained at E = 10–20 keV and transferred to 60 keV overestimate $\Delta Q/\Delta x$ since multiple scattering is frequent at low energies. For relativistic energies $E > 60$ keV, $\Delta Q/\Delta x$ becomes proportional to β^{-2} according to the Bethe formula (5.100).

The heat generated is proportional to the mass thickness of electron-transparent films. There is a stronger increase for greater thicknesses due to multiple scattering and to the increase of the inelastic scattering probability with decreasing electron energy [11.24]. When bulk specimens such as specimen-grid bars or diaphragms are irradiated, the generation of heat P per unit time saturates for thicknesses of the order of the electron range at a value

$$P = fP_0 = fIU, \tag{11.3}$$

where $P_0 = IU$ denotes the total beam power. Some electrons are lost by backscattering, with the result that only a fraction $f = 0.7$ for copper contributes to P.

11.1.3 Calculation of Specimen Temperature

The specimen temperature becomes stationary when the heat generated is equal to the heat dissipated by radiation and thermal conduction. This problem of thermal conduction can be solved only for simple geometries such as a circular hole covered with a uniform foil [11.22] or for rod-shaped specimens (needles) [11.14].

The power dissipated by radiation can be estimated using the Stefan–Boltzmann law

$$P_{\mathrm{rad}} = SA\sigma(T^4 - T_0^4), \tag{11.4}$$

where S denotes the specimen surface area (both surfaces must be counted for foils), A the absorptive power of the blackbody radiation, which is the same as the emissivity (Kirchhoff's law), σ is the Stefan–Boltzmann constant, and T is the temperature of the specimen and T_0 that of the surroundings. The absorptive power A is equal to unity only for a blackbody and is of the order of 0.01–0.05 for bulk metals. For thin transparent films, A is still smaller: $A = 9 \times 10^{-4}$ for a 10 nm collodion film, for example [11.28]. The influence of radiation loss can therefore be neglected for thin-foil specimens and has to be considered only if the heat dissipation by thermal conduction is reduced by the presence of a large self-supporting area and/or if the temperature is high because the radiation loss increases with the fourth power of T in (11.4). This can be seen from the following example [11.10]. An SiO foil was placed over a 400 μm-diameter hole, and indicator spots of Ge (melting point $T_{\mathrm{m}} = 958°\mathrm{C}$) and In (156°C) were deposited near the center. The current densities necessary for melting were in the ratio 9.3 : 1. A ratio 6.9 : 1 would be expected for pure thermal conduction and 82 : 1 for pure radiation loss. The increase of the rate of dissipation of heat in Ge relative to the value for pure conduction is just detectable, whereas the same experiment with a 200 μm diaphragm gave the ratio expected for pure conduction. The irradiation of small particles with a diameter of the order of micrometers on a supporting film of low thermal conductivity produces another extreme case. The temperature then increases to a value at which the radiation loss becomes dominant, with the result that the current density required to melt the particles (e.g., small crystals of alkali halogenides) is proportional to T_{m}^4 [11.16].

The heat transfer by conduction through an area S caused by a gradient ∇T of the specimen temperature is described by

$$P_{\rm c} = -\lambda S \nabla T, \qquad (11.5)$$

where λ is the thermal conductivity. Two extreme irradiation conditions for a foil over a diaphragm of radius R will now be discussed: uniform illumination and highly localized (small-area) illumination.

Uniform Illumination. The whole foil is irradiated with a uniform current density j (A m^{-2}) by means of the strongly defocused condenser-lens system. The thermal power generated inside a circle of radius r and area πr^2 with the foil center at $r = 0$ has to be transferred by thermal conduction through an area $S = 2\pi r t$ (t: foil thickness). This results in the equilibrium relation

Power dissipated = power transferred by thermal conduction

$$\pi r^2 \frac{j}{e}\frac{\Delta Q}{\Delta x}\rho t = -\lambda 2\pi r t \frac{{\rm d}T}{{\rm d}r} \quad \rightarrow \quad \frac{{\rm d}T}{{\rm d}r} = -\frac{j\rho}{2e\lambda}\frac{\Delta Q}{\Delta x} r. \qquad (11.6)$$

The stationary temperature distribution $T(r)$ of (11.6) has a parabolic form (Fig. 11.2, curve a) with the maximum temperature

$$T_{\rm max} = T_0 + \frac{j\rho}{4e\lambda}\frac{\Delta Q}{\Delta x}R^2 \qquad (11.7)$$

at the center ($r = 0$) of the diaphragm.

The rise of the temperature with time can be determined from the time-dependent equation of heat conduction

$$\lambda \Delta T(r,\tau) + \frac{j\rho}{e}\frac{\Delta Q}{\Delta x} = c_p \rho \frac{\partial T}{\partial \tau}, \qquad (11.8)$$

c_p being the specific heat of the foil. The solution of (11.8) for the boundary condition $T = T_0$ at $r = R$ can be expressed as an eigenfunction series [11.27]

$$T(r,\tau) = T_0 + \sum_{n=1}^{\infty} b_n J_0(a_n r)\left[1 - \exp\left(\frac{\lambda a_n^2 \tau}{c_p \rho}\right)\right]. \qquad (11.9)$$

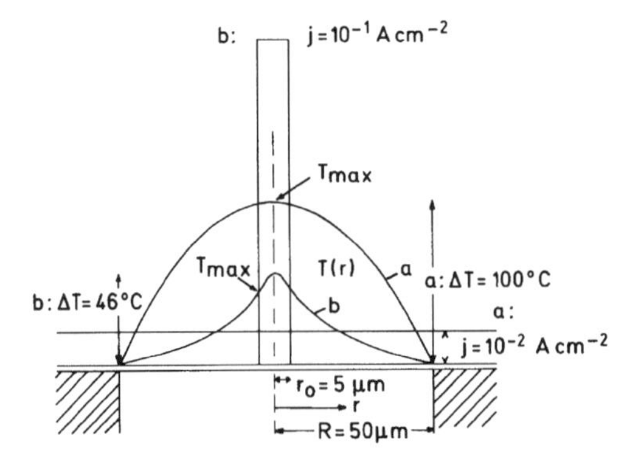

Fig. 11.2. Temperature distribution in a formvar film irradiated with 60 keV electrons for (curve *a*) uniform current density j and (curve *b*) small-area illumination.

Table 11.2. Thermal conductivity λ, density ρ, specific heat c_p, and time constant τ_1 for $R = 50\ \mu$m.

Substance	λ [J/K cm s]	ρ [g/cm^3]	c [J/g K]	τ_1 [ms]
Carbon film	1.5×10^{-2} [11.29]	2.0	–	–
Formvar film	2.4×10^{-3} [11.29]	1.2	2.0	4.3
Glass (SiO)	10^{-2}	2.2	0.8	0.8
Metal (Cu)	4	8.9	0.36	3.5×10^{-3}

The coefficients a_n are determined by the condition that at the boundary $r = R$ the temperature stays constant. This can be achieved only if $x_n = a_n R$ is the nth zero of the Bessel function $J_0(x)$. The quantity $\tau_n = (c_p\rho)/(\lambda a_n^2)$ describes the decay time of the nth term. The largest decay time $\tau_1 = c_p\rho/(a_1^2\lambda) = c_p\rho R^2/(5.78\lambda)$.

Inserting realistic values, we find that the temperature of an electron microscope specimen rises so rapidly that the stationary value will be attained immediately after the irradiation conditions are changed (see Table 11.2). However, the thermal conductivity of organic specimens can be altered by radiation damage and normally increases with increasing irradiation as a result of cross-linking. This causes a decrease of temperature with increasing irradiation time and constant illumination conditions. This damage process requires a much higher electron dose $j\tau$ than does the mass loss of organic films, for example (Sect. 11.2). The thermal conductivity of pure carbon and SiO films is not affected by irradiation [11.10].

Small-Area Illumination. Equation (11.6) has to be modified when a smaller area is irradiated by using a more strongly focused condenser lens. If only a small area of radius r_0 is irradiated with a current density j ($r_0 \ll R$), the heat-transfer equation becomes, for $r \geq r_0$,

$$\pi r_0^2 \frac{j}{e}\frac{\Delta Q}{\Delta x}\rho t = -\lambda 2\pi r t \frac{\mathrm{d}T}{\mathrm{d}r} \quad \rightarrow \quad \frac{\mathrm{d}T}{\mathrm{d}t} = -\frac{j\rho}{2e\lambda}\frac{\Delta Q}{\Delta x}\frac{r_0^2}{r}\,. \tag{11.10}$$

With $T = T_0$ for $r = R$, the stationary temperature becomes

$$T(r,\infty) = T_0 + \frac{j\rho}{2e\lambda}\frac{\Delta Q}{\Delta x}\, r_0^2 \ln\frac{R}{r} \quad \text{for} \quad r \geq r_0. \tag{11.11}$$

Inside the irradiated area (that is, for $r \leq r_0$), the temperature distribution is again parabolic, provided that the current density j inside the radius r_0 is uniform. At $r = r_0$, the solution must take the value $T(r_0,\infty)$ given by (11.11). However, the small increase of temperature from $r = r_0$ to $r = 0$ can be neglected in comparison with the increase from $r = R$ to $r = r_0$, and the temperature in the foil center becomes

$$T_{\max} = T_0 + \frac{j\rho}{2e\lambda}\frac{\Delta Q}{\Delta x}\, r_0^2 \ln\frac{R}{r_0}\,. \tag{11.12}$$

Table 11.3. Rise of temperature ΔT in the center of a circular diaphragm ($R = 50$ μm) covered with a supporting film and irradiated with 100 keV electrons.

	Uniform illumination	Small-area illumination
Substance	$R = 50\ \mu$m, $j = 100$ A m^{-2}	$r_0 = 0.5\ \mu$m, $j = 10^4$ A m^{-2}
Formvar	62°C	6°C
Glass (SiO)	27°C	2.5°C
Metal (Cu)	0.3°C	0.03°C

This is a much smaller temperature rise $\Delta T = T_{\max} - T_0$ than that predicted by (11.7) even for larger current densities. (Thus for a tenfold increase of current density in Fig. 11.2, ΔT reaches only half the value for uniform illumination; see also the numerical examples of Table 11.3.)

For more exact calculations, the current density distribution within the electron beam, typically Gaussian, has to be considered [11.10, 11.13]. If the electron beam hits the diaphragm or the specimen-grid bars, the temperature T_0 in (11.7) can be increased sharply owing to the greater generation of heat (11.3) in bulk material. In practice, therefore, irradiation of bulk parts of the specimen support should be kept as low as possible to limit specimen heating. The estimated values in the last column of Table 11.3 show that under these conditions, the rise of temperature due to electron irradiation can be kept small.

Whereas an increase of specimen temperature of a few kelvins has no significant effect with the specimen at room temperature (300 K), this can become a large relative increase at liquid-helium temperature (4 K). The radiation damage depends very sensitively on temperature (Sect. 11.2.3), and most of the discrepancies in experimental results at low temperatures may be attributed to a temperature rise [11.30]. The thermal conductivities of most substances decrease by one or more orders of magnitude when the temperature is decreased from 300 K to 4 K [11.31]. It is very important that there be a good thermal conductivity of the supporting film (carbon), good thermal contact between film and grid and support, and good cryoshielding of parts of the microscope at high temperature. The special problems of examining frozen-hydrated biological specimens in a cold stage are discussed in [11.32, 11.33].

The temperature increase caused by a moving electron probe is of interest for scanning transmission electron microscopy [11.34, 11.27].

11.2 Radiation Damage of Organic Specimens

11.2.1 Elementary Damage Processes in Organic Specimens

Radiation damage in organic material is caused by all kinds of ionizing radiation. The damage depends on the energy dissipated per unit volume, which is proportional to the number of incident electrons $n = j\tau/e$ per unit area, where τ is the irradiation time in seconds. The incident charge density $q = j\tau = en$ ($\mathrm{C\ m^{-2}}$) can thus be used to compare different irradiation conditions. The quantity q is called the *electron dose*, although, in radiation chemistry, dose is defined as energy dissipated per unit mass [measured in grays: 1 gray (Gy) = 1 J/kg]. A dose of 1 $\mathrm{C\ m^{-2}}$ corresponds to a number of electrons $n = 6 \times 10^{18}\ \mathrm{m^{-2}} = 6\ \mathrm{nm^{-2}}$ and at $E = 100$ keV to a transferred energy density $n\rho|\Delta E/\Delta x| = 1.4 \times 10^{27}\ \mathrm{eV\ m^{-3}} \simeq 2.2 \times 10^{8}\ \mathrm{J\ m^{-3}} = 2.2 \times 10^{5}$ Gy if we insert the values $\rho = 1\ \mathrm{g\ cm^{-3}}$ and $|\Delta E/\Delta x| = 2.4\ \mathrm{eV\ \mu g^{-1}\ cm^{2}}$ that correspond to carbon (Fig. 11.1) and are converted to 100 keV. This last value is given by the Bethe formula (5.100) and (11.2). Most of this energy density is consumed in ionization processes. From (5.100), we see that this contribution decreases as v^{-2} with increasing electron energy (see also Sect. 11.2.3).

Current densities of the order of $j = 100\ \mathrm{A\ m^{-2}}$ are necessary at magnifications $M \simeq 10\,000$, which corresponds to an energy density of 160 $\mathrm{eV\ nm^{-3}}$ in 1 s. High-resolution micrographs require an electron dose of $q = 5 \times 10^{3}$ $\mathrm{C\ m^{-2}}$ to expose a photographic emulsion with a density $S = 1$, which means an energy density of $10^{4}\ \mathrm{eV\ nm^{-3}}$! Table 11.4 contains a scale of physical and biological damage effects (see reviews in [1.43, 11.35–11.45]). The energy dissipated in an electron microscope specimen after a brief irradiation therefore corresponds to conditions that, outside a microscope, occur only near the center of a nuclear explosion! Results from radiation chemistry, where the energy densities are much lower, of the order of $10^{6} - 10^{7}\ \mathrm{J\ m^{-3}}$ [11.35, 11.36, 11.46–11.52], are relevant only to the very early stage of radiation damage.

The primary damage process is inelastic scattering, which causes molecular excitation or ionization or collective molecular excitations (similar to plasmon excitations of a free-electron gas). The energy dissipated is either converted to molecular vibrations (heat) or causes bond scission; for example, a loss of hydrogen and production of radicals (R: residual organic molecule, *: molecular excitation, •: radical)

$$\mathrm{RH}^{*} \;\rightarrow\; \mathrm{R}^{\bullet} + \mathrm{H}^{\bullet}$$

(the bonding energy of H to C atoms is of the order of 4 eV) or a bond break in a carbon chain:

$$\mathrm{R{-}CH_2{-}CH_2{-}CH_2^{*}{-}R} \;\rightarrow\; \mathrm{R{-}CH_3 + CH_2{=}CH{-}R.}$$

C–H bonds break more frequently in aliphatic chains than in aromatic compounds owing to the spread of energy dissipation by the π-electron system of benzene rings [11.53]. Bond scission also leads to the production of low-molecular-weight molecules and radicals. These primary processes are unaffected by temperature and cannot be avoided by specimen cooling.

A quantum-mechanical calculation has been performed for radicals produced by cleaving hydrogen from the DNA bases adenine, guanine, cytosine,

Table 11.4. Scale of radiation-damage processes with 100 keV electrons.

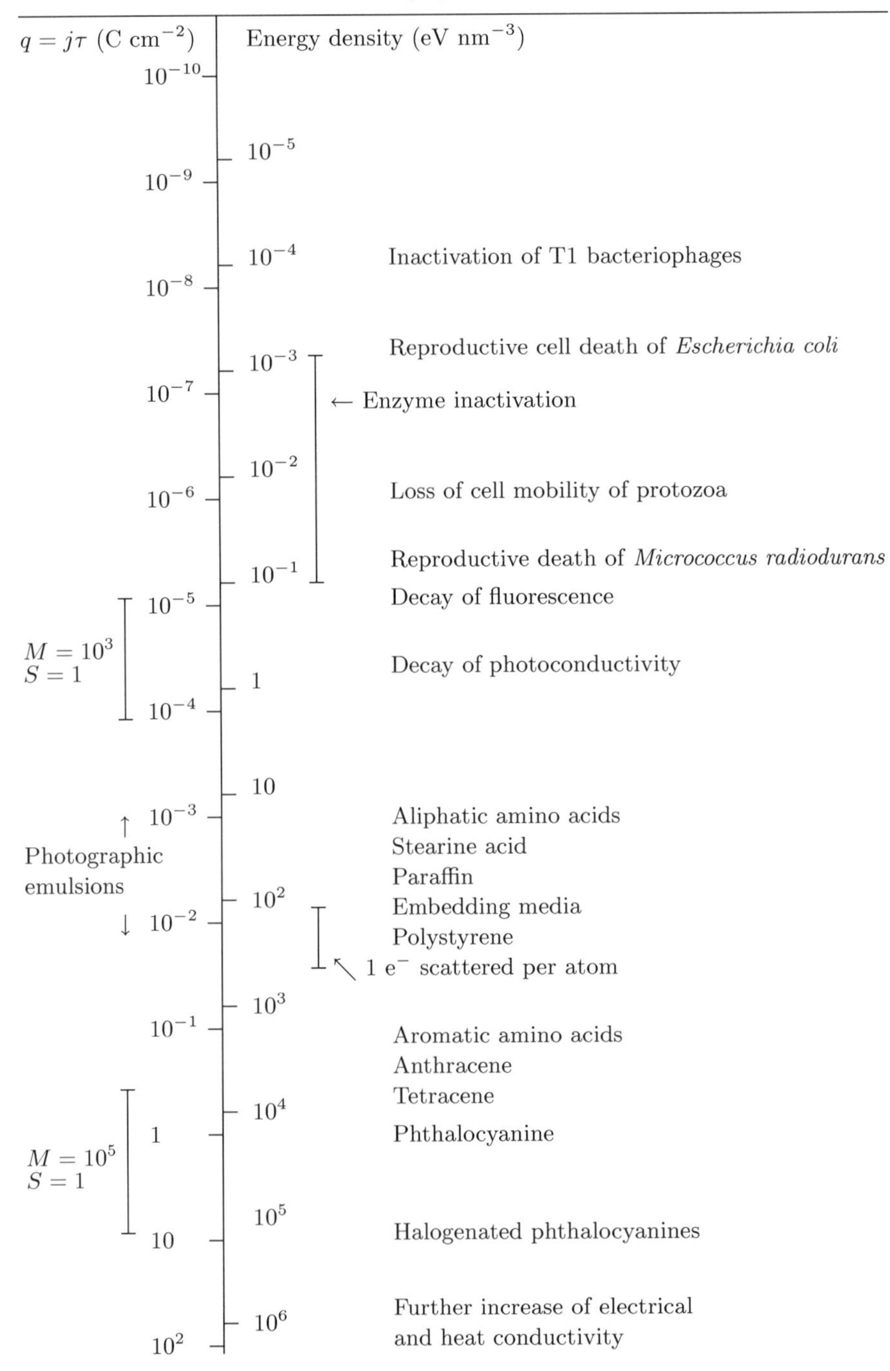

and thymine [11.54]. These results can be used to discuss the change of the fine structure of the carbon edge in electron energy-loss spectra [11.43].

The loss of hydrogen atoms would not noticeably affect the molecular and crystal architecture, but the scission of carbon chains and side groups and the secondary processes cause great damage. Secondary processes are, for example, the diffusion of hydrogen atoms to other molecules and the formation of additional radicals at unsaturated bonds or a hydrogen molecule:

$$\mathrm{R\text{-}H + H^\bullet \rightarrow R\text{-}H_2^\bullet} \qquad \text{or} \qquad \mathrm{R\text{-}H + H^\bullet \rightarrow R^\bullet + H_2}.$$

Other typical secondary processes are cross-linkings of molecular chains, such as

$$\begin{array}{ccc} \mathrm{R - CH_2 = CH - R} & & \mathrm{R - CH_2 - CH - R} \\ & \rightarrow & \qquad\quad | \qquad\qquad +\mathrm{H^\bullet}\,. \\ \mathrm{R - CH_2 - CH_2^* - R} & & \mathrm{R - CH_2 - CH - R} \end{array}$$

There may also be reactions with radicals and thermal diffusion and evaporation of fragmented atoms and molecules of low atomic weight, such as H_2, CH_4, CO_2, and NH_3. The generation of H_2 by electron irradiation was demonstrated by mass spectroscopy of H_2 released from specimens irradiated at low temperature (10 K) and warmed to room temperature [11.55]. These processes, which involve loss of mass, can be diminished by specimen cooling. The resulting cage or frozen-in effect is the most important way of reducing secondary processes at low temperatures (Sect. 11.2.3). In some cases, the probability that the scission products recombine or cross-link before leaving the specimen can be increased [11.56]. A typical consequence of secondary processes is the collapse of crystal structure and molecular architecture due to mass loss. The ultimate stage is a cross-linked, carbon-rich, polymerized cinder. The continuing increase of the thermal and electrical conductivities at very high doses $>10^4$ C m^{-2} indicates that the carbonization of the material proceeds and that the rearrangement of atoms and molecules does not stop [11.35].

The decay curves of fluorine in PTFE and chlorine in PVC measured by x-ray microanalysis saturate at $1\text{–}5 \times 10^6$ C m^{-2}, and the amount of Os from OsO_4 stain acting as a reagent for double bonds shows a maximum at 5×10^5 C m^{-2} [11.58].

The mean number of specific reactions (e.g. bond rupture, cross-linking, or disappearance of original molecules) that are produced by an energy loss of 100 eV is known in radiation chemistry as the G value. Typical values are shown in Table 11.5: $G_{-\mathrm{M}}$ corresponds to the disappearance of molecules, $G_{\mathrm{H_2}}$ to the appearance of H_2, etc. These figures show that $G_{-\mathrm{M}}$ decreases in the following order: unsaturated and saturated hydrocarbons, ethers, aldehydes, carboxylic acids, and aromatic compounds. The very low value of $G_{\mathrm{H_2}}$ for aromatics is of special interest, as is the higher value for saturated hydrocarbons, which is of nearly the same order as the $G_{-\mathrm{M}}$ value. Thus, for

Table 11.5. Principal bonds broken by radiation damage in pure organic compounds and some G values [11.47, 11.57].

Compounds and sites of attack	$G_{-\mathrm{M}}$	$G_{\mathrm{H_2}}$	$G_{\mathrm{CO_2}}$	G_{CO}	$G_{\mathrm{H_2O}}$	$G_{\mathrm{CH_4}}$
Hydrocarbons,						
saturated						
C–H, C–C	6–9	3.8–5.6				0.2–0.7
unsaturated						
C–H, C–C	11–10	0.8–1.2				0.13
aromatic						
C–H, side chain C–C	0.2–1	0.01–0.18				
Alcohols						
HC–OH, C–COH	3–6	3.5–4.5		0.04–0.23	0.3–0.9	
Ethers						
C–H, C–OR	7	2.0–3.6		0.06–0.13		
Aldehydes and ketones						
C–H, C–C=O	7	0.8–1.2		0.6–1.6		0.1–2.6
Esters						
C–H, O=C–OR	4	0.5–0.9	0.3–1.6	0.15–1.6		0.4–2.0
Carboxylic acids						
C–H, C–COOH	5	0.5–2.3	0.5–4.0	0.1–0.5	0.1–2.2	0.5–1.4
Amino acids	$G_{-\mathrm{M}}$	$G_{\mathrm{H_2}}$	$G_{\mathrm{CO_2}}$	$G_{\mathrm{NH_3}}$		
Leucine						
C–H, C–NH_2, C–COOH	14	0.5	2.8	5.1		
Valine						
C–H, C–NH_2, C–COOH	8	0.2	0.6	4.1		

a high-resolution micrograph, $G = 1$ with 10^4 eV nm^{-3} provokes 10^2 nm^{-3} damage processes! This estimate gives an idea of the demands that electron microscopists make on their organic specimens.

11.2.2 Quantitative Methods of Measuring Damage Effects

Our knowledge of the individual damage processes is very poor. Even for a particular molecule, the scale of primary and secondary processes is very broad and complex, and it is impossible to describe the whole damage process in detail. For practical electron microscopy, damage processes that can be observed in the final image or in a diffraction pattern and that can be used directly for quantitative measurement of damage are of greatest interest [11.59].

(a) Loss of Mass. The transmission T of an amorphous specimen layer decreases as $\exp(-x/x_k)$ (6.6) up to mass thicknesses $x = \rho t = 30$–$50\ \mu\mathrm{g\ cm}^{-2}$ at $E = 100$ keV. The quantity $-\log_{10}T$ is proportional to x (Fig. 6.1) and can be used to measure the loss of mass during irradiation [11.16, 11.61, 11.62, 11.63,

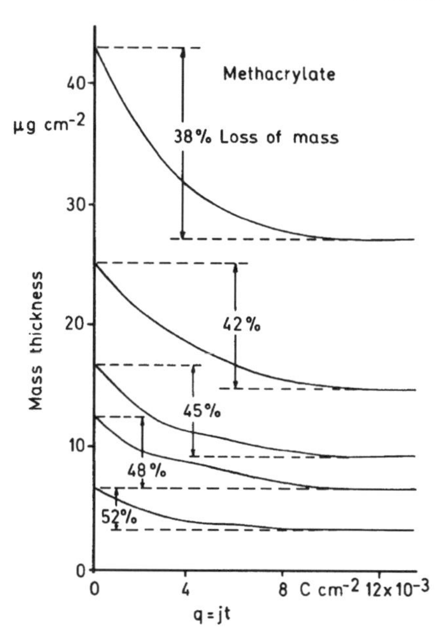

Fig. 11.3. Mass thickness of polymethacrylate foils (methyl : butyl = 20 : 80) of different initial mass thicknesses as a function of electron dose $q = j\tau$ obtained from measurements of electron transmisssion ($E = 60$ keV).

11.64]. For high-molecular-weight polymers, collodion, formvar, methacrylate, epon, and other embedding media, the mass thickness shows an approximately exponential decrease down to a residual value (Fig. 11.3) that for many substances corresponds roughly to the carbon content. This confirms that many noncarbon atoms can leave the specimen after bond scission. A fraction of the carbon atoms can also leave the specimen as volatile fragments, and some of the noncarbon atoms will be bound more strongly. It is necessary to measure the transmission with low current densities, of the order of $j = 1$ A m^{-2}, because the dose that corresponds to the terminal mass loss is of the order of $q_{\max} = j\tau = 100$ C m^{-2} for polymers. The percentage of mass lost by polymethacrylate films shows a systematic decrease with increasing film thickness (Fig. 11.3). This is by no means the rule, however, and protective evaporated carbon films (see also Sect. 11.3.3e) do not decrease the mass loss for all substances. The mass loss and other damage effects are proportional to the dose so long as the irradiation conditions do not cause appreciable specimen heating. Some substances, such as methacrylate, lose more mass when the temperature is increased by a higher current density [11.65]. The elastic dark-field signal of a scanning transmission electron microscope can be used to measure the mass loss of TMV, HPI layers, or freeze-dried phages [6.33].

In practice, it is prudent to preirradiate and stabilize a biological section with a low current density to avoid any increase of mass loss by specimen heating. For high resolution, the tendency is to irradiate the specimen with a dose as low as possible (Sects. 6.6.5 and 11.2.4).

The mass lost by organic single or polycrystalline films cannot be determined by measurement of the transmission because they show diffraction contrast, which depends on the accidental excitation error and film thickness. Normally, a crystalline film shows a greater averaged transmission T than an amorphous film of equal mass thickness [6.1], and the transmission may even decrease in the early stage of irradiation [11.66].

Another way of determining the mass loss is to weigh the specimen before and after irradiation, but for such experiments a reasonably large mass (10–20 μm films of about 1 cm^2 in area) is necessary [11.35, 11.62, 11.67]. The change of optical density nt (n: refractive index, t: specimen thickness) can also be measured by interferometric methods or by analyzing interference colors [11.68, 11.69].

(b) Fading of Electron-Diffraction Intensities. Damage to single molecules also distorts the crystal lattice, which results in a decrease of the electron-diffraction intensity of Debye–Scherrer rings or single-crystal spots [11.70, 11.71] and in some cases in a shift and broadening of the reflections as well [11.72].

Most evaporated films of organic compounds are crystalline. This method is therefore widely used to investigate the influence of chemical structure on resistance to radiation damage and the dependence on electron energy and specimen temperature (Sect. 11.2.3). The dose necessary for complete disordering of the lattice is also of interest for estimation of the irradiation conditions in which diffraction contrast can be observed or lattice planes in organic crystals resolved. Figure 11.4 shows examples of decreasing intensity of Debye–Scherrer rings with increasing charge density. Whereas the lattice of aliphatic compounds is destroyed at low $q_{max} \simeq 10 - 100$ C m^{-2} after irradiation with 100 keV electrons, the extrapolated terminal dose $q_{max} \simeq 1000$ C m^{-2} for tetracene in Fig. 11.4, for example, increases considerably if the compounds contain benzene rings (aromatic compounds). The cross section for the damage of aromatic compounds corresponds to the ionization cross

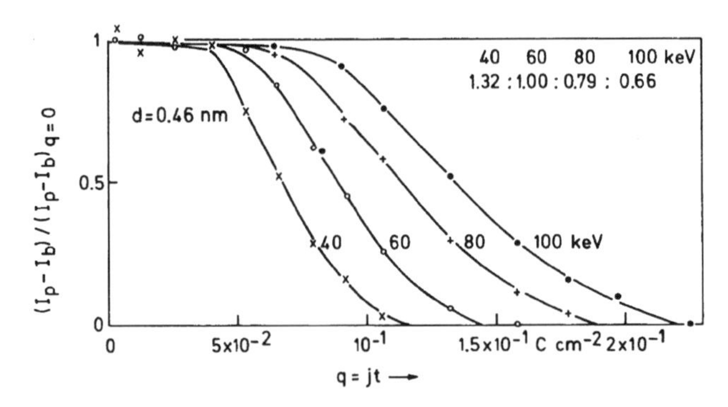

Fig. 11.4. Decrease of the electron-diffraction intensity of the $d = 0.46$ nm lattice planes in tetracene films with increasing electron dose $q = j\tau$ for different electron energies $E = 40$–100 keV (I_p: peak intensity of the Debye–Scherrer ring, I_b: background intensity of the diffraction pattern).

section of the carbon K shell, and indeed measurement of the dose needed to destroy molecules shows that the probability of damage decreases rapidly for electron energies below 2 keV as the carbon edge is approached [11.73].

Crystals of molecules of phthalocyanine and its metal derivatives, which were used by Menter [9.115] to resolve lattice-plane fringes of the order of 1.2 nm for the first time, have a very large value of $q_{max} = 1–2 \times 10^4$ C m^{-2} for these lattice planes. However, a larger lattice-plane spacing is less affected by lattice distortions than a smaller one. If we assume that the damage can be described by frozen-in lattice vibrations, although this is a rather crude model, the first decrease of intensity can be described in terms of a Debye–Waller factor (7.95) [11.74],

$$I_g = I_{g0} \exp(-4\pi^2 \langle u^2 \rangle g^2). \tag{11.13}$$

This equation can be used to estimate an average displacement $\langle u^2 \rangle$. When the dose q required to obtain the same value of $\langle u^2 \rangle = (0.05 \text{ nm})^2$ is calculated, the values for phthalocyanines are found to be no larger than for other aromatic compounds. Only the halogen-substituted phthalocyanines (e.g., $CuCl_{16}C_{32}N_8$) show an exceptionally large resistance with $q_{max} = 1.5–4 \times 10^5$ C m^{-2} (see also [11.75]). They can be used for high-resolution study of crystal lattices [11.76, 11.77], and the damage effects can be observed directly in the image at molecular dimensions [11.78]. The collapse of phthalocyanine crystals is initiated locally; thus parts of the crystal with exposed edges damage preferentially relative to the perfect bulk crystal. For further models describing the fading of electron-diffraction patterns, see [11.56, 11.79, 11.80, 11.81].

The decrease of the diffraction intensities with increasing dose has to be interpreted with care because the shape of the fading curve can depend strongly on the film structure; for example, the texture in polycrystalline films. So-called *latent doses* (ranges of q in which the fading curve shows very little decrease, see Fig. 11.3 for example) can be artifacts caused by the specimen structure and by the transition between the conditions in which the dynamical and kinematical theories of electron diffraction are applicable. However, the extrapolated terminal doses q_{max} (Table 11.6) are in agreement with the results of other methods.

Spectroscopic Methods. Changes in the structure of molecules can be detected by their influence on the photon-absorption and electron energy-loss spectra. The absorption spectrum of evaporated dye films can be measured inside one irradiated mesh of the supporting grid (0.1×0.1 mm^2) [11.35, 11.86]. The terminal dose for the damage, as indicated by the absorption maxima (Fig. 11.5), coincides with the value given by electron diffraction (Table 11.6). Photoconductivity and cathodoluminescence are very sensitive to radiation damage; the photoconductivity of phthalocyanine, for example, is lost for $q_{max} \simeq 0.5$ C m^{-2} [11.35].

Infrared absorption spectra need a larger specimen area, and the specimen has to be irradiated in the final image plane. The absorption maxima in the far infrared caused by vibrations and rotations of the molecules disappear first,

Table 11.6. The electron dose q needed at 300 K for complete destruction of organic compounds, measured by different methods (ML: mass loss, C: contrast, ED: electron diffraction, LA: light absorption, EELS: electron energy-loss spectroscopy, XRMA: x-ray microanalysis).

Substance	q (C m^{-2})	E (keV)	Method	Refs.
Amino acids				
Glycine	15	60	ED	[11.70]
l-Valine	15	80	ED	[11.82]
Leucine	15 – 20	60	ED	[11.70]
Aliphatic hydrocarbons				
Stearic acid	20 – 30	60	ED	[11.70]
Paraffin	30 – 50	60	ED, C	[11.70]
	100	100	ED, C	[11.66]
Polymethacrylate	80 – 100	60	C	[11.60]
Polyoxymethylene	70 – 80	80/100	ED	[11.71, 11.83]
Polyethylene	70 – 100	100	ED	[11.71, 11.83, 11.84]
Nylon 6	120	100	ED	[11.71]
Polyvinylformol	100	75	ML	[11.62]
Polyamide	150 – 200	75	ML	[11.62]
Polyester	200	75	ML	[11.62]
Fluorinated ethylene polymer	50 – 100	75	ML	[11.62]
Tetrafluorinated ethylene polymer	100 – 150	75	ML	[11.62]
Gelatin	100	75	ML	[11.62]
Bases of nucleic acids				
Adenosine	100	80	ED	[11.82]
Adenine	5000	100	ED	[11.74]
Cytosine	3000	20	EELS	[11.85]
	400	20	ED	[11.85]
Guanine	6000	20	EELS	[11.85]
Uracil	$1 - 2 \times 10^3$	100	ED	[11.74]
Aromatic compounds and dyes				
Anthracene	600 – 800	60	ED	[11.70]
Tetracene	2000	60	ED	[11.66, 11.70]
Pentacene + Tetracene	300 – 500	60	LA	[11.86]
Coronene	500	100	ED	[11.45]
Indigo	150	60	ED	[11.70]
	5000	60	LA	[11.86]
Cloranile	1000	100	ED	[11.74]
Bromanile	$3 - 5 \times 10^3$	100	ED	[11.74]
Hexabromobenzene	$5\text{–}7 \times 10^4$	100	ED	[11.74]
Phthalocyanine (d=0.6 nm)	1000	100	ED	[11.70]
Cu-Phthalocyanine (d=0.6 nm)	4000	100	ED	[11.74]
(d=1.2 nm)	$1\text{–}2 \times 10^4$	60	ED	[11.70]
	$2\text{–}3 \times 10^4$	60	LA	[11.86]
$CuCl_{16}$-Phthalocyanine	$2.5\text{–}3.5 \times 10^5$	100	ED	[11.87]
PTVE (F loss)	2×10^6	100	XRMA	[11.58]
PVC (Cl loss)	$4\text{–}5 \times 10^6$	100	XRMA	[11.58]

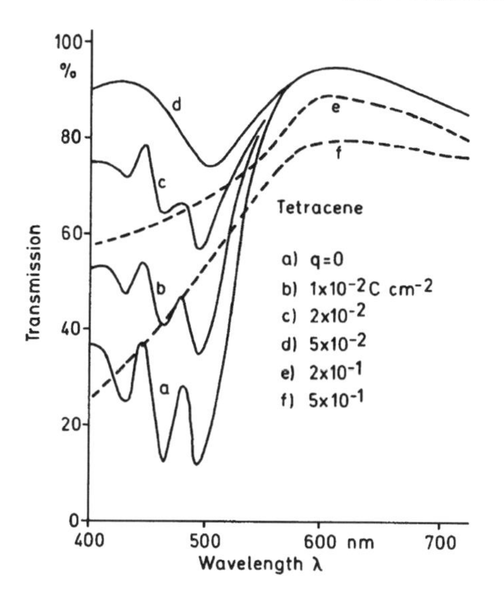

Fig. 11.5. Light-absorption spectra of a 560 nm thick tetracene film for various electron doses q (E = 60 keV) [11.86].

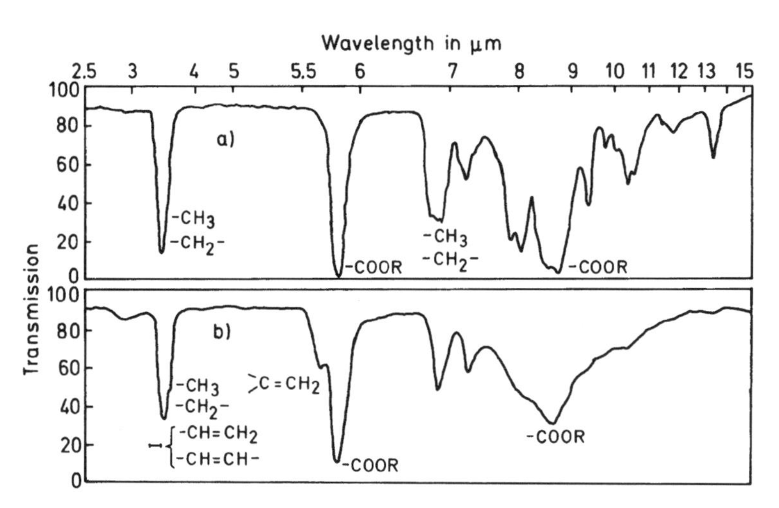

Fig. 11.6. Infrared absorption spectra of a polymethacrylate foil (14 μm thick) (**a**) not irradiated and (**b**) irradiated with a dose of q = 0.4 C m^{-2} at E = 60 keV [11.35].

at relatively low doses (Fig. 11.6) [11.35, 11.62, 11.67]. This indicates that cross-linking and scission of molecules occur. Absorption maxima in the near infrared can be attributed to typical groups such as $-CH_3$, $-CH_2-$, $=CH_2-$, $=C-$, $-COOH$, and $OH-$. Figure 11.6 shows the decrease of the maxima and the arrival of new maxima beside the $-COOR$ maximum, which can be attributed to $>C=CH_3$ and confirms that double bonds are generated by the loss of H atoms.

Infrared spectra recorded during a single transmission need 5–10 μm thick specimens (the electron range for 100 keV electrons is about 100 μm). (With modern infrared spectrometers, much thinner and smaller areas can be investigated.) The infrared technique can be applied to monomolecular layers by using the multiple internal reflection technique [11.89, 11.88]. An amino acid analyzer has also been used to investigate the radiation products in detail. This technique was applied to catalase, for example, and some 0.2–0.4 mg of material were required [11.90]. Irradiation with $n = 0$–100 electrons per nm^2 produces a more or less rapid decay of the amino acids Asx, Glx, Arg, His, Lys, Thr, Ser, Met, Cys, Pro, and Tyr (e.g., –40% for Lys); the amounts of Val, Leu, Ile, and Phe remain nearly constant, whereas the amounts of Gly, Ala, and Abu increase (e.g., +25% for Ala). This is an indication of possible transformations of amino acids, confirmed by the appearance of Abu (α-amino butyric acid), which does not occur in the native, unirradiated catalase. These experiments show that transformations are important in irradiated homo- and heteropolypeptides and that secondary and tertiary chemical reactions can be observed that are not found when single amino acids are irradiated.

The application of electron energy-loss spectroscopy (EELS) to the study of radiation damage has the advantage of providing direct elemental analysis of C, N, or O atoms, among others, and the decrease of their contribution to the EEL spectrum can be followed during electron irradiation [11.91, 11.92, 11.93, 11.94] (Fig. 11.7). Energy- or wavelength-dispersive x-ray microanalysis can also be used to investigate the loss of heavy atoms, such as F and Cl [11.58] (Table 11.6) or S and P [11.95], whereas the decrease of the background intensity indicates the loss of mass [11.96].

Investigation of irradiated organic films with a laser microprobe mass analyzer (LAMMA) [11.97] shows that the $(M+H)^+$ peak in the positive ion spectrum decreases with increasing dose of electron preirradiation. The termi-

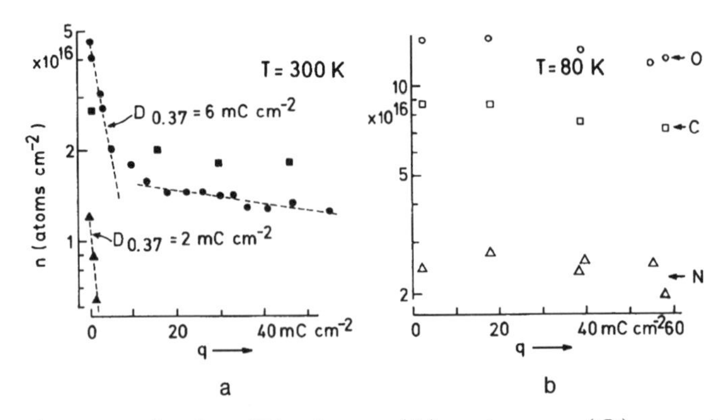

Fig. 11.7. Amounts of carbon (□), nitrogen (△), and oxygen (○) per unit area of a thin collodion film, depending on the incident electron dose q at (**a**) $T = 300$ K and (**b**) $T = 80$ K (liquid-nitrogen cooled), measured by electron energy-loss spectroscopy at $E = 80$ keV [11.92].

nal doses agree with those found by electron-diffraction fading. Some smaller molecular fragments first increase and are therefore scission products. A disadvantage of this method is that the mass spectrum already contains molecular fragments in the unirradiated state. Other sensitive methods are electron-spin resonance [11.51] and nuclear magnetic resonance [11.98].

A further method for the investigation of radiation damage is inelastic-tunneling spectroscopy [11.99]. This technique has been used to reveal the vibrational modes of organic compounds that are included in the insulating layer of a metal-insulator-metal (MIM) tunnel diode. The COH functional group in β-D fructose is disrupted and the –C=C– bond increases, for example.

11.2.3 Methods of Reducing Radiation Damage

(a) High-Voltage Electron Microscopy. One way of reducing the damage rate is to use high-voltage electron microscopy (HVEM). The reduction in energy dissipation with increasing energy, $|dE/ds| \propto v^{-2}$, that results from the variations of the inelastic cross section (Sect. 5.2.2) and the Bethe loss formula (5.100) has been confirmed by experiments [11.71, 11.84, 11.87, 11.100, 11.101] that use the electron-diffraction method. A gain of two in the terminal dose q_{max} that causes complete fading of the electron-diffraction pattern can be obtained if E is increased from 100 to 200 keV, but when E is further increased to 1 MeV, the gain is only about three. We have to keep in mind, however, that the image contrast for weakly scattering objects is reduced by the same factor.

(b) Cryoprotection. Another possibility is to reduce the specimen temperature. Siegel [11.66] reported experiments at 300, 70, and 4 K on paraffin and tetracene with gains of two and four, respectively. If, however, the temperature of the specimen initially irradiated at 4 K is raised to 300 K without further irradiation, the fading of the electron-diffraction pattern is the same as would have been observed if the specimen had been irradiated with the same electron dose at 300 K.

Measurements of the number of C, O, and N atoms by EELS (Fig. 11.7) also show that mass loss is strongly reduced at liquid-nitrogen temperature [11.92, 11.93]. Mass loss has also been observed by quite a different method [11.102]; ^{14}C-labeled T4 phages and *E. coli* bacteria were irradiated at 4 K and 300 K. The residual ^{14}C content was measured by the autoradiographic method (deposition of a photographic emulsion on the specimen grid and development of the grains exposed by the β-emission of ^{14}C). Exposures up to 10^4 C m^{-2} show no significant loss when irradiated at 4 K, whereas the loss at 300 K is of the order of 30%.

All experiments confirm that the primary process of ionization and bond rupture is not influenced by the temperature, whereas the secondary processes of loss of mass and crystallinity require migration and diffusion of the reaction products, which decrease as the specimen temperature is reduced. This confirms that the mobility of fragments is strongly reduced at 4 K. Radiation-induced recombination processes cannot be excluded, but these do not necessarily recreate the original structure. However, the degree of improvement varies from one substance to another. Discrepancies among the reported gains on cooling to liquid-helium temperature can be attributed to uncertainties about the exact specimen temperature.

All electron-diffraction experiments [11.44, 11.56, 11.66, 11.82, 11.83, 11.103, 11.104] show that the critical doses (1/e fading or terminal dose) are improved by a factor of at most 5 to 7, and reported gains larger than 100 [11.105] have not been confirmed. Also, the terminal doses at 4 K are not noticeably better than those at 80 K. Arrhenius plots of the logarithm of extinction dose against reciprocal temperature reveal two activation energies for the degradation, the higher activation energy dying out at $\simeq$ 80 K [11.112].

(c) Hydrated Organic Specimens. Biological material is normally observed in the dehydrated state. For ultramicrotomy, the specimens are dried, usually in a series of baths of increasing concentration of the dehydrating agent (e.g., alcohol), after which the intermediate dehydrating fluid is replaced by a resin, which can be polymerized. Material not prepared in this way loses water in the vacuum of the microscope. There is, however, an ever-increasing interest in observing biological material in the native state. One way of achieving this is to use an environmental cell in which a partial pressure of water is maintained [11.106]; alternatively, the specimen may be frozen and cryosections cut by means of a cryoultramicrotome [11.107, 11.108, 11.109].

It is known from the radiation chemistry of aqueous systems that the G value for the formation of H and OH radicals and H_2O_2 is high [$G(-H_2O)$ $\simeq$4.5 for the liquid state] and that these products cause strong secondary reactions with the biological material. However, this G value decreases to 3.4 in ice at 263 K, to 1.0 at 195 K, and to 0.5 at 73 K; this can be attributed to an increase of the molecular recombination of the water molecule fragments. Model experiments in different states have been done with catalase crystals. The outer diffraction spots that correspond to higher resolution faded away first. The electron dose required for complete fading was 300 electrons per nm^2 for frozen catalase at $\simeq$150 K, which is ten times greater than the value for wet catalase at 300 K [11.44]. No significant difference in the dose for 1/e fading, which is about 200–300 electrons per nm^2, has been found for crystalline and vitreous ice even when the specimen temperature is decreased from 110 K to 4 K [11.44]. This confirms that, at low temperatures, the effect of radiolysis in the presence of water can be neglected, and the damage rate shows the same magnitude in the frozen and dried states. After complete fading of the diffraction pattern, voids and bubbles are formed in frozen-hydrated catalase, which do not appear in thin, dried samples or inside pure ice. These bubbles can be detected at doses larger than 1000 electrons per nm^2 at 110 K and 20 000 electrons per nm^2 at 4 K [11.110]. Embedded organic material may show shrinkage or swelling effects that are different from those seen in the absence of ice [11.111].

(d) Elemental Substitution. The replacement of the H atoms by Cl or Br in benzene or phthalocyanine, for example, increases the terminal dose of the electron-diffraction pattern by more than one order of magnitude [11.69, 11.74, 11.87] (Table 11.6). Terminal doses considerably larger than 10^4 C m^{-2} allow crystal-structure imaging of these substances with high resolution [11.77, 11.87]. This can be explained by the reduced mobility of halogen atoms (cage effect) and the ability to recombine. Halogens are reagents for carbon double bonds, for example.

(e) Conductive Coatings. Coating with a thin evaporated layer of carbon can reduce the mass loss, especially in substances with a high fraction of volatile scission products, such as polymethacrylate [11.16]. Gold-sandwiched coronene crystals were found to have increased radiation resistance by a factor of 5 [11.113], and carbon encapsulation increases the terminal dose by a factor of 3–12 [11.114]. However, these results are not the rule and must be checked.

11.2.4 Radiation Damage and High Resolution

High resolution is possible only by elastic scattering. As shown in Table 11.4, a mean electron dose of the order of 100 C m^{-2} is necessary at $E = 100$ keV for one elastic-scattering process per atom. The fraction of scattered electrons that can be used to provide image contrast depends on the operating mode used (e.g., bright- or dark-field TEM or STEM). Exposure of a photographic emulsion to a density $S = 1$ at $M = 100\,000$ needs a dose of 10^3–10^4 C m^{-2}.

In order to estimate the minimum dose $q_{\min}$, we consider that $n_0 = j\tau = q/e$ electrons are incident per unit area. A fraction f contributes to the image background (e.g., $f = 1 - \epsilon$, $\epsilon \ll 1$ for the bright-field or $f = 10^{-3} - 10^{-2}$ for the dark-field TEM mode). The number $N = nd^2 = fn_0d^2$ forms the image element of area d^2. We assume that the image contrast C is caused by a difference Δn in the number of electrons,

$$C = \frac{\Delta n}{n} = \frac{\Delta n}{fn_0}\,, \tag{11.14}$$

and hence

$$\Delta N = \Delta n\, d^2 = fn_0d^2C. \tag{11.15}$$

The shot noise of the background signal (Poisson statistics) is $N^{1/2}$. For the signal to be significant, the signal-to-noise ratio κ must be larger than 3–5; this is known as the Rose condition [11.115]:

$$\frac{\text{Signal}}{\text{Noise}} = \frac{\Delta N}{N^{1/2}} = Cd(fn_0)^{1/2} = Cd(fq/e)^{1/2} > \kappa. \tag{11.16}$$

Solving for q, we find that

$$q_{\min} = \frac{e\kappa^2}{fd^2C^2} \tag{11.17}$$

is the minimum dose with which a specimen detail of area d^2 can be detected. As a numerical example, assuming a bright field image ($f \simeq 1$), $\kappa = 5$, $d = 1$ nm, $C = 1\%$, we obtain $q_{\min} = 400$ C m^{-2}.

These examples demonstrate that the squares of d and C in the denominator of (11.17) can have a considerable influence on $q_{\min}$ and that the minimum dose is of the same order of magnitude as the dose required for the exposure of a photographic emulsion.

These doses are so high that severe damage to biological specimens is inevitable, especially because taking a micrograph consists of three steps – searching, focusing, and recording – and the first two need a larger dose than the third. Minimum exposure techniques have therefore been developed in which the grid is scanned at low magnification with a strongly reduced electron dose; the microscope is focused on a different specimen region, after which the beam is switched to the area of interest by means of deflection coils and a shutter is opened for exposure [11.116, 11.117, 11.118, 11.119].

Another way of further decreasing the electron dose involves the low-exposure averaging technique, in which the noise is reduced by averaging over a large number of identical structures. This technique is therefore particularly suitable for periodic specimens [6.207, 11.120]. If R denotes the number of repeated unit cells, the Rose equation (11.16) becomes $Cd(fRq/e)^{1/2} > \kappa$ and the minimum dose $q_{\min}$ is then reduced by a factor $1/R$. The fog level of the photographic emulsion becomes a serious limitation for low exposure. Nuclear track emulsions and nowadays CCD arrays have been found better than the emulsions normally used at 100 keV [11.120, 11.121]. With the aid of cross-correlation methods, the technique can also be employed for non-periodic specimens (see also Sect. 6.6.4 and Fig. 6.35) [11.122]. From a series of low-exposure micrographs with an increasing number of electrons per nm^2, the evolution of the damage with time can be followed either by calculating crystal-structure projections, by Fourier transforming the measured intensities of a selected-area electron-diffraction pattern with assumptions about the phase based on crystal symmetry [11.123], or by averaging low-dose exposures with cross-correlation methods [11.124].

11.3 Radiation Damage of Inorganic Specimens

11.3.1 Damage by Electron Excitation

Electron excitations in metals and in most covalent semiconductors are reversible and cause no damage. Only electron–nucleus collisions (knock-on processes, Sect. 11.3.2) can cause atomic displacements. However, it should be recalled that the formation of defect clusters can also be observed during prolonged irradiation in a 100 keV TEM. This damage may be attributable to negative ions accelerated between cathode and anode [11.125].

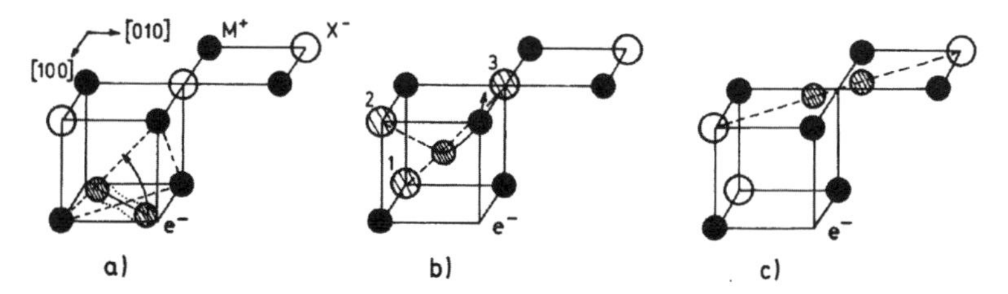

Fig. 11.8. Radiolysis sequence in alkali halides having the NaCl structure: (**a**) formation of a Cl_2^- bond; (**b**) intermediate state leading to the final state (**c**) consisting of an interstitial Cl_2^- cation (H center) and an anion vacancy with a trapped electron (F center).

In ionic crystals (e.g., alkali halides), the most important excitations are:

1. inner shell ionization,
2. plasmon losses as collective oscillations of the valence electrons,
3. ionization of valence electrons, and
4. creation of locally bound electron–hole pairs (excitons).

The probability of inner-shell ionization (1) is low, and the energy is lost in x-ray or Auger-electron emission. Plasmons (2) are the most probable excitations but are much too delocalized to cause any localized transfer of energy; they may, however, decay into more localized single-electron excitations of exciton character. Mobile electrons (3) excited into higher states of the conduction band recombine with the less mobile holes (of large effective mass) via intermediate exciton states. These secondary and primary excitons (4) are responsible for the radiolysis. Figure 11.8 shows as an example a possible radiolytic sequence in NaCl [11.126]. A localized hole behaves like a chlorine atom and, in some exciton states with energy around 7 eV, this neutralized anion is tightly localized in a Cl_2^- bond between neighboring anions while the excited electron stays in hydrogen-like orbitals of large diameter near the surrounding Na^+ cations (Fig. 11.8a). The Cl_2^- moves by hole tunneling and interstitial propagation (Fig. 11.8b,c), which results in formation of a Frenkel pair that consists of an interstitial (H center) and an anion vacancy with a trapped electron (F center).

Similar processes also cause radiolysis in other alkali halides and alkali-earth fluorides (CaF_2, MgF_2). In MgO, however, radiolysis cannot be observed because the displacement energy for the ions is greater than the available energy of the excitons.

Radiolysis is strongly dependent on temperature if the transition to a Frenkel pair requires an activation energy, which is of the order of 0.1 eV in KI, NaCl, and NaBr. This causes a strong decrease of radiolysis at low temperatures, which allows defects to be observed in specimens cooled below 50 K [11.127, 11.128].

Other processes consist of either recombination of the Frenkel pairs or defect accumulation of interstitial halogen atoms, resulting in interstitial dislocation loops even at 50 K. The kinetics of these processes depends on the mobility of the defects (temperature) and their concentration. As shown in Sect. 11.2.1, the energy dissipated by electron bombardment in a transmission electron microscope is very much greater than in irradiation experiments with x-rays or UV quanta. Results obtained with these radiations are therefore not directly comparable with those observed during electron irradiation in a transmission electron microscope.

Radiolysis and secondary processes are not proportional to the dose $q = j\tau$, unlike damage in organic material. Some secondary processes may not be observed at low current densities but only at greater j, for which the rate of Frenkel pair formation is greater. At high temperatures, the enhanced mobility of anion vacancies and halogen molecules leads to the formation of colloidal metal inclusions and halogen bubbles, whereas the dislocation loops grow and form a dense dislocation network [11.129, 11.130]. Thus, in CaF_2, voids about 10 nm in diameter condense into a superlattice [11.131, 11.132]. Loss of interlayer sodium has been observed in mica [11.133].

Irradiation of quartz causes a radiolytic transformation into an amorphous state (vitrification). The mechanism is not understood in detail, but rupture of Si–O bonds causes rotations of $[SiO_4]$ tetrahedra in the silicate structure [11.134]. Amorphization has also been observed in natural zeolites, for example [11.135].

Some decomposition products can be identified by electron diffraction, and degradation by electron-beam heating may not lead to the same result as thermal decomposition, owing to the additional action of excitation processes. For example, dolomite ($Ca_{0.5}Mg_{0.5}CO_3$) decomposes in different stages to CaO and MgO [11.136].

High-resolution electron microscopy in either the crystal-structure or the surface-profile imaging modes (Sects. 9.6.3 and 9.7.3) needs very high current densities ($1–4 \times 10^5$A m^{-2}), and electron-beam-induced radiolysis and decomposition effects can be observed, especially by surface-profile imaging [11.137].

Damage effects can also be analyzed by EELS experiments. For example, in LiCl, an O K edge appears as a result of oxidation of Li, and a loss of Cl can be observed after irradiating NaCl [11.138]. The ELNES of hydroxyborate minerals (rhodizite, colemanite, howlite) indicates that a BO_4 to BO_3 transformation occurs, leading to a structure damage [11.139]; a reduction of Mn^{4+} in manganese oxides has been revealed by parallel EELS [11.140].

11.3.2 Radiation Damage by Knock-On Collisions

Besides the damage produced by ionization (Sects. 11.2 and 11.3.1) and specimen heating (Sect. 11.1), radiation damage by knock-on collisions has to be considered in HVEM; for reviews, see [11.141–11.145].

It was shown in Sect. 5.1.2 that, during an elastic collision between a beam electron and a nucleus, the energy transferred can become greater than the

Table 11.7. Mean displacement energy E_{d} and electron thereshold energy E_{th} for direct knock-on of atoms (for carbon, $E_{\mathrm{d}} = 5$ eV corresponds to molecules and 10 eV for graphite). Displacement cross sections σ_{d} and maximum energy transfer $E_{\max}$ for a head-on impact ($\theta = 180°$) at $E = 1$MeV.

Element	C	Si	Cu	Mo	Au
E_{d} (eV)	5 (10)	13	19	33	33
E_{th} (keV)	27.2 (54.4)	145	400	810	1300
σ_{d} (10^{-24}cm^2)	89 (43.6)	69	59	7.2	–
$E_{\max}$ (eV)	366	155	68	45	22

mean (polycrystalline) value of the displacement energy E_{d} (Table 11.7). As a result of such a displacement, a Frenkel pair that consists of a vacancy and an interstitial atom is produced. Similarly, an atom can be pushed into a neighboring vacancy, or an interstitial atom can be moved to another interstitial site. Neighboring knocked-on atoms can transfer momentum to a mobile (thermally activated) atom. These processes result in radiation-induced or radiation-enhanced diffusion. The displacement energies are greater than for similar processes caused by thermal activation because an atom is pushed to an interstitial position across the saddle point of neighboring atoms so quickly that the lattice cannot relax by exciting lattice vibrations.

With increasing electron energy E, the energy transfer ΔE first becomes greater than E_{d} for a scattering angle $\theta = 180°$ at the threshold energy E_{th}. With increasing atomic mass A, the values of E_{th} become greater (Table 11.7) but the cross sections σ_{d} for displacement increase more rapidly for greater A at high energies (Fig. 5.3). The threshold energy also depends on the direction of the knock-on momentum and is least in the close-packed directions $\langle 110 \rangle$, $\langle 100 \rangle$ and, $\langle 11\bar{2}0 \rangle$ for fcc, bcc, and hexagonal close-packed metals, respectively. Thus, for Cu, E_{d} is 20 eV near $\langle 110 \rangle$ and increases to 45 eV near $\langle 111 \rangle$ but can be as low as 10 eV for a narrow angular region about 10° away from $\langle 110 \rangle$ [11.146, 11.147, 11.148].

The concentration of displacements $c_{\mathrm{d}} = n_{\mathrm{d}}/n$ (n: number of atoms per unit volume) is proportional to the current density j_{n} at the nuclei and the irradiation time τ:

$$c_{\mathrm{d}} = \sigma_{\mathrm{d}} j_{\mathrm{n}} \tau / e. \tag{11.18}$$

Irradiation of copper for 1 min with a 5 μm spot and a current of 0.2 μA gives $c_{\mathrm{d}} = 0.25\%$ at $E = 600$ keV and $c_{\mathrm{d}} = 1.25\%$ at $E = 1$ MeV. However, unlike the case of ionization damage, the cross sections σ_{d} are so small that the damage can be avoided even when working at high resolution. Alternatively, the damage can be exacerbated by increasing the current density and the irradiation time, and a high-voltage electron microscope becomes a powerful tool for investigating radiation-damage effects in situ because the production of Frenkel pairs can be three to four orders of magnitude greater than with electron accelerators or in a nuclear reactor. Very high electron energies will be useful for studying damage in materials of high atomic number (e.g., 2.5 MeV for Au [11.149]).

The dependence of c_d on the current density j_n at the nuclei enhances the production rate up to a factor of 4 if the Bloch-wave intensity is greatest at the nuclei. There is thus a sensitive dependence of c_d on crystal orientation, with a maximum near the Bragg position, where c_d becomes proportional to the intensity of excess Kikuchi bands (Fig. 7.20b) [11.147, 11.150].

Defect clusters are observable by TEM only as secondary damage products. The diffusion of vacancies and interstitials depends strongly on temperature. Whereas interstitials are highly mobile during irradiation at all temperatures, vacancies need higher temperatures (room temperature in Cu, for example) to become mobile. A decrease of specimen temperature to 4 K can be used to stabilize a defect structure obtained at high temperature, or the accumulation of defects generated at 4 K can be observed as the temperature is increased.

If the thermally activated diffusion of point defects does not lead to a recombination of Frenkel pairs, then defect clusters, interstitial-dislocation loops, stacking-fault tetrahedra, or voids may be formed. Surfaces, dislocations, and grain boundaries are sinks for point defects, which decrease the defect concentration over distances of 100–150 μm.

In alloys, atomic displacements can cause radiation enhanced diffusion; in Al-Cu or Al-Zn alloys, for example, radiation-enhanced precipitation may result. Disordering can be observed in ordered alloys (e.g., Ni_3Mn), but ordering may occur simultaneously as a result of radiation-enhanced diffusion (e.g., Fe-Ni, Au_4Mn).

11.4 Contamination

11.4.1 Origin and Sources of Contamination

Radiation damage of adsorbed hydrocarbon molecules on the specimen surface causes a carbon-rich, polymerized film to form; this grows on electron-irradiated areas of the specimen by cross-linking. In competition with this contamination, reactions with activated, adsorbed H_2O, O_2 or N_2 molecules cause etching of carbonaceous material. Depending on specimen preparation, partial pressures, specimen temperature, and irradiation conditions, growth of either sign (positive for contamination and negative for etching) may prevail. It is not easy to work at equilibrium (zero growth) (see [11.152] for a review).

Various sources of hydrocarbon molecules are

1. adsorbed layers on the specimen, introduced during preparation or by atmospheric deposition,
2. vacuum oil from the rotary and diffusion pumps, and
3. grease and rubber O-rings and adsorbed layers (e.g., fingermarks) on the microscope walls.

These contributions to contamination can be kept small by taking the following precautions.

Even pure specimens become contaminated by hydrocarbon molecules if exposed to air for a period of a day. This type of contamination cannot be reduced by the liquid-nitrogen-cooled anticontamination blades (or cold finger) inside the microscope. It is certainly the most important source of contamination, especially when small electron probes $<0.1\,\mu$m are used. Contamination of this kind will be introduced even with the specimen under ultrahigh vacuum conditions. Washing the specimen cartridge and the specimen in methyl alcohol is the simplest way of eliminating this contaminant [11.153].

The partial pressure of vacuum-oil molecules can be reduced by using oil of low vapor pressure and a good baffle between the diffusion pump and the column, and by switching over from the rotary to the diffusion pump at a pressure of about 10 Pa to avoid back streaming of the rotary-pump oil. The best solution is to use an oil-free turbomolecular pump.

Grease must be avoided or be used very sparingly. Vacuum leaks can never be cured by heavy greasing but only by careful polishing of the sealing surfaces. Viton rings should be used in preference to rubber. All surfaces should be washed with methyl alcohol, which evaporates completely in air. Finger-marks should be avoided by wearing gloves when opening the microscope.

The etching process is affected by the composition of the residual gas and by the partial pressure of water that comes from photographic materials. Emulsions should be pre-evacuated in the presence of water-absorbing material (e.g., P_2O_5) and raised to atmospheric pressure for only a very short time [4.126]. Photographic emulsions are a very uncertain source of water vapor.

11.4.2 Methods for Decreasing Contamination

The precautions mentioned in the preceding section are essential if the contamination rate is to be effectively reduced. Further improvements can be obtained by the following methods.

(1) Specimen Heating to 200–300°C increases the desorption of hydrocarbon molecules and decreases the contamination [11.155]. When small electron probes are employed, it is, however, necessary to heat the specimen initially to 400–500°C.

(2) Specimen Cooling [11.156]. The specimen cartridge and additional components act as a cryoshield and decrease the adsorption of hydrogen molecules by decreasing their partial pressure near the specimen. At low temperatures, the contamination changes over to etching of carbon or organic material (Fig. 11.9a). This is attributed to a radiation-induced chemical reaction with adsorbed residual gas molecules (H_2O, N_2, O_2, CO, H_2), resulting in volatile compounds with carbon. The observed spread of experimental results in Fig. 11.9 is a consequence of the variable composition of the residual gas. When metal or inorganic specimens are cooled below –100°C, there is no

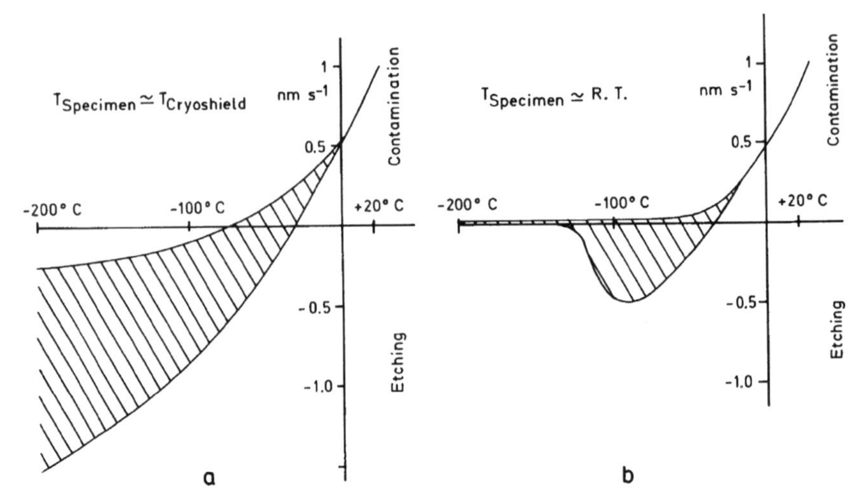

Fig. 11.9. Rate of carbon deposition by contamination (positive sign) and removal by etching (negative sign) as a function of temperature T for (**a**) specimen cooling with the specimen and the cryoshield at the same temperature and (**b**) the anticontamination blades at the temperature T and the specimen at room temperature [11.157, 11.158].

contamination. If the cryoshield is not sufficient, an ice film may grow on the specimen, but this will vanish if a small electron probe is being employed.

(3) Anticontamination Blades cooled with liquid nitrogen surround the specimen and act as cryoshields, but the specimen is still at room temperature. Cooling first causes a decrease of the partial pressure of hydrocarbons, and etching of the specimen predominates (Fig. 11.9b) [11.157, 11.158]. Further cooling also reduces the partial pressure of the residual gas molecules that are responsible for etching. At low temperatures of the blades, contamination or etching continues at a very low rate. The precautions of Sect. 11.4.1, together with anticontamination blades, are mostly sufficient for routine use of a transmission electron microscope with illuminated areas larger than one micrometer in diameter. A much higher contamination rate is, however, observed when small electron probes are employed for STEM, microdiffraction, or microanalysis. The main source of trouble is then the surface diffusion of adsorbed hydrocarbon molecules to the irradiated area (see Sect. 11.4.3).

11.4.3 Dependence of Contamination on Irradiation Conditions

Three types of irradiation conditions may be distinguished.

(1) Uniform Irradiation of a Reasonably Large Area (Fig. 11.10a). The growth rate of contamination is proportional to the current density and the irradiation time or proportional to the charge density $q = j\tau$. The growth

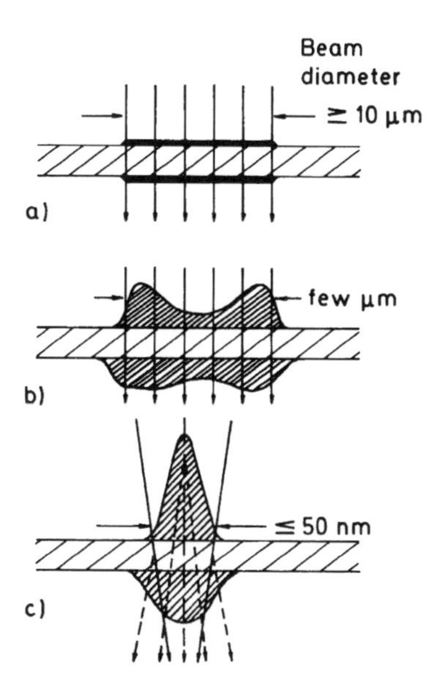

Fig. 11.10. Buildup of (**a**) a uniform contamination layer by uniform irradiation of a larger area, (**b**) a contamination ring by uniform irradiation of a small area, and (**c**) a contamination needle by irradiation with a small electron probe.

rate in the central part of the irradiated area is mainly determined by adsorption of molecules from the gas phase. Adsorbed hydrocarbons introduced into the microscope with the specimen are damaged in the first stage of irradiation and are fixed by cross-linking.

(2) Irradiation of a Small Area with Uniform Current Density (Fig. 11.10b). If the diameter of the area illuminated is reduced to a few micrometers by imaging the condenser diaphragm on the specimen, an annular contamination spot can be observed with more contamination at the periphery of the irradiated area than at the center. This phenomenon can be explained by surface diffusion of adsorbed hydrocarbons, which are cross-linked when they are struck by the electron beam and have little opportunity to diffuse to the center of the illuminated area. The contamination rate is higher than that found with uniform illumination of a larger area because the whole foil is a source, which supplies molecules by surface diffusion. There is no reason to assume that the electric fields caused by secondary electron emission have any influence, as claimed in [11.159]. The model calculations of Müller [11.160] discussed below in detail are fully adequate to explain the generation of contamination rings (see the end of this section).

(3) Irradiation with an Electron Probe (Fig. 11.10c). The diameter of the annular zone of higher contamination rate (described above) decreases if the electron-probe diameter is less than 0.1 μm, and a needle-shaped contamination spot is formed. This spot can be broadened at its bottom by multiple scattering of electrons at the top of the contamination needle. The contamination cone on the bottom of the foil is also normally broadened by multiple scattering [11.161]. It is very difficult to combat this type of contamination.

An excellent vacuum, washing the specimen with methyl alcohol, and specimen heating or cooling are necessary precautions. A cold finger, which helps to decrease the partial pressure and the contamination under irradiation condition 1, is normally not sufficient for this illumination mode. It has been found that the growth of contamination needles increases the longer the specimen stays in the microscope vacuum. If a larger area (grid mesh) is subsequently irradiated, the contamination rate again decreases, which indicates that the adsorbed molecules from the vacuum are fixed by irradiation and cannot diffuse further to the irradiated spot.

The higher contamination rate with small electron probes can be exploited for microwriting [11.160, 11.162]. The local foil thickness can be determined by tilting the irradiated specimen through 45° to separate the top and bottom cones (needles) of contamination [10.44].

All of these processes and different illumination conditions can be described by a single differential equation [11.160]. The number k of molecules that are cross-linked to the surface per unit time and per unit area is proportional to the density n of the adsorbed molecules and to the current density j:

$$k = n\sigma \frac{j}{e} \,. \tag{11.19}$$

The number ν of molecules incident on a unit area in a unit time can be calculated from the partial pressure p:

$$\nu = \frac{p}{(2\pi mkT)^{1/2}} \,. \tag{11.20}$$

These molecules are adsorbed and desorbed again with a time constant τ_0. The equilibrium concentration will be

$$n_\infty = \nu\tau_0 . \tag{11.21}$$

Four mechanisms can change the concentration n:

(1) adsorption of molecules	$(\partial n/\partial\tau)_1 = \nu$.
(2) desorption	$(\partial n/\partial\tau)_2 = -n/\tau_0$.
(3) diffusion	$(\partial n/\partial\tau)_3 = \lambda\nabla^2 n$.
(4) contamination	$(\partial n/\partial\tau)_4 = -(j/e)\sigma n$.

This may be expressed by the partial differential equation

$$\frac{\partial n}{\partial\tau} = \nu - \frac{n}{\tau_0} + \lambda\nabla^2 n - \frac{j}{e}\sigma n . \tag{11.22}$$

Applying the equilibrium condition $\partial n/\partial\tau = 0$ and imposing rotational symmetry, we obtain

$$\nu - \frac{n}{\tau_0} + \lambda\left(\frac{\mathrm{d}^2 n}{\mathrm{d}r^2} + \frac{1}{r}\frac{\mathrm{d}n}{\mathrm{d}r}\right) - \frac{j}{e}\sigma n = 0 . \tag{11.23}$$

With the abbreviations $\rho = (\tau_0\lambda)^{1/2}$, $\alpha = \sigma/e\lambda$, $\nu/\lambda = n_\infty/\rho^2$, this equation can be written as

$$\frac{\mathrm{d}^2 n}{\mathrm{d}r^2} + \frac{1}{r}\frac{\mathrm{d}n}{\mathrm{d}r} - \frac{1}{\rho^2}n(r) - \alpha j(r)n(r) + \frac{n_\infty}{\rho^2} = 0. \tag{11.24}$$

If uniform irradiation j_0 is assumed in the area $r \leq R$, the equation can be solved in this inner area,

$$n_2 = n_\infty \left[\frac{\rho_0^2}{\rho^2} + C_2 \mathrm{I}_0\left(\frac{r}{\rho_0}\right)\right] \quad \text{with} \quad \rho_0 = \rho(1 + \alpha^2\rho^2 j_0)^{-1/2}, \tag{11.25}$$

and in the outer, nonirradiated area $r \geq R$,

$$n_1 = n_\infty \left[1 - C_1 \mathrm{K}_0\left(\frac{r}{\rho}\right)\right]. \tag{11.26}$$

The constants C_1 and C_2 can be determined from the boundary conditions at $r = R$: $n_1(R) = n_2(R)$ and $\mathrm{d}n_1(R)/\mathrm{d}r = \mathrm{d}n_2(R)/\mathrm{d}r$. I_0 and K_0 are the modified Bessel (Hankel) functions. An example of the resulting distribution $n(r)$ and the thickness $t_c = \sigma n(r) j\tau/e$ of the contamination layer is plotted in Fig. 11.11. It shows how the irradiated area acts as a sink for hydrocarbons. The contamination rate is proportional to $n(r)$, and the contamination ring discussed earlier is predicted. On increasing the current density j by a factor of 2 (a $\rightarrow$ b) and (c $\rightarrow$ d), the density $n(r)$ decreases to approximately half the value observed at low j. This results in a growth of contamination of thickness t_c, which will be proportional only to the irradiation time τ $[n(r)j = \text{const}]$ and not to the charge density $q = j\tau$. For constant j, increasing the radius R of the irradiated area (a $\rightarrow$ c) and (b $\rightarrow$ d) results in a decrease of $n(r)$ that is approximately proportional to r^{-2}. The proportionality of $n(r)$ to $(jR^2)^{-1} = I_\mathrm{p}^{-1}$ results in the formation of a constant total mass of contamination per unit time. There is thus a saturation effect caused by the delayed diffusion of hydrocarbon molecules from the unirradiated part of the foil.

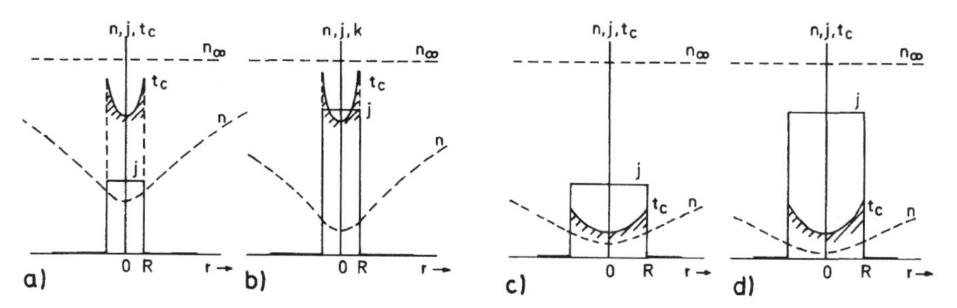

Fig. 11.11. Equilibrium of the concentration $n(r)$ of adsorbed organic molecules. The irradiated area $r \leq R$ acts as a sink for mobile molecules, so that a contamination layer of thickness $t_c \propto n(r)j\tau$ is formed (n_∞ is the equilibrium concentration at a large distance). The irradiated area shows in (**c**) and (**d**) twice the diameter $2R$ as in (**a**) and (**b**), and the current density in (**b**) and (**d**) is a factor of 2 larger than in (**a**) and (**c**) [11.160].

Scanning a larger area in the STEM mode will result in a uniform contamination layer for fast scans when $n \simeq n_\infty$. For slow scans, the discussion of the contamination rate is more complicated because the decrease of n is asymmetrical and the contamination trails behind the moving electron probe. This nonstationary case can be calculated by solving of the time-dependent equation (11.23) numerically [11.163].

References

Chapter 1

Transmission Electron Microscopy (for electron optics, see [2.4–2.7])

General

[1.1] C. Magnan (ed.): *Traité de Microscopie Electronique*, Vols. 1, 2 (Hermann, Paris, 1961)
[1.2] D.H. Kay: *Techniques for Electron Microscopy*, 2nd ed. (Blackwell, Oxford, 1965)
[1.3] V.E. Cosslett, R. Barer (eds.): *Advances in Optical and Electron Microscopy*, (Academic, London, 1966–1994)
[1.4] J. Picht, J. Heydenreich: *Einführung in die Elektronenmikroskopie* (VEB Verlag Technik, Berlin, 1966)
[1.5] L. Reimer: *Elektronenmikroskopische Untersuchungs- und Präparationsmethoden*, 2. Aufl. (Springer, Berlin, 1967)
[1.6] G. Schimmel: *Elektronenmikroskopische Methodik* (Springer, Berlin, 1969)
[1.7] M. von Heimendahl: *Einführung in die Elektronenmikroskopie* (Vieweg, Braunschweig, 1970)
[1.8] D.L. Misell: *Image Analysis, Enhancement and Interpretation* (North-Holland, Amsterdam, 1970)
[1.9] P. Grivet: *Electron Optics II: Instruments* (translated by P.W. Hawkes) (Pergamon, Oxford, 1972)
[1.10] P.W. Hawkes: *Electron Optics and Electron Microscopy* (Taylor & Francis, London, 1972)
[1.11] P.W. Hawkes (ed.): *Image Processing and Computer-Aided Design in Electron Optics* (Academic, London, 1973)
[1.12] B.M. Siegel, D.R. Beaman (eds.): *Physical Aspects of Electron Microscopy and Microbeam Analysis* (Wiley, New York, 1975)
[1.13] W.O. Saxton: *Computer Techniques for Image Processing in Electron Microscopy* (Academic, New York, 1978)
[1.14] E. Ruska: *Die frühe Entwicklung der Elektronenlinsen und der Elektronenmikroskopie.* (Barth, Leipzig, 1979); *The Early Development of Electron Lenses and Electron Microscopy* (Hirzel, Stuttgart, 1980)

[1.15] P.W. Hawkes (ed.): *Computer Processing of Electron Microscope Images* Topics in Current Physics, Vol. 13 (Springer, Berlin, 1980)
[1.16] S. Wischnitzer: *Introduction to Electron Microscopy*, 3rd edn. (Pergamon, New York, 1981)
[1.17] J.M. Cowley: *Diffraction Physics*, 2nd ed. (North-Holland, Amsterdam, 1981)
[1.18] D. Chescoe, P. Goodhew: *The Operation of the Transmission Electron Microscope* (Oxford University Press, Oxford, 1984)
[1.19] W. Krakow, M.A. O'Keefe: *Computer Simulation of Electron Microscope Diffraction and Images* (The Minerals, Metals and Materials Society, Warrendale, PA, 1989)
[1.20] J.C. Russ: *Computer-Assisted Microscopy* (Plenum, New York, 1990)
[1.21] P.W. Hawkes: *Électrons et Microscopes: Vers les nanosciences* (CNRS Editions, Paris, 1995)
[1.22] D.B. Williams, C.B. Carter: *Transmission Electron Microscopy* (Plenum, New York, 1996)
[1.23] H. Alexander: *Physikalische Grundlagen der Elektronenmikroskopie* (Teubner, Stuttgart, 1997)
[1.24] C. Colliex: *La microscopie électronique* (Presses Universitaires de France, Paris, 1998)
[1.25] M. DeGraef: *Introduction to Conventional Transmission Electron Microscopy* (Cambridge Univ. Press, Cambridge 2003)

Materials Science

[1.26] P.B. Hirsch, A. Howie, R.B. Nicholson, D.W. Pashley, M.J. Whelan: *Electron Microscopy of Thin Crystals* (Butterworths, London, 1965), 2nd ed. (Krieger, New York, 1977)
[1.27] F. Hornbogen: *Durchstrahlungselektronenmikroskopie fester Stoffe* (Chemie, Weinheim, 1971)
[1.28] U. Valdrè: *Electron Microscopy in Materials Science* (Academic, New York, 1971)
[1.29] B. Jouffrey (ed.): *Méthodes et Techniques Nouvelles d'Observation en Métallurgie Physique* (Société Française du Microscopie Electronique, Paris, 1972)
[1.30] M.H. Loretto, R.E. Smellman: *Defect Analysis in Electron Microscopy* (Chapman and Hall, London, 1975)
[1.31] U. Valdrè, E. Ruedl (eds.): *Electron Microscopy in Materials Science*, Part I–IV (Commission of the European Communities, Brussels, 1975)
[1.32] P.J. Grundy, G.A. Jones: *Electron Microscopy in the Study of Materials* (Edwards Arnold, London, 1976)
[1.33] H.R. Wenk (ed.): *Electron Microscopy in Mineralogy* (Springer, Berlin, 1976)
[1.34] S. Amelinckx, R. Gevers, G. Remaut, J. Van Landuyt: *Modern Diffraction and Imaging Techniques in Materials Science* (North-Holland, Amsterdam 1970) 2nd ed. in 2 Vols., published in 1978
[1.35] G. Thomas, M.J. Goringe: *Transmission Electron Microscopy of Metals* (Wiley, New York, 1979)
[1.36] D.B. Williams: *Practical Analytical Electron Microscopy in Materials Science* (Chemie International, Weinheim, 1979)

[1.37] O. Brümmer, J. Heydenreich, K.H. Krebs. H.G. Schneider (eds.): *Handbuch der Festkörperanalyse mit Elektronen, Ionen und Röntgenstrahlen* (Vieweg, Braunschweig, 1980)
[1.38] H. Bethge, J. Heydenreich (eds.): *Elektronenmikroskopie in der Festkörperphysik* (Springer, Berlin, 1982); *Electron Microscopy in Solid State Physics* (EL Series, Amsterdam, 1987)
[1.39] B. Jouffrey, A. Bourret, C. Colliex (eds.): *Microscopie Electronique en Science des Matériaux* (Editions du CNRS, Paris, 1983)
[1.40] W. Krakow, D.A. Smith, L.W. Hobbs (eds.): *Electron Microscopy of Materials* (North-Holland, Amsterdam, 1984)
[1.41] P.G. Merli, M.V. Antisari (eds.): *Electron Microscopy in Materials Science* (World Scientific, Singapore, 1992)
[1.42] P.R. Buseck (ed.): *Minerals and Reactions at the Atomic Scale: Transmission Electron Microscopy*, Reviews in Mineralogy, Vol. 27 (Mineralogical Society of America, Washington, 1992)
[1.43] B. Fultz and J. Howe: *Transmission Electron Microscopy and Diffractometry of Materials* (Springer, Berlin, 2001)

Biology

[1.44] A.M. Glauert (ed.): *Practical Methods in Electron Microscopy* (North-Holland, Amsterdam, 1972) series continues
[1.45] J.K. Koehler (ed.): *Advanced Techniques in Biological Electron Microscopy* (Springer, Berlin, 1973); *Specific Ultrastructural Problems* (Springer, Berlin, 1978)
[1.46] G.A. Meek: *Practical Electron Microscopy for Biologists*, 2nd ed. (Wiley, London, 1976)
[1.47] W. Hoppe, R. Mason (eds.): *Unconventional electron microscopy for molecular structure determination.* Advances in Structure Research by Diffraction Methods, Vol. 7 (Vieweg, Braunschweig, 1979) 1
[1.48] W. Baumeister, W. Vogell (eds.): *Electron Microscopy at Molecular Dimensions* (Springer, Berlin, 1980)
[1.49] M.A. Hayat: *Principles and Techniques of Electron Microscopy: Biological Applications*, Vol. 1, 2nd ed. (University Park Press, Baltimore, 1980)
[1.50] J.D. Griffiths (ed.): *Electron Microscopy in Biology* (Wiley, New York, 1981) series continued
[1.51] R.H. Lange, J. Blödorn: *Das Elektronenmikroskop, TEM+REM* (Thieme, Stuttgart, 1981)
[1.52] J.N. Turner (ed.): *Three-Dimensional Ultrastructure in Biology*, Methods in Cell Biology, Vol. 22 (Academic, New York, 1981)
[1.53] H. Plattner, H.P. Zingsheim: *Elektronenmikroskopische Methodik in der Zell- und Molekularbiologie* (Fischer, Stuttgart, 1987)
[1.54] R.A. Steinbrecht, K. Zierold: *Cryotechniques in Biological Electron Microscopy* (Springer, Berlin, 1987)
[1.55] P.W. Hawkes, U. Valdre (eds.): *Biophysical Electron Microscopy* (Academic, London, 1990)

International Conferences on Electron Microscopy

[1.56] R. Ross (ed.): *3rd International Conference on Electron Microscopy*, London 1954 (Royal Microscopic Society, London, 1956)

[1.57] W. Bargmann, G. Möllenstedt, H. Niehrs, D. Peters, E. Ruska, C. Wolpers: *Vierter Internationaler Kongreß für Elektronenmikroskopie*, Berlin 1958, Vols. 1, 2 (Springer, Berlin, 1960)

[1.58] S.S. Breese (ed.): *Electron Microscopy 1962*, Philadelphia, Vols. 1, 2 (Academic, New York, 1962)

[1.59] R. Uyeda (ed.): *Electron Microscopy 1966*, Kyoto, Vols. 1, 2 (Maruyen, Tokyo, 1966)

[1.60] P. Favard (ed.): *Microscopie Electronique 1970*, Grenoble, Vols. 1–3 (Société Française de Microscopie Electronique, Paris, 1970)

[1.61] J.V. Sanders, D.J. Goodchild (eds.): *Electron Microscopy 1974*, Canberra, Vols. 1, 2 (Australian Academy of Sciences, Canberra, 1974)

[1.62] J.M. Sturgess (ed.): *Electron Microscopy 1978*, Toronto, Vols. 1–3 (Microscopical Society of Canada, Toronto, 1978)

[1.63] The Congress Organizing Committee: *Electron Microscopy 1982*, Hamburg, Vols. 1–3 (Deutsche Gesellschaft für Elektronenmikroskopie, Frankfurt, 1982)

[1.64] T. Imura, S. Maruse, T. Suzuki (eds.): *Electron Microscopy 1986*, Kyoto, Vols. 1–4 (Japanese Society of Electron Microscopy, Tokyo, 1986)

[1.65] L.D. Peachey, D.B. Williams (eds.): *Electron Microscopy 1990*, Seattle, Vols. 1–4 (San Francisco Press, San Francisco, 1990)

[1.66] B. Jouffrey, C. Colliex (eds.): *Electron Microscopy 1994*, Paris, Vols. 1–5 (Les Editions de Physique, Les Ulis, 1994)

[1.67] H.A. Calderán-Beanvides, M.J. Yacamán (eds.): *Electron Microscopy 1998*, Cancun, Vols. 1–4 (Institute of Physics, Bristol, 1998)

[1.68] J. Engelbrecht, T. Sevell, M. Witcomb, R. Cross, P. Richards (eds.): *15th International Congress on Electron Microscopy* Vol. 1-3 (Microscopy Society of South Africa, 2002)

High-Resolution Electron Microscopy (HREM)

[1.69] J.C.H. Spence: *Experimental High-Resolution Electron Microscopy*, 2nd edn. (Oxford University Press, Oxford, 1988)

[1.70] P. Buseck, J. Cowley, L. Eyring (eds.): *High-Resolution Transmisson Electron Microscopy and Associated Techniques* (Oxford University Press, Oxford, 1988)

[1.71] S. Horiuchi: *Fundamentals of High-Resolution Transmission Electron Microscopy* (North-Holland, Amsterdam, 1994)

[1.72] H. Lichte: Electron image plane off-axis holography of atomic structures, in *Advances in Optical and Electron Microscopy*, Vol. 12, ed. by R. Barer, V.E. Cosslett (Academic, London, 1991) p. 25

[1.73] A. Tonomura: *Electron Holography*, Springer Series in Optical Sciences, Vol. 70 (Springer, Berlin, 1993)

[1.74] M. Lentzen, B. Jahnen, C.L. Jia, A. Thust, K. Tillmann, K. Urban: High-resolution imaging with an aberration-corrected transmission electron microscope. Ultramicroscopy **92**, 233 (2002)

Analytical Electron Microscopy

[1.75] J.J. Hren, J.I. Goldstein, D.C. Joy: *Introduction to Analytical Electron Microscopy* (Plenum, New York, 1979)

[1.76] J.R. Fryer: *The Chemical Applications of Transmission Electron Microscopy* (Academic, London, 1979)

[1.77] D.C. Joy, A.D. Romig, J.I. Goldstein (eds.): *Principles of Analytical Electron Microscopy* (Plenum, New York, 1986)

[1.78] R.F. Egerton: *Electron Energy-Loss Spectroscopy in the Electron Microscope*, 2nd ed. (Plenum, New York, 1996)

[1.79] M.M. Disko, C.C. Ahn, B. Fultz (eds.): *Transmission Electron Energy-loss Spectrometry in Materials Science* (The Minerals, Metals and Materials Society, Warrendale, PA, 1992)

[1.80] P. Goodman (ed.): *Fifty Years of Electron Diffraction* (Reidel, Dordrecht, 1981)

[1.81] J.C.H. Spence, J.M. Zuo: *Electron Microdiffraction* (Plenum, New York, 1992)

[1.82] J.M. Cowley (ed.): *Electron Diffraction Techniques* (Oxford University Press, Oxford) Vol. 1 (1992), Vol. 2 (1993)

[1.83] J.P. Morniroli: *Large Angle Convergent-Beam Diffraction (LACBED)* (Société Française des Microscopies, Paris 2002)

[1.84] L. Reimer (ed.): *Energy-Filtering Transmission Electron Microscopy*, Springer Series in Optical Sciences, Vol. 71 (Springer, Berlin, 1995)

High-Voltage Electron Microscopy (HVEM)

[1.85] G. Dupouy: Electron microscopy at very high voltages, in*Advances in Optical and Electron Microscopy*, Vol. 2, ed. by R. Barer, V.E. Cosslett (Academic, London, 1966)

[1.86] P.R. Swann (ed.): Proceedings of the Symposium on HVEM 1972, published in J. Microsc. **97**, Parts 1 and 2, 1–269 (1973)

[1.87] P.R. Swann, C.J. Humphreys, M.J. Goringe (eds.): *High Voltage Electron Microscopy* (Academic, London, 1974)

[1.88] B. Jouffrey, P. Favard (eds.): *Microscopie Electronique à Haute Tension* (Société Française de Microscopie Electronique, Paris, 1976)

[1.89] T. Imura, H. Hashimoto (eds.): *High Voltage Electron Microscopy* (Japanese Society of Electron Microscopy, Kyoto, 1977)

[1.90] P. Brederoo, J. Van Landuyt (eds.): *Electron Microscopy 1980*, Vol. 4: High Voltage (Seventh European Congress on Electron Microscopy Foundation, Leiden, 1980)

[1.91] H. Fujita, K. Ura, H. Mori (eds.): New directions and future aspects of HVEM. Ultramicroscopy **39**, 1–408 (1991)

Scanning Transmission Electron Microscocy

[1.92] A.V. Crewe: The current state of high resolution scanning electron microscopy. Q. Rev. Biophys. **3**, 137 (1970)

[1.93] A.V. Crewe: Scanning transmission electron microscopy. J. Microsc. **100**, 247 (1974)

[1.94] C. Colliex, C. Mory: Quantitative aspects of scanning transmission electron microscopy, in *Quantitative Electron Microscopy*, ed. by J.N. Chapman, A.J. Craven (Scottish Universities Summer School in Physics, Edinburgh, 1984) p. 149

[1.95] P.E. Batson, N. Dellby, O.L. Krivanek: Sub-ångstrom resolution using aberration corrected electron optics. Nature **418**, 617 (2002)

Emission Electron Microscopy

[1.96] H. Düker: Emissions-Elektronenmikroskope. Acta Phys. Austriaca **18**, 232 (1964)

[1.97] G. Möllenstedt, F. Lenz: Electron emission microscopy. Adv. Electron. Electron Phys. **18**, 251 (1963)

[1.98] L. Wegmann: Photoemissions-Elektronenmikroskopie, in *Handbuch der zerstörungsfreien Materialprüfung*, ed. by E.A.W. Müller (Oldenbourg, München, 1969) R.31, 1

[1.99] L. Wegmann: The photoemission electron microscope, its technique and applications. J. Microsc. **96**, 1 (1972)

[1.100] H. Bethge, M. Klaus: Photo-electron emission microscopy of work function changes. Ultramicroscopy **11**, 207 (1983)

[1.101] O.H. Griffith, G.F. Rempfer: Photoelectron imaging: photoelectron microscopy and related techniques, in *Advances in Optical and Electron Microscopy*, Vol. 10, ed. by R. Barer, V.E. Cosslett (Academic, London, 1987) p. 269

[1.102] O.H. Griffith, W. Engel (eds.): Emission microscopy and related techniques. Ultramicroscopy **36**, 1–274 (1991)

[1.103] S. Grund, W. Engel, P. Teufel: Photoelektronen-Emissionsmikroskop und Immunofluoreszenz. J. Ultrastruct. Res. **50**, 284 (1975)

[1.104] R. Fink, M.R. Weiss, E. Umbach, D. Preikszas, H. Rose, R. Spehr, P. Hartel, W. Engel, R. Degenhardt, R. Wichtendahl, H. Kuhlenbeck, W. Erlebach, K. Ihmann, R. Schlögl, H.-J. Freund, A.M. Bradshaw, G. Lilienkamp, Th. Schmidt, E. Bauer, G. Benner: SMART: a planned ultrahigh-resolution spectromicroscope for BESSY II. J. Electron Spectr. and Related Phenomena **84**, 231 (1997)

[1.105] W. Święch, G.H. Fecher, Ch. Ziethen, O. Schmidt, G. Schönhense, K. Grzelakowski, C.M. Schneider, R. Frömter, H.P. Oepen, J. Kirschner: Recent progress in photoemission microscopy with emphasis on chemical and magnetic sensitivity. J. Electron Spectrosc. Relat. Phenom. **84**, 171 (1997)

Reflection and Mirror Electron Microscopy

[1.106] E. Ruska, H.O. Müller: Über Fortschritte bei der Abbildung elektronenbestrahlter Oberflächen. Z. Phys. **116**, 366 (1940)

[1.107] V.E. Cosslett, D. Jones: A reflexion electron microscope. J. Sci. Instr. **32**, 86 (1955)

[1.108] C. Fert: Observation directe de la surface d'un échantillon massif en microscopie électronique, in [Ref. 1.1, Vol. 1, p. 277]

[1.109] Z.L. Wang: *Reflection Electron Microscopy and Spectrometry for Surface Analysis* (Cambridge University Press, Cambridge, 1996)

[1.110] A.B. Bok, J.B. Le Poole, J. Roos, H. Delang, H. Bethge, J. Heydenreich, M.E. Barnett: Mirror electron microscopy, in *Advances in Optical and Electron Microscopy*, Vol. 4, ed. by R. Barer, V.E. Cosslett (Academic, London, 1971) p. 161

[1.111] E. Bauer: Low energy electron microscopy, in *Chemistry and Physics of Solid Surfaces VIII*, ed. by R. Vanselov, R. Howe, Springer Series in Surface Science, Vol. 22 (Springer, Berlin, 1990) p. 267

[1.112] G.V. Spivak, V.P. Ivannikov, A.E. Luk'yanov, E.I. Rau: Development of scanning mirror electron microscopy for quantitative evaluation of electric microfields. J. Microsc. Spectrosc. Electron. **3**, 89 (1978)

[1.113] J. Witzani, E.M. Hörl: Scanning electron mirror microscopy. Scanning **4**, 53 (1980)

Scanning Electron Microscopy

[1.114] C.W. Oatley: *Scanning Electron Microscopy I. The Instrument* (Cambridge University Press, Cambridge, 1972)

[1.115] D.B. Holt, M.D. Muir, P.R. Grant, I.M. Boswarva (eds.): *Quantitative Scanning Electron Microscopy* (Academic, London, 1974)

[1.116] O.C. Wells: *Scanning Electron Microscopy* (McGraw-Hill, New York, 1974)

[1.117] J.I. Goldstein, H. Yakowitz: *Practical Scanning Electron Microscopy* (Plenum, New York, 1975)

[1.118] L. Reimer, G. Pfefferkorn: *Raster-Elektronenmikroskopie*, 2. Aufl. (Springer, Berlin, 1977)

[1.119] F. Maurice, L. Meny, R. Tixier (eds.): *Microanalyse et Microscopie Electronique a Balayage* (Les Editions de Physique, Orsay, 1978) [English transl.: *Microanalysis and Scanning Electron Microscopy*, 1979]

[1.120] J.I. Goldstein, D.E. Newbury, P. Echlin, D.C. Joy, C. Fiori, E. Lifshin: *Scanning Electron Microscopy and X-Ray Microanalysis* (Plenum, New York, 1981)

[1.121] B.L. Gabriel: *Biological Scanning Electron Microscopy* (Van Nostrand Reinhold, New York, 1982)

[1.122] L. Reimer: *Scanning Electron Microscopy, Physics of Image Formation and Microanalysis*, Springer Series in Optical Sciences, Vol. 45, 2nd ed. (Springer, Berlin, 1998)

[1.123] L. Reimer: *Image Formation in Low-Voltage Scanning Electron Microscopy*, Tutorial Texts in Optical Engineering, Vol. TT 12 (SPIE, Bellingham, WA, 1993)

X-Ray Microanalysis

[1.124] S.B. Reed: *Electron Microprobe Analysis* (Cambridge University Press, Cambridge, 1975)

[1.125] J.A. Chandler: *X-Ray Microanalysis in the Electron Microscope* (North-Holland, Amsterdam, 1977)

[1.126] D.A. Erasmus (ed.): *Electron Probe Microanalysis in Biology* (Chapman and Hall, London, 1978)

[1.127] P. Lechene, R.E. Warner (eds.): *Microbeam Analysis in Biology* (Academic, New York, 1979)

[1.128] M.A. Hayat (ed.): *X-Ray Microanalysis in Biology* (Univ. Park Press, Baltimore, 1980)
[1.129] T.E. Hutchinson, A.B. Somlyo: *Microprobe Analysis of Biological Systems* (Academic, New York, 1981)
[1.130] K.F.J. Heinrich: *Electron Beam X-Ray Microanalysis* (Van Nostrand, New York, 1981)
[1.131] A.J. Morgan: *X-Ray Microanalysis in Electron Microscopy for Biologists* (Oxford University Press, Oxford, 1984)
[1.132] P. Echlin: *Analysis of Organic and Biological Surfaces* (Wiley, New York 1884)

Scanning-Probe Microscopy

[1.133] R.J. Behm, N. Garcia, H. Rohrer (eds.): *Scanning Tunneling Microscopy and Related Methods*, NATO ASI Series E: Applied Sciences, Vol. 184 (Kluwer, Dordrecht, 1990)
[1.134] H.J. Güntherodt, R. Wiesendanger (eds.): *Scanning Tunneling Electron Microscopy I-III*, 2nd ed., Springer Series in Surface Science, Vols. 20, 28 and 29 (Springer, Berlin) Vol. 20 (1994), Vol. 28 (1995), and Vol. 29 (1996)
[1.135] D.A. Bonnell (ed.): *Scanning Tunneling Microscopy and Spectroscopy* (VCH Publishers, New York, 1993)
[1.136] R. Wiesendanger: *Scanning Probe Microscopy and Spectroscopy* (Cambridge University Press, Cambridge, 1994)
[1.137] R. Wiesendanger: *Scanning Probe Microscopy* (Springer, Berlin, 1998)

Chapter 2

[2.1] J. Kessler: *Polarized Electrons*, 2nd ed., Springer Series on Atoms and Plasmas, Vol. 1 (Springer, Berlin, 1985)
[2.2] W. Glaser: *Grundlagen der Elektronenoptik* (Springer, Wien, 1952)
[2.3] W. Glaser: Elektronen- und Ionenoptik, in *Encyclopedia of physics*, Vol. 33, ed. by S. Flügge (Springer, Berlin, 1956)
[2.4] A. Septier: *Focusing of Charged Particles* (Academic, New York, 1967)
[2.5] P. Grivet: *Electron Optics*, Pt. 1: Optics, Pt. 2: Instruments, 2nd ed. (translated by P.W. Hawkes) (Pergamon, Oxford, 1972)
[2.6] P.W. Hawkes (ed.): *Properties of Magnetic Lenses*, Topics in Applied Physics, Vol. 18 (Springer, Berlin, 1982)
[2.7] P.W. Hawkes, E. Kasper: *Principles of Electron Optics*, Vol. 1: Basic Geometrical Optics, Vol. 2: Applied Geometrical Optics (Academic, London, 1989)
[2.8] J.D. Jackson: *Classical Electrodynamics* (Wiley, New York, 1999)
[2.9] W. Glaser: Strenge Berechnung magnetischer Linsen der Feldform $H = H_0/[1+(z/a)^2]$. Z. Phy. **117**, 285 (1941)
[2.10] J. Dosse: Strenge Berechnung magnetischer Linsen mit unsymmetrischer Feldform nach $H = H_0/[1+(z/a)^2]$. Z. Phys. **117**, 316 (1941)
[2.11] E. Ruska: Über die Auflösungsgrenzen des Durchstrahlungs-Elektronenmikroskops. Optik **22**, 319 (1965)
[2.12] W.D. Riecke: Objective lens design for TEM. A review of the present state of the art, in *Electron Microscopy 1972* ed. by V.E. Cosslett (IoP, London, 1972) p. 98

[2.13] W. Kamminga: Properties of magnetic lenses with highly saturated pole pieces. Optik **45**, 39; and **46**, 226 (1976)
[2.14] I. Dietrich: *Superconducting Electron-Optical Devices* (Plenum, New York, 1976)
[2.15] P.W. Hawkes, U. Valdre: Superconductivity and electron microscopy. J. Phys. E **10**, 309 (1977)
[2.16] G. Lefranc, E. Knapek, I. Dietrich: Superconducting lens design. Ultramicroscopy **10**, 111 (1982)
[2.17] W.D. Riecke: Practical lens design, in [2.5] p. 164
[2.18] J.P. Adriaanse: High T_c superconductors and magnetic electron lenses, in *Advances in Optical and Electron Microscopy*, Vol. 14, ed. by T. Mulvey, C.J.R. Sheppard (Academic, London, 1994) p. 1
[2.19] T. Mulvey, M.J. Wallington: Electron lenses. Rep. Prog. Phys. **36**, 347 (1973)
[2.20] T. Mulvey: Unconventional lens design, in [2.5] p. 359
[2.21] P.A. Sturrock: Perturbation characteristic functions and their application to electron optics. Proc. Roy. Soc. (London) A **210**, 269 (1952)
[2.22] H. Rose: Hamiltonian magnetic optics. Nucl. Instr. Meth. Phys. Res. A **258**, 374 (1987)
[2.23] C.E. Hall: Method of measuring spherical aberration of an electron microscope objective. J. Appl. Phys. **20**, 631 (1949)
[2.24] K. Heinemann: In-situ measurement of objective lens data of a high resolution electron microscope. Optik **34**, 113 (1971)
[2.25] T.F. Budinger, R.M. Glaeser: Measurement of focus and spherical aberration on electron microscope objective lens. Ultramicroscopy **2**, 31 (1976)
[2.26] W. Glaser, H. Grümm: Die Kaustikfläche von Elektronenlinsen. Optik **7**, 96 (1950)
[2.27] S. Leisegang: Zum Astigmatismus von Elektronenlinsen. Optik **10**, 5 (1953)
[2.28] J. Dosse: Über optische Kenngrößen starker Elektronenlinsen. Z. Phys. **117**, 722 (1941)
[2.29] V.E. Cosslett: Energy loss and chromatic aberration in electron microscopy. Z. Angew. Phys. **27**, 138 (1969)
[2.30] L. Reimer, P. Gentsch: Superposition of chromatic error and beam broadening in TEM of thick carbon and organic specimens. Ultramicroscopy **1**, 1 (1975)
[2.31] S. Katagiri: Experimental investigation of chromatic aberration in the electron microscope. Rev. Sci. Instr. **26**, 870 (1955)
[2.32] O. Rang: Der elektronenoptische Stigmator, ein Korrektiv für astigmatische Elektronenlinsen. Optik **5**, 518 (1949)
[2.33] O. Scherzer: Z. Phys. **101**, 593 (1936)
[2.34] O. Scherzer: Sphärische und chromatische Korrektur von Elektronenlinsen. Optik **2**, 114 (1947)
[2.35] A. Septier: Lentille quadrupolaire magnéto-électrique corrigée de l'aberration chromatique. Aberration d'ouverture de ce type de lentilles. C.R. Acad. Sci. Paris **256**, 2325 (1963)
[2.36] H. Rose: Über den sphärischen und den chromatischen Fehler unrunder Elektronenlinsen. Optik **25**, 587 (1967)
[2.37] H. Rose: Elektronenoptische Aplanate. Optik **34**, 285 (1971)
[2.38] H. Koops, G. Kuck, O. Scherzer: Erprobung eines elektrostatischen Achromators. Optik **48**, 225 (1977)

[2.39] H. Rose: Outline of a spherically corrected semiaplanatic medium-voltage TEM. Optik **85**, 19 (1990)
[2.40] M. Haider, G. Braunshausen, E. Schwan: State of development of a C_s corrected high resolution 200 kV TEM, in [Ref. 1.61, Vol. 1, p. 195]
[2.41] C. L. Jia, M. Lentzen, K. Urban: Atomic-resolution imaging of oxygen in perovskite ceramics. Science **299**, 870 (2003)
[2.42] O.L. Krivanek, N. Dellby, A.R. Lupini: Towards sub-Å electron beams. Ultramicroscopy **78**, 1 (1999)
[2.43] H. Rose: Outline of an ultracorrector compensating for all primary chromatic and geometrical aberrations of charged particle lenses. Nucl. Instr. Methods **A 519**, 12 (2004)
[2.44] F. Zemlin, K. Weiss, P. Schiske, W. Kunath, K.H. Hermann: Coma-free alignment of high resolution electron microscopes with the aid of optical diffractograms. Ultramicroscopy **3**, 49 (1978)
[2.45] F. Zemlin: A practical procedure for alignment of a high resolution electron microscope. Ultramicroscopy **4**, 241 (1979)
[2.46] K. Ishizuka: Coma-free alignment of a high resolution electron microscope with three-fold astigmatism. Ultramicroscopy **55**, 407 (1978)
[2.47] D.J. Smith, W.A. Saxton, M.A. O'Keefe, G.J. Wood, W.M. Stobbs: The importance of beam alignment and crystal tilt in high resolution electron microscopy. Ultramicroscopy **11**, 263 (1983)
[2.48] W.A. Saxton, D.J. Smith, S.J. Erasmus: Procedures for focusing, stigmating and alignment in high resolution electron microscopy. J. Microsc. **130**, 187 (1983)
[2.49] A.J. Koster, A. van den Bos, K.D. van der Mast: An autofocus method for a TEM. Ultramicroscopy **21**, 209 (1987)
[2.50] A.J. Koster, W.J. de Ruijter, A. van den Bos, K.D. van der Mast: Autotuning of a TEM using minimum electron dose. Ultramicroscopy **27**, 251 (1989)
[2.51] A.J. Koster, W.J. de Ruijter: Practical autoalignment of transmission electron microscopes. Ultramicroscopy **40**, 89 (1992)

Chapter 3

[3.1] L. Schiff: *Quantum Mechanics* (McGraw-Hill, New York, 1968)
[3.2] L.D. Landau, E. M. Lifshitz *Quantum Mechanics* (Pergamon, Oxford, 1965)
[3.3] A. Messiah: *Quantum Mechanics*, Vol. I (North-Holland, Amsterdam, 1970)
[3.4] Y. Aharonov, D. Bohm: Significance of electromagnetic potentials in the quantum theory. Phys. Rev. **115**, 485 (1959)
[3.5] R.G. Chambers: Shift of an electron interference pattern by enclosed magnetic flux. Phys. Rev. Lett. **5**, 3 (1960)
[3.6] H.A. Fowler, L. Marton, J.A. Simpson, J.A. Suddeth: Electron interferometer studies of iron whiskers. J. Appl. Phys. **32**, 1153 (1961)
[3.7] H. Boersch, H. Hamisch, K. Grohmann, D. Wohlleben: Experimenteller Nachweis der Phasenschiebung von Elektronenwellen durch das magnetische Vektorpotential. Z. Phys. **165**, 79 (1961)
[3.8] W. Bayh: Messung der kontinuierlichen Phasenschiebung von Elektronenwellen im kraftfeldfreien Raum durch das magnetische Vektorpotential einer Wolframwendel. Z. Phys. **169**, 492 (1962)

[3.9] G. Schaal, C. Jönsson, E.F. Krimmel: Weitgetrennte kohärente Elektronen-Wellenzüge und Messung des Magnetflusses $\Phi_0 = h/e$. Optik **24**, 529 (1967)
[3.10] B. Lischke: Bestimmung des Fluxoidquants in supraleitenden Hohlzylindern. Z. Phys. **237**, 469 (1970) and **239**, 360 (1970)
[3.11] N. Osakabe, T. Matsuda, T. Kawasaki, J. Endo, A. Tonomura, S. Yano, H. Yameda: Experimental confirmation of the Aharonov–Bohm effect using a toroidal magnetic field confined by a superconductor. Phys. Rev. **A34**, 815 (1986)
[3.12] C. Jönsson, H. Hoffmann, G. Möllenstedt: Messung des mittleren inneren Potentials von Be im Elektronen-Interferometer. Phys. Kondens. Mater. **3**, 193 (1963)
[3.13] M. Keller: Ein Biprisma-Interferometer für Elektronenwellen und seine Anwendung. Z. Phys. **164**, 274 (1961)
[3.14] R. Buhl: Interferenzmikroskopie mit Elektronenwellen. Z. Phys. **155**, 395 (1959)
[3.15] H. Hoffmann, C. Jönsson: Elektroneninterferometrische Bestimmung der mittleren inneren Potentiale von Al, Cu und Ge unter Verwendung eines neuen Präparationsverfahrens. Z. Phys. **182**, 360 (1965)
[3.16] K.H. Gaukler, R. Schwarzer: Verbessertes Verfahren zur Bestimmung des mittleren inneren Potentials aus Reflexions-Kikuchi-Diagrammen. Optik **33**, 215 (1971)
[3.17] Z.G. Pinsker: *Electron Diffraction* (Butterworths, London, 1953)
[3.18] K. Molière, H. Niehrs: Interferenzbrechung von Elektronenstrahlen. I. Zur Theorie der Elektroneninterferenzen an parallelepipedischen Kristallen. Z. Phys. **137**, 445 (1954)
[3.19] H.J. Altenheim, K. Molière: Interferenzbrechung von Elektronenstrahlen. II. Die Feinstruktur der Interferenzen von Magnesiumoxid-Kristallen. Z. Phys. **139**, 103 (1954)
[3.20] D.K. Saldin, J.C.H. Spence: On the mean inner potential in high- and low-energy electron diffraction. Ultramicroscopy **55**, 397 (1994)
[3.21] G. Möllenstedt, H. Düker: Beobachtungen und Messungen an Biprisma-Interferenzen mit Elektronenwellen. Z. Phys. **145**, 377 (1956)
[3.22] H. Düker: Lichtstarke Interferenzen mit einem Biprisma für Elektronenwellen. Z. Naturforsch. A **10**, 256 (1955)
[3.23] G. Möllenstedt, G. Wohland: Direct interferometric measurement of the coherence length of an electron wave packet using a Wien filter, in *Electron Microscopy, 1980*, Vol. 1, ed. by P. Brederoo, G. Boom (Seventh European Congress on Electron Microscopy Foundation, Leiden, 1980) p. 28
[3.24] P.W. Hawkes: Coherence in electron optics, in *Advances in Optical and Electron Microscopy*, Vol. 7, ed. by R. Barer, V.E. Cosslett (Academic, London, 1978) p. 101
[3.25] H.A. Ferweda: Coherence of illumination in electron microscopy, in *Imaging Processes and Coherence in Physics*, ed. by M. Schlenker, M. Fink, J.P. Goedgebuer, C. Malgrange, J.C. Vienot, R.H. Wade, Lecture Notes in Physics, Vol. 112 (Springer, Berlin, 1980) p. 85
[3.26] H. Boersch: Fresnelsche Beugung im Elektronenmikroskop. Phys. Z. **44**, 202 (1943)
[3.27] K.J. Hanszen, B. Morgenstern: Fresnelsche Beugungssäume im elektronenmikroskopischen Bild einer Objektkante und Kontrastübertragungsdaten des Mikroskopobjektivs. Optik **24**, 442 (1967)

[3.28] L. Reimer, H. Rüberg: Der Einfluß von Objektparametern auf die Intensitätsverteilung in elektronenoptischen Fresnelsäumen an Kanten. Optik **40**, 29 (1974)
[3.29] A.R. Wilson, L.A. Bursill, A.E.C. Spargo: Fresnel diffraction effects on high resolution images: Effect of spherical aberration on the Fresnel fringe. Optik **52**, 313 (1979)
[3.30] J.N. Ness, W.M. Stobbs, T.F. Page: A TEM Fresnel-diffraction-based method for characterizing thin grain-boundary and interfacial films. Philos. Mag. A **54**, 679 (1986)
[3.31] P.W. Hawkes: Note on the sign of the wave aberration in electron optics. Optik **46**, 357 (1976) and **48**, 253 (1977)
[3.32] M.E.C. MacLachlan: The sign of the wave aberration in electron optics. Optik **47**, 363 (1977)
[3.33] O. Scherzer: The theoretical resolution limit of the electron microscope. J. Appl. Phys. **20**, 20 (1949)
[3.34] K.J. Hanszen: Generalisierte Angaben über die Phasenkontrast- und Amplitudenkontrast-Übertragungsfunktion für elektronenmikroskopische Objektive. Z. Angew. Phys. **20**, 427 (1966)
[3.35] K.J. Hanszen: The optical transfer theory of the electron microscope: Fundamental principles and applications, in *Advances in Optical and Electron Microscopy*, Vol. 4, ed. by R. Barer, V.E. Cosslett (Academic, London, 1971) p. 1

Chapter 4

[4.1] H. Ibach, H. Lüth: *Solid-State Physics* (Springer, Berlin, 1995)
[4.2] M.E. Haine, P.A. Einstein, P.H. Borcherds: Resistance bias characteristics for the electron microscope gun. Brit. J. Appl. Phys. **9**, 482 (1958)
[4.3] M. Takeguchi, C. Hanqing, Y. Kimura, T. Ando, R. Shimizu: Development of Zr-O/W(100) thermal field emission TEM. Optik **92**, 83 (1992)
[4.4] L.W. Swanson, L.C. Crouser: Total-energy distribution of field-emitted electrons and single-plane work functions for tungsten. Phys. Rev. **163**, 622 (1967)
[4.5] H. Boersch: Experimentelle Bestimmung der Energieverteilung in thermisch ausgelösten Elektronenstrahlen. Z. Phys. **139**, 115 (1954)
[4.6] K.H. Loeffler: Energy-spread generation in electron-optical instruments. Z. Angew. Phys. **27**, 145 (1969)
[4.7] R.W. Ditchfield, M.J. Whelan: Energy broadening of the electron beam in the electron microscope. Optik **48**, 163 (1977)
[4.8] H. Rose, R. Spehr: On the theory of the Boersch effect. Optik **57**, 339 (1980)
[4.9] H. Rose, R. Spehr: Energy broadening in high-density electron and ion beams: The Boersch effect, in *Applied Charged Particle Optics*, ed. by A. Septier (Academic, New York, 1983) Pt.C, p. 479
[4.10] M. Troyon, P. Zinzindohoué: Energy spread of different thermionic electron sources, in [Ref. 1.59, Vol. 1, p. 273]
[4.11] P. Zinzindohoué, M. Troyon: Energy spread of different field emission electron beams, in [Ref. 1.59, Vol. 1, p. 271]
[4.12] D.B. Langmuir: Theoretical limitations of cathode-ray tubes. Proc. IRE **25**, 977 (1937)

[4.13] J. Dosse: Theoretische und experimentelle Untersuchungen über Elektronenstrahler. Z. Phys. **115**, 530 (1940)

[4.14] A.N. Broers: Electron gun using long-life LaB_6 cathode. J. Appl. Phys. **38**, 1991 (1967)

[4.15] A.N. Broers: Some experimental and estimated characteristics of the LaB_6 rod cathode electron gun. J. Phys. E **2**, 273 (1969)

[4.16] S.D. Ferris, D.C. Joy, H.J. Leamy, C.K. Crawford: A directly heated LaB_6 electron source, in *Scanning Electron Microscopy 1975*, ed. by O. Johari (IIT Research Institute, Chicago, 1976) p. 11

[4.17] S. Nakagawa, T. Yanaka: A high stable electron probe obtained with LaB_6 cathode electron gun, in *Scanning Electron Microscopy 1975*, ed. by O. Johari (IIT Research Institute, Chicago, 1976) p. 19

[4.18] C.K. Crawford: Mounting methods and operating characteristics for LaB_6 cathodes, in *Scanning Electron Microscopy 1979 I*, ed. by O. Johari (SEM/AMF O'Hare, Chicago, 1979) p. 19

[4.19] P.H. Schmidt, D.C. Joy, L.D. Longinotti, H.J. Leamy, S.D. Ferris, Z. Fisk: Anisotropy of thermionic electron emission values of LaB_6 single-crystal emitter cathodes. Appl. Phys. Lett. **29**, 400 (1976)

[4.20] M.E. Haine, P.A. Einstein: Characteristics of the hot cathode electron microscope gun. Br. J. Appl. Phys. **3**, 40 (1952)

[4.21] R. Lauer: Characteristics of triode electron guns, in *Advances in Optical and Electron Microscopy*, Vol. 8, ed. by R. Barer, V.E. Cosslett (Academic, London, 1982) p. 137

[4.22] D.W. Tuggle, J.Z. Li, L.W. Swanson: Point cathodes for use in virtual source electron optics. J. Microsc. **140**, 293 (1985)

[4.23] D.W. Tuggle, L.W. Swanson: Emission characteristics of the ZrO/W thermal field electron source. J. Vac. Sci. Techn. B **3**, 220 (1985)

[4.24] E. Kasper: Field electron emission systems, in *Advances in Optical and Electron Microscopy*, Vol. 8, ed. by R. Barer, V.E. Cosslett (Academic, London, 1982) p. 207

[4.25] A.V. Crewe, D.N. Eggenberger, J. Wall, L.M. Welter: Electron gun using a field-emission source. Rev. Sci. Instr. **39**, 576 (1968)

[4.26] E. Munro: Design of electrostatic lenses for field-emission guns, in *Electron Microscopy 1972* ed. by V.E. Cosslett (IoP, London, 1972) p. 22

[4.27] D. Kern. D. Kurz, R. Speidel: Elektronenoptische Eigenschaften eines Strahlerzeugungssystemes mit Feldemissionskathode. Optik **52**, 61 (1978)

[4.28] G.H.N. Riddle: Electrostatic einzel lenses with reduced spherical aberration for use in field-emisssion gun. J. Vac. Sci. Technol. **15**, 857 (1978)

[4.29] J. Orloff, L.W. Swanson: An asymmetric lens for field-emission microprobe applications. J. Appl. Phys. **50**, 2494 (1979)

[4.30] F.H. Plomp, L. Veneklasen, B. Siegel: Development of a field emission electron source for an electron microscope, in *Electron Microscopy 1968*, Vol. 1., ed. by D.S. Bocciarelli (Tipografia Poliglotta Vaticana, Rome, 1968) p. 141

[4.31] L.H. Veneklasen, B.M. Siegel: A field emission illuminating system for transmission microscopy, in [Ref. 1.55, Vol. 2, p. 87]

[4.32] T. Someya, T. Goto, Y. Harada, M. Watanabe: Development of field emission electron gun for high resolution 100 kV electron microscope, in *Electron Microscopy 1972* ed. by V.E. Cosslett (IoP, London, 1972) p. 20

[4.33] W. Engel, W. Kunath, S. Krause: Properties of three electrode accelerating lenses for field emission guns, in [Ref. 1.56, Vol. 1, p. 118]

[4.34] J.R.A. Cleaver: Field emission electron gun system incorporating single-pole magnetic lenses. Optik **52**, 293 (1979)

[4.35] M. Troyon: A magnetic field emission electron probe forming system, in *Electron Microscopy 1980*, Vol. 1, ed. by P. Brederoo, G. Boom (Seventh European Congress on Electron Microscopy Foundation, Leiden, 1980) p. 56

[4.36] W.D. Riecke: Zur Zentrierung des magnetischen Elektronenmikroskopes. Optik **24**, 397 (1966)

[4.37] W.D. Riecke: Instrument operation for microscopy and microdiffraction, in [Ref. 1.27, Pt.1, p. 19]

[4.38] V.E. Cosslett: Probe size and probe current in the STEM. Optik **36**, 85 (1972)

[4.39] L.H. Veneklasen: Some general considerations concerning the optics of the field emission illumination system. Optik **36**, 410 (1972)

[4.40] J.R.A. Cleaver, K.C.A. Smith: Two-lens probe forming systems employing field emission guns, in *Scanning Electron Microscopy 1973*, ed. by O. Johari (ITT Research Institute, Chicago, 1973) p. 49

[4.41] G. Benner, W. Probst: Köhler illumination in the TEM: Fundamentals and advantages. J. Microsc. **174**, 133 (1994)

[4.42] M. Müller, Th. Koller: Preparation of aluminium oxide films for high resolution electron microscopy. Optik **35**, 287 (1972)

[4.43] D. Dorignac, M.E.C. MacLachlan, B. Jouffrey: Low-noise boron supports for high resolution electron microscopy. Ultramicroscopy **4**, 85 (1979)

[4.44] S. Iijima: Thin graphite supporting films for high resolution electron microscopy. Micron **8**, 41 (1977)

[4.45] W. Baumeister, M.H. Hahn: Suppression of lattice periods in vermiculite single crystal supports for high resolution electron microscopy. J. Microsc. **101**, 111 (1974)

[4.46] U. Valdre, M.J. Goringe: *Electron Microscopy in Materials Science* (Academic, New York, 1971) p. 207

[4.47] U. Valdre: General considerations on specimen stages, in *Electron Microscopy 1972* ed. by V.E. Cosslett (IoP, London, 1972) p. 317

[4.48] J.A. Venables: In-situ experiments in electron microscopes, in *Electron Microscopy 1972* ed. by V.E. Cosslett (IoP, London, 1972) p. 344

[4.49] U. Messerschmidt, M. Bartsch: High-temperature straining stage for in situ experiments in the high-voltage electron microscope. Ultramicroscopy **56**, 163 (1994)

[4.50] H.G. Heide: Principle of a TEM specimen device to meet highest requirements: Specimen temperature 5–300 K, cryo transfer, condensation protection, specimen tilt, stage stability for highest resolution. Ultramicroscopy **6**, 115 (1981)

[4.51] H.G. Heide: Design and operation of cold stages. Ultramicroscopy **10**, 125 (1982)

[4.52] J.E. Eades: A helium-cooled specimen stage for electron microscopy. J. Phys. E **15**, 184 (1982)

[4.53] D.F. Parsons, V.R. Matricardi, J. Subjeck, I. Uydess, G. Wray: High-voltage electron microscopy of wet whole cancer and normal cells: Visualization of cytoplasmatic structure and surface projections. Biochim. Biophys. Acta **290**, 110 (1972)

[4.54] J. Stabenow: Herstellung dünnwandiger Objektivaperturblenden für die Elektronenmikroskopie. Naturwissenschaften **54**, 163 (1967)

[4.55] J. Kala, J. Podbrdsky: Thin foil apertures with very small openings for electron microscopy. J. Phys. E **4**, 609 (1971)

[4.56] E. Schabtach: A method for the fabrication of thin foil apertures for electron microscopy. J. Microsc. **101**, 121 (1974)

[4.57] C.F. Oster, D.C. Skillman: Determination and control of electron microscopic magnification, in [Ref. 1.53, p. EE-3]

[4.58] G.F. Bahr, E. Zeitler: The determination of magnification in the electron microscope. Lab. Invest. **14**, 880 (1965)

[4.59] P.F. Elbers, J. Pieters: Accurate magnification determination in the Siemens Elmiskop I, in *Electron Microscopy 1964*, Vol. A, ed. by M. Titlbach (Czechoslovak Academy of Sciences, Prague, 1964) p. 123

[4.60] P.J. Wilbrandt: A simple concept for better alignment and simplified operation of a TEM. Ultramicroscopy **52**, 193 (1993)

[4.61] J. Porstendörfer, J. Heyder: Elektronenmikroskopische Untersuchungen an Latex-Teilchen. Optik **35**, 73 (1972)

[4.62] W.C.T. Dowell: Die Bestimmung der Vergrößerung des Elektronenmikroskops mittels Elektroneninterferenz. Optik **21**, 26 (1964)

[4.63] R. Luftig: An accurate measurement of the catalase crystal period and its use as an internal marker for electron microscopy. J. Ultrastruct. Res. **20**, 91 (1967)

[4.64] N.G. Wrigley: The lattice spacing of crystalline catalase as an internal standard of length in electron microscopy. J. Ultrastruct. Res. **24**, 454 (1968)

[4.65] J. McCaffrey, J.M. Baribeau: TEM calibration sample for all magnification, camera constant, and image-diffraction pattern rotation calibrations, in [Ref. 1.61, Vol. 1, p. 265]

[4.66] J.B. LePoole, P. Stam: An objective method for focusing, in [Ref. 1.51, p. 666]

[4.67] H. Koike, K. Ueno, M. Suzuki: Scanning device combined with conventional electron microscope, in *Proceedings of the 29th Annual Meeting of EMSA* (Claitor, Baton Rouge, LA, 1971) p. 28

[4.68] L. Reimer, P. Hagemann: The use of transmitted and backscattered electrons in the scanning mode of a TEM, in *Developments in Electron Microscopy and Analysis*, ed. by D.L. Misell (IoP, London, 1977) p. 135

[4.69] S.J. Pennycook, L.M. Brown, A.J. Craven: Observation of cathodoluminescence at single dislocations by STEM. Philos. Mag. A **41**, 589 (1980)

[4.70] N. Yamamoto, J.C.H. Spence, D. Fathy: Cathodoluminescence and polarization studies from individual dislocations in diamonds. Philos. Mag. B **49**, 609 (1984)

[4.71] S.J. Pennycook, A. Howie: Study of single electron excitations by electron microscopy. Philos. Mag. A **41**, 809 (1980)

[4.72] P.M. Petroff, D.V. Lang, J.L. Strudel, R.A. Logan: STEM techniques for simultaneous electronic analysis and observation of defects in semiconductors, in *Scanning Electron Microscopy 1978/I*, ed. by O. Johari (SEM/AMF O'Hare, Chicago, 1978) p. 325

[4.73] M.J. Leamy: Charge collection scanning electron microscopy. J. Appl. Phys. **53**, R51 (1982)

[4.74] H. Blumtritt, R. Gleichmann, J. Heydenreich, J. Johansen: Combined scanning (EBIC) and transmission electron microscopic investigations of dislocations in semiconductors. Phys. Status Solidi A **55**, 1517 (1977)
[4.75] T.G. Sparrow, U. Valdre: Application of STEM to semiconductor devices. Philos. Mag. **36**, 1517 (1977)
[4.76] P.M. Petroff, D.V. Lang: A new spectroscopic technique for imaging the spatial distribution of nonradiative defects in a STEM. Appl. Phys. Lett. **31**, 60 (1977)
[4.77] A.V. Crewe, J. Wall, L.M. Welter: A high-resolution STEM. J. Appl. Phys. **39**, 5861 (1968)
[4.78] A.V. Crewe, J. Wall: Contrast in a high-resolution STEM. Optik **30**, 461 (1970)
[4.79] A.V. Crewe, M. Isaacson, D. Johnson: A high-resolution electron spectrometer for use in transmission scanning electron microscopy. Rev. Sci. Instr. **42**, 411 (1971)
[4.80] A.V. Crewe: Production of electron probes using a field emission source, in *Progress in Optics*, Vol. 11 (North-Holland, Amsterdam, 1973) p. 225
[4.81] J.M. Cowley: Image contrast in a transmission scanning electron microscope. Appl. Phys. Lett. **15**, 58 (1969)
[4.82] E. Zeitler, M.G.R. Thomson: Scanning transmission electron microscopy. Optik **31**, 258 and Optik **31**, 359 (1970)
[4.83] C. Colliex, A.J. Craven, C.J. Wilson: Fresnel fringes in STEM. Ultramicroscopy **2**, 327 (1977)
[4.84] D.C. Joy, D.M. Maher, A.G. Cullis: The nature of defocus fringes in STEM. J. Microsc. **108**, 185 (1976)
[4.85] A.V. Crewe, M. Isaacson, D. Johnson: A high resolution electron spectrometer for use in TEM. Rev. Sci. Instr. **42**, 411 (1971)
[4.86] R.F. Egerton: A simple electron spectrometer for energy analysis in TEM. Ultramicroscopy **3**, 39 (1978)
[4.87] R.F. Egerton: Design of an aberration-corrected electron spectrometer for the TEM. Optik **57**, 229 (1980)
[4.88] H. Shuman: Correction of the second-order aberrations of uniform field magnetic sectors. Ultramicroscopy **5**, 45 (1980)
[4.89] R.F. Egerton: The use of electron lenses between a TEM specimen and an electron spectrometer. Optik **56**, 363 (1980)
[4.90] D.E. Johnson: Pre-spectrometer optics in CTEM/TEM. Ultramicroscopy **5**, 163 (1980)
[4.91] A.W. Blackstock, R.D. Birkhoff, M. Slater: Electron accelerator and high resolution analyser. Rev. Sci. Instr. **26**, 274 (1955)
[4.92] J. Lohff: Charakteristische Energieverluste bei der Streuung mittelschneller Elektronen an Aluminium-Oberflächen. Z. Phys. **171**, 442 (1963)
[4.93] A.V. Crewe, J. Wall, L.M. Welter: A high resolution scanning transmission electron microscope. J. Appl. Phys. **39**, 5861 (1968)
[4.94] W.H.J. Anderson, J.B. LePoole: A double Wien filter as a high resolution, high transmission electron energy analyser. J. Phys. E **3**, 121 (1970)
[4.95] G.H. Curtis, J. Silcox: A Wien filter for use as an energy analyzer with an electron microscope. Rev. Sci. Instr. **42**, 630 (1971)
[4.96] H. Boersch, J. Geiger, W. Stickel: Das Auflösungsvermögen des elektrostatischen-magnetischen Energieanalysators für schnelle Elektronen. Z. Phys. **180**, 415 (1964)

[4.97] P.E. Batson: Prospects for high-resolution EELS experiments with the STEM. Ultramicroscopy **18**, 125 (1985)

[4.98] M. Terauchi, R. Kuzuo, F. Satoh, M. Tanaka, K. Tsuno, J. Ohyama: Performance of a new high-resolution electron energy-loss spectroscopy microscope. Microsc. Microanal. Microstruct. **2**, 351 (1991)

[4.99] M. Terauchi, M. Tanaka, K. Tsuno, M. Ishida: Development of a high resolution electron energy-loss spectroscopy microscope. J. Microscopy **194**, 203 (1999)

[4.100] P.C. Tiemeijer: Operation modes of a TEM monochromator. Inst. Phys. Conf. Ser. **161**, 191 (1999)

[4.101] H.W. Mook, P. Kruit: Construction and characterization of the fringe field monochromator for a field emission gun. Ultramicroscopy **81**, 129 (2000)

[4.102] S. Uhlemann, M. Haider: Experimental set-up of a purely electrostatic monochromator for high resolution and analytical purposes of a 200 kV TEM. Microsc. Microanal. **8**, (Suppl. 2), 584 (2002)

[4.103] O.L. Krivanek, A.J. Gubbens, N. Dellby: Developments in EELS instrumentation for spectroscopy and imaging. Microsc. Microanal. Microstruct. **2**, 315 (1991)

[4.104] O.L. Krivanek, S.L. Friedman, A.J. Gubbens, B. Kraus: An imaging filter for biological applications. Ultramicroscopy **59**, 267 (1995)

[4.105] H.A. Brink, M.M.G. Barfels, R.P. Burgner, B.N. Edwards: A sub-50meV spectrometer and energy filter for use in combination with 200 kV monochromated (S)TEMs. Ultramicroscopy **96**, 367 (2003)

[4.106] A.J. Gubbens, B. Kraus, O.L. Krivanek, P.E. Mooney: An imaging filter for high voltage electron microscopy. Ultramicroscopy **59**, 255 (1995)

[4.107] R. Castaing, L. Henry: Filtrage magnétique des vitesses en microscope électronique. C.R. Acad. Sci. Paris **225**, 76 (1962)

[4.108] R. Castaing: Quelques application du filtrage magnétique des vitesses en microscopie électronique. Z. Angew. Phys. **27**, 171 (1969)

[4.109] R.M. Henkelman, F.P. Ottensmeyer: An energy filter for biological electron microscopy. J. Microsc. **102**, 79 (1979)

[4.110] W. Egle, A. Rilk, F.P. Ottensmeyer: A new analytical TEM with imaging electron energy loss spectrometer, in *Electron Microscopy 1984*, Vol. I, ed. by A. Csanády et al., (Motesz, Budapest, 1984) p. 63

[4.111] L. Reimer, I. Fromm, R. Rennekamp: Operation modes of electron imaging and electron energy-loss spectroscopy in a TEM. Ultramicroscopy **24**, 339 (1988)

[4.112] L. Reimer: Energy-filtering transmission electron microscopy. Adv. Electron. Electron Phys. **81**, 43 (1991)

[4.113] G. Zanchi, J.Ph. Pérez, J. Sevely: Adaptation of a magnetic filtering device in a one megavolt electron microscope. Optik **43**, 495 (1975)

[4.114] J.Ph. Pérez, J. Sirvin, A. Séguéla, J.C. Lacaze: Étude, au premier ordre, d'une sytéme dispersif, magnétique symmétric, de type alpha. J. Physique **45**, C 2, Suppl. 2, 171 (1984)

[4.115] H. Rose, E. Plies: Entwurf eines fehlerarmen magnetischen Energie-Analysators. Optik **40**, 336 (1974)

[4.116] S. Lanio: High-resolution imaging magnetic energy filter with simple structure. Optik **73**, 99 (1986)

[4.117] H. Rose, W. Pejas: Optimisation of imaging magnetic energy filters free of second-order aberration. Optik **54**, 235 (1979)

[4.118] A. Lanio, H. Rose, D. Krahl: Test and and improved design of a corrected imaging magnetic filter. Optik **73**, 56 (1986)

[4.119] D. Krahl, H. Rose: Electron optics of imaging energy filters, in *Energy-Filtering Transmission Electron Microscopy*, ed. by L. Reimer, Springer Series in Optical Sciences Vol. 71 (Springer, Berlin, 1995) p. 43

[4.120] S. Uhlemann, H. Rose: The MANDOLINE-filter – a new high-performance imaging filter for sub-eV EFTEM. Optik **96**, 163 (1994)

[4.121] E. Essers, B. Huber: Experimental Setup and Verification of the MANDOLINE Filter. Microsc. Microanal. **9** (Suppl. 3) 20 (2003)

[4.122] L. Reimer: in Ref. [1.84] p. 347

[4.123] C. Colliex, M. Tencé, E. Lefèvre, C. Mory, H. Gu, D. Bouchet, C. Jeanguillaume: Electron energy loss spectrometry mapping. Microchim. Acta **114**, 71 (1994)

[4.124] L. Reimer: Electron spectroscopic imaging, in [Ref. 1.74, p. 347]

[4.125] V.E. Cosslett, G.L. Jones, R.A. Camps: Image viewing and recording in high voltage electron microscopy, in [Ref 1.78 p. 147]

[4.126] H.G. Heide: Zur Vorevakuierung von Photomaterial für Elektronenmikroskopie. Z. Angew. Phys. **19**, 348 (1965)

[4.127] H. Frieser, H. Klein: Die Eigenschaften photographischer Schichten bei Elektronenbestrahlung. Z. Angew. Phys. **10**, 337 (1958)

[4.128] H. Frieser, H. Klein, E. Zeitler: Das Verhalten photographischer Schichten bei Elektronenbestrahlung. Z. Angew. Phys. **11**, 190 (1959)

[4.129] R.C. Valentine: The response of photographic emulsions to electrons, in *Advances in Optical and Electron Microscopy*, Vol. 1, ed. by R. Barer, V.E. Cosslett (Academic, London, 1966) p. 180

[4.130] R.E. Burge, D.F. Garrard: The resolution of photographic emulsions for electrons in the energy range 7–60 keV. J. Phys. E **1**, 715 (1968)

[4.131] R.E. Burge, D.F. Garrard, M.T. Browne: The response of photographic emulsions to electrons in the energy range 7–60 keV. J. Phys. E **1**, 707 (1968)

[4.132] G.C. Farnell, R.B. Flint: The response of photographic materials to electrons with particular reference to electron micrography. J. Microsc. **97**, 271 (1973)

[4.133] W. Lippert: Erfahrungen mit der photographischen Methode bei der Massendickebestimmung im Elektronenmikroskop. Optik **29**, 372 (1969)

[4.134] N. Mori, T. Oikawa, T. Katoh, J. Miyahara, Y. Harada: Application of the "imaging plate" to TEM image recording. Ultramicroscopy **25**, 195 (1988)

[4.135] D. Shindo, K. Hiraga, T. Oku: Quantification in high-resolution electron microscopy with the imaging plate. Ultramicroscopy **39**, 50 (1991)

[4.136] N. Ogura, K. Yoshida, Y. Kojima, H. Saito: Development of the 25 micron pixel imaging plate system for TEM, in [Ref. 1.61, Vol. 1, p. 219]

[4.137] K.H. Herrmann, D. Krahl: Electronic image recording in conventional electron microscopy, in *Advances in Optical and Electron Microscopy*, Vol. 8, ed. by R. Barer, V.E. Cosslett (Academic, London, 1984) p. 1

[4.138] Z. Tang, R. Ho, Z. Xu, Z. Shao, A.P. Somlyo: A high-sensitivity CCD system for parallel EELS. J. Microsc. **175**, 100 (1994)

[4.139] H. Shuman: Parallel recording of electron energy-loss spectra. Ultramicroscopy **6**, 163 (1981)

[4.140] P.T.E. Roberts, J.N. Chapman, A.M. MacLeod: A CCD-based image recording system for the CTEM. Ultramicroscopy **8**, 385 (1982)

[4.141] R.F. Egerton: Parallel-recording systems for electron energy-loss spectroscopy. J. Electron Microsc. Technol. **1**, 37 (1984)
[4.142] I. Daberkow, K.H. Herrmann, L. Liu, W.D. Rau: Performance of electron image converters with YAG single crystal screens and CCD sensors. Ultramicroscopy **38**, 215 (1991)
[4.143] S. Kujawa, D. Krahl: Performance of a low-noise CCD camera adapted to a TEM. Ultramicroscopy **46**, 395 (1992)
[4.144] O.L. Krivanek, P.E. Mooney: Applications of slow-scan CCD cameras in TEM. Ultramicroscopy **49**, 95 (1993)
[4.145] K. H. Herrmann, L. Liu: Performance of image converters using slow-scan CCDs in MeV electron microscopy. Optik **92**, 48 (1992)
[4.146] G.Y. Fan, M.H. Ellisman: High-sensitivity lens-coupled slow-scan CCD camera for TEM. Ultramicroscopy **52**, 21 (1993)
[4.147] D.A. Gedcke, J.B. Ayers, P.B. DeNee: A solid state backscattered electron detector capable of operating at TV scan rates, in *Scanning Electron Microscopy 1978/I*, ed. by O. Johari (SEM/AMF O'Hare, Chicago, 1978) p. 581
[4.148] M. Kikuchi, S. Takashima: Multi-purpose backscattered electron detector, in [Ref. 1.57, Vol. 1, p. 82]
[4.149] J. Pawley: Performance of STEM scintillation materials, in *Scanning Electron Microscopy 1974*, ed. by O. Johari (IIT Research Institute, Chicago, 1974) p. 28
[4.150] W. Baumann, A. Niemitz, L. Reimer, B. Volbert: Preparation of P-47 scintillators for STEM. J. Microsc. **122**, 181 (1981)
[4.151] R. Autrata, P. Walther, S. Kriz, M. Müller: A BSE scintillation detector in the STEM. Scanning **8**, 3 (1986)

Chapter 5

[5.1] S. Gasiorowicz: *Quantum Physics* (Wiley, New York, 1974)
[5.2] G. Molière: Theorie der Streuung schneller geladener Teilchen. Z. Naturforsch. A **2**, 133 (1947)
[5.3] R.J. Glauber: High-energy collision theory, in *Lectures in Theoretical Physics*, Vol. 1, ed. by W.E. Brittin, G. Dunham (Interscience, New York, 1959) p. 315
[5.4] W.J. Byatt: Analytical representation of Hartree potentials and electron scattering. Phys. Rev. **104**, 1298 (1956)
[5.5] T. Tietz: Über den Mottschen Polarisationseffekt bei der Streuung mittelschneller Elektronen. Nuovo Cimento **36**, 1365 (1965)
[5.6] H.L. Cox, R.A. Bonham: Elastic electron scattering amplitudes for neutral atoms calculated using the partial wave method at 10, 40, 70 and 100 kV for $Z = 1$ to $Z = 54$. J. Chem. Phys. **47**, 2599 (1967)
[5.7] H. Raith: Komplexe Atomstreuamplituden für die elastische Elektronenstreuung an Festkörperatomen. Acta Cryst. A **24**, 85 (1968)
[5.8] L. Reimer, K.H. Sommer: Messungen und Berechnungen zum elektronenmikroskopischen Streukontrast für 17 bis 1200 keV-Elektronen. Z. Naturforsch. A **23**, 1569 (1968)
[5.9] E. Zeitler, H. Olsen: Screening effects in elastic electron scattering. Phys. Rev. A **136**, 1546 (1964); Complex scattering amplitudes in elastic electron scattering. Phys. Rev. A **162**, 1439 (1967)

[5.10] J. Haase: Berechnung der komplexen Streufaktoren für schnelle Elektronen unter Verwendung von Hartree-Fock-Atompotentialen. Z. Naturforsch. A **23**, 1000 (1968)

[5.11] F. Lenz: Zur Streuung mittelschneller Elektronen in kleinste Winkel. Z. Naturforsch. A **9**, 185 (1954)

[5.12] F. Arnal, J.L. Balladore, G. Soum, P. Verdier: Calculations of the cross sections of electron interaction with matter. Ultramicroscopy **2**, 305 (1977)

[5.13] R.E. Burge, G.H. Smith: A new calculation of electron scattering cross sections and a theoretical discussion of image contrast in the electron microscope. Proc. Phys. Soc. **79**, 673 (1962)

[5.14] J.A. Ibers, B.K. Vainshtein: Scattering amplitudes for electrons, in *International Tables for X-Ray Crystallography*, Vol. 3, ed. by K. Lonsdale (Kynoch, Birmingham, 1962)

[5.15] P.A. Doyle, P.S. Turner: Relativistic Hartree-Fock x-ray and electron scattering factors. Acta Cryst. A **24**, 390 (1968)

[5.16] P.A. Doyle, J.M. Cowley: in *International Tables for X-Ray Crystallography*, Vol. 4, ed. by K. Lonsdale (Kynoch, Birmingham, 1974)

[5.17] J. Geiger: Zur Streuung von Elektronen am Einzelatom, in [Ref. 1.53, p. AA-12]

[5.18] F. Salvat, R. Mayol: Elastic scattering of electrons and positrons by atoms. Schrödinger and Pauli partial wave analysis. Comp. Phys. Commun. **74**, 358 (1993)

[5.19] J.P. Langmore, J. Wall, M. Isaacson: The collection of scattered electrons in dark field electron microscopy. I. Elastic scattering. Optik **38**, 335 (1973); II. Inelastic scattering. Optik **39**, 359 (1974)

[5.20] J. Geiger, K. Wittmaack: Wirkungsquerschnitte für die Anregung von Molekülschwingungen durch schnelle Elektronen. Z. Phys. **187**, 433 (1965)

[5.21] H. Boersch, J. Geiger, A. Bohg: Wechselwirkung von Elektronen mit Gitterschwingungen in NH_4Cl und NH_4Br. Z. Phys. **227**, 141 (1969)

[5.22] B. Schröder, J. Geiger: Electron-spectrometric study of amorphous Ge and Si in the two-phonon region. Phys. Rev. Lett. **28**, 301 (1972)

[5.23] H. Raether: *Solid State Excitations by Electrons*, Springer Tracts in Modern Physics, Vol. 38 (Springer, Berlin, 1965) p. 84

[5.24] R.D. Leapman, V.E. Cosslett: Energy loss spectrometry of inner shell excitations, in *Electron Microscopy 1976*, Vol. 1, ed. by D.G. Brandon (Tal International, Jerusalem, 1976) p. 431

[5.25] M. Isaacson: Interaction of 25 keV electrons with the nucleic acid bases adenine, thymine and uracil. J. Chem. Phys. **56**, 1803 (1972)

[5.26] H. Bethe: Zur Theorie des Durchganges schneller Korpuskularstrahlen durch Materie. Ann. Phys. **5**, 325 (1930)

[5.27] M. Inokuti: Inelastic collisions of fast charged particles with atoms and molecules – The Bethe theory revisited. Rev. Mod. Phys. **43**, 297 (1971)

[5.28] B.G. Williams, T.G. Sparrow, R.F. Egerton: Electron Compton scattering from solids. Proc. Roy. Soc. London A **393**, 409 (1984)

[5.29] L. Reimer, R. Rennekamp: Imaging and recording of multiple scattering effects by angular-resolved EELS. Ultramicroscopy **28**, 258 (1989)

[5.30] B.G. Williams: *Compton Scattering: The Investigation of Electron Momentum Distribution* (McGraw-Hill, London 1977)

[5.31] M.J. Cooper: Compton scattering and electron momentum distribution. Rep. Progr. Phys. **48**, 415 (1985)

[5.32] P. Schattschneider, P. Jonas, M. Mändl: Electron Compton scattering on solids – a feasibility experiment on a PEELS system. Microsc. Microanal. Microstruct. **2**, 367 (1991)
[5.33] P. Jonas, P. Schattschneider: The experimental conditions for Compton scattering in the electron microscope. Phys. Cond. Matter **5**, 7173 (1993)
[5.34] P. Schattschneider, A. Exner: Progress in electron Compton scattering. Ultramicroscopy **59**, 241 (1995)
[5.35] H. Koppe: Der Streuquerschnitt von Atomen für unelastische Streuung schneller Elektronen. Z. Phys. **124**, 658 (1948)
[5.36] R.F. Egerton: Measurement of inelastic/elastic scattering ratio for fast electrons and its use in the study of radiation damage. Phys. Status Solidi A **37**, 663 (1976)
[5.37] L. Reimer, M. Ross-Messemer: Contrast in the electron spectroscopic imaging mode of a TEM. II. Z-ratio, structure-sensitive and phase contrast. J. Microsc. **159**, 143 (1990)
[5.38] J. Geiger: *Elektronen und Festkörper* (Vieweg, Braunschweig, 1968)
[5.39] J. Daniels, C. von Festenberg, H. Raether, K. Zeppenfeld: *Optical Constants of Solids by Electron Spectroscopy*, Springer Tracts in Modern Physics, Vol. 54 (Springer, Berlin, 1970)
[5.40] P.M. Platzman, P.A. Wolff: *Waves and Their Interactions in Solids* Solid State Physics Supplement 13 (Academic, New York, 1973)
[5.41] H. Raether: *Excitation of Plasmons and Interband Transitions by Electrons*, Springer Tracts Mod. Phys., Vol. 88 (Springer, Berlin, 1980)
[5.42] P. Schattschneider, B. Jouffrey: Plasmons and related excitations, in *Energy-Filtering Transmission Electron Microscopy*, ed. by L. Reimer, Springer Series in Optical Sciences Vol. 71 (Springer, Berlin, 1995) p. 151
[5.43] L.D. Landau, E.M. Lifschitz: *Lehrbuch der theoretischen Physik*, Vol. 8, Elektrodynamik der Kontinua, (Akademie, Berlin, 1990)
[5.44] H. Boersch, J. Geiger, H. Hellwig, H. Michel: Energieverluste von Elektronen in Metallen in verschiedenen Aggregatzuständen. Messungen an Al und Hg. Z. Phys. **169**, 252 (1962)
[5.45] E. Petri, A. Otte: Direct nonvertical interband and intraband transitions in Al. Phys. Rev. Lett. **34**, 1238 (1975)
[5.46] C.H. Chen, J. Silcox: Direct nonvertical interband transitions of large wavevectors in aluminum. Phys. Rev. B **16**, 1246 (1977)
[5.47] D. Pines: Collective energy losses in solids. Rev. Mod. Phys. **28**, 184 (1956)
[5.48] H. Kohl: Spatially Sensitive Electron Energy Loss Spectroscopy. in: *Fundamental Electron and Ion Beam Interactions with Solids for Microscopy, Microanalysis and Microsltihography* J. Schou. P. Kruit, and D. Newbury (eds.) Scanning Microsc. Suppl. 4 (1989) 17
[5.49] D. Bohm, D. Pines: A collective description of electron interactions. III. Coulomb interactions in a degenerate electron gas. Phys. Rev. **92**, 609 (1953)
[5.50] B. Rafferty, S.J. Pennycook, L.M. Brown: Zero loss peak deconvolution for bandgap EEL spectra. J. Electron Microsc. **49**, 517 (2000)
[5.51] S. Pokrant, M. Cheynet, S. Jullian, R. Pantel: Chemical analysis of nanometric dielectric layers using spatially resolved VEELS. Ultramicroscopy **104**, 233 (2005)
[5.52] W. Sigle, L. Gu, V. Srot, C. Koch, P. van Aken: Low-loss EELS with monochromatized electrons. Microsc. Microanal. **13** Supplement 3, 54 (2007)

[5.53] J. Lindhard: On the properties of gas of charged particles. Danske Vidensk. Selsk. Mat.-Fys. Medd. **28**, 1 (1954)

[5.54] R.W. Ditchfield, A.G. Cullis: Identification of impurity particles in epitaxially grown Si films using combined electron microscopy and energy analysis, in [Ref. 1.55, Vol. 2, p. 125]

[5.55] R.F. Cook: Electron energy loss spectroscopy of glass, in [Ref. 1.55, Vol. 2, p. 127]

[5.56] J. Hainfeld, M. Isaacson: The use of electron energy loss spectroscopy for studying membrane architecture. Ultramicroscopy **3**, 87 (1978)

[5.57] G. Meyer: Über die Abhängigkeit der charakteristischen Energieverluste von Temperatur und Streuwinkel. Z. Phys. **148**, 61 (1957)

[5.58] L.B. Leder, L. Marton: Temperature dependence of the characteristic energy loss of electrons in Al. Phys. Rev. **112**, 341 (1958)

[5.59] D.R. Spalding, A.J.F. Metherell: Plasmon losses in Al-Mg alloys. Philos. Mag. **18**, 41 (1968)

[5.60] S.L. Cundy, A.J.F. Metherell, M.J. Whelan, P.N.T. Unwin, R.B. Nicolson: Studies of segregation and the initial stages of precipitation at grain boundaries in Al-7wt% Mg alloy with an energy analysing microscope. Proc. Roy. Soc. London A **307**, 267 (1968)

[5.61] D.R. Spalding, R.E. Villagrana, G.A. Chadwick: A study of copper distribution in lamellar, Al-$CuAl_2$ eutectics using an energy analysing microscope. Philos. Mag. **20**, 471 (1969)

[5.62] R.F. Cook, S.L. Cundy: Plasmon energy losses in Al-Zn alloys. Philos. Mag. **20**, 665 (1969)

[5.63] G. Hibbert, J.W. Eddington: Experimental errors in combined electron microscopy and energy analysis. J. Phys. D **5**, 1780 (1972)

[5.64] G. Hibbert, J.W. Eddington: Superposition effects in the energy analysing electron microscope. Philos. Mag. **26**, 1071 (1972)

[5.65] R.F. Cook, A. Howie: Effect of elastic constraints on electron energy loss measurements in inhomogeneous alloy. Philos. Mag. **20**, 641 (1969)

[5.66] D.R. Spalding: Electron microscope evidence of plasmon-dislocation interactions. Philos. Mag. **34**, 1073 (1976)

[5.67] I. Fromm, L. Reimer, R. Rennekamp: Investigation and use of plasmon losses in energy-filtering TEM. J. Microsc. **166**, 257 (1992)

[5.68] H. Watanabe: Experimental evidence for the collective nature of the characteristic energy loss of electrons in solids. J. Phys. Soc. Jpn. **11**, 112 (1956)

[5.69] C. Kunz: Die Winkelverteilung der charakteristischen Energieverluste von Elektronen, gemessen am 15 eV-Al-Verlust und am 17 eV-Si-Verlust. Phys. Status Solidi **1**, 441 (1961)

[5.70] C. Kunz: Über die Winkelabhängigkeit der charakteristischen Energieverluste an Al, Si, Ag. Z. Phys. **167**, 53 (1962)

[5.71] H. Boersch, H. Miessner, W. Raith: Untersuchungen zur Winkelabhängigkeit des 14, 7 eV-Energieverlustes von Elektronen in Al. Z. Phys. **168**, 404 (1962)

[5.72] J. Geiger: Winkelverteilung der Energieverluste mittelschneller Elektronen in Antimon. Z. Naturforsch. A **17**, 696 (1962)

[5.73] P. Schmüser: Anregung von Volumen- und Oberflächen-Plasmaschwingungen in Al und Mg durch mittelschnelle Elektronen. Z. Phys. **180**, 105 (1964)

[5.74] T. Kloos: Plasmaschwingungen in Al, Mg, Li, Na und K angeregt durch schnelle Elektronen. Z. Phys. **265**, 225 (1973)

[5.75] S.E. Schnatterly: Inelastic electron scattering spectroscopy, Sol. State Phys. **34**, 275 (1979)
[5.76] M. Creuzburg, H. Raether: On the behaviour of nuclear energy losses of electrons in alkali halides. Solid State Commun. **2**, 175 (1964)
[5.77] K. Zeppenfeld: Anistropie der Plasmaschwingungen in Graphit. Z. Phys. **211**, 391 (1968)
[5.78] C.H. Chen, J. Silcox: Detection of optical surface guided modes in thin graphite films by high-energy electron scattering. Phys. Rev. Lett. **35**, 390 (1975)
[5.79] M. Urner-Wille, H. Raether: Anisotropy of the 15 eV plasmon dispersion in Al. Phys. Lett. A **58**, 265 (1976)
[5.80] R.A. Ferrell: Characteristic energy loss of electrons passing through metal foils. Phys. Rev. **107**, 450 (1968)
[5.81] J. Sevely, J.Ph. Perez, B. Jouffrey: Energy losses of electrons through Al and C films from 300 keV up to 1200 keV, in [Ref. 1.77, p. 32]
[5.82] L. Marton, J.A. Simpson, H.A. Fowler, N. Swanson: Plural scattering of 20 keV electrons in Al. Phys. Rev. **126**, 182 (1962)
[5.83] M. Creuzburg, H. Dimigen: Energieanalyse im Elektroneninterferenzbild von Si-Einkristallen. Z. Phys. **174**, 24 (1963)
[5.84] R.E. Burge, D.L. Misell: Electron energy loss spectra of evaporated carbon films. Philos. Mag. **18**, 251 (1968)
[5.85] R.E. Burge, D.L. Misell: Convolution effects in electron energy-loss spectra recorded by electron transmission. J. Phys. C **2**, 1397 (1969)
[5.86] L. Reimer: Calculation of the angular and energy distribution of multiple scattered electron using Fourier transforms. Ultramicroscopy **31**, 169 (1989)
[5.87] R.H. Ritchie: Plasma losses by fast electrons in thin films. Phys. Rev. **106**, 874 (1957)
[5.88] E. Kröger: Berechnung der Energieverluste schneller Elektronen in dünnen Schichten mit Retardierung. Z. Physik **216**, 115 (1968)
[5.89] E. Kröger: Transition radiation, Cerenkov radiation and energy losses of relativistic charged particles traversing thin foils at oblique incidence. Z. Phys. **235**, 403 (1970)
[5.90] H. Boersch, J. Geiger, A. Imbusch, H. Niedrig: High resolution investigation of the energy losses of 30 keV electrons in Al foils of various thicknesses. Phys. Lett. **22**, 146 (1966)
[5.91] R.B. Pettit, J. Silcox, R. Vincent: Measurement of surface-plasmon dispersion in oxidized Al films. Phys. Rev. B **11**, 1306 (1975)
[5.92] T. Kloos: Zur Dispersion der Oberflächenplasmaverluste an reinen und oxydierten Al-Oberflächen. Z. Phys. **208**, 77 (1968)
[5.93] E.A. Stern, R.A. Ferrell: Surface plasma oscillations of a degenerate electron gas. Phys. Rev. **120**, 130 (1960)
[5.94] M. Creuzburg: Unsymmetrie in der Intensitätsverteilung charakteristischer Energieverluste. Z. Naturforsch. A **18**, 101 (1963); Über die Winkelabhängigkeit und ihre Unsymmetrie von Energieverlusten an Si und Ge. Z. Phys. **174**, 511 (1963)
[5.95] C. von Festenberg, E. Kröger: Retardation effects for the electron energy loss probability in GaP and Si. Phys. Lett. A **26**, 339 (1968)
[5.96] C.H. Chen, J. Silcox, R. Vincent: Electron energy loss in silicon: Bulk and surface plasmons and Čerenkov radiation. Phys. Rev. B **12**, 64 (1975)

[5.97] L.D. Marks: Observation of the image force for fast electrons near a MgO surface. Solid State Commun. **43** 727 (1982)
[5.98] A. Howie, R.H. Milne: Electron energy loss spectrum and reflection images from surfaces. J. Microsc. **136**, 279 (1984)
[5.99] P.E. Batson: Surface plasmon coupling in clusters of small spheres. Phys. Rev. Lett. **49**, 936 (1982)
[5.100] D. Ugarte, C. Colliex, P. Trebbia: Surface-plasmon and Interface-plasmon modes on small semiconducting spheres. Phys. Rev. B **45**, 4332 (1992)
[5.101] J. Nelayah, M. Kociak, O. Stphan, F.J. Garcie de Abajo, M. Tenc, L. Henrard, D. Taverna, I. Pastroiza-Santos, L.M. Liz-Marzan, C. Colliex: Mapping surface plasmons on a single metallic nanoparticle. Nature Physics **31** 348 (2007)
[5.102] C.C. Ahn, O.L. Krivanek: EELS Atlas. Copies from Gatan Inc., 780 Commonwealth Drive, Warrendale, PA 15086 (1983)
[5.103] L. Reimer, U. Zepke, J. Moesch, St. Schulze-Hillert, M. Ross-Messemer, W. Probst, E. Weimer: *EELS Spectroscopy: A Reference Handbook of Standard Data for Identification and Interpretaton of Electron Energy Loss Spectra and for Generation of Electron Spectroscopic Images* (Carl Zeiss, Electron Optics Division, Oberkochen, 1992)
[5.104] C. Colliex: Electron energy loss spectroscopy in the electron microscope, in *Advances in Optical and Electron Microscopy*, Vol. 9, ed. by R. Barer, V.E. Cosslett (Academic, London, 1984) p. 65
[5.105] F. Hofer: Inner-shell ionization, in *Energy-Filtering Transmission Electron Microscopy*, ed. by L. Reimer, Springer Series in Optical Sciences, Vol. 71 (Springer, Berlin, 1995) p. 225
[5.106] N.J. Zaluzec, T. Schober, D.G. Westlake: Application of EELS to the study of metal–hydrogen systems, in *Proceedings of the 39th Annual Meeting of EMSA*, ed. by G.W. Bailey (Claitor, Baton Rouge, LA, 1981) p. 194
[5.107] O.T. Woo, G.J.C. Carpenter: EELS characterization of zirconium hydrides. Microsc. Microanal. Microstruct. **3**, 35 (1992)
[5.108] G.J. Thomas: Study of hydrogen and helium in metals by EELS, in *Analytical Electron Microscopy 1981*, ed. by R.H. Geiss (San Francisco Press, San Francisco, 1981) p. 195
[5.109] R. Manzke, M. Campagna: Study of He bubbles in Al by electron energy loss spectroscopy. Sol. State Commun. **39**, 313 (1981)
[5.110] J. Fink: Recent development in energy-loss spectroscopy. Adv. Electron. Electron Phys. **75**, 121 (1989)
[5.111] D.H. Madison, E. Merzbacher: Theory of charged-particle excitation, in *Atomic Inner-Shell Processes*, Vol. 1, ed. by B. Crasemann (Academic, New York, 1975) p. 1
[5.112] R.F. Egerton: K-shell ionization cross-sections for use in microanalysis. Ultramicroscopy **4**, 169 (1979)
[5.113] W.J. Veigele: Photon cross-sections from 0.1 keV to 1 MeV for elements Z = 1 to Z = 94. Atomic Data Tables **5**, 51 (1973)
[5.114] T.F. Malis, J.M. Titchmarsh: A 'k factor' approach to EELS analysis, in *Electron Microscopy and Analysis 1985*, ed. by G.J. Tatlock (IoP, Bristol, 1985) p. 181
[5.115] B.H. Choi, E. Merzbacher, G.S. Khandelwale: Tables of Born approximation calculations of L-subshell ionization by simple heavy charged particles. Atomic Data Tables **5**, 291 (1973)

[5.116] R.F. Egerton: SIGMAL: A program for calculating L-shell ionization cross-sections, in *Proceedings of the 39th Annual Meeting of EMSA*, ed. by G.W. Bailey (Claitor, Baton Rouge, LA, 1981) p. 198
[5.117] B.P. Luo, E. Zeitler: M-shell cross-sections for fast electron inelastic collisions based on photoabsorption data. J. Electron Spectrosc. Relat. Phenom. **57**, 285 (1991)
[5.118] R.D. Leapman, P. Rez, D.F. Mayers: K, L and M shell generalized oscillator strengths and ionization cross-sections for fast electron collisions. J. Chem. Phys. **72**, 1232 (1980)
[5.119] R. Knippelmeyer, P. Wahlbring, H. Kohl: Relativistic cross-sections for use in microanalysis. Ultramicroscopy **68**, 25 (1996)
[5.120] P. Rez: Cross sections for energy-loss spectrometry. Ultramicroscopy **9**, 283 (1982)
[5.121] D. Rez, P. Rez: The contribution of discrete transitions to integrated inner shell ionization cross-sections. Microsc. Microanal. Microstruct. **3**, 433 (1992)
[5.122] F. Hofer, P. Golob: Quantification of EELS with K and L shell ionization cross-sections. Micron and Microscopica Acta **19**, 73 (1988)
[5.123] F. Hofer, P. Golob, A. Brunegger: EELS quantification of the elements Sr to W by means of M_{45} edges. Ultramicroscopy **25**, 81 (1988)
[5.124] F. Hofer: EELS quantification of the elements Ba to Tm by means of N_{45} edges. J. Microsc. **156**, 279 (1989)
[5.125] P. Wilhelm, F. Hofer: EELS microanalysis of the elements Ca to Cu using M_{23} edges, in Electron Microscopy 1992, Vol. 1, ed. by A. Rios et al. (University of Granada, Granada, 1992) p. 281
[5.126] M. Grande, C.C. Ahn: Deconvolution and quantification of energy loss transition metal oxide spectra, Inst. Phys. Conf. Ser. Vol. 68 (IoP, Bristol, 1984) p. 123
[5.127] P.A. Crozier, J.N. Chapman, A.J. Craven, J.M. Titchmarsh: On the determination of inner-shell cross-sections from NiO using EELS. J. Microsc. **148**, 279 (1987)
[5.128] R.F. Egerton: A simple parametrization scheme for inner-shell cross-sections, in *Proceedings of the 46th. Annual Meeting of EMSA*, ed. by G.W. Bailey (San Francisco Press, San Francisco, CA, 1988) p. 532
[5.129] F. Hofer: Determination of inner-shell cross-sections for EELS quantification. Microsc. Microanal. Microstruct. **2**, 215 (1991)
[5.130] G.J. Auchterlonie, D.R. McKenzie, D.J.H. Cockayne: Using ELNES with parallel EELS for differentiating between a-Si:X thin films. Ultramicroscopy **31**, 217 (1989)
[5.131] R.D. Leapman, L.A. Grunes, P.L. Fejes: Study of the L_{23} edges in the 3d transition metals and their oxides by EELS with comparison to theory. Phys. Rev. B **26**, 614 (1982)
[5.132] C. Colliex, B. Jouffrey: Diffusion inélastique des électrons dans un solide par excitation de niveaux atomiques profonds. Philos. Mag. **25**, 491 (1972)
[5.133] D.E. Johnson: The interaction of 25 keV electrons with guanine and cytosine. Radiation Res. **49**, 63 (1972)
[5.134] R.D. Leapman, V.E. Cosslett: Extended fine structure above the x-ray edge in electron energy loss spectra. J. Phys. D **9**, L29 (1976)
[5.135] R.D. Leapman, J. Silcox: Orientation dependence of core edges in electron energy loss spectra. Phys. Rev. Lett. **42**, 1361 (1979)

[5.136] R.D. Leapman, P.L. Fejes, J. Silcox: Orientation dependence from anisotropic materials determined by inelastic scattering of fast electrons. Phys. Rev. B **28**, 2361 (1983)
[5.137] R. Brydson, H. Sauer, W. Engel, J.M. Thomas, E. Zeitler: Coordination fingerprints in electron loss near-edge structure: Determination of the local site symmetry of Al and Be in ultrafine minerals. J. Chem. Soc. Chem. Commun. **15**, 1010 (1989)
[5.138] R. Brydson, H. Sauer, W. Engel, E. Zeitler: EELS as a fingerprint of the chemical coordination of light elements. Microsc. Microanal. Microstruct. **2**, 159 (1991)
[5.139] D.F. Blake: The nature and origin of interstellar diamond. Nature **332**, 611 (1988)
[5.140] P.A. van Aken, B.H. Liebscher, V.J. Styrsa: Quantitative determination of iron oxidation states in minerals using Fe $L_2$3-edge electron energy-loss near-edge structure spectroscopy. Physics Chem. Minerals **25**, 323 (1998)
[5.141] P. Rez, J. Bruley, P. Brohan, M. Payne, L.A.J. Garvie: Review of methods for calculating near edge structure. Ultramicroscopy **59**, 159 (1965)
[5.142] T.L. Loucks: *Augmented Plane Wave Method* (Benjamin, New York, 1967)
[5.143] J.E. Muller, O. Jepsen, J.W. Wilkins: X-ray absorption spectra: K-edges of 3d metals, edges of 3d and 4d metals and M-edges of palladium. Solid State Commun. **42**, 365 (1982)
[5.144] P. Blaha, K. Schwarz: Electron densities in TiC, TiN, and TiO derived from energy band calculations. Int. J. Quantum Chem. **23**, 1535 (1983)
[5.145] X. Weng, P. Rez, O.F. Sankey: Pseudo-atomic-orbital band theory applied to electron energy loss near-edge structures. Phys. Rev. B **40**, 5694 (1989)
[5.146] X. Weng, H. Ma, P. Rez: Carbon K-shell near-edge structure: Multiple scattering and band theory calculations. Phys. Rev. B **40**, 4175 (1989)
[5.147] J. Pflüger, J. Fink, K. Schwarz: Electronic structure of unoccupied states of stoichiometric ZrN, NbC and NbN as determined by high resolution EELS. Solid State Commun. **55**, 675 (1985)
[5.148] D.D. Vvendensky, D.K. Salin, J.B. Pendry: An update of DLXANES, the calculation of x-ray absorption near-edge structure. Comp. Phys. Commun. **40**, 421 (1986)
[5.149] R. Brydson, D.D. Vvendensky, W. Engel, H. Sauer, B.G. Williams, E. Zeitler, J.M. Thomas: Chemical information from ELNES. Core hole effects in the Be and B K-edges in rhodizite. J. Phys. Chem. **92**, 962 (1988)
[5.150] R. Brydson, B.G. Williams, H. Sauer, W. Engel, R. Schlögl, M. Muller, E. Zeitler, J.M. Thomas: EELS and the crystal chemistry of rhodizite. Part 2: Near-edge structure. J. Chem. Soc. Faraday Trans. **84**, 631 (1988)
[5.151] P. Rez, X. Weng: Multiple scattering approach to oxygen K near-edge structure in EELS of alkaline earths. Phys. Rev. B **39**, 7405 (1989)
[5.152] K.M. Krishnan: Iron L_{23} near-edge structure studies. Ultramicroscopy **32**, 309 (1990)
[5.153] B. Thole, G. van der Laan: Branching ratio in x-ray absorption spectroscopy. Phys. Rev. B **38**, 3158 (1988)
[5.154] D.E. Johnson, S. Csillag, E.A. Stern: Analytical electron microscopy using extended energy-loss fine structure (EXELFS), in *Scanning Electron Microscopy 1981/I* (SEM/AMF O'Hare, Chicago, 1981) p. 105
[5.155] K. Kambe, D. Krahl, K.H. Herrmann: Extended fine structure in electron energy-loss spectra of MgO crystallites. Ultramicroscopy **6**, 157 (1981)

[5.156] B.M. Kincaid, A.E. Meixner, P.M. Platzman: Carbon K edge in graphite measured using electron energy-loss spectroscopy. Phys. Rev. Lett. **40**, 1296 (1978)
[5.157] V. Serin, G. Zanchi, J. Sévely: EXELFS as a structural tool for studies of low Z-elements. Microsc. Microanal. Microstruct. **3**, 201 (1992)
[5.158] G. Hug, G. Blanche, M. Jaouen, A.M. Flank, J.J. Rehr: Simulation of the extended fine structure of K-shell edges in intermetallic ordered alloys. Ultramicroscopy **59**, 121 (1995)
[5.159] M. Qian, M. Sarikaya, E.A. Stern: Development of EXELFS technique for high accuracy structural information. Ultramicroscopy **59**, 137 (1995)
[5.160] Z.W. Yuan, S. Csillag, M.A. Tafreshi, C. Colliex: High spatial resolution EXELFS investigations of silicon dioxide compounds. Ultramicroscopy **59**, 149 (1995)
[5.161] D.E. Sayers, E.A. Stern, F.W. Lytle: New technique for investigating noncrystalline structures: Fourier analysis of the extended x-ray absorption fine structure. Phys. Rev. Lett. **27**, 1204 (1971)
[5.162] C. Hèbert, P. Schattschneider: A proposal for dichroic experiments in the electron microscope. Ultramicroscopy **96**, 463 (2003)
[5.163] G. Schütz, W. Wagner, W. Wilhelm, P. Kienle: Absorption of circularily polarized x rays in iron. Phys. Rev. Lett. **58**, 737 (1987)
[5.164] P. Schattschneider, S. Rubino, C. Hébert, S. Rusz, J. Kunes, P. Novák, E. Carlino, M. Fabrizioli, P. Panaccione, G. Rossi: Detection of magnetic circular dichroism using a transmission electron microscope. Nature **441**, 486 (2006)
[5.165] M.M. Disko, O.L. Krivanek, P. Rez: Orientation dependent extended fine structure in electron energy-loss spectra. Phys. Rev. B **25**, 4252 (1982)
[5.166] H.G. Badde, H. Kappert, L. Reimer: Wellenoptische Theorie des Ripple-Kontrastes in der Lorentzmikroskopie. Z. Angew. Phys. **30**, 83 (1970)
[5.167] R.E. Burge, D.L. Misell, J.W. Smart: The small-angle scattering of electrons in thin films of evaporated carbon. J. Phys. C **3**, 1661 (1970)
[5.168] D.L. Misell, R.E. Burge: Convolution, deconvolution and small-angle plural electron scattering. J. Phys. C **2**, 61 (1969)
[5.169] R.A. Crick, D.L. Misell: A theoretical consideration of some defects in electron optical images. A formulation of the problem for the incoherent case. J. Phys. D **4**, 1 (1971)
[5.170] L. Landau: On the energy loss of fast electrons by ionization. J. Phys. USSR **8**, 201 (1944)
[5.171] O. Blunck, S. Leisegang: Zum Energieverlust schneller Elektronen in dünnen Schichten. Z. Phys. **128**, 500 (1950)
[5.172] H.D. Maccabee, D.G. Papworth: Correction to Landau's energy loss formula. Phys. Lett. A **30**, 241 (1969)
[5.173] L. Reimer, K. Brockmann, U. Rhein: Energy losses of 20–40 keV electrons in 150–650 μg cm^{-2} metal films. J. Phys. D **11**, 2151 (1978)
[5.174] L. Reimer, P. Gentsch: Superposition of chromatic error and beam broadening in TEM of thick carbon and organic specimens. Ultramicroscopy **1**, 1 (1975)
[5.175] P. Gentsch, H. Gilde, L. Reimer: Measurement of the top–bottom effect in STEM of thick anorphous specimens. J. Microsc. **100**, 81 (1974)
[5.176] L. Reimer, M. Ross-Messemer: Top–bottom effect in energy-selecting TEM. Ultramicroscopy **21**, 385 (1987)

[5.177] K. Jost, J. Kessler: Die Ortsverteilung mittelschneller Elektronen bei Mehrfachstreuung. Z. Phys. **176**, 126 (1963)

[5.178] T. Groves: Thick specimens in the CEM and STEM: Resolution and image formation. Ultramicroscopy **1**, 15 and 170 (1975)

[5.179] H. Rose: The influence of plural scattering on the limit of resolution in electron microscopy. Ultramicroscopy **1**, 167 (1975)

[5.180] L. Reimer, H. Gilde, K.H. Sommer: Die Verbreiterung eines Elektronenstrahles (17–1200 keV) durch Mehrfachstreuung. Optik **30**, 590 (1970)

[5.181] W. Bothe: Die Streuung von Elektronen in schrägen Folien. Sitzungsber. Heidelb. Akad. Wiss. **7** , 307 (1951)

[5.182] H.W. Thümmel: *Durchgang von Elektronen- und Betastrahlung durch Materieschichten.* (Akademie, Berlin, 1974)

[5.183] J.I. Goldstein, J.L. Costley, G.W. Lorimer, S.J.B. Reed: Quantitative x-ray analysis in the electron microscope, in *Scanning Electron Microscopy 1977/I*, ed. by O. Johari (IIT Research Institute, Chicago, 1977) p. 315

[5.184] T. Just, H. Niedrig, H. Yersin: Schichtdickenbestimmung mittels Elektronen-Rückstreuung. Z. Angew. Phys. **25**, 89 (1986)

[5.185] H. Niedrig, P. Sieber: Rückstreuung mittelschneller Elektronen an dünnen Schichten. Z. Angew. Phys. **31**, 27 (1971)

[5.186] F.J. Hohn, H. Niedrig: Elektronenrückstreuung an dünnen Metall- und Isolatorschichten. Optik **35**, 290 (1972)

[5.187] L. Reimer, B. Lödding: Calculation and tabulation of Mott cross-sections for large-angle scattering. Scanning **6**, 128 (1984)

[5.188] L. Reimer, E.R. Krefting: The effect of scattering models on the results of Monte Carlo calculations, in *Use of Monte Carlo Calculations in Electron Probe Microanalysis and Scanning Electron Microscopy*, ed. by K.F.J. Heinrich, D.E. Newbury, H. Jakowitz, NBS Special Publ. 460 (U.S. Government Printing Office, Washington, 1976) p. 45

[5.189] L. Reimer, D. Stelter: Fortran 77 Monte Carlo program for minicomputers using Mott cross-sections. Scanning **8**, 265 (1986)

[5.190] H. Seiler: Einige aktuelle Probleme der Sekundärelektronenemission. Z. Angew. Phys. **22**, 249 (1967)

[5.191] H. Seiler: Secondary electron emission in the SEM. J. Appl. Phys. **54**, R1 (1983)

[5.192] H. Drescher, L. Reimer, H. Seidel: Rückstreukoeffizient und Sekundärelektronenausbeute von 10–100 keV-Elektronen und Beziehungen zur Raster-Elektronenmikroskopie. Z. Angew. Phys. **29**, 331 (1970)

[5.193] L. Reimer, H. Drescher: Secondary electron emission of 10–100 keV electrons from transparent films of Al and Au. J. Phys. D **10**, 805 (1977)

Chapter 6

[6.1] L. Reimer: Deutung der Kontrastunterschiede von amorphen und kristallinen Objekten in der Elektronenmikroskopie. Z. Angew. Phys. **22**, 287 (1967)

[6.2] W. Lippert: Über die "elektronenmikroskopische Durchlässigkeit" dünner Schichten. Optik **13**, 506 (1956)

[6.3] L. Reimer: Zur Elektronenabsorption dünner Metallaufdampfschichten im Elektronenmikroskop. Z. Angew. Phys. **9**, 34 (1957)

[6.4] L. Reimer: Messung der Abhängigkeit des elektronenmikroskopischen Bildkontrastes von Ordnungszahl, Strahlspannung und Aperturblende. Z. Angew. Phys. **13**, 432 (1961)

[6.5] L. Reimer, K.H. Sommer: Messungen und Berechnungen zum elektronenmikroskopischen Streukontrast für 17–1200 keV Elektronen. Z. Naturforsch. A **23**, 1569 (1968)

[6.6] E. Zeitler, G.F. Bahr: Contributions to the quantitative interpretation of electron microscope pictures. Exp. Cell Res. **12**, 44 (1957)

[6.7] W. Lippert: Bemerkungen zur elektronenmikroskopischen Dickenmessung von Kohleschichten. Z. Naturforsch. B **17**, 335 (1962)

[6.8] W. Schwertfeger: Zur Kleinwinkelstreuung von mittelschnellen Elektronen beim Durchgang durch amorphe Festkörperschichten. Dissertation, Universität Tübingen (1974)

[6.9] G. Dupouy, F. Ferrier, P. Verdier: Amélioration du contraste des images d'objets amorphes minces en microscopie électronique. J. Microscopie **5**, 655 (1966)

[6.10] R.F. Whiting, F.P. Ottensmeyer: Heavy atoms in model compounds and nucleic acids by dark field TEM. J. Mol. Biol. **67**, 173 (1972)

[6.11] J. Dubochet, M. Ducommun, M. Zollinger, E. Kellenberger: A new preparation method for dark-field electron microscopy of biomacromolecules. J. Ultrastruct. Res. **35**, 147 (1971)

[6.12] G.J. Brakenhoff, N. Nanninga, J. Pieters: Relative mass determination from dark-field electron micrographs, with an application to ribosomes. J. Ultrastruct. Res. **41**, 238 (1972)

[6.13] W. Krakow, L.A. Howland: A method for producing hollow cone illumination electronically in the conventional transmission microscope. Ultramicroscopy **2**, 53 (1976)

[6.14] L. Reimer, M. Ross-Messemer: Contrast in the electron spectroscopic imaging mode of a TEM. I. Influence of zero-loss filtering on scattering contrast. J. Microsc. **155**, 169 (1989)

[6.15] R. Bauer: Electron spectroscopic imaging: An advanced technique for imaging and analysis in TEM, in *Methods in Microbiology*, Vol. 20, ed. by F. Mayer (Academic, London, 1988) p. 113

[6.16] R. Bauer, U. Hezel, D. Kurz: High-resolution imaging of thick biological specimens with an imaging electron energy loss spectrometer. Optik **77**, 171 (1987)

[6.17] H.J. Wagner: Contrast tuning by electron spectroscopic imaging of half-micrometer-thick sections of nervous tissue. Ultramicroscopy **32**, 42 (1990)

[6.18] C. Colliex, C. Mory, A.L. Olins, D.E. Olins, M. Tencé: Energy-filtered STEM imaging of thick biological sections. J. Microsc. **153**, 1 (1989)

[6.19] L. Reimer, R. Rennekamp, I. Fromm, M. Langenfeld: Contrast in the electron spectroscopic imaging mode of a TEM: IV. Thick specimens imaged by the most-probable energy loss. J. Microsc. **162**, 3 (1991)

[6.20] E. Zeitler, M.G.R. Thomson: Scanning transmission electron microscopy. Optik **31**, 258 and 359 (1970)

[6.21] L. Reimer, P. Gentsch, P. Hagemann: Anwendung eines Rasterzusatzes zu einem TEM. I. Grundlagen und Abbildung amorpher Objekte. Optik **43**, 431 (1975)

[6.22] E. Carlemalm, E. Kellenberger: The reproducible observation of unstained embedded cellular material in thin sections: Visualisation of an integral membrane protein by a new mode of imaging for STEM. EMBO J. **1**, 63 (1982)
[6.23] R. Reichelt, E. Carlemalm, A. Engel: Quantitative contrast evaluation for different STEM imaging modes, in *Scanning Electron Microscopy 1984/III* (SEM/AMF O'Hare, Chicago, 1984) p. 1011
[6.24] R.F. Egerton: Thickness dependence of the STEM ratio image. Ultramicroscopy **10**, 297 (1982)
[6.25] P.J. Andree, J.E. Mellema, R.W.H. Ruignek: Discrimination of heavy and light elements in a specimen by use of STEM. Ultramicroscopy **17**, 237 (1985)
[6.26] W. Tichelaar, C. Ferguson, J.C. Olivo, K.R. Leonard, M. Haider: A novel method of Z-contrast imaging in STEM applied to double-labelling. J. Microsc. **175**, 10 (1994)
[6.27] C.E. Hall: Electron densitometry of stained virus particles. J. Biophys. Biochem. Cytol. **1**, 1 (1955)
[6.28] E. Krüger-Thiemer: Ein Verfahren für elektronenmikroskopische Massendickemessungen an nichtkristallinen Objekten. Z. Wiss. Mikr. **62**, 444 (1955)
[6.29] N.R. Silvester, R.E. Burge: A quantitative estimation of the uptake of two new electron stains by the cytoplasmic membrane of rat sperm. J. Biophys. Biochem. Cytol. **6**, 179 (1959)
[6.30] L. Reimer, P. Hagemann: Recording of mass thickness in STEM. Ultramicroscopy **2**, 297 (1977)
[6.31] M.K. Lamvik: Electron microscopic mass determination using photographic isodensity techniques. Ultramicroscopy **1**, 187 (1976)
[6.32] A. Engel: Molecular weight determination by STEM. Ultramicroscopy **3**, 273 (1978)
[6.33] S.A. Müller, K.N. Goldie, R. Bürki, R. Häring, A. Engel: Factors influencing the precision of quantitative STEM. Ultramicroscopy **46**, 317 (1992)
[6.34] J. Trachtenberg, K.R. Leonard, W. Tichelaar: Radial mass density functions of vitrified helical specimens determined by STEM: Their potential use as substitutes for equatorial data. Ultramicroscopy **45**, 307 (1992)
[6.35] P.W.J. Linders, P. Hagemann: Mass determination of the biological specimens using backscattered electrons. Ultramicroscopy **11**, 13 (1983)
[6.36] E. Zeitler, G.F. Bahr: A photometric procedure for weight determination of submicroscopic particles. J. Appl. Phys. **33**, 847 (1962)
[6.37] G.F. Bahr, E. Zeitler: The determination of dry mass in populations of isolated particles. Lab. Invest. **14**, 955 (1965)
[6.38] F.S. Sjöstrand: The importance of high resolution electron microscopy in tissue cell ultrastructure research. Sci. Tools **2**, 25 (1955)
[6.39] B. von Borries, F. Lenz: Über die Entstehung des Kontrastes im elektronenmikroskopischen Bild, in *Electron Microscopy*, Proceedings of the Stockholm Conference 1956, ed. by F.J. Sjöstrand, J. Rhodin (Almqvist and Wiksells, Stockholm, 1957) p. 60
[6.40] F. Thon: Zur Defokussierungsabhängigkeit des Phasenkontrastes bei der elektronenmikroskopischen Abbildung. Z. Naturforsch. A **21**, 476 (1966)

[6.41] F. Lenz, W. Scheffels: Das Zusammenwirken von Phasen- und Amplitudenkontrast in der elektronenmikroskopischen Abbildung. Z. Naturforsch. A **13**, 226 (1958)

[6.42] W. Krakow, K.H. Downing, B.M. Siegel: The use of tilted specimens to obtain the contrast transfer characteristics of an electron microscope imaging system. Optik **40**, 1 (1974)

[6.43] A. Howie, O.L. Krivanek, M.L. Rudee: Interpretation of electron micrographs and diffraction patterns of amorphous materials. Philos. Mag. **27**, 235 (1973)

[6.44] G.J. Brakenhoff: On the sub-nanometre structure visible in high-resolution dark-field electron microscopy. J. Microsc. **100**, 283 (1974)

[6.45] A. Oberlin, M. Oberlin, M. Maubois: Study of thin amorphous and crystalline carbon films by electron microscopy. Philos. Mag. **32**, 833 (1975)

[6.46] L. Reimer, H. Gilde: Scattering theory and image formation in the electron microscope, in [Ref. 1.11, p. 138]

[6.47] L. Albert, R. Schneider, H. Fischer: Elektronenmikroskopische Sichtbarmachung von $\leq 10\mathring{A}$ großen Fremdstoffeinschlüssen in elektrolytisch abgeschiedenen Nickelschichten mittels Phasenkontrast durch Defokussierung. Z. Naturforsch. A **19**, 1120 (1964)

[6.48] M. Rühle, M. Wilkens: Defocusing contrast of cavities, in *Electron Microscopy 1972* ed. by V.E. Cosslett (IoP, London, 1972) p. 146

[6.49] L. Reimer, H. Gilde: Electron optical phase contrast of small gold particles. Optik **41**, 524 (1975)

[6.50] P. Hirsch, L. Reimer: Influence of zero-loss filtering on electron optical phase contrast. J. Microsc. **174**, 143 (1994)

[6.51] R. Knippelmeyer, A. Thesing, H. Kohl: Determination of the contrast transfer function by analysing diffractograms of thin amorphous foils. Z. Metallkunde **94**, 282 (2003)

[6.52] C.B. Eisenhandler, B.M. Siegel: Imaging of single atoms with the electron microscope by phase contrast. J. Appl. Phys. **37**, 1613 (1966)

[6.53] R. Langer, W. Hoppe: Die Erhöhung von Auflösung und Kontrast im Elektronenmikroskop mit Zonenkorrekturplatten: Optik **24**, 470 (1966); **25**, 413 and 507 (1967)

[6.54] L. Reimer: Elektronenoptischer Phasenkontrast. Z. Naturforsch. A **24**, 377 (1969)

[6.55] H. Niehrs: Optimale Abbildungsbedingungen und Bildintensitätsverlauf bei einer Elektronenmikroskopie von Atomen. Optik **30**, 273 (1969); **31**, 51 (1970)

[6.56] D.L. Misell: Image formation in the electron microscope. J. Phys. A **4**, 782 and 798 (1971)

[6.57] D.L. Misell: Image resolution and image contrast in the electron microscope. J. Phys. A **6**, 62, 205, and 218 (1973)

[6.58] T. Kobayashi, L. Reimer: Computation of electron microscopical images of single organic molecules. Optik **43**, 237 (1975)

[6.59] W. Chiu, R.M. Glaeser: Single atom image contrast: conventional dark-field and bright-field electron microscopy. J. Microsc. **103**, 33 (1975)

[6.60] H. Hoch: Dunkelfeldabbildung von schwachen Phasenobjekten im Elektronenmikroskop. Optik **47**, 65 (1977)

[6.61] W. Krakow: Computer experiments for tilted dark-field imaging. Ultramicroscopy **1**, 203 (1976)

[6.62] K.J. Hanszen: Problems of image interpretation in electron microscopy with linear and nonlinear transfer. Z. Angew. Phys. **27**, 125 (1969)
[6.63] H. Hashimoto, A. Kumao, K. Hino, H. Yotsumoto, A. Ono: Images of Th atoms in TEM. Jpn. J. Appl. Phys. **10**, 1115 (1971)
[6.64] R.M. Henkelman, F.P. Ottensmeyer: Visualization of single heavy atoms by dark field electron microscopy. Proc. Nat. Acad. Sci. USA **68**, 3000 (1971)
[6.65] F.P. Ottensmeyer, E.E. Schmidt, T. Jack, J. Powell: Molecular architecture: the optical treatment of dark field electron micrographs of atoms. J. Ultrastruct. Res. **40**, 546 (1972)
[6.66] F. Thon, D. Willasch: Imaging of heavy atoms in dark field electron microscopy using hollow cone illumination. Optik **36**, 55 (1972)
[6.67] K.J. Hanszen: The relevance of dark field illumination in conventional and scanning TEM. PTB-Bericht A Ph-7 (Physikalisch-Technische Bundesanstalt, Braunschweig, 1974)
[6.68] D.L. Misell: Image resolution in high voltage electron microscopy. J. Phys. D **6**, 1409 (1973)
[6.69] H. Formanek, M. Müller, M.H. Hahn, T. Koller: Visualization of single heavy atoms with the electron microscope. Naturwissenschaften **58**, 339 (1971)
[6.70] J.R. Parsons, H.M. Johnson, C.W. Hoelke, R.R. Hosbons: Imaging of uranium atoms with the electron microscope by phase contrast. Philos. Mag. **27**, 1359 (1973)
[6.71] W. Baumeister, M.H. Hahn: Electron microscopy of monomolecular layers of thorium atoms. Nature **241**, 445 (1973)
[6.72] S. Iijima: Observation of single and clusters of atoms in bright field electron microscopy. Optik **48**, 193 (1977)
[6.73] E.B. Prestridge, D.J.C. Yates: Imaging the rhodium atom with a conventional high resolution electron microscope. Nature **234**, 345 (1971)
[6.74] D. Dorignac, B. Jouffrey: Atomic resolution at 3 MV, in [Ref. 1.79, p. 143]
[6.75] D. Dorignac, B. Jouffrey: Iron single atom images, in *Electron Microscopy 1980*, Vol. 1, ed. by P. Brederoo, G. Boom (Seventh European Congress on Electron Microscopy Foundation, Leiden, 1980) p. 112
[6.76] M. Retsky: Observed single atom elastic cross sections in a scanning electron microscope. Optik **41**, 127 (1974)
[6.77] M. Isaacson, J.P. Langmore, H. Rose: Determination of the non-localization of the inelastic scattering of electrons by electron microscopy. Optik **41**, 92 (1974)
[6.78] A.V. Crewe, J.P. Langmore, M.S. Isaacson: Resolution and contrast in the STEM, in [Ref. 1.12, p. 47]
[6.79] M. Isaacson, M. Utlaut, D. Kopf: Analog computer processing of STEM images, in [Ref. 1.15, p. 257]
[6.80] A.V. Crewe, J. Langmore, M. Issacson, M. Retsky: Understanding single atoms in STEM, in [Ref. 1.56, Vol. 1, p. 260]
[6.81] M.S. Isaacson, J. Langmore, N.W. Parker, D. Kopf, M. Utlaut: The study of adsorption and diffusion of heavy atoms on light element substrates by means of the atomic resolution STEM. Ultramicroscopy **1**, 359 (1976)
[6.82] J.S. Wall, J.F. Hainfeld, J.W. Bittner: Preliminary measurements of uranium atom motion on carbon films at low temperatures. Ultramicroscopy **3**, 81 (1978)

[6.83] K.J. Hanszen, B. Morgenstern, K.J. Rosenbruch: Aussagen der optischen Übertragungstheorie über Auflösung und Kontrast im elektronenmikroskopischen Bild. Z. Angew. Phys. **16**, 477 (1964)
[6.84] K.J. Hanszen, B. Morgenstern: Die Phasenkontrast und Amplitudenkontrast-Übertragung des elektronenmikroskopischen Objektivs. Z. Angew. Phys. **19**, 215 (1965)
[6.85] K.J. Hanszen: Contrast transfer and image processing, in [Ref. 1.11, p. 16]
[6.86] M. Foschepoth, H. Kohl: Amplitude contrast – a way to obtain directly interpretable high-resolution images in a spherical aberration corrected transmission electron microscope. Phys. Stat. Sol. a **166**, 357 (1998)
[6.87] K.J. Hanszen, L. Trepte: Der Einfluß von Strom- und Spannungsschwankungen sowie der Energiebreite der Strahlelektronen auf Kontrastübertragung und Auflösung des Elektronenmikroskopes. Optik **32**, 519 (1971)
[6.88] K.J. Hanszen, L. Trepte: Die Kontrastübertragung im Elektronenmikroskop bei partiell kohärenter Beleuchtung. Optik **33**, 166 and 182 (1971)
[6.89] J. Frank: The envelope of electron microscopic transfer functions for partially coherent illumination. Optik **38**, 519 (1973)
[6.90] R.H. Wade, J. Frank: Electron microscope transfer functions for partially coherent axial illumination and chromatic defocus spread. Optik **49**, 81 (1977)
[6.91] W.O. Saxton: Spatial coherence in axial high resolution conventional electron microscopy. Optik **49**, 51 (1977)
[6.92] H. Yoshida, A. Ohshita, H. Tomita: Determination of spatial and temporal coherence functions from a single astigmatic image. Jpn. J. Appl. Phys. **20**, 2427 (1981)
[6.93] P.W. Hawkes: Coherence in electron optics, in *Advances in Optical and Electron Microscopy*, Vol. 7, ed. by R. Barer, V.E. Cosslett (Academic, London, 1978) p. 101
[6.94] W. Hoppe, D. Köstler, D. Typke, N. Hunsmann: Kontrastübertragung für die Hellfeld-Bildrekonstruktion mit gekippter Beleuchtung in der Elektronenmikroskopie. Optik **42**, 43 (1975)
[6.95] K.H. Downing: Note on transfer function in electron microscopy with tilted illumination. Optik **43**, 199 (1975)
[6.96] S.C. McFarlane: The imaging of amorphous specimens in a tilted-beam electron microscope. J. Phys. C **8**, 2819 (1975)
[6.97] R.H. Wade: Concerning tilted beam electron microscope transfer functions. Optik **45**, 87 (1976)
[6.98] P.W. Hawkes: Electron microscope transfer functions in closed form with tilted illumination. Optik **55**, 207 (1980)
[6.99] W. Krakow: Calculation and observation of atomic structure for tilted beam dark-field microscopy, in *Development of Electron Microscopy and Analysis*, ed. by J.A. Venables (Academic, London, 1976) p. 261
[6.100] W. Hoppe, Towards three-dimensional electron microscopy at atomic resolution. Naturwissenschaften **61**, 239 (1974)
[6.101] W. Kunath: Signal-to-noise enhancement by superposition of bright-field images obtained under different illumination tilts. Ultramicroscopy **4**, 3 (1979)

[6.102] W. Kunath, F. Zemlin, K. Weiss: Apodization in phase-contrast electron microscopy realised with hollow-cone illumination. Ultramicroscopy **16**, 123 (1985)
[6.103] O. Scherzer: Zur Theorie der Abbildung einzelner Atome in dicken Objekten. Optik **38**, 387 (1973)
[6.104] W.O. Saxton, W.K. Jenkins, L.A. Freeman, D.J. Smith: TEM observations using bright field hollow cone illumination. Optik **49**, 505 (1978)
[6.105] H. Rose: Nonstandard imaging methods in electron microscopy. Ultramicroscopy **2**, 251 (1977)
[6.106] J. Fertig, H. Rose: On the theory of image formation in the electron microscope. Optik **54**, 165 (1979)
[6.107] R. Eusemann, H. Rose: Optimum bright-field of strong scatterers in CTEM and STEM. Ultramicroscopy **9**, 85 (1982)
[6.108] C. Dinges, H. Kohl, H. Rose: High-resolution imaging of crystalline objects by hollow-cone illumination. Ultramicroscopy **55**, 91 (1994)
[6.109] H. Rose: Phase contrast in STEM. Optik **39**, 416 (1974)
[6.110] N.H. Dekkers, H. de Lang: Differential phase contrast in STEM. Optik **41**, 452 (1974)
[6.111] W.C. Stewart: On differential phase contrast with an extended illumination source. J. Opt. Soc. Am. **66**, 813 (1976)
[6.112] H. Rose: Image formation by inelastically scattered electrons in electron microscopy. Optik **45**, 139 (1976)
[6.113] P.W. Hawkes: Half-plane apertures in TEM, split detectors in STEM and ptychography. J. Optique (Paris) **9**, 235 (1978)
[6.114] G.R. Morrison, J.N. Chapman: STEM imaging with a quadrant detector, in *Electron Microscopy 1981*, ed. by M.J. Goringe (IoP, London, 1981) p. 329
[6.115] I. Daberkow, K.H. Herrmann, F. Lenz: A configurable angle-resolving detector for STEM. Ultramicroscopy **50**, 75 (1993)
[6.116] M. Haider, A. Epstein, P. Jarron, C. Boulin: A versatile, software configurable multichannel STEM detector for angle-resolved imaging. Ultramicroscopy **54**, 41 (1994)
[6.117] M. Hammel, H. Rose: Optimum rotationally symmetric detector configurations for phase-contrast imaging in STEM. Ultramicroscopy **58**, 403 (1995)
[6.118] H. Kohl, H. Rose: Theory of image formation by inelastically scattered electrons in the electron microscope. Adv. Electron. Electron Phys. **65**, 173 (1985)
[6.119] J.M. Martin, J.L. Mansot, M. Hallouis: Energy filtered electron microscopy of overbased reverse micelles. Ultramicroscopy **30**, 321 (1989)
[6.120] A.J. Craven, C. Colliex: The effect of energy loss on phase contrast, in *Developments in Electron Microscopy and Analysis 1977*, ed. by D.L. Misell (IoP, Bristol, 1977) p. 271
[6.121] P. Hirsch, L. Reimer: Influence of zero-loss filtering on electron optical phase contrast. J. Microsc. **174**, 143 (1994)
[6.122] W. Hoppe: Ein neuer Weg zur Erhöhung des Auflösungsvermögens des Elektronenmikroskopes. Naturwissenschaften **48**, 736 (1961)
[6.123] F. Lenz: Zonenplatten zur Öffnungsfehlerkorrektur und zur Kontrasterhöhung. Z. Phys. **172**, 498 (1963)
[6.124] K.H. Müller: Phasenplatten für Elektronenmikroskope. Optik **45**, 73 (1976)
[6.125] D. Willasch: High resolution electron microscopy with profiled phase plates. Optik **44**, 17 (1975)

[6.126] E. Majorovits, B. Barton, K. Schultheiß, F. Pèrez-Willard, D. Gerthsen, R.R. Schröder: Optimizing phase contrast in transmission electron microscopy with an electrostatic (Boersch) phase plate. Ultramicroscopy **107**, 213 (2007)

[6.127] R. Cambie, K.H. Downing, D. Typke, R.M. Glaeser, J. Jin: Design of a microsfabricated, two-electrode phase-contrast element suitable for electron microscopy. Ultramicroscopy **107**, 329 (2007)

[6.128] L. Reimer, H.G. Badde, E. Drewes, H. Gilde, H. Kappert, H.J. Höhling, D.B. von Bassewitz, A. Rössner: Laserbeugung an elektronenmikroskopischen Aufnahmen. Forschungsberichte Landes Nordrhein Westfalen Nr. 2314 (Westdeutscher Verlag, Opladen, 1973)

[6.129] J.R. Berger, D. Harker: Optical diffractometer for production of Fourier transforms of electron micrographs. Rev. Sci. Instr. **38**, 292 (1967)

[6.130] T. Isshiki, K. Nishio, H. Saijo, M. Shiojiri: Real-time Fourier transformation of electron microscopy images on liquid crystal display panel by optical diffraction, in [Ref. 1.61, Vol. 1, p. 263]

[6.131] O.L. Krivanek: A method of determining the coefficient of spherical aberration from a single electron micrograph. Optik **45**, 97 (1976)

[6.132] L. Reimer, H.G. Heine, R.A. Ajeian: Optimalbedingungen für den Beugungsnachweis von Defokussierungsstrukturen in elektronenmikroskopischen Aufnahmen. Z. Naturforsch. A **24**, 1846 (1969)

[6.133] L. Reimer, H. Kappert: Bestimmung der Domänenwanddicke aus defokussierten elektronenoptischen Aufnahmen von ferromagnetischen Schichten. Z. Angew. Phys. **26**, 58 (1969)

[6.134] J. Frank: Nachweis von Objektbewegungen im lichtoptischen Diffraktogramm von elektronenmikroskopischen Aufnahmen. Optik **30**, 171 (1969)

[6.135] J. Frank: Observation of the relative phases of electron microscopic phase contrast zones with the aid of the optical diffractometer. Optik **35**, 608 (1972)

[6.136] L. Reimer, B. Volbert, P. Bracker: Quality control of SEM micrographs by laser diffractometry. Scanning **1**, 233 (1978)

[6.137] D. Gabor: Microscopy by reconstructed wavefronts. Proc. Roy. Soc. (London) A **197**, 454 (1949); Proc. Phys. Soc. B **64**, 449 (1950)

[6.138] K.J. Hanszen: Holographische Rekonstruktionsverfahren in der Elektronenmikroskopie und ihre kontrastübertragungstheoretische Deutung. Optik **32**, 74 (1970)

[6.139] A. Tonomura, A. Fukuhara, H. Watanabe, T. Komoda: Optical reconstruction of image from Fraunhofer electron hologram. Jpn. J. Appl. Phys. **7**, 295 (1968)

[6.140] J. Munch: Experimental electron holography. Optik **43**, 79 (1975)

[6.141] K.J. Hanszen, G. Ade, R. Lauer: Genauere Angaben über sphärische Längsaberration, Verzeichnung in der Pupillenebene und über die Wellenaberration von Elektronenlinsen. Optik **35**, 567 (1972)

[6.142] K.J. Hanszen: Neuere theoretische Erkenntnisse und praktische Erfahrungen über die holographische Rekonstruktion elektronenmikroskopischer Aufnahmen, PTB-Bericht A Ph-4 (Physikalisch-Technische Bundesanstalt, Braunschweig, 1973)

[6.143] G. Ade: Erweiterung der Kontrastübertragungstheorie auf nicht-isoplanatische Abbildungen. Optik **50**, 143 (1978)

[6.144] A. Lohmann: Optische Einseitenbandübertragung angewandt auf das Gabor-Mikroskop. Opt. Acta **3**, 97 (1956)
[6.145] K.J. Hanszen: Einseitenband-Holographie. Z. Naturforsch. A **24**, 1849 (1969)
[6.146] W. Hoppe, R. Langer, F. Thon: Verfahren zur Rekonstruktion komplexer Bildfunktionen in der Elektronenmikroskopie. Optik **30**, 538 (1970)
[6.147] W. Hoppe: Zur Abbildung komplexer Bildfunktionen in der Elektronenmikroskopie. Z. Naturforsch. A **26**, 1155 (1971)
[6.148] K.H. Downing: Compensation of lens aberrations by single-sideband holography, in *Proceedings of the 30th Annual EMSA Meeting* (Claitor, Baton Rouge, LA, 1972) p. 562
[6.149] P. Sieber: High resolution electron microscopy with heated apertures and reconstruction of single-sideband micrographs, in [Ref. 1.56, Vol. 1, p. 274]
[6.150] K.H. Downing, B.M. Siegel: Discrimination of heavy and light components in electron microscopy using single-sideband holographic techniques. Optik **42**, 155 (1975)
[6.151] E.N. Leith, J. Upatnieks: Reconstructed wavefronts and communication theory. J. Opt. Soc. Am. **52**, 1123 (1962)
[6.152] G. Möllenstedt, H. Wahl: Elektronenholographie und Rekonstruktion mit Laserlicht. Naturwissenschaften **55**, 340 (1968)
[6.153] H. Lichte: Electron holography approaching atomic resolution. Ultramicroscopy **20**, 293 (1986)
[6.154] E. Völkl, H. Lichte: Electron holograms for sub-Ångstrøm point resolution. Ultramicroscopy **32**, 177 (1990)
[6.155] Th. Leuthner, H. Lichte, H.H. Herrmann: STEM holography using the electron biprism. Phys. Status Solidi a **116**, 113 (1989)
[6.156] H. Banzhof, K.H. Herrmann: Reflection electron holography. Ultramicroscopy **48**, 475 (1993)
[6.157] N. Osakabe, T. Matsuda, J. Endo, A. Tonomura: Reflection electron holographic observation of surface displacement field. Ultramicroscopy **48**, 483 (1993)
[6.158] A. Tonomura, J. Endo, T. Matsuda: An application of electron holography to interference microscopy. Optik **53**, 143 (1979)
[6.159] J. Endo, T. Matsuda, A. Tonomura: Interference electron microscopy by means of holography. Jpn. J. Appl. Phys. **18**, 2291 (1979)
[6.160] A. Tonomura, T. Matsuda, J. Endo, T. Arii, K. Mihama: Direct observation of fine structure of magnetic domain walls by electron holography. Phys. Rev. Lett. **44**, 1430 (1980)
[6.161] N. Osakabe, K. Yoshida, Y. Horiuchi, T. Matsuda, H. Tanabe, T. Okuwaki, J. Endo, H. Fujiwara, A. Tonomura: Observation of recorded magnetization pattern by electron holography. Appl. Phys. Lett. **42**, 746 (1983)
[6.162] K.J. Hanszen, R. Lauer, G. Ade: Discussions of the possibilities and limitations of in-line and off-axis holography in electron microscopy. PTB-Bericht A Ph-15 (Physikalisch-Technische Bundesanstalt, Braunschweig, 1980)
[6.163] K.H. Hanszen: Methods of off-axis holography and investigations of the phase structure in crystals. J. Phys. D **19**, 373 (1986)
[6.164] K.J. Hanszen: Holography in electron microscopy. Adv. Electron. Electron Phys. **59**, 1 (1982)
[6.165] A. Tonomura, L.F. Allard, G. Pozzi, D.C. Joy, Y.A. Ono (eds.): *Electron Holography* (Elsevier, Amsterdam, 1995)

[6.166] K.J. Hanszen: Lichtoptische Anordnungen mit Laser-Lichtquellen als Hilfsmittel für die Elektronenmikroskopie, in *Electron Microcopy 1968*, Vol. 1, ed. by D.S. Bocciarelli (Tipografia Poliglotta Vaticana, Rome, 1968) p. 153
[6.167] K.J. Hanszen: Holographische Rekonstruktionsverfahren in der Elektronenmikroskopie und ihre kontrastübertragungstheoretische Deutung. Optik **32**, 74 (1970)
[6.168] A. Maréchal, P. Croce: Un filtre de fréquences spatiales pour l'amélioration du contraste des images optiques. C. R. Acad. Sci. Paris **237**, 607 (1953)
[6.169] M.H. Hahn: Eine optische Ortsfrequenzfilter- und Korrelationsanlage für elektronenmikroskopische Aufnahmen. Optik **35**, 326 (1972)
[6.170] G.W. Stroke, M. Halioua, F. Thon, D. Willasch: Image improvement in high resolution electron microscopy using holographic image deconvolution. Optik **41**, 319 (1974)
[6.171] R.E. Burge, R.F. Scott: Binary filters for high resolution electron microscopy. Optik **43**, 53 (1975); ibid. **44**, 159 (1976)
[6.172] S. Boseck, H. Hager: Beseitigung des spatialen Rauschens in elektronenmikroskopischen Aufnahmen durch lichtoptische Filterung. Optik **28**, 602 (1968)
[6.173] S. Boseck, R. Lange: Ausschöpfung des Informationsgehaltes von elektronenmikroskopischen Aufnahmen biologischer Objekte mit Hilfe des Abbéschen Beugungsapparates, gezeigt am Beispiel kristallartiger Strukturen. Z. Wiss. Mikr. **70**, 66 (1970)
[6.174] J.B. Bancroft, G.J. Hills, R. Markham: A study of the self-assembly process in a small spherical virus. Virology **31**, 354 (1967)
[6.175] A. Klug, D.J. deRosier: Optical filtering of electron micrographs: reconstruction of one-sided images. Nature **212**, 29 (1966)
[6.176] C.A. Taylor, J.K. Ranniko: Problems in the use of selective optical spatial filtering to obtain enhanced information from electron micrographs. J. Microsc. **100**, 307 (1974)
[6.177] R. Markham, J.H. Hitchborn, G.J. Hills, S. Frey: The anatomy of tobacco mosaic virus. Virology **22**, 342 (1964)
[6.178] R.C. Warren, R.M. Hicks: A simple method of linear integration for resolving structures in periodic lattices. J. Ultrastruct. Res. **36**, 861 (1971)
[6.179] R. Markham, S. Frey, G.J. Hills: Methods for the enhancement of image detail and accentuation of structure in electron microscopy. Virology **20**, 88 (1963)
[6.180] P.W. Hawkes: Processing electron images, in *Quantitative Electron Microscopy*, ed. by J.N. Chapman, A.J. Craven (Scottish Universities Summer School, Edinburgh, 1984) p. 351
[6.181] R. Hegerl: A brief survey of software packages for image processing in biological electron microscopy. Ultramicroscopy **46**, 417 (1992)
[6.182] D.L. Misell: The phase problem in electron microscopy, in *Advances in Optical and Electron Microscopy*, Vol. 7, ed. by R. Barer, V.E. Cosslett (Academic, London, 1978) p. 185
[6.183] W.O. Saxton: Computer techniques for image processing in electron microscopy. Adv. Electron. Electron Phys. Suppl. **10**, 289 (1978)
[6.184] W.O. Saxton: Recovery of specimen information for strongly scattering objects, in [Ref. 1.15, p. 35]
[6.185] R.W. Gerchberg, W.O. Saxton: Phase determination from image and diffraction plane pictures in the electron microscope. Optik **34**, 275 (1971)

[6.186] R.W. Gerchberg, W.O. Saxton: A practical algorithm for the determination of phase from image and diffraction plane picture. Optik **35**, 237 (1972)

[6.187] J. Frank: A remark on phase determination in electron microscopy. Optik **38**, 582 (1973)

[6.188] R.W. Gerchberg: Holography without fringes in the electron microscope. Nature **240**, 404 (1972)

[6.189] J.N. Chapman: The application of iterative techniques to the investigation of strong phase objects in the electron microscope. Philos. Mag. **32**, 527 and 541 (1975)

[6.190] D.L. Misell: An examination of an iterative method for the solution of the phase problem in optics and electron optics. J. Phys. D **6**, 2200 and 2217 (1973)

[6.191] P. Schiske: Phase determination from a focal series and the corresponding diffraction pattern in electron microscopy for strongly scattering objects. J. Phys. D **8**, 1372 (1975)

[6.192] W.O. Saxton: What is the focus variation method? Is it new? Is it direct? Ultramicroscopy **55**, 171 (1994)

[6.193] D. Van Dyck, M. Op de Beeck: A new approach to object wavefunction reconstruction in electron microscopy. Optik **93**, 103 (1993)

[6.194] W.O. Saxton, W.M. Stobbs: BF/DF image subtraction for image linearization. In *Electron Microscopy 1984*, Vol. 1, ed. by Á. Csanády, P. Röhlich, D. Szabó (MOTESZ, Budapest, 1984) p. 287

[6.195] P. van Toorn, A.M.J. Huiser, H.A. Ferwerda: Proposals for solving the phase retrieval problem for semi-weak objects from noisy electron micrographs. Optik **51**, 309 (1978)

[6.196] R. Langer, J. Frank, A. Feltynowski, W. Hoppe: Anwendung des Bilddifferenzverfahrens auf die Untersuchung von Strukturänderungen dünner Kohlefolien bei Elektronenbestrahlung. Ber. Bunsenges. Phys. Chem. **74**, 1120 (1970)

[6.197] J. Frank: Two-dimensional correlation functions in electron microscope image analysis, in *Electron Microscopy 1972* ed. by V.E. Cosslett (IoP, London 1972) p. 622

[6.198] M. van Heel, M. Schatz, E. Orlova: Correlation functions revisited. Ultramicroscopy **46**, 307 (1992)

[6.199] W. Hoppe, R. Langer, J. Frank, A. Feltynowski: Bilddifferenzverfahren in der Elektronenmikroskopie. Naturwissenschaften **56**, 267 (1969)

[6.200] J. Frank, P. Bu ler, R. Langer, W. Hoppe: Einige Erfahrungen mit der rechnerischen Analyse und Synthese von elektronenmikroskopischen Bildern hoher Auflösung. Ber. Bunsenges. Phys. Chem. **74**, 1105 (1970)

[6.201] T.A. Welton: A computational critique of an algorithm for image enhancement in bright field electron microscopy. Adv. Electron. Electron Phys. **48** 37 (1978)

[6.202] R.A. Crowther, L.A. Amos: Harmonic analysis of electron microscope images with rotational symmetry. J. Mol. Biol. **60**, 123 (1971)

[6.203] B.K. Jap, M. Zulauf, T. Scheybani, A. Hefti, W. Baumeister, U. Aebi, A. Engel: 2D crystallization: from art to science. Ultramicroscopy **46**, 45 (1992)

[6.204] R. Dürr: Displacement field analysis: calculation of distortion measures from displacement maps. Ultramicroscopy **38**, 135 (1991)

[6.205] W.O. Saxton, R. Dürr, W. Baumeister: From lattice distortion to molecular distortion: characterising and exploiting crystal deformation. Ultramicroscopy **46**, 287 (1992)

[6.206] H.P. Erikson, A. Klug: Measurements and compensation of defocusing and aberrations by Fourier processing of electron micrographs. Philos. Trans. B **261**, 105 (1971)

[6.207] A.M. Kuo, R.M. Glaeser: Development of methodology for low exposure, high resolution electron microscopy of biological specimens. Ultramicroscopy **1**, 53 (1975)

[6.208] P.N.T. Unwin, R. Henderson: Molecular structure determination by electron microscopy of unstained crystalline specimens. J. Mol. Biol. **94**, 425 (1975)

[6.209] H. Gross, Th. Müller, I. Wildhaber, H. Winkler: High resolution metal replication, quantified by image processing of periodic test specimens. Ultramicroscopy **16**, 287 (1985)

[6.210] I. Wildhaber, H. Gross, H. Moor: Comparative studies of very thin shadowing films produced by atom beam sputtering and electron beam evaporation. Ultramicroscopy **16**, 321 (1985)

[6.211] W.O. Saxton, J. Frank: Motif detection in quantum noise-limited electron micrographs by cross-correlation. Ultramicroscopy **2**, 219 (1976)

[6.212] J. Frank: Averaging of low exposure electron micrographs of nonperiodic objects. Ultramicroscopy **1**, 159 (1975)

[6.213] J. Frank: Optimal use of image formation using signal detection and averaging techniques. Ann. New York Acad. Sci. **306**, 112 (1978)

[6.214] J. Frank, W. Goldfarb, D. Eisenberg, T.S. Baker: Reconstruction of glutamine synthease using computer averaging. Ultramicroscopy **3**, 283 (1978)

[6.215] J. Frank, A. Verschoor, M. Boublik: Computer averaging of electron micrographs of 40S ribosomal subunits. Science **214**, 1356 (1981)

[6.216] M. van Heel: Detection of objects in quantum-noise-limited images. Ultramicroscopy **7**, 331 (1982)

[6.217] M. van Heel, J. Frank: Use of multivariate statistics in analysing the images of biological macromolecules. Ultramicroscopy **6**, 187 (1981)

[6.218] J. Frank: The role of multivariate image analysis in solving the architecture of the *Limulus polyphemus* hemocyanin molecule. Ultramicroscopy **13**, 153 (1984)

[6.219] M. van Heel: Multivariate statistical classification of noisy images (randomly oriented biological macromolecules). Ultramicroscopy **13**, 165 (1984)

[6.220] J.G. Helmcke: Theorie und Praxis der elektronenmikroskopischen Stereoaufnahme. Optik **11**, 201 (1954); **12**, 253 (1955)

[6.221] R.I. Garrod, J.F. Nankivell: Some remarks on the accuracy obtainable in electron stereomicroscopy. Optik **16**, 27 (1959)

[6.222] R.A. Crowther, D.J. deRosier, A. Klug: The reconstruction of a three-dimensional structure from projections and its application to electron microscopy. Proc. Roy. Soc. London A **317**, 319 (1970)

[6.223] G.N. Ramachandran, A.V. Lakshminarayanan: Three-dimensional reconstruction from radiographs and electron micrographs. Proc. Nat. Acad. Sci. USA **68**, 2236 (1971)

[6.224] M. van Heel, W. Keegstra: IMAGIC: a fast, flexible and friendly image analysis software system. Ultramicroscopy **7**, 113 (1981)

[6.225] R.A. Crowther, A. Klug: ART and science or conditions for three-dimensional structure from projections and its application to electron microscopy. J. Theor. Biol. **32**, 199 (1971)
[6.226] R. Gordon, R. Bender, G.T. Herman: Algebraic reconstruction techniques (ART) for three-dimensional electron microscopy and x-ray photography. J. Theor. Biol. **29**, 471 (1970)
[6.227] P.F.C. Gilbert: The reconstruction of a three-dimensional structure from projections and its application to electron microscopy. II Direct methods. Proc. Roy. Soc. London B **182**, 89 (1972)
[6.228] E. Zeitler: The reconstruction of objects from their projections. Optik **39**, 396 (1974)
[6.229] W. Hoppe, H.J. Schramm, M. Sturm, N. Hunsmann, J. Gaßmann: Three-dimensional electron microscopy of individual biological objects. Z. Naturforsch. A **31**, 645, 1370, and 1380 (1976)
[6.230] M. Zwick, E. Zeitler: Image reconstruction from projections. Optik **38**, 550 (1973)
[6.231] M. Carazo, J. Carrascosa: Information recovery in missing angular data cases: an approach by the convex projections method in three dimensions. J. Microsc. **145**, 23 (1987)
[6.232] M. Carazo, J. Carrascosa: Restoration of direct Fourier three-dimensional reconstruction of crystalline specimens by the method of convex projections. J. Microsc. **145**, 159 (1987)
[6.233] M.I. Sezan: An overview of convex projections theory and its application to image recovery problems. Ultramicroscopy **40**, 55 (1992)
[6.234] D.P. Barnard, J.N. Turner, J. Frank, B.F. McEwen: A 360° single-axis tilt stage for the HVEM. J. Microsc. **167**, 39 (1992)
[6.235] M. Radermacher, T. Wagenknecht, A. Verschoor, J. Frank: Three-dimensional reconstruction from single-exposure, random conical tilt series applied to the 50S ribosomal subunit of *Escherichia coli*. J. Microsc. **146**, 113 (1987)
[6.236] J. Frank (ed.): *Electron Tomography* (Plenum, New York, 1992)
[6.237] J. Frank, M. Rademacher: Three-dimensional reconstruction of single particles negatively stained or in vitreous ice. Ultramicroscopy **46**, 241 (1992)
[6.238] M. Schatz, M. van Heel: Invariant recognition of molecular projections in vitreous ice. Ultramicroscopy **45**, 15 (1992)
[6.239] K. Dierksen, D. Typke, R. Hegerl, A.J. Koster, W. Baumeister: Towards automatic electron tomography. Ultramicroscopy **40**, 71 (1992)
[6.240] K. Dierksen, D. Typke, R. Hegerl, W. Baumeister: Towards automatic tomography. II. Implementation of autofocus and low-dose procedures. Ultramicroscopy **49**, 109 (1993)
[6.241] P.A. Midgley, M. Weyland: 3D electron microscopy in the physical sciences: the development of Z-contrast and EFTEM tomography. Ultramicroscopy **96**, 413 (2003)
[6.242] I. Arslan, J.R. Tong, P.A. Midgley: Reducing the missing wedge: high-resolution dual axis tomography of inorganic materials. Ultramicroscopy **106**, 994 (2006)
[6.243] J.N. Chapman: The investigation of magnetic domain structures in thin foils by electron microscopy. J. Phys. D **17**, 623 (1984)
[6.244] P.J. Grundy, R.S. Tebble: Lorentz electron microscopy. Adv. Phys. **17**, 153 (1968)

[6.245] R.H. Wade: Lorentz microscopy or electron phase microscopy of magnetic objects, in *Advances in Optical and Electron Microscopy*, Vol. 5, ed. by R. Barer, V.E. Cosslett (Academic, London, 1973) p. 239
[6.246] J.P. Jakubovics: Lorentz microscopy and application (TEM and SEM), in Ref. 1.31, Part IV, p. 1303
[6.247] J. Zweck, M. Schneider, M. Sessner, T. Uhlig, M. Heumann. Lorentz electron microscopic observation of micromagnetic configurations in nanostructured materials. Adv. Solid State Physics **41**, 533 (2001)
[6.248] K. Tsuno, T. Taoka: Magnetic-field-free objective lens around the specimen for observing fine structure of ferromagnetic materials in TEM. Jpn. J. Appl. Phys. **22**, 1041 (1983)
[6.249] E. Fuchs: Magnetische Strukturen in dünnen ferromagnetischen Schichten, untersucht mit dem Elektronenmikroskop. Z. Angew. Phys. **14**, 203 (1962)
[6.250] K. Schaffernicht: Messung der Magnetisierungsverteilungen in dünnen Eisenschichten durch die Ablenkung von Elektronen. Z. Angew. Phys. **15**, 275 (1963)
[6.251] D.H. Warrington, J.M. Rodgers, R.S. Tebble: The use of ferromagnetic domain structure to determine the thickness of iron films in TEM. Philos. Mag. **7**, 1783 (1962)
[6.252] R.H. Wade: Electron diffraction from a magnetic phase grating. Phys. Status Solidi **19**, 847 (1967)
[6.253] M.J. Goringe, J.P. Jakubovics: Electron diffraction from periodic magnetic fields. Philos. Mag. **15**, 393 (1967)
[6.254] H. Boersch, H. Raith: Elektronenmikroskopische Abbildung Weißscher Bezirke in dünnen ferromagnetischen Schichten. Naturwissenschaften **46**, 574 (1959)
[6.255] H.W. Fuller, M.E. Hale: Domains in thin magnetic films observed by electron microscopy. J. Appl. Phys. **31**, 1699 (1960)
[6.256] J. Podbrdsky: High resolution in-focus Lorentz electron microscopy. J. Microsc. **101**, 231 (1974)
[6.257] W. Rollwagen, Ch. Schwink: Die Empfindlichkeit einfacher elektronenoptischer Schlierenanordnungen. Optik **10**, 525 (1953)
[6.258] Ch. Schwink: Über neue quantitative Verfahren der elektronenoptischen Schattenmethode. Optik **12**, 481 (1955)
[6.259] A.G. Cullis, D.M. Maher: High-resolution topographical imaging by direct TEM. Philos. Mag. **30**, 447 (1974)
[6.260] H.W. Fuller, M.E. Hale: Determination of magnetization distribution in thin films using electron microscopy. J. Appl. Phys. **31**, 238 (1960)
[6.261] H. Boersch, H. Hamisch, D. Wohlleben, K. Grohmann: Antiparallele Weißsche Bezirke als Biprisma für Elektroneninterferenzen. Z. Phys. **159**, 397 (1960)
[6.262] D. Wohlleben: Diffraction effects in Lorentz microscopy. J. Appl. Phys. **38**, 3341 (1967)
[6.263] L. Reimer, H. Kappert: Elektronen-Kleinwinkelstreuung und Bildkontrast in defokussierten Aufnahmen magnetischer Bereichsgrenzen. Z. Angew. Phys. **27**, 165 (1969)
[6.264] R.H. Wade: The determination of domain wall thickness in ferromagnetic films by electron microscopy. Proc. Phys. Soc. **79**, 1237 (1962)

[6.265] T. Suzuki, A. Hubert: Determination of ferromagnetic domain wall widths by means of high voltage Lorentz microscopy. Phys. Status Solidi **35**, K5 (1970)
[6.266] T. Suzuki, M. Wilkens: Lorentz-electron microscopy of ferromagnetic specimens at high voltages. Phys. Status Solidi A **3**, 43 (1970)
[6.267] H. Gong, J.N. Chapman: On the use of divergent wall images in the Fresnel mode of Lorentz microscopy for the measurement of the widths of very narrow domain walls. J. Magnetism Magn. Mat. **67**, 4 (1987)
[6.268] D.S. Hothersall: The investigation of domain walls in thin sections of iron by the electron interference method. Philos. Mag. **20**, 89 (1969)
[6.269] D.C. Hothersall: Electron images of domain walls in Co foils. Philos. Mag. **24**, 241 (1971)
[6.270] D.C. Hothersall: Electron images of two-dimensional domain walls. Phys. Status Solidi B **51**, 529 (1972)
[6.271] P. Schwellinger: The analysis of magnetic domain wall structures in the transition region of Néel and Bloch walls by Lorentz microscopy. Phys. Status Solidi A **36**, 335 (1976)
[6.272] C.G. Harrison, K.D. Leaver: A second domain wall parameter measurable by Lorentz microscopy. Phys. Status Solidi A **12**, 413 (1972)
[6.273] R. Ajeian, H. Kappert, L. Reimer: Fraunhofer-Beugung an Lorentz-mikroskopischen Aufnahmen des Magnetisierungs-Ripple. Z. Angew. Phys. **30**, 80 (1970)
[6.274] T. Susuki: Investigations into ripple wavelength in evaporated thin films by Lorentz microscopy. Phys. Status Solidi **37**, 101 (1970)
[6.275] C. Mory, C. Colliex: Inelastic effects in Lorentz microscopy. Philos. Mag. **33**, 97 (1976)
[6.276] M. Blackman, A.E. Curzon, A.T. Pawlowicz: Use of an electron beam for detecting superconducting domains of lead in its intermediate state. Nature **200**, 157 (1963)
[6.277] G. Pozzi, U. Valdre: Study of electron shadow patterns of the intermediate state of superconducting lead. Philos. Mag. **23**, 745 (1971)
[6.278] J.P. Jacubovics: The effect of magnetic domain structure on Bragg reflection in TEM. Philos. Mag. **10**, 277 (1964)
[6.279] J.N. Chapman, E.H. Darlington: The application of STEM to the study of thin ferromagnetic films. J. Phys. E **7**, 181 (1974)
[6.280] J.N. Chapman, E.M. Waddell, P.E. Batson, R.P. Ferrier: The Fresnel mode of Lorentz microscopy using a STEM. Ultramicroscopy **4**, 283 (1979)
[6.281] J.N. Chapman, P.E. Batson, E.M. Waddell, R.P. Ferrier: The direct determination of magnetic domain wall profiles by differential phase contrast electron microscopy. Ultramicroscopy **3**, 203 (1978)
[6.282] J.N. Chapman, P. Ploessl, D.M. Donnet: Differential phase contrast microscopy of magnetic materials. Ultramicroscopy **47**, 331 (1992)
[6.283] A. Olivei: Holography and interferometry in electron Lorentz microscopy. Optik **30**, 27 (1969)
[6.284] A. Olivei: Magnetic inhomogeneities and holographic methods in electron Lorentz microscopy. Optik **33**, 93 (1971)
[6.285] M.S. Cohen, K.J. Harte: Domain wall profiles in magnetic films. J. Appl. Phys. **40**, 3597 (1969)

[6.286] J.E. Bonevich, K. Harada, T. Matsuda, H. Kasai, T. Yoshida, G. Pozzi, A. Tonomura: Electron holography observation of vortex lattices in a superconductor. Phys. Rev. Lett. **70**, 2952 (1993)

[6.287] V.I. Petrov, G.V. Spivak: On a stroboscopic Lorentz microscope. Z. Angew. Phys. **27**, 188 (1969)

[6.288] O. Bostanjoglo, Th. Rosin: Resonance oscillations of magnetic domain walls and Bloch lines observed by stroboscopic electron microscopy. Phys. Status Solidi A **57**, 561 (1980)

[6.289] H. Mahl, W. Weitsch: Nachweis von fluktuierenden Ladungen in isolierenden Filmen bei Elektronenbestrahlung. Optik **17**, 107 (1960)

[6.290] H. Mahl, W. Weitsch: Versuche zur Beseitigung von Aufladungen auf Durchstrahlungsobjekten durch zusätzliche Bestrahlung mit langsamen Elektronen. Z. Naturforsch. A **17**, 146 (1962)

[6.291] G.H. Curtis, R.P. Ferrier: The electric charging of electron microscopical specimens. J. Phys D **2**, 1035 (1969)

[6.292] L. Reimer: Aufladung kleiner Teilchen im Elektronenmikroskop. Z. Naturforsch. A **20**, 151 (1965)

[6.293] V. Drahos, J. Komrska, M. Lenc: Shadow images of charged spherical particles, in *Electron Microscopy 1968*, Vol. 1, ed. by D.S. Bocciarelli (Tipografia Poliglotta Vaticana, Rome, 1968) p. 157

[6.294] C. Jönsson, H. Hoffmann: Der Einfluß von Aufladungen auf die Stromdichteverteilung im Elektronenschattenbild dünner Folien. Optik **21**, 432 (1964)

[6.295] H. Pfisterer, E. Fuchs, W. Liesk: Elektronenmikroskopische Abbildung ferroelektrischer Domänen in dünnen $BaTiO_3$-Einkristallschichten. Naturwissenschaften **49**, 178 (1962)

[6.296] H. Blank, S. Amelinckx: Direct observation of ferroelectric domains in $BaTiO_3$ by means of the electron microscope. Appl. Phys. Lett. **2**, 140 (1963)

[6.297] E. Fuchs, W. Liesk: Elektronenmikroskopische Beobachtung von Domänenkonfigurationen und von Umpolarisationsvorgängen in dünnen $BaTiO_3$-Einkristallen. J. Phys. Chem. Solidi **25**, 845 (1964)

[6.298] R. Ayroles, J. Torres, J. Aubree, C. Roucau, M. Tanaka: Electron-microscope observation of structure domains in the ferroelectric phase of lead phosphate. Appl. Phys. Lett. **34**, 4 (1979)

[6.299] C. Manolikas, S. Amelinckx: Phase transitions in ferroelastic lead orthovanadate as observed by means of electron microscopy and electron diffraction. Phys. Status Solidi A **60**, 607 (1980)

[6.300] M. Tanaka, G. Honjo: Electron optical studies of $BaTiO_3$ single crystal films. J. Phys. Soc. Jpn. **19**, 954 (1964)

[6.301] J.M. Titchmarsh, G.R. Booker: The imaging of electric field regions associated with p-n junctions, in *Electron Microscopy 1972* ed. by V.E. Cosslett (IoP, London, 1972) p. 540

[6.302] P.G. Merli, G.F. Missiroli, G. Pozzi: TEM observations of p-n junctions. Phys. Status Solidi A **30**, 699 (1975)

[6.303] C. Capiluppi, P.G. Merli, G. Pozzi, I. Vecchi: Out-of-focus observations of p-n junctions by high-voltage microscopy. Phys. Status Solidi A **35**, 165 (1976)

[6.304] S. Frabboni, G. Matencci, G. Pozzi: Electron holographic observation of the electrostatic field associated with thin reverse-biased p-n junctions. Phys. Rev. Lett. **55**, 2196 (1985)

[6.305] G. Matteucci, G.F. Missiroli, G. Pozzi: Electron holography of electrostatic fields. J. Electron Microscopy **45**, 19 (1996)

Chapter 7

[7.1] C. Kittel: *Introduction to Solid State Physics* (Wiley, New York, 1976)

[7.2] N.W. Ashcroft, N.D. Mermin: *Solid State Physics* (Holt-Saunders, New York, 1976)

[7.3] M.J. Buerger: *Contemporary Crystallography* (McGraw-Hill, New York, 1970)

[7.4] C.G. Darwin: The theory of x-ray diffraction. Philos. Mag. **27**, 315 and 675 (1914)

[7.5] Z.G. Pinsker: *Dynamical Scattering of X-rays in Crystals*, Springer Series in Solid-State Science, Vol. 3 (Springer, Berlin, 1978)

[7.6] A. Howie, M.J. Whelan: Diffraction contrast of electron microscopic images of lattice defects. Proc. Roy. Soc. A **263**, 217 (1961); **267**, 206 (1962)

[7.7] H. Bethe: Theorie der Beugung von Elektronen an Kristallen. Ann. Phys. **87**, 55 (1928)

[7.8] G. Thomas, E. Levine: Increase of extinction distance with temperature in Si. Phys. Status Solidi **11**, 81 (1965)

[7.9] A. Howie, U. Valdre: Temperature dependence of the extinction distance in electron diffraction. Philos. Mag. **15**, 777 (1967)

[7.10] L. Sturkey: The use of electron-diffraction intensities in structure determination. Acta Cryst. **10**, 858 (1957)

[7.11] H. Niehrs: Die Formulierung der Elektronenbeugung mittels einer Streumatrix und ihre praktische Verwendbarkeit. Z. Naturforsch. A **14**, 504 (1959)

[7.12] F. Fujimoto: Dynamical theory of electron diffraction in Laue-case. J. Phys. Soc. Jpn. **14**, 1558 (1959); **15**, 859 and 1022 (1960)

[7.13] A.J.F. Metherell: Diffraction of electrons by perfect crystals, in Ref. [1.31], Vol. 2, 397

[7.14] C.J. Humphreys, R.M. Fisher: Bloch wave notation in many-beam electron diffraction theory. Acta Cryst. A **27**, 42 (1971)

[7.15] J.P. Spencer, C.J. Humphreys: Electron diffraction from tilted specimens and its application to SEM, in [Ref. 1.78, p. 310]

[7.16] L.E. Thomas, C.G. Shirley, J.S. Lally, R.M. Fisher: The critical voltage effect and its applications, in *High Voltage Electron Microscopy* (Academic, London, 1974) p. 38

[7.17] P.B. Hirsch, A. Howie, R.B. Nicholson, D.W. Pashley, M.J. Whelan: *Electron Microscopy of Thin Crystals* (Butterworths, London, 1965)

[7.18] A. Howie: Inelastic scattering of electrons by crystals. Proc. Roy. Soc. A **271**, 268 (1963)

[7.19] H. Yoshioka: Effect of inelastic waves on electron diffraction. J. Phys. Soc. Jpn. **12**, 618 (1957)

[7.20] Z.L. Wang: *Elastic and Inelastic Scattering in Electron Diffraction and Imaging* (Plenum, New York, 1995)

[7.21] G. Radi: Unelastische Streuung in der dynamischen Theorie der Elektronenbeugung. Z. Phys. **212**, 146 (1968)

[7.22] R. Serneels, D. Haentjens, R. Gevers: Extension of the Yoshioka theory of inelastic electron scattering in crystals. Philos. Mag. A **42**, 1 (1980)

[7.23] C.J. Humphreys, M.J. Whelan: Inelastic scattering of fast electrons by crystals. Philos. Mag. **20**, 165 (1969)

[7.24] A. Weickenmeier, H. Kohl: Computation of the atomic inner-shell cross-sections for fast electrons in crystals. Philos. Mag. B **60**, 467 (1989)

[7.25] A. Weickenmeier, E. Quandt, H. Kohl, H. Rose, H. Niedrig: Computation and measurement of characteristic energy-loss large-angle convergent-beam patterns of molybdenum selenide. Ultramicroscopy **49**, 210 (1993)

[7.26] C.R. Hall, P.B. Hirsch: Effect of thermal diffuse scattering on propagation of high energy electrons through crystals. Proc. Roy. Soc. A **286**, 158 (1965)

[7.27] P. Rez, C.J. Humphreys, M.J. Whelan: The distribution of intensity in electron diffraction patterns due to phonon scattering. Philos. Mag. **35**, 81 (1977)

[7.28] Y. Kainuma: The theory of Kikuchi pattern. Acta Cryst. **8**, 247 (1955)

[7.29] R.G. Blake, A. Jostsons, P.M. Kelly, J.G. Napier: The determination of extinction distances and anomalous absorption coefficients by STEM. Philos. Mag. A **37**, 1 (1978)

[7.30] J.W. Steeds: Many-beam diffraction effects in gold and measurements of absorption parameters by fitting computer graphs. Phys. Status Solidi **38**, 203 (1970)

[7.31] D.M. Bird, Q.A. King: Absorptive factors for high-energy electron diffraction. Acta Cryst. A **46**, 202 (1990)

[7.32] C.J. Allen, C.J. Rossouw: Absorptive potentials due to ionization and thermal diffuse scattering by fast electrons in crystals. Phys. Rev. B **42**, 11644 (1990)

[7.33] A. Weickenmeier, H. Kohl: Computation of absorptive form factors for high-energy electron diffraction. Acta Cryst. A **47**, 590 (1991)

[7.34] A. Weickenmeier, H. Kohl: The influence of anisotropic thermal vibrations on absorptive form factors for high-energy electron diffraction. Acta Cryst. A **54**, 283 (1998)

[7.35] W. Coene, D. Van Dyck: Inelastic scattering of high-energy electrons in real space. Ultramicroscopy **33**, 261 (1990)

[7.36] H. Hashimoto: Energy dependence of extinction distance and transmission power for electron waves in crystals. J. Appl. Phys. **35**, 277 (1964)

[7.37] A. Mazel, R. Ayroles: Étude de la distance d'extinction et du coefficient d'absorption des electrons dans des échantillons d'aluminium pour des tensions comprises 50 et 1200 kilovolts. J. Microscopie **7**, 793 (1968)

[7.38] G. Dupouy, F. Perrier, R. Uyeda, R. Ayroles, A. Mazel: Mesure de coefficient d'absorption des électrons accélérés sons des tensions comprises 100 et 1200 kV. J. Microscopie **4**, 429 (1965)

[7.39] G. Radi: Complex lattice potentials in electron diffraction calculated for a number of crystals. Acta Cryst. A **26**, 41 (1970)

[7.40] P.A. Doyle: Absorption coefficients for Al 111 systematics: Theory and comparison with experiment. Acta Cryst. A **26**, 133 (1970)

[7.41] L. Reimer, M. Wächter: Complex Fourier coefficients of the crystal lattice potential, in *Electron Microscopy 1980*, Vol. 3, ed. by P. Brederoo, G. Boom (Seventh European Congress Electron Microscopy Foundation, Leiden, 1980) p. 192

[7.42] G. Meyer-Ehmsen: Untersuchungen zur normalen und anomalen Absorption von Elektronen in Si- und Ge-Einkristallen bei verschiedenen Temperaturen. Z. Phys. **218**, 352 (1969)
[7.43] M.J. Goringe: Temperature dependence of the absorption of fast electrons in Cu. Philos. Mag. **14**, 93 (1966)
[7.44] M.J. Goringe, M.J. Whelan: The absorption of fast electrons in crystals, in [Ref. 1.54, Vol. 1, p. 49]
[7.45] D. Renard, P. Croce, M. Gandais, M. Sauvin: Etude expérimentale de l'absorption des électrons dans l'or. Phys. Status Solidi B **47**, 411 (1971)
[7.46] H.G. Badde, L. Reimer: Measurement of complex structure potentials in Au and PbTe by convergent electron diffraction, in *Electron Microscopy 1972* (IoP, London, 1972) p. 440
[7.47] P. Goodman, G. Lehmpfuhl: Electron diffraction study of MgO h00-systematic interactions. Acta Cryst. **22**, 14 (1967)
[7.48] K.G. Gaukler, K. Graff: Struktur- und Absorptionspotentiale von KCl und NaCl aus Beugungsaufnahmen in konvergentem Elektronenbündel. Z. Phys. **232**, 190 (1970)
[7.49] H. Hashimoto, A. Howie, M.J. Whelan: Anomalous electron absorption effects in metal foils. Proc. Roy. Soc. A **269**, 80 (1962)
[7.50] P. Hagemann, L. Reimer: An experimental proof of the dependent Bloch wave model by large angle scattering from thin crystals. Philos. Mag. **40**, 367 (1979)
[7.51] M.V. Berry: Diffraction in crystals at high energies. J. Phys. C **4**, 697 (1971)
[7.52] M.V. Berry, K.E. Mount: Semiclassical approximations in wave mechanics. Rep. Progr. Phys. **35**, 315 (1972)
[7.53] K. Kambe, G. Lehmpfuhl, F. Fujimoto: Interpretation of electron channelling by the dynamical theory of electron diffraction. Z. Naturforsch. A **29**, 1034 (1974)
[7.54] F. Nagata, A. Fukuhara: 222 electron reflection from Al and systematic interaction. Jpn. J. Appl. Phys. **6**, 1233 (1967)
[7.55] R. Uyeda: Dynamical effects in high voltage electron diffraction. Acta Cryst. A **24**, 175 (1968)
[7.56] J.S. Lally, C.J. Humphreys, A.J.F. Metherell, R.M. Fisher: The critical voltage effect in high voltage electron microscopy. Philos. Mag. **25**, 321 (1972)
[7.57] L.E. Thomas: Kikuchi patterns in HVEM. Philos. Mag. A **26**, 1447 (1972)
[7.58] A.F. Moodie, J.R. Sellar, D. Imeson, C.J. Humphreys: Convergent beam diffraction in the high voltage electron microscope, in [Ref. 1.80, p. 191]
[7.59] J.R. Sellar, D. Imeson, C.J. Humphreys: Experimental and theoretical study of the convergent-beam critical voltage effect in high voltage electron diffraction, in *Electron Microscopy 1980*, Vol. 1, ed. by P. Brederoo, G. Boom (Seventh European Congress on Electron Microscopy Foundation, Leiden, 1980) p. 120
[7.60] T. Arii, R. Uyeda: Vanishing voltages of the second order reflections in electron diffraction. Jpn. J. Appl. Phys. **8**, 621 (1969)
[7.61] T. Arii, R. Uyeda, O. Terasaki, D. Watanabe: Accurate determination of atomic scattering factors of fcc and hcp metals by high voltage electron diffraction. Acta Cryst. A **29**, 295 (1973)
[7.62] A. Fukuhara, A. Yanagisawa: Vanishing of 222 Kikuchi line from Ag crystal. Jpn. J. Appl. Phys. **8**, 1166 (1969)

[7.63] M. Fujimoto, O. Terasaki, D. Watanabe: Determination of atomic scattering factors of V and Cr by means of vanishing Kikuchi line method. Phys. Lett. A **41**, 159 (1972)

[7.64] A. Rocher, B. Jouffrey: Contribution á l'étude des tensions critiques dans le Cu et Al. C.R. Acad. Sci. Paris B **275**, 133 (1972)

[7.65] D. Watanabe, R. Uyeda, A. Fukuhara: Determination of the atom form factor by high voltage electron diffraction. Acta Cryst. A **25**, 138 (1969)

[7.66] E.A. Hewet, C.J. Humphreys: Si(111) and Ge(111) and (200) scattering factors determined from critical voltage measurements, in [Ref. 1.78, p. 52]

[7.67] E.P. Butler: Application of the critical voltage effect to the study of compositional changes in Ni-Au alloys. Philos. Mag. **26**, 33 (1972)

[7.68] I.P. Jones, E.G. Tapetado: The dependence of electron distribution and atom vibration in hcp metals on the c/a ratio: An investigation using the critical voltage technique, in [Ref. 1.78, p. 48]

[7.69] K. Kuroda, Y. Tomokiyo, T. Eguchi: Temperature dependence of critical voltages in Cu-based alloys, in *Electron Microscopy 1980*, Vol. 4, ed. by P. Brederoo, J. Van Landuyt (Seventh European Congress on Electron Microscopy Foundation, Leiden, 1980) p. 112

[7.70] C.G. Shirley, R.M. Fisher: Application of the critical voltage effect to alloy studies, in *Electron Microscopy 1980*, Vol. 4, ed. by P. Brederoo, J. Van Landuyt (Seventh European Congress on Electron Microscopy Foundation, Leiden, 1980) p. 88

[7.71] J. Gjønnes, R. Høier: The application on non-systematic many-beam dynamic effects to structure-factor determination. Acta Cryst. A **27**, 313 (1971)

[7.72] J. Tafto, J. Gjønnes: The intersecting Kikuchi line technique: critical voltage at any voltage. Ultramicroscopy **17**, 329 (1985)

[7.73] R. Leonhardt, H. Richter, W. Rossteutscher: Elektronenbeugungsuntersuchungen zur Struktur dünner nichtkristalliner Schichten. Z. Phys. **165**, 121 (1961)

[7.74] D.J.H. Cockayne, D.R. McKenzie: Electron diffraction analysis of polycrystalline and amorphous thin films. Acta Cryst. A **44**, 870 (1988)

[7.75] D. Cockayne, D. McKenzie: Structural studies of amorphous and polycrystalline materials using energy filtered RDF analysis, in *Electron Microscopy 1992*, Vol. 1, ed. by A. Rios et al. (University of Granada, 1992) p. 179

[7.76] M. Horstmann, G. Meyer: Messung der elastischen Elektronenbeugungsintensitäten polykristalliner Al-Schichten. Acta Cryst. **15**, 271 (1962)

[7.77] M. Blackman: On the intensities of electron diffraction rings. Proc. Roy. Soc. A **173**, 68 (1939)

[7.78] M. Horstmann, G. Meyer: Eine Gegenfeldanordung zur Messung von Energie- und Winkelverteilungen gestreuter Elektronen. Z. Phys. **159**, 563 (1960)

[7.79] P.H. Denbigh, C.W.B. Grigson: Scanning electron diffraction with energy analysers. J. Sci. Instr. **42**, 395 (1965)

[7.80] M.F. Tompsett: Review: scanning high-energy electron diffraction in materials science. J. Mater. Sci. **7**, 1069 (1972)

[7.81] L. Reimer, I. Fromm, I. Naundorf: Electron spectroscopic diffraction. Ultramicroscopy **32**, 80 (1990)

[7.82] J. Mayer, C. Deininger, L. Reimer: Electron spectroscopic diffraction, in *Energy-Filtering Transmission Electron Microscopy*, Springer Series in Optical Sciences, Vol. 71, ed. by L. Reimer (Springer, Berlin, 1995) p. 291

[7.83] A.A. Maradudin, E.W. Montroll, G.H. Weiss, I.P. Ipatova: *Theory of Lattice Dynamics in the Harmonic Approximation.* Solid State Physics Supplement 3 (Academic, New York, 1971)

[7.84] C.J. Humphreys, P.B. Hirsch: Absorption parameters in electron diffraction. Philos. Mag. **18**, 115 (1968)

[7.85] C.R. Hall: The scattering of high energy electrons by the thermal vibrations of crystals. Philos. Mag. **12**, 815 (1965)

[7.86] G. Honjo, S. Kodera, N. Kitamura: Diffuse streak diffraction patterns from single crystals. J. Phys. Soc. Jpn. **19**, 351 (1964)

[7.87] K. Komatsu, K. Teramoto: Diffuse streak patterns from various crystals in x-ray and electron diffraction. J. Phys. Soc. Jpn. **21**, 1152 (1966)

[7.88] N. Kitamura: Temperature dependence of diffuse streaks in single crystal Si electron diffraction patterns. J. Appl. Phys. **37**, 2187 (1966)

[7.89] H.P. Herbst, G. Jeschke: Diffuse streak-patterns from PbJ_2- and Bi-single crystals and their temperature dependence, in *Electron Microscopy 1968*, Vol. 1, ed. by D.S. Bocciarelli (Tipografia Poliglotta Vaticana, Rome, 1968) p. 293

[7.90] E.M. Hörl: Thermisch-diffuse Elektronenstreuung in As-, Sb- und Bi-Kristallen. Optik **27**, 99 (1968)

[7.91] Z.L. Wang: Dynamics of thermal diffuse scattering in high-energy electron diffraction and imaging: theory and experiments. Philos. Mag. B **65**, 559 (1992)

[7.92] M. Horstmann: Einfluß der Kristalltemperatur auf die Intensitäten dynamischer Elektroneninterferenzen. Z. Phys. **183**, 375 (1965)

[7.93] M. Horstmann, G. Meyer: Messung der Elektronenbeugungs-Intensitäten polykristalliner Al-Schichten bei tiefer Temperatur und Vergleich mit der dynamischen Theorie. Z. Phys. **182**, 380 (1965)

[7.94] M. Horstmann: Messung der thermisch diffusen Elektronenstreuung in polykristallinen Al-Schichten. Z. Phys. **188**, 412 (1965)

[7.95] W. Zechnall: Temperaturabhängigkeit des Streuuntergrundes im Elektroneninterferenzdiagramm polykristalliner Ag-Schichten. Z. Phys. **229**, 62 (1969)

[7.96] J. Hansen-Schmidt, M. Horstmann: Temperaturabhängigkeit der Streuabsorption schneller Elektronen in polykristallinen Au-Schichten. Z. Naturforsch. A **20**, 1239 (1965)

[7.97] H. Boersch, O. Bostanjoglo, H. Niedrig: Temperaturabhängigkeit der Transparenz dünner Schichten für schnelle Elektronen. Z. Phys. **180**, 407 (1964)

[7.98] W. Glaeser, H. Niedrig: Temperature dependence of dynamical electron diffraction intensities of polycrystalline foils. J. Appl. Phys. **37**, 4303 (1966)

[7.99] W.W. Albrecht, H. Niedrig: Temperature dependence of dynamical electron diffraction intensities of polycrystalline foils II. J. Appl. Phys. **39**, 3166 (1968)

[7.100] G. Jeschke, D. Willasch: Temperaturabhängigkeit der anomalen Elektronenabsorption von Bi-Einkristallen. Z. Phys. **238**, 421 (1970)

[7.101] C.R. Hall: On the thickness dependence of Kikuchi band contrast. Philos. Mag. **22**, 63 (1970)

[7.102] H. Boersch: Über Bänder bei Elektronenbeugung. Phys. Z. **38**, 1000 (1937)

[7.103] H. Pfister: Elektroneninterferenzen an Bleijodid bei Durchstrahlung im konvergenten Bündel. Ann. Phys. **11**, 239 (1953)

[7.104] M. Komura, S. Kojima, T. Ichinokawa: Contrast reversal of Kikuchi bands in transmission electron diffraction. J. Phys. Soc. Jpn. **33**, 1415 (1972)

[7.105] S. Takagi: On the temperature diffuse scattering of electrons. J. Phys. Soc. Jpn. **13**, 287 (1958)

[7.106] J. Gjønnes: The influence of Bragg scattering on inelastic and other forms of diffuse scattering of electrons. Acta Cryst. **20**, 240 (1966)

[7.107] K. Ishida: Inelastic scattering of fast electrons by crystals. J. Phys. Soc. Jpn. **28**, 450 (1970); **30**, 1439 (1971)

[7.108] K. Okamoto, T. Ichinokawa, Y.H. Ohtsuki: Kikuchi patterns and inelastic scattering. J. Phys. Soc. Jpn. **30**, 1690 (1971)

[7.109] R. Høier: Multiple scattering and dynamical effects in diffuse electron scattering. Acta Cryst. A **29**, 663 (1973)

[7.110] R. Serneels, C. van Roost, G. Knuyt: Kikuchi patterns in transmission electron diffraction. Philos. Mag. A **45**, 677 (1982)

[7.111] L. Reimer, U. Heilers, G. Saliger: Kikuchi band contrast in diffraction patterns recorded by transmitted and backscattered electrons. Scanning **8**, 101 (1986)

[7.112] J.G. Philip, M.J. Whelan, R.F. Egerton: The contribution of inelastically scattered electrons to the diffraction pattern and images of a crystalline specimen, in [Ref. 1.56, Vol. 1, p. 276]

[7.113] L. Reimer, I. Fromm, Ch. Hülk, R. Rennekamp: Energy-filtering transmission electron microscopy in materials science. Microsc. Microanal. Microstruct. **3**, 141 (1992)

Chapter 8

[8.1] J.B. Le Poole: Ein neues Elektronenmikroskop mit stetig regelbarer Vergrö ßerung. Philips Techn. Rundsch. **9**, 33 (1947)

[8.2] M.E. Haine, R.S. Page, R.G. Garfitt: A three-stage electron microscope with stereographic dark field and electron diffraction capabilities. J. Appl. Phys. **21**, 173 (1950)

[8.3] W.D. Riecke, E. Ruska: Über ein Elektronenmikroskop mit Einrichtungen für Feinbereichsbeugung und Dunkelfeldabbildung durch Einzelreflex. Z. Wiss. Mikr. **63**, 288 (1957)

[8.4] A.W. Agar: Accuracy of selected-area microdiffraction in the electron microscope. Br. J. Appl. Phys. **11**, 185 (1960)

[8.5] W.D. Riecke: Über die Genauigkeit der Übereinstimmung von ausgewähltem und beugendem Bereich bei der Feinbereichs-Elektronenbeugung im LePooleschen Strahlengang. Optik **18**, 278 (1961)

[8.6] W.D. Riecke: Verzeichung und Auflösung der im LePooleschen Strahlengang aufgenommenen Beugungsdiagramme. Optik **18**, 373 (1961)

[8.7] W.C.T. Dowell: Fehler von Beugungsdiagrammen, die mittels Elektronenlinsen erzeugt und abgebildet sind. Optik **20**, 581 (1963)

[8.8] J.C. Lodder, K.G. van der Berg: A method for accurately determining lattice parameters using electron diffraction in a commercial electron microscope. J. Microsc. **100**, 93 (1974)

[8.9] F. Fujimoto, K. Komaki, S. Takagi, H. Koike: Diffraction patterns obtained by scanning electron microscopy. Z. Naturforsch. A **27**, 441 (1972)

[8.10] A.G. Pogany, P.S. Turner: Reciprocity in electron diffraction and microscopy. Acta Cryst. A **24**, 103 (1968)

[8.11] M.N. Thompson: A scanning transmission microscope: some techniques and applications, in *Scanning Electron Microscopy: Systems and Applications*, ed. by W.C. Nixon (IoP, London, 1973) p. 176

[8.12] D.M. Maher: Scanning electron diffraction in TEM and SEM operating in the transmission mode, in *Scanning Electron Microscopy 1974*, ed. by O. Johari (IIT Research Institute, Chicago, 1974) p. 176

[8.13] K.J. van Oostrum, A. Leenhouts, A. Jore: A new scanning micro-diffraction technique. Appl. Phys. Lett. **23**, 283 (1973)

[8.14] R.H. Geiss: STEM electron diffraction from 30Å diameter areas, in *Developments in Electron Microscopy and Analysis 1975*, ed. by J.A. Venables (Academic, London, 1976) p. 61

[8.15] J.P. Chevalier, A.J. Craven: Microdiffraction, application to short range order in a quenched copper-platinum alloy. Philos. Mag. **36**, 67 (1977)

[8.16] L.M. Brown, A.J. Craven, L.G.P. Jones, A. Griffith, W.M. Stobbs, C.J. Wilson: Application of high resolution STEM to material science, in *Scanning Electron Microscopy 1976/I*, ed. by O. Johari (IIT Research Inst, Chicago, 1976) p. 353

[8.17] H. von Harrach, C.E. Lyman, G.E. Verney, D.C. Joy, G.R. Booker: Performance of the Oxford field-emission STEM, in *Developments in Electron Microscopy and Analysis*, ed. by J.A. Venables (Academic, London 1976) p. 7

[8.18] L. Reimer: Electron diffraction methods in TEM, STEM and SEM. Scanning **2**, 3 (1979)

[8.19] W. Kossel, G. Möllenstedt: Elektroneninterferenzen im konvergenten Bündel. Naturwissenschaften **26**, 660 (1938)

[8.20] W. Kossel, G. Möllenstedt: Dynamische Anomalie von Elektroneninterferenzen. Ann. Phys. **42**, 287 (1942)

[8.21] P. Goodman, G. Lehmpfuhl: Elektronenbeugungsuntersuchungen im konvergenten Bündel mit dem Siemens Elmiskop I. Z. Naturforsch. A **20**, 110 (1965)

[8.22] H. Raith: Elektronenbeugung im konvergenten Bündel an gekühlten Präparaten mit dem Siemens Elmiskop I. Z. Naturforsch. A **20**, 855 (1965)

[8.23] W.J. Vine, R. Vincent, P. Spellward, J.W. Steeds: Observation of phase contrast in CBED patterns. Ultramicroscopy **41**, 423 (1992)

[8.24] K. Tsuda, M. Terauchi, M. Tanaka, T. Kaneyama, T. Honda: Observation of coherent CBED patterns from small lattice spacings and stacking faults, in [Ref. 1.61, Vol. 1, p. 865]

[8.25] N. Tanaka, M. Egi, K. Kimoto: Coherent CBED of PbTe/MgO double layers, in [Ref. 1.61, Vol. 1, p. 869]

[8.26] P.A. Midgley, M. Saunders, R. Vincent, J.W. Steeds: Energy-filtered convergent-beam diffraction: examples and future prospects. Ultramicroscopy **59**, 1 (1995)

[8.27] D.J.H. Cockayne, P. Goodman, J.C. Mills, A.F. Moodie: Design and generation of an electron diffraction camera for the study of small crystalline regions. Rev. Sci. Instr. **38**, 1097 (1967)

[8.28] J.M. Cowley, D.J. Smith, G.A. Sussex: Application of a high voltage STEM, in *Scanning Electron Microscopy 1970*, ed. by O. Johari (IIT Research Institute, Chicago, 1970) p. 11
[8.29] P. Goodman: Observation of background contrast in convergent beam patterns. Acta Cryst. A **28**, 92 (1972)
[8.30] M. Tanaka, R. Saito, K. Ueno, Y. Harada: Large-angle CBED. J. Electron Microsc. **29**, 408 (1980)
[8.31] I.K. Jordan, C.J. Rossouw, R. Vincent: Effects of energy filtering in LACBED patterns. Ultramicroscopy **35**, 237 (1991)
[8.32] C. van Essen: SEM channelling pattern from 2 μm selected areas, in [Ref. 1.55, Vol. 1, p. 237]
[8.33] R.J. Woolf, D.C. Joy, J.M. Titchmarsh: Scanning transmission electron diffraction in the SEM, in *Electron Microscopy 1972* ed. by V.E. Cosslett (IoP, London 1972) p. 498
[8.34] A.J. Craven: Specimen orientation in STEM, in *Developments in Electron Microscopy and Analysis 1977*, ed. by D.L. Misell (IoP, London, 1977) p. 311
[8.35] G. Möllenstedt, H.R. Meyer: Strahlengang zur Strukturanalyse von Einkristallen durch Elektronen-Transmissions-Doppelwinkelabrasterung. Optik **42**, 487 (1975)
[8.36] J.A. Eades: Another way to form zone-axis patterns, in *Electron Microscopy and Analysis 1980*, ed. by T. Mulvey (IoP, London, 1980) p. 9
[8.37] M. Tanaka, R. Saito, K. Ueno, Y. Harada: Large-angle convergent-beam electron diffraction. J. Electron Microsc. **29**, 408 (1980)
[8.38] H. Mahl, W. Weitsch: Kleinwinkelbeugung mit Elektronenstrahlen. Naturwissenschaften **47**, 301 (1960); Z. Naturforsch. A **15**, 1051 (1960)
[8.39] R.P. Ferrier: Small angle electron diffraction in the electron microscope, in *Advances in Optical and Electron Microscopy*, Vol. 2, ed. by R. Barer, V.E. Cosslett (Academic, London, 1969) p. 155
[8.40] R.H. Wade, J. Silcox: Small angle electron scattering from vacuum condensed metallic films. Phys. Status Solidi **19**, 57 and 63 (1967)
[8.41] J. Smart, R.E. Burge: Small-angle electron diffraction patterns of assemblies of spheres and viruses. Nature **205**, 1296 (1965)
[8.42] V. Drahos, A. Delong: Low-angle electron diffraction from defined specimen area, in [Ref. 1.55, Vol. 2, p. 147]
[8.43] R.T. Murray, R.P. Ferrier: Biological applications of electron diffraction. J. Ultrastruct. Res. **21**, 361 (1967)
[8.44] G.A. Bassett, A. Keller: Low-angle scattering in an electron microscope applied to polymers. Philos. Mag. **9**, 817 (1964)
[8.45] L. Reimer, K. Freking: Versuch einer quantitativen Erfassung der Textur von Au-Aufdampfschichten. Z. Phys. **184**, 119 (1965)
[8.46] R. Anderson, M.J. Carr, V.L. Himes, A.D. Mighell: The EISI index – a search/match tool for electron diffraction phase analysis, in [Ref. 1.60, Vol. 2, p. 514]
[8.47] C.W. Grigson: Improved scanning electron diffraction system. Rev. Sci. Instr. **36**, 1587 (1965)
[8.48] F.C.S.M. Totthill, W.C. Nixon, C.W.B. Grigson: Ultra-high vacuum modification of an AEI EM6 electron microscope for studies of nucleation in evaporated films, in *Electron Microscopy 1968*, Vol. 1, ed. by. D.S. Bocciarelli (Tipografia Poliglotta Vaticana, Rome, 1968) p. 229

[8.49] A.M. MacLeod, J.N. Chapman: A digital scanning and recording system for spot electron diffraction patterns. J. Phys. E **10**, 37 (1977)

[8.50] W. Riecke, F. Stöcklein: Eine Objektkammer mit universell beweglichem Präparattisch für Elektronenbeugungsuntersuchungen. Z. Phys. **156**, 163 (1959)

[8.51] J.M. Cowley: Surface energies and surface structure of small crystals studied by use of a STEM instrument. Surf. Sci. **114**, 587 (1982)

[8.52] C. Elibol, J.H. Ou, G.G. Hembree, J.M. Cowley: Improved instrument for medium energy electron diffraction and microscopy of surfaces. Rev. Sci. Instr. **56**, 1215 (1985)

[8.53] J.A. Venables, C.J. Harland: Electron back-scattering patterns – a new technique for obtaining crystal information in the SEM. Philos. Mag. **27**, 1193 (1973)

[8.54] M.N. Alam, M. Blackman, D.W. Pashley: High-angle Kikuchi patterns. Proc. Roy. Soc. London A **221**, 224 (1954)

[8.55] L. Reimer, W. Pöpper, B. Volbert: Contrast reversals in the Kikuchi bands of backscattered and transmitted electron diffraction patterns, in *Developments in Electron Microscopy and Analysis 1977*, ed. by D.L. Misell (IoP, London, 1977) p. 259

[8.56] D.G. Coates: Kikuchi-like reflection patterns obtained with the SEM. Philos. Mag. **16**, 1179 (1967)

[8.57] G.R. Booker: Scanning electron microscopy: Electron channelling effects, in [Ref. 1.30, p. 613]

[8.58] L. Reimer: Electron specimen interactions in SEM, in *Developments in Electron Microscopy and Analysis*, ed. by J.A. Venables (Academic, London, 1976) p. 86

[8.59] J.W. Steeds, G.J. Tatlock, J. Hamson: Real space crystallography. Nature **241**, 435 (1973)

[8.60] G.J. Tatlock, J.W. Steeds: Real space crystallography in molybdenite. Nature **246**, 126 (1973)

[8.61] J.W. Steeds, P.M. Jones, G.M. Rackham, M.D. Shannon: Crystallographic information from zone axis patterns, in *Developments in Electron Microscopy and Analysis*, ed. by J.A. Venables (Academic, London, 1976) p. 351

[8.62] J.W. Steeds, P.M. Jones, J.E. Loveluck, K.E. Cooke: The dependence of zone axis patterns on string integrals or the number of bound states in high energy electron diffraction. Philos. Mag. **36**, 309 (1977)

[8.63] M.D. Shannon, J.W. Steeds: On the relationship between projected crystal potential and the form of certain zone axis patterns in high energy electron diffraction. Philos. Mag. **36**, 279 (1977)

[8.64] W. Witt: Zur absoluten Präzisionsbestimmung von Gitterkonstanten mit Elektroneninterferenzen am Beispiel von Thallium-(I)-Chlorid. Z. Naturforsch. A **19**, 1363 (1964)

[8.65] J.M. Corbett, F.W. Boswell: Use of thin single crystals as reference standards for precision electron diffraction. J. Appl. Phys. **37**, 2016 (1966)

[8.66] A.L. MacKay: Calibration of diffraction patterns taken in the electron microscope. J. Phys. E **3**, 248 (1970)

[8.67] J.T. Jubb, E.E. Laufer: The beam-tilt device of an electron microscope as an internal diffraction standard. J. Phys. E **9**, 871 (1976)

[8.68] E.E. Laufer, J.T. Jubb, K.S. Milliken: The use of the beam tilt circuitry of an electron microscope for rapid determination of lattice constants. J. Phys. E **8**, 671 (1975)

[8.69] H. König: Gitterkonstantenbestimmung im Elektronenmikroskop. Naturwissenschaften **33**, 343 (1946)

[8.70] F.W.C. Boswell: A standard substance for precise electron diffraction measurements. Phys. Rev. **80**, 91 (1950)

[8.71] C. Lu, E.W. Malmberg: ZnO smoke as a reference standard in electron wavelength calibration. Rev. Sci. Instr. **14**, 271 (1943)

[8.72] R. Rühle: Über Gesetzmäßigkeiten in Texturaufnahmen von Elektronenbeugungsbildern. Optik **7**, 279 (1950)

[8.74] B.K. Vainshtein: *Structure Analysis by Electron Diffraction* (Pergamon, Oxford, 1964)

[8.75] J.A. Gard: *Interpretation of Electron Micrographs and Diffraction Patterns: The Electron Optical Investigation of Clays.* (Mineralogical Society, London, 1971)

[8.76] J.M. Cowley: Crystal structure determination by electron diffraction. Prog. Mater. Sci. **13**, 267 (1966)

[8.77] S. Nagakura: A method for correcting the primary exctinction effect in electron diffraction. Acta Cryst. **10**, 601 (1957)

[8.78] B.K. Vainshtein, A.N. Lobacher: Dynamic scattering and its use in structural electron diffraction studies. Sov. Phys. Cryst. **6**, 609 (1961)

[8.79] J.M. Cowley: Structure analysis of single crystals by electron diffraction. Acta Cryst. **6**, 516, 522, and 846 (1953)

[8.80] J.M. Cowley: The theoretical basis for electron diffraction structure analysis, in [Ref. 1.53, Vol. 1, p. JJ-1]

[8.81] S. Fujime, D. Watanabe, S. Ogawa: On forbidden reflection spots and unexpected streaks appearing in electron diffraction patterns from hexagonal Co. J. Phys. Soc. Jpn. **19**, 711 (1964)

[8.82] R. Vincent, P.A. Midgley: Double conical beam-rocking system For measurement of integrated electron diffraction intensities. Ultramicroscopy **53**, 271 (1994)

[8.83] Ultramicroscopy 107 (2007)

[8.84] J.F. Brown, D. Clark: The use of the three-stage electron microscope in crystal-structure analysis. Acta Cryst. **5**, 615 (1952)

[8.85] J.A. Gard: The use of stereoscopic tilt device of the electron microscope in unit-cell determinations. Br. J. Appl. Phys. **7**, 361 (1956)

[8.86] J.A. Gard: Interpretation of electron diffraction patterns, in [Ref. 1.29, p. 52]

[8.87] R.R. Dayal, J.A. Gard, F.P. Glasser: Crystal data on $FeAlO_3$. Acta Cryst. **18**, 574 (1965)

[8.88] J.A. Gard, J.M. Bennet: A goniometric specimen stage, and its use in crystallography, in [Ref. 1.54, Vol. 1, p. 593]

[8.89] G. Cliff, J.A. Gard, G.W. Lorimer, H.F.W. Taylor: Tacharanite. Mineral. Mag. **40**, 113 (1975)

[8.90] S. Kuwabara: Accurate determination of hydrogen positions in NH_4Cl by electron diffraction. J. Phys. Soc. Jpn. **14**, 1205 (1959)

[8.91] V.V. Udalova, Z.G. Pinsker: Electron diffraction study of the structure of ammonium sulfate. Sov. Phys. – Cryst. **8**, 433 (1963)

[8.92] J.A. Gard, H.F.W. Taylor, L.W. Staples: Studies in crystal structure using electron diffraction of single crystals, in [Ref. 1.52, Vol. 1, p. 449]
[8.93] H.M. Otte, J. Dash, H.F. Schaake: Electron microscopy and diffraction of thin films. Interpretation and correlation of images and diffraction patterns. Phys. Status Solidi **5**, 527 (1964)
[8.94] C. Laird, E. Eichen, W.R. Bitler: Accuracy in the use of electron diffraction spot patterns for determining crystal orientations. J. Appl. Phys. **37**, 2225 (1966)
[8.95] K. Lücke, H. Perlwitz, W. Pitsch: Elektronenmikroskopische Bestimmung der Orientierungsverteilung der Kristallite in gewalztem Kupfer. Phys. Status Solidi **7**, 733 (1964)
[8.96] F. Haessner, U. Jakubowski, N. Wilkens: Anwendung elektronenmikroskopischer Feinbereichsbeugung zur Ermittlung der Walztextur von Kupfer. Phys. Status Solidi **7**, 701 (1964)
[8.97] P.L. Ryder, W. Pitsch: The uniqueness of orientation determination by selected area electron diffraction. Philos. Mag. **15**, 437 (1967)
[8.98] P.L. Ryder, W. Pitsch: On the accuracy of orientation determination by selected area diffraction. Philos. Mag. **18**, 807 (1968)
[8.99] D.J. Mazey, R.S. Barnes, A. Howie: On interstitial dislocation loops in aluminium bombarded with alpha-particles. Philos. Mag. **7**, 1861 (1962)
[8.100] M.H. Loretto, L.M. Clarebrough, P. Humble: Nature of dislocation loops in quenched Al. Philos. Mag. **13**, 953 (1966)
[8.101] M. von Heimendahl: Determination of metal foil thickness and orientation in electron microscopy. J. Appl. Phys. **35**, 457 (1964)
[8.102] S.S. Sheinin, C.D. Cann: The determination of orientation from Kikuchi patterns. Phys. Status Solidi **11**, K1 (1965)
[8.103] R. Bonnet, F. Durand: Precise determination of the relative orientation of two crystals from the analysis of two Kikuchi patterns. Phys. Status Solidi A **27**, 543 (1975)
[8.104] W. Griem, P. Schwaab, U. Stockhofe: Behandlung von Epitaxie-Fragen bei der Elektronenbeugung mit Hilfe der Datenverarbeitung. Arch. Eisenhüttenwesen **43**, 509 (1972)
[8.105] W. Griem, P. Schwaab: Behandlung von gesetzmäßigen Verwachsungen nicht-kubischer und teilkohärenter Phasen bei der Elektronenbeugung. Arch. Eisenhüttenwesen **44**, 677 (1973)
[8.106] R. Bonnet, E.E. Laufer: Precise determination of the relative orientation of two crystals from the analysis of spot diffraction patterns. Phys. Status Solidi A **40**, 599 (1977)
[8.107] M.D. Drazin, M.H. Otte: The systematic determination of crystallographic orientations from three octahedral traces on a plane surface. Phys. Status Solidi **3**, 824 (1963)
[8.108] A.G. Crocker, M. Bevis: The determination of the orientation and thickness of thin foils from transmission electron micrographs. Phys. Status Solidi **6**, 151 (1964)
[8.109] G. Thomas: *Transmission Electron Microscopy of Metals* (Wiley, New York, 1962)
[8.110] A. Baltz: Rotation of image and selected area diffraction patterns in the RCA EMU3 electron microscope. Rev. Sci. Instr. **33**, 246 (1962)
[8.111] P. Delavignette: Determination of some instrumental constants of the electron microscope Philips EM 200. J. Sci. Instr. **40**, 461 (1963)

[8.112] H. Raether: Reflexion von schnellen Elektronen an Einkristallen, Z. Phys. **78**, 527 (1932)
[8.113] R.D. Heidenreich: Theory of the forbidden (222) electron reflection in the diamond structure. Phys. Rev. **77**, 271 (1950)
[8.114] M. Takagi, S. Morimoto: The forbidden 222 electron diffraction from Ge. J. Phys. Soc. Jpn. **18**, 819 (1963)
[8.115] H. Göttsche: Zur Struktur dünner Ag-Schichten. Z. Phys. **134**, 517 (1953)
[8.116] W. Pitsch: Kristallographische Eigenschaften von Eisennitrid-Ausscheidungen im Ferrit. Arch. Eisenhüttenwesen **32**, 493 and 573 (1961)
[8.117] S. Ogawa, D. Watanabe, H. Watanabe, T. Komoda: The direct observation of the long period of the ordered alloy CuAu(II) by means of electron microscope, in [Ref. 1.52, Vol. 1, p. 334]
[8.118] S. Ogawa: On the antiphase domain structures in ordered alloys. J. Phys. Soc. Jpn. **17**, Suppl. B-II, 253 (1962)
[8.119] P. Goodman: A practical method of three-dimensional space group analysis using CBED. Acta Cryst. A **31**, 804 (1975)
[8.120] B.F. Buxton, J.A. Eades, J.W. Steeds, G.M. Rackham: The symmetry of electron diffraction zone axis patterns. Philos. Trans. Roy. Soc. London A **281**, 171 (1976)
[8.121] P.E. Champness: Convergent beam electron diffraction. Mineral. Mag. **51**, 33 (1987)
[8.122] M. Tanaka: Symmetry analysis. J. Electron Microsc. Techn. **13**, 27 (1989)
[8.123] J.W. Steeds, J.P. Morniroli: Selected area electron diffraction (SAED) and convergent beam electron diffraction (CBED), in *Minerals and Reactions at the Atomic Scale: Transmission Electron Microscopy*, Reviews in Mineralogy, Vol. 27, ed. by P.R. Buseck (Mineralogical Society of America, Washington, 1992) p. 37
[8.124] J. Gjønnes, A.F. Moodie: Extinction conditions in the dynamical theory of electron diffraction. Acta Cryst. **19**, 65 (1965)
[8.125] J.W. Steeds, R. Vincent: Use of high-symmetry zone axes in electron diffraction in determining crystal point and space groups. J. Appl. Cryst. **16**, 317 (1983)
[8.126] I. Ackermann: Beobachtungen an dynamischen Interferenzerscheinungen im konvergenten Elektronenbündel. Ann. Phys. **2**, 19 and 41 (1948)
[8.127] P.M. Kelly, A. Jostsons, R.G. Blake, J.G. Napier: The determination of foil thickness by STEM. Phys. Status Solidi A **31**, 771 (1975)
[8.128] S.M. Allen: Foil thickness measurements from convergent beam diffraction patterns. Philos. Mag. A **43**, 325 (1981)
[8.129] R.G. Blake, A. Jostsons, P.M. Kelly, J.G. Napier: The determination of extinction distances and anomalous absorption coefficients by STEM. Philos. Mag. A **37**, 1 (1978)
[8.130] J.M. Zuo, J.C.H. Spence, M. O'Keefe: Bonding in GaAs. Phys. Rev. Lett. **61**, 353 (1988)
[8.131] K. Kambe: Study of simultaneous reflexion in electron diffraction by crystals. J. Phys. Soc. Jpn. **12**, 13 (1957)
[8.132] J. Gjønnes, R. Høier: The application of non-systematic many-beam dynamical effects to structure-factor determination. Acta Cryst. A **27**, 313 (1971)
[8.133] K. Marthinsen, R. Høier: Many-beam dynamic effects and phase information in electron channelling patterns. Acta Cryst. A **42**, 484 (1986)

[8.134] D.M. Bird, R. James, A.R. Preston: Direct measurement of crystallographic phase by electron diffraction. Phys. Rev. Lett. **59**, 1216 (1989)
[8.135] J.M. Cowley: The determination of structure factors from dynamical effects in electron diffraction. Acta Cryst. A **25**, 129 (1969)
[8.136] J.M. Zuo, K. Gjønnes, J.C.H. Spence: FORTRAN source listing for simulating three-dimensional convergent beam patterns with absorption by the Bloch wave method. J. Electron Microsc. Techn. **12**, 29 (1989)
[8.137] J.M. Zuo, J.C.H. Spence: Automated structure factor measurement by convergent-beam electron diffraction. Ultramicroscopy **35**, 185 (1991)
[8.138] R. Høier, L.N. Bakken, K. Marthinsen, R. Holmestad: Structure factor determination in non-centrosymmetric crystals by a 2-dimensional CBED-based multi-parameter refinement method. Ultramicroscopy **49**, 159 (1993)
[8.139] C. Deininger, G. Necker, J. Mayer: Determination of structure factors, lattice strains and accelerating voltage by energy-filtered CBED. Ultramicroscopy **54**, 15 (1994)
[8.140] J.M. Zuo. J.C.H. Spence, J. Downs, J. Mayer: Measurement of individual structure-factor phases with tenth-degree accuracy: The 00.2 reflection in BeO studied by electron and x-ray diffraction. Acta Cryst. A **49**, 422 (1993)
[8.141] J.W. Steeds, K.K. Fung: Application of convergent beam electron microscopy in materials science, in [Ref. 1.57, p. 620]
[8.142] J.W. Steeds: Convergent beam electron diffraction, in [Ref. 1.66, p. 387]
[8.143] R. Ayer: Determination of unit cell. J. Electron Microsc. Techn. **13**, 16 (1989)
[8.144] D.M. Bird: Theory of zone axis electron diffraction. J. Electron Microsc. Techn. **13**, 77 (1989)
[8.145] P.M. Jones, G.M. Rackham, J.W. Steeds: Higher order Laue zone effects in electron diffraction and their use in lattice parameter determination. Proc. Roy. Soc. London A **354**, 197 (1977)
[8.146] B.F. Buxton: Bloch waves and higher order Laue zone effects in high energy electron diffraction. Proc. Roy. Soc. (London) A **350**, 335 (1976)
[8.147] J.W. Steeds: Information about the crystal potential from zone axis patterns, in *Electron Microscopy 1980*, Vol. 4, ed. by P. Brederoo, J. Van Landuyt (Seventh European Congress on Electron Microscopy Foundation, Leiden, 1980) p. 96
[8.148] J.R. Baker, S. McKernan: Structure factor information from HOLZ beam intensities in convergent-beam HEED, in *Electron Microscopy and Analysis 1981*, ed. by M.J. Goringe (IoP, London, 1982) p. 283
[8.149] G.M. Rackham, P.M. Jones, J.W. Steeds: Upper layer diffraction effects in zone axis patterns, in [Ref. 1.55, Vol. 1, p. 336 and 355]
[8.150] J.E. Loveluck, J.W. Steeds: Crystallography of lithium tantalate and quartz, in *Developments in Electron Microscopy and Analysis 1977*, ed. by D.L. Misell (IoP, London, 1977) p. 293
[8.151] G.M. Rackham, J.W. Steeds: Convergent beam observation near boundaries and interfaces, in *Developments in Electron Microscopy and Analysis*, ed. by J.A. Venables (Academic, London, 1976) p. 457
[8.152] P. Goodman: A practical method for three-dimensional space-group analysis using convergent beam electron diffraction. Acta Cryst. A **31**, 804 (1975)
[8.153] P. Goodman: The symmetry of electron diffraction zone axis patterns. Philos. Mag. A **281**, 171 (1976)

[8.154] J.W. Steeds: Electron crystallography, in *Quantitative Electron Microscopy*, ed. by J.N. Chapman, A.J. Craven (Scottish Universities Sommer School in Physics, Edinburgh, 1984) p. 49

[8.155] T. Okuyama, S. Matsumura, N. Kuwano, K. Oki: Dynamical diffraction effects on higher-order Laue zone lines in CBED patterns of semiconductors. Ultramicroscopy **31**, 309 (1989)

[8.156] Y.P. Lin, D.M. Bird, R. Vincent: Errors and correction terms for HOLZ line simulations. Ultramicroscopy **27**, 233 (1989)

[8.157] J.M. Zuo: Perturbation theory in high-energy transmission electron diffraction. Acta Cryst. A **47**, 87 (1991)

[8.158] J.M. Zuo: Automated lattice parameter measurement from HOLZ lines and their use for the measurement of oxygen content in $YBa_2Cu_3O_{7-\delta}$. Ultramicroscopy **41**, 211 (1992)

[8.159] Y. Kondo, Y. Harada: New electron diffraction technique to obtain HOLZ patterns using hollow-cone illumination, in *Electron Microscopy 1984*, Vol. 1, ed. by Á. Csanády. P. Röhlich, D. Szabó (MOTESZ, Budapest, 1984) p. 337

[8.160] M. Tanaka, M. Terauchi: Whole pattern in CBED using the hollow-cone method. J. Electron Microsc. **34**, 52 (1985)

[8.161] M. Tanaka, M. Terauchi, T. Kaneyama: *Convergent-Beam Electron Diffraction II* (JEOL, Tokyo, 1988)

[8.162] D. Cherns, A.R. Preston: Convergent beam diffraction studies of interfaces, defects and multilayers. J. Electron Microsc. Techn. **13**, 111 (1989)

[8.163] P. Cordier, J.P. Morniroli: Characterization of crystal defects in quartz by LACBED. Philos. Mag. **72**, 1421 (1995)

[8.164] A.R. Preston, D. Cherns: Measurement of strain in convergent beam electron diffraction. in Institute of Physics Conference Series 78 (IoP, London, 1986) p. 41

[8.165] D. Cherns, C.J. Kiely, A.R. Preston: Electron diffraction studies of strain in epitaxial bicrystals and multilayers. Ultramicroscopy **24**, 355 (1988)

[8.166] R. Vincent, J. Wang, D. Cherns, S.J. Bailey, A.R. Preston, J.W. Steeds: Diffraction and imaging from semiconductor multilayers. in Institute of Physics Conference Series **90** (IoP, London, 1987) p. 233

[8.167] C.J. Rossouw, M. Al-Khafaji, D. Cherns, J.W. Steeds, R. Touaitia: A treatment of dynamical diffraction to multiply layered structures. Ultramicroscopy **35**, 229 (1991)

[8.168] D. Cherns, R. Touaitia, A.R. Preston, C.J. Rossouw, D.L. Houghton: CBED studies of strain in Si/SiGe superlattices. Philos. Mag. A **64**, 597 (1991)

[8.169] D. Cherns, I.K. Jordan, R. Vincent: CBED from AlGaAs/GaAs single quantum wells. Philos. Mag. Lett. **58**, 45 (1988)

[8.170] Y. Atici, D. Cherns: Observation of crystal distortions in SiGe/Si superlattice using a new application of LACBED. Ultramicroscopy **58**, 435 (1995)

Chapter 9

[9.1] A.J.F. Metherell, M.J. Whelan: Measurement of absorption of fast electrons in single crystal films of Al. Philos. Mag. **15**, 755 (1967)

[9.2] A. Iijima: Intensity of fast electron transmitted through thick single crystals. J. Phys. Soc. Jpn. **35** 213 (1973)

[9.3] L. Reimer: Contrast in amorphous and crystalline objects. Lab. Invest. **14**, 939 (1965)
[9.4] L. Reimer: Deutung der Kontrastunterschiede von amorphen und kristallinen Objekten in der Elektronenmikroskopie. Z. Angew. Phys. **22**, 287 (1967)
[9.5] G. Dupouy, F. Ferrier, R. Uyeda, R. Ayroles, A. Mazel: Mesure du coefficient d'absorption des électrons accélérés sous des tensions comprises entre 100 et 1200 kV. J. Microscopie **4**, 429 (1965)
[9.6] A. Mazel, R. Ayroles: Etude dans des cristaux d'oxyde de magnesium des distances d'extinction correspondant à diverses réflexions systématiques, in [Ref. 1.55, Vol. 1., p. 99]
[9.7] G. Möllenstedt: Elektronenmikroskopische Sichtbarmachung von Hohlstellen in Einkristall-Lamellen. Optik **10**, 72 (1953)
[9.8] L. Reimer: Elektronenoptische Untersuchung zur Zwillingsbildung in Silber-Aufdampfschichten. Optik **16**, 30 (1959)
[9.9] P. Rao: Separation and identification of phases with through-focus dark-field electron microscopy. Philos. Mag. **32**, 755 (1975)
[9.10] G.M. Michal, R. Sinclair: A quantitative assessment of the capabilities of 2 1/2 D microscopy for analysing crystalline solids. Philos. Mag. A **42**, 691 (1980)
[9.11] T. Mitsuishi, H. Nagasaki, R. Uyeda: A new type of interference fringes observed in electron microscopy of crystalline substances. Proc. Imp. Acad. Jpn. **27**, 86 (1951)
[9.12] G.A. Bassett, J.W. Menter, D.W. Pashley: Moiré patterns of electron micrographs and their application to the study of dislocations in metals. Proc. Roy. Soc. London A **246**, 345 (1958)
[9.13] O. Rang: Zur geometrischen Theorie der Moiré-Muster auf Elektronenbildern übereinanderliegender Einkristalle. Z. Krist. **114**, 98 (1960)
[9.14] H. Hashimoto, M. Mannani, T. Naiki: Dynamical theory of electron diffraction for the electron microscopical image of crystal lattices. Philos. Trans. Roy. Soc. London A **253**, 459 (1961)
[9.15] R. Gevers: Dynamical theory of moiré fringe patterns. Philos. Mag. **7**, 1681 (1962)
[9.16] J. Demmy: Aussagen des Verdrehungsmoirés über Gitterfehler. Z. Naturforsch. A **15**, 194 (1960)
[9.17] J.W. Matthews, W.M. Stobbs: Measurement of the lattice displacement across a coincidence grain boundary. Philos. Mag. **36**, 373 (1977)
[9.18] J. Matthews, W.M. Stobbs: Geometric factors in fcc and bcc metal-on-metal epitaxy. Philos. Mag. **36** 1331 (1977)
[9.19] K. Takayanaki, K. Yagi, K. Kobayashi, G. Honjo: Technique for routine UHV in situ electron microscopy of growth processes of epitaxial thin films. J. Phys. E **11**, 441 (1978)
[9.20] L. Reimer: Contrast in the different modes of SEM, in *Scanning Electron Microscopy: Systems and Applications 1973* (IoP, London, 1973) p. 120
[9.21] G.R. Booker, D.C. Joy, J.P. Spencer, H. von Harrach: Contrast effects from crystalline material using STEM, in *Scanning Electron Microscopy 1974*, ed. by O. Johari (IIT Research Institute, Chicago, 1974) p. 225
[9.22] D.M. Maher, D.C. Joy: The formation and interpretation of defect images from crystalline materials in a STEM. Ultramicroscopy **1**, 239 (1976)

[9.23] L. Reimer, P. Hagemann: STEM of crystalline specimens, in *Scanning Electron Microscopy 1976/I*, ed. by O. Johari (IIT Research Institute, Chicago, 1976) p. 321
[9.24] L. Reimer, P. Hagemann: Anwendung eines Rasterzusatzes zu einem TEM. II. Abbildung kristalliner Objekte. Optik **47**, 325 (1977)
[9.25] T. Yamamoto, H. Nishizawa: Imaging of crystalline substances in STEM. Phys. Status Solidi A **28**, 237 (1975)
[9.26] C.J. Humphreys, M.J. Whelan: Inelastic scattering of fast electrons by crystals. Philos. Mag. **20**, 165 (1969)
[9.27] R. Castaing, P. Henoc, L. Henry, M. Natta: Degre de coherence de la diffusion électronique par interaction électron-phonon. C.R. Acad. Sci. Paris **265**, 1293 (1967)
[9.28] S.L. Cundy, A.J.F. Metherell, M.J. Whelan: Contrast preserved by elastic and quasi-elastic scattering of fast electrons near Bragg beams. Philos. Mag. **15**, 623 (1967)
[9.29] S.L. Cundy, A. Howie, U. Valdre: Preservation of electron microscope image contrast after inelastic scattering. Philos. Mag. **20**, 147 (1969)
[9.30] S. Kuwubara, T. Uefuji: Variation of electron microscopic thickness fringes of Al single crystals with energy loss. J. Phys. Soc. Jpn. **38**, 1090 (1975)
[9.31] A.J. Craven, J.M. Gibson, A. Howie, D.R. Spalding: Study of single-electron excitations by electron microscopy. I. Image contrast from delocalized excitations. Philos. Mag. A **38**, 519 (1978)
[9.32] A. Bakenfelder, I. Fromm, L. Reimer, R. Rennekamp: Contrast in the electron spectroscopic imaging mode of a TEM. III. Bragg contrast of crystalline specimens. J. Microsc. **159**, 161 (1990)
[9.33] P.H. Duval, L. Henry: Calcul d'influence de la diffusion inélastique des électron sur les images de monocristaux. Philos. Mag. A **35**,1381 (1977)
[9.34] S. Doniach, C. Sommers: Coherence of inelastically scattered fast electrons in crystals of finite thickness. Philos. Mag. A **51**, 419 (1985)
[9.35] W.M. Stobbs, A.J. Bourdillon: Current applications of electron energy loss spectroscopy. Ultramicroscopy **9**, 303 (1982)
[9.36] G. Lehmpfuhl, D. Krahl, M. Swoboda: Electron microscopic channelling imaging of thick specimens with medium-energy electrons in an energy-filter microscope. Ultramicroscopy **31**, 161 (1989)
[9.37] J. Mayer: Electron spectroscopic imaging and diffraction applications in materials science, in *Proceedings of the 50th Annual Meeting of EMSA* (San Francisco Press, San Francisco, CA, 1991) p. 616
[9.38] H. Hashimoto: High voltage TEM – contrast theory, in [Ref. 1.78, p. 9]
[9.39] C.J. Humphreys, L.E. Thomas, J.S. Lally, R.M. Fisher: Maximising the penetration in HVEM. Philos. Mag. **23**, 87 (1971)
[9.40] A. Rocher, R. Ayroles, A. Mazel, C. Mory, B. Jouffrey: Electron penetration in Al, Cu, and MgO at high voltages up to 3 MV, in [Ref. 1.79, p. 436]
[9.41] C.J. Humphreys, J.S. Lally: Aspects of Bloch-wave channelling in high-voltage electron microscopy. J. Appl. Phys. **41**, 232 (1970)
[9.42] M.S. Spring: Electron channelling at high energies. Phys. Lett. A **31**, 421 (1970)
[9.43] R. Uyeda, M. Nonoyama: The observation of thick specimens by HVEM. Jpn. J. Appl. Phys. **6**, 557 (1967)
[9.44] G. Thomas: Electron microscopy at high voltages. Philos. Mag. **17**, 1097 (1968)

[9.45] G. Thomas, J.C. Lacaze: Transmission electron microscopy at 2.5 MeV. J. Microscopie **97**, 301 (1973)

[9.46] H. Fujita, T. Tabata: Voltage dependence of the maximum observable thickness by electron microscopy up to 3MV. Jpn. J. Appl. Phys. **12**, 471 (1973)

[9.47] H. Fujita, T. Tabata, K. Yoshida, N. Sumida, S. Katagiri: Some applications of an ultra-high voltage electron microscope in materials science. Jpn. J. Appl. Phys. **11**, 1522 (1972)

[9.48] M.J. Whelan: An outline of the theory of diffraction contrast observed at dislocations and other defects in thin crystals examined by TEM. J. Inst. Met. **87**, 392 (1959)

[9.49] P.B. Hirsch, A. Howie, M.J. Whelan: A kinematical theory of diffraction contrast of electron transmission microscope images of dislocations and other defects. Philos. Trans. Roy. Soc. London A **252**, 499 (1960)

[9.50] R. Gevers: On the dynamical theory of electron transmission microscope images of dislocations and stacking faults. Phys. Status Solidi **3**, 415 (1963)

[9.51] R. Gevers: On the dynamical theory of different types of electron microscopic transmission fringe patterns. Phys. Status Solidi **3**, 1672 (1963)

[9.52] A. Howie, M.J. Whelan: Diffraction contrast of electron microscope images of crystal lattice defects. Proc. Roy. Soc. London A **263**, 217 (1961); ibid. **267**, 206 (1962)

[9.53] C.J. Ball: A relation between dark field electron micrographs of lattice defects. Philos. Mag. **9**, 541 (1964)

[9.54] A. Howie: Inelastic scattering of electrons by crystals. Proc. Roy. Soc. London A **271**, 268 (1963)

[9.55] M. Wilkens: Zur Theorie des Kontrastes von elektronenmikroskopisch abgebildeten Gitterfehlern. Phys. Status Solidi **5**, 175 (1964)

[9.56] M. Wilkens: Streuung von Blochwellen schneller Elektronen in Kristallen mit Gitterbaufehlern. Phys. Status Solidi **6**, 939 (1964)

[9.57] M. Wilkens, M. Rühle: Black–white contrast figures from small dislocation loops. Phys. Status Solidi B **49**, 749 (1972)

[9.58] J. van Landuyt, R. Gevers, S. Amelinckx: Fringe patterns at anti-phase boundaries with $\alpha = \pi$ observed in the electron microscope. Phys. Status Solidi **7**, 519 (1964)

[9.59] C.M. Drum, M.J. Whelan: Diffraction contrast effects from stacking faults with phase-angle π. Philos. Mag. **11**, 295 (1965)

[9.60] M.J. Whelan, P.B. Hirsch: Electron diffraction from crystals containing stacking faults. Philos. Mag. **2**, 1121 and 1303 (1957)

[9.61] A. Art, R. Gevers, S. Amelinckx: The determination of the type of stacking faults in fcc alloys by means of contrast effects in the electron microscope. Phys. Status Solidi **3**, 697 (1963)

[9.62] R. Gevers, A. Art, S. Amelinckx: Electron microscopic images of single and intersecting stacking faults in thick foils. Phys. Status Solidi **3**, 1563 (1963)

[9.63] M.J. Marcinkowski: Theory and direct observation of antiphase boundaries and dislocations in superlattices, in *Electron Microscopy and Strength of Crystals*, ed. by G. Thomas, J. Washburn (Interscience, New York, 1963) p. 333

[9.64] S. Amelinckx: The study of planar interfaces by means of electron microscopy, in [Ref. 1.30, p. 257]

[9.65] S. Amelinckx, J. van Landuyt: Contrast effects at planar interfaces, in [Ref. 1.29, p. 68]

[9.66] R. Serneels, M. Snykers, P. Delavignette, R. Gevers, S. Amelinckx: Friedel's law in electron diffraction as applied to the study of domain structure in non-centrosymmetrical crystals. Phys. Status Solidi B **58**, 277 (1973)

[9.67] O. van der Biest, G. Thomas: Identification of enantiomorphism in crystals by electron microscopy. Acta Cryst. A **31**, 70 (1975)

[9.68] A.J. Morton: Inversion anti-phase domains in Cu-rich γ-brasses. Phys. Status Solidi A **31**, 661 (1975)

[9.69] R. Portier, D. Gratias, M. Fayard: Electron microscopy study of enantiomorphic ordered structures. Philos. Mag. **36**, 421 (1977)

[9.70] R. Gevers, P. Delavignette, H. Blank, S. Amelinckx: Electron microscope transmission images of coherent domain boundaries. Phys. Status Solidi **4**, 383 (1964)

[9.71] R. Gevers, P. Delavignette, H. Blank, J. van Landuyt, S. Amelinckx: Electron microscope transmission images of coherent domain boundaries. Phys. Status Solidi **5**, 595 (1964)

[9.72] R. Gevers, J. van Landuyt, S. Amelinckx: Intensity profiles for fringe patterns due to planar interfaces as observed by electron microscopy. Phys. Status Solidi **11**, 689 (1965)

[9.73] J. van Landuyt, R. Gevers, S. Amelinckx: Dynamical theory of the images of microtwins as observed in the electron microscope. Phys. Status Solidi **9**, 135 (1965)

[9.74] P. Delavignette, S. Amelinckx: Electron microscopic observation of antiferromagnetic domain walls in NiO. Appl. Phys. Lett. **2**, 236 (1963)

[9.75] A.J. Ardell: Diffraction contrast at planar interfaces of large coherent precipitates. Philos. Mag. **16**, 147 (1967)

[9.76] S.S. Sheinin, J.M. Corbett: Application of the multi-beam dynamical theory to crystals containing twins. Phys. Status Solidi A **38**, 675 (1976)

[9.77] J. van Landuyt, R. Gevers, S. Amelinckx: Diffraction contrast from small voids as observed by electron microscopy. Phys. Status Solidi **10**, 319 (1965)

[9.78] C.M. Wayman: Martensitic transformations, in [Ref. 1.30, p. 187]

[9.79] R. Gevers: On the kinematical theory of diffraction contrast of electron transmission microscope images of perfect dislocations of mixed type. Philos. Mag. **7**, 651 (1962)

[9.80] W.J. Turnstall, P.B. Hirsch, J. Steeds: Effects of surface stress relaxations on the electron microscope images of dislocations normal to thin metal foils. Philos. Mag. **9**, 99 (1964)

[9.81] M. Wilkens, M. Rühle, F. Häussermann: On the nature of the long-range dislocation contrast in electron transmission micrographs. Phys. Status Solidi **22**, 689 (1967)

[9.82] P. Delavignette, S. Amelinckx: Dislocation nets in bismuth and antimony tellurides. Philos. Mag. **5**, 729 (1960)

[9.83] A.K. Head: The computer generation of electron microscope pictures of dislocations. Aust. J. Phys. **20**, 557 (1967)

[9.84] P. Humble: Computed electron micrographs for tilted foils containing dislocations and stacking faults. Aust. J. Phys. **21**, 325 (1968)

[9.85] A.K. Head, P. Humble, L.M. Clarebrough, A.J. Morton, C.T. Forwood: Computed electron micrographs and defect identification, in *Defects in Crystalline Solids*, Vol. 7, ed. by S. Amelinckx, R. Gevers, J. Nihoul (North-Holland, Amsterdam, 1973)

[9.86] A.R. Thölén: A rapid method for obtaining electron microscope contrast maps of various lattice defects. Philos. Mag. **22**, 175 (1970)

[9.87] D.J.H. Cockayne, I.L.F. Ray, M.J. Whelan: Investigations of dislocation strain fields using weak beams. Philos. Mag. **20**, 1265 (1969)

[9.88] D.J.H. Cockayne: A theoretical analysis of the weak-beam method of electron microscopy. Z. Naturforsch. A **27**, 452 (1972)

[9.89] D.J.H. Cockayne: The principles and practice of the weak-beam method of electron microscopy. J. Microsc. **98**, 116 (1973)

[9.90] I.L.F. Ray, D.J.H. Cockayne: The dissociation of dislocations in silicon. Proc. Roy. Soc. London A **325**, 543 (1971)

[9.91] A. Howie, Z.S. Basinski: Approximation of the dynamical theory of diffraction contrast. Philos. Mag. **17**, 1039 (1968)

[9.92] C.J. Humphreys, R.A. Drummond: The column approximation and high-resolution imaging of defects, in *Electron Microscopy 1976*, Vol. 1, ed. by D.G. Brandon (Tal International, Jerusalem, 1976) p. 142

[9.93] D.J.H. Cockayne, M.L. Jenkins, I.L.F. Ray: The measurement of stacking fault energies of pure face-centred cubic metals. Philos. Mag. **24**, 1383 (1971)

[9.94] M.L. Jenkins: Measurement of the stacking-fault energy of gold using the weak-beam technique of electron microscopy. Philos. Mag. **26**, 747 (1972)

[9.95] C.B. Carter, S.M. Holmes: The stacking-fault energy of nickel. Philos. Mag. **35**, 1161 (1977)

[9.96] C.G. Rhodes, A.W. Thomson: The composition dependence of stacking fault energy in austenitic stainless steel. Metall. Trans. A **8**, 1901 (1977)

[9.97] A. Gomez, D.J.H. Cockayne, P.B. Hirsch, V. Vitek: Dissociation of near-screw dislocations in Ge and Si. Philos. Mag. **31**, 105 (1975)

[9.98] G.W. Groves, M.J. Whelan: The determination of the sense of the Burgers vector of a dislocation from its electron microscope images. Philos. Mag. **7**, 1603 (1962)

[9.99] R. Siems, P. Delavignette, S. Amelinckx: Die direkte Messung von Stapelfehlerenergien. Z. Phys. **165**, 502 (1961)

[9.100] M.H. Loretto, L.K. France: The influence of the degree of the deviation from the Bragg condition on the visibility of dislocations in copper. Philos. Mag. **19**, 141 (1969)

[9.101] K. Marukawa: A new method of Burgers vector identification from electron microscope images. Philos. Mag. A **40**, 303 (1979)

[9.102] M.F. Ashby, L.M. Brown: On diffraction contrast from inclusions. Philos. Mag. **8**, 1649 (1963)

[9.103] M.F. Ashby, L.M. Brown: Diffraction contrast from spherically symmetrical coherency strains. Philos. Mag. **8**, 1083 (1963)

[9.104] M.J. Makin, A.D. Whapham, F.J. Minter: The formation of dislocation loops in copper during neutron irradiation. Philos. Mag. **7**, 285 (1962)

[9.105] U. Essmann, M. Wilkens: Elektronenmikroskopische Kontrastexperimente an Fehlstellenagglomeraten in neutronen-bestrahltem Kupfer. Phys. Status Solidi **4**, K53 (1964)

[9.106] M. Wilkens: Identification of small defect clusters in particle-irradiated crystals by means of TEM, in [Ref. 1.30, p. 233]

[9.107] M. Rühle, M. Wilkens, U. Essmann: Zur Deutung der elektronenmikroskopischen Kontrasterscheinungen an Fehlstellenagglomeraten in neutronenbestrahltem Kupfer. Phys. Status Solidi **11**, 819 (1965)

[9.108] M. Rühle: Elektronenmikroskopie kleiner Fehlstellenagglomerate in bestrahlten Metallen. Phys. Status Solidi **19**, 263 and 279 (1967)
[9.109] M. Rühle, M. Wilkens: Small vacancy dislocation loops in neutron-irradiated copper. Philos. Mag. **15**, 1075 (1967)
[9.110] K.H. Katerbau: The contrast of dynamical images of small lattice defects in the electron microscope. Phys. Status Solidi A **38**, 463 (1976)
[9.111] B.L. Eyre, D.M. Maher, R.C. Perrin: Electron microscope image contrast from small dislocation loops. J. Phys. F **7**, 1359 and 1371 (1978)
[9.112] M.L. Jenkins, K.H. Katerbau, M. Wilkens: TEM studies of displacement cascades in Cu_3Au. Philos. Mag. **34**, 1141 (1976)
[9.113] M. Wilkens, M.L. Jenkins, K.H. Katerbau: TEM diffraction contrast of lattice defects causing strain contrast and structure factor contrast simultaneously. Phys. Status Solidi A **39**, 103 (1977)
[9.114] M. Rühle, M. Wilkens: Defocusing contrast of cavities. Cryst. Lattice Defects **6**, 129 (1975)
[9.115] J.W. Menter: the direct study by electron microscopy of crystal lattices and their imperfections. Proc. Roy. Soc. London **236**, 119 (1956)
[9.116] R. Scholz, H. Bethge: High resolution study of 20°[001] tilt boundaries in gold, in *Electron Microscopy, 1980*, Vol. 1, ed. by J. Brederoo, G. Boom (Seventh European Congress on Electron Microscopy Foundation, Leiden, 1980) p. 238
[9.117] T. Komoda: On the resolution of the lattice imaging in the electron microscope. Optik **21**, 93 (1964)
[9.118] R. Sinclair: Microanalysis by lattice imaging, in [Ref. 1.66, p. 507]
[9.119] R. Sinclair, R. Gronsky, G. Thomas: Optical diffraction from lattice images of alloys. Acta Metall. **24**, 789 (1976)
[9.120] C.K. Wu, R. Sinclair, G. Thomas: Lattice imaging and optical microanalysis of a Cu-Ni-Cr spinoidal alloy. Metall. Trans. A **9**, 381 (1978)
[9.121] W.C.T. Dowell: Das elektronenmikroskopische Bild von Netzebenenscharen und sein Kontrast. Optik **20**, 535 (1963)
[9.122] R. Sinclair, J. Dutkiewicz: Lattice imaging of the B19 ordering transformation and interfacial structure in Mg_3Cd. Acta Metall. **25**, 235 (1977)
[9.123] V.A. Phillips: Lattice resolution measurements of strain fields at Guinier–Preston zones in Al-3.0% Cu. Acta Metall. **21**, 219 (1973)
[9.124] T. Komoda: Electron microscopic observation of crystal lattices on the level with atomic dimensions. Jpn. J. Appl. Phys. **5**, 603 (1966)
[9.125] S. Iijima, S. Kimura, M. Goto: High resolution microscopy of nonstoichiometric $Nb_{22}O_{54}$ crystals: point defects and structural defects. Acta Cryst. A **30**, 251 (1974)
[9.126] J.G. Allpress, J.V. Sanders: The direct observation of the structure of real crystals by lattice imaging. J. Appl. Cryst. **6**, 165 (1973)
[9.127] P.L. Fejes, S. Iijima, J.M. Cowley: Periodicity in thickness of electron microscope crystal-lattice images. Acta Cryst. A **29**, 710 (1973)
[9.128] F. Phillipp, R. Höschen, M. Osaki, G. Möbus, M. Rühle: New high-voltage atomic resolution microscope approaching 1 Å point resolution installed in Stuttgart. Ultramicroscopy **56**, 1 (1994)
[9.129] M. Haider, S. Uhlemann, E. Schwan, H. Rose, B. Kabius, K. Urban: Electron microscopy enhanced. Nature **392**, 768 (1998)
[9.130] H. Rose: Correction of aberrations, a promising means for improving the spatial and energy resolution of energy-filtering electron microscopes. Ultramicroscopy **56**, 11 (1994)

[9.131] G. Benner, M. Matijevic, A. Orchowski, B. Schindler, M. Haider, P. Hartel: State of the First Aberration-Corrected, Monochromatized 200 kV FEG-TEM. Microsc. Microanal. **9** (Suppl. 2), 938 (2003)

[9.132] C.L. Jia, K. Urban: Atomic-resolution measurement of oxygen concentration in oxide materials. Science **303**, 2001 (2004)

[9.133] C.L Jia, M. Lentzen, K. Urban: High-resolution transmission electron microscopy using negative spherical aberration. Microsc. Microanal. **10**, 174 (2004)

[9.134] J.M. Cowley, S. Iijima: The direct imaging of crystal structures, in [Ref. 1.29], p. 123

[9.135] J.M. Cowley, A.F. Moodie: The scattering of electrons by atoms and crystals. I. A new theoretical approach. Acta Cryst. **10**, 698 (1957)

[9.136] D.F. Lynch: Out-of-zone effects in dynamical electron diffraction intensities from Au. Acta Cryst. A **27**, 399 (1971)

[9.137] P. Goodman, A.F. Moodie: Numerical evaluation of n-beam wave functions in electron scattering by the multi-slice method. Acta Cryst. A **30**, 280 (1974)

[9.138] P.A. Stadelmann: EMS – a software package for electron diffraction analysis and HREM image simulation in materials science. Ultramicroscopy **21**, 131 (1987)

[9.139] E.J. Kirkland: *Advanced Computing in Electron Microscopy* (Plenum, New York, 1998)

[9.140] D. van Dyck, W. Coene: The real space method for the dynamical electron diffraction calculations in HREM. Ultramicroscopy **15**, 29 and 41 (1984)

[9.141] D. van Dyck: Image calculations in high-resolution electron microscopy: problems, progress and prospects. Adv. Electron. Electron Phys. **65**, 295 (1985)

[9.142] A. Thust, K. Urban: Quantitative high-speed matching of HREM images. Ultramicroscopy **45**, 23 (1992)

[9.143] W.E. King, G.H. Campbell: Quantitative HREM using non-linear least-squares methods. Ultramicroscopy **56**, 46 (1994)

[9.144] G. Möbus, M. Rühle: Structure determination of metal-ceramic interfaces by numerical contrast evaluation of HRTEM micrographs. Ultramicroscopy **56**, 54 (1994)

[9.145] W. Coene, A.G. Janssen, M. Op de Beeck, D. van Dyck: Phase retrieval through focus variation for ultra-resolution in field-emission TEM. Phys. Rev. Lett. **69**, 3743 (1992)

[9.146] R. Bierwolf, M. Hohenstein: Premise-free reconstruction of the exit-surface wave function in HRTEM. Ultramicroscopy **56**, 32 (1994)

[9.147] A. Thust, M. Lentzen, K. Urban: Non-linear reconstruction of the exit plane wave function from periodic HREM images. Ultramicroscopy **53**, 101 (1994)

[9.148] A. Ourmazd, D.W. Taylor, J. Cunningham, C.W. Tu: Chemical mapping of semiconductor interfaces at near-atomic resolution. Phys. Rev. Lett. **62**, 933 (1989)

[9.149] P. Schwander, C. Kisielowski, M. Seibt, F.H. Baumann, Y. Kim, A. Ourmazd: Mapping projected potential, interfacial roughness and composition in general crystalline solids by quantitative TEM. Phys. Rev. Lett. **71**, 4150 (1993)

[9.150] D. Stenkamp, W. Jäger: Compositional and structural characterization of Si_xGe_{1-x} alloys and heterostructures by HRTEM. Ultramicroscopy **50**, 321 (1993)
[9.151] M. Hohenstein: Reconstruction of the exit surface wave function from experimental HRTEM micrographs. Ultramicroscopy **35**, 119 (1991)
[9.152] M. Hohenstein: Single sideband imaging in HREM. Appl. Phys. A **54**, 485 (1992)
[9.153] A. Orchowski, W.D. Rau, H. Lichte: Electron holography surmounts resolution limit of electron microscopy. Phys. Rev. Lett. **74**, 399 (1995)
[9.154] P. Bayle, T. Deutsch, B. Gilles, F. Lançon, A. Marty, J. Thibault: Quantitative analysis of the deformation and chemical profiles of strained multilayers. Ultramicroscopy **56**, 94 (1994)
[9.155] M.I. Buckett, K.L. Merkle: Determination of grain boundary volume expansion by HREM. Ultramicroscopy **56**, 71 (1994)
[9.156] W.O. Saxton, D.J. Smith: The determination of atomic positions in high-resolution electron micrographs. Ultramicroscopy **18**, 39 (1985)
[9.157] D. Hofmann, F. Ernst: Quantitative HRTEM of the incoherent $\Sigma 3(211)$ boundary in Cu. Ultramicroscopy **53**, 205 (1994)
[9.158] D.J. Smith, L.A. Bursill, G.J. Wood: Non-anomalous high-resolution imaging of crystalline materials. Ultramicroscopy **16**, 19 (1985)
[9.159] Y. Wang, Li Jihong, Ye Hengqiang: Structure analysis of the $\Sigma = 7$ $((12\bar{3}0)/[0001]21.8^\circ)$ grain boundary in α-Ti. Philos. Mag. A **73**, 213 (1996)
[9.160] A. Olsen, J.C.H. Spence: Distinguishing dissociated glide and shuffle set dislocations by HREM. Philos. Mag. A **43**, 945 (1981)
[9.161] A. Bourret, J. Desseaux, A. Renault: Core structure of the Lomer dislocation in Ge and Si. Philos. Mag. A **45**, 1 (1982)
[9.162] M. Tanaka, B. Jouffrey: Dissociated dislocations in GaAs observed in HREM. Philos. Mag. A **50**, 733 (1984)
[9.163] M. Ichimura, J. Narayami: Atomistic study of partial misfit dislocations in Ge/Si(001) heterostructures. Philos. Mag. A **73**, 767 (1996)
[9.164] J.M. Penission, A. Bourret: High resolution study of [011] low-angle tilt boundaries in Al. Philos. Mag. A **40**, 811 (1979)
[9.165] A. de Crecy, A. Bourret, S. Naka, A. Lasalmonie: High resolution determination of the core structure of $1/3\langle 11\bar{2}0\rangle\{10\bar{1}0\}$ edge dislocations in titanium. Philos. Mag. A **47**, 245 (1983)
[9.166] M.J. Mills, P. Stadelmann: A study of the structure of Lomer and 60° dislocations in Al using HRTEM. Philos. Mag. A **60**, 355 (1989)
[9.167] J.C.H. Spence, J.M. Cowley: Lattice imaging in STEM. Optik **50**, 129 (1978)
[9.168] S.J. Pennycook: Z-contrast STEM for materials science. Ultramicroscopy **30**, 58 (1988)
[9.169] S.J. Pennycook, D.E. Jesson: High-resolution Z-contrast imaging of crystals. Ultramicroscopy **37**, 14 (1991)
[9.170] Z.L. Wang, J.M. Cowley: Simulating high-angle annular dark-field STEM images including inelastic thermal diffuse scattering. Ultramicroscopy **31**, 437 (1989)
[9.171] Z.L. Wang, J.M. Cowley: Dynamic theory of high-angle annular-dark-field STEM lattice images for a Ge/Si interface. Ultramicroscopy **32**, 275 (1990)
[9.172] R.F. Loane, P. Xu, J. Silcox: Incoherent imaging of zone axis crystals with ADF STEM. Ultramicroscopy **40**, 121 (1992)

[9.173] J. Silcox, P. Xue, F. Loane: Resolution limits in annular dark field STEM. Ultramicroscopy **47**, 173 (1989)

[9.174] S. Mader: Elektronenmikroskopische Untersuchung der Gleitlinienbildung auf Cu-Einkristallen. Z. Phys. **149**, 73 (1957)

[9.175] L. Reimer, C. Schulte: Elektronenmikroskopische Oberflächenabdrücke und ihr Auflösungsvermögen. Naturwissenschaften **53**, 489 (1966)

[9.176] G.A. Bassett: A new technique for decoration of cleavage and slip steps on ionic crystal surfaces. Philos. Mag. **3**, 1042 (1958)

[9.177] H. Bethge, K.W. Keller: Über die Abbildung von Versetzungen durch Abdampfstrukturen auf NaCl-Kristallen. Z. Naturforsch. A **15**, 271 (1960)

[9.178] H. Bethge, K.W. Keller, N. Stenzel: Zur elektronenmikroskopischen Sichtbarmachung unterschiedlicher Bindungsenergien und Adsorptionseigenschaften an Lamellenstufen auf NaCl-Kristallen. Naturwissenschaften **49**, 152 (1962)

[9.179] K. Kambe, G. Lehmpfuhl: Weak-beam technique for electron microscopic observation of atomic steps on thin single-crystal surfaces. Optik **42**, 187 (1975)

[9.180] D. Cherns: Direct resolution of surface atomic steps by TEM. Philos. Mag. **30**, 549 (1974)

[9.181] G. Lehmpfuhl, T. Takayanagi: Electron microscopic contrast of atomic steps on fcc metal crystal surfaces. Ultramicroscopy **6**, 195 (1981)

[9.182] M. Klaua, H. Bethge: Imaging of atomic steps on ultrathin Au films by TEM. Ultramicroscopy **11**, 125 (1983)

[9.183] A.F. Moodie, C.E. Warble: The investigation of primary step growth in MgO by direct TEM. Philos. Mag. **16**, 891 (1967)

[9.184] G. Lehmpfuhl, C.E. Warble: Direct electron microscope imaging of surface topography by diffraction and phase contrast. Ultramicroscopy **19**, 135 (1986)

[9.185] K. Takayanagi, Y. Tanishiro, M. Takahashi, S. Takahashi: Structure analysis of Si(111) 7×7 by UHV transmission electron diffraction and microscopy. J. Vac. Sci. Technol. A **3**, 1502 (1985)

[9.186] J.C.H. Spence: High energy transmission electron diffraction and imaging studies of the Si(111) 7×7 surface structure. Ultramicroscopy **11**, 117 (1983)

[9.187] G. Binnig, H. Rohrer: Surface imaging by scanning tunneling microscopy. Ultramicroscopy **11**, 157 (1983)

[9.188] E. Ruska, H.O. Müller: Über Fortschritte bei der Abbildung elektronenbestrahlter Oberflächen. Z. Phys. **116**, 366 (1940)

[9.189] B. von Borries: Sublichtmikroskopische Auflösung bei der Abbildung von Oberflächen im Übermikroskop. Z. Phys. **116**, 370 (1940)

[9.190] Ch. Fert: Observation directe des surfaces en microscopie électronique par réflexion. Optik **13**, 378 (1956)

[9.191] V.E. Cosslett, D. Jones: A reflexion electron microscope. J. Sci. Instr. **32**, 86 (1955)

[9.192] J.C.H. Spence: Energy-filtered reflection electron microscopy, in [Ref. 1.74, p. 401]

[9.193] L.M. Peng, K.H. Ino (guest eds.): Reflection electron microscopy, special issue of Ultramicroscopy **48**, No.4, 367–490 (1993)

[9.194] N. Osakabe, Y. Tanishiro, K. Yagi, G. Honjo: Observation of Si(111) 7×7 – 1×1 transition by REM. Surf. Sci. **97**, 393 (1980)

[9.195] N. Osakabe, Y. Tanishiro, K. Yagi, G. Honjo: Direct observation of the phase transition between the (7×7) and (1×1) structures of clean (111) silicon surfaces. Surf. Sci. **109**, 353 (1981)

[9.196] K. Takayanagi, Y. Tanishiro, K. Kobayashi, K. Akijama, K. Yagi: Surface structures observed by high resolution UHV electron microscopy at atomic level. Jpn. J. Appl. Phys. **26**, L597 (1987)

[9.197] T. Hsu, L.M. Peng: Experimental studies of steps in REM. Ultramicroscopy **22**, 217 (1987)

[9.198] Y. Uchida, G. Lehmpfuhl: Observation of double contours of monoatomic steps in REM. Ultramicroscopy **23**, 53 (1987)

[9.199] J.M. Cowley, L.M. Peng: The image contrast of surface steps in REM. Ultramicroscopy **16**, 59 (1985)

[9.200] N. Osakabe, Y. Tanishiro, K. Yagi, G. Honjo: Image contrast of dislocations and atomic steps on (111) silicon surface in REM. Surf. Sci. **102**, 424 (1981)

[9.201] L.M. Peng, J.M. Cowley: Dynamical diffraction calculations for RHEED and REM. Acta Cryst. A **42**, 545 (1986)

[9.202] Z.L. Wang, J. Liu, P. Lu, J.M. Cowley: Electron resonance reflections from perfect crystals and crystal surfaces with steps. Ultramicroscopy **27**, 101 (1989)

[9.203] N. Osakabe, Y. Tanishiro, K. Yagi, G. Honjo: Reflection electron microscopy of clean and gold deposited (111) silicon surfaces. Surf. Sci. **97**, 393 (1980)

[9.204] T. Hsu, J.M. Cowley: Reflexion electron microscopy (REM) of fcc metals. Ultramicroscopy **11**, 239 (1983)

[9.205] T. Tanji, J.M. Cowley: Interactions of electron beams with surfaces of MgO crystals. Ultramicroscopy **17**, 287 (1985)

[9.206] L.D. Marks, D.J. Smith: Direct surface imaging in small metal particles. Nature **303**, 316 (1983)

[9.207] L.D. Marks, D.J. Smith: Direct atomic imaging of solid surfaces. Surf. Sci. **143**, 495 (1984)

[9.208] L.D. Marks: Direct atomic imaging of solid surfaces. Surf. Sci. **139**, 281 (1984)

[9.209] N. Ikarashi, K. Kobayashi, H. Koike, H. Hasegawa, K. Yagi: Profile and plane-view imaging of reconstructed surface structures of gold. Ultramicroscopy **26**, 195 (1988)

[9.210] Y. Takai, Y. Taniguchi, T. Ikuta, R. Shimizu: Spherical-aberration-free observation of profile images of the Au(001) surface by defocus-modulation image processing. Ultramicroscopy **54**, 250 (1994)

[9.211] J.M. Gibson, M.L. Mcdonald, F.C. Unterwald: Direct imaging of a novel silicon surface reconstruction. Phys. Rev. Lett. **55**, 1765 (1985)

[9.212] L.D. Marks, V. Heine, D.J. Smith: Direct observation of elastic and plastic deformation at Au (111) surfaces. Phys. Rev. Lett. **52**, 656 (1984)

[9.213] J.L. Hutchison, N.A. Briscoe: Surface profile imaging of spinel catalyst particles. Ultramicroscopy **18**, 435 (1985)

[9.214] T. Isshiki, M. Toyoshima, M. Tsujikawa, H. Saijo: Contrast of adsorbed or removal atoms in surface profile images by high and ultra-high-resolution electron microscopy. Ultramicroscopy **41**, 201 (1992)

[9.215] R.L. Wallenberg, J.O. Bovin, D.J. Smith: Atom hopping on small gold particles imaged by high-resolution electron microscopy. Naturwissenschaften **72**, 539 (1985)

[9.216] S. Iijima, T. Ishihashi: Structural instability of ultrafine particles in metals. Phys. Rev. Lett. **56**, 616 (1986)

[9.217] D.J. Smith, L.A. Bursill, D.A. Jefferson: Atomic images of oxide surfaces. Surf. Sci. **175**, 673 (1986)

Chapter 10

[10.1] L.D. Landau, E.M. Lifschitz: *Lehrbuch der theoretischen Physik*, Vol. 2, Klassische Feldtheorie (Akademie, Berlin, 1997)

[10.2] S.T. Stephenson: The continous x-ray spectrum, in *Handbuch der Physik*, Vol. 30, ed. by S. Flügge (Springer, Berlin 1957) p. 337

[10.3] N.A. Dyson: *X-Rays in Atomic and Nuclear Physics* (Longman, London, 1973)

[10.4] H.H. Kramers: On the theory of x-ray absorption and of the continuous x-ray spectrum. Philos. Mag. **46**, 836 (1923)

[10.5] R.H. Pratt, H.K. Tseng, C.M. Lee, L. Kissel: Bremsstrahlung energy spectra from electrons of kinetic energy 1 keV$\leq T_1 \leq$2000 keV incident on neutral atoms 2$\leq Z \leq$92. Atomic Data Nucl. Data Tables **20**, 175 (1977)

[10.6] H.K. Tseng, C.M. Lee: Electron bremsstrahlung angular distributions in the 1–500 keV energy range. Phys. Rev. A **19**, 187 (1979)

[10.7] P. Bernsen, L. Reimer: Total rate imaging with x-rays in a SEM, in *Scanning Electron Microscopy 1984/IV* (SEM/AMF O'Hare, Chicago, 1984)) p. 1707

[10.8] A. Sommerfeld: Über die Beugung und Bremsung der Elektronen. Ann. Phys. **11**, 257 (1931)

[10.9] P. Kirkpatrick, L. Wiedmann: Theoretical continuous x-ray energy and polarization. Phys. Rev. **67**, 321 (1945)

[10.10] N.F. Mott, H.S.W. Massey: *The Theory of Atomic Collisions*, 3rd ed. (Oxford University Press, Oxford, 1965)

[10.11] C.R. Worthington, S.G. Tomlin: The intensity of emission of characteristic x-radiation. Proc. Phys. Soc. A **69**, 401 (1956)

[10.12] M. Gryzinski: Classical theory of atomic collisions. I. Theory of inelastic collisions. Phys. Rev. A **138**, 336 (1965)

[10.13] C.J. Powell: Cross-sections for ionization of inner-shell electrons by electrons. Rev. Mod. Phys. **48**, 33 (1976)

[10.14] C.J. Powell: Inelastic scattering of electrons, in *Electron Beam Interactions with Solids*, ed. by D.F. Kyser et al. (SEM Inc./ AMF O'Hare, Chicago, 1982) p. 19.

[10.15] J.W. Motz, R.C. Placious: K-ionization cross sections for relativistic electrons. Phys. Rev. A **136**, 662 (1964)

[10.16] J.A. Bearden: X-ray wavelengths. Rev. Mod. Phys. **39**, 78 (1967)

[10.17] J.A. Bearden, A.F. Burr: Reevaluation of x-ray atomic energy levels. Rev. Mod. Phys. **39**, 125 (1967)

[10.18] R.W. Fink, R.C. Jopson, H. Mark, C.D. Swift: Atomic fluorescence yields. Rev. Mod. Phys. **38**, 513 (1966)

[10.19] W. Bambynek, B. Crasemann, R.W. Fink, H.U. Freund, H. Mark, C.D. Swift, R.E. Price, R.V. Rao: X-ray fluorescence yields, Auger and Coster–Kronig transition probabilities. Rev. Mod. Phys. **44**, 716 (1972)

[10.20] H.U. Freund: Recent experimental values for K-shell fluorescence yields. X-ray Spectrom. **4**, 90 (1975)

[10.21] P. Kruit, J.A. Venables: High-spatial-resolution surface-sensitive electron spectroscopy using a magnetic parallelizer. Ultramicroscopy **25**, 183 (1988)

[10.22] T.K. Kelly: Mass absorption coefficients and their relevance in electron probe microanalysis. Trans. Inst. Min. Metall. B **75**, 59 (1966)

[10.23] K.F.J. Heinrich: X-ray absorption uncertainty, in *The Electron Microprobe*, ed. by T.D. McKinley, K.F.J. Heinrich, D.B. Wittry (Wiley, New York, 1966)

[10.24] J.W. Mayer, E. Rimini: *Ion Beam Handbook for Materials Analysis* (Academic, New York, 1977)

[10.25] J.H. Hubble: Photon mass attenuation and energy absorption coefficients from 1 keV to 20 keV. Int. J. Appl. Radiat. Isot. **33**, 1269 (1982)

[10.26] E.B. Saloman, J.H. Hubbell, J.H. Scofield: X-ray attenuation cross sections for energies 100 eV to 100 keV and elements Z = 1 to Z = 92. Atomic Data Nucl. Data Tables **38**, 1 (1988)

[10.27] C.J. Cooke, P. Duncumb: Performance analysis of a combined microscope and electron probe microanalyser 'EMMA', in *5th International Congress on X-Ray Optics and Microanalysis*, ed. by G. Möllenstedt, K.H. Gaukler (Springer, Berlin, 1969) p. 245

[10.28] C.J. Cooke, I.K. Openshaw: Combined high resolution electron microscopy and x-ray microanalysis, in [Ref. 1.55, Vol. 1, p. 175]

[10.29] D.A. Gedcke: The Si(Li) x-ray energy analysis system: operating principles and performance. X-ray Spectrom. **1**, 129 (1972)

[10.30] D.A. Gedcke: The Si(Li) x-ray spectrometer for x-ray microanalysis, in [Ref. 1.105, p. 403]

[10.31] C.E. Fiori, D.E. Newbury: Artifacts observed in energy-dispersive x-ray spectrometry in the SEM, in *Scanning Electron Microscopy 1978/I* (SEM/AMF O'Hare, Chicago, 1978) p. 401

[10.32] T.J. White, D.R. Cousens, G.J. Auchterlonie: Preliminary characterization of an intrinsic germanium detector on a 400-keV microscope. J. Microsc. **162**, 379 (1991)

[10.33] P.J. Staham, J.V.P. Long, G. White, K. Kandiah: Quantitative analysis with an energy-dispersive detector using a pulsed electron probe and active signal processing. X-ray spectrometry **3**, 153 (1974)

[10.34] C.E. Lyman, J.I. Goldstein, D.B. Williams, D.W. Ackland, S. von Harrach, A.W. Nicholls. P.J. Staham: High-performance x-ray detection in a new analytical electron microscope. J. Microsc. **176**, 85 (1994)

[10.35] T.A. Hall: Reduction of background due to backscattered electrons in energy dispersive x-ray microanalysis. J. Microsc. **110**, 103 (1977)

[10.36] B. Neumann, L. Reimer: A permanent magnet system for electron deflection in front of an energy dispersive x-ray spectrometer. Scanning **1**, 130 (1978)

[10.37] N.C. Barbi, A.O. Sandborg, J.C. Russ, C.E. Soderquist: Light element analysis on the SEM using a windowless energy dispersive x-ray spectrometer, in *Scanning Electron Microscopy 1974*, ed. by O. Johari (IIT Research Institute, Chicago, 1974) p. 289

[10.38] J.C. Russ: Procedures for quantitative ultralight element energy dispersive x-ray analysis, in *Scanning Electron Microscopy 1977/I*, ed. by O. Johari (IIT Research Institute, Chicago, 1977) p. 289

[10.39] C.E. Lyman, D.B. Williams, J.I. Goldstein: X-ray detectors and spectrometers. Ultramicroscopy **28**, 137 (1988)

[10.40] R. Schmidt, M. Feller-Kniepmeier: Investigation of system-induced background radiation using a 0–160 keV high-purity germanium detector. Ultramicroscopy **34**, 229 (1990)

[10.41] F. Hofer, W. Grogger, P. Golob: Detector strategy in x-ray microanalysis, in *Analytical TEM in Materials Science*, Proceedings of the Autumn School 1993, ed. by J. Heydenreich, W. Neumann (Max-Planck Institut für Mikrostrukturphysik, Halle, 1993), p. 50

[10.42] M.J. Jacobs, J. Baborovska: Quantitative microanalysis of thin foils with a combined electron microscope–microanalyser (EMMA-3), in *Electron Microscopy 1972* ed. by V.E. Cosslett (IoP, London, 1972) p. 136

[10.43] D.A. Wollman, K.D. Irwin, G.C. Hilton, L.L. Dulcie, D.E. Newbury, J.M. Martinis: High-resolution, energy-dispersive microcalorimeter spectrometer for x-ray microanalysis. J. Microscopy **188** 196 (1997)

[10.44] G.W. Lorimer, F. Cliff, J.N. Clark: Determination of the thickness and spatial resolution for the quantitative analysis of thin foils, in *Developments in Electron Microscopy and Analysis*, ed. by J.A. Venables (Academic, London, 1976) p. 153

[10.45] R. König: Quantitative microanalysis of thin foils, in [Ref. 1.29, p. 526]

[10.46] J.I. Goldstein, J.L. Costley, G.W. Lorimer, S.J.B. Reed: Quantitative x-ray analysis in the electron microscope, in *Scanning Electron Microscopy 1977/I*, ed. by O. Johari (IIT Research Institute, Chicago, 1977) p. 315

[10.47] Z. Horita, T. Sano, M. Nemoto: A new form of the extrapolation method for absorption correction in quantitative x-ray microanalysis with the analytical electron microscope. Ultramicroscopy **35**, 27 (1991)

[10.48] J. Philibert, R. Tixier: Electron probe microanalysis of TEM specimens, in [Ref. 1.12, p. 333]

[10.49] G.W. Lorimer, S.A. Al-Salman, G. Cliff: The quantitative analysis of thin specimens: Effects of absorption, fluorescence and beam spreading, in *Developments in Electron Microscopy and Analysis 1977*, ed. by D.L. Misell (IoP, London, 1977) p. 369

[10.50] I.M. Anderson, J. Bentley, C.B. Carter: The secondary fluorescence correction for x-ray microanalysis in the analytical electron microscope. J. Microsc. **178**, 226 (1995)

[10.51] C.R. Hall: On the production of characteristic x-rays in thin metal crystals. Proc. Roy. Soc. London A **295**, 140 (1966)

[10.52] D. Cherns, A. Howie, M.H. Jacobs: Characteristic x-ray production in thin crystals. Z. Naturforsch. A **28**, 565 (1973)

[10.53] B. Neumann, L. Reimer: Anisotropic x-ray generation in thin and bulk single crystals. J. Phys. D **13**, 1737 (1980)

[10.54] A.J. Bourdillon, P.G. Self, W.M. Stobbs: Crystallographic orientation effects in energy dispersive x-ray analysis. Philos. Mag. A **44**, 1335 (1981)

[10.55] I. Hashimoto, E. Wakai, H. Yamaguchi: Dependence of the x-ray detection orientation on Cliff–Lorimer factor for quantitative microanalysis in an electron microscope. Ultramicroscopy **32**, 121 (1990)

[10.56] J.C.H. Spence, J. Taftø: ALCHEMI: a new technique for locating atoms in small crystals. J. Microsc. **130**, 147 (1983)

[10.57] S.J. Pennycook: Delocalization corrections for electron channeling analysis. Ultramicroscopy **26**, 239 (1988)

[10.58] C.J. Rossouw, P.S. Turner, T.J. White, A.J. O'Connor: Statistical analysis of electron channelling microanalytical data for the determination of site occupancies of impurities. Philos. Mag. Lett. **60**, 225 (1989)

[10.59] W. Qian, B. Tötdal, R. Hoier, J.C.H. Spence: Channelling effects on oxygen-characteristic x-ray emission and their use as a reference site for ALCHEMI. Ultramicroscopy **41**, 147 (1992)

[10.60] J. Taftø, Z. Liliental: Studies of the cation atom distribution in $ZnCr_xFe_{2-x}O_4$ spinels using the channeling effect in electron induced x-ray emission. J. Appl. Cryst. **15**, 260 (1992)

[10.61] L.J. Allen, T.W. Josefsson, C.J. Rossouw: Interaction delocalization in characteristic x-ray emission from light elements. Ultramicroscopy **55**, 258 (1994)

[10.62] W. Nüchter, W. Sigle: Electron channelling: a method in real-space crystallography and a comparison with the ALCHEMI. Philos. Mag. A **71**, 165 (1995)

[10.63] J. Taftø, O.L. Krivanek: Site-specific valence determination by EELS. Phys. Rev. Lett. **48**, 560 (1982)

[10.64] J. Taftø, G. Lehmpfuhl: Direction dependence in EELS from single crystals. Ultramicroscopy **7**, 287 (1982)

[10.65] J. Bentley, N.J. Zaluzec, E.A. Kenik, R.W. Carpenter: Optimization of an analytical electron microscope for x-ray microanalysis, in *Scanning Electron Microscopy 1979/II*, ed. by O. Johari (SEM/AMF O'Hare, Chicago, 1979) p. 581

[10.66] J. Philibert, R. Tixier: Electron penetration and the atomic number correction in electron probe microanalysis. J. Phys. D **1**, 685 (1968)

[10.67] M.J. Nasir: Quantitative analysis on thin films in EMMA-4 using block standards, in *Electron Microscopy 1972* ed. by V.E. Cosslett (IoP, London, 1972) p. 142

[10.68] G. Cliff, G.W. Lorimer: Quantitative analysis of thin metal foils using EMMA-4 – the ratio technique, in *Electron Microscopy 1972* ed. by V.E. Cosslett (IoP, London, 1972) p. 140

[10.69] G. Cliff, G.W. Lorimer: The quantitative analysis of thin specimens. J. Microsc. **103**, 203 (1975)

[10.70] M.N. Thompson, P. Doig, J.W. Edington, P.E.J. Flewitt: The influence of specimen thickness on x-ray count rates in STEM microanalysis. Philos. Mag. **35**, 1537 (1977)

[10.71] J.K. Park, A.J. Ardell: Solute-enriched surface layers and x-ray microanalysis of thin foils of a commercial aluminium alloy. J. Microsc. **165**, 301 (1992)

[10.72] P. Schwaab: Quantitative energy-dispersive x-ray microanalysis of thin metal specimens using the STEM. Scanning **9**, 1 (1987)

[10.73] R. Gauvin, G. L'Espérance: Determination of the C_{nl} parameter in the Bethe formula for the ionization cross-section by use of Cliff–Lorimer k_{ab} factors obtained at different accelerating voltages. J. Microsc. **163**, 295 (1991)

[10.74] T.P. Schreiber, A.M. Wims: A quantitative x-ray microanalysis thin film method using K-, L- and M-lines. Ultramicroscopy **6**, 323 (1981)

[10.75] J.E. Wood, D.C. Williams, J.I. Goldstein: Experimental and theoretical determination of $k_{\mathrm{a,Fe}}$ factors for quantitative x-ray microanalysis in the analytical electron microscope. J. Microsc. **133**, 255 (1984)

[10.76] D.B. Williams, J.R. Michael, J.I. Goldstein, A.D. Romig: Definition of the spatial resolution of x-ray microanalysis in thin foils. Ultramicroscopy **47**, 121 (1992)

[10.77] C.E. Lyman, P.E. Manning, D.J. Duquette, E. Hall: STEM microanalysis of duplex stainless steel weld metal, in *Scanning Electron Microscopy 1978/I*, ed. by O. Johari (SEM/AMF O'Hare, Chicago, 1978) p. 213

[10.78] D.B. Williams, J.I. Goldstein: STEM/x-ray microanalysis across α/γ interfaces in FeNi meteorites, in [Ref. 1.57, p. 416]

[10.79] N.J. Long: Digital x-ray mapping on an HB501 STEM, a new approach for the analysis of interfaces. Ultramicroscopy **34**, 81 (1990)

[10.80] A.M. Ritter, W.G. Morris, M.F. Henry: Factors affecting the measurement of composition profiles in STEM, in *Scanning Electron Microscopy 1979/I*, ed. by O. Johari (SEM/AMF O'Hare, Chicago, 1979) p. 121

[10.81] T.A. Hall: The microprobe assay of chemical elements, in *Physical Techniques in Biological Research*, Vol. 1, Pt.A, ed. by G. Oster (Academic, New York, 1971) p. 157

[10.82] T.A. Hall, H.C. Anderson, T. Appleton: The use of thin specimens for x-ray microanalysis in biology. J. Microsc. **99**, 177 (1973)

[10.83] T.A. Hall, B.L. Gupta: EDS quantitation and application to biology, in [Ref. 1.66, p. 169]

[10.84] A. Warley: Standards for the application of x-ray microanalysis to biological specimens. J. Microsc. **157**, 135 (1990)

[10.85] A. Patak, A. Wright, A.T. Marshall: Evaluation of several common standards for the x-ray microanalysis of thin biological sections. J. Microsc. **170**, 265 (1993)

[10.86] H. Shuman, A.V. Somlyo, A.P. Somlyo: Quantitative electron probe microanalysis of biological thin sections: methods and validity. Ultramicroscopy **1**, 317 (1976)

[10.87] G.M. Roomans: Standards for x-ray microanalysis of biological specimens, in *Scanning Electron Microscopy 1979/II*, (SEM/AMF O'Hare, Chicago, 1979) p. 649

[10.88] F.F. Ingram, M.J. Ingram: Electron microprobe calibration for measurements of intracellular water, in *Scanning Electron Microscopy 1979/II* (SEM/AMF O'Hare, Chicago, 1979) p. 649

[10.89] G.M. Roomans, H.L.M. van Gaal: Organometallic and organometalloid compounds as standards for microprobe analysis of epoxy resin embedded tissue. J. Microsc. **109**, 235 (1977)

[10.90] N. Roos, T. Barnard: Aminoplastic standards for quantitative x-ray microanalysis of thin sections of plastic embedded biological material. Ultramicroscopy **15**, 277 (1984)

[10.91] A.J. Morgan, C. Winters: Practical notes on the production of thin aminoplastic standards for quantitative x-ray microanalysis. Micron Microsc. Acta **20**, 209 (1989)

[10.92] W.C. De Bruijn, M.I. Cleton-Soeteman: Application of Chelex standard beads in integrated morphometrical and x-ray microanalysis, in *Scanning Electron Microscopy 1985/II* (SEM/AMF O'Hare, Chicago, 1985) p. 715

[10.93] A. Dörge, R. Rick, K. Gehring, K. Thurau: Preparation of frozen-dried cryosections for quantitative x-ray microanalysis of electrolytes in biological soft tissue. Pflügers Arch. **373**, 85 (1978)

[10.94] T. von Zglinicki, M. Bimmler, W. Krause: Estimation of organelle water fractions from frozen-dried cryosections. J. Microsc. **146**, 67 (1987)

[10.95] K.E. Tvedt, G. Kopstad, J. Halgunset, O.A. Haugen: Rapid freezing of small biopsies and standards for cryosectioning and x-ray microanalysis. Am. J. Clin. Pathol. **92**, 51 (1989)

[10.96] D.C. Joy, D.M. Maher: The electron energy-loss spectrum: facts and artifacts, in *Scanning Electron Microscopy 1980/I* (SEM/AMF, Chicago, 1980) p. 25

[10.97] A.J. Craven, T.W. Buggy: Correcting electron energy loss spectra for artefacts introduced by serial collection. J. Microsc. **136**, 227 (1984)

[10.98] O.L. Krivanek, C.C. Ahn, R.B. Keeney: Parallel detection electron spectrometer using quadrupole lenses. Ultramicroscopy **22**, 103 (1987)

[10.99] A.J. Gubbens, O.L. Krivanek: Applications of post-column imaging filter in biology and materials science. Ultramicroscopy **51**, 146 (1993)

[10.100] R.F. Egerton, Y.Y. Yang, S.C. Cheng: Characterization and use of the Gatan parallel-recording electron energy-loss spectrometer. Ultramicroscopy **48**, 239 (1993)

[10.101] D.W. Johnson: A Fourier series method for numerical Kramers–Kronig analysis. J. Phys. A **8**, 490 (1975)

[10.102] R.F. Egerton, S.C. Cheng: Thickness measurement by EELS, in *Proceedings of the 43rd Annual Meeting of EMSA*, (San Francisco Press, San Francisco, CA, 1985) p. 389

[10.103] D.R. Liu, D.B. Williams: Influence of some practical factors on background extrapolation in EELS quantification. J. Microsc. **156**, 201 (1987)

[10.104] T. Pun, J.R. Ellis, M. Eden: Weighted least squares estimation of background in EELS imaging. J. Microsc. **137**, 93 (1985)

[10.105] C. Colliex, C. Jeanguillaume, P. Trebbia: Quantitative local microanalysis with EELS, in [Ref. 1.119, p. 251]

[10.106] M. Unser, J.R. Ellis, T. Oun, M. Eden: Optimal background estimation in EELS. J. Microsc. **145**, 245 (1987)

[10.107] C.W. Sorber, G.A.M. Ketelaars, E.S. Gelsema, J.F. Jongkind, W.C. De Bruijn: Quantitative analysis of electron energy-loss spectra from ultrathin-sectioned biological material. J. Microsc. **162**, 23 (1991)

[10.108] A.L.D. Beckers, E.S. Gelsema, W.C. De Bruijn: An efficient method for calculating the least-squares background fit in EELS. J. Microsc. **171**, 87 (1993)

[10.109] J. Bentley, G.L. Lehmann, P.S. Sklad: Improved background fitting for EELS, in [Ref. 1.58, Vol. 1, p. 585]

[10.110] H. Shuman, P. Kruit: Quantitative data-processing of parallel recorded EELS with low signal to background. Rev. Sci. Instr. **56**, 231 (1985)

[10.111] M.K. Kundmann: Analysis of semiconductor EELS in the low-loss regime, in *Microbeam Analysis 1986*, ed. by A.D. Romig, W.F. Chambers (San Francisco Press, San Francisco, CA, 1986) p. 417

[10.112] C.P. Scott, A.J. Craven, C.J. Gilmore, A.W. Bowen: Background fitting in the low-loss region of electron energy-loss spectra, in [Ref. 1.60, Vol. 2, p. 56]

[10.113] J.D. Steele, J.M. Titchmarsh, J.N. Chapman, J.H. Paterson: A single stage process for quantifying electron energy-loss spectra. Ultramicroscopy **17**, 273 (1985)

[10.114] R.D. Leapman, C.R. Swyt: Separation of overlapping core edges in electron energy loss spectra by multiple least squares fitting. Ultramicroscopy **26**, 393 (1988)

[10.115] D.W. Johnson, J.C.H. Spence: Determination of the single-scattering probability distribution from plural-scattering data. J. Phys. D **7**, 771 (1974)

[10.116] C.R. Swyt, R.D. Leapman: Plural scattering in electron energy-loss microanalysis, in *Scanning Electron Microscopy 1982/I*, (SEM/AMF, Chicago, 1982) p. 73

[10.117] R.F. Egerton, P.A. Crozier: The use of Fourier techniques in EELS, in *Sanning Electron Microscopy*, Suppl.2 (SEM/AMF, Chicago, 1988) p. 245

[10.118] C.R. Bradley, M.L. Wroge, P.C. Gibbons: How to remove multiple scattering from core-excitation spectra. Ultramicroscopy **16**, 95 (1985); **19**, 317 (1986); **21**, 305 (1987)

[10.119] D.S. Su, P. Schattschneider: Numerical aspects of the deconvolution of angle-integrated electron energy-loss spectra. J. Microsc. **167**, 63 (1992)

[10.120] D.S. Su, P. Schattschneider: Deconvolution of angle-resolved electron energy-loss spectra. Philos. Mag. A **65**, 1127 (1992)

[10.121] K. Wong, R.F. Egerton: Correction for the effects of elastic scattering in core-loss quantification. J. Microsc. **178**, 198 (1995)

[10.122] M. Isaacson, D. Johnson: The microanalysis of light elements using transmitted energy loss electrons. Ultramicroscopy **1**, 33 (1975)

[10.123] R.F. Egerton, M.J. Whelan: High resolution microanalysis of light elements by electron energy loss spectrometry, in [Ref. 1.56, Vol. 1, p. 384]

[10.124] J. Sévely, J.Ph. Pérez, B. Jouffrey: Energy loss of electrons through Al and carbon films from 300 keV up to 1200 keV, in [Ref. 1.78, p. 32]

[10.125] R.F. Egerton: Formulae for light-element microanalysis by electron energy-loss spectrometry. Ultramicroscopy **3**, 243 (1978)

[10.126] P.G. Self, P.R. Buseck: Low-energy limit to channelling effects in the inelastic scattering of fast electrons. Philos. Mag. A **48**, L21 (1983)

[10.127] J. Taftø, O.L. Krivanek: Characteristic energy-losses from channeled 100 keV electrons. Nucl. Instr. Methods **194**, 153 (1982)

[10.128] R.F. Egerton, C.J. Rossouw, M.J. Whelan: Progress towards a method for quantitative microanalysis of light elements by electron energy-loss spectrometry, in *Developments in Electron Microscopy and Analysis*, ed. by J.A. Venables (Academic, London, 1976) p. 129

[10.129] C.E. Fiori, R.D. Leapman, C.R. Swyt, S.B. Andrews: Quantitative x-ray mapping of biological cryosections. Ultramicroscopy **24**, 237 (1988)

[10.130] D.E. Johnson, K. Izutsu, M. Cantino, J. Wong: High spatial resolution spectroscopy in the elemental microanalysis and imaging of biological systems. Ultramicroscopy **24**, 221 (1988)

[10.131] A. LeFurgey, S.D. Davilla, D.A. Kopf, J.R. Sommer, P. Ingram: Real-time quantitative elemental analysis and mapping: microchemical imaging in cell physiology. J. Microsc. **165**, 191 (1992)

[10.132] A. Berger, J. Mayer, H. Kohl: Detection limits in elemental distribution images produced by energy filtering TEM: case study of grain boundaries in Si_3N_4. Ultramicroscopy **55**, 101 (1994)

[10.133] C. Colliex: An illustrated review of various factors governing the high spatial resolution capabilities in EELS microanalysis. Ultramicroscopy **18**, 131 (1985)

[10.134] R.D. Leapman: STEM elemental mapping by electron energy-loss spectroscopy. Ann. New York Acad. Sci. **483**, 326 (1986)

[10.135] H. Shuman, C.F. Chang, E.L. Bahe, A.P. Somlyo: Electron energy-loss spectroscopy: quantitation and imaging. Ann. New York Acad. Sci. **483**, 295 (1986)

[10.136] R.H. Barckhaus, H.J. Höhling, I. Fromm, P. Hirsch, L. Reimer: Electron spectroscopic diffraction and imaging of the early and mature stages of calcium phosphate formation in the epiphyseal growth plate. J. Microsc. **162**, 155 (1991)

[10.137] H. Lehmann, U. Kunz, A. Jacob: A simplified preparation procedure of plant material for elemental analysis by ESI and EELS techniques. J. Microsc. **162**, 77 (1991)

[10.138] W. Probst, E. Zellmann, R. Bauer: Electron spectroscopic imaging of frozen-hydrated sections. Ultramicroscopy **28**, 312 (1989)

[10.139] R.R. Schröder: Zero-loss energy-filtered imaging of frozen-hydrated proteins: model calculations and implications for future developments. J. Microsc. **166**, 389 (1992)

[10.140] M. Creuzburg: Entstehung von Alkalimetallen bei der Elektronenbestrahlung von Alkalihalogeniden. Z. Phys. **194**, 211 (1966)

[10.141] P.A. Crozier, J.N. Chapman, A.J. Craven, J.M. Titchmarsh: Some factors affecting the accuracy of EELS in determining elemental concentrations in thin films, in *Analytical Electron Microscopy 1984*, ed. by D.B. Williams, D.C. Joy (San Francisco Press, San Francisco, 1984) p. 79

[10.142] T.O. Ziebold: Precision and sensitivity in electron microprobe analysis. Anal. Chem. **39**, 858 (1967)

[10.143] D.C. Joy, D.M. Maher: Sensitivity limits for thin specimen x-ray analysis, in *Scanning Electron Microscopy 1977/I* (IIT Research Institute, Chicago, IL, 1977) p. 325

[10.144] R.D. Leapman: EELS quantitative analysis, in [Ref. 1.70, p. 47]

[10.145] D.C. Joy, D.M. Maher: Electron energy loss spectroscopy: detectable limits for elemental analysis. Ultramicroscopy **5**, 333 (1980)

[10.146] R.D. Leapman, N.W. Rizzo: Towards single atom analysis of biological structures. Ultramicoscopy **78**, 251 (1999)

[10.147] K. Suenaga, M. Tencé, C. Mory, C. Colliex, H. Kato, T. Okazaki, H. Shinohara, K. Hirahara, S. Bandow, S. Iijima: Element-selective single atom imaging. Science **290**, 2280 (2000)

[10.148] K.M. Adamson-Sharpe, F.P. Ottensmeyer: Spatial resolution and detection sensitivity in microanalysis by electron energy loss selected imaging. J. Microsc. **122**, 309 (1981)

[10.149] D.P. Bazett-Jones, F.P. Ottensmeyer: DNA organization in nucleosomes. Can. J. Biochem. **60**, 364 (1982)

[10.150] F.P. Ottensmeyer, D.W. Andrews, A.L. Arsenault, Y.M. Heng, G.T. Simon, G.C. Weatherly: Elemental imaging by electron energy loss microscopy. Scanning **10**, 227 (1988)

[10.151] U. Plate, H.J. Höhling, L. Reimer, R.H. Barckhaus, R. Wienecke, H.P. Wiesmann, A. Boyde: Analysis of the calcium distribution in predentine by EELS and of the early crystal formation in dentine by ESI and ESD. J. Microsc. **166**, 329 (1992)

[10.152] C. Colliex: An illustrated review of various factors governing the high spatial resolution capabilities in EELS microanalysis. Ultramicroscopy **18**, 131 (1985)

[10.153] A. Berger, H. Kohl: Optimum imaging parameters for elemental mapping in an energy filtering TEM. Optik **92**, 175 (1993)

[10.154] G. Kothleitner, F. Hofer: Optimization of the signal to noise ratio in EFTEM elemental maps with regard to different ionization types. Micron **29**, 349 (1998)

[10.155] B. Gralla, A. Thesing, H. Kohl: Optimization of the positions and the width of energy windows for the recording of EFTEM elemental maps. *Proceedings of the MC 2005* (PSI, Villingen, 2005)

[10.156] O.L. Krivanek, M.K. Kundman, K. Kimoto: J. Microscopy **180**, 277 (1995)

[10.157] R.F. Egerton, K. Wong: Some practical consequences of the Lorentzian angular distribution of inelastic scattering. Ultramicroscopy **59**, 169 (1995)

[10.158] R.F. Egerton, P.A. Crozier: The effect of lens aberrations on the spatial resolution of an energy-filtered TEM image. Micron **28**, 117 (1997)

[10.159] R.H. Ritchie: Quantal aspects of the spatial resolution of energy-loss measurements in electron microscopy. Philos. Mag. A **44**, 931 (1981)

[10.160] H. Kohl: Image formation by inelastically scattered electrons: image of a surface plasmon. Ultramicroscopy **11**, 53 (1983)

[10.161] A. Berger, H. Kohl: Elemental mapping using an imaging energy filter: image formation and resolution limits. Microsc. Microanal. Microstruct. **3** 159 (1992)

[10.162] H. Kohl, A. Berger: The resolution limit for elemental mapping in energy-filtering TEM. Ultramicroscopy **59**, 191 (1995)

[10.163] D.A. Muller, J. Silcox: Delocalization in inelastic scattering. Ultramicroscopy **59**, 195 (1995)

[10.164] K. Kimoto, T. Hirano, K. Usami, H. Hoshiya: High spatial resolution elemental mapping of multilayers using a field emission electron microscope equipped with an imaging filter, Jpn. J. Appl. Phys. **33** L1642 (1994)

[10.165] W. Jäger, J. Mayer: Energy-filtered TEM of Si_mGe_n superlattices and Si-Ge heterostructures. I. Experimental results. Ultramicoscopy **59**, 33 (1995)

[10.166] W. Mader, B. Freitag: Element specific imaging with high lateral resolution: an experimental study on layer structures. J. Microscopy **194**, 42 (1999)

[10.167] P. Stallknecht, H. Kohl: Computation and interpretation of contrast in crystal lattice images formed by inelastically scattered electrons in a transmission electron microscope. Ultramicroscopy **66**, 261 (1996)

[10.168] T. Navidi-Kasmai, H. Kohl: Computation of contrasts in atomic resolution electron spectroscopic images of planar defects in crystalline specimens. Ultramicroscopy **81**, 223 (2000)

[10.169] L.J. Allen, S.D. Findlay, M.P. Oxley, C. Witte, N.J. Zaluzec: Modelling high-resolution electron microscopy based on core-loss spectroscopy. Ultramicroscopy **106**, 1001 (2006)

Chapter 11

[11.1] I.G. Stojanowa, E.M. Belawzewa: Experimentelle Untersuchung der thermischen Einwirkung des Elektronenstrahls auf das Objekt im Elektronenmikroskop, in [Ref. 1.52, Vol. 1, p. 100]

[11.2] M. Watanabe, T. Someya, Y. Nagahama: Temperature rise of specimen due to electron irradiation, in [Ref. 1.53, Vol. 1, p. A-8]

[11.3] D.D. Thornburg, C.M. Wayman: Specimen temperature increases during transmission electron microscopy. Phys. Status Solidi A **15**, 449 (1973)

[11.4] L. Reimer, R. Christenhusz, J. Ficker: Messung der Objekttemperatur im Elektronenmikroskop mittels Elektronenbeugung. Naturwissenschaften **47**, 464 (1960)

[11.5] M. Fukamachi, T. Kikuchi: Application of the critical voltage effect to the measurement of temperature increase of metal foils during the observation with HVEM. Jpn. J. Appl. Phys. **14**, 587 (1975)

[11.6] A. Winkelmann: Messung der Temperaturerhöhung der Objekte bei Elektronen-Interferenzen. Z. Angew. Phys. **8**, 218 (1956)

[11.7] E. Gütter, H. Mahl: Einfluß einer periodischen Objektbeleuchtung auf die elektronenmikroskopische Abbildung. Optik **17**, 233 (1960)

[11.8] G. Honjo, N. Kitamura, K. Shimaoka, K. Mihama: Low temperature specimen method for electron diffraction and electron microscopy. J. Phys. Soc. Jpn. **11**, 527 (1956)

[11.9] L. Reimer, R. Christenhusz: Reversible Temperaturindikatoren in Form von Aufdampfschichten zur Ermittlung der Objekttemperatur im Elektronenmikroskop. Naturwissenschaften **48**, 619 (1961)

[11.10] L. Reimer, R. Christenhusz: Experimenteller Beitrag zur Objekterwärmung im Elektronenmikroskop. Z. Angew. Phys. **14**, 601 (1962)

[11.11] S. Yamaguchi: Über die Temperaturerhöhung der Objekte im Elektronenmikroskop. Z. Angew. Phys. **8**, 221 (1956)

[11.12] S. Leisegang: Zur Erwärmung elektronenmikroskopischer Objekte bei kleinem Strahlquerschnitt, in [Ref. 1.51, p. 176]

[11.13] B. Gale, K.F. Hale: Heating of metal foils in an electron microscope. Brit. J. Appl. Phys. **12**, 115 (1961)

[11.14] K. Kanaya: The temperature distribution along a rod-specimen in the electron microscope. J. Electron Microsc. **4**, 1 (1956)

[11.15] P. Balk, J. Ross Colvin: Note on an indirect measurement of object temperature in electron microscopy. Kolloid Z. **176**, 141 (1961)

[11.16] L. Reimer: Zur Zersetzung anorganischer Kristalle im Elektronenmikroskop. Z. Naturforsch. A **14**, 759 (1959)

[11.17] L. Reimer: Ein experimenteller Beitrag zur Thermokraft dünner Schichten. Z. Naturforsch. A **12**, 525 (1957)

[11.18] D. Thornburg, C.M. Wayman: Thermoelectric power of vacuum evaporated Au-Ni thin film thermocouples. J. Appl. Phys. **40**, 3007 (1969)

[11.19] G.R. Piercy, R.W. Gilbert, L.M. Howe: A liquid helium cooled finger for the Siemens electron microscope. J. Sci. Instr. **40**, 487 (1963)

[11.20] G.M. Parkinson, W. Jones, J.M. Thomas: Electron microscopy at liquid helium temperatures, in [Ref. 1.43]

[11.21] S. Kritzinger, E. Ronander: Local beam heating in metallic electron microscope specimens. J. Microsc. **102**, 117 (1974)

[11.22] K. Kanaya: The temperature distribution of specimens on thin substrates supported on a circular opening in the electron microscope. J. Electron Microsc. **3**, 1 (1955)

[11.23] L. Reimer, R. Christenhusz: Determination of specimen temperature. Lab. Invest. **14**, 1158 (1965)

[11.24] R. Christenhusz, L. Reimer: Schichtdickenabhängigkeit der Wärmeerzeugung durch Elektronen -bestrahlung im Energiebereich zwischen 9 und 100 keV. Z. Angew. Phys. **23**, 397 (1967)

[11.25] S. Leisegang: Elektronenmikroskope, in *Handbuch der Physik*, Vol. 33 (Springer, Berlin, 1956) p. 396

[11.26] V.E. Cosslett, R.N. Thomas: Multiple scattering of 5–30 keV electrons in evaporated metal films. II. Range-energy relations. Br. J. Appl. Phys. **15**, 1283 (1964)

[11.27] H. Kohl, H. Rose, H. Schnabl: Dose-rate effect at low temperatures in FBEM and STEM due to object-heating. Optik **58**, 11 (1981)

[11.28] A. Brockes: Zur Objekterwärmung im Elektronenmikroskop. Kolloid Z. **158**, 1 (1958)

[11.29] R. Christenhusz, L. Reimer: Wärmeleitfähigkeit elektronenmikroskopischer Trägerfolien. Naturwissenschaften **55**, 439 (1968)

[11.30] E. Knapek, J. Dubochet: Beam damage to organic material is considerably reduced in cryo-electron microscopy. J. Mol. Biol. **141**, 147 (1980)

[11.31] I. Dietrich, F. Fox, H.G. Heide, E. Knapek, R. Weyl: Radiation damage due to knock-on processes on carbon foils cooled to liquid helium temperature. Ultramicroscopy **3**, 185 (1978)

[11.32] Y. Talmon, E.L. Thomas: Temperature rise and sublimation of water from thin frozen hydrated specimens in cold stage microscopy, in *Scanning Electron Microscopy 1977/I*, ed. by O. Johari (IIT Research Institute, Chicago, IL, 1977) p. 265

[11.33] Y. Talmon, E.L. Thomas: Beam heating of a moderately thick cold stage specimen in the SEM/STEM. J. Microsc. **111**, 151 (1977)

[11.34] L.G. Pittaway: The temperature distribution in the foil and semi-infinite targets bombarded by an electron beam. Brit. J. Appl. Phys. **15**, 967 (1964)

[11.35] L. Reimer: Irradiation changes in organic and inorganic objects. Lab. Invest. **14**, 1082 (1965)

[11.36] L. Reimer: Review of the radiation damage problem of organic specimens in electron microscopy, in [Ref. 1.12, p. 231]

[11.37] K. Stenn, G.F. Bahr: Specimen damage caused by the beam of the transmission electron microscope, a correlative consideration. J. Ultrastruct. Res. **31**, 526 (1970)

[11.38] D.T. Grubb, A. Keller: Beam-induced radiation damage in polymers and its effect on the image formed in the electron microscope, in *Electron Microscopy 1972* ed. by V.E. Cosslett (IoP, London, 1972) p. 554

[11.39] R.M. Glaeser: Radiation damage and biological electron microscopy, in [Ref. 1.12, p. 205]

[11.40] E. Zeitler (ed.). *Cryomicroscopy and Radiation Damage* (North-Holland, Amsterdam, 1982), published also in Ultramicroscopy **10**, 1–178 (1982); further conference report in Ultramicroscopy **14**, 163–315 (1984)

[11.41] M.S. Isaacson: Inelastic scattering and beam damage of biological molecules, in [Ref. 1.12, p. 247]

[11.42] D.F. Parsons: Radiation damage in biological materials, in [Ref. 1.12, p. 259]

[11.43] M.S. Isaacson: Specimen damage in the electron microscope, in *Principles and Techniques of Electron Microscopy*, Vol. 7, ed. by M.A. Hayat (Van Nostrand-Reinhold, New York, 1977) p. 1

[11.44] R.M. Glaeser, K.A. Taylor: Radiation damage relative to transmission electron microscopy of biological specimens at low temperature: a review. J. Microsc. **112**, 127 (1978)
[11.45] V.E. Cosslett: Radiation damage in the high resolution electron microscopy of biological materials: a review. J. Microsc. **113**, 113 (1978)
[11.46] Z.M. Bacq, P. Alexander: *Fundamentals of Radiobiology* (Pergamon, Oxford, 1961)
[11.47] R.D. Bolt, J.G. Carroll (eds.): *Radiation Effects on Organic Materials* (Academic, New York, 1963)
[11.48] A. Charlesby: *Atomic Radiation and Polymers* (Pergamon, Oxford, 1960)
[11.49] A.J. Swallow: *Radiation Chemistry of Organic Compounds* (Pergamon, Oxford, 1960)
[11.50] H. Dertinger, H. Jung: *Molekulare Strahlenbiologie* (Springer, Berlin, 1968)
[11.51] H.C. Box: Cryoprotection of irradiated specimens, in [Ref. 1.12, p. 279]
[11.52] J. Hüttermann: Solid-state radiation chemistry of DNA and its constituents. Ultramicroscopy **10**, 25 (1982)
[11.53] R. Spehr, H. Schnabl: Zur Deutung der unterschiedlichen Strahlen-Empfindlichkeit organischer Moleküle: Z. Naturforsch. A **28**, 1729 (1973)
[11.54] H. Schnabl: Does removal of hydrogen change the electron energy-loss spectra of DNA bases? Ultramicroscopy **5**, 147 (1980)
[11.55] G.M. Parkinson, M.J. Goringe, W. Jones, W. Rees, J.M. Thomas, J.O. Williams: Electron induced damage in organic molecular crystals: Some observations and theoretical considerations, in *Development in Electron Microscopy and Analysis*, ed. by J.A. Venables (Academic, London, 1976) p. 315
[11.56] L. Reimer, J. Spruth: Interpretation of the fading of diffraction patterns from organic substances irradiated with 100 keV electrons at 10–300 K. Ultramicroscopy **10**, 199 (1982)
[11.57] T. Gejvall, G. Löfroth: Radiation induced degradation of some crystalline amino acids. Radiat. Eff. **25**, 187 (1975)
[11.58] J. Vesely: Electron beam damage of amorphous synthetic polymers. Ultramicroscopy **14**, 279 (1984)
[11.59] L. Reimer: Methods of detection of radiation damage in electron microscopy. Ultramicroscopy **14**, 291 (1984)
[11.60] L. Reimer: Quantitative Untersuchung zur Massenabnahme von Einbettungsmitteln (Methacrylat, Vestopal und Araldit) unter Elektronenbeschuß. Z. Naturforsch. B **14**, 566 (1959)
[11.61] W. Lippert: Über thermisch bedingte Veränderungen an dünnen Folien im Elektronenmikroskop. Z. Naturforsch. A **15**, 612 (1960)
[11.62] G.F. Bahr, F.B. Johnson, E. Zeitler: The elementary composition of organic objects after electron irradiation. Lab. Invest. **14**, 1115 (1965)
[11.63] K. Ramamurti, A.V. Crewe, M.S. Isaacson: Low temperature mass loss of l-phenylalanine and l-tryptophan upon electron irradiation. Ultramicroscopy **1**, 156 (1975)
[11.64] R. Freeman, K.R. Leonard: Comparative mass measurement of biological macromolecules by STEM. J. Microsc. **122**, 175 (1981)
[11.65] W. Lippert: Über Massendickeveränderungen bei Kunststoffen im Elektronenmikroskop. Optik **19**, 145 (1962)

[11.66] G. Siegel: Der Einfluß tiefer Temperaturen auf die Strahlenschädigung von organischen Kristallen durch 100 keV-Elektronen. Z. Naturforsch. A **27**, 325 (1972)

[11.67] A. Brockes: Über Veränderungen des Aufbaus organischer Folien durch Elektronen-Bestrahlung. Z. Phys. **149**, 353 (1957)

[11.68] V.E. Cosslett: The effect of the electron beam on thin sections, in *Proceedings of the European Regional Conference on Electron Microscopy*, Vol. 2, ed. by A.L. Houwink, B.J. Spit (Nederlandse Vereniging voor Electronenmicroscopie, Delft, 1960) p. 678

[11.69] L. Reimer: Interferenzfarben von Methacrylatschnitten und ihre Veränderung unter Elektronenbeschuß . Photogr. Wiss. **9**, 25(1960)

[11.70] L. Reimer: Veränderungen organischer Kristalle unter Beschuß mit 60 keV Elektronen im Elektronenmikroskop. Z. Naturforsch. A **15**, 405 (1960)

[11.71] K. Kobayashi, K. Sakaoku: Irradiation changes in organic polymers at various accelerating voltages. Lab. Invest. **14**, 1097 (1965)

[11.72] H. Orth, E.W. Fischer: Änderungen der Gitterstruktur hochpolymerer Einkristalle durch Bestrahlung im Elektronenmikroskop. Makromol. Chem. **88**, 188 (1965)

[11.73] A. Howie, F.J. Rocca, U. Valdre: Electron beam ionization damage processes in p-therphenyl. Philos. Mag. B **52**, 751 (1982)

[11.74] L. Reimer, J. Spruth: Information about radiation damage of organic molecules by electron diffraction. J. Microsc. Spectr. Electron. **3**, 579 (1978)

[11.75] W.R.K. Clark, J.N. Chapman, A.M. MacLeod, R.P. Ferrier: Radiation damage mechanism in copper phthalocyanine and its chlorinated derivatives. Ultramicroscopy **5**, 195 (1980)

[11.76] N. Uyeda, T. Kobayashi, E. Suito, Y. Harada, M. Watanabe: Molecular image resolution in electron microscopy. J. Appl. Phys. **43**, 5181 (1972)

[11.77] T. Kobayashi, Y. Fujiyoshi, K. Ishizuka, N. Uyeda: Structure determination and atom identification on polyhalogenated molecule, in *Electron Microscopy 1980*, Vol. 4, ed. by P. Brederoo, J. van Landuyt (Seventh European Congress on Electron Microscopy Foundation, Leiden, 1980) p. 158

[11.78] Y. Murata: Studies of radiation damage mechanisms by optical diffraction analysis and high resolution image, in [Ref. 1.57, Vol. 3, p. 49]

[11.79] J.R. Fryer: Radiation damage in organic crystalline films. Ultramicroscopy **14**, 277 (1984)

[11.80] D.J. Smith, J.R. Fryer, R.A. Camps: Radiation damage and structure studies: halogenated phthalocyanines. Ultramicroscopy **19**, 279 (1986)

[11.81] D. van Dyck, M. Wilkens: Aspects of electron diffraction from radiation-damaged crystals. Ultramicroscopy **14**, 237 (1984)

[11.82] R.M. Glaeser: Limitations to significant information in biological electron microscopy as a result of radiation damage. J. Ultrastruct. Res. **36**, 466 (1971)

[11.83] D.T. Grubb, G.W. Groves: Rate of damage of polymer crystals in the electron microscope: dependence on temperature and beam voltage. Philos. Mag. **24**, 815 (1971)

[11.84] L.E. Thomas, C.J. Humphreys, W.R. Duff, D.T. Grubb: Radiation damage of polymers in the million volt electron microscope. Radiat. Eff. **3**, 89 (1970)

[11.85] A.V Crewe, M. Isaacson, D. Jonson: Electron beam damage in biological molecules, in *Proceedings of the 28th Annual Meeting of EMSA* (Claitor, Baton Rouge, LA, 1970) p. 264

[11.86] L. Reimer: Veränderungen organischer Farbstoffe im Elektronenmikroskop. Z. Naturforsch. B **16**, 166 (1961)

[11.87] N. Uyeda, T. Kobayashi, M. Ohara, M. Watanabe, T. Taoka, Y. Harada: Reduced radiation damage of halogenated copper-phthalocyanine, in *Electron Microscopy 1972* ed. by V.E. Cosslett (IoP, London, 1972) p. 566

[11.88] W. Baumeister, U.P. Fringeli, M. Hahn, F. Kopp, J. Seredynski: Radiation damage in tripalmitin layers studied by means of infrared spectroscopy and electron microscopy. Biophys. J. **16**, 791 (1976)

[11.89] W. Baumeister, J. Seredynski: Radiation damage to proteins: changes on the primary and secondary structure level, in [Ref. 1.57, Vol. 3, p. 40]

[11.90] W. Baumeister, M. Hahn, J. Seredynski, L.M. Herbertz: Radiation damage of proteins in the solid state: changes of amino acid composition in catalase. Ultramicroscopy **1**, 377 (1976)

[11.91] M. Isaacson: Electron beam induced damage of organic solids: implications for analytical electron microscopy. Ultramicroscopy **4**, 193 (1979)

[11.92] R.F. Egerton: Chemical measurements of radiation damage in organic samples at and below room temperature. Ultramicroscopy **5**, 521 (1980)

[11.93] R.F. Egerton: Organic mass loss at 100 K and 300 K. J. Microsc. **126**, 95 (1982)

[11.94] M. Misra, R.F. Egerton: Assessment of electron irradiation damage to biomolecules by electron diffraction and EELS. Ultramicroscopy **15**, 337 (1984)

[11.95] H. Shuman, A.V. Somlyo, P. Somlyo: Quantitative electron probe microanalysis of biological thin sections: methods and validity. Ultramicroscopy **1**, 317 (1976)

[11.96] T.A. Hall, B.L. Gupta: Beam-induced loss of organic mass under electron microscope conditions. J. Microsc. **100**, 177 (1974)

[11.97] P. Bernsen, L. Reimer, P.F. Schmidt: Investigation of electron irradiation damage of evaporated organic films by laser microprobe mass analysis. Ultramicroscopy **7**, 197 (1981)

[11.98] S.H. Faraj, S.M. Salih: Spectroscopy of electron irradiated polymers in electron microscopy. Radiat. Eff. **55**, 149 (1981)

[11.99] P.K. Haasma, M. Parikh: A tunneling spectroscope study of molecular degradation due to electron irradiation. Science **188**, 1304 (1975)

[11.100] M.J. Richardson, K. Thomas: Aspects of HVEM of polymers, in *Electron Microscopy 1972* ed. by V.E. Cosslett (IoP, London, 1972) p. 562

[11.101] S.M. Salih, V.E. Cosslett: Some factors influencing radiation damage in inorganic substances, in [Ref. 1.56, Vol. 2, p. 670]

[11.102] J. Dubochet: Carbon loss during irradiation of T4 bacteriophages and E. coli bacteria in electron microscopes. J. Ultrastruct. Res. **52**, 276 (1975)

[11.103] K.H. Downing: Temperature dependence of the critical electron exposure for hydrocarbon monolayers. Ultramicroscopy **11**, 229 (1983)

[11.104] L. Zuppiroli, N. Housseau, L. Fooro, J.P. Guillot, J. Pelissier: Fading of the Bragg spots in irradiated organic conductors: temperature and composition effects. Ultramicroscopy **19**, 325 (1986)

[11.105] E. Knapek: Properties of organic specimens and their supports at 4 K under irradiation in an electron microscope. Ultramicroscopy **10**, 71 (1982)

[11.106] D.F. Parsons, V.R. Matricardi, R.C. Moretz, J.N. Turner: Electron microscopy and diffraction of wet unstained and unfixed biological objects. Adv. Biol. Med. Phys. **15**, 161 (1974)

[11.107] H.G. Heide, S. Grund: Eine Tiefkühlkette zum Überführen von wasserhaltigen biologischen Objekten ins Elektronenmikroskop. J. Ultrastruct. Res. **48**, 259 (1974)

[11.108] K.A. Taylor, R.M. Glaeser: Electron microscopy of frozen hydrated biological specimens. J. Ultrastruct. Res. **55**, 448 (1976)

[11.109] T.E. Hutchinson, D.E. Johnson, A.P. Mackenzie: Instrumentation for direct observation of frozen hydrated specimens in the electron microscope. Ultramicroscopy **3**, 315 (1978)

[11.110] J. Lepault, J. Dubochet: Beam damage and frozen-hydrated specimens, in [Ref. 1.59, Vol. 1, p. 25]

[11.111] Y. Talmon: Radiation damage to organic inclusions in ice. Ultramicroscopy **14**, 305 (1984)

[11.112] J.R. Fryer, C.H. McConnell: Effect of temperature on radiation damage to aromatic organic molecules. Ultramicroscopy **40**, 163 (1992)

[11.113] S.M. Salih, V.E. Cosslett: Reduction in electron irradiation damage to organic compounds by conducting coatings. Philos. Mag. **30**, 225 (1974)

[11.114] J.R. Fryer, F. Holland: The reduction of radiation damage in the electron microscope. Ultramicroscopy **11**, 67 (1983)

[11.115] A. Rose: Television pickup tubes and the problem of noise. Adv. Electron. **1**, 131 (1948)

[11.116] R.C. Williams, H.W. Fischer: Electron microscopy of tobacco mosaic virus under conditions of minimal beam exposure. J. Mol. Biol. **52**, 121 (1970)

[11.117] M. Ohtsuki, E. Zeitler: Minimal beam exposure with a field emission source. Ultramicroscopy **1**, 163 (1975)

[11.118] K.H. Herrmann, J. Menadue, H.T. Pearce-Percy: The design of compact deflection coils and their application to a minimum exposure system, in *Electron Microscopy 1976*, Vol. 1, ed. by D.G. Brandon (Tal International, Jerusalem, 1976) p. 342

[11.119] Y. Fujiyoshi, T. Kobayashi, K. Ishizuka, N. Uyeda, Y. Ishida, Y. Harada: A new method for optimal-resolution electron microscopy of radiation-sensitive specimens. Ultramicroscopy **5**, 459 (1980)

[11.120] S.B. Hayward, R.M. Glaeser: Radiation damage of purple membrane at low temperatures. Ultramicroscopy **4**, 201 (1979)

[11.121] W. Chiu, R.M. Glaeser: Evaluation of photographic emulsion for low-exposure imaging, in [Ref. 1.43, p. 194]

[11.122] M. Kessel, J. Frank, W. Goldfarb: Low-dose microscopy of individual macromolecules, in [Ref. 1.43, p. 154]

[11.123] D.L. Dorset, F. Zemlin: Structural changes in electron irradiated paraffin crystals at <15 K and their relevances to lattice imaging experiments. Ultramicroscopy **17**, 229 (1985)

[11.124] W. Kunath, K. Weiss, H. Sack-Kongehl, M. Kessel, E. Zeitler: Time-resolved low-dose microscopy of glutamine synthease molecules. Ultramicrocopy **13**, 241 (1984)

[11.125] D.W. Pashley, A.E.B. Presland: Ion damage to metal films inside an electron microscope. Philos. Mag. **6**, 1003 (1961)

[11.126] M.N. Kabler, T.T. Williams: Vacancy-interstitial pairs production via electron-hole recombination in halide crystals. Phys. Rev. B **18**, 1948 (1978)

[11.127] H. Strunk: High voltage transmission electron microscope of the dislocation arrangement in plastically deformed NaCl crystals, in [Ref. 1.78, p. 285]
[11.128] L.W. Hobbs, A.E. Hughes, D. Pooley: A study of interstitial clusters in irradiated alkali halides using direct electron microscopy. Proc. Roy. Soc. London A **332**, 167 (1973)
[11.129] L.W. Hobbs: Radiation effects in the electron microscopy of beam-sensitive inorganic solids, in *Developments in Electron Microscopy and Analysis*, ed. by J.A. Venables (Academic, London, 1976) p. 287
[11.130] L.W. Hobbs: Radiation damage in electron microscopy of inorganic solids. Ultramicroscopy **3**, 381 (1979)
[11.131] T. Evans: Decomposition of calcium fluoride and strontium fluoride in the electron microscope. Philos. Mag. **8**, 1235 (1963)
[11.132] L.E. Murr: TEM study of crystal defects in natural fluorite. Phys. Status Solidi A **22**, 239 (1974)
[11.133] J.H. Ahn, D.R. Feacor, E.J. Essene: Cation-diffusion-induced characteristic beam damage in TEM images of micas. Ultramicroscopy **19**, 375 (1986)
[11.134] R.D. Baeta, K.H.G. Ashbee: Electron irradiation damage in synthetic quartz, in *Developments in Electron Microscopy and Analysis*, ed. by J. A. Venables (Academic, London, 1976) p. 307
[11.135] Y. Yokota, H. Hashimoto, T. Yanaguchi: Electron beam irradiation of natural zeolites at low and room temperature. Ultramicroscopy **54**, 207 (1994)
[11.136] E.D. Kater: Mechanism of decomposition of dolomite in the electron microscope. Ultramicroscopy **18**, 241 (1985)
[11.137] D.J. Smith: Atomic resolution studies of surface structure and reactions, in [Ref. 1.59, Vol. 2, p. 929]
[11.138] R.F. Egerton, P.A. Crozier, P. Rice: EELS and chemical damage. Ultramicroscopy **23**, 305 (1987)
[11.139] H. Sauer, R. Brydson, P.N. Rowley, W. Engel, J.M. Thomas: Determination of coordinations and coordination-specific site occupancies by EELS: an investigation of boron-oxygen compounds. Ultramicroscopy **49**, 198 (1993)
[11.140] L.A.J. Garvie, A.J. Craven: Electron-beam-induced reduction of Mn^{4+} in manganese oxides as revealed by parallel EELS. Ultramicroscopy **59**, 83 (1994)
[11.141] M.J. Makin: Atom displacement radiation damage in electron microscopes, in [Ref. 1.57, Vol. 3, p. 330]
[11.142] M. Wilkens, K. Urban: Studies of radiation damage in crystalline materials by means of high voltage electron microscopy, in [Ref. 1.78, p. 332]
[11.143] K. Urban: Radiation damage in inorganic materials in the electron microscope, in *Electron Microscopy 1980*, Vol. 4, ed. by P. Brederoo, J. van Landuyt (Seventh European Congress on Electron Microscopy Foundation, Leiden 1980) p. 188
[11.144] V.E. Cosslett: Radiation damage by electrons, with special reference to the knock-on process, in *Electron Microscopy and Analysis 1979*, ed. by T. Mulvey (IoP, London, 1979) p. 177
[11.145] M. Kiritani, T. Yoshiie, E. Ishida, S. Kojima, Y. Satoh: In-situ electron radiation damage study of materials by HVEM, in [Ref. 1.59, Vol. 2, p. 1089]
[11.146] M. Wilkens: Radiation damage in crystalline materials, displacement cross sections and threshold energy surfaces, in [Ref. 1.80, p. 475]

[11.147] N. Yoshida, K. Urban: A study of the anisotropy of the displacement threshold energy in copper by means of a new high-resolution technique, in [Ref. 1.80, p. 493]
[11.148] W.E. King, K.L. Merkle, M. Meshii: Study of the anisotropy of the threshold energy in copper using in-situ electrical resistivity measurements in the HVEM, in [Ref. 1.81, p. 212]
[11.149] M.O. Ruault: In situ study of radiation damage in thin foils of gold by HVEM. Philos. Mag. **36**, 835 (1977)
[11.150] L.E. Thomas: The diffraction dependence of electron damage in a HVEM. Radiat. Eff. **5**, 183 (1970)
[11.151] N. Yoshida, K. Urban: Electron diffraction channelling and its effect on displacement damage formation, in [Ref. 1.80, p. 485]
[11.152] J.J. Hren: Barriers of AEM: contamination and etching, in [Ref. 1.66, p. 481]
[11.153] L. Reimer, M. Wächter: Contribution to the contamination problem in TEM. Ultramicroscopy **3**, 169 (1978)
[11.155] A.E. Ennos, The sources of electron-induced contamination in the electron microscope. Br. J. Appl. Phys. **5**, 27 (1954)
[11.156] S. Leisegang: Über Versuche in einer stark gekühlten Objektpatrone, in [Ref. 1.51, p. 184]
[11.157] H.G. Heide: Die Objektverschmutzung im Elektronenmikroskop und das Problem der Strahlenschädigung durch Kohlenstoffabbau. Z. Angew. Phys. **15**, 116 (1963)
[11.158] H.G. Heide: Die Objektraumkühlung im Elektronenmikroskop. Z. Angew. Phys. **17**, 73 (1964)
[11.159] J.T. Fourie: The controlling parameter in contamination of specimens in electron microscopes. Optik **44**, 11 (1975)
[11.160] K.H. Müller: Elektronen-Mikroschreiber mit geschwindigkeitsgesteuerter Strahlführung. Optik **33**, 296 (1971)
[11.161] G. Love, V.D. Scott, N.M.T. Dennis, L. Laurenson: Sources of contaminants in electron optical equipment. Scanning **4**, 32 (1981)
[11.162] M.T. Browne, P. Charalambous, R.E. Burge: Uses of contamination in STEM projection electron lithography, in *Developments in Electron Microscopy and Analysis*, ed. by M.J. Goringe (IoP, London, 1981) p. 47
[11.163] P. Hirsch, M. Kässens, M. Püttmann, L. Reimer: Contamination in a SEM and the influence of specimen cooling. Scanning **16**, 101 (1994)

Index

G value, 469, 478
Z-contrast, 113
Ω-Filter, 121
α-fringes, 384
(A.S.T.M. Index), 345
(Gjønnes–Moodie lines), 352
(Grigson mode), 339

Aberration correction, 40, 404
absorption correction, 433, 435
absorptive power of radiation, 463
Acceleration voltage, 7, 17, 99, 107, 356, 357, 404
acceptance angle, 445–447
Achromatic circle, 234
Achromatic image plane, 120, 121
Adjustment, 92
Aharonov-Bohm effect, 48
Airy distribution, 70, 76, 94, 233
ALCHEMI, 434
Alignment, 43, 92, 107, 134, 249, 254
Alloy composition, 316
amino acid analyzer, 476
amorphization, 482
amorphous diffraction, 360
amorphous layer, 407
Amorphous specimen contrast, 196
Amorphous specimen diffraction, 317, 318
Amplitude-phase diagram (APD), 57, 292, 373, 374, 378, 392
Analytical Electron Microscopy, 5
analytical sensitivity, 453
angle-resolved EELS, 7, 126, 159, 168, 443
Angular momentum, 24, 152, 174
anisotropic aberration, 31, 38
annular Aperture, 213
annular detector, 337
Anode, 17, 85
Anomalous absorption, 306, 312, 325, 376, 380, 390, 434
anomalous electron transmission, 337
Anomalous transmission, 99, 309, 312, 335, 362, 369, 370, 372, 382
anticomplementary, 382
Anticontamination blade, 100, 101, 485, 486
antiferromagnetic domain, 384
antiphase boundaries, 351, 378, 383
Aperture illumination, 61
arrhenius plot, 478
Astigmatism, 31, 34, 40, 73, 92, 103, 104, 239, 244, 255, 331
atomic displacement, 409, 480, 484
atomic number correction, 432, 433
Auger electron, 155, 424, 427
Auger electron microanalysis, 14, 183
autoalignment, 43
Autobiasing, 87
Autocorrelation, 160, 254
Autofocusing, 250, 262
autoradiography, 477
Autotuning, 250, 404
Averaging by Fourier filtering, 255

Averaging of periodic structures, 255, 261
Averaging rotational symmetry, 256

Back focal plane, 29, 96, 110
Backscattered electron coefficient, 193, 210
backscattered electrons (BSE), 13, 110, 138, 139, 192, 337, 429
Backscattering, 313, 324, 432
Backscattering coefficient, 139
Barber's rule, 116
Bend contour, 268, 311, 315, 341, 358–360, 362, 363, 369, 370, 388, 401
Bethe approximation, 313
Bethe dynamic potential, 360
Bethe dynamical potential, 313, 314
Bethe formula, 187, 467
Bethe loss, 187, 194, 462, 477
Bethe range, 192
Bethe ridge, 7, 159
Bethe sum rule, 158, 440
Bethe surface, 158, 159
Binomial distribution, 132, 136
Biological sections, 8, 98, 100, 196, 203, 207, 209, 210, 259, 421, 452
Biomacromolecules, 249, 256
Biprism fringes, 246, 267, 268
Bloch wave, 294, 295, 297, 434
Bloch wave absorption, 306
Bloch wave absorption parameter, 308, 311, 313, 321–323
Bloch wave boundary condition, 297, 298, 306
Bloch wave dependent and independent model, 311
Bloch wave excitation amplitude, 297, 305, 309
Bloch wave field, 297, 337, 370, 371
Bloch wave intensity, 312
Bloch wave probability density, 308, 313
Bloch wave type, 303
bloch-wave, 376
bloch-wave absorption, 397, 399
Bloch-wave channeling, 342, 372
bloch-wave excitation amplitude, 375, 376, 382, 383
bloch-wave method, 382, 383, 390, 405
bloch-wave type, 383
Blocking, 313
Boersch effect, 39, 82, 89
Bohr model, 423
Bohr radius, 147, 148
Boltzmann constant, 78
bond rupture, 469
Bormann effect, 434
Born approximation, 146, 148, 150, 151, 153, 200, 294
Bragg angle, 106, 284, 330, 364, 460
Bragg condition, 283, 285, 287, 292, 300, 342, 348
Bragg reflection, 98, 213, 306, 330, 335, 338, 342, 349, 359, 363, 369, 400
Bragg reflection amplitude, 291, 383
Bragg reflection intensity, 300
Bragg spot diameter, 333
Bragg transmission intensity, 300
Bright-field, 104, 196, 209, 219, 224, 329
Brillouin zone, 302
Burgers vector, 100, 357, 359, 386, 389, 394

cage effect, 479
Camera length, 104, 331, 339, 366, 395
Canonical momentum, 47
Castaing-Henry imaging energy filter, 121
Catalase, 108
Cathode, 17, 84, 86
Cathode lifetime, 79, 88
Cathode temperature, 78
Cathode-ray tube, 110
Cathodoluminescence, 13, 110, 127
Caustic, 76, 92, 336
cavity, 362, 366, 384
CBED, 333, 335, 342, 352–355
CCD, 1, 339, 353
Central beam stop, 218, 224
Centrifugal potential barrier, 176
Čerenkov radiation, 173
Channeling effects, 112, 194, 313
Channeling pattern, 194, 311
Charge density, 127, 129, 201, 467
distributions, 353
Charge-collection efficiency, 138
Charge-coupled-device (CCD), 107, 134
charge-sensitive amplifier, 428

Charging, 332, 339, 340, 363
chemical etching, 370
Chromatic Aberration, 42
Chromatic Aberration C_c, 7, 94, 98, 103, 105, 106, 110, 120, 155, 188–190, 204, 370, 402
cleavage, 370
Cliff–Lorimer ratio, 179, 434, 435
Close-packed structure, 278
Coherence, 90, 230, 234, 400, 404, 406
Coherence condition, 55
Coherence length, 55
Coherence partial, 55
Coherence spatial, 55
Coherence temporal, 55
coherent Precipitates, 396
Cold finger, 100
Colloidal gold, 217, 238
Column approximation, 289, 373, 393
coma, 31, 37
Coma-free alignment, 32, 38, 43, 93, 241, 404
Complex scattering amplitude, 148, 152, 197, 200, 220
composition profile, 408
Compton angle, 446
Compton effect, 424
Compton scattering, 7, 160, 171, 326
condensation of gases, 461
Condenser lens, 90
condenser-objective Lens, 96, 109
Contamination, 90, 103, 108, 113, 238, 332, 333, 356, 362, 418, 435, 484
Contamination mark, 269
contamination needle, 488
contamination rate, 489
contamination ring, 489, 490
Contrast thickness, 198, 200, 205, 210, 445
Contrast transfer in STEM, 236, 237
Contrast tuning, 125, 206
contrast-transfer envelope, 402
Contrast-transfer function, 3, 4, 100, 105, 106, 130, 131, 213, 218, 220, 222, 228, 251, 404, 406
Contrast-transfer function envelope, 231, 232
Convergent beam electron diffraction (CBED), 6, 7, 110, 329, 333, 341, 362, 412
Convolution, 64, 185, 191
Core polarizability, 167
Cornu spiral, 58, 61
correction of aberrations, 40, 404
Coulomb energy, 147, 152, 157, 178
Coulomb energy (potential), 285, 304, 347
Coulomb force, 51
Coulomb interactions, 82
Coulomb potential, 347
Critical voltage, 313, 321, 342, 460
Cross-correlation, 43, 249, 254, 257, 407, 480
cross-linking, 465, 469, 475, 484, 487
Crossover, 82, 86, 91, 96
cryo-methods, 437, 453
cryo-shield, 466
cryosection, 478
crystal bending, 333
Crystal boundaries, 357
Crystal orientation, 100, 343, 347, 349
crystal potential, 354
Crystal Structure, 345, 346, 349, 352
crystal symmetry, 342
crystal thicknesses, 346, 354
Crystal-structure imaging, 253, 402, 479
crystalline arrays of Biomacromolecules, 256
Crystallography, 274
Current density, 78
cyclotron frequency, 24

damage effects, 467
Dark-field mode, 93, 106, 201, 213, 218, 224, 350, 351, 359, 399, 412
Dark-field tilt method, 201, 224
Debye temperature, 317, 322
Debye–Scherrer ring, 342, 343, 460, 472
Debye–Waller factor, 183, 294, 319, 321, 322, 355, 356, 473
Debye-Scherrer ring broadening, 289
Debye-Scherrer ring intensity, 318
deconvolution, 438, 440, 442, 443, 452
decoration technique, 412
dedicated scanning transmission electron microscope, 124

dedicated Scanning transmission electron microscopy (STEM), 207, 210, 225, 237
dedicated STEM, 112
Deep-level transient spectroscopy, 112
defect cluster, 398, 480, 484
deflection coil, 362
Defocus series, 34, 239, 256, 407
Defocusing, 3, 71, 109, 249, 254
Delocalization, 226
Depletion layer, 112, 138
Depth of focus, 109
Depth of image, 108
Detection limit, 456
Detection quantum efficiency, 132, 134, 136
Detector annular, 225, 227
detector aperture, 110, 114, 208, 336, 368
Detector half-plane, 269
detector Noise, 132
Detector quadrant, 269
Diaphragm annular, 202
Diaphragm half-plane, 233, 254
Dichroism, 183
dielectric, 439
Dielectric function, 164
Dielectric theory, 6, 163
differential Inelastic scattering, 156
differential Inelastic scattering cross section, 163, 197
differential Phase contrast, 237, 269
diffraction, 331
Diffraction Contrast, 100, 104, 268, 360, 369, 384, 400, 415, 472
diffraction error, 32
Diffraction lens, 104
diffuse streaks, 350
Digital image processing, 134
Dipole matrix element, 158
dipole radiation, 419
Dislocation, 111, 357–359, 363, 367, 369, 371, 372, 375, 385
dislocation dipole, 393
Dislocation loop, 101, 388, 398, 461, 482
dislocation residual contrast, 396, 400
Dislocation screw, 386, 388, 390, 392, 394
dislocation simulated image, 390
Disorder, 183
Dispersion surface, 297, 302, 303, 305, 308, 311, 313, 314, 355, 357
displacement vector, 358, 376, 378, 379, 383, 384, 386, 397
dissociated dislocation, 383, 391, 393, 411
distortion, 31, 36, 331, 341–343
Drude model, 164
Duane–Hunt law, 419
Dynamical n-beam theory, 311
dynamical interaction, 354
Dynamical theory of electron diffraction, 98, 197, 292, 333, 359, 360, 366, 390
Dynamical theory of electron diffraction fundamental equations, 295, 298, 306, 313

edge contour (fringe), 382, 384, 401
edge dislocation, 387, 388, 392, 394
Edge spread function, 136
effective number of electrons, 439, 440
Eigenvalue, 296, 298, 300, 306, 375, 405
Eigenvalue problem, 294, 296, 311
Eigenvector, 300, 375, 405
Eigenvector orthogonality relation, 297, 312
Eigenvector periodicity condition, 297
eikonal method, 32
elastic anisotropy, 396
Elastic scattering, 141
Elastic scattering amplitude, 146, 148, 183
Elastic scattering characteristing angle θ_0, 200, 225
Elastic scattering differential cross section, 142
Elastic scattering mean free path, 141, 184, 197, 200
Electric field, 17, 18, 78, 86, 89
Electrometer, 140
Electron Acceleration, 17
electron channeling pattern, 13, 337, 341
Electron charge density distribution, 151
Electron deflection, 20, 49
electron distribution image, 448

electron dose, 4, 467, 479
Electron emission, 78
Electron energy-loss spectroscopy (EELS), 5, 9, 424, 434, 437, 454, 476, 477
Electron energy-loss spectroscopy (EELS) ratio method, 446
electron energy-loss spectroscopy (EELS) sensitivity, 454
electron energy-loss spectrum background, 447, 449
electron energy-loss spectrum background fitting, 441
electron energy-loss spectrum background subtraction, 441
Electron energy-loss spectrum delayed edge, 178
Electron gun, 1, 78
Electron interferometry, 52
electron kinetic energy, 19
electron mass, 20
Electron momentum, 20, 143
Electron probe, 3, 12, 90, 93, 96, 109, 110, 113
Electron probe broadening, 99
Electron probe diameter, 94, 95, 97, 332, 435
Electron probe optimum aperture, 94
Electron range, 129, 138, 192, 193, 463
electron rest energy, 20
Electron spectrometer, 9, 115
Electron spectrometer double-stigmatic focusing, 117
Electron spectrometer second-order aberration, 118, 122, 123
Electron spectroscopic diffraction, 159, 168, 169, 326, 327, 353
Electron spectroscopic imaging, 99, 121, 125, 173, 448, 449, 453, 456
electron spectroscopic imaging series, 449
Electron spin, 20, 174
electron spin resonance, 477
Electron Trajectories, 22, 24
electron velocity, 20
Electron wavelength, 3, 20, 45
Electron–nucleus collision, 143
Electron–phonon scattering, 145, 185, 304, 350, 369
electron-backscattering pattern, 13, 340
Electron-beam-induced current (EBIC), 110
Electron-energy loss spectrum, 174
Electron-energy-loss spectroscopy, 90, 93, 107, 110, 113, 115, 120, 121
Electron-energy-loss spectroscopy diffraction mode, 126
electron-hole pair, 112, 138, 193, 428, 429
Electron-optical refractive index, 51, 146, 148, 405
Electron-phonon scattering, 303, 306, 322, 326
electron-probe aperture, 94, 114
Electron-probe broadening, 188
electron-probe current, 427
Electron-probe stationary, 332, 335
Electron-spectroscopic diffraction, 121, 125
Electropolishing, 1, 99, 360
Electrostatic fields, 270
Element distribution image, 7, 125, 456
elemental map, 456
Embedding, 203, 452
Emission current density, 89
emission electron microscope, 10
enantiomorphic phase, 384
Energy dissipation, 165, 193
energy filtering, 339, 353, 369
Energy loss near-edge structure (ELNES), 179
Energy spread, 55, 77, 81, 89, 90, 95, 218, 230, 232, 234, 402, 404, 406, 442
energy-dispersive plane, 120
energy-dispersive X-ray detector, 5
Energy-filtering, 334, 335
energy-filtering transmission electron microscopy, 3, 99, 113, 124
energy-loss near-edge structure (ELNES), 6, 442, 445
Environmental experiment, 8, 101
epitaxy, 365, 367
Equipotentials, 85
Ewald sphere, 283, 284, 287, 288, 295, 298, 299, 311, 314, 343–345, 350, 354, 393, 395, 414

excitation, 333, 335, 336, 355, 357, 359, 360, 362, 368, 369, 372, 375, 379, 388, 391, 394, 403
Excitation error, 268, 287, 292, 293, 301, 352
Excitation point, 297, 298, 303
exciton, 481
Exit wave function, 249, 253, 405
Extended energy-loss fine structure (EXELFS), 6, 182, 442, 445
extinctiion rule, 378
Extinction distance, 290, 293, 294, 301, 314, 321, 360, 377, 379, 382, 390, 397, 412
extinction rules, 343, 345, 351
Extractor electrode, 88
extrinsic stacking fault, 378

fading of electron diffraction, 473, 477
Fano factor, 429
Faraday cage, 139, 184, 194, 339
Fast Fourier transform (FFT), 239
fault matrix, 376, 377
Faunhofer diffraction, 289
Fermi distribution, 78, 81
Fermi energy, 78
Fermi level, 78, 80, 154, 159, 166, 174, 180, 182, 194
Fermi surface, 302
Ferritin, 215, 455
Ferroelectric domain, 270, 384
fiber axis, 343, 344
Fiber plate, 139
fiber texture, 343, 344
fibre plate, 438
Field curvature, 244
Field emission, 80
Field emission gun, 89, 90, 92, 95, 113, 246
field-effect transistor, 428
field-emission gun, 4, 332
fluorescence correction, 433
Fluorescent screen, 91, 126
Flux quantum, 267
Fluxon criterion, 267
focal length, 330
Focus series, 215
Focusing, 109
forbidden reflection, 349, 352, 411, 412, 414
Formvar film, 99
forward-scattered electrons, 337, 341
Fourier filter, 251
Fourier sum, 68
Fourier synthesis, 345, 346
Fourier tranform, 218
Fourier transform, 3, 70, 74, 232, 281, 439
Fourier transform convolution theorem, 68, 76, 241, 251, 406, 443
Fourier transform translation theorem, 67, 252, 263
Fourier-log method, 442, 444
Fourier-ratio method, 441, 442
Fowler–Nordheim formula, 81
Fowler-Nordhein plot, 80
Fraunhofer Diffraction, 61, 146, 213
Fraunhofer Holography, 243, 244
Freeze-drying, 257
Freeze-fracturing, 257
frenkel defect, 398
Frenkel pair, 481
Fresnel biprism, 53
Fresnel diffraction, 55, 61, 242, 262, 289, 293, 406
Fresnel Fringes, 59, 61, 90, 233
Fresnel Holography, 243
Fresnel integral, 59
Fresnel mode, 184
Fresnel zone, 57, 290
Fresnel-zone lens, 242
Friedel's law, 376
Front focal plane, 96
fundamental reflections, 383

Gaussian distribution, 83, 94, 230
Gaussian function, 191
Gaussian image plane, 243
generalized Oscillator strength (GOS), 158, 177
Gerchberg–Saxton algorithm, 253, 269
Goniometer, 100, 346, 360
grain boundary, 411, 484
Granularity, 212, 214, 239
Grease, 484
Green's function, 304

Guinier–Preston zone, 350, 398, 402
Gun brightness, 82, 91, 94, 404

Hamiltonian, 303
Hartree–Fock method, 147, 151, 153
Hartree–Slater wave function, 178
heat generation, 461
Heisenberg uncertainty relation, 55
heterostructures, 408
High resolution microscopy, 3, 9, 97, 104–106, 400–416
High Voltage Electron Microscopy, 7
high-angle annular-dark-field method (HAADF), 412
High-order Laue zone (HOLZ), 6, 329, 346, 354
High-order Laue zone (HOLZ) line, 328
High-order Laue zone (HOLZ) pattern, 284, 326, 342, 354
High-resolution electron microscopy, 3
High-voltage electron microscopy, 1, 101, 112, 127, 135, 145, 190, 315, 360, 369, 372, 390, 477, 483
Hollow-beam, 86
Hollow-cone illumination, 2, 4, 93, 105, 202, 214, 225, 233, 234, 237
Holography, 4, 90, 241, 408
HOLZ line, 352, 355
HOLZ rings, 355
Honeycomb pattern, 135
Howie–Whelan equations, 293, 299, 306, 311, 373, 376, 390, 392
Huygens principle, 114
Hydrogenic model, 178
Hysteresis, 107

Ice embedding, 249
illumination aperture, 4, 42, 96, 104, 109, 110, 196, 230, 244, 331, 333, 367, 406
Illumination system, 96
Image drift, 241
Image EELS, 126
image matching, 407
image plate, 1
Image recording, 126
Image restoration, 238, 249, 252
image rotation, 348, 394
image shift, 407
image simulation, 407
Image-reconstruction, 215
imaginary Fourier coefficient, 306
imaginary Lattice potential, 306, 322
Imaging energy filter, 115, 119, 121, 123, 124
Imaging modes, 104
Imaging plate, 131
Imaging prism spectrometer, 124
Imaging system, 103
Impact parameter, 142
improvement Contrast-transfer funktion, 238
in situ experiments, 100, 371, 483
incoherent precipitate, 396
inelastic cross-section, 370
Inelastic scattering, 153, 369
Inelastic scattering angular distribution, 237
Inelastic scattering characteristic angle θ_E, 158, 237
Inelastic scattering in crystals, 302
Inelastic scattering localization, 3, 225
Inelastic scattering mean free path, 184
inelastic-to-elastic total cross sections, 205, 225
infrared absorption, 473
Inner potential, 248, 265, 295
Inner-shell cross section, 177, 178
Inner-shell cross section parametrization, 179
Inner-shell ionization, 6, 171, 174, 305, 306, 313, 369, 481
insibility criterion (g*b=0), 396
integrated Oscillator strength, 179
Interband excitation (transition), 154, 163, 166, 303, 305
interband transitions, 6
interface, 407
Intermediate lens, 103, 104
interstitial (site), 145, 398, 483
Intraband excitation (transition), 154, 163, 303, 305, 369
invisibility criterion, 357
Ion-beam etching, 1, 99, 370
Ionization depth distribution, 193
Ionization edge, 175
Ionization energy, 154, 174
Isoplanatic patch, 244

Köhler illumination, 97
Kikuchi, 337, 340
Kikuchi band, 7, 305, 321, 323, 325, 327, 335, 337, 353, 484
Kikuchi diagram (pattern), 311
Kikuchi line, 6, 305, 315, 321, 323, 325, 328, 332, 340, 348, 349, 355, 363, 393, 394
Kikuchi line intersecting technique, 317
Kinematical theory of electron diffraction, 210, 283, 294, 295, 298, 300, 302, 319, 373, 379, 380, 385, 387
Kirchhoff diffraction, 55
Kirchhoff's law, 463
knock-of collision, 480, 482
Knock-on collision, 145
Kossel bands, 335
Kossel cone, 323, 325, 349, 394
Kossel pattern, 332, 334, 336
Kramers–Kronig relation, 167

LACBED, 6, 329, 333, 335, 336, 338, 352, 357, 358, 396
Landau theory, 186, 207
Lanthanum hexaboride (LaB_6) cathode, 79, 84
large-angle convergent-beam (LACBED), 335
large-angle convergent-beam diffraction (LACBED), 335
large-angle convergent-beam electron diffraction (LACBED), 396
large-angle Electron diffraction (LACBED), 326
laser (Fraunhofer) diffraction, 339
Laser diffraction, 232, 238
laser microprobe mass analyser (LAMMA), 476
latent electron dose, 473
Lattice amplitude, 285
Lattice defect, 99, 100, 111, 112, 259, 357, 359, 365, 367, 369
lattice defect symmetry rule, 376
lattice displacement, 367
lattice distortion, 333
Lattice fringes, 100
lattice parameters, 357
lattice planar faults, 376–378
Lattice plane, 277
Lattice plane imaging, 98, 108, 113
Lattice potential, 306, 315, 319, 322, 352
Lattice-plane distance (spacing), 275, 281, 473
lattice-plane fringes, 400, 402, 416
lattice-plane spacing, 342, 345
lattice-potential coefficients, 346
Laue equations (conditions), 281, 283
Laue zone, 282, 284, 326, 354, 412, 414
Laue zone (HOLZ), 346
layer structures, 354
Light pipe, 111, 139
light-optical Reconstruction, 246
Lindhard model, 166
Liquid crystal display, 239
Logarithmic amplifier, 210
long-rang strain cantrast, 398
Lorentz force, 262, 268
Lorentz force microscopy, 262
Lorentz microscopy, 84, 90, 91, 107, 113, 339
Lorentz microscopy Foucault mode, 264, 265, 269, 271
Lorentz microscopy Fresnel mode, 265, 269–271
Lorentz microscopy holographic mode, 269, 271
Lorentz microscopy STEM mode, 269
Lorentz microscopy stroboscopy, 270
Lorentz model, 166
low-dose exposure, 249, 257, 480
low-energy electron diffraction (LEED), 340, 416
Low-light-level camera, 134
low-voltage scanning electron microscopy (LVSEM), 13

Mach–Zehnder interferometer, 247, 248, 269
Magnetic domain wall, 263, 264
Magnetic domain wall thickness, 267
Magnetic scalar potential, 22
magnetic stray fields, 12
Magnetic vector potential, 47, 49, 248, 262
Magnetization ripple, 263, 268
Magnification, 107, 400
magnification calibration, 402

Magnification standard, 108
Main transfer band, 229, 232
martensitic transformation, 384
masking function, 75
Mass density distribution, 259, 260
mass loss, 5, 469, 471, 477, 479
mass spectroscopy, 469
Mass thickness, 113, 196, 197, 201, 202, 209, 370, 470
Mass–thickness, 143
matrix method, 375
matrix strain-field contrast, 398
Maxwell–Boltzmann distribution, 81, 82, 86, 231
Mean energy, 187
Mean ionization energy, 162, 186, 197
micro-writing, 488
Microcalorimeter, 431
Microorganism, 209
microtwin, 363, 370
Miller indices, 277, 279, 319, 363
minimum-exposure technique, 480
Minority carrier, 112
Mirror charge, 78
Mirror Electron Microscopy, 11
mirror lines, 352
misfit, 397, 409
Missing cone, 260
moiré contrast, 397
Moiré fringes, 135, 365, 384, 397
Molecule excitation, 153
Monochromator, 119, 154
Monte Carlo simulation, 136, 188, 191, 194
most-probable Energy loss, 186, 187
Most-probable-loss imaging, 125, 207
Motiv detection, 257
Mott cross section, 152, 194
Muffin-tin model, 148, 152, 200, 285
multi-beam image (MBI), 369, 381, 390
Multichannel STEM detector, 237
multilayers, 358
multiple diffraction, 345
multiple internal reflection, 476
Multiple scattering, 99, 158, 171, 184, 202, 462, 487
Multiple scattering angular distribution, 184
Multiple scattering energy distribution, 186
Multiple-beam interferometry, 210
Multiple-scattering integral, 185, 191
Multislice method, 294, 405–407, 418
Multivariate statistical analysis, 257
Mutual-correlation function, 255
Myelin lamellae, 215

Negative staining, 98, 204, 257
Newton's law, 164
non-centrosymmetric crystals, 353, 384
nuclear magnetic resonance, 477
nuclear track emulsion, 480

Objective aperture, 98, 104, 113, 196, 208, 359, 368
Objective diaphragm, 98, 100, 104, 198, 330, 360, 362, 366, 396, 400
Objective lens, 2, 104
Objective prefield lens, 96
Optical analog filtering, 250
Optical constants, 163
optical density, 472
Optical diffractometry, 70, 250, 257
optical Oscillator strength, 158
Optical path difference, 146
Optical theorem, 152, 221
Ordered Alloys, 350, 383
Organic Particle, 204
orientation contrast, 397

p-n junction, 112, 138
Parallax, 259
parallel recording electron energy-loss spectroscopy (EELS), 437
parallel-recording EELS, 121, 124
partial coherence, 3
partial dislocation, 383, 390, 394
partial Inner-shell cross section, 178
Partial pressure, 100, 101, 478, 484–486, 488
Partial wave analysis, 151, 197
Pauli's principle, 174
Peltier cooling, 134
Pendellösung, 301, 310, 312, 335, 346, 353, 355, 360, 372
Phase contrast, 2, 55, 73, 90, 98, 105, 113, 114, 184, 204, 211, 262, 398, 415, 417

Phase contrast of inelastic scattering, 224, 237
Phase plate, 238
phase shift, 3, 12, 56, 57, 146, 152, 183, 211, 271, 397, 405, 407, 409
phase shift by magnetic fields, 52, 262
phase transition, 460
phase-grating approximation, 405
Phonon excitation, 153, 156
Photoabsorption, 178
Photoelectron, 139
Photographic density, 127, 479
Photographic emulsion, 127, 129, 131, 192, 479, 480, 485
Photomultiplier, 139
plane Wave, 46
Plasmon, 6, 7, 154, 165
Plasmon cut-off angle, 169
Plasmon dispersion, 168, 169, 172
Plasmon frequency, 165
Plasmon loss, 163, 369, 481
Plasmon loss anisotropy, 169
Plasmon loss differential cross section, 169, 173
Plasmon loss imaging, 125, 237
Plasmon loss mean free path, 170
Plasmon loss of alloys, 168
plastic Scintillator, 138
point defect, 156, 371
point groups, 352
Point-spread function, 76, 136, 218, 251, 255
Poisson distribution, 133, 136, 370
Poisson distribution (statistics), 171, 185, 186, 479
Poisson's ratio, 386, 398
Polystyrene spheres, 107, 188, 189
Potential barrier, 80, 81, 88
Potential energy, 78
powder scintillator, 134, 138
Precipitate, 101, 289, 349, 350, 367, 384, 396, 398
Preionization peak, 180
principal planes, 73
Probability density, 49
probe aperture, 208, 332, 336
Projected angular distribution, 191
Projected lateral distribution, 191
Pseudo-topographic contrast, 233, 265
Pupil function, 218, 251

Quadrant detector, 113
Quantum Efficiency, 132, 139
Quantum number, 174
Quantum-mechanical tunneling effect, 80

Radial density distribution, 183, 186
radiation chemistry, 467, 469, 478
Radiation damage, 4, 9, 90, 101, 108, 126, 139, 145, 156, 188, 204, 209, 225, 227, 249, 257, 260, 465
radiation damage by negative ions, 480
radiation damage inorganic specimens, 480
radiation damage measurements, 470
radiation damage organic specimens, 466
radiation damage reduction, 477
radiation-enhanced diffusion, 483, 484
radiation-induced reaction, 485
radical, 467, 478
radiolysis, 478, 481
Random conical tilting, 262
Random-phase approximation, 166
Reciprocal lattice, 279, 282, 284
Reciprocal lattice needles, 288
reciprocal lattice points, 340, 347, 349, 350
Reciprocal lattice vector, 280, 282
reciprocal-lattice vectors, 347
Reconstruction of off-axis hologram, 246
Reconstruction of phase and amplitude, 246
reconstruction surface layer, 415, 417
Reduced aperture, 70, 220
Reduced coordinates, 73, 228
Reduced defocus, 73, 220
Reflection Electron Microscope, 416
Reflection Electron Microscopy, 11, 340
Reflection electron microscopy (REM), 246
Reflection High-Energy Electron Diffraction, 11, 340, 417
relation of scattering contrast Phase contrast, 220

Resolution, 1, 3, 188, 220, 241, 456
resolution limit, 456
retarding field Electron spectrometer, 118
retarding-field filter, 339
RHEED, 52, 344
Richardson's law, 78
Rocking, 93, 109, 110, 194, 324, 336
rocking beam, 331, 336
Rocking curve, 301, 309, 311, 312, 315, 335, 352, 357, 358, 362, 369
rocking probe, 337
rocking-beam, 331
Rose condition, 479
Rutherford cross section, 152, 192, 193, 201, 337

SAED, 331, 349
sampling theorem, 70
scalar magnetic potential, 22
Scanning coils, 113, 331
Scanning electron diffraction, 339
Scanning Electron Microscopy, 12, 110, 417
Scanning transmission electron microscopy (STEM), 73, 94, 95, 99, 109, 112, 124, 189, 210, 246, 340, 367, 411, 471, 479, 486
scanning tunneling microscope, 14
Scanning tunnelling electron microscopy (STEM), 3
Scanning-probe Microscopy, 14, 15
Scattering amplitude, 146, 148, 217
scattering amplitudes for x-rays, 347
Scattering contrast, 2, 104, 195, 220, 221, 396
Scattering contrast in STEM mode, 208
scattering matrix, 376
Scherzer defocus, 3
Scherzer focus, 220, 222
Scherzer formula, 73
Schottky barrier, 112
Schottky effect, 79
Schottky emission, 79, 81
Schottky emission gun, 84, 88, 89, 246
Schottky gun, 4
Schottky plot, 80
Schrödinger equation, 53, 146, 177, 375, 393
Scintillation detector, 138, 192, 193, 210
Scintilator-photomultiplier detector, 210
scintillator–photomultiplier combination, 339
Scintillator-photomultiplier detector, 110, 139
scission, 467, 469, 471, 477
Screening, 147, 150, 174, 178
screw dislocation, 386, 388, 390, 392, 394
Secondary electron, 110, 270, 340, 487
Secondary-electron yield, 139
sector field Electron spectrometer, 116
Selected-Area Electron Diffraction (SAED), 6, 104, 329, 362
selection error, 330, 332
Selection rule, 156, 178
selector diaphragm, 329, 331, 335, 363
Semiconductor, 193
Semiconductor detector, 110, 138, 192, 210, 337
serial recording electron energy-loss spectroscopy (EELS), 437
serial-recording EELS, 122
Shadow-casting film, 202, 257
shot Noise, 96, 132, 133, 139, 479
Sigle-sideband holography, 233
SIGMA programs, 178
Signal-to-noise ratio, 95, 96, 110, 131, 132, 134, 208, 255, 256
Single atom imaging, 113, 115, 221
Single atom imaging in STEM, 225
Single atom imaging in TEM, 221
Single sideband holography, 238
single-sideband Holography, 243, 244, 254, 265
Single-sideband transfer, 233
SIT camera, 134, 239
Small-Angle Electron Diffraction, 91, 338
space group, 342, 346, 352, 357
spartial Coherence, 3, 4, 105
Spatial distribution, 188
Spatial frequency, 62, 70, 73, 75, 104, 105, 109, 129, 212, 218, 228, 238, 243
Spatially resolved EELS, 126
Specimen annealing, 8, 101

Specimen cartridge, 100
Specimen charging, 270
Specimen cooling, 101, 333, 355, 467, 469, 488
Specimen deformation, 101
Specimen drift, 90
Specimen grid, 99
Specimen Heating, 100, 188, 333, 459, 461, 462, 471, 482, 488
Specimen height, 107
Specimen manipulation, 100
Specimen mounting, 99
specimen Reconstruction, 249
Specimen rotation, 100
Specimen straining, 100
specimen temperature, 461, 463, 464, 472, 477, 478, 484
specimen temperature calculation, 461
specimen temperature measurement, 459
Specimen thickness, 98, 99
Specimen tilt, 100
Spherical aberration, 3, 94, 96, 103, 107, 115, 330, 336
Spherical aberration C_s, 212, 223, 239, 241, 249, 255, 405
spherical wave, 49
Spin-orbit interaction, 174
Spin-orbit splitting, 176
spinodal alloy, 401
Stacking fault, 101, 112, 348, 350, 358, 372, 375, 378–380, 383, 390, 395, 396
stacking fault contrast, 395, 397
stacking fault energy, 396, 411, 461
stacking fault fringes, 370
stacking fault tetrahedra, 398, 484
Staining, 203, 257
Stationary Electron Probe, 332, 335
Stefan–Boltzmann law, 463
STEM, 73, 340
STEM Phase contrast, 237
Stereo pair, 8, 100, 399
stereo viewer, 365
Stereology, 207
Stereometry, 258
Stigmator, 92, 103, 113
strain, 356, 358, 398
strain contrast, 397, 417
strain field, 375, 391, 397–399, 417
strain relaxation, 386
Streaks, 326
Stroboscopy, 252, 270
Structure amplitude, 281, 282, 285, 286, 289, 290, 294, 315, 349, 353, 354, 378, 383, 414
structure factor, 346, 353, 354
structure factor contrast, 397, 400
Structure-sensitive contrast, 125, 206
Superconducting transition, 268
Superconducting vortices, 246
Superlattice reflection, 326, 351, 383, 400
Superlattice Structure, 350
superstructure reflection, 416
Supporting film, 100, 201, 204, 209, 212, 214, 225, 226, 270, 412
surface dislocation, 418
surface Plasmon losses, 156, 171
Surface replica, 99, 107, 196, 202, 259, 417
surface step, 12, 412
Surface-charge wave, 171
surface-profile imaging, 418, 482
Systematic row, 311, 314, 346, 402

temporal coherence, 3, 4
terminal electron dose, 473
terminal mass loss, 471
Texture, 320, 343–345, 473
Theorem of reciprocity, 113, 114, 208, 225, 305, 331, 336, 341, 362, 411
thermal conductivity, 461, 463, 464, 466
Thermal diffuse scattering, 185, 210, 306, 321
Thermal expansion, 167, 294, 460
Thermionic electron gun, 84, 118, 332
Thermionic emission, 78
thermocouple, 460, 462
thickness fringes, 370
thickness-defocus tableau, 408
thin-film relaxations, 356
thin-foil Diaphragm, 265
Thomas–Fermi model, 147
Three-beam case, 311, 314
three-dimensional Reconstruction, 258
threefold astigmatism, 32, 43
Tilt coils, 343

Tilt parameter, 299, 301, 311, 360, 379, 401
Tilt series, 260, 262
tilt stage, 360
Tilted-beam illumination, 233, 402
Time constant, 138, 140
Time-resolved EELS, 126
Tomography, 5, 100, 249, 259
Top–bottom effect, 189, 190, 206
total Elastic cross section, 143, 148, 152, 155
total Elastic scattering, 197
total Inelastic cross section, 155, 162, 163, 197
total Inelastic scattering cross-section, 197
traces of structures, 348
Transfer gap, 229, 233, 238, 239, 244, 249, 253
transfer matrix, 348
Translation vector, 274
Transmission, 151
Transmission polycrystalline films, 201, 210
Transmission amorphous films, 196
transmission crystalline specimen, 370
Triode gun, 82, 85
Tunneling effect, 80, 89
Two-beam case, 292, 293, 298, 309, 353, 375–377, 391, 400

Unit cell, 274

vacancy, 481, 483
Vacancy cluster, 215, 399
Vacuum, 85, 89, 112
Viewing screen, 109
Virtual source, 91, 92
vitrification, 482
Voltage center, 92

Wave aberration, 2, 70, 71, 109, 212, 219, 234, 239, 400–402, 404, 407
Wave amplitude, 46, 142
Wave function, 49, 156, 303
Wave optical imaging, 73
Wave packet, 49, 50
Wave-optical calculation, 216
Wavefront, 49
Wavelength, 45
wavelength-dispersive X-ray detector, 5
weak-beam method, 375, 377, 386, 391, 393, 394, 396, 411, 415
Wehnelt electrode, 82, 85, 86, 88
Wentzel (potential) atomic model, 147, 151, 153, 162, 200, 201
White line, 176, 177, 179, 181
Wien filter, 55, 116
Wiener optimum filter, 256
Wiener spectrum, 255
WKB method, 46, 146, 150, 152, 197, 200
WKB-approximation, 46
Wobbling, 109
Work function, 78, 89, 194

X-ray Compton scattering, 160
x-ray continuum, 8
X-ray crystallography, 352, 353
x-ray microanalyser, 5, 8
x-ray microanalysis, 14, 90, 93, 107, 110, 113, 469, 476
X-ray scattering amplitude, 150, 162

Y-modulation, 210
YAG single crystal Scintillator, 134, 138
young fringes, 241

Z-contrast, 209
Zernike phase plate, 211, 238
Zero-loss filtering, 7, 125, 190, 199, 205, 207, 238, 267, 318, 320, 326, 335, 353, 370, 404
Zone axis, 282, 284, 346, 352, 414
Zone plate, 238, 250
zone-axis pattern (ZAP), 338, 341, 362

Springer Series in

OPTICAL SCIENCES

Volume 1

1 **Solid-State Laser Engineering**
By W. Koechner, 5th revised and updated ed. 1999, 472 figs., 55 tabs., XII, 746 pages

Published titles since volume 100

100 **Quantum Interference and Coherence**
Theory and Experiments
By Z. Ficek and S. Swain, 2005, 178 figs., XV, 418 pages

101 **Polarization Optics in Telecommunications**
By J. Damask, 2005, 110 figs., XVI, 528 pages

102 **Lidar**
Range-Resolved Optical Remote Sensing of the Atmosphere
By C. Weitkamp (Ed.), 161 figs., XX, 416 pages

103 **Optical Fiber Fusion Splicing**
By A.D. Yablon, 2005, 137 figs., XIII, 306 pages

104 **Optoelectronics of Molecules and Polymers**
By A. Moliton, 2005, 229 figs., 592 pages

105 **Solid-State Random Lasers**
By M. Noginov, 2005, 131 figs., XII, 238 pages

106 **Coherent Sources of XUV Radiation**
Soft X-Ray Lasers and High-Order Harmonic Generation
By P. Jaeglé, 2006, 332 figs., XIII, 416 pages

107 **Optical Frequency-Modulated Continuous-Wave (FMCW) Interferometry**
By J. Zheng, 2005, 137 figs., XVIII, 254 pages

108 **Laser Resonators and Beam Propagation**
Fundamentals, Advanced Concepts and Applications
By N. Hodgson and H. Weber, 2005, 587 figs., XXVI, 794 pages

109 **Progress in Nano-Electro Optics IV**
Characterization of Nano-Optical Materials and Optical Near-Field Interactions
By M. Ohtsu (Ed.), 2005, 123 figs., XIV, 206 pages

110 **Kramers–Kronig Relations in Optical Materials Research**
By V. Lucarini, J.J. Saarinen, K.-E. Peiponen, E.M. Vartiainen, 2005,
37 figs., X, 162 pages

111 **Semiconductor Lasers**
Stability, Instability and Chaos
By J. Ohtsubo, 2005, 169 figs., XII, 438 pages

112 **Photovoltaic Solar Energy Generation**
By A. Goetzberger and V.U. Hoffmann, 2005, 139 figs., XII, 234 pages

113 **Photorefractive Materials and Their Applications 1**
Basic Effects
By P. Günter and J.P. Huignard, 2006, 169 figs., XIV, 421 pages

114 **Photorefractive Materials and Their Applications 2**
Materials
By P. Günter and J.P. Huignard, 2006, 370 figs., XVII, 640 pages

115 **Photorefractive Materials and Their Applications 3**
Applications
By P. Günter and J.P. Huignard, 2007, 316 figs., X, 366 pages

116 **Spatial Filtering Velocimetry**
Fundamentals and Applications
By Y. Aizu and T. Asakura, 2006, 112 figs., XII, 212 pages

Springer Series in
OPTICAL SCIENCES

117 **Progress in Nano-Electro-Optics V**
Nanophotonic Fabrications, Devices, Systems, and Their Theoretical Bases
By M. Ohtsu (Ed.), 2006, 122 figs., XIV, 188 pages

118 **Mid-infrared Semiconductor Optoelectronics**
By A. Krier (Ed.), 2006, 443 figs., XVIII, 751 pages

119 **Optical Interconnects**
The Silicon Approach
By L. Pavesi and G. Guillot (Eds.), 2006, 265 figs., XXII, 389 pages

120 **Relativistic Nonlinear Electrodynamics**
Interaction of Charged Particles with Strong and Super Strong Laser Fields
By H.K. Avetissian, 2006, 23 figs., XIII, 333 pages

121 **Thermal Processes Using Attosecond Laser Pulses**
When Time Matters
By M. Kozlowski and J. Marciak-Kozlowska, 2006, 46 figs., XII, 217 pages

122 **Modeling and Analysis of Transient Processes in Open Resonant Structures**
New Methods and Techniques
By Y.K. Sirenko, N.P. Yashina, and S. Ström, 2007, 110 figs., XIV, 353 pages

123 **Wavelength Filters in Fibre Optics**
By H. Venghaus (Ed.), 2006, 210 figs., XXIV, 454 pages

124 **Light Scattering by Systems of Particles**
Null-Field Method with Discrete Sources: Theory and Programs
By A. Doicu, T. Wriedt, and Y.A. Eremin, 2006, 123 figs., XIII, 324 pages

125 **Electromagnetic and Optical Pulse Propagation 1**
Spectral Representations in Temporally Dispersive Media
By K.E. Oughstun, 2007, 74 figs., XX, 456 pages

126 **Quantum Well Infrared Photodetectors**
Physics and Applications
By H. Schneider and H.C. Liu, 2007, 153 figs., XVI, 250 pages

127 **Integrated Ring Resonators**
The Compendium
By D.G. Rabus, 2007, 243 figs., XVI, 258 pages

128 **High Power Diode Lasers**
Technology and Applications
By F. Bachmann, P. Loosen, and R. Poprawe (Eds.) 2007, approx. 535 figs., VI, 546 pages

129 **Laser Ablation and its Applications**
By C.R. Phipps (Ed.) 2007, approx. 300 figs., XX, 586 pages

130 **Concentrator Photovoltaics**
By A. Luque and V. Andreev (Eds.) 2007, approx. 242 figs., XV, 298 pages

131 **Surface Plasmon Nanophotonics**
By M.L. Brongersma and P.G. Kik (Eds.) 2007, approx. X, 298 pages

132 **Ultrafast Optics V**
By S. Watanabe and K. Midorikawa (Eds.) 2007, approx. 303 figs., XII, 520 pages

133 **Frontiers in Surface Nanophotonics**
Principles and Applications
By D.L. Andrews and Z. Gaburro (Eds.) 2007, approx. 89 figs., X, 250 pages

134 **Strong Field Laser Physics**
By T. Brabec, 2007, approx. 150 figs., XV, 500 pages

135 **Optical Nonlinearities in Chalcogenide Glasses and their Applications**
By A. Zakery and S.R. Elliott, 2007, approx. 102 figs., IX, 199 pages

Printed by Printforce, the Netherlands